European ☰ Business School
International University · Schloss Reichartshausen

The Innovation for Development Report 2009–2010

Strengthening Innovation for the Prosperity of Nations

Augusto López-Claros, Editor
Director, EFD–Global Consulting Network

**The Innovation for Development Report
2009–2010:** Strengthening Innovation for the
Prosperity of Nations

European Business School

Augusto López-Claros
Editor

The terms country and nation as used in this report do not in
all cases refer to a territorial entity that is a state as understood by international law and practice. The term covers
well-defined, geographically self-contained economic areas
that may not be states, but for which statistical data are maintained on a separate and independent basis.

Copyright © 2010
Augusto López-Claros

All rights reserved. No reproduction, copy or transmission of
this publication may be made without written permission.

No paragraph of this publication may be reproduced, copied
or transmitted save with written permission or in accordance
with the provisions of the Copyright, Designs and Patents
Act 1988, or under the terms of any licence permitting limited copying issued by the Copyright Licensing Agency, 90
Tottenham Court Road, London W1T 4LP.

Any person who does any unauthorized act in relation to this
publication may be liable to criminal prosecution and civil
claims for damages.

The authors have asserted their rights to be identified as
the authors of this work in accordance with the Copyright,
Designs and Patents Act 1988.

First published 2010 by
PALGRAVE MACMILLAN
Houndmills, Basingstoke, Hampshire RG21 6XS and
175 Fifth Avenue, New York, N.Y. 10010

PALGRAVE MACMILLAN is the global academic imprint
of the Palgrave Macmillan division of St. Martin's Press, LLC
and of Palgrave Macmillan Ltd. Macmillan® is a registered
trademark in the United States, United Kingdom and other
countries. Palgrave is a registered trademark in the European
Union and other countries.

ISBN 9780230239661

This book is printed on paper suitable for recycling and made
from fully managed and sustained forest sources.

A catalogue record for this book is available from the British
Library. A catalogue record for this book is available from the
Library of Congress.

10 9 8 7 6 5 4 3 2 1
19 18 17 16 15 14 13 12 11 10

Printed and bound in Great Britain by Hobbs the Printer,
Totton, Hampshire.

Contents

Preface

Professor Christopher Jahns,
President, European Business School

Seen in the context of the current financial and economic crisis, at least two reasons why many may have seen a diminished role for innovation stand out. First, many question how it is possible to generate R&D-driven, technological innovation, given the shortage of funds and the apparent unwillingness or reluctance of banks to lend, even to solvent companies. Second, the entire notion of innovation may have suffered in the public imagination. Financial "innovations" have been at the centre of the crisis itself—derivatives, structured products, financial engineering driven to perfection, yet often incomprehensible even to many bankers and their managers, and seen as value-destroyers by the general public.

However, it is particularly in times like these that innovation—done the right way—should play a more prominent role in companies and nations at large. Innovation is a powerful value-*driver*, which can be utilized by governments and companies to improve their competitive advantage and emerge from the crisis in a stronger position, better able to face the challenges of our increasingly complex global marketplace.

Our understanding of what drives national prosperity has evolved over time. Natural resources, population growth, industrialization, geography, climate, and military might have all played a role in the past. We also know that the relative importance of these drivers has shifted over time, and that in recent decades more importance has been given to the coherence and quality of policies and the development of supporting institutions. A relative newcomer to this debate—identified as perhaps one of the most important modern engines of productivity and growth—has been the innovation excellence of a country; that is, its industries, researchers, developers, creative thinkers, politicians, lecturers, managers, and clusters.

The "discovery" of innovation as a driver of prosperity is not only an indication of the rising social welfare awareness of nations. It also constitutes an unexpected shift in direction towards a more equitable world and a fairer division of the fruits of global prosperity at this critical stage in the evolution of capitalism. The prosperity of a country no longer depends solely on raw materials, capital, and other structural endowments, as the current unequal distribution suggests, but increasingly mirrors a nation's innovation strength, as reflected in the quality of its governance and the strength of its institutions. More critical today is the extent to which societies allocate resources for the development of human capacities through education and training, and succeed in promoting social inclusion, gender equity, and environmental sustainability.

Commendable as the insight about the importance of innovation for prosperity may be, the challenge is to make this realization practical. How does a country boost its own innovation potential? What constitutes a nourishing innovation climate? How can it be created, cultivated, and allowed to flourish? If the innovation capacity of a country developed in a particular way in the past, how will it develop over the next few years when changes have taken place in the global environment? In which direction is the innovation capacity of a country moving? Has innovation resulted from greater attention to training the labor force in new skills and from investment in human capital? Or has greater attention been paid to creating a more transparent regulatory environment, one in which the state sees its role as the setter of sensible rules rather than mindless bureaucracy? What are the tradeoffs and how quickly can changes in direction be made? Is the pace of change consistent with shifts in the environmental dynamic?

In contrast, what are innovation climate "killers" and how can countries avoid them? These are all important policy questions that must be probed. The answers to these questions entail more than the usual sweeping recommendations which often come from the mouths of policymakers. Answers are needed that will allow every country to assess its strengths and weaknesses and gauge where it stands in comparison to other countries, to see more clearly which factors contribute to its current ranking, and to understand how it can improve innovation through a combination of more efficient allocation of scarce resources and more coherent policies based on scientifically established strategies and roadmaps. *The Innovation for Development Report* provides an excellent foundation for intelligent debate on these central questions, for the saying that "You can't manage and improve what you can't measure" applies equally well in this context.

Unquestionably, nations have come to recognize the importance of innovation for productivity growth. Indeed, the global race for excellence in innovation is on. However, desire alone is not sufficient to "win the race." It is the support for the culture of innovation, adequate structures and effective and efficient processes, which make the difference between will

and reality. I am convinced that leaving decisions of how to develop the right atmosphere, structures and processes to the free market alone is at best inefficient, and at worse dangerous. The free market is not only blind to fair and equal distribution in theory and practice; it is also blind in other ways. It only innovates what is economically advantageous or profitable. Sometimes it does not innovate at all, or not enough of what is absolutely necessary for future generations—witness our well-justified, present-day worries about the dangers of climate change and the extent to which environmental problems reflect serious market failures. It is the responsibility of scholars, researchers, and managers to remedy these failures. For these reasons, we feel it is worthwhile collaborating in the preparation of a report which addresses these fundamental questions.

In its broad view of innovation, this *Report* goes far beyond R&D as the main source of (product) innovation, beyond conventional process and business models, and beyond the private sector, which is so often seen as the main locus of innovation. The *Report* provides a macro perspective of the entire value-chain of innovation and innovation management, and includes analysis of the many factors which underpin a modern conception of innovation. Furthermore, as the title indicates, innovation is seen as an important driver of development in its broadest sense. It is here that the strengths of the *Report's* Editor, Augusto López-Claros, are of greatest value. His many years of service with such institutions as the International Monetary Fund and the World Economic Forum, have given him an understanding of innovation from a truly global perspective. Together with a team of eminent contributors from diverse backgrounds, institutions and countries, he has put together a volume that delivers both a range of insightful perspectives on many dimensions of innovation and that offers the Innovation Capacity Index, a tool for assessing the extent to which nations have succeeded in developing a climate that will nourish the potential for innovation. The Index allows policymakers and entrepreneurs around the world to examine the broad range of country-specific factors which underlie innovation capacity, creating a quantified intellectual framework for formulating and implementing better policies for the creation of an environment supportive of innovation.

The *Report* contains three thematic sections. Part 1 features the Innovation Capacity Index, a methodological tool that examines those factors, policies, and institutions that critically affect innovation in a large number of countries. Part 2 offers the contributions of a number of leading experts, who deal with different dimensions of innovation, and address such questions as: What is the effect of increasing access to information and communication technologies on a country's economic growth? What is the role of good governance in fostering a culture of innovation? What are the myths and the realities of knowledge-led productivity growth? Does the US patent system strengthen or weaken innovation and progress? How do emerging markets innovate? Finally, Part 3 presents innovation profiles for those 68 important countries, which account for the lion's share of world output.

The European Business School (EBS), with its strong ties to government, industry and entrepreneurs around the globe, is honoured to support this study. Our mission is to educate students to become responsible leaders, to inspire each individual with the vision of sustainable business through innovation. With our international profile and clear focus on emerging markets, EBS already offers an inside view on many countries to students, stakeholders, and the scientific community.

In supporting this publication we are making a contribution to further discussion on how to analyze, measure, and encourage innovation on various levels of society. I am confident that this publication will provide the basis for the fruitful exchange of ideas between countries, governments, policymakers, entrepreneurs, managers, and academics in our mutual striving to create a better world for the generations to come.

Executive Summary

Augusto López-Claros and Yasmina N. Mata

The first chapter, "**The Innovation Capacity Index: Factors, Policies, and Institutions Driving Country Innovation**," by authors Augusto López-Claros and Yasmina N. Mata, begins with a glimpse at some of the little-known history of innovation, long before the industrial revolution. We learn that the invention of eyeglasses not only extended productive working life, but spawned the invention of precision instruments, laying the foundation for later articulated machines with fitted parts. The clock permitted the ordering of life in cities, but gave rise to the very notion of productivity, leading to Adam Smith's insight that wealth and prosperity depend directly on the "productive powers of labor." As the authors show, the varied paths followed by different nations in their approach to innovation and scientific discovery determined their ability to capitalize on their innovations and buttress their development and technological potential. They explain how, despite the priceless inventions they bequeathed to the world—printing, paper, the compass, gunpowder, porcelain, silk, the use of coal and coke for smelting iron, and the numerous inroads into scientific research which far surpassed what was known in Europe in their day—the totalitarian nature of the regimes in the Arab world and China stifled the possibilities for further development. With the coming of the Renaissance and the establishment of scientific societies and formal programs of scientific enquiry, Europe imposed fewer constraints on innovators, leading inexorably to the industrial revolution and the culture of innovation and research which we now see as powerful engines of economic and social development.

There is no doubt that, in recent years, progress in the dissemination of knowledge and the use of information and communications technologies (ICT) has become increasingly widespread and has resulted in improved productivity. As the authors make clear, the traditional sources of power and influence, such as territory, resources, raw manpower, and military might—for centuries the chief determinants of nations' prosperity—are far less important today, and have given way to a world in which successful development is not only increasingly linked to sound policies, good governance, and effective management of scarce financial resources, but, most important, to the ability of societies to release and harness the latent creative capacities of their populations. Successful countries today are not necessarily large geographically, neither are they richly endowed with natural resources, or able to project military power beyond their borders. More and more, the countries to look to are those which have managed to expand opportunities for their populations through the full exploitation of the opportunities afforded by the world economy through international trade, foreign investment, the adoption of new technologies, macroeconomic stability, and high rates of saving.

In building the **Innovation Capacity Index** (ICI), the authors draw on a sound theoretical framework and the best available data to correlate the wide-ranging set of relevant factors, policies, and institutional characteristics which play a central role in boosting a nation's capacity for innovation. In its 2009 edition, the ICI covers 131 countries and identifies over 60 factors that are seen to have a bearing on a country's ability to create an environment that encourages innovation, such as a nation's institutional environment, human capital endowment, the presence of social inclusion, the regulatory and legal framework, the infrastructure for research and development, and the adoption and use of information and communication technologies, among others. Fully 90 percent of the variables used in the construction of the Index are "hard"—i.e., measuring directly some underlying factor, such as the budget deficit, expenditure in education, or cumbersome regulations, etc.—and, therefore, not dependent on a survey instrument.

The authors explain in detail the construction of the Index, which explicitly incorporates the notion that, while there are many factors which influence countries' innovation capacity, their relative importance varies, depending on the stage of a country's development and the particular political regime in which policies are being implemented. These differing stages of development are closely correlated with rising economic prosperity and per capita income. But, the authors also take the view, anchored in empirical observation, that democracies tend to do better than authoritarian regimes at encouraging the creation of friendly environments for innovation. These notions are reflected in the weight distribution assigned to the different pillars of the Index according to countries' per capital income and political regime classification. Those pillars which have more to do with people, institutions, and social networks are shown to be foundations for the pillars dealing with means and other enabling factors. The weight distribution encourages achievements in the last set of pillars in countries where the institutional and human resource foundations are well laid, whereas the reverse obtains for achievements in these same areas, in countries where these foundations are lacking.

The ICI is offered as a policy tool to promote dialogue for examining more closely the broad range of policies and institutions which foster an environment conducive to innovation. The methodologies developed offer country-specific policy prescriptions, based on nations' stages of development, and the nature of their political regimes. The authors have constructed the Index on the foundation of the large body of work which sees indexes—with all their limitations—as working tools to generate debate on key policy issues, and to track progress over time in the evolution of those factors which help explain national performance. The Innovation Capacity Index rankings 2009–2010 are presented in Table 1. This year's printed edition of the *Innovation for Development Report* includes the individual innovation profiles of 68 countries, accounting for the lion's share of world output. The remaining 63 can be found at the dedicated Website: www.innovationfordevelopmentreport.org

Following a detailed description of the constituent elements of the Index and its construction, the authors highlight the uses to which the ICI can be deployed, and examine in some depth the innovation capacity of five countries: Sweden, Chile, India, Russia, and Taiwan, brief descriptions of which follow:

Sweden (ICI rank 1) is the ICI's top performing country in 2009, serving as a benchmark for other countries. The authors point to Sweden's important presence in the global economy and to elements in its approach to innovation, which are of particular relevance not only to other industrialized countries, but to many middle-income countries with aspirations to join the league of top innovators. Sweden is impressive not only in combining open and transparent government, universal social protections, and high levels of competitiveness and productivity—making it one of the most innovative economies in the world—but equally so in the extent to which the country's excellent policy framework has turned the private sector into the main engine of innovation.

Chile is presented as an interesting case, proving that sound policies and good institutions are not the result, but rather the

Table 1. Innovation Capacity Index rankings 2009–2010*

Country	ICI rank	ICI score
Sweden	1	82.2
Finland	2	77.8
United States	3	77.5
Switzerland	4	77.0
Netherlands	5	76.6
Singapore	6	76.5
Canada	7	74.8
United Kingdom	8	74.6
Norway	9	73.5
New Zealand	10	73.4
Luxembourg	11	73.3
Denmark	11	73.3
Taiwan	13	72.9
Iceland	14	72.6
Japan	15	72.1
Hong Kong SAR	16	71.3
Australia	17	71.2
Ireland	18	70.5
Korea, Republic of	19	70.0
Germany	20	68.8
Israel	21	68.2
Belgium	22	67.6
Austria	23	66.7
France	24	65.4
Estonia, Republic of	25	62.7
Lithuania, Republic of	26	60.7
Latvia, Republic of	27	60.5
Spain	28	60.3
Chile	29	59.4
Italy	30	59.1
Slovenia, Republic of	31	58.6
Czech Republic	32	58.0
Bulgaria	33	57.7
Malaysia	34	57.3
Portugal	35	57.2
Bahrain, Kingdom of	36	56.6
United Arab Emirates	37	56.2
Croatia, Republic of	38	56.0
Slovak Republic	39	55.8
Poland	40	55.7
Hungary	41	55.6
Georgia	42	55.1
Thailand	43	54.6
Jordan	44	53.9
Qatar	45	53.8

Country	ICI rank	ICI score
South Africa	46	53.3
Macedonia, FYR	47	53.1
Romania	47	53.1
Uruguay	49	52.8
Russian Federation	49	52.8
Mauritius	49	52.8
Malta	52	52.4
Cyprus	53	52.3
Ukraine	54	52.0
Saudi Arabia	55	51.9
Tunisia	56	51.8
Kazakhstan, Republic of	57	51.6
Costa Rica	58	51.5
Turkey	59	50.8
Peru	60	50.6
Mexico	61	50.5
Oman	62	50.2
Greece	62	50.2
Kuwait	64	50.1
China, People's Republic of	65	49.5
Argentina	66	49.2
Botswana	67	49.1
Panama	68	48.9
Trinidad and Tobago	69	48.7
Bosnia and Herzegovina	70	48.3
El Salvador	70	48.3
Colombia	72	48.0
Namibia	73	47.5
Azerbaijan, Republic of	74	47.3
Philippines	75	47.0
Algeria	76	46.7
Ghana	77	46.6
Vietnam	78	46.4
Dominican Republic	79	46.3
Egypt, Arab Republic of	79	46.3
Jamaica	81	46.2
Honduras	82	46.0
Lebanon	83	45.8
Iran, Islamic Republic of	84	45.7
India	85	45.6
Sri Lanka	86	45.5
Brazil	87	45.2
Indonesia	88	44.9
Guatemala	89	44.5
Paraguay	90	44.3

Country	ICI rank	ICI score
Ecuador	91	44.2
Tanzania	92	43.7
Nicaragua	93	43.4
Madagascar	93	43.4
Morocco	95	43.3
Kenya	95	43.3
Pakistan	97	42.7
Belize	98	42.1
Zambia	99	41.8
Bolivia	100	41.5
Papua New Guinea	101	41.3
Venezuela	102	40.9
Nepal	103	40.3
Nigeria	104	40.2
Suriname	105	40.1
Bangladesh	106	39.8
Syrian Arab Republic	107	39.4
Mozambique, Republic of	108	39.1
Uganda	109	38.3
Cameroon	109	38.3
Senegal	111	38.1
Cambodia	112	37.5
Malawi	112	37.5
Ethiopia	114	37.3
Mauritania	115	37.1
Lao PDR	116	36.8
Yemen, Republic of	117	35.1
Sudan	118	35.0
Iraq	119	34.2
Mali	120	33.8
Angola	121	33.4
Rwanda	122	33.3
Congo, Republic of	123	33.0
Côte d'Ivoire	124	32.4
Zimbabwe	125	31.8
Niger	126	30.6
Togo	127	30.1
Guinea	128	29.1
Haiti	129	28.7
Chad	130	25.6
Afghanistan, Islamic Republic of	131	24.0

*All rankings and scores are after rounding.

engines for, the creation of wealth and prosperity. Chile's performance (ICI rank 29) is far ahead of any other country in Latin America, and in many critical areas it is already ahead of the European Union average. A mix of sound macroeconomic management—including one of the most virtuous fiscal policies in the world—institutional reforms, and the opening of its economy to the benefits of free trade, foreign investment, and international competition, have combined to create a reliable engine of high growth and poverty reduction. The authorities have also sought to implement micro-policies aimed at enhancing the efficiency of public services through various electronic platforms, and facilitating the use of ICTs more generally. Chile is well poised to catch up with the richer members of the EU.

India is acknowledged as one of the world's fastest-growing economies and has aspirations to be a global player in the field of technological innovation. Its economic performance over the past two decades has been impressive, and has turned it into the world's fourth largest economy. India has not only a long political tradition of democracy and rule of law, but also favorable demographics, with a growing working age population which, if properly educated, could spur rising productivity and growth. But the authors deal also with India's disadvantages, including high illiteracy, a poorly developed infrastructure, a festering fiscal deficit problem, and a highly bureaucratic regulatory framework, all of which seriously discourage entrepreneurship and innovation. While its ranking in the ICI (85) is not high, they indicate that there is wide scope for the implementation of better policies, including institutional reforms, which might allow India to scale up in the rankings.

Russia (ICI rank 49), despite its well-established tradition of solid contributions to basic science, is shown to be lagging far behind its true potential for innovation performance. In previous decades a leader in space exploration, nuclear technology, and aviation, it has had a difficult transition from the inefficiencies of bureaucratic central planning to the challenges of a market economy. The authors describe how the commodity boom of the past five years has increased Russia's economic dependence on energy and other raw materials exports, and how the country's unfriendly business environment hinders entrepreneurship and the incubation of new ideas and approaches to new products or process creation. They

point also to corruption, the lack of independence of judges and courts, and the gradual return to authoritarian forms of governance as factors which do not bode well for the creation of an environment conducive to various forms of innovation. However, they conclude that there is no intrinsic reason why a country with such rich human and natural resources and distinguished history of scientific innovation should not be able to catch up with the best of the world's innovators.

Taiwan (ICI rank 13) is offered as the most impressive example in the post-World War II period of the consequences of high growth and the policies that underpin it. That a country should be able to increase its income per capita from under US$200 in 1952 to close to US$17,000 in 2007 is nothing short of astounding. Taiwan's success is attributed to two factors: first, its success in achieving high growth, while taking full advantage of the benefits of international trade and investment and the acquisition of new technologies, and second, in avoiding the errors that have inhibited development in so many other countries. While acknowledging Taiwan's rapid transformation in less than a half century from a simple agrarian society in the earliest stage of development into a global technology powerhouse and world leader in the production of ICT equipment, the authors suggest that Taiwan's challenge in coming years will be to find creative ways to cooperate with China—an emerging technology power in her own right, with a much lower cost structure—and to move closer to the best performers in the ICI.

Other dimensions of innovation

Laura Altinger

The chapter entitled **"Technology and Innovation for Addressing Climate Change: Delivering on the Promise,"** by Laura Altinger, first provides an overview of the current thinking about climate change and then an analysis of the promising role for innovation in global efforts to reduce greenhouse gases, in adapting technologies to decrease the vulnerability of those most likely to suffer the most serious impact of climate change, and in moving our economies onto green and sustainable growth trajectories. While key technologies already exist with great potential for limiting GHG emissions, leading up to 2050, the technologies that will be relied upon to make the substantial cuts in GHG emissions required to keep the planet safe will target energy efficiency in all key sectors: carbon capture and sequestration for power generation and industry, nuclear power, biofuels, wind, electric and plug-in vehicles, and hydrogen fuel cells. The author describes the roles to be played by the private sector in developing technologies, and by the public sector in addressing important market failures, using such mechanisms as carbon markets and regulatory regimes to provide investment incentives for alternative environmental technologies, to facilitate research, and ensure an adequate pool of human capital. She proposes that governments boost investment, supplemented by carbon taxes, in supportive infrastructure for new technologies or green fiscal stimulus spending aimed at environmental technology development. Finally, she discusses the importance of legally binding commitments to regulate countries already generating significant GHG emissions, improving risk management in smaller developing countries, removing barriers to trade and investment, and support by developed countries for the transfer of clean technology in regions of the world where it is most needed.

Sarah Box and Ester Basri

In their chapter entitled **"International Mobility of the Highly Skilled: Impact and Policy Approaches,"** Sarah Box and Ester Basri discuss the international mobility of highly skilled people, with a particular focus on "human resources for science and technology" (HRST), the group of skilled individuals, such as scientists, engineers, and researchers, who play an important role in stimulating innovative activity. The mobility of such skilled people, including human resources in science and technology, has become a central aspect of globalization, with talented migrants playing an important role in shaping the skilled labor forces of many countries and influencing the creation and diffusion of knowledge. The authors present selected data to describe the broad patterns of mobility of highly skilled people, the importance of mobility for the transfer of knowledge, the effects on receiving and sending countries, focusing in particular on the potential impacts of "brain gain," "brain circulation," and the diaspora. Finally, they outline the policy responses of selected OECD countries regarding mobility of HRST. Their central message is that mobility of the highly skilled has the potential to ben-

efit the migrant, the receiving country, and the sending country, but that the policy environment plays an important role in whether this mobility can lift innovative performance. The evidence on return migration and brain circulation, beneficial brain drain and diasporas suggests that there are a variety of mechanisms by which migrants can continue to contribute to knowledge creation and innovation in their home countries.

Simon Commander

The article entitled **"How Do Emerging Markets Innovate? Evidence from Brazil and India"** by Simon Commander examines the productivity effects of ICT adoption and use in two of the emerging market economies, Brazil and India. The author takes off from the past studies on the varied productivity and growth consequences of ICT adoption across countries and regions and inquires into the factors behind such variation. ICT has been adopted and managed in different ways in different parts of the world, and, not surprisingly, associated organizational dimensions of the new technology appear to play an important role in explaining differences in outcomes. The author's research, based on interviews with 1,000 manufacturing firms in the two countries, describes the factors explaining the pace of ICT adoption—including policy and financing constraints—and the consequences of that adoption. The results show that that there are differences not only in the timing of adoption and the patterns of ICT use across the two countries, but also within the countries themselves. Larger sized firms and foreign ownership tend to be associated with higher adoption, and in both countries, associated with a higher share of educated workers and a change in the skill mix. The Brazilian firms are shown, on average, to have adopted more ICT than their Indian counterparts, and to have used it more intensively. However, firms operating in Indian states with good institutional arrangements tend to have adoption rates similar to those in Brazil. There is clear evidence that high returns in productivity have resulted from investment in organizational change and improvement in the quality of infrastructure arrangements.

Alexander Ebner and Florian Täube

In their article **"Dynamics and Challenges of Innovation in Germany,"** authors Alexander Ebner and Florian Täube analyze the historical underpinnings and current challenges facing innovation in Germany. They first review the conceptual frameworks for assessing innovation dynamism and survey the relevant institutional components of the German economy, discussing the trade regime, competition law, labor relations, the financial system, and entrepreneurship policies. They then highlight the basic features of the German innovation system, in particular pointing to factors such as education and training, R&D, and university-industry relations. From its renowned position as a "social market economy," combining technological innovativeness, international openness, and industrial competitiveness with an extensive welfare system, Germany has become an institutional "hybrid" moving in the direction of greater entrepreneurial spirit. The innovation system, based in a rather bureaucratic, bank-dominated economy, has been seriously challenged by globalization, technological change, demographic pressures, persistent unemployment, and the burdens of reunification. They contend that the current situation requires an urgent institutional response, in the form of increased venture capital, high-growth stock markets, more flexible regulatory measures, the removal of hindrances to innovation in small and medium enterprises, and increased public support for R&D. According to Ebner and Täube, the German service sector lags behind in knowledge-intensive services to create employment in the current world economy and there is a significant lack of human capital in high-tech industries which could be addressed by creating more attractive conditions in the research and educational systems.

Anil. K. Gupta

"Grassroots Green Innovations for Inclusive Sustainable Development," the title of the article by Anil K. Gupta, is an exploration of open, user-driven innovation, as exemplified by individuals in India. Long reliant on internal R&D for innovation, large companies are often constrained in their ability to identify and meet the needs of what he calls "excluded clients." The ability of corporations to influence the lives of common people with a variety of products and services has not increased in the recent past. The author presents the model of the "Honey Bee Network," which offers new thinking to help the formal sector learn from grassroots innovators and traditional knowledge-holders, enabling them to solve problems in an affordable, accountable, and accessible manner. Using intriguing examples of grassroots innovations by users—a bi-

cycle that generates energy from bumpy roads, a peanut pod-collecting device, an organic pesticide, and more—the author stresses the importance of reorganizing consumption and production relationships, minimizing investment in wasteful packaging, creating frugal design and development processes, allowing communities to take creative ownership in order to solve serious local problems, creating global markets of grassroots products, and redesigning supply chains. Gupta stresses the importance of including grassroots innovators in the benefits of their creativity when products are marketed, and giving them full credit for their work, through patenting and intellectual property rights.

Markus Haacker

"Quantifying the Impact of ICTs on Growth in Developing Economies" is the title of the article by Markus Haacker, in which he sets out to quantify the growth impact of technological advances in ICTs across the developing world. The author analyzes existing data on the production of ICT equipment, and builds a dataset covering the absorption of ICT equipment in a cross-section of developing economies. His analysis suggests that the direct growth impact of technological advances in the production of ICT equipment plays a subordinate role in the developing world. However, advances in ICTs do affect economic growth across the developing countries, as lower prices of ICT-equipment result in ICT-related capital deepening. Haacker finds that the growth impact of ICTs across the countries covered has increased from 0.19 percent annually in 1991–1995 to 0.26 percent annually in 2000–2006, and that there appears to be a greater growth impact in low-middle-income countries than in low-income countries, reflecting higher rates of ICT-related investment. He concludes that the sources of these growth increments are divided evenly between capital deepening related to IT and to communications equipment and that, while investment in communications equipment has been roughly twice as high as investment in IT equipment, the rate of technological progress regarding IT equipment has been higher.

Alan Hughes

In his article "Innovation Policy as Cargo Cult: Myth and Reality in Knowledge-Led Productivity Growth," author Alan Hughes compares the mid-20th century Melanesian "cargo cults" to the danger he perceives that the evolution of innovation policy structures which copy perceived cultural and structural characteristics of the US innovation system will also fail to deliver the "goods," viz. economic well-being through improved productivity. Hughes describes these "ritual structures" as increased R&D expenditure, the commercialization of science, and the promotion of an entrepreneurial culture based on the subsidization of risk-taking in venture capital investment. In questioning the emphasis on R&D-intensive high-technology spin-offs, he contends that they have been exaggerated to the neglect of other key factors in the innovation system which must be considered. After considering these factors (e.g., diffusion and use of ICT as a general-purpose technology, the role of performance transformation of existing firms as compared to new entrants in driving productivity, and the role of universities in the creation of human capital, role of public procurement policy, among others), he concludes by arguing that the crafting of any specific national innovation policy requires a careful consideration of its own structural features and particular opportunities and challenges. In order for the innovation "cargo" to be delivered, space must be created in institutional mechanisms for the practical utilization of scientific advances, focused problem solving, and the recognition and potential exploitation of commercial opportunities.

Josh Lerner and Adam B. Jaffe

In their article **"The US Patent System: Does It Strengthen or Weaken Innovation and Progress?"** authors Lerner and Jaffe ask the critical question: whether it strengthens or weakens innovation and progress. In order for technological innovation to create broad social benefits, to enable us "to live differently from our grandparents," institutions must create incentives for individuals and firms to invest money in a financially rewarding process. After a brief excursion into the history of the "passionate debates" over patenting in Britain and the Netherlands, the authors describe the workings of the US Patent Office (PTO), the process by which patent applications are examined for utility, novelty, and non-obviousness, and, how, if granted, patents are intended to ensure intellectual property rights and protect inventors from the risks of infringement. They then outline three ways in which patents fail to protect and how they are wielded to retard innovation. By

analyzing the 60 largest countries (by total economic activity) in 2000, they show how patent systems changed from 1859 to 1990, significantly constraining the discretion of government officials, increasing the length of patent shelf life, but not solving the universal dilemma of patent validity. Changes to US patent law and policy between 1982 and 1990 have resulted in a decline in rigor with which the standards of novelty and non-obviousness are applied, pushing under-qualified, underpaid, and overworked examiners, using "flawed and obsolete tools" to resolve cases quickly. Coupled with the explosion in patent litigation, the deterioration in the examination standards of the overworked PTO has resulted in thousands of noxious "patent weeds" which threaten the innovation garden. The authors end with concrete recommendations for reform, with the aim of achieving a better balance between rapid approval of good applications and reliable rejection of bad ones, without dramatically increasing expenditures.

Daniel Kaufmann

The article **"Good Governance for Sustained Growth and Development,"** by Daniel Kaufmann, provides more solid evidence for judging the effects of governance on development, and the effectiveness of strategies to improve it. Kaufmann's research disproves the common assumption that becoming rich is a precondition for a country to "afford good governance," to have a competent government bureaucracy, sound rule of law, and an environment in which corruption is not condoned. Contrary to popular belief, he says, corruption is not the direct result of low income, and good governance is not a "luxury good." His studies point to better governance as being the *cause* of higher economic growth and improved development, and not the reverse. By introducing the notion of "state capture" and "legal corruption," the author explains why traditional definitions and views of the investment climate—usually focused on the public sector—have tended to underestimate the importance of governance factors and why they do not accurately reflect what enterprises themselves report as being of greatest significance for their operation. He ends his chapter by showing how moves toward transparency, gender equality, freedom of expression, and public participation result not only in better socio-economic and human development indicators, but in higher competitiveness, less corruption, and the fostering of a culture of innovation. Reforms in such areas have proven to be net *savers* of public resources, obviating the necessity for excessive regulations or rules. Kaufmann makes strong recommendations to the international aid community to rethink strategies and embrace more fully good governance approaches.

Mohsen Khalil and Ellen Olafsen

In their article **"Enabling Innovation and Entrepreneurship through Business Incubation,"** the authors demonstrate how a globally applied mechanism of "business incubation" is facilitating the creation of an innovation and entrepreneurship ecosystem that encourages entrepreneurs who are willing and eager to take the risk of bringing new ideas to the market, and helps them turn the potential of their ideas and ambitions into real social and economic value. Basing their research on the international network of "infoDev," they provide the reader with examples of a wide range of innovations in products, services and business models which have been brought to market by developing country entrepreneurs: from biogas cooking stoves in Rwanda, to improved honey production methods in the Ukraine, to improved bus services through telephony and tracking in Brazil. They discuss in some detail the various methods countries can use to spur innovation domestically and adapt imported innovations, in order to couple such innovations with viable entrepreneurship. In addition to the more commonly encountered factors which foster innovation, they stress an often-overlooked characteristic called "cultural capital," which refers to the level of tolerance for risk and the interpersonal trust that exists in a given society, affecting both decisions to start a business and the entrepreneur's ability to grow it. The authors describe the intricate web of interactions making up the innovation "ecosystem" and the linkages between all stakeholders in the system. They then illustrate how business incubators assist early-stage enterprises to become competitive and grow by interacting with all the actors in the ecosystem, either directly or indirectly through the enterprises they serve, and help them to meet the challenges their clients face, whether related to regulations, finance, labor, or infrastructure.

Hernán Rincón

In his article **"Innovation and Social Development in Latin America,"** author Hernán Rincón shares his view of the current financial crisis and its effects on developing economies from the perspective of the CEO of Microsoft Latin America. He takes issue with the notion that "developing" necessarily implies moving from a low- to a high-income economic model, and to the idea that countries are going to return to the seemingly "stable" high-income regime that prevailed before the current crisis. For Rincón, economic development does not necessarily dovetail perfectly with economic growth—as measured by GDP—any more than GDP satisfactorily measures economic well-being. Referring to the current situation as a challenge of recalibration rather than recession, Rincón uses the example of the photographic industry which failed to foresee the onslaught of the digital camera, and stresses the critical importance of people-centered training, re-training, R&D, and IT applications in managing the shift to new business models. He points to the success of Brazil's San Luis Digital project in providing an entire province with free wireless Internet access, and Mexico's Housing Funds Institute, which gave partners access to instant messaging, video and audio conference capabilities, vastly simplifying communications, and increasing productivity. Rincón is optimistic that Latin America is now better positioned to face the downturn and adapt to new conditions, both because of the continent's past experience in surviving severe economic, political, and social upheavals, and because of recent stronger economic policies which lowered inflation and interest rates, improved public finances, reduced external debt, and substantially increased foreign exchange reserves. Rincón concludes by describing Microsoft's "Unlimited Potential" program to promote sustained social and economic opportunity for the world's five billion people, who have yet to benefit from technological advances, by transforming education, fostering local innovation, and creating jobs and opportunities.

Andrew Stirling

"From Enlightenment to Enablement: Opening up Choices for Innovation," by Andrew Stirling, provides insight into the "knowledge society" and the widespread notion that scientific and technological progress is linear and cumulative, that every possible or feasible path will be realized. Rather, Stirling writes, "whether deliberately, blindly, or unconsciously," societies pursue only a restricted subset of diverse possibilities, in which certain pathways for change are "closed down," while others are "opened up." The factors driving choice are determined by whether power is exercised deliberately and democratically, and whether public policy is open, inclusive, and accountable in dealing with links between technological risk, scientific uncertainty, social values, political priorities, and economic interests. Stirling analyzes the relationships between social and technological progress, on the one hand, and public participation and responsible precaution, on the other, and asks what are the most appropriate and practical ways, under different conditions, to "get the best out of specialist expertise," while "engaging stakeholders, learning from different experiences, and empowering the least privileged groups in society." Stirling analyzes the vulnerability of society from technology (biological, environmental, etc.), and its intriguing opposite: the risks for technology from society, such as when wise, feasible choices are foreclosed because of "market lock-in," prejudice, or the needs, preferences, values, and interests of restricted groups. After a discussion of the governance of these vulnerabilities, the author examines some of the unfounded assumptions about knowledge itself: that every marketable innovation is socially acceptable, or that the knowledge responsible for an innovation also encompasses its consequences, and reminds us that even apparently complete knowledge may be indeterminate in its implications, that facts and values are not necessarily interdependent. The article ends with a description of the "precautionary principle" which acknowledges both the potential for irreversible harm and the impossibility of scientific certainty, and opens up "directions for choice."

Part 1 The Innovation Capacity Index

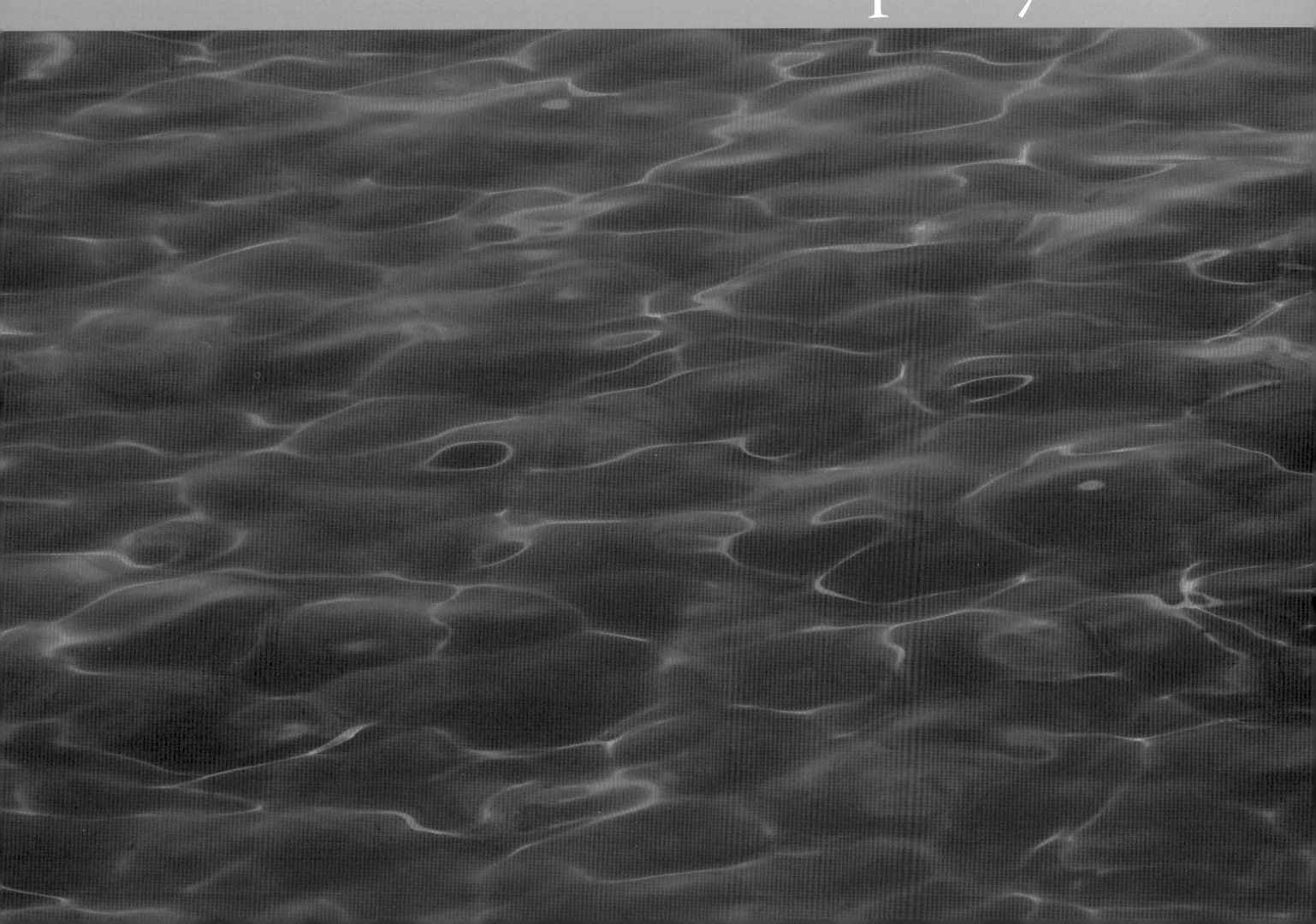

Chapter 1.1

The Innovation Capacity Index: Factors, Policies, and Institutions Driving Country Innovation

Augusto López-Claros,
EFD–Global Consulting Network

Yasmina N. Mata,
Consultant

Introduction

The relative importance of various drivers of economic growth and prosperity has evolved over time and, for a growing number of countries, innovation, in its many dimensions, is emerging now as a leading factor.[1] This chapter discusses the role of innovation in enhancing the development process. In particular, it features the Innovation Capacity Index, a methodological tool that examines a broad array of factors, policies, and institutions that have a bearing on strengthening innovation in a large number of countries, including their institutional environment, their human capital endowment, the presence of social inclusion, the regulatory and legal framework, the infrastructure for research and development, and the adoption and use of information and communication technologies, among others. The primary aim is to offer a didactic tool for policy dialogue on various dimensions of innovation. As will be shown, the methodologies developed allow the formulation of policy prescriptions that are country-specific, based on a nation's stage of development, and the nature of its political regime.

This chapter is divided as follows: Section 1 presents a brief historical overview of the role of innovation in economic and social development, with particular emphasis on its role in boosting factor productivity. In Section 2, we examine some of the factors which appear to be essential for the creation of an environment that will encourage innovation and the types of initiatives that will contribute in some way to boosting productivity and, hence, economic growth. Implicit in Section 2 is the idea that as countries have managed to sort out some of the more basic building blocks of development (macroeconomic stability, reasonably working institutions, and the creation of predictable mechanisms for social protection), they have had to give increasing attention to the role of technology and innovation as the primary engines of productivity growth.[2] The content of this section, which draws on insights

---

[1] For their insightful comments on particular dimensions of this project, the authors would like to thank Sergei Alexashenko, Farshad Arjomandi, Neil Buckley, Arthur Lyon Dahl, Yegor Gaidar, Evgeny Gavrilenkov, Pablo Guidotti, David S. Hong, Jui-Bin Hung, Natalia Ivanova, Jason Kao, Wang Kong, Shyh-Nan Kao, Yao Chung Liao, David Lin, Ricardo López Murphy, Alexander Pumpiansky, Beatriz Nofal, Anne Pringle, Hernán Rincón, Eduardo Rodriguez Veltze, José María Valdepeñas, Armida Sanchez, Sergei Vasilyev, Ignacio Walker, Stanley Wang, Randy T. M. Yen, and Mikhail Zadornov. The authors remain solely responsible for its contents.

[2] In this respect, our approach and arguments have some of the flavor found in Rostow (1960) and Porter (1990) and of their analysis and discussion of the central themes of the stages of economic growth. A thoughtful application of these concepts can also be found in Sala-i-Martin and Artadi (2004).

in economic theory and practice accumulated during the past half century, will be central to determining the major building blocks of the Innovation Capacity Index (ICI) featured later in the chapter. Section 3 presents a brief overview of international benchmarking as a means of enhancing analysis and policy dialogue in a number of important areas. Against the background of this discussion and the vast international experience acquired thus far with benchmarking exercises, Section 4 goes on to present the Innovation Capacity Index and to discuss various dimensions of its architecture. Section 5 presents the main results of the ICI for 2009, with particular reference to a handful of countries: Sweden, Chile, India, Russia, and Taiwan, which are seen as exhibiting some especially interesting features, or as suggesting patterns that may be of broader interest. Finally, we present our main conclusions and discuss the way forward.

1. Innovation: A brief historical overview

David Landes (1998) gives several examples of scientific innovation in Europe of the Middle Ages which contributed to substantially enhancing labor productivity. Eyeglasses significantly lengthened the working life of skilled workers. He notes that a medieval craftsman of 40 years of age could realistically expect—provided he could see well—to work for another 20 years, a development made possible by the invention of spectacles, which greatly boosted the productivity of toolmakers, weavers, metalworkers, scribes, and others who depended on their eyesight to do fine work. The first eyeglasses appeared in Pisa around the end of the 13th century. Although these early spectacles were initially not particularly accurate, by the middle of the 15th century, "Florentines at least understood that visual acuity declines with age and so made the convex lenses in five-year strengths and the concave in two, enabling users to buy in batches and change with time" (p. 47). More significantly, eyeglasses not only prolonged the productive working life of large numbers of people, but, in doing so—in a fascinating case of reverse causation—they also encouraged the invention of a whole battery of new precision instruments (e.g., gauges and micrometers), which could not have been invented, had work-

ers not been able to see particularly well, thus laying "the basis for articulated machines with fitted parts" (p. 47).[3]

The mechanical clock is characterized by Landes as "the greatest achievement of medieval mechanical ingenuity" (p. 49), both for its revolutionary conception (the first instance of a digital as opposed to an analog device) but, equally important, because it permitted the ordering of life in the cities in ways that had a major impact on productivity. "Indeed, the very notion of productivity is a by-product of the clock: once one can relate performance to uniform time units, work is never the same" (p. 49–50). It was the invention of the mechanical clock which in turn led to one of Adam Smith's seminal insights: wealth and prosperity depend directly—to use Smith's language—on the "productive powers of labor."[4]

Printing was a Chinese invention in the ninth century, but it did not take off in a major way until it made its way to Europe several centuries later. Landes notes that "much publication depended on government initiative, and the Confucian mandarinate discouraged dissent and new ideas (p. 51)." In Europe, in contrast, written manuscripts had been much in demand for centuries before Gutenberg printed the first Bible in 1452–55, and after the arrival of movable type had led to an explosion of printed materials. In Italy alone, more than 2 million books were printed before 1501. Other cultures, however, took longer to accept this new technological innovation. According to Landes, the Muslim countries found the idea of a printed Koran unacceptable, leaving the operation of printing presses in Istanbul to Jews and Christians, but not Muslims. Indians, likewise, did not adopt the new technology until the early 19th century when the first printing presses made their appearance. In Europe, in sharp contrast, not even the Church was able to restrain the new technology and all its uses. In all of these examples, one sees innovations spreading gradually, sometimes over several decades, "diffusing across countries and regions as people moved up learning curves and gained efficiency through practicing and improving the new techniques" (Goldstone, 1996).

An interesting question concerns the factors that may help create an environment that nurtures the capacity for inno-

---

[3] Indeed, Adam Smith himself had noted this feedback mechanism in his *Wealth of Nations*. "This great increase in the quantity of work, which, in consequence of the division of labor, the same number of people are capable of performing, is owing to three different circumstances; first, to the increase of dexterity in every particular workman; secondly, to the saving of the time which is commonly lost in passing from one species of work to another; and lastly, to the invention of the great number of machines which facilitate and abridge labor, and enable one man to do the work of many" (p. 7). But then he adds that "the invention of all those machines by which labor is so much facilitated and abridged" is itself the result of the improvements in productivity made possible by the division of labor.

[4] Smith, 1994, p. 5.

vation. Two cultures that showed great promise of playing a leading role in advancing the cause of scientific discovery and innovation were those of Islam and China, and it is instructive to say a few words about each. There seems to be little doubt that in the 400-year period to 1100, as noted by Landes, "Islamic science and technology far surpassed those of Europe, which needed to recover its heritage and do so to some extent through contacts with Muslims in such frontier areas as Spain. Islam was Europe's teacher." [5,6]

Gradually, after the year 1100, Islamic science came to a standstill as the faith was taken over by zealots, and the emphasis within the community shifted to one of conformity and obedience to its rulers, itself facilitated by the non-separation of the church and the state. Not surprisingly, "native springs of invention seem to have dried up."[7]

The case of China is equally fascinating because, at a time when Europe was a backwater of scientific enquiry, Chinese inventions—printing, paper, the compass, gunpowder, porcelain, silk, the use of coal and coke for smelting iron—suggested the existence of great technological potential. Why China failed to realize this potential and in the next several centuries fell hopelessly behind Europe is an intriguing question. Several explanations have been put forward by sinologists, among which the role of the state figures prominently. At one level, the lack of a well-defined framework for property rights and the absence of a free market seem to have been lethal. "The Chinese state was always interfering with private enterprise—taking over lucrative activities, prohibiting others, manipulating prices, exacting bribes, curtailing private enrichment."[8] During the Ming dynasty (1368–1644), serious attempts were made to shut down all trade with the outside world, efforts which in turn led to the proliferation of smuggling, rent-seeking, corruption, and violence.

The sinologist Etienne Balazs puts the blame for China's still-born technological prowess squarely on the emergence of totalitarian control:

The word 'totalitarian' has a modern ring to it, but it serves well to describe the scholar-officials' state if it is understood to mean that the state has *complete control over all activities* (emphasis in original), absolute domination at all levels…Nothing escaped official regimentation. Trade, mining, building, ritual, music, schools, in fact the whole of public life and a great deal of private life as well, were subjected to it…A final totalitarian characteristic was the state's tendency to clamp down immediately on any form of private enterprise (and this in the long run kills not only initiative but even the slightest attempts at innovation), or, if it did not succeed in putting a stop to it in time, to take over and nationalize it…Most probably the main inhibiting cause was the intellectual climate of Confucianist orthodoxy, not at all favorable for any form of trial or experiment, for innovations of any kind, or for the free play of the mind. The bureaucracy was perfectly satisfied with traditional techniques. Since these satisfied its practical needs, there was nothing to stimulate any attempt to go beyond the concrete and the immediate.[9]

At least one author has suggested that an additional factor in explaining the abortive nature of China's technological potential stemmed from the confinement of women to the home, which severely restricted the employment of women outside of the household and limited the supply of workers to labor-intensive industries, such as textiles.[10]

Potential innovators in Europe were considerably less subject to such constraints. What was more important: Europe had entered an era of free enterprise. "Innovation worked and paid, and rulers and vested interests were limited in their ability to prevent or discourage innovation. Success bred imitation and emulation."[11] It led to the establishment of scientific societies and formal programs of scientific enquiry and, in time, created a culture of innovation and research which saw

---

[5] Landes, p. 54.

[6] An early example of Islamic innovation is provided by Sells (1999): "At the time Muhammad was reciting the first Qur'anic revelations to a skeptical audience in the town of Mecca, several developments were leading to a transformation of Arabia's place in the world. One was a technological revolution. Sometime around the period of Muhammad's life, the Bedouin developed a new kind of camel saddle that allowed their camels to carry previously unimagined weight. Camels, which had been used largely for milk and transport of individuals and small loads, became the center of a transportation revolution. Within a hundred years, the Hellenistic and Roman worlds of transport and commerce, based on donkey carts and the upkeep of roads, were replaced by camel caravans. And the Bedouin in Arabia, who had been traders with and raiders of the established civilizations, were to control the vehicle of trade and commerce in the Western world: the dromedary camel." (Sells, 1999, p. 7).

[7] Landes, p. 55.

[8] Idem, p. 56.

[9] Balazs, Etienne, 1964, pp. 13–27.

[10] Goldstone, 1996. He further states: "In northwest Europe, with its pattern of late marriages and nuclear families, there existed a stage in the life course of most women—between puberty in their early teens and marriage in their mid-twenties—when they were available for labor and routinely performed work for wages outside their natal households. *No such stage existed in the life course of Chinese women, at least from the Ming through the end of the Imperial era (to 1911)* (emphasis in original). This would have posed a great obstacle to the creation of textile factories along the lines of their development in Europe and North America at any time in China's late Imperial history" (p. 3).

[11] Landes, p. 59.

the progress of science and technology as powerful engines of economic and social development.[12]

2. Factors, policies and institutions fostering innovation[13]

The broader context

Development as a *global* objective for improving the economic well-being of ordinary people is a relatively recent concept. It was first embodied in the UN Charter, which said: "the United Nations shall promote higher standards of living, full employment, and conditions of economic and social progress and development."[14] While this may be the first instance of a specific commitment on the part of the international community to promoting "development," the UN Charter does not itself define what are to be the defining elements of economic and social progress. In time—at least among practicing economists in academia and policymakers in government—it was interpreted to imply improved economic opportunity through increased production of goods and services in ever more efficient ways or, to use economic jargon, capital formation and rising productivity. The implicit assumption was that growth would lead to rising living standards, increases in longevity, reduced mortality, improved nutrition and literacy, and so on.

Between 1950 and 2007, world GDP/capita expanded at an annual average rate of 2.1 percent and this expansion—although with considerable variation over different regions of the world [15]—was associated with a remarkable evolution in three key indicators of human welfare. In particular, in the near half-century between 1960 and 2007

- Infant mortality fell from 140 to 44 per 1000 live births;
- Average life expectancy at birth rose from 43 to 66 years;
- Illiteracy (percent of adults) fell from 53 to 18 percent.

It is perhaps equally impressive that there was a sharp drop in the incidence of poverty. Data from a comprehensive study done at the World Bank shows that, between 1981 and 2001, the globalization phase of the 20th century, the share of the world's population living in extreme poverty fell from 40.4 percent to 21.1 percent.[16] While this still left about 1.1 billion people living under harsh conditions,[17] the existence of a positive trend was undeniable and, against the low expectations of the late 1940s, was a welcome development. As noted by Richard Cooper,

> performance in the period 1950–2000 can only be described as fantastic in terms of the perspective of 1950, in the literal sense that if someone had forecast what actually happened he would have been dismissed by contemporaries as living in a world of fantasy…There is, to be sure, much work to be done, since too many people still live in poverty. But it is also necessary to note success when there has been success, to avoid drawing erroneous conclusions. [18]

---

[12] For an excellent overview of innovation in the financial world, from the early days of money lending in Venice in the 14th century, through the gradual emergence of credit and currency markets under the Medici, to the appearance of bond, insurance, and real estate markets elsewhere in Europe, see Ferguson, 2008.

[13] There have been some attempts to define "innovation." For the OECD, for instance, innovation is "the implementation of a new or significantly improved product (good or service), or process, a new marketing method, or a new organizational method in business practices, workplace organization or external relations." (OECD and European Communities, 2005, p. 46). We are sympathetic to the view that any definition is likely to be constraining and is unlikely to apply and be meaningful when seen in the context of several thousand years of recorded history. In the context of this study, we think of innovation as the creative use of knowledge to allow individuals (and, by extension, corporations and nation-states) "to go farther, faster, deeper and cheaper" (Friedman, 1999). In most instances, innovation will involve a rise in factor productivity and, hence, other things being equal, living standards.

[14] "Charter of the United Nations and Statute of the International Court of Justice" available at: http://www.un.org/aboutun/charter/ This is not to suggest that individual countries, particularly during the period of empire building which began in the 15th century and stretched to the second half of the 20th were not, in some fashion, committed to the development of those lands and peoples under their control. According to Landes (1998), even the East India Company recognized the need—for the company's sake—to protect the welfare of those it saw as having fallen under its care. "India was compared to a landed estate where the interests of tenant and landlord were the same" (p. 163). (See also Landes' fuller discussion of colonialism on pages 422–441).

[15] For instance, Asia grew at 3.4 percent, but sub-Saharan Africa at 1.0 percent. Other regions include Western Europe (2.8 percent), Latin America (1.6 percent), Eastern Europe (2 percent), former USSR (1 percent), U.S., Canada, and Australia (2.2 percent). For a comprehensive set of economic and social indicators see, for instance, World Bank, 2008b.

[16] See Chen and Ravallion, 2008.

[17] Poverty is defined by the World Bank as living on less than US$2 per day; for extreme poverty the threshold is lowered to US$1 per day. The number of people living in extreme poverty in 1981 was 1.5 billion, or 400 million more than in 2001. Nevertheless, while accepting these figures, Joseph Stiglitz makes the valid point that "life for people this poor is brutal," with malnutrition endemic, life expectancy well below the global average, and medical care scarce or non-existent. (Stiglitz, 2006, p. 10).

[18] Cooper, 2004b, p. 39. Many critics of development practices during the past half century will tend to focus on the unfinished agenda, the fact that, notwithstanding the gains made during this period, there is still too much poverty in the world and that this poverty coexists uncomfortably with rising income disparities. Some of these critics call into question the very approach to development taken by such institutions as the World Bank and the International Monetary Fund and the aid agencies of the large donor countries, which also happen to be the largest shareholders of these two development organizations. Often, calls are made for "a new development model," although it is not spelled out what that development model should consist of and, equally important, whether such calls have any practical, conceptual, and political underpinnings. For a particularly incisive, well thought out, non-dogmatic, and unusually pragmatic analysis of the problems of the 58 poorest countries in the world and what the international community can do about it, see Collier, 2007.

The observation that economic growth had been the main engine of poverty reduction and other improvements in human welfare led many to ask themselves what could be done to accelerate growth everywhere, particularly in Africa, where the incidence of poverty actually rose during this period.[19, 20] The question acquired particular urgency among policymakers in the developing world, given the pressing needs to continue to make progress in improving living standards, against the background of rising expectations among their respective populations.

These numbers led notable economists like William Easterly (2002) to say that a key priority for policymakers should be "to discover the means by which poor countries in the tropics could become rich like the rich countries in Europe and North America." In a highly influential book published in 2002, he said that he cared about economic growth because "it makes the lives of poor people better…[and] frees the poor from hunger and disease." He then proceeded to show that growth improves infant mortality, and that, for instance, in Africa 500,000 deaths could have been averted if growth in the decade of the 1980s had been 1.5 percent higher.

The above insights, in turn, have led to a remarkable re-examination among professional economists and policymakers about the relative importance of various factors in creating the conditions for sustainable growth, including the role of institutions, education and social inclusion, the quality of governance, of macroeconomic management, of public administration, the presence of economic opportunities, and the increasingly crucial role of technology and innovation in enhancing the efficiency of the development process.[21]

An increasingly important factor in explaining rising prosperity and economic efficiency concerns the agility with which an economy adopts existing technologies to enhance the productivity of its industries. As countries have made considerable progress in improving their institutional and macroeconomic framework, attention turned to other drivers of productivity, and, without doubt, technology and innovation have been at the top of the list. Economic output is no longer just a function of capital and labor but, increasingly, of knowledge and the acquisition of *new* knowledge.

Why are these issues critical? Because technological differences have been shown to explain much of the variation in productivity between countries. In fact, the relative importance of technology adoption and innovation for rising productivity has been increasing in recent years, as progress in the dissemination of knowledge and the increasing use of information and communications technologies (ICT) have become increasingly widespread. For example, the strong productivity growth recorded in the United States since 1995 has been linked to the improved performance of industries which have used the latest technologies intensively to transform key elements of their operations. This has been particularly the case with wholesaling, retailing, and financial services. High-tech producers such as Microsoft, with well-established traditions of heavy spending in research and development, are enabling those sectors of the economy using the latest information technologies to improve their productivity performance and thus contributing to an overall boost to productivity growth.[22]

The central questions which follow from this discussion are: What are the factors, policies and institutions which are conducive to the creation of an economic and social environment that boosts the capacity for innovation? What is their relative importance? How do they interact with each other? How successful have countries been in identifying and adopting them? Let us now consider some high-priority areas.

Education and social inclusion

Social inclusion refers to the arrangements in place for education and health care which influence the individual's freedom to live better. We want this for two reasons: First, because, as pointed out by Nobel Laureate Amartya Sen (1999), a

---

[19] A report prepared by the United Nations Development Program for the 2008 United Nations General Assembly shows that the Millennium Development Goal of halving world poverty between 2000 and 2015 is within reach, largely because between 1990 and 2005, China brought some 475 million people out of poverty, compared to an *increase* of some 100 million during the same period in sub-Saharan Africa. (See *The London Financial Times*, "Number of poor rises in developing countries—China bucks trend, UN figures show; targets for 2015 still within reach," September 12, 2008).

[20] In Latin America, using US$1 as poverty line, the reduction was from 9.7 to 9.5 percent. Using a US$2 poverty line, it was from 26.9 to 24.5 percent. In sub-Saharan Africa, the corresponding figures are an increase from 41.6 to 46.4 percent for the $1 line, and 73.3 to 76.6 percent, for the $2 line. In millions of people, the figures are: for Latin America: 35.6m to 49.8m for the $1 line and 98.9m to 128.2m for $2 line. For sub-Saharan Africa: 163.6m to 312.7m for the $1 line and 287.9m to 516m for the $2 line. Idem, p. 56.

[21] Indeed, this debate has intensified in the past year as a result of the ongoing international financial crisis and the soul-searching it has precipitated. Robert Shiller (2009), a leading observer of financial markets, who issued repeated warnings about the real estate bubble in the United States, thinks that "capitalist economies, left to their own devices, without the balancing of governments, are essentially unstable." Nobel Laureate Amartya Sen (2009) recently wrote that "the question that arises most forcefully now is not so much about the end of capitalism as about the nature of capitalism and the need for change."

[22] See for example the chapter by Alan Hughes: "Innovation Policy As Cargo Cult: Myth and Reality in Knowledge-Led Productivity Growth" in this volume.

healthy life prevents morbidity and premature mortality. But, perhaps just as importantly, because education and good public health allow for more effective participation in the economic and political life of the nation. Illiteracy, for instance, can be a major barrier to participation in economic activities and the use of, and access to, technological innovations. Lack of such basic skills severely limits the possibilities of citizens to participate in the development process, to be gainfully employed, to be well-informed judges of government policies and politicians, and to avoid falling prey to the manipulations of demagogues—as we have seen in recent years in various corners of the world. From a business perspective, as noted by Porter (1990),

> . . . the quality of human resources must be steadily rising if a nation's economy is to upgrade. Not only does achieving higher productivity require more skilled managers and employees, but improving human resources in other nations sets a rising standard even to maintain current competitive positions. [23]

Notwithstanding the progress achieved in reducing levels of illiteracy noted above, much work remains to be done. According to UNESCO, almost 40 percent of India's population—well over 400 million people—still cannot read or write, representing a staggering burden for Indian society. Furthermore, an undue focus on enrolment rates has disguised important differences in the quality of education and in the particular approach taken by governments and the private sector to improving the educational system and its supporting institutions. Education and training are emerging as key drivers of productivity growth. As the global economy has become more complex, it is now evident that in order to compete and maintain a presence in global markets, it is essential to boost the human capital endowments of the labor force, whose members must have access to new knowledge, be continually trained in new processes, and in the operation of the latest technologies. Porter provides useful insights in his discussion of the role of education in contributing to an upgrading of an economy's productive apparatus. We find his emphasis on high educational standards (which the state must take the lead in setting) to be well placed, as are his calls for an

educational system that delivers education and training with a fair degree of practical orientation. Equally worth noting is his additional emphasis on the need to strengthen technical and vocational education, to facilitate interactions between educational institutions and firms, to empower the former to deliver graduates with good grounding on the needs of the business community, and on immigration policies that allow the movement of workers with specialized skills. [24]

As coverage of primary education has expanded rapidly in the developing world, higher education has gained importance. Thus, countries which have invested heavily in creating a well-developed infrastructure for tertiary education have reaped enormous benefits in terms of growth. Education has been a particularly important driver in the development of the capacity for technological innovation, as the experience of Japan, Finland, Sweden, Korea, Taiwan, and Israel clearly shows. [25] Without doubt, today's globalizing economy requires economies to create troops of well-educated workers, who are able to adapt rapidly to their changing environment. Conversely, as revealed by many innovation surveys, the absence of skilled personnel greatly hampers innovation (OECD, 2000).

Governments in many regions have made considerable progress in expanding social opportunities to their populations. The trend has definitely been in the right direction. However, the speed of progress has been at times adversely affected by the lack of a long-established tradition of fiscal discipline. Disorderly fiscal management has more often been the rule rather than the exception in much of the developing world, and this has curtailed the ability of governments to be more proactive in investing in education, public health, and infrastructure. [26] We will come back to this issue later, when we discuss the importance of macroeconomic management.

Institutions

According to Acemoglu, Johnson, and Robinson (2004), by institutions we mean the rules that establish the terms under which economic agents interact with each other in society and that also determine the incentives for such interactions.

---

[23] Porter, 1990, p. 628.

[24] Porter, 1990, p. 628–630.

[25] On the role of education in the emergence of Israel as an ICT power, see López-Claros and Mia, 2006.

[26] The notable exception is, of course, Chile, and the data demonstrate this quite explicitly. According to a report in the weekly *The Economist*, "poverty has fallen further, faster, in Chile than anywhere else in Latin America. Sustained economic growth and job creation since the mid-1980s are the main explanation, though it helps that poorer Chileans are having fewer children than in the past." The data show that while poverty rates in Latin America fell from about 48 percent to 39 percent between 1990 and 2006, the drop in Chile over the same period, from 38 percent to 13 percent, was far more dramatic. The authors add: "Chile has a chance of all but abolishing poverty in the next few years." ("Chile: Destitute No More." *The Economist*. 18 August, 2007).

The institutional framework has a crucial bearing on growth and development. It plays a central role in the ways societies distribute the benefits and bear the burdens of development strategies and policies. Indeed, it is the case that "without property rights, individuals will not have the incentive to invest in physical or human capital or adopt more efficient technologies… Societies with economic institutions that facilitate and encourage factor accumulation, innovation and the efficient allocation of resources will prosper." [27]

It is of fundamental importance the extent to which governments are accountable to their respective populations. Investors care deeply whether judges and courts are reasonably independent, or whether they are subject to undue interference or, far worse, are for sale to the highest bidder. Do businesses have to pay bribes to settle their tax obligations? Are they under pressure to hire private security outfits because police services are unreliable or, in some cases, indistinguishable from, or even working with, criminal organizations? Are governments biased in their decisions, or are they even-handed in their relations with the business community, playing more the role of impartial formulators of transparent rules, rather than meddling arbiters? Are public resources being allocated to education and essential infrastructure, or spent on wasteful and unproductive projects or schemes, including the maintenance of military establishments?

Needless to say, laying a sound institutional foundation is far from an easy task. Nor is it a process which produces results quickly, as is often the case with purely macroeconomic measures—an interest rate hike here, a tax cut there. Attempts at institutional reform often run up against strong opposition, as they often challenge powerful and deeply entrenched vested interests. [28] Some of the institutional factors that come to mind are respect for property rights; the ethics of government behavior and the incidence of corruption; the independence of the judiciary; the extent to which the government gives the private sector freedom to operate or engages in interventionist discretionary practices; the levels of government inefficiency reflected in the waste of public resources; a heavy regulatory burden; and the ability to provide an environment for economic activity characterized by adequate levels of public safety, to name a few.

Governance

Over the last few decades, there has been a noticeable (and most welcome) broadening of the debate as to what constitutes successful economic development. One element of this concerns the role of government in general and, more to the point, the exercise of political authority in a society for the purpose of managing its resources. Governance is the term that is now used in the development community to underscore the fundamental role of the *quality* of government in this process. Because this is so fundamental for successful development, let us briefly examine a few basic elements:

Accountability. The exercise of power must be guided by the need to improve the standard of living and well-being of the population. Adequate safeguards must be introduced to prevent the emergence of situations where ruling elites use political power for personal gain rather than public benefit. Democracy and political pluralism should facilitate this task which, at a minimum, involves the periodic legitimization of governments through popular choice, in such a way that gives adequate voice to the opposition, making politicians more responsive to the needs of society. The issue of accountability is closely linked to that of participatory development. Unless people feel that they have a say about those who rule them, they cannot be expected to fully support the government's development strategies and policies. Without such public support, even well-designed plans will in the end amount to very little. Sen (1999) convincingly argues that those countries in which governments operate in an environment of political legitimacy tend to be much better at allowing the formation of vital understandings and beliefs among the population that directly impinge upon aspects of the development process— for example, the notion that female education, employment, and ownership rights exert powerful influences on women's ability to control their environment and improve their condition and thus better contribute to national prosperity.

Transparency. Societies operate better on some presumption of trust. Here, we refer to the need for openness, the freedom to deal with one another under what Sen calls guarantees of disclosure and honesty. This is tremendously im-

---

27 Acemoglu et al., p. 2.
28 My years as an economist at the International Monetary Fund, including several years as Resident Representative in Russia during the 1990s, persuaded me that well-meaning governments will always find it easier to frame economic policies in purely macroeconomic terms. It is far easier to agree to an interest rate hike or some other budgetary measure than to get on with the far more difficult task of improving the legal framework for property rights, which, of necessity, may well take a decade or longer. Part of the ineffectiveness of the organization over the past couple of decades is linked, in part, to this macro short-term bias. Of course, a short-term macro bias, de facto, becomes a permanent one, with deleterious effects for the evolution of the country: viz. the rapid descent of Russia into the bottom ranks of the most corrupt countries in the world.

portant for preventing corruption, and financial and other abuses. Experience has shown that where there is trust, citizens and businesses pay their taxes. This, in turn, enables the government to formulate policies to achieve various social ends—for example, to dramatically increase access to the internet in the schools—because the resources are available to invest in these areas. As societies see the fruits of these efforts, trust in the government is reinforced and the country enters into what one can call a "virtuous cycle" of development. Of course, "vicious cycles" are also possible, and we have seen these in many parts of the world more often than we would care to remember.

Daniel Kaufmann (2003) and a number of other researchers have shown the central importance of the establishment of an institutional environment characterized by openness and transparency in the management of public resources. Corruption poisons the development process. It leads to resource misallocation, as funds are no longer directed toward their most productive ends, but are instead captured for private gain. It undermines the credibility of those who are perceived as being its beneficiaries (e.g., public officials, government ministers, and business leaders) and thus sharply limits their ability to gain public support for economic and other reforms. Work done at the World Bank has shown that the benefits for income per capita associated with improvements in governance are very large—"an estimated 400 percent improvement in per capita income associated with an improvement in governance by one standard deviation."[29]

Lack of transparency in the workings of the global financial system has been very much at the heart of the present crisis. Had the authorities been more effective in monitoring the explosive growth of increasingly sophisticated and opaque financial instruments—the so-called "weapons of financial mass destruction," to use the term coined by Warren Buffett—it is quite conceivable that the current crisis might not have been so severe in its intensity. Sen notes that societies operate better under some presumption of trust and that, therefore, they will benefit from greater openness. In a *Financial Times* article of 11 March 2009, entitled "Adam Smith's market never stood alone," Sen observes, "the far-reaching consequences of mistrust and lack of confidence in others, which have contributed to generating this crisis and are making a recovery so very difficult, would not have puzzled him."

Justice. Closely linked to the issue of accountability is the need for the rule of law, the notion that the rules which govern a society—and hence those that regulate economic activity—are applicable to all. There is increasing recognition that without a reasonably objective, efficient, and predictable judicial system and legal framework, accountability will have no legal underpinnings, and the goals of good governance will be undermined. As regards the economy in particular, it has long been recognized that the absence of an adequate legal framework and judicial system will increase business costs, discourage investment, and introduce an element of uncertainty into economic activity which will be detrimental to the development process.

From the above discussion, it is clear that these various elements of good governance: accountability, transparency, and justice, are not independent of one another. Interactions are inevitable and conflicts can arise in the short run. Participatory processes implemented in an environment of political pluralism and openness may add an element of unpredictability to the decision-making process. It may take much longer to forge the necessary consensus around a particular strategy. But this does not detract from their intrinsic value and the overriding need to pursue them as essential ingredients of good governance.

The potential benefits of an approach to development that seeks to incorporate the above mutually reinforcing elements should not be underestimated. To take an example: in an environment of accountability and political legitimacy, people will be far more likely to become active participants in the economy. A broadly shared sense of entitlement to economic transactions will then become an engine of economic growth. A growing economy will boost private incomes and enable the state to collect taxes out of which it will be able to finance expenditures, including in vitally important social areas, such as education, research, and development. Higher levels of spending on education and health care have been shown to be associated with reductions in infant mortality and a fall in birth rates. Female literacy and improved schooling have profound effects on women's fertility behavior, with resulting widespread implications for the environment, the pressures on which are often linked to rapid population growth. Conversely, it is possible to attribute the often disappointing fruits of economic development in many countries during the last half century to the absence of the above building blocks.

---

[29] Kaufmann, 2003, p. 146.

Indeed, neglect of these building blocks of good governance will make it difficult to create an environment that will release people's creative potential, so vital for the construction of a culture of innovation. One cannot help agreeing with Easterly (2002) when he observes that in such a country "skilled people opt for activities that distribute income rather than create growth."[30]

The macroeconomic environment

However important the role of governance, education, and social inclusion are for enhancing countries' capacity for innovation, a stable financial environment is essential for the successful implementation of broad-based reforms and the establishment of a macroeconomic environment supportive of private sector activity. Countries should pursue prudent fiscal policies that allow adequate levels of private sector credit, while limiting the growth of total credit to levels consistent with non-inflationary growth in the money supply and a viable external position. Cautious fiscal and monetary policies that contribute to low inflation rates and a more stable domestic environment also contribute strongly to business confidence and the willingness of domestic and foreign investors to undertake investment projects. In this way, government economic policies that reduce inflation and encourage macroeconomic stability have played a critical role in fostering economic growth and, more generally, in creating an environment that will foster innovation.

Clearly, fiscal policy should give priority to public sector expenditures that contribute directly to growth, such as outlays for human capital and spending in essential infrastructure, as against, for instance, the maintenance of large military establishments, or other unproductive expenditures. One element of this is the quality of public administration itself, which has many dimensions: policy coordination and responsiveness, service delivery and operational efficiency, merit and ethics, pay adequacy and management of the wage bill, among others.

Although not a "macroeconomic stability" issue per se, the question of a country's integration with the global economy has acquired growing importance over the past decade, particularly in the context of discussion about the interactions between the process of globalization and economic development. In an increasingly interdependent world economy, a more outward-looking orientation has become an essential element of successful economic reforms. In addition to the well-known gains from international trade, it is clear that relative openness and strong links with the world economy impose on domestic producers the valuable discipline of international competition and provide opportunities for new exports. An open orientation can also attract much needed capital and expertise, thus enhancing the prospects for growth through increased efficiency and productivity. Greater integration with the world economy also serves as an important channel for absorbing technological advances from abroad, including improvements in management practice and positive effects on the build-up of human capital that derive from being able to tap into global systems of knowledge, as is evident from the experience of many outward-oriented economies that have developed strong export sectors based on new manufacturing industries.

Economic opportunities

These refer to the chances that individuals have to utilize economic resources for the purpose of consumption, production, or exchange. Freedom to enter markets can make a significant contribution to development. Indeed, not an inconsiderable share of the progress made in India and China in the past 20 years reflects a reorientation of policies which significantly relaxed the barriers to entry to goods, labor, and financial markets.

For several years now the World Bank has published the *Doing Business Report* (DBR), an excellent compendium of business regulation in 181 countries. The picture that emerges from that study for a large number of countries is not a pleasant one.[31] Recently, the scope of the DBR has expanded significantly, such that now, in addition to the usual indicators on opening a new business (number of procedures needed, time taken, cost), one can also look at such things as: which countries make it easy to pay taxes, or to get licenses; where is it easier or more difficult to enforce contracts; who regulates property registration most closely; where are investors provided the greatest protection; or which countries have the most restrictive labor legislation, making it very difficult, for instance, to adjust the size of the payroll.

The data in Table 1 eloquently highlight the extent to which many countries *discourage* the development of entrepreneurship and, hence, the capacity for innovation of their own private sectors. And it is clear from the data that these are

[30] Easterly, 2002, p. 8.

[31] The *Doing Business Report* is available free of charge, at: www.worldbank.org

Table 1. *Doing Business Report*: **An international perspective on regulation**

	Brazil	India	China	Russian Federation	Venezuela	Greece	New Zealand
Ease of doing business*	125	122	83	120	174	96	2
Starting a business*	127	121	151	65	142	133	1
Number of procedures	18	13	14	8	16	15	1
Time (days)	152	30	40	29	141	19	1
Dealing with construction permits*	108	136	176	180	96	45	2
Employing workers*	121	89	111	101	180	133	14
Registering property*	111	105	30	49	92	101	3
Time (days)	42	45	29	52	47	22	2
Protecting investors*	70	38	88	88	170	150	1
Paying taxes*	145	169	132	134	177	62	12
Enforcing contracts*	100	180	18	18	71	85	11
Time (days)	616	1420	406	281	510	819	216
Closing a business*	127	140	62	89	149	41	17
Time (years)	4	10	1.7	3.8	4	2	1.3

* Rank from 181 countries

Source: 2009 Doing Business Report

problems existing not only in developing countries.

The sobering irony of the DBR is that those countries with the greatest need for entrepreneurship and private sector development are those that generally create the greatest obstacles for the creation of new enterprises, or that otherwise intervene in ways that retard the emergence of entrepreneurial capacities which are so central to the development of an enabling environment for innovation. Here, the critical factor is political will. Red tape, excessive regulation, and bureaucracy are self-imposed evils, which are potentially amenable to speedy elimination.

Other factors

The list of other factors which contribute to create an enabling environment for innovation is long. Without additional comment, let us quickly add a few more:

- What is the legal basis for secure property (including intellectual) and contract rights?
- What are the overall patterns of revenue mobilization, both as regards tax structure and equity?
- Is there timely and accurate accounting and reporting?
- What is the structure and level of sophistication of the financial sector, and of the policies and regulations that af-

fect it? Is the financial sector deep enough to allow reasonably free access to finance and the emergence of venture capital?

- Is the trade regime unduly restrictive, or it is reasonably open, encouraging competition and gains in efficiency?
- What are the levels of spending in education, both in absolute terms (percent of GDP) and in relative terms (as percent of total government expenditure)?
- How freely are women able to engage in the labor market, and how well are they represented in decision-making bodies, whether in parliament, cabinet, or the board room?
- Is there an adequate safety net to provide workers with some degree of financial security in times of economic stress?
- Is regulation of the labor market appropriate, or does it provide perverse incentives for both employers and workers?
- What is the level of expenditure in research and development?
- What is level of expenditure in information and communication technologies?
- What is the proportion of university students enrolled in science and engineering?

- How prevalent is knowledge of English?
- What are the penetration rates of the latest technologies?
- How effective is the government in providing information and public services for the people, and is this done through an electronic platform?
- Are public procurement policies and systems open and transparent and do they encourage the adoption of new technologies and reward innovation?
- To what extent do environmental policies foster the protection and sustainable use of natural resources and the management of pollution?
- What is the degree of collaboration between industry and the universities? Do they work independently from each other, or do they consult and give each other feedback?
- Where they exist, are government tax incentives well targeted, limited in duration, and applied transparently, or do they distort the incentives system?
- Do government immigration policies encourage the arrival of skilled workers and other highly qualified professionals?
- Is there public funding for long-term research?

3. Measuring innovation: Composite indicators

"What we measure affects what we do. We will never have perfect measures—and we need different measures for different purposes."

— Joseph Stiglitz[32]

A composite indicator can be thought of as the result of aggregating a set of statistical data in order to measure the overall performance of a certain phenomenon or issue (e.g., environmental sustainability, gender equity, competitiveness, etc.) that is directly or indirectly affected by its components. This definition highlights at least two key areas that influence the development of an effective indicator: a) choosing a proper data set, and b) the method of aggregation. However, there seems to be broad consensus that such indicators will be more credible if their construction is underpinned by a sound theoretical framework that enlightens in a plausible way the choice of variables and the ways in which these are combined. There has been wide debate with respect to the usefulness of these types of measures. The debate has been limited not only to technical aspects and methodological questions, but also to subjective perceptions of the public at large and, more specifically, to whether their advantages outweigh their potential disadvantages. It is not our intention to enter into this debate. Suffice it to say that the past decade has seen a remarkable increase in the number of credible organizations that have opted for the development of composite indicators, scoring mechanisms, and associated rankings.

The *Handbook on Constructing Composite Indicators* by the OECD and the European Commission Joint Research Centre (EC JRC) lists some of their main advantages and disadvantages (Table 2). Some of the functionalities implied are: i) support for decision-makers, since such indicators may allow more considered judgements as to various policy options available; ii) the ability to assess progress over time and to make meaningful international comparisons; and iii) contribute to public debate and the promotion of greater accountability. According to the *Handbook*, the two main criteria for evaluating composite indicators are ease of interpretation and the transparency of the methodology used. In other words, synthesis and construction. In view of the disadvantages, perhaps one of the main conclusions of this analysis is that composite indicators must be used with caution and as useful complements to other information and analysis, including well-informed judgements and common sense.

As a source of information, composite indicators can influence policymaking from a variety of perspectives. For instance, composite indicators can be useful for quantifying and outlining numerical goals and benchmarks. International benchmarking as a means of providing incentives for "changing behavior" has a well-established record. For example, the *Human Development Index* (HDI)[33] rankings have encouraged many countries to invest in preparing better and updated statistical series. The practice of synthesizing large volumes of information into a scoring system which can be translated into an index and an associated set of rankings can provide considerable value-added, particularly where efforts have been made to identify the critical factors deemed to affect the dependent variable. For instance, Transparency International (TI) has been associated with the *Corruption Perceptions Index* (CPI) since 1993. Despite occasional criticism—mainly from countries which do not wish attention to be drawn to a broad range of institutional

---

[32] Stiglitz, 2009, p. 28.
[33] Available at: http://www.undp.org

Table 2. Advantages and disadvantages of composite indicators

Advantages	Disadvantages
• Can summarize complex, multi-dimensional realities with a view to supporting decisionmakers; • Are easier to interpret than a battery of many separate indicators; • Can assess progress of countries over time; • Reduce the visible size of a set of indicators without dropping the underlying information base, thus making it possible to include more information within the existing size limit; • Place issues of country performance and progress at the centre of the policy arena; • Facilitate communication with general public (i.e., citizens, media) and promote accountability; • Help to construct/underpin narratives for lay and literate audiences; • Enable users to compare complex dimensions effectively.	• May send misleading policy messages if poorly constructed or misinterpreted; • May invite simplistic policy conclusions; • May be misused, e.g., to support a desired policy, if the construction process is not transparent and/or lacks sound statistical or conceptual principles; • The selection of indicators and weights could be the subject of political dispute; • May disguise serious failings in some dimensions and increase the difficulty of identifying proper remedial action, if the construction process is not transparent; • May lead to inappropriate policies if dimensions of performance that are difficult to measure are ignored.

Source: OECD and European Community Joint Research Centre, *Handbook on constructing composite indicators: Methodology and user guide,* 2008.

weaknesses[34]—the CPI has come to be accepted by civil society, the business community, and the media as a valuable tool, providing relevant data about the prevalence of corruption and corrupt practices in a large number of countries.

Composite indicators can also contribute to developing a common discourse and values when framing a problem in the light of public debate. Indexes and the associated rankings are useful benchmarking tools to focus public attention on a particular set of policy issues. When supported by detailed data, they can provide valuable information about underlying strengths and weaknesses, which can then become a catalyst for enhanced policy debate and efforts to improve particular areas of deficiency. For instance, the *Human Development Index* is an alternative measure of human welfare that captures a social dimension not existing in conventional GDP measures. The United Nations Development Program also publishes gender-related indices which attempt to assess the extent to which countries have succeeded in empowering women and reducing gender disparities.[35]

Finally, they can also help to highlight priority areas for policy reform and existing areas of achievement. For instance, the World Bank has developed the *Country Policy and Institutional Assessments*, a rating system that captures a broad array of factors affecting the policy environment in a large number of developing countries. The CPIA encompass such concepts as the quality of public sector management, the extent to which authorities have improved the policy framework through various structural policies aimed at enhancing resource use, as well as various elements of social policy, including aspects of social protection and poverty reduction, among others.[36]

The International Monetary Fund has published the *Trade Restrictiveness Index*, which nicely captures tariff and non-tariff barriers to trade. As noted by the IMF at the time of its release, "the index was constructed to provide a baseline of each country's overall trade policy stance" and "to provide policy handles for discussions with national authorities."[37]

The Innovation Capacity Index was built against the background of this large body of work which sees indexes—with all their limitations—as working tools to generate debate on key policy issues, and to track progress over time in the evolution of

---

[34] For a recent example, see "Transparency Group Fears for Staff in Bosnia" (*Financial Times*, 22 July 2008) in which it is reported that "The New York-based Human Rights Watch last week condemned [Prime Minister] Dodik's 'campaign of intimidation' against TI."

[35] See, for instance, the UNDP's *Gender Empowerment Measure* (GEM) and the *Gender-related Development Index* (GDI), both at www.undp.org

[36] According to the World Bank "The CPIA consists of a set of criteria representing the different policy and institutional dimensions of an effective poverty reduction and growth strategy. The criteria have evolved over time, reflecting lessons learned and mirroring the evolution of the development paradigm. In 1998, the criteria were substantially revised and coverage was expanded to include governance and social policies. The number of criteria was set at 20 (where it remained until 2004), and the ratings scale was changed from a 5- to a 6-point scale. To strengthen the comparability of country scores, specifically across regions, the ratings process was revised to include the benchmarking step." (World Bank, 2005, available at: www.worldbank.org).

[37] International Monetary Fund, 2005, available at: www.imf.org

those factors which help explain national performance. A well-designed composite indicator could thus provide a useful frame of reference for evaluation, the effectiveness of which will be enhanced if greater attention is placed on ways to improve national performance than on the relative rankings themselves.

4. The Innovation Capacity Index

"I have no data yet. It is a capital mistake to theorize before one has data. Insensibly one begins to twist facts to suit theories, instead of theories to suit facts."

— Sherlock Holmes, in "A Scandal in Bohemia"
(Arthur Conan Doyle, 1891)

The construction of the ICI was a response to three interrelated questions: What are the factors, policies, and institutions which are conducive to the creation of an economic and social environment that boosts the capacity for innovation? What is their relative importance, how do they interact with each other, and how are they dependent on a country's given stage of development and political system? Can we develop a methodology that will suggest, on a country-specific basis, the priority areas for strengthening the capacity for innovation? These three questions, in turn, suggested a work agenda that would involve two distinct components: first, a comprehensive assessment and identification of the factors that play a role in boosting the capacity for innovation; and second, the need to incorporate in the measurement of innovation capacity the country's stage of development—as captured by its income per capita—and the nature of its political regime. These, in turn, would lead to the development of a methodological tool that would allow policymakers to track progress in a country's capacity for innovation, both in relation to other countries and with respect to its own history. The result was the construction of the Innovation Capacity Index (ICI), which in its 2009 edition covers 131 countries and identifies over 60 factors that are seen to have a bearing on a country's ability to create an environment that will encourage innovation. The ICI is not the first attempt at the complex task of measuring innovation. There are several examples of innovation analy-

ses consisting of "scoreboards" of non-aggregated indicators, variables, and/or benchmarks, which track the performance of a particular region, nation, or groups of nations, including, for instance, the *Oregon Innovation Index*,[38] the *Mississippi Innovation Index*,[39] the *Index of the Massachussetts Innovation Economy*,[40] and the *OECD Science, Technology and Industry scoreboard*.[41] The composite indicator approach that generates cross-country rankings allowing international comparisons on the basis of comparable data is less common. Among these one may find the following examples:

- *Summary Innovation Index*.[42] Part of the *European Innovation Scoreboard*, created to examine the strengths and weaknesses and convergence in innovation of the European member states and their gap with respect to the U.S. and Japan. It measures innovation from an input/output perspective. Sample inputs include: tertiary education, ICT penetration, R&D and ICT expenditures, and small and medium-sized firm policies. Sample outputs include: high-tech exports and employment, sales of new market products, and patents and trademarks;

- *Innovation Index*.[43] Created to measure US innovative capacity with respect to other OECD countries over a 25-year period. Indicators include: personnel employed in R&D, expenditures on R&D, openness to international trade and investment, strength of protection for intellectual property, share of GDP spent on secondary and tertiary education, share of total R&D expenditure funded by private industry, and share of total R&D outlays carried out by universities;

- *National Innovative Capacity Index*.[44] Research derived from the US Innovation Index described above was expanded to cover other countries, using data from the World Economic Forum's 2001 Executive Opinion Survey (EOS). Qualitative measures were selected from the survey to construct different subindexes around the main areas of patents and number of scientists and engineers, including concepts such as intellectual property protection, market sophistication, quality of scientific research institutions, and venture capital availability. This work was further expanded in 2003 to cover 78 countries, by aggregat-

---

[38] Oregon Innovation Council, 2007. Available at:. http://www.oregoninc.org/
[39] Mississippi Technology Alliance. Available at: http://www.innovationindex.ms/
[40] Massachussetts Technology Collaborative, 2008. Available at: http://www.masstech.org/
[41] Available at: http://www.sourceoecd.org/scoreboard
[42] European Innovation Scoreboard, 2007.
[43] Porter and Stern, 1999.
[44] Porter and Stern, 2002.

ing science and engineering manpower, innovation policy, the cluster innovation environment, innovation linkages, and company operations and strategy subindexes;[45]

- *Global Innovation Index.*[46] Created by INSEAD in collaboration with the Confederation of Indian Industries, groups over 90 indicators combining quantitative data with a large number of indicators drawn from the World Economic Forum's (WEF) Executive Opinion Survey.

The ICI is an attempt to extend and build upon the work done by others in a number of specific ways. It is worthwhile to mention at least three areas in which the work underlying the construction of the ICI makes this a novel and, in our view, a far-reaching policy instrument. We discuss these in turn.

A. Overwhelming use of hard data

The ICI makes overwhelming use of hard data indicators. A full 90 percent of the variables used in the construction of the Index can be regarded as hard, that is, measuring directly some underlying factor (e.g., the budget deficit, expenditure in education, cumbersome regulations, etc.), and, therefore, not dependent on some survey instrument capturing (typically), business *perceptions*. This is not to suggest that there is no place for surveys in the construction of indexes. However, over the past decade or so, we have seen considerable improvement in the ability of various international organizations to develop indicators for a large number of countries that capture factors that had previously not been easily measured. An excellent example of this is the work done at the World Bank on business regulation and obstacles to the creation of new enterprises. Most of the concepts captured in the *Doing Business Report* published by the World Bank were in the past "measured" only through some opinion survey, such as the one carried out annually by the World Economic Forum. Many of these concepts, however, are now available through the comprehensive field work done by the Bank to examine the actual—as opposed to *perceived*—obstacles faced by the business community in a large number of countries. While this may perhaps be the best example, it is by no means the only one. In recent years, the International Tele-

communications Union has broadened the scope of the variables which they track that attempt to capture various indicators of the breadth and use of the latest technologies. As noted earlier, the IMF has compiled a measure of trade openness and the World Bank has put together at least two impressive scoring mechanisms: one is the Worldwide Governance Indicators which capture a large number of governance and rule-of-law measures; the second is the Country Policy and Institutional Assessment (CPIA), which examines various elements of a country's policy environment, such as the quality of public administration, the efficiency of the financial sector, and so on.[47] All of these have been used in the construction of the ICI.

B. Explicit incorporation of a "stages-of-development" theoretical framework

The construction of the Index explicitly incorporates the notion that while there are many factors which will have a bearing on countries' innovation capacity, the relative importance of these will vary depending on their stage of development and the particular political regime against which policies are being implemented. As regards the stages of development, our work is close in spirit to that done by Porter (1990), who divides countries and their respective industries into three broad categories: factor-driven, investment-driven, and innovation-driven. These categories, in turn, are highly correlated with rising economic prosperity, as captured by the growth of per capita income. Porter highlights some of the features of each of these stages and it will be useful to provide here a brief summary.

Factor-driven

Countries are in this stage when they derive advantages from basic factors of production, such as natural resources, plentiful and inexpensive labor, and, in some cases, a benign climate which may create favorable conditions for agriculture. These factors may impose some constraints on the kinds of industries that can develop and, thus, may limit a country's presence in the global economy. At the factor-driven stage, countries will compete on the basis of price advantage, and technologies will

[45] The authors limit themselves to the use of survey data, as these are "the only alternative because there are no quantitative data at all available on most of the areas measured, much less for a meaningful number of countries, so that Survey data are the only alternative." (Porter and Stern, 2003, p. 96).

[46] INSEAD, Global Innovation Index 2008–2009. Available at: http://elab.insead.edu

[47] One area where we are likely to continue to rely on survey instruments is the measurement of corruption. Transparency International's *Corruption Perceptions Index* is survey-based, and it is unlikely that, due to the nature of this problem, we will be able to dispense with opinion surveys any time soon. In such cases, we are firmly of the view that it is far better to use surveys—with all their limitations—than to fail to measure, however inadequately, the problem in question. There is no doubt whatever that TI has succeeded well in calling the attention of the international community to a serious problem, which has a grievous impact on development and, in the case of our subject, the development of the capacity for innovation.

usually be adopted from other countries, as opposed to created from within. Typically, human capital resources will not be particularly well developed, a feature that will constrain a country's ability to innovate and to see sustained productivity growth. Because countries will be largely price-takers in international markets, they will be vulnerable to business cycle fluctuations, exchange rate movements, or other external shocks that may lead to sharp changes in the terms of trade. At this stage, countries will have institutions in the early stages of development and one may see high levels of corruption, weaknesses in the legal framework and the rule of law, relatively low levels in the quality of the public administration and, as a result, a poor macroeconomic situation, characterized, for instance, by high inflation or loose public finances. In light of these observations, for nations in the factor-driven stage, the focus of policies should be the achievement of macroeconomic stability and the establishment and improvement of the basic institutions underpinning the modern market economy. To the extent that policies are not geared to these ends, nations may get stuck at this stage for decades, if not, in fact, much longer.

Investment-driven

At this stage, we witness heavy investment aimed at modernizing the economy's infrastructure. According to Porter, firms will invest to "construct modern, efficient, and often large-scale facilities equipped with the best technology available on global markets."[48] Technologies and processes discovered or developed elsewhere will not simply be adopted but may also be improved upon. The range of technologies imported from abroad may also widen to include not only basic ones, but also the most sophisticated. The main underlying theme of this stage is the willingness of firms to invest to upgrade factors to enhance productivity growth. This may include improvements in education and training, which create a pool of skilled workers who are able to assimilate and improve upon imported technologies or, in any case, adapt them to local conditions. Cost factors are still important and economies operating at this level are not immune from shifts in the global business cycle (or the exchange rate). But at this stage, investment aimed at a more efficient use of resources will often bring about a diversification in the economy's sources of wealth creation, and, thus, the emergence of a greater degree of resilience to changes in the terms of trade. As a result of the above, one may also see a fairly sustained increase in wages and labor costs. At this stage, the focus of policies broadens somewhat. While macrostability and institutional development are still important, these policies must be supplemented by policies aimed at further structural reforms, increasingly formulated in a medium-term framework. At this stage, for instance, governments may focus on fiscal sustainability issues and may implement pension reform to establish a sounder financial basis for the social security system, may aim to significantly improve the infrastructure for higher education, and find ways to change the nature of public administration so that it plays a more supportive role for private sector development.

Innovation-driven

Consumers in countries operating at this stage of development have high levels of income per capita, sophisticated and demanding tastes, and, on average, higher levels of education than at the factor-driven or investment-driven stages, all of which create a demand for improvement and innovation. At this stage, firms may continue to use and improve existing technologies, but, increasingly, they create them. "Favorable demand conditions, a supplier base, specialized factors, and the presence of related industries in the nation allow firms to innovate and to sustain innovation."[49] This stage may also see countries essentially ceding to nations in earlier stages of development those industries that are less-sophisticated, or where demand is highly price-sensitive. Firms operating in innovation-driven countries will have their own marketing and supply networks and will have, in many cases, established recognizable brands. They will also become important investors abroad and become truly global players, not only in terms of markets for sale and sources of inputs, but also in terms of sources of funding, labor supply and the location of production. This stage also sees a further upgrade in the training of the labor force and the emergence of highly-skilled workers with specialized know-how and able to command high wages. The role of public policy at the innovation stage is more subdued than at the previous two stages. Governments—overwhelmingly in the context of democratic institutions and processes—are called upon to preserve the gains made over the previous decades in terms of macro management and in-

---

[48] Porter, 1990, p. 548.
[49] Porter, 1990, p. 554.

stitutional development. Above all, governments are expected to do no harm to the policy environment, and the prospect that they can always be voted out of office generally tends to explain a certain level of policy stability. In these countries "the impetus to innovate, the skills to do so, and the signals that guide its directions must come largely from the private sector."[50, 51]

The above stages are not meant to be interpreted in a rigid way. It may be possible, for instance, for a country to be in the factor-driven stage, while some of its industries, in specialized niche sectors, may be operating at a higher stage of development. Neither should countries be seen as steadily and gradually progressing from the factor-driven to the innovation-driven stage. Korea, Singapore, and Taiwan are examples of economies that have made the transition to the innovation stage in a relatively short span of time; indeed, Taiwan has made the transition from an agricultural economy with low income per capita to a prosperous global industrial ICT powerhouse in less than 40 years, an impressive achievement. By way of contrast, Argen-

tina was a G10 power in the first part of the 20th century and had the best scientific and higher education infrastructure in Latin America by the 1950s, but has since *regressed*, in the wake of decades of economic mismanagement, to an economy with all the characteristics of the factor-driven stage.[52] This regression was caused, in particular, by an undue reliance on exports of primary commodities as the primary source of economic growth, high levels of corruption and, in an unusual turn, the gradual disappearance of reliable statistics, as a result of authoritarian, state-sponsored tampering and manipulation.[53]

In all cases, as should be evident, the role of policy matters enormously for how quickly and efficiently countries are able to make the transition through these three stages. Table 3 presents World Bank data on average income per capita for 2007, on the basis of which countries are classified as being high-income, upper-middle and lower-middle-income, and low-income. One may apply Porter's stages-of-development framework to suggest that low-income countries are at the factor-driven stage,

Table 3. Average GNI per capita, current US dollars, 2007 (World Bank Atlas Method)

High-income	GNI per capita > $11,456		Average: $34,907
Full democracies	Flawed democracies	Hybrid regimes	Authoritarian regimes
$40,066	$16,292	$32,040	$34,362
Upper-middle-income	GNI per capita: $3,706–$11,455		Average: $6,662
Full democracies	Flawed democracies	Hybrid regimes	Authoritarian regimes
$5,797	$6,790	$7,168	$5,060
Lower-middle-income	GNI per capita: $936–$3,705		Average: $2,374
Full democracies	Flawed democracies	Hybrid regimes	Authoritarian regimes
–	$2,328	$2,849	$2,288
Low-income	GNI per capita < $935		Average: $536
Full democracies	Flawed democracies	Hybrid regimes	Authoritarian regimes
–	$850	$501	$555

Source: World Bank.

---

[50] Porter, 1990, p. 555. Porter also identifies a "wealth-driven" stage which, in essence, is one of decline, where "the motivations of investors, managers, and individuals shift in ways that undermine sustained investment and innovation, and hence upgrading…and where malaise and an eroding sense of purpose may set in." It is conceivable that countries may enter periods of decline, and it is certainly the case that industries may also do so, partly through the failure of managers to anticipate technological change. But there is nothing to suggest that the entire collectivity of nations will go through a period of decadence and decline. The more likely scenario would appear to be one where nations gradually progress through the three stages identified above. Although some may remain in a given stage for a very long time—perhaps lasting even many decades, if not longer—a few may see temporary regression (e.g., Argentina and many of the poorest nations in Africa which can degrade to failed states). But the majority find themselves in a path of gradual forward, though at times uneven, progress.

[51] For an application of Porter's stages-of-development approach to the measurement of competitiveness see Sala-i-Martin and Artadi, 2004.

[52] Argentina remains to this day the only country in Latin America to have earned three Nobel prizes in science, with the awards going to Messrs. Houssay (Physiology or Medicine), Leloir (Chemistry), and Milstein (Physiology or Medicine).

[53] See, for instance, "Hocus-pocus: The real world consequences of producing unreal inflation figures." *The Economist*, 14 June 2008, p. 56. A more recent assessment by *The Economist*, commenting on mid-term elections, suggests that inflation figures are worth little because: "Mr Kischner put stooges in the statistics office and they massage the numbers." (See: "A chance to change course," 20 June 2009)

middle-income countries would have moved to the investment-driven stage, and high-income countries would have entered the innovation-driven stage. While there will be exceptions to this categorization (e.g., a rich oil exporter in the Gulf region), we find that, in general, countries broadly possess the characteristics identified by Porter for each of the levels of income. A further sobering feature of this table is the relatively huge income gaps across the various categories: for instance, from an average of US$6,662 for upper-middle-income to US$34,907 for high-income, or from US$2,374 for lower-middle-income to US$536 for low-income, displaying well known, large, and growing, income disparities.

C. The nature of a country's political regime matters for innovation

The above theoretical (and practical) considerations, as explained further below, have had a direct bearing on the choice of weights for the various factors which have been used to construct the Innovation Capacity Index. In addition to the embedding of a formal stages-of-development framework into the determination of key elements of the Index structure, we have also seen the benefits of establishing a further distinguishing criterion for nations: namely the type of political regime under which policies are implemented. For these purposes we have used the four categories developed in The Economist's Democracy Index: *full democracies, flawed democracies, hybrid regimes,* and *authoritarian regimes.* There is ample empirical evidence suggesting that democracies are much better at creating the sorts of conditions in a country that are conducive to the nurturing of creativity and independence of thought that are so essential for innovation. Therefore, our work attaches to the nature of a country's political regime a significance that is not captured by purely looking at the level of income per capita as a proxy for the country's stage of development.

The question of the relationship between democracy and development has been amply debated in the economics and political science literature. Without entering into this debate—which is outside the scope of this paper—there is overwhelming empirical support for the thesis that, for instance, poor democracies do much better than poor autocracies, arguably the most relevant comparison to cast light on this sub-

ject.[54] Siegle, Weinstein, and Halperin (2004) look at annual data drawn from the World Bank's *World Development Indicators* for the period 1960–2003 to show that the median per capita growth rates of poor democracies have been 50 percent higher than those of autocracies.[55] Citizens in poor democracies live, on average, nine years longer than in low-income autocracies, have a 40 percent higher chance of attending secondary school, will enjoy higher levels of agricultural productivity, and much lower infant mortality rates.

The latter statistic is particularly relevant as it reflects, in turn, better prenatal care for pregnant women, higher levels of nutrition, higher quality drinking water, and more opportunities for the education of girls. It turns out that poor democracies are also far better than poor autocracies in avoiding severe economic contractions—annual drops of 10 percent or higher in real GDP. "Seventy percent of autocracies have experienced at least one such episode since 1980, whereas only 5 of the 80 worst examples of economic contraction over the last 40 years have occurred in democracies."[56] In a nutshell: "poor democracies outperform authoritarian countries because their institutions enable power to be shared and because they encourage openness and adaptability. ... An integral virtue of democracies, therefore, is that they provide a sphere of private space, which, protected by law, nurtures inventiveness, independent action, and civic activity. ... Democracies are open: they spur the flow of information. ... The free flow of ideas, every bit as much as the flow of goods, fosters efficient, customized, and effective policies."[57]

Index structure and formulation

In constructing the Index, we have tried to strike a balance between reasonably broad coverage of those factors which affect the capacity for innovation, on the one hand, and a certain degree of economy, on the other, as there is, in principle, a potentially large number of variables which could conceivably have a bearing on a nation's ability to innovate. Once these factors had been identified, an early priority was to organize them in a sensible way, bringing similar variables—for instance, those pertaining to a country's human capital endowment—under one category or pillar. Obviously, there is no unique way to do this, nor is there a "magic" number of pillars that may be used.

---

[54] To compare like with like; it makes no sense to compare, for instance, high-income democracies with poor autocracies.

[55] Indeed, the true gap is probably larger, because the data excludes figures for Cuba, North Korea, and Somalia, among the worst-performing authoritarian regimes.

[56] Siegle et al., 2004, p. 60.

[57] Siegle et al., pp. 63–64.

We feel comfortable with the following formulation which identifies five pillars:

1. Institutional environment
2. Human capital, training and social inclusion
3. Regulatory and legal framework
4. Research and development
5. Adoption and use of information and communication technologies

which, a priori, theoretical, or empirical considerations might suggest are relevant. This was the case, for instance, with knowledge of the English language. English being the most widely used language of science and technology, global finance, and the Internet, common sense would suggest that, other things being equal, knowledge of English would have a tangible impact on boosting a nation's capacity to innovate. But there appear to be no data on English literacy for the large number of countries

Figure 1. **The Innovation Capacity Index**

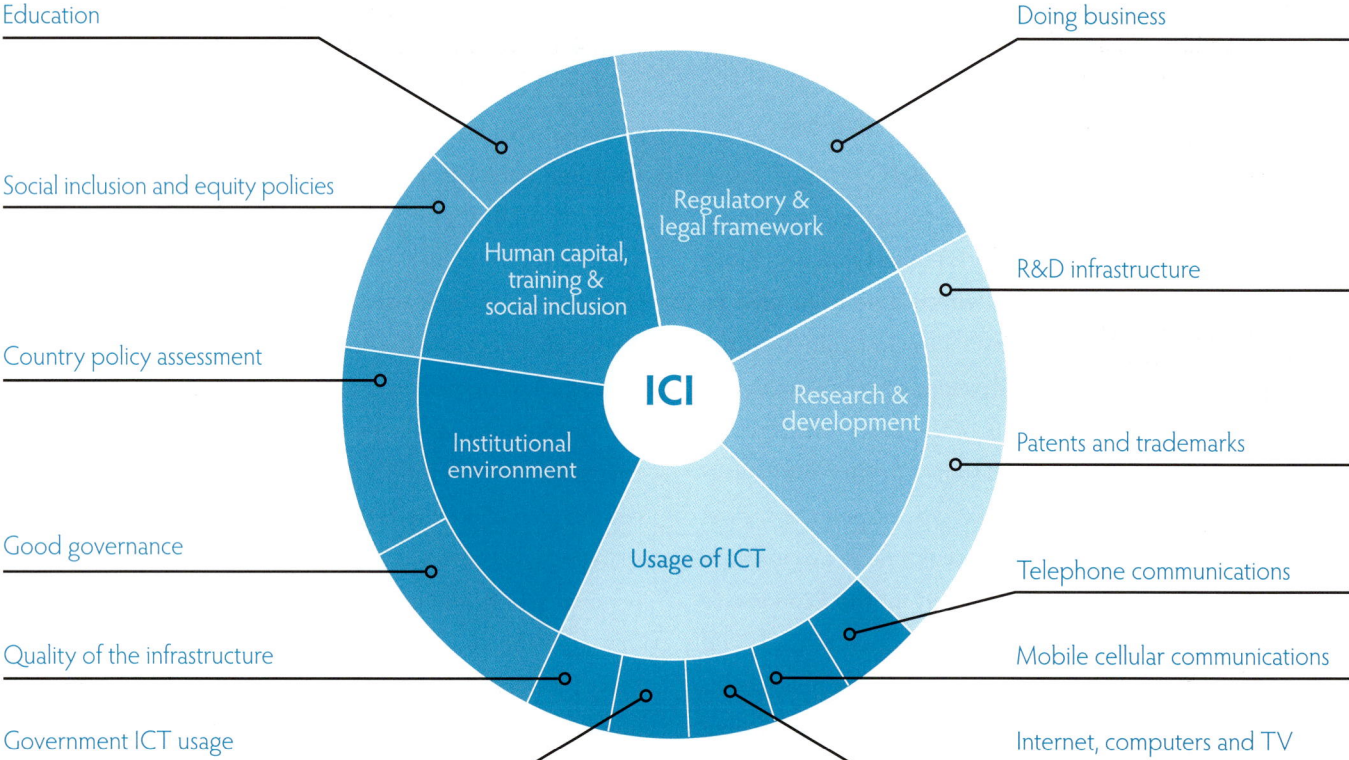

A more detailed representation can be seen in Figure 1 and in Box 1.

The choice of pillars and variables is based on the theoretical and empirical considerations discussed in detail in Section 2. It is worthwhile at this point to make several additional remarks to cast some light on some methodological issues which arose in the construction of the ICI.

Missing variables

One constraint faced by researchers in the construction of such indexes is the lack of reliable or internationally comparable data. The absence of data may prevent the inclusion of some variables

that figure in this study. However, since these omissions were mostly exceptional, we were not greatly hampered by lack of data, a fact partly to be attributed to the progress that has been made over the past decade in quantifying a growing number of previously "soft" variables.[58]

Data sources

Because a key virtue of an index is its ability to make meaningful international comparisons, we have gone to sources which compile the data on a comparable basis, using a common methodology. These include: the International Telecommunication Union, which provides the most up-to-date

---

58 See the Technical Note (at: www.innovationfordevelopmentreport.org) which addresses the issue of how we deal with missing data for individual indicators for a small set of countries. The Technical Note also touches upon other data issues, including normalization, weighing and aggregation, and sensitivity analysis.

Box 1. Structure of the Innovation Capacity Index (ICI)

The ICI is built upon five pillars composed of a total of 61 variables. For synthetic purposes only, the variables are grouped into conceptual subsections, which may be thought of as subindexes. The ICI ranks countries according to their overall performance and also provides scores by pillars and subindexes which give a general idea of performance in those areas. Variable definitions are presented in the Appendix.

1st Pillar: Institutional environment

A. Good governance
- 1.01 Voice and accountability
- 1.02 Political stability
- 1.03 Government effectiveness
- 1.04 Rule of law
- 1.05 Property rights framework
- 1.06 Transparency and judicial independence
- 1.07 *Corruption Perceptions Index* (TI)

B. Country policy assessment
1. Public sector management
   - 1.08 Quality of budgetary and financial management
   - 1.09 Quality of public administration
2. Structural policies
   - 1.10 Financial sector efficiency
   - 1.11 Trade openness
   - 1.12 Foreign direct investment gross inflows (as % of GDP)
3. Macroeconomy
   - 1.13 Debt levels
   - 1.14 Fiscal balance
   - 1.15 Macro stability

2nd Pillar: Human capital, training and social inclusion

A. Education
- 2.01 Adult literacy rate (% aged 15 and older)
- 2.02 Secondary gross enrolment ratio (%)
- 2.03 Tertiary gross enrolment ratio (%)
- 2.04 Expenditure in education (as % of GDP)

B. Social inclusion and equity policies
- 2.05 Gender Equity
- 2.06 Environmental sustainability
- 2.07 Health worker density
- 2.08 Inequality measure: ratio of richest 20% to poorest 20%

3rd Pillar: Regulatory and legal framework

A. Doing business
1. Starting a business
   - 3.01 Number of procedures
   - 3.02 Time (days)
   - 3.03 Cost (as % of income per capita)
2. Ease of employing workers
   - 3.04 Ease of employing workers
3. Paying taxes
   - 3.05 Paying taxes
4. Protecting investors
   - 3.06 Strength of investor protection
5. Registering property
   - 3.07 Number of procedures
   - 3.08 Time (days)
   - 3.09 Cost (as % of property value)

4th Pillar: Research and development

A. R&D infrastructure
- 4.01 Research and development expenditure (as % of GDP)
- 4.02 Information and communication technology expenditure (as % of GDP)
- 4.03 R&D worker density
- 4.04 Students in science and engineering (as % of tertiary students)
- 4.05 Scientific and technical journal articles (per million people)
- 4.06 Schools connected to the internet (%)

B. Patents and trademarks
- 4.07 Patents granted to residents (per million people)
- 4.08 Trademark applications filed by residents (per million people)
- 4.09 Receipts of royalty and license fees (US$ per person)
- 4.10 Payments of royalty and license fees (US$ per person)

5th Pillar: Adoption and use of information and communication technologies

A. Telephone communications
- 5.01 Main (fixed) telephone lines per 100 inhabitants
- 5.02 Waiting list for main (fixed) lines per 1000 inhabitants
- 5.03 Business connection charge (as % of GDP/capita)
- 5.04 Business monthly subscription (as % of GDP/capita)
- 5.05 Residential connection charge (as % of GDP/capita)
- 5.06 Residential monthly subscription (as % of GDP/capita)

B. Mobile cellular communications
- 5.07 Subscribers per 100 inhabitants
- 5.08 Prepaid subscribers per 100 inhabitants
- 5.09 Population coverage (%)
- 5.10 Connection charge (as % of GDP/capita)

C. Internet, computers and TV
- 5.11 Total fixed internet subscribers per 100 inhabitants
- 5.12 Total fixed broadband subscribers per 100 inhabitants
- 5.13 Internet users per 100 inhabitants
- 5.14 Personal computers per 100 inhabitants
- 5.15 Television receivers per 100 inhabitants

D. Government ICT usage
- 5.16 E-government readiness index

E. Quality of the infrastructure
- 5.17 Electrification rate (%)
- 5.18 Electric power transmission and distribution losses (as % of output)
- 5.19 Roads paved (as % of total roads)

and complete database of ICT and telecommunication statistics;[59] the World Bank's *World Development Indicators* (WDI), which makes available data on some 800 indicators covering different dimensions of economic and social development; [60] the World Bank/International Finance Corporation's *Doing Business Report* (DBR), which contains objective measures of business regulations and their enforcement across 181 economies;[61] the United Nations Development Programme's Human Development Report (HDR), with its ample database on critical issues for human development worldwide;[62] and the World Economic Outlook (WEO), the main instrument for the IMF's global surveillance activities,[63] among others.

Country categories

For operational and analytical purposes, countries were divided into two different categories by income level and political system, according to the following criteria:

Income levels: Gross National Income (GNI) per capita based on the World Bank 2007 country classifications:[64]
High-income: GNI per capita > $11,456
Upper-middle-income: GNI per capita: $3,706 – 11,455
Lower-middle-income: GNI per capita: $936 – 3,705
Low-income: GNI per capita < $935

Average incomes per capita for each country grouping are shown in Table 3.

Political systems: The Economist Intelligence Unit's *Index of Democracy* 2008 [65] analyzes electoral process and pluralism, prevalence of civil liberties, the functioning of government, issues of political participation, and political culture, and classifies countries as:
Full democracies: scores 8–10
Flawed democracies: scores 6–7.9
Hybrid regimes: scores 4–5.9
Authoritarian regimes: scores < 4

The 131 countries included in the ICI may thus be presented as shown in Table 4.

Weights

We have given considerable thought to the issue of how to weight the five pillars of the Index across the 131 countries. In choosing the weights, our starting point has been the theoretical considerations put forward by Rostow (1960) and Porter (1990, as highlighted in the section above), which we find intuitively appealing and in conformity with extensive empirical observation over the post-World War II period, particularly in the context of the work carried out by organizations such as the World Bank and the International Monetary Fund. Such work suggests that the relative importance of factors affecting innovation will be a function of a country's stage of development. Countries in earlier stages—Rostow called them "traditional societies" but, as in Porter, we may think of them as countries with relatively under-developed institutions and human capital, which act as constraints on the level of attainable output per capita—will need to prioritize those areas which are essential prerequisites for the next stage.[66] Thus, before it can join the group of nations doing innovation, a low-income country in sub-Saharan Africa will need to focus reform efforts and resources in developing the institutional infrastructure and in building up its human resource endowments. At the other end of the development spectrum, an innovator such as Sweden—already endowed with efficiently working institutions and with a highly skilled labor force—will have to focus its energies on improving those factors which more directly sustain and further boost an established capacity for innovation, for example, ensuring that the system of higher education is able to provide training immediately relevant for industry, or ensuring that the government makes further improvements in the regulatory environment and provides the incentives that underpin the creation of new businesses.[67] An alternative way to see this is to say that those pillars which more fundamentally have to do with people, institutions, and

---

[59] International Telecommunication Union (ITU), available at: http://www.itu.int

[60] World Bank, 2008b, available at: http://www.worldbank.org

[61] World Bank, 2008a, available at: http://www.doingbusiness.org

[62] United Nations Development Programme (UNDP), available at: http://www.undp.org

[63] International Monetary Fund (IMF), 2009a, available at: http://www.imf.org

[64] Available at: http://www.worldbank.org

[65] The Economist Intelligence Unit's *Index of Democracy*, available at: http://www.eiu.com

[66] This is how Rostow (1960) expressed it: "The second stage of growth embraces societies in the process of transition; that is, the period when the preconditions for take-off are developed; for it takes time to transform a traditional society in the ways necessary for it to exploit the fruits of modern science, to fend off diminishing returns, and thus to enjoy the blessings and choices opened up by the march of compound interest" (p. 6).

Table 4. ICI Country clusters according to income level and political regime

High-income: GNI per capita > US$11,456			
Full democracies	**Flawed democracies**	**Hybrid regimes**	**Authoritarian regimes**
Australia Korea, Republic of Austria Luxembourg Belgium Malta Canada Netherlands Czech Republic New Zealand Denmark Norway Finland Portugal France Slovenia, Republic of Germany Spain Greece Sweden Iceland Switzerland Ireland United Kingdom Italy United States Japan	Cyprus Estonia, Republic of Hungary Israel Slovak Republic Taiwan Trinidad and Tobago	Hong Kong SAR Singapore	Bahrain, Kingdom of Kuwait Oman Qatar Saudi Arabia United Arab Emirates

Upper-middle-income: GNI per capita: US$3,706–US$11,455			
Full democracies	**Flawed democracies**	**Hybrid regimes**	**Authoritarian regimes**
Costa Rica Mauritius Uruguay	Argentina Lithuania, Republic of Belize Malaysia Botswana Mexico Brazil Panama Bulgaria Poland Chile Romania Croatia, Republic of South Africa Jamaica Suriname Latvia, Republic of	Lebanon Russian Federation Turkey Venezuela	Kazakhstan, Republic of

Lower-middle-income: GNI per capita: US$936–US$3,705			
Full democracies	**Flawed democracies**	**Hybrid regimes**	**Authoritarian regimes**
	Bolivia Namibia Colombia Nicaragua Dominican Republic Paraguay El Salvador Peru Guatemala Philippines Honduras Sri Lanka India Thailand Indonesia Ukraine Macedonia, FYR	Bosnia and Herzegovina Ecuador Georgia Iraq	Algeria Angola Azerbaijan, Republic of Cameroon China, People's Republic of Congo, Republic of Egypt, Arab Republic of Iran, Islamic Republic of Jordan Morocco Sudan Syrian Arab Republic Tunisia

Low-income: GNI per capita < US$935			
Full democracies	**Flawed democracies**	**Hybrid regimes**	**Authoritarian regimes**
	Papua New Guinea	Bangladesh Mozambique, Cambodia Republic of Ethiopia Nepal Ghana Pakistan Haiti Senegal Kenya Tanzania Madagascar Uganda Malawi Zambia Mali	Afghanistan, Islamic Niger Republic of Nigeria Chad Rwanda Côte d'Ivoire Togo Guinea Vietnam Lao PDR Yemen, Republic of Mauritania Zimbabwe

social networks (pillars 1 and 2) are seen as the foundations for the pillars which deal with means and other enabling factors (pillars 3, 4, and 5). Innovation would be the last frontier, provided that the foundations of governance and human resources are well on their way to being broadly secured.

These theoretical considerations have been further complemented by extensive data analysis which is described in greater detail in a Technical Note (available at www.innovationfordevelopmentreport.org). Nevertheless, it is useful to provide here the gist of that analysis, which largely corroborates the above observations derived from the work of Rostow and Porter. A first step was to determine the influence of the three country categories chosen (income levels, type of political regime, and geographical location[68]) on the raw index scores. This was achieved in two stages: first, we obtained a set of raw pillar and index scores without imposing any prior organizational principle on the data with respect to a country's level of income, its political regime, or its geographical location; second, we used statistical techniques developed by Pavlidis and Noble (2001) to create a template for a correlation analysis with respect to numerical values assigned to each category;[69] that is, income levels were given a number from 1 to 4, from lowest to highest income, and political regimes from 1 to 4, from least democratic to most democratic, and so on, thus generating three category data sets. In this way the raw index and pillar scores were used as templates and compared with the category data, in order to find if there was a correlation between the different categories and scores. Only those correlations with p-values equal or lower than 0.05 were deemed significant.[70] According to these tests (see Figure 2), the two main categories with the greatest influence on the index and pillar scores were income levels followed by political regime. In the age of globalization, geographic location appears to play a role of declining importance. This created

16 possible country clusters based on four income categories and four different types of political regime (Table 4). The final weight allocation is shown in Table 5.

5. Innovation Capacity Index rankings 2009–2010

The results for this year's rankings for the 131 countries covered by the Innovation Capacity Index are presented in Table 6. Table 7 presents a more detailed version of the results, identifying individual pillar scores and ranks and the corresponding scores and ranks for the subindexes that make up the various pillar components, such as "good governance" and "country policy assessment" for pillar 1, on a country's institutional environment. Table 8, on the other hand, present Index ranks and scores for the various country clusters, depending on each country's income per capita (e.g., stage of development) and political regime. This Table is useful, as it addresses the occasional criticism against rankings involving a relatively large number of countries, namely, that they force comparisons between markedly different sets of countries, possibly at very different stages of development or having other important structural differences. From this Table one can see, for instance, that although Jordan has a rank of 44 in the ICI, it is first among lower-middle-income authoritarian regimes, ahead of Tunisia and China. Likewise, Ghana's rank of 77 among all 131 countries highlights a large number of weaknesses across all the pillars of the ICI, but the country does much better when the comparator group includes only low-income countries with either a hybrid or an authoritarian regime.

While these tables provide a good overview of the main results, we direct the attention of the reader to the inno-

---

[67] Again, Rostow provides useful insights: "This is the stage in which an economy demonstrates that it has the technological and entrepreneurial skills to produce not everything, but anything that it chooses to produce. It may lack (like contemporary Sweden and Switzerland, for example) the raw materials or other supply conditions required to produce a given type of output economically; but its dependence is a matter of economic choice or political priority rather than a technological or institutional necessity" (Rostow, op. cit., p. 10).

[68] The choice of geographic location was not induced by any sense of geographic determinism, that is, the notion, as discussed by Diamond (1999), that differences across countries and cultures are largely determined by climate, fauna, and flora. Rather, the idea was in keeping with Diamond's sensible observation that "all human societies contain inventive people. It's just that some environments provide more starting materials, and more favorable conditions for utilizing inventions, than do other environments" (p. 408).

[69] Pavlidis and Noble, 2001. In this paper, the authors demonstrated the ease and feasibility of using this type of correlation analysis when dealing with large data sets, and applied in their case to array expression patterns of DNA. They note that the advantages of template matching (that is, using a set of data as a pattern in order to find correlations with other data sets) are that this feature selection method is simple, can be used to differentiate between any number of categories, and permits rankings according to different levels of differentiation. In fact, the large data set generated by our study was managed and analyzed with the aid of a free open-source DNA microarray analysis suite, the Multiexperiment Viewer, developed at the Institute for Genomic Research (TIGR) in California. For more information see: Saeed et al., 2003. Available at: http://www.tm4.org/mev.html

[70] The p-value determines to what extent the different correlations obtained were due to chance. It is a probability value that varies from 0 to 1. A significance level of 0.05 indicates that the there is only a 5 percent probability that the correlation value was determined purely by chance.

vation profiles contained in Part 3 of the Report, which provide additional information on individual country performance. Part 3 includes profiles for a total of 68 countries, with the remaining 63 innovation profiles available at: www.innovationfordevelopmentreport.org

To highlight the type of analysis which is made possible through the Innovation Capacity Index we discuss here this year's results for Sweden, Chile, India, Russia, and Taiwan. These countries are interesting for a variety of reasons: Sweden, because it is this year's top-ranked nation and provides an impressive benchmark against which to assess other countries' performance. Chile is not only the highest-ranked country in Latin America, but is far ahead (20 places) from the next best performer in the region, Uruguay. What are the factors that account for this significantly better performance, which puts Chile at levels above the EU average? India is arguably the country with one of the highest potentials to become a leading center for technological innovation. Yet, it is a country whose innovation potential is saddled by major shortcomings in education and human capital accumulation, inadequate infrastructure, mind-boggling levels of bureaucracy and red

tape, an unreformed budget, and correspondingly high levels of public debt. Russia is a country with an impressive human capital endowment which during the time of the Soviet Union had made substantial inroads in such areas as space exploration, nuclear power, and basic sciences. Yet today, it is operating well below its capacity, largely confined to acquiring advanced technologies from abroad, and not providing any homegrown innovations. What are the challenges which are now preventing the development of its latent innovation capacities and what is needed for a better interaction between private sector strategies and public sector policies that will release the country's untapped potential? Taiwan, as noted earlier, has made impressive progress over the past decades in transforming itself into a leading player in the ICT industry, and the ingredients for its success are well worth studying, as key components of Taiwan's strategy have international relevance.

Figure 2. Correlation coefficients (R in %) of the different country category groups with respect to raw index and pillar scores*

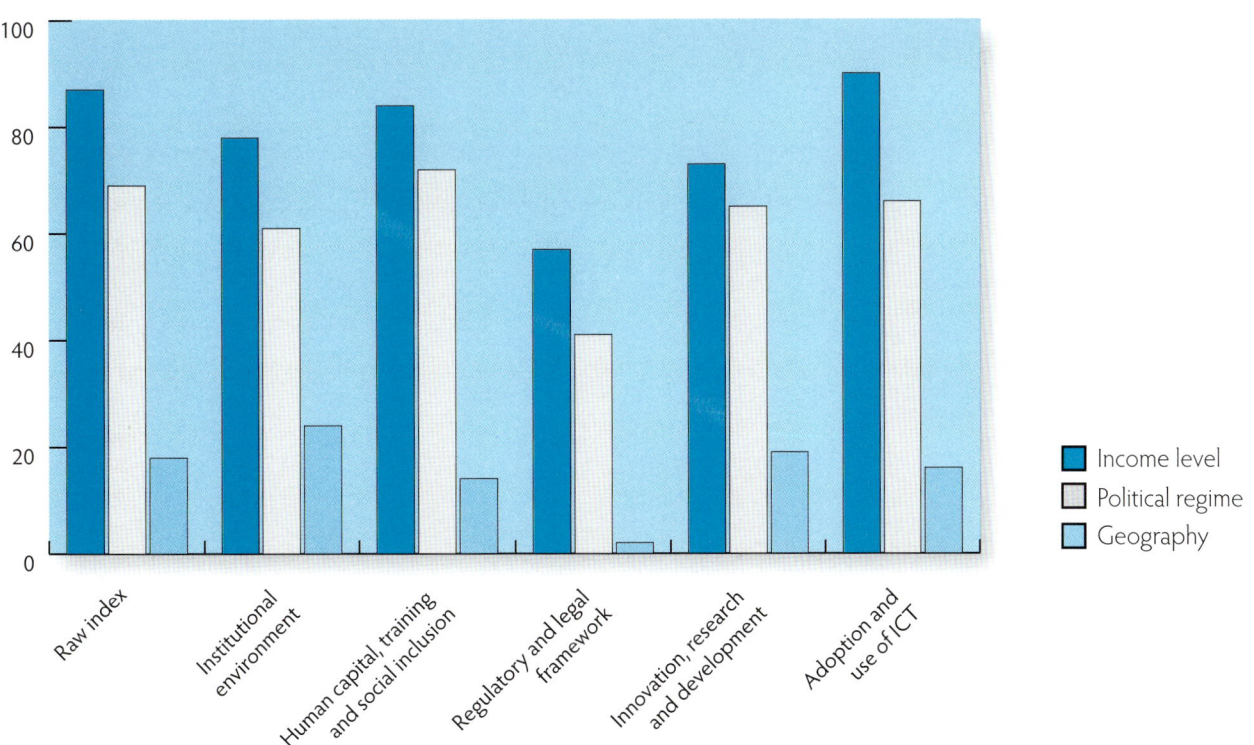

Income level
Political regime
Geography

* Pillars 2, 3 and 5 with respect to geography showed p-values above 0.05. These were 0.12, 0.07 and 0.85 respectively.

Table 5. **Weighting of pillars in the Innovation Capacity Index (in percent)**

	High-income: GNI per capita > US$11,456			
	Full democracies	Flawed democracies	Hybrid regimes	Authoritarian regimes
Institutional environment	10	15	20	20
Human capital, training and social inclusion	10	15	20	20
Regulatory and legal framework	20	20	20	20
Research and development	30	25	20	20
Adoption and use of ICT	30	25	20	20
Total	100	100	100	100

	Upper-middle-income: GNI per capita: US$3,706–US$11,455			
	Full democracies	Flawed democracies	Hybrid regimes	Authoritarian regimes
Institutional environment	25	25	25	25
Human capital, training and social inclusion	25	25	25	25
Regulatory and legal framework	20	20	20	20
Research and development	15	15	15	15
Adoption and use of ICT	15	15	15	15
Total	100	100	100	100

	Lower-middle-income: GNI per capita: US$936–US$3,705			
	Full democracies	Flawed democracies	Hybrid regimes	Authoritarian regimes
Institutional environment	-	30	30	30
Human capital, training and social inclusion	-	30	30	30
Regulatory and legal framework	-	20	20	20
Research and development	-	10	10	10
Adoption and use of ICT	-	10	10	10
Total	-	100	100	100

	Low-income: GNI per capita < US$935			
	Full democracies	Flawed democracies	Hybrid regimes	Authoritarian regimes
Institutional environment	-	30	30	30
Human capital, training and social inclusion	-	30	30	30
Regulatory and legal framework	-	20	20	20
Research and development	-	10	10	10
Adoption and use of ICT	-	10	10	10
Total	-	100	100	100

Table 6. Innovation Capacity Index rankings 2009–2010*

Country	ICI rank	ICI score	Country	ICI rank	ICI score	Country	ICI rank	ICI score
Sweden	1	82.2	South Africa	46	53.3	Ecuador	91	44.2
Finland	2	77.8	Macedonia, FYR	47	53.1	Tanzania	92	43.7
United States	3	77.5	Romania	47	53.1	Nicaragua	93	43.4
Switzerland	4	77.0	Uruguay	49	52.8	Madagascar	93	43.4
Netherlands	5	76.6	Russian Federation	49	52.8	Morocco	95	43.3
Singapore	6	76.5	Mauritius	49	52.8	Kenya	95	43.3
Canada	7	74.8	Malta	52	52.4	Pakistan	97	42.7
United Kingdom	8	74.6	Cyprus	53	52.3	Belize	98	42.1
Norway	9	73.5	Ukraine	54	52.0	Zambia	99	41.8
New Zealand	10	73.4	Saudi Arabia	55	51.9	Bolivia	100	41.5
Luxembourg	11	73.3	Tunisia	56	51.8	Papua New Guinea	101	41.3
Denmark	11	73.3	Kazakhstan, Republic of	57	51.6	Venezuela	102	40.9
Taiwan	13	72.9	Costa Rica	58	51.5	Nepal	103	40.3
Iceland	14	72.6	Turkey	59	50.8	Nigeria	104	40.2
Japan	15	72.1	Peru	60	50.6	Suriname	105	40.1
Hong Kong SAR	16	71.3	Mexico	61	50.5	Bangladesh	106	39.8
Australia	17	71.2	Oman	62	50.2	Syrian Arab Republic	107	39.4
Ireland	18	70.5	Greece	62	50.2	Mozambique, Republic of	108	39.1
Korea, Republic of	19	70.0	Kuwait	64	50.1	Uganda	109	38.3
Germany	20	68.8	China, People's Republic of	65	49.5	Cameroon	109	38.3
Israel	21	68.2	Argentina	66	49.2	Senegal	111	38.1
Belgium	22	67.6	Botswana	67	49.1	Cambodia	112	37.5
Austria	23	66.7	Panama	68	48.9	Malawi	112	37.5
France	24	65.4	Trinidad and Tobago	69	48.7	Ethiopia	114	37.3
Estonia, Republic of	25	62.7	Bosnia and Herzegovina	70	48.3	Mauritania	115	37.1
Lithuania, Republic of	26	60.7	El Salvador	70	48.3	Lao PDR	116	36.8
Latvia, Republic of	27	60.5	Colombia	72	48.0	Yemen, Republic of	117	35.1
Spain	28	60.3	Namibia	73	47.5	Sudan	118	35.0
Chile	29	59.4	Azerbaijan, Republic of	74	47.3	Iraq	119	34.2
Italy	30	59.1	Philippines	75	47.0	Mali	120	33.8
Slovenia, Republic of	31	58.6	Algeria	76	46.7	Angola	121	33.4
Czech Republic	32	58.0	Ghana	77	46.6	Rwanda	122	33.3
Bulgaria	33	57.7	Vietnam	78	46.4	Congo, Republic of	123	33.0
Malaysia	34	57.3	Dominican Republic	79	46.3	Côte d'Ivoire	124	32.4
Portugal	35	57.2	Egypt, Arab Republic of	79	46.3	Zimbabwe	125	31.8
Bahrain, Kingdom of	36	56.6	Jamaica	81	46.2	Niger	126	30.6
United Arab Emirates	37	56.2	Honduras	82	46.0	Togo	127	30.1
Croatia, Republic of	38	56.0	Lebanon	83	45.8	Guinea	128	29.1
Slovak Republic	39	55.8	Iran, Islamic Republic of	84	45.7	Haiti	129	28.7
Poland	40	55.7	India	85	45.6	Chad	130	25.6
Hungary	41	55.6	Sri Lanka	86	45.5	Afghanistan, Islamic Republic of	131	24.0
Georgia	42	55.1	Brazil	87	45.2			
Thailand	43	54.6	Indonesia	88	44.9			
Jordan	44	53.9	Guatemala	89	44.5			
Qatar	45	53.8	Paraguay	90	44.3			

*All rankings and scores are after rounding.

Table 7. Innovation Capacity Index 2009–2019: Pillar rankings*

COUNTRY	Pillar 1: Institutional environment						Pillar 2: Human capital, training, and social inclusion			
	Pillar		Good governance		Country policy assessment		Pillar		Education	
	RANKING	SCORE	RANKING	SCORE	RANKING	SCORE	RANKING	SCORE	RANKING	SCORE
Afghanistan, Islamic Republic of	127	26.6	130	14.3	116	40.3	131	12.7	129	16.1
Algeria	70	46.5	102	33.0	30	59.9	82	51.4	66	59.5
Angola	104	38.2	121	24.5	58	53.5	128	27.8	121	26.9
Argentina	104	38.2	81	38.8	125	37.6	41	63.6	30	70.4
Australia	11	79.4	8	88.7	12	70.2	9	79.9	11	79.1
Austria	15	71.7	13	86.6	45	56.7	22	74.0	33	68.9
Azerbaijan, Republic of	96	40.9	112	28.5	59	53.3	58	59.0	67	59.4
Bahrain, Kingdom of	27	64.1	49	55.5	9	72.7	42	62.9	45	67.3
Bangladesh	111	37.0	115	27.8	95	46.3	104	41.6	107	36.8
Belgium	20	69.9	18	79.2	26	60.5	8	81.7	18	75.7
Belize	95	41.2	66	43.5	123	38.1	91	47.7	92	48.1
Bolivia	85	43.2	92	35.5	77	51.0	88	49.7	51	65.5
Bosnia and Herzegovina	68	47.1	82	38.3	40	57.0	47	61.0	79	56.2
Botswana	26	65.3	34	63.9	18	66.7	93	46.9	91	48.9
Brazil	81	44.5	65	43.6	97	45.4	71	53.0	59	62.6
Bulgaria	47	53.7	60	46.2	24	61.1	35	67.8	32	69.0
Cambodia	115	36.2	119	26.6	96	45.8	112	39.7	111	33.3
Cameroon	100	40.1	111	28.7	72	51.4	111	40.0	109	35.2
Canada	14	74.3	11	88.1	26	60.5	11	79.1	12	77.5
Chad	125	30.1	127	16.5	100	45.1	129	20.8	131	14.4
Chile	19	70.2	25	71.0	14	69.5	63	56.5	47	67.0
China, People's Republic of	64	49.1	77	40.5	37	57.7	87	50.3	89	50.6
Colombia	93	41.4	78	40.1	105	42.8	76	51.9	60	61.9
Congo, Republic of	116	35.2	121	24.5	93	47.2	110	40.1	106	37.5
Costa Rica	44	56.7	40	59.7	55	53.7	52	60.4	72	58.3
Côte d'Ivoire	126	29.0	126	18.7	120	39.3	118	35.7	114	31.1
Croatia, Republic of	50	52.1	52	52.6	70	51.6	32	68.2	36	68.6
Cyprus	28	64.0	24	71.2	41	56.9	43	62.7	48	66.6
Czech Republic	43	56.8	37	61.5	66	52.1	25	71.9	28	71.1
Denmark	4	83.7	1	93.6	5	73.7	5	83.3	9	79.6
Dominican Republic	99	40.3	78	40.1	115	40.5	64	56.0	74	57.3
Ecuador	108	37.6	117	27.6	91	47.6	73	52.7	88	50.7
Egypt, Arab Republic of	106	37.9	88	36.1	118	39.7	75	52.3	68	58.7
El Salvador	74	45.8	63	44.2	92	47.4	78	51.8	62	60.9
Estonia, Republic of	16	70.6	22	71.7	14	69.5	18	75.7	3	83.6
Ethiopia	119	34.7	105	29.6	118	39.7	109	40.3	122	26.2
Finland	6	81.6	5	92.6	11	70.7	3	83.9	7	80.1
France	25	65.4	21	75.7	52	55.2	20	75.3	29	70.5
Georgia	50	52.1	68	42.7	21	62.5	50	60.5	37	68.5
Germany	17	70.5	14	84.7	48	56.4	14	77.0	38	68.4
Ghana	61	49.8	55	50.3	82	49.4	99	43.8	100	41.6
Greece	60	49.9	46	57.5	107	42.4	15	76.6	2	84.2
Guatemala	89	42.3	91	35.6	86	49.0	97	45.1	108	36.3
Guinea	130	23.9	125	19.0	130	28.8	122	31.6	127	21.1
Haiti	124	31.0	124	20.9	109	42.1	122	31.6	93	47.5
Honduras	76	45.5	89	35.8	52	55.2	81	51.5	84	53.3
Hong Kong SAR	3	84.4	16	83.0	1	85.7	46	61.4	55	64.0
Hungary	58	51.0	38	61.1	112	41.0	30	69.5	31	70.0
Iceland	1	85.6	6	92.3	3	78.9	2	86.7	1	90.6
India	72	46.3	64	43.8	89	48.8	94	45.9	99	41.9
Indonesia	81	44.5	100	33.3	50	55.8	85	50.9	81	55.1
Iran, Islamic Republic of	101	39.8	114	28.2	63	52.7	86	50.4	78	56.5
Iraq	129	24.9	131	13.7	120	39.3	113	39.4	96	45.1
Ireland	13	75.5	15	84.1	18	66.7	12	78.0	27	72.3
Israel	37	58.3	35	63.0	57	53.6	36	67.3	52	65.1
Italy	69	47.0	47	56.3	124	37.8	19	75.5	16	76.5
Jamaica	110	37.5	61	44.6	129	29.6	72	52.9	86	51.5
Japan	35	59.2	20	77.6	114	40.7	29	70.6	22	73.8
Jordan	48	53.6	50	54.2	61	52.9	50	60.5	61	61.2
Kazakhstan, Republic of	66	47.8	98	34.0	22	61.6	40	64.5	42	67.8
Kenya	98	40.8	103	31.5	78	50.1	98	44.9	97	44.2
Korea, Republic of	31	62.9	30	65.7	29	60.1	33	68.1	5	81.4
Kuwait	53	51.6	53	51.7	71	51.5	62	56.6	82	54.5
Lao PDR	120	34.1	118	27.2	110	41.9	100	43.1	111	33.3
Latvia, Republic of	35	59.2	42	59.4	34	58.9	28	70.8	24	73.0
Lebanon	123	32.7	101	33.2	128	32.1	55	60.0	39	68.0

Table 7. Innovation Capacity Index 2009–2019: Pillar rankings* (cont'd.)

| | Pillar 1: Institutional environment | | | | | | Pillar 2: Human capital, training, and social inclusion | | | |
| | Pillar | | Good governance | | Country policy assessment | | Pillar | | Education | |
COUNTRY	RANKING	SCORE	RANKING	SCORE	RANKING	SCORE	RANKING	SCORE	RANKING	SCORE
Lithuania, Republic of	37	58.3	43	58.8	36	57.8	24	72.9	12	77.5
Luxembourg	2	84.5	9	88.3	2	80.7	21	74.8	17	75.8
Macedonia, FYR	73	46.1	70	42.5	78	50.1	44	62.0	73	58.1
Madagascar	57	51.1	67	42.9	33	59.3	108	40.5	102	40.2
Malawi	111	37.0	84	37.5	126	36.5	117	36.9	116	30.0
Malaysia	37	58.3	45	58.0	35	58.7	68	54.7	76	56.9
Mali	85	43.2	74	41.8	101	44.7	126	30.1	126	21.7
Malta	23	66.0	23	71.4	26	60.5	49	60.6	39	68.0
Mauritania	103	38.8	95	34.8	105	42.8	119	35.4	123	25.4
Mauritius	46	56.1	33	64.2	90	47.9	69	53.4	77	56.6
Mexico	75	45.6	71	42.3	87	48.9	61	57.0	70	58.5
Morocco	92	41.5	72	42.2	113	40.9	105	41.5	104	39.3
Mozambique, Republic of	80	44.6	86	37.1	66	52.1	127	29.8	128	20.5
Namibia	49	52.9	48	55.8	80	50.0	92	47.2	90	50.5
Nepal	113	36.8	106	29.5	102	44.2	102	41.7	115	30.1
Netherlands	12	77.2	7	89.3	20	65.1	6	82.0	19	75.2
New Zealand	5	82.1	4	92.7	10	71.4	10	79.2	9	79.6
Nicaragua	90	42.2	95	34.8	82	49.4	84	51.0	93	47.5
Niger	84	43.6	94	34.9	64	52.4	130	20.3	130	15.0
Nigeria	87	42.8	109	28.9	46	56.6	107	40.8	101	40.6
Norway	9	80.6	12	87.6	6	73.5	1	88.9	4	82.4
Oman	34	59.9	44	58.6	23	61.3	66	55.4	43	67.5
Pakistan	108	37.6	120	26.1	84	49.1	106	40.9	113	32.4
Panama	53	51.6	57	46.9	49	56.3	70	53.1	68	58.7
Papua New Guinea	93	41.4	104	30.9	68	51.9	114	38.1	118	29.9
Paraguay	114	36.7	108	29.4	104	44.0	89	49.1	49	66.4
Peru	77	45.4	80	39.0	69	51.8	57	59.9	41	67.9
Philippines	102	39.5	99	33.9	99	45.2	48	60.9	56	63.7
Poland	53	51.6	51	54.0	84	49.1	39	66.1	46	67.2
Portugal	33	60.2	27	69.2	75	51.2	27	70.9	35	68.8
Qatar	21	69.6	32	64.4	4	74.8	90	49.0	71	58.4
Romania	64	49.1	59	46.8	73	51.3	55	60.0	58	63.2
Russian Federation	79	45.2	106	29.5	25	60.8	37	66.7	43	67.5
Rwanda	96	40.9	84	37.5	102	44.2	124	31.5	124	22.6
Saudi Arabia	42	57.6	72	42.2	8	72.9	67	55.0	15	77.0
Senegal	67	47.3	75	41.7	62	52.8	120	34.5	125	22.1
Singapore	8	80.8	9	88.3	7	73.4	33	68.1	33	68.9
Slovak Republic	44	56.7	39	59.8	55	53.7	31	68.8	50	65.6
Slovenia, Republic of	30	63.6	26	69.9	39	57.2	26	71.0	26	72.9
South Africa	37	58.3	40	59.7	41	56.9	78	51.8	63	60.5
Spain	28	64.0	29	68.2	31	59.8	13	77.1	22	73.8
Sri Lanka	107	37.8	87	36.6	122	39.1	73	52.7	65	59.9
Sudan	128	25.3	128	16.1	127	34.4	116	37.0	103	40.0
Suriname	71	46.4	62	44.3	87	48.9	76	51.9	80	56.1
Sweden	6	81.6	2	93.4	13	69.7	4	83.4	21	74.3
Switzerland	9	80.6	3	93.3	17	67.8	7	81.9	14	77.3
Syrian Arab Republic	118	34.8	110	28.8	111	41.5	96	45.3	87	51.3
Taiwan	32	60.7	31	64.8	47	56.5	23	73.9	6	81.0
Tanzania	63	49.2	76	40.9	38	57.4	95	45.5	116	30.0
Thailand	62	49.7	69	42.6	43	56.8	59	58.5	54	64.6
Togo	122	34.0	115	27.8	117	40.2	125	31.4	119	29.1
Trinidad and Tobago	52	51.9	57	46.9	43	56.8	45	61.8	83	53.7
Tunisia	56	51.3	54	51.2	73	51.3	65	55.9	84	53.3
Turkey	59	50.7	56	47.6	54	53.9	82	51.4	75	57.0
Uganda	88	42.7	97	34.2	76	51.1	101	42.7	120	27.6
Ukraine	83	44.4	89	35.8	60	53.0	38	66.5	20	75.1
United Arab Emirates	24	65.5	36	61.8	16	69.2	53	60.3	64	60.4
United Kingdom	17	70.5	17	81.3	32	59.7	15	76.6	24	73.0
United States	22	66.6	19	77.8	51	55.5	17	76.5	8	80.0
Uruguay	41	57.9	28	68.7	94	47.1	54	60.2	53	65.0
Venezuela	120	34.1	123	22.7	97	45.4	60	57.2	57	63.5
Vietnam	91	42.1	93	35.1	81	49.8	80	51.7	95	45.4
Yemen, Republic of	117	35.0	113	28.4	107	42.4	121	33.6	105	38.2
Zambia	78	45.3	82	38.3	64	52.4	115	37.6	110	33.9
Zimbabwe	131	14.2	129	16.0	131	12.2	102	41.7	98	42.3

Table 7. Innovation Capacity Index 2009–2019: Pillar rankings* (cont'd.)

COUNTRY	Pillar 2: Human capital, training, and social inclusion		Pillar 3: Regulatory and legal framework			
	Social inclusion and equity policies		Pillar		Doing business	
	RANKING	SCORE	RANKING	SCORE	RANKING	SCORE
Afghanistan, Islamic Republic of	131	2.4	114	50.6	114	50.6
Algeria	91	46.1	99	57.6	99	57.6
Angola	127	28.6	124	44.0	124	44.0
Argentina	49	59.0	88	61.3	88	61.3
Australia	11	80.3	10	81.4	10	81.4
Austria	17	77.3	49	69.3	49	69.3
Azerbaijan, Republic of	51	58.8	76	63.4	76	63.4
Bahrain, Kingdom of	52	58.4	13	80.3	13	80.3
Bangladesh	100	44.8	94	59.1	94	59.1
Belgium	6	85.6	26	74.4	26	74.4
Belize	87	47.4	71	64.5	71	64.5
Bolivia	115	39.2	119	46.6	119	46.6
Bosnia and Herzegovina	38	65.7	109	54.4	109	54.4
Botswana	95	45.5	60	66.8	60	66.8
Brazil	88	46.6	114	50.6	114	50.6
Bulgaria	36	66.9	38	71.4	38	71.4
Cambodia	101	44.1	103	57.1	103	57.1
Cameroon	102	43.2	118	49.1	118	49.1
Canada	12	80.1	3	88.8	3	88.8
Chad	128	26.0	125	41.8	125	41.8
Chile	77	49.6	23	75.4	23	75.4
China, People's Republic of	75	50.1	58	67.3	58	67.3
Colombia	98	45.2	56	67.6	56	67.6
Congo, Republic of	105	42.2	129	39.3	129	39.3
Costa Rica	42	61.8	93	59.3	93	59.3
Côte d'Ivoire	116	38.8	117	49.9	117	49.9
Croatia, Republic of	33	68.0	82	62.2	82	62.2
Cyprus	46	59.6	ND	ND	ND	ND
Czech Republic	21	72.5	54	68.0	54	68.0
Denmark	5	85.7	9	81.7	9	81.7
Dominican Republic	61	55.1	80	62.9	80	62.9
Ecuador	64	54.0	96	58.2	96	58.2
Egypt, Arab Republic of	83	48.0	68	65.8	68	65.8
El Salvador	93	45.7	69	65.6	69	65.6
Estonia, Republic of	25	70.4	18	77.3	18	77.3
Ethiopia	76	49.7	84	62.1	84	62.1
Finland	4	86.4	19	77.2	19	77.2
France	16	78.4	64	66.3	64	66.3
Georgia	60	55.3	12	80.5	12	80.5
Germany	9	82.8	44	70.4	44	70.4
Ghana	96	45.3	42	70.5	42	70.5
Greece	23	71.5	110	54.1	110	54.1
Guatemala	72	50.9	66	66.1	66	66.1
Guinea	117	38.6	120	46.2	120	46.2
Haiti	130	21.0	128	40.9	128	40.9
Honduras	74	50.4	100	57.5	100	57.5
Hong Kong SAR	47	59.3	4	88.4	4	88.4
Hungary	29	69.2	57	67.4	57	67.4
Iceland	8	84.3	16	78.7	16	78.7
India	81	48.6	79	63.1	79	63.1
Indonesia	82	48.1	96	58.2	96	58.2
Iran, Islamic Republic of	90	46.4	95	58.9	95	58.9
Iraq	125	30.9	87	61.6	87	61.6
Ireland	10	81.8	7	83.8	7	83.8
Israel	31	68.7	21	76.7	21	76.7
Italy	18	74.8	40	70.6	40	70.6
Jamaica	65	53.7	45	70.3	45	70.3
Japan	32	68.5	17	77.7	17	77.7
Jordan	44	60.1	80	62.9	80	62.9
Kazakhstan, Republic of	41	62.2	31	73.8	31	73.8
Kenya	96	45.3	77	63.3	77	63.3
Korea, Republic of	48	59.2	53	68.3	53	68.3
Kuwait	53	58.3	30	73.9	30	73.9
Lao PDR	77	49.6	111	51.3	111	51.3
Latvia, Republic of	28	69.4	32	73.6	32	73.6
Lebanon	66	53.6	67	66.0	67	66.0

Table 7. Innovation Capacity Index 2009–2019: Pillar rankings* (cont'd.)

| | Pillar 2: Human capital, training, and social inclusion | | Pillar 3: Regulatory and legal framework | | | |
| | Social inclusion and equity policies | | Pillar | | Doing business | |
COUNTRY	RANKING	SCORE	RANKING	SCORE	RANKING	SCORE
Lithuania, Republic of	26	69.8	29	74.1	29	74.1
Luxembourg	19	74.3	71	64.5	71	64.5
Macedonia, FYR	40	64.6	61	66.6	61	66.6
Madagascar	111	40.8	73	64.4	73	64.4
Malawi	107	41.5	92	60.4	92	60.4
Malaysia	68	53.2	14	80.1	14	80.1
Mali	121	35.7	113	51.0	113	51.0
Malta	68	53.2	ND	ND	ND	ND
Mauritania	106	42.1	102	57.2	102	57.2
Mauritius	73	50.8	25	75.0	25	75.0
Mexico	57	55.9	47	69.9	47	69.9
Morocco	103	42.9	89	61.1	89	61.1
Mozambique, Republic of	123	34.5	65	66.2	65	66.2
Namibia	99	45.0	78	63.2	78	63.2
Nepal	79	49.4	52	68.4	52	68.4
Netherlands	3	86.6	28	74.2	28	74.2
New Zealand	14	78.9	1	96.2	1	96.2
Nicaragua	67	53.3	91	60.5	91	60.5
Niger	129	23.9	111	51.3	111	51.3
Nigeria	109	40.9	105	56.2	105	56.2
Norway	1	93.2	8	81.9	8	81.9
Oman	93	45.7	24	75.3	24	75.3
Pakistan	88	46.6	50	69.1	50	69.1
Panama	79	49.4	74	64.2	74	64.2
Papua New Guinea	109	40.9	35	72.8	35	72.8
Paraguay	118	37.6	63	66.4	63	66.4
Peru	62	54.5	46	70.1	46	70.1
Philippines	50	58.9	96	58.2	96	58.2
Poland	39	65.3	59	66.9	59	66.9
Portugal	22	72.3	34	73.1	34	73.1
Qatar	114	39.6	27	74.3	27	74.3
Romania	54	57.8	69	65.6	69	65.6
Russian Federation	37	66.1	48	69.8	48	69.8
Rwanda	119	37.5	121	46.1	121	46.1
Saudi Arabia	120	37.4	15	79.4	15	79.4
Senegal	104	42.8	123	45.1	123	45.1
Singapore	34	67.4	2	89.9	2	89.9
Slovak Republic	24	70.9	36	72.5	36	72.5
Slovenia, Republic of	27	69.7	85	62.0	85	62.0
South Africa	92	46.0	21	76.7	21	76.7
Spain	13	79.2	61	66.6	61	66.6
Sri Lanka	84	47.8	55	67.8	55	67.8
Sudan	122	34.6	75	63.5	75	63.5
Suriname	84	47.8	126	41.2	126	41.2
Sweden	2	89.5	11	80.8	11	80.8
Switzerland	7	85.0	37	72.2	37	72.2
Syrian Arab Republic	112	40.5	106	56.1	106	56.1
Taiwan	29	69.2	39	71.1	39	71.1
Tanzania	59	55.8	101	57.3	101	57.3
Thailand	63	54.4	20	77.1	20	77.1
Togo	124	33.3	126	41.2	126	41.2
Trinidad and Tobago	35	67.3	40	70.6	40	70.6
Tunisia	55	57.7	82	62.2	82	62.2
Turkey	86	47.6	33	73.3	33	73.3
Uganda	71	52.8	114	50.6	114	50.6
Ukraine	43	60.8	108	55.4	108	55.4
United Arab Emirates	44	60.1	42	70.5	42	70.5
United Kingdom	14	78.9	5	87.3	5	87.3
United States	20	74.2	5	87.3	5	87.3
Uruguay	56	57.1	85	62.0	85	62.0
Venezuela	70	53.1	122	45.2	122	45.2
Vietnam	57	55.9	89	61.1	89	61.1
Yemen, Republic of	126	30.5	107	55.9	107	55.9
Zambia	113	40.0	51	68.5	51	68.5
Zimbabwe	108	41.4	104	56.3	104	56.3

Table 7. Innovation Capacity Index 2009–2019: Pillar rankings* (cont'd.)

| COUNTRY | Pillar 4: Research and development | | | | | | Pillar 5: Adoption and use of information and communication technologies | | | |
| | Pillar | | R&D infrastructure | | Patents and trademarks | | Pillar | | Telephone communications | |
	RANKING	SCORE	RANKING	SCORE	RANKING	SCORE	RANKING	SCORE	RANKING	SCORE
Afghanistan, Islamic Republic of	129	0.0	127	0.0	119	0.0	124	20.8	123	44.9
Algeria	79	10.8	84	18.0	88	0.7	76	47.6	77	77.8
Angola	62	15.2	73	22.5	95	0.4	101	33.1	81	76.3
Argentina	46	20.0	70	23.7	30	16.3	49	56.5	45	86.4
Australia	15	51.2	15	58.4	17	41.0	18	78.9	16	94.1
Austria	19	48.6	13	59.1	22	33.9	17	79.0	24	92.4
Azerbaijan, Republic of	113	2.0	113	3.8	85	0.9	83	44.3	87	74.6
Bahrain, Kingdom of	80	10.6	68	24.6	53	5.0	40	65.2	35	89.2
Bangladesh	65	13.8	46	30.9	106	0.2	112	30.2	115	54.5
Belgium	17	50.0	20	53.2	15	45.6	23	75.3	20	93.3
Belize	103	4.9	97	11.5	66	2.2	89	41.9	79	77.2
Bolivia	95	7.1	91	13.2	85	0.9	97	36.3	109	63.1
Bosnia and Herzegovina	117	1.0	121	0.3	78	1.2	73	49.3	74	78.2
Botswana	87	8.3	94	12.9	92	0.6	86	43.3	66	79.7
Brazil	53	17.8	62	27.2	54	4.7	60	53.4	64	81.0
Bulgaria	35	24.4	37	35.1	39	9.4	44	62.5	54	84.6
Cambodia	96	6.5	100	10.0	92	0.6	117	26.8	116	53.3
Cameroon	64	14.4	39	33.6	119	0.0	111	30.5	113	57.4
Canada	12	54.7	14	58.8	11	48.9	8	84.4	3	98.2
Chad	129	0.0	127	0.0	119	0.0	126	19.8	122	48.3
Chile	33	25.4	45	31.3	28	17.1	47	58.7	50	85.6
China, People's Republic of	55	16.9	54	29.5	56	4.4	79	45.5	114	55.1
Colombia	68	13.0	77	21.1	72	1.8	68	51.4	58	84.5
Congo, Republic of	102	5.7	112	4.1	119	0.0	125	20.0	128	31.5
Costa Rica	61	15.5	76	21.2	40	8.6	59	53.8	36	88.9
Côte d'Ivoire	125	0.2	121	0.3	112	0.1	113	29.8	111	59.5
Croatia, Republic of	39	22.8	40	33.4	43	7.9	35	66.9	28	91.1
Cyprus	37	23.8	53	29.6	29	16.9	33	67.5	30	90.6
Czech Republic	26	36.1	21	52.6	36	13.0	30	68.9	40	87.4
Denmark	23	45.9	11	65.6	26	18.4	3	88.2	12	95.2
Dominican Republic	116	1.1	126	0.1	77	1.3	75	47.7	72	78.5
Ecuador	91	8.0	96	11.9	58	4.2	77	47.3	65	79.8
Egypt, Arab Republic of	72	12.7	72	23.0	95	0.4	74	47.9	71	79.0
El Salvador	92	7.7	95	12.7	64	2.8	69	51.2	46	86.3
Estonia, Republic of	32	27.8	31	41.0	38	11.9	25	73.2	34	89.9
Ethiopia	104	4.8	103	8.8	119	0.0	127	19.4	120	49.7
Finland	3	74.3	3	81.2	7	64.7	20	78.4	30	90.6
France	21	46.6	17	55.3	21	36.2	16	80.3	7	96.2
Georgia	89	8.2	86	17.0	78	1.2	85	44.0	93	70.8
Germany	14	52.0	12	59.8	16	42.6	13	81.4	2	98.4
Ghana	84	8.4	74	22.2	112	0.1	98	35.3	94	70.7
Greece	34	25.3	33	37.4	41	8.3	43	63.7	9	96.0
Guatemala	99	6.1	101	9.9	66	2.2	81	44.8	90	74.1
Guinea	80	10.6	59	28.3	119	0.0	123	21.0	117	52.6
Haiti	125	0.2	127	0.0	106	0.2	129	17.3	127	31.8
Honduras	55	16.9	43	32.0	72	1.8	96	36.6	102	67.3
Hong Kong SAR	24	40.0	30	42.1	20	37.1	11	82.4	10	95.7
Hungary	30	29.9	32	39.0	27	17.2	38	66.3	38	87.8
Iceland	15	51.2	5	70.3	24	28.2	14	81.1	4	98.1
India	69	12.9	71	23.1	88	0.7	93	40.5	73	78.4
Indonesia	107	4.5	105	7.8	83	1.1	88	42.0	84	76.0
Iran, Islamic Republic of	59	16.5	58	28.4	66	2.2	67	51.9	63	81.7
Iraq	120	0.4	121	0.3	95	0.4	118	25.9	131	6.0
Ireland	18	49.0	24	49.7	12	47.9	18	78.9	15	94.2
Israel	6	66.5	1	83.5	14	46.1	29	69.7	24	92.4
Italy	28	31.8	28	44.1	33	14.6	21	77.3	22	92.9
Jamaica	84	8.4	92	13.0	63	2.9	53	55.5	105	63.8
Japan	4	69.0	9	66.7	4	72.2	22	76.4	32	90.3
Jordan	54	17.5	55	29.1	78	1.2	61	53.0	80	77.1
Kazakhstan, Republic of	109	3.6	110	5.7	69	2.0	56	54.7	69	79.4
Kenya	57	16.8	27	44.4	102	0.3	107	32.2	108	63.3
Korea, Republic of	10	61.1	7	68.9	10	50.1	10	83.1	14	94.5
Kuwait	98	6.3	98	11.1	95	0.4	46	62.2	43	86.5
Lao PDR	111	3.0	106	7.5	112	0.1	110	31.1	92	71.3
Latvia, Republic of	40	21.1	41	32.5	44	7.4	34	67.3	52	84.9
Lebanon	73	12.2	59	28.3	106	0.2	70	50.2	99	68.0

Table 7. Innovation Capacity Index 2009–2019: Pillar rankings* (cont'd.)

COUNTRY	Pillar 4: Research and development						Pillar 5: Adoption and use of information and communication technologies			
	Pillar		R&D infrastructure		Patents and trademarks		Pillar		Telephone communications	
	RANKING	SCORE	RANKING	SCORE	RANKING	SCORE	RANKING	SCORE	RANKING	SCORE
Lithuania, Republic of	42	20.3	42	32.4	50	5.8	36	66.8	47	86.2
Luxembourg	8	62.1	26	45.1	2	83.3	7	86.2	8	96.1
Macedonia, FYR	51	18.1	47	30.7	62	3.0	52	55.6	76	77.9
Madagascar	96	6.5	90	14.4	102	0.3	121	23.2	118	51.1
Malawi	122	0.3	118	0.7	106	0.2	108	31.8	91	71.4
Malaysia	40	21.1	49	30.3	41	8.3	39	65.8	54	84.6
Mali	129	0.0	127	0.0	119	0.0	130	15.9	125	34.4
Malta	31	28.9	61	28.2	23	29.9	31	68.7	17	93.4
Mauritania	112	2.5	106	7.5	119	0.0	109	31.3	96	68.9
Mauritius	75	11.5	85	17.8	59	3.9	48	58.2	39	87.5
Mexico	47	19.5	50	30.2	54	4.7	62	52.6	60	83.0
Morocco	57	16.8	44	31.8	72	1.8	80	45.2	98	68.2
Mozambique, Republic of	82	10.5	87	16.5	95	0.4	119	24.7	124	43.2
Namibia	101	5.8	89	14.5	88	0.7	87	42.1	69	79.4
Nepal	119	0.5	116	0.8	95	0.4	114	29.7	104	64.1
Netherlands	11	60.2	19	53.9	5	71.1	1	92.6	17	93.4
New Zealand	22	46.1	22	52.0	18	37.9	15	80.8	27	91.8
Nicaragua	115	1.4	115	0.9	70	1.9	105	32.5	121	49.6
Niger	127	0.1	125	0.2	119	0.0	131	11.1	129	24.3
Nigeria	106	4.6	99	10.4	106	0.2	100	34.0	99	68.0
Norway	20	47.1	18	54.1	19	37.4	6	86.9	21	93.1
Oman	87	8.3	64	26.9	85	0.9	65	52.0	54	84.6
Pakistan	74	12.1	75	21.8	94	0.5	91	41.2	81	76.3
Panama	59	16.5	80	19.5	37	12.9	72	49.4	59	83.2
Papua New Guinea	120	0.4	116	0.8	112	0.1	116	28.2	86	74.7
Paraguay	84	8.4	114	1.4	34	14.0	81	44.8	85	75.5
Peru	100	6.0	104	8.0	61	3.7	84	44.1	88	74.4
Philippines	76	11.3	82	18.4	78	1.2	92	41.1	112	58.4
Poland	36	24.0	34	36.2	46	6.9	45	62.3	43	86.5
Portugal	29	30.7	29	43.1	35	13.3	32	67.7	33	90.1
Qatar	77	11.1	79	20.5	72	1.8	41	65.0	37	88.7
Romania	47	19.5	56	29.0	48	6.2	41	65.0	42	86.6
Russian Federation	52	18.0	52	29.7	60	3.8	55	54.8	78	77.7
Rwanda	127	0.1	121	0.3	119	0.0	120	23.9	119	51.0
Saudi Arabia	77	11.1	57	28.9	95	0.4	50	56.4	51	85.3
Senegal	70	12.8	67	25.6	112	0.1	104	32.6	103	65.9
Singapore	9	62.0	8	67.6	9	55.2	12	81.7	17	93.4
Slovak Republic	38	23.6	36	35.5	47	6.8	37	66.4	47	86.2
Slovenia, Republic of	27	35.7	23	50.2	31	15.5	24	73.4	26	92.0
South Africa	43	20.2	47	30.7	52	5.4	71	49.6	68	79.5
Spain	25	36.7	25	49.2	25	19.1	26	73.0	23	92.5
Sri Lanka	94	7.4	92	13.0	72	1.8	93	40.5	110	62.0
Sudan	114	1.6	111	4.2	119	0.0	99	34.3	83	76.1
Suriname	89	8.2	69	24.1	70	1.9	95	40.2	75	78.0
Sweden	2	75.6	2	82.4	6	66.1	2	89.6	6	97.2
Switzerland	7	66.2	6	69.1	8	61.2	5	88.0	1	99.4
Syrian Arab Republic	122	0.3	120	0.4	102	0.3	90	41.4	97	68.3
Taiwan	1	82.7	4	76.9	1	100.0	27	71.3	5	97.6
Tanzania	83	8.6	51	30.1	119	0.0	115	29.1	107	63.6
Thailand	70	12.8	83	18.2	48	6.2	57	54.2	66	79.7
Togo	108	4.4	108	7.2	112	0.1	128	18.2	130	20.5
Trinidad and Tobago	63	14.8	62	27.2	65	2.3	54	55.3	52	84.9
Tunisia	43	20.2	38	33.8	83	1.1	65	52.0	54	84.6
Turkey	50	18.4	66	26.4	45	7.1	64	52.2	49	86.1
Uganda	110	3.5	109	7.0	119	0.0	122	22.6	126	34.1
Ukraine	43	20.2	34	36.2	57	4.3	51	55.8	89	74.3
United Arab Emirates	67	13.6	64	26.9	102	0.3	28	70.9	29	90.9
United Kingdom	13	53.2	16	57.3	13	47.4	3	88.2	13	94.8
United States	5	68.8	10	66.3	3	72.4	9	83.4	11	95.3
Uruguay	49	18.5	78	21.0	31	15.5	58	54.0	41	86.7
Venezuela	93	7.5	102	9.5	51	5.6	63	52.5	62	82.1
Vietnam	66	13.7	81	19.0	78	1.2	78	46.9	61	82.8
Yemen, Republic of	118	0.6	127	0.0	88	0.7	103	32.8	95	70.0
Zambia	122	0.3	119	0.6	106	0.2	106	32.3	101	67.8
Zimbabwe	105	4.7	88	16.3	112	0.1	102	32.9	105	63.8

Table 7. Innovation Capacity Index 2009–2019: Pillar rankings* (cont'd.)

	Pillar 5: Adoption and use of information and communication technologies							
	Mobile cellular communications		Internet, computers, and TV		Government ICT usage		Quality of the infrastructure	
COUNTRY	RANKING	SCORE	RANKING	SCORE	RANKING	SCORE	RANKING	SCORE
Afghanistan, Islamic Republic of	125	26.8	117	1.7	122	20.5	117	23.7
Algeria	56	76.8	92	6.8	91	35.2	45	82.8
Angola	119	36.5	112	2.2	96	33.3	111	34.4
Argentina	39	84.7	57	20.0	39	58.4	66	67.2
Australia	44	83.2	17	66.7	8	81.1	55	77.1
Austria	32	86.0	21	62.0	16	74.3	6	98.3
Azerbaijan, Republic of	80	63.4	84	9.9	77	46.1	65	67.4
Bahrain, Kingdom of	1	99.9	47	27.6	42	57.2	26	91.3
Bangladesh	106	46.9	111	2.3	104	29.4	102	43.7
Belgium	41	84.4	24	56.5	24	67.8	24	91.4
Belize	91	56.4	67	16.5	84	41.0	121	17.0
Bolivia	100	49.8	102	4.1	66	48.7	93	48.7
Bosnia and Herzegovina	68	73.4	62	17.9	80	45.1	74	62.1
Botswana	58	76.6	99	4.3	89	36.5	92	49.8
Brazil	69	73.0	54	23.5	45	56.8	79	58.8
Bulgaria	15	90.6	44	30.5	42	57.2	23	91.6
Cambodia	99	50.7	129	0.4	103	29.9	125	13.1
Cameroon	107	46.6	113	1.9	110	27.3	101	44.1
Canada	78	64.6	3	87.2	7	81.7	55	77.1
Chad	121	32.3	129	0.4	129	10.5	130	0.8
Chile	45	82.3	49	25.9	40	58.2	61	72.3
China, People's Republic of	89	58.0	66	16.9	61	50.2	29	91.0
Colombia	66	74.9	67	16.5	50	53.2	85	56.5
Congo, Republic of	103	47.9	118	1.4	109	27.4	128	8.3
Costa Rica	94	53.1	48	26.2	55	51.4	63	69.9
Côte d'Ivoire	105	47.0	116	1.8	125	18.5	103	43.0
Croatia, Republic of	11	91.6	35	40.3	46	56.5	49	80.8
Cyprus	17	90.2	32	41.5	35	60.2	50	80.6
Czech Republic	12	91.4	33	41.1	25	67.0	11	96.8
Denmark	70	72.6	4	84.1	2	91.3	2	99.0
Dominican Republic	74	69.5	83	11.0	64	49.4	67	65.8
Ecuador	54	77.6	74	13.2	68	48.4	104	42.8
Egypt, Arab Republic of	83	60.6	85	9.3	71	47.7	42	84.8
El Salvador	27	87.8	87	8.9	63	49.7	75	60.2
Estonia, Republic of	24	88.1	22	58.8	13	76.0	90	53.7
Ethiopia	131	2.9	128	0.5	124	18.6	107	38.0
Finland	59	76.4	14	69.3	15	74.9	35	87.4
France	59	76.4	19	65.3	9	80.4	7	98.2
Georgia	65	75.0	79	12.0	78	46.0	83	57.6
Germany	29	86.4	19	65.3	22	71.4	4	98.7
Ghana	97	51.7	113	1.9	101	30.0	94	48.3
Greece	25	87.9	53	23.8	42	57.2	19	93.1
Guatemala	71	71.6	93	6.4	81	42.8	64	67.9
Guinea	123	27.7	126	0.6	127	14.0	127	9.8
Haiti	130	8.0	97	4.8	120	21.0	113	31.0
Honduras	96	52.5	100	4.2	86	40.5	98	47.2
Hong Kong SAR	19	89.8	15	68.5	ND	ND	30	90.6
Hungary	21	89.1	30	42.8	30	64.9	70	64.1
Iceland	30	86.2	10	71.8	21	71.8	54	77.4
India	110	45.6	85	9.3	87	38.1	89	53.8
Indonesia	88	58.1	96	5.0	83	41.1	70	64.1
Iran, Islamic Republic of	90	57.9	50	24.3	85	40.7	52	78.2
Iraq	93	53.9	119	1.3	112	26.9	33	88.9
Ireland	8	95.0	23	57.6	19	73.0	13	96.6
Israel	48	81.9	38	37.9	17	73.9	2	99.0
Italy	1	99.9	26	53.7	26	66.8	8	97.3
Jamaica	16	90.4	45	28.9	75	46.8	48	81.5
Japan	77	67.2	18	66.0	11	77.0	24	91.4
Jordan	56	76.8	74	13.2	48	54.8	19	93.1
Kazakhstan, Republic of	79	63.8	64	17.4	72	47.4	41	85.1
Kenya	109	46.0	100	4.2	92	34.7	112	33.3
Korea, Republic of	12	91.4	13	69.7	6	83.2	22	91.8
Kuwait	18	89.9	46	27.9	54	52.0	31	89.8
Lao PDR	108	46.2	113	1.9	115	23.8	123	14.4
Latvia, Republic of	25	87.9	29	44.7	36	59.4	37	86.5
Lebanon	85	59.9	55	22.4	68	48.4	36	86.8

Table 7. Innovation Capacity Index 2009–2019: Pillar rankings* (cont'd.)

| | Pillar 5: Adoption and use of information and communication technologies | | | | | | | |
| | Mobile cellular communications | | Internet, computers, and TV | | Government ICT usage | | Quality of the infrastructure | |
COUNTRY	RANKING	SCORE	RANKING	SCORE	RANKING	SCORE	RANKING	SCORE
Lithuania, Republic of	4	97.7	39	36.5	28	66.2	44	83.7
Luxembourg	9	94.1	9	75.2	14	75.1	1	99.8
Macedonia, FYR	52	79.1	43	30.9	66	48.7	73	62.8
Madagascar	122	31.1	122	0.9	99	30.7	124	13.3
Malawi	104	47.6	125	0.7	108	28.8	115	26.0
Malaysia	47	82.2	37	38.7	34	60.6	21	92.2
Mali	128	12.4	124	0.8	126	15.9	100	44.6
Malta	23	88.2	34	40.7	29	65.8	39	85.6
Mauritania	95	52.9	110	2.6	123	20.3	126	11.3
Mauritius	55	77.0	50	24.3	59	50.9	10	97.0
Mexico	61	76.2	59	19.0	37	58.9	86	55.8
Morocco	64	75.3	77	12.5	104	29.4	60	72.8
Mozambique, Republic of	118	41.4	121	1.1	113	25.6	110	35.6
Namibia	86	59.7	76	13.0	95	34.5	106	39.1
Nepal	124	27.4	122	0.9	110	27.3	91	52.7
Netherlands	30	86.2	1	95.9	5	86.3	17	95.6
New Zealand	28	86.6	11	71.6	17	73.9	39	85.6
Nicaragua	92	55.8	103	4.0	88	36.7	95	48.2
Niger	127	13.2	131	0.3	128	11.4	119	20.6
Nigeria	102	48.2	103	4.0	100	30.6	105	40.3
Norway	49	81.6	5	83.4	3	89.2	32	89.7
Oman	40	84.5	71	14.9	74	46.9	75	60.2
Pakistan	81	62.3	94	5.6	97	31.6	78	59.3
Panama	67	74.4	81	11.3	73	47.2	68	64.9
Papua New Guinea	126	24.6	106	3.6	121	20.8	129	3.5
Paraguay	84	60.4	90	7.6	76	46.5	53	77.8
Peru	116	41.6	63	17.7	53	52.5	81	58.0
Philippines	72	69.8	89	8.2	62	50.0	82	57.8
Poland	22	88.9	42	31.0	33	61.3	51	78.6
Portugal	5	97.3	40	32.6	31	64.8	26	91.3
Qatar	6	96.7	40	32.6	51	53.1	43	84.3
Romania	34	85.3	28	46.8	49	53.8	80	58.4
Russian Federation	12	91.4	58	19.9	56	51.2	57	75.1
Rwanda	120	32.5	126	0.6	104	29.4	120	19.0
Saudi Arabia	20	89.2	56	20.3	64	49.4	62	70.7
Senegal	101	49.2	105	3.9	114	25.3	108	36.8
Singapore	33	85.6	12	70.4	23	70.1	5	98.6
Slovak Republic	37	84.9	31	42.0	37	58.9	28	91.2
Slovenia, Republic of	51	79.4	25	55.2	26	66.8	12	96.7
South Africa	45	82.3	82	11.1	56	51.2	75	60.2
Spain	35	85.1	27	47.2	20	72.3	15	95.8
Sri Lanka	82	62.2	98	4.6	82	42.4	58	74.5
Sudan	129	11.9	73	13.9	117	21.9	97	47.3
Suriname	76	68.5	87	8.9	92	34.7	114	26.3
Sweden	36	85.0	2	91.0	1	91.6	59	74.2
Switzerland	43	83.4	6	83.3	12	76.3	9	97.2
Syrian Arab Republic	87	59.4	78	12.1	90	36.1	84	56.8
Taiwan	63	75.6	16	67.7	ND	ND	ND	ND
Tanzania	113	43.2	120	1.2	107	29.3	116	24.7
Thailand	53	78.3	72	14.1	60	50.3	15	95.8
Togo	116	41.6	106	3.6	117	21.9	118	22.3
Trinidad and Tobago	62	75.7	60	18.5	51	53.1	47	81.6
Tunisia	42	83.7	79	12.0	94	34.6	46	82.3
Turkey	50	81.2	65	17.1	70	48.3	87	55.6
Uganda	111	45.4	108	3.3	98	31.3	122	16.0
Ukraine	7	95.4	61	18.1	41	57.3	34	88.7
United Arab Emirates	3	98.6	36	40.2	32	63.0	18	94.5
United Kingdom	10	93.0	7	81.9	10	78.7	13	96.6
United States	73	69.6	8	79.8	4	86.4	38	86.3
Uruguay	74	69.5	52	24.0	46	56.5	88	55.4
Venezuela	37	84.9	70	15.4	58	51.0	72	63.2
Vietnam	98	51.1	69	15.6	79	45.6	69	64.5
Yemen, Republic of	112	44.8	94	5.6	119	21.4	109	35.7
Zambia	115	42.8	109	2.7	116	22.7	99	45.8
Zimbabwe	114	42.9	91	7.2	101	30.0	95	48.2

Table 8. Innovation Capacity Index 2009–2010: Country clusters: Index scores and rankings*

High-income: GNI per capita > US$11,456

Full democracies	Within group rank	Overall ICI rank	ICI score		Czech Republic	24	32	58.0
Sweden	1	1	82.2		Portugal	25	35	57.2
Finland	2	2	77.8		Malta	26	52	52.4
United States	3	3	77.5		Greece	27	62	50.2
Switzerland	4	4	77.0		**Flawed democracies**	**Within group rank**	**Overall ICI rank**	**ICI score**
Netherlands	5	5	76.6		Taiwan	1	13	72.9
Canada	6	7	74.8		Israel	2	21	68.2
United Kingdom	7	8	74.6		Estonia	3	25	62.7
Norway	8	9	73.5		Slovak Republic	4	39	55.8
New Zealand	9	10	73.4		Hungary	5	41	55.6
Luxembourg	10	11	73.3		Cyprus	6	53	52.3
Denmark	10	11	73.3		Trinidad and Tobago	7	69	48.7
Iceland	12	14	72.6		**Hybrid regimes**	**Within group rank**	**Overall ICI rank**	**ICI score**
Japan	13	15	72.1		Singapore	1	6	76.5
Australia	14	17	71.2		Hong Kong SAR	2	16	71.3
Ireland	15	18	70.5		**Authoritarian regimes**	**Within group rank**	**Overall ICI rank**	**ICI score**
Korea, Republic of	16	19	70.0		Bahrain, Kingdom of	1	36	56.6
Germany	17	20	68.8		United Arab Emirates	2	37	56.2
Belgium	18	22	67.6		Qatar	3	45	53.8
Austria	19	23	66.7		Saudi Arabia	4	55	51.9
France	20	24	65.4		Oman	5	62	50.2
Spain	21	28	60.3		Kuwait	6	64	50.1
Italy	22	30	59.1					
Slovenia, Republic of	23	31	58.6					

Upper-middle-income: GNI per capita: US$3,706–US$11,455

Full democracies	Within group rank	Overall ICI rank	ICI score		Flawed democracies	Within group rank	Overall ICI rank	ICI score
Uruguay	1	49	52.8		Lithuania, Republic of	1	26	60.7
Mauritius	2	49	52.8		Latvia, Republic of	2	27	60.5
Costa Rica	3	58	51.5		Chile	3	29	59.4
Hybrid regimes	**Within group rank**	**Overall ICI rank**	**ICI score**		Bulgaria	4	33	57.7
					Malaysia	5	34	57.3
Russian Federation	1	49	52.8		Croatia, Republic of	6	38	56.0
Turkey	2	59	50.8		Poland	7	40	55.7
Lebanon	3	83	45.8		South Africa	8	46	53.3
Venezuela	4	102	40.9		Romania	9	47	53.1
Authoritarian regimes	**Within group rank**	**Overall ICI rank**	**ICI score**		Mexico	10	61	50.5
					Argentina	11	66	49.2
Kazakhstan, Republic of	NA	57	51.6		Botswana	12	67	49.1
					Panama	13	68	48.9
					Jamaica	14	81	46.2
					Brazil	15	87	45.2
					Belize	16	98	42.1
					Suriname	17	105	40.1

*All rankings and scores are after rounding.

Lower-middle-income: GNI per capita: US$936–US$3,705

Flawed democracies	Within group rank	Overall ICI rank	ICI score	Hybrid regimes	Within group rank	Overall ICI rank	ICI score
Thailand	1	43	54.6	Georgia	1	42	55.1
Macedonia, FYR	2	47	53.1	Bosnia and Herzegovina	2	70	48.3
Ukraine	3	54	52.0				
Peru	4	60	50.6	Ecuador	3	91	44.2
El Salvador	5	70	48.3	Iraq	4	119	34.2
Colombia	6	72	48.0	Authoritarian regimes	Within group rank	Overall ICI rank	ICI score
Namibia	7	73	47.5	Jordan	1	44	53.9
Philippines	8	75	47.0	Tunisia	2	56	51.8
Dominican Republic	9	79	46.3	China, People's Republic of	3	65	49.5
Honduras	10	82	46.0				
India	11	85	45.6	Azerbaijan	4	74	47.3
Sri Lanka	12	86	45.5	Algeria	5	76	46.7
Indonesia	13	88	44.9	Egypt, Arab Republic of	6	79	46.3
Guatemala	14	89	44.5	Iran, Islamic Republic of	7	84	45.7
Paraguay	15	90	44.3	Morocco	8	95	43.3
Nicaragua	16	93	43.4	Syrian Arab Republic	9	107	39.4
Bolivia	17	100	41.5	Cameroon	10	109	38.3
				Sudan	11	118	35.0
				Angola	12	121	33.4
				Congo, Republic of	13	123	33.0

Low-income: GNI per capita < US$935

Flawed democracies	Within group rank	Overall ICI rank	ICI score	Hybrid regimes	Within group rank	Overall ICI rank	ICI score
Papua New Guinea	NA	101	41.3	Ghana	1	77	46.6
Authoritarian regimes	Within group rank	Overall ICI rank	ICI score	Tanzania	2	92	43.7
				Madagascar	3	93	43.4
Vietnam	1	78	46.4	Kenya	4	95	43.3
Nigeria	2	104	40.2	Pakistan	5	97	42.7
Mauritania	3	115	37.1	Zambia	6	99	41.8
Lao PDR	4	116	36.8	Nepal	7	103	40.3
Yemen, Republic of	5	117	35.1	Bangladesh	8	106	39.8
Rwanda	6	122	33.3	Mozambique, Republic of	9	108	39.1
Côte d'Ivoire	7	124	32.4	Uganda	10	109	38.3
Zimbabwe	8	125	31.8	Senegal	11	111	38.1
Niger	9	126	30.6	Cambodia	12	112	37.5
Togo	10	127	30.1	Malawi	12	112	37.5
Guinea	11	128	29.1	Ethiopia	14	114	37.3
Chad	12	130	25.6	Mali	15	120	33.8
Afghanistan, Islamic Republic of	13	131	24.0	Haiti	16	129	28.7

*All rankings and scores are after rounding.

Sweden: Why is its innovation outlook so bright?

An impressive performance

Sweden is the top ranked country in the 2009 edition of the Innovation Capacity Index because it does exceptionally well in all the areas captured by the Index. Figure 3 shows Sweden's relative performance with respect to other high-income countries in 10 of the indicators used in the estimation of the Index. As can be seen, Sweden is an exceptionally good performer, very often placing in the top ranks in those areas identified as being particularly important to assessing innovation capacity. Indeed, Sweden has a rank of number one among 131 countries in transparency and judicial independence, corruption perceptions, gender equity, e-government readiness, personal computer penetration rates, receipts of royalties and license fees, as well as the "doing business" indicators for the time and number of procedures required to register property. It has a rank of 2 in scientific and technical journal articles per capita, environmental sustainability, and research and development expenditure in relation to GDP, where it is second only to Israel. There are 12 other indicators in which Sweden has a top 8 rank, including the quality of its public administration, the effectiveness of its government, rule of law, the more egalitarian distribution of national income, Internet penetration rates, as well as other indicators of good governance. Table 9 shows Sweden's pillar ranks in the ICI.

Sweden's rank is richly deserved. It is a country that has had an extremely virtuous fiscal policy for the past decade, running budget surpluses with the aim of saving resources to deal with the long-term effects of population aging, but also generating, in the short-term, substantial resources to invest heavily in knowledge and training, to earn a top position in terms of labor productivity growth among high income countries. On a per capita basis, Sweden has the largest university system in the world. According to the OECD, "Swedish research is, in relation to the size of its population, leading in the world in terms of scientific output, measured by the number of publications in internationally acknowledged scientific journals."[71] Sweden is also a leader in terms of patent registration.

Openness and transparency

Sweden has in impressive record of openness and transparency in government. It has put in place comprehensive safety nets which provide security to vulnerable groups in the population. It has thus been able, during periods of economic stress—such as in the context of the 2008–09 world financial crisis—to shelter its population from the effects of the global economic slowdown. Since it also has levels of public debt that are well below those prevailing among competitor countries, Sweden has greater flexibility when it is time to provide fiscal stimulus.

Women in Sweden have access to a wider spectrum of educational, political, and work opportunities and enjoy a higher standard of living than women in other parts of the world. They also have achieved the highest echelons of political power and have an important presence in the business world. Sweden is also an egalitarian society with a more even income distribution than most countries in the OECD, and, thus, a strong sense of solidarity and stable labor relations. The country has also achieved an enviable record in terms of caring for the environment; it ranks second in the world in the Environmental Sustainability Index.

Sweden's public sector is highly qualified and enjoys unusually high degrees of credibility with the business community and civil society. Although the country has high tax rates, there is no evidence that this has discouraged entrepreneurship and innovation. More likely than not, this reflects the fact that the relatively high levels of revenue collection are then reinvested in the economy at large in education, infrastructure development and modernization, public health, and other components of the safety net, as well as training and other productivity-enhancing initiatives, all of which are directly beneficial to the private sector. Having an honest public administration—as demonstrated by Sweden's privileged and consistently high rankings in Transparency International's *Corruption Perceptions Index*—suggest that what matters is not whether tax rates are high or not, but rather whether the government uses the taxes collected in ways that will be productive and that will boost its credibility with economic agents.

A leader in ICT

The government has also played an important catalytic role in encouraging the use of the entire spectrum of information and communication technologies, as made clear by the very high penetration rates of mobile phones, computers, broadband, and the Internet. Not only does the government spend gener-

---

[71] OECD and European Communities, 2005, p. 189.

Figure 3. Sweden: Significant indicators above income group average

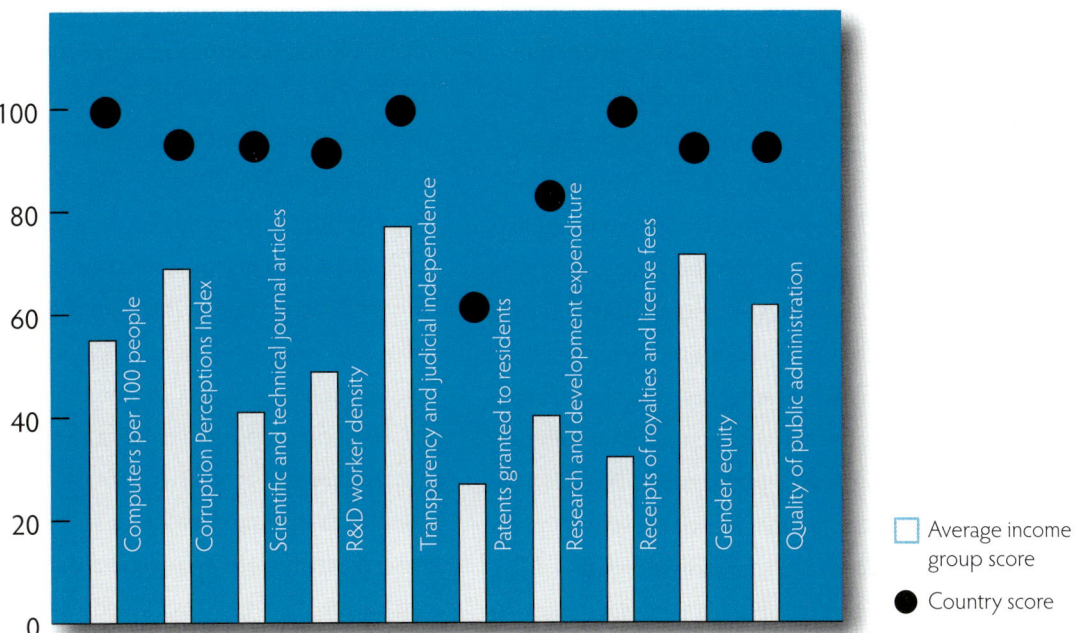

ously in research and development (particularly through institutions of higher education), but the Swedish business sector has also been a driving force in R&D spending, particularly in the telecommunications and pharmaceutical sectors. Sweden has benefited from an economy that, according to the OECD, is dominated by public-private partnerships between manufacturing groups that allocate considerable resources to R&D on the one hand, and public agencies and companies, on the other. This has led, in turn, to the emergence of a manufacturing sector that spans "all of the high-technology and medium high-technology industries" (OECD, 2005, p. 190).

A virtuous cycle of development

Sweden is likely to retain a privileged position in future editions of the Innovation Capacity Index. A combination of solid institutions, good policies and a public administration strongly committed to the idea of building upon past

achievements has pushed the country into what one might call a *virtuous cycle of development*. Successive governments have implemented policies whose primary motivation has been the public good. This in turn has transformed the business community and civil society into active, well informed participants in the shaping of public policies. Just as citizens and corporations pay their taxes because the benefits of doing so are tangible and transparent, governments have been empowered to focus their energies and talents in devising innovative ways to improve the quality of governance. Sweden and its Nordic neighbors provide a useful template for other countries to examine, and, where feasible, to emulate. There is much in their approach to development—combining key elements of modern capitalism without some of its excesses, with a strong commitment to social policies that are fundamentally egalitarian in nature—that is worthy of close examination and study.

Table 9. Sweden: ICI pillar rankings

	Rank	Score
Overall position	1	82.2
1. Institutional environment	6	81.6
2. Human capital, training, and social inclusion	4	83.4
3. Regulatory and legal framework	11	80.8
4. Research and development	2	75.6
5. Usage of information and communication technologies	2	89.6

Chile: Catching up with the top performers

The best innovation capacity in Latin America

With a rank of 29 among the 131 countries included in the ICI, Chile is by far the best performing country in Latin America. Indeed, it has a rank a full 20 places ahead of Uruguay (49), the next best performer (Table 10). As may be seen in Table 6, Chile is firmly positioned among 12 members of the European Union, with some slightly ahead (e.g., Belgium, Austria, France, and Spain), and others slightly behind (Italy, Slovenia, the Czech Republic, and Portugal). Chile has the highest rank among countries with a broadly similar level of income per capita, with only Malaysia (34) exhibiting a similar performance. As shown in Table 10, Chile has a rank of 1 in Latin America in several important indicators including government effectiveness, rule of law, absence of corruption, the fiscal balance (as a proxy indicator for the strength of macroeconomic policies), the number of schools connected to the Internet, the ease of paying taxes, broadband penetration rates, reliability of electricity generation, and a top 5 rank in a much larger set of indicators.

Chile's strong performance in the Innovation Capacity Index is the result of a combination of several factors, two of which have played a central role and are, therefore, desirable to highlight: first, the gradual build-up of an institutional environment that has been broadly supportive of private sector development; and second, the introduction of a range of policies that have explicitly sought to enhance the role of high technologies in promoting gains in factor productivity. It will be useful to present here a brief overview of both.

Chile ranks 23rd among 180 countries in Transparency International's *Corruption Perceptions Index 2008*, tied with France (23) and ahead of Spain (28), Portugal (32), and far ahead of Korea (40), Italy (55), Mexico (72), Brazil (80), and Argentina (109). In fact, the 22 countries with a better score than Chile are all high-income countries, as defined by the World Bank. In the ICI's own Good Governance subindex—which also includes measures of voice and accountability, political stability, government effectiveness, rule of law, the property rights framework, and transparency and judicial independence—and in the Country Policy Assessment subindex, which captures various measures of the quality of public sector policies, Chile ranks 25 and 14 respectively, out of 131 countries in 2009 (see Table 7).

Legitimizing market reforms

Market reforms in Chile have been legitimized in the eyes of the public because they have benefited the population in tangible ways, for instance, by increases in per capita income or, as noted earlier, sustained reductions in poverty levels. This contrasts sharply with other countries in the region, where the motivations for public policy have more often been a mixture of dubious ideology or some confusion about public ends and private benefits among the ruling elites. In addition, on those occasions when flaws in the public administration in Chile have emerged, the authorities' response has been swift and effective. For example, Chile today has a demanding campaign contributions law that is tougher than those found in the statutes of many high-income democracies. Furthermore, the authorities have generally been very good about generating a broad consensus for their policies, which ensures sustainability in the policy environment. Successive governments over the past 19 years, following the country's return to democracy, have been fairly successful in setting in motion processes of consultation, to elicit the views of various sectors in society, such as opposition political parties, trade unions, and various organizations of civil society. This has resulted in greater understanding on the part of the population, and elicited their commitment to the often painful measures that accompany the implementation of various economic adjustment measures. This approach has also led to a more equitable distribution of the costs of adjustment and contributed to political stability.

A solid macro environment

Together with the Nordics, Chile is part of a small group of countries in which the political process has resulted in broad-based support for fiscal discipline, where safeguards have been introduced, which effectively insulate the budget from the short-term horizon of politicians, and from the diverse demands placed upon it by economic agents in a pluralistic democracy. The net effect has been a virtuous fiscal policy, which has contributed to a sustained reduction in the levels of public debt, from close to 90 percent of GDP in the mid-1980s, to less than 7 percent of GDP in 2008. We find no example, either among industrialized countries or in the developing world, with as sustained a downward adjustment in debt levels as in Chile. In fact, quite the opposite is the case: the vast majority of OECD members have higher levels of public debt today than 10 years ago. Indeed, according to the IMF, against the background of the global financial crisis and the fiscal stimulus measures that have been taken to address the effects of the crisis, public debt in the advanced economies will rise from 75 percent of GDP in 2008 to 110 percent of GDP in 2014.

Chile's policies have, in contrast, greatly reduced the debt-servicing burden of the public debt in Chile, contributed to sharply lower interest rates, and to the highest credit ratings in Latin America. Indeed, in 2009 Chile was the only country to have actually seen a rise in its credit ratings, at a time of massive ratings downgrades worldwide, affecting corporations and sovereign debt issuers alike. A lower debt burden has, of course, allowed spending to rise in other areas, including education and public health, and is very much behind the progress made in reducing the incidence of poverty, which fell from 38.6 percent in 1990 to 13 percent in 2006.[72]

Moreover, as noted above, not only has Chile done much to establish a clear, transparent framework for public policies, also involving a solid legal and regulatory framework—it has a ranking of 23 in the third pillar of the ICI, which captures several indicators measuring various obstacles to private sector activity—but the government has also played a leading role in promoting other innovation-friendly policies which have nicely complemented those aimed at improving the institutional climate.

Good innovation policies

The government has shown remarkable commitment to e-government, to increasing efficiency in public management, to diminishing the transaction and coordination costs between public entities, to facilitating innovation and creativity in management, to increasing the public value of services, improving government transparency and, more generally, to enhancing the quality of the services provided by the government to civil society.[73] Three areas in which this has been done in a particularly effective way, providing best practice, are those reforms introduced at the Internal Revenue Service and through the electronic platforms ChileCompra and Trámite Fácil. At the IRS, e-government has boosted direct interactions with tax payers and greatly facilitated tax compliance. Close to 100 percent of Chilean tax-payers now pay income taxes through the Internet, and the Chilean IRS is acknowledged to be one of the most modern, efficient, high-quality taxation administrations in the

world, setting high international standards for tax compliance.

ChileCompra was launched in 2000 and is a public electronic system for purchasing and hiring, based on an Internet platform. It has earned a worldwide reputation for excellence, transparency and efficiency. It serves companies, public organizations, and citizens, and is by far the largest business-to-business site in Chile, involving over 1000 purchasing organizations which invoiced well in excess of US$2 billion in transactions by 2005. It has also been a catalyst for the use of the Internet throughout the country. Trámite Fácil is a government site coordinating the work of over 240 government agencies and bodies, and taking care of a broad range of processes online, including birth certificates, identity documents, pension fund payments, trademarks/patents, housing subsidies, university credits, and so on. The government's efforts to integrate the Chilean school system with the Internet have been no less successful, and have involved heavy infrastructure investments, the training of over 90,000 teachers in the basics of ICTs, digital literacy campaigns, encouraging the study of English and several novel public/private partnerships aimed at bringing to the classroom the latest technologies and know-how.

Some challenges ahead to boost innovation capacity

The authorities in Chile have shown remarkable leadership, as well, in identifying the key challenges ahead to strengthening the role of ICTs in improving productivity and in boosting the innovation capacities of the public and private sectors and civil society. In this respect, they feel that it is necessary to expand and intensify the integration of digital technologies in the educational curriculum and to improve the education and training of highly qualified workers (see Table 11 showing the OECD's Program for International Student Assessment (PISA) results for Chile and other countries). It is also necessary, in their view, to enhance connectivity, especially among the lowest four-fifths of the income distribution, by overcoming unequal income distribution, restrictions facing micro- and small companies, and connectivity problems in rural and remote regions. They would also like to encourage the development by the private sector of computer packages for low-income households and micro-companies so that they can access the Internet more cheaply and effectively, and

72 For a discussion of the institutional framework in place for the implementation of fiscal policy in Chile, including the targeting of a surplus in the government balance since 2000, as well as other progress made in the implementation of a sound institutional framework, see López-Claros (2004).
73 For a comprehensive discussion of these issues see Alvarez Voullième et al., 2006.

Table 10. The Innovation Capacity Index: Chile and Latin America

Selected variables								
Innovation Capacity Index			Government effectiveness			Rule of law		
Score	Rank* (131)	Region Rank	Score	Rank* (131)	Region Rank	Score	Rank* (131)	Region Rank
Chile	59.4	29	1	70.8	25	1	79.2	23	1
Uruguay	52.8	49	2	55.0	41	2	62.1	45	2
Costa Rica	51.5	58	3	50.7	47	3	60.8	47	3
Peru	50.6	60	4	30.3	87	13	32.2	98	14
Mexico	50.5	61	5	44.1	59	6	35.5	86	11
Argentina	49.2	66	6	37.7	72	11	37.0	79	8
Panama	48.9	68	7	47.1	53	5	44.8	66	4
Trinidad and Tobago	48.7	69	8	50.1	50	4	44.3	67	5
El Salvador	48.3	70	9	35.5	75	12	33.0	94	13
Colombia	48.0	72	10	41.9	63	8	35.7	85	10
Dominican Republic	46.3	79	11	29.8	90	14	36.2	83	9
Jamaica	46.2	81	12	43.9	60	7	34.2	89	12
Honduras	46.0	82	13	27.0	93	15	28.6	106	16
Brazil	45.2	87	14	38.0	70	10	38.9	73	7
Guatemala	44.5	89	15	26.7	95	16	22.3	119	20
Paraguay	44.3	90	16	20.2	112	18	25.6	114	18
Ecuador	44.2	91	17	15.5	120	21	23.9	116	19
Nicaragua	43.4	93	18	18.8	117	20	28.9	103	15
Bolivia	41.5	100	19	20.8	111	17	26.0	111	17
Venezuela	40.9	102	20	19.6	115	19	13.1	127	22
Suriname	40.1	105	21	40.4	66	9	43.8	68	6
Haiti	28.7	129	22	8.5	124	22	14.4	124	21
Memorandum items:								
Finland	77.8	2	-	88.6	8	-	96.4	8	-
New Zealand	73.4	10	-	87.6	10	-	97.6	5	-
Ireland	70.5	18	-	81.8	17	-	94.0	14	-
Spain	60.3	28	-	65.5	31	-	77.9	24	-
Portugal	57.2	35	-	62.6	33	-	73.6	27	-

* Ranks after rounding to one decimal point.

to continue government subsidies for rural and remote areas and low-income communities and microcompanies. Priority is also being given to increasing R&D in the use of ICTs to stimulate competitiveness of the main export sectors, to rectify limitations in the legal system, to provide an appropriate institutional framework to stimulate/encourage e-trade, e-government, and use of ICTs, and to assure public trust in electronic operations and platforms. Finally, priority is also being given to facilitating the takeoff of the ICT industry by improving virtuous cycles of cooperation between institutions of higher education and the business community. This is seen as essential for narrowing the skills gap that exists today between Chile and the average in the OECD, made evident by the results of the PISA tests (Table 11).

India: Priority areas for boosting innovation capacity

Viewed in a long-term perspective, India's recent economic

Table 10. The Innovation Capacity Index: Chile and Latin America (cont'd.)

	Corruption Perceptions Index			Fiscal balance			Paying taxes		
Selected variables	Score	Rank* (131)	Region Rank	Score	Rank* (131)	Region Rank	Score	Rank* (131)	Region Rank
Chile	69.0	21	1	62.3	9	1	84.2	17	1
Uruguay	69.0	21	1	30.3	65	15	63.1	91	10
Costa Rica	51.0	40	3	38.1	37	5	59.3	99	12
Peru	36.0	61	6	30.7	62	14	76.6	39	3
Mexico	36.0	61	6	29.4	71	17	63.4	89	9
Argentina	29.0	87	16	32.0	57	12	45.4	120	18
Panama	34.0	72	11	37.1	39	7	53.1	112	15
Trinidad and Tobago	36.0	61	6	56.3	16	2	75.2	43	4
El Salvador	39.0	56	4	21.8	105	21	62.5	92	11
Colombia	38.0	59	5	19.4	113	22	43.5	122	19
Dominican Republic	30.0	82	14	29.3	72	18	55.8	109	14
Jamaica	31.0	79	12	35.0	45	8	49.6	118	17
Honduras	26.0	98	17	29.7	68	16	58.8	102	13
Brazil	35.0	68	10	30.8	61	13	42.8	123	20
Guatemala	31.0	79	12	26.5	85	19	68.6	74	7
Paraguay	24.0	106	19	38.3	35	4	71.2	69	6
Ecuador	20.0	116	20	34.1	49	10	74.9	45	5
Nicaragua	25.0	103	18	34.0	51	11	52.7	115	16
Bolivia	30.0	82	14	37.5	38	6	36.1	127	22
Venezuela	19.0	120	21	41.8	25	3	38.8	124	21
Suriname	36.0	61	6	23.4	99	20	83.7	19	2
Haiti	14.0	130	22	34.5	47	9	66.8	82	8
Memorandum items:									
Finland	90.0	5	-	48.2	20	-	74.0	52	-
New Zealand	93.0	1	-	51.4	18	-	87.7	12	-
Ireland	77.0	16	-	43.8	24	-	89.3	9	-
Spain	65.0	26	-	40.9	26	-	72.9	61	-
Portugal	61.0	29	-	19.2	116	-	78.2	36	-

* Ranks after rounding to one decimal point.

performance has been quite impressive. According to the OECD, GDP per capita has accelerated from 1.2 percent in the 30-year period to 1980 to 7.5 percent currently, a growth rate, which, if sustained, would double income per capita in a decade. This is clearly an important achievement that has brought with it a substantial reduction in the incidence of poverty, from 36 percent in 1994 to some 27 percent by 2005.[74]

Inevitably, the global financial crisis has contributed to a deceleration of India's economic growth in 2008 and 2009, and the emergence of other problems, such as a substantial widening of the budget deficit (see below). However, assuming this to be a temporary phenomenon, the key question for Indian economic policy for the foreseeable future will be what policies will allow it to sustain or, indeed, accelerate its growth performance over the next decade. Just as China has benefited from a massive process of urbanization in the past two decades which has contributed in an important way to its high economic growth rates, India has a similar structural fea-

---

[74] This progress notwithstanding, China has grown more quickly than India over the same period and, consequently, has seen much faster reduction in poverty levels, regardless of the poverty line chosen. China has much lower infant mortality, higher life expectancy, and lower illiteracy rates than India.

Table 10. The Innovation Capacity Index: Chile and Latin America (cont'd.)

Selected variables	Environmental sustainability			Total fixed broadband subscribers per 100 inhabitants			E-government readiness index		
	Score	Rank* (131)	Region Rank	Score	Rank* (131)	Region Rank*	Score	Rank* (131)	Region Rank
Chile	83.4	28	4	19.8	42	1	58.2	40	3
Uruguay	82.3	35	8	13.6	51	3	56.5	46	5
Costa Rica	90.5	5	1	8.1	56	8	51.4	55	9
Peru	78.1	55	13	5.6	64	11	52.5	53	8
Mexico	79.8	43	11	11.8	52	4	58.9	37	1
Argentina	81.8	37	9	18.1	43	2	58.4	39	2
Panama	83.1	29	5	2.8	77	15	47.2	73	15
Trinidad and Tobago	70.4	81	19	3.2	73	14	53.1	51	7
El Salvador	77.2	60	15	3.6	72	13	49.7	63	11
Colombia	88.3	9	2	7.2	57	9	53.2	50	6
Dominican Republic	83.0	32	6	4.3	68	12	49.4	64	12
Jamaica	79.1	51	12	8.2	55	7	46.8	75	16
Honduras	75.4	68	17	0.0	108	21	40.5	86	19
Brazil	82.7	33	7	11.6	53	5	56.8	45	4
Guatemala	76.7	64	16	0.6	91	20	42.8	81	18
Paraguay	77.7	59	14	2.2	81	16	46.5	76	17
Ecuador	84.4	22	3	6.6	60	10	48.4	68	14
Nicaragua	73.4	72	18	0.9	87	19	36.7	88	20
Bolivia	64.7	96	20	1.0	86	18	48.7	66	13
Venezuela	80.0	42	10	8.5	54	6	51.0	58	10
Suriname	-	-	-	1.6	84	17	34.7	92	21
Haiti	60.7	104	21	0.0	108	21	21.0	120	22
Memorandum items:									
Finland	91.4	4	-	91.7	4	-	74.9	15	-
New Zealand	88.9	7	-	44.4	28	-	73.9	17	-
Ireland	82.7	33	-	45.1	27	-	73.0	19	-
Spain	83.1	29	-	49.4	25	-	72.3	20	-
Portugal	85.8	18	-	41.6	30	-	64.8	31	-

* Ranks after rounding to one decimal point.

ture: favorable demographics, which is likely to fuel growth. For the next 20 years, the share of the working age population will rise, and India will have to find ways to bring its masses of young people into the mainstream by spending on education and improving the quality of its educational institutions, in order to boost the productivity of its young, particularly the poor.

There has also been a significant improvement in recent years in the quality of India's policy environment and the degree of sophistication of its private sector. In those areas in which the government has decided to open up participation to the private sector—telecommunications, civil aviation—the response has been impressive. According to the OECD, India's telecommunications sector has become the third largest in the world. In contrast, in electricity generation, where public enterprises are still dominant, shortages are common, and there is a serious problem of non-payment due to "poor management of distribution enterprises and a failure to eradicate theft" (OECD,

Table 11. The Innovation Capacity Index and PISA scores: Latin America

| | Innovation Capacity Index | | | PISA (Program for International Student Assessment)* | | | | | |
| | | | | Science | | Reading | | Mathematics | |
	Score	Rank** (131)	Region Rank	Score	Upper and Lower Ranks*** (57)	Score	Upper and Lower Ranks*** (57)	Score	Upper and Lower Ranks*** (57)
Chile	59.4	29	1	438	40-42	442	37-40	411	44-48
Uruguay	52.8	49	2	428	42-45	413	41-44	427	42-43
Costa Rica	51.5	58	3	-	-	-	-	-	-
Peru	50.6	60	4	-	-	-	-	-	-
Mexico	50.5	61	5	410	48-49	410	41-44	406	46-48
Argentina	49.2	66	6	391	50-55	374	51-53	381	50-53
Panama	48.9	68	7	-	-	-	-	-	-
Trinidad and Tobago	48.7	69	8	-	-	-	-	-	-
El Salvador	48.3	70	9	-	-	-	-	-	-
Colombia	48.0	72	10	388	50-55	385	48-53	370	52-55
Dominican Republic	46.3	79	11	-	-	-	-	-	-
Jamaica	46.2	81	12	-	-	-	-	-	-
Honduras	46.0	82	13	-	-	-	-	-	-
Brazil	45.2	87	14	390	50-54	393	46-51	370	53-55
Guatemala	44.5	89	15	-	-	-	-	-	-
Paraguay	44.3	90	16	-	-	-	-	-	-
Ecuador	44.2	91	17	-	-	-	-	-	-
Nicaragua	43.4	93	18	-	-	-	-	-	-
Bolivia	41.5	100	19	-	-	-	-	-	-
Venezuela	40.9	102	20	-	-	-	-	-	-
Suriname	40.1	105	21	-	-	-	-	-	-
Haiti	28.7	129	22	-	-	-	-	-	-
Memorandum items:									
Finland	77.8	2	-	563	1-1	547	2-2	548	1-4
New Zealand	73.4	10	-	530	3-9	521	4-6	522	8-13
Ireland	70.5	18	-	508	15-22	517	5-8	501	17-23
Spain	60.3	28	-	488	26-34	461	34-36	480	31-34
Portugal	57.2	35	-	474	35-38	472	29-34	466	35-38

* *PISA 2006: Science Competencies for Tomorrow's World.* Executive Summary. OECD, 2007.

** Ranks after rounding to one decimal point.

*** Rankings for all participating countries. On the basis of the samples of students assessed by PISA, it is not always possible to say with confidence which of two countries with similar performance has a higher mean score for the whole population. However, it is possible to give a range of possible rankings within which each country falls.

2007). There would thus appear to be wide scope for gains in efficiency in resource allocation in India, with corresponding gains in productivity and economic growth.

India does not do well in the Innovation Capacity Index, with an overall ranking of 85 among 131 countries (Table 6). Looking at the various pillars of the ICI, India's worst ranking (94) corresponds to human capital, training, and social inclusion, followed by adoption and use of information and communication technologies (93) (see Table 7). To boost its capacity for innovation, policymakers in India will have to address a number of important weaknesses, of which the most important are discussed below. Figure 4 presents the ICI's top priorities for policy reform for India.

Education and labor market

India continues to have high illiteracy rates—its rank in the ICI on this particular indicator is 110—suggesting that illiteracy still afflicts several hundred million people, not surprisingly a serious blight on innovation capacity. School enrolment rates remain low by international standards, with its rank for secondary school level an unimpressive 94. The scope for improvement in girls' education is especially intense—the ICI attaches to India a rank of 89 on the gender equity index. Given the wide range of positive payoffs associated with improvements in girls' education and, more generally, gender equity, much more will have to be done over the longer term to integrate women into the economy, the educational system, and India's political establishment. India will also have to educate and train its young poor, to enable them to join the labor force with usable skills, particularly in those sectors with potential comparative advantage. There is every expectation that world demand for outsourcing will rise in coming years, reflecting the continued shift of backroom operations associated with further reductions in the cost of communications. For India to be able to take full advantage of these opportunities, it will have to improve the level of skills and training of its workforce. In this respect, it is particularly worrying to see that India suffers from huge inefficiencies in its labor market, with laws governing regular employment contracts much stricter than in many emerging markets, and in virtually all members of the OECD. As noted by the OECD, one major reason for this is "the requirement to obtain government permission to lay off just one worker from manufacturing plants with more than 100 workers." Not surprisingly, a rigid labor market will prevent India from deriving the full benefit of its comparative advantage in labor-intensive industries.

A serious fiscal deficit problem

For many years now India has had a serious problem with its public finances. Essentially, it has been running deficits of some 6-10 percent of GDP for the past decade, among the highest in the world. This problem has many dimensions and it is worthwhile to highlight several here. First, India's public debt level, at 83 percent of GDP in 2009, is already very high by international standards; indeed, it is larger than that of Brazil and Argentina, twice that of Turkey, four times that of China, and well over ten times larger than that of Russia, as well as of most OECD countries. Second, with total revenue collection in the neighborhood of 18 percent of GDP (again, extremely low by international standards) due to its very narrow revenue base—the central government collects no more than about 11 percentage points of GDP in taxes—the revenue-to-debt ratio is among the lowest in the world.

In an attempt to bring about some measure of medium-term fiscal adjustment, the government brought into force in 2003 a Fiscal Responsibility Budget Management Act (FRBMA) which established a path of deficit reduction through 2009. The high economic growth rates during the period 2004–07 boosted government revenue and some progress was made in reducing the deficit, but the 2008 financial crisis and the need

Figure 4. India: Top priorities for policy reform

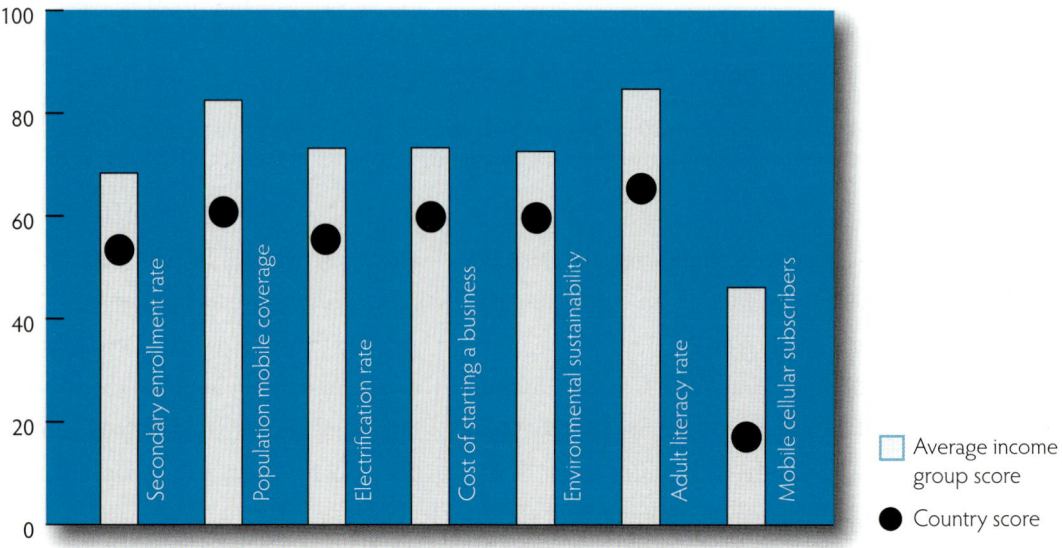

to respond to the weakening of economic activity through fiscal stimulus means that the deficit in 2009 will be back to some 10 percent of GDP. In any case, the law has generally applied to the central government only, whereas, in fact, a large share of the deficit problem is with the states. Moreover, it does not contain a medium-term debt target that might act as a binding constraint on the public finances. The law also does not establish any penalties or sanctions for departures from the path of fiscal adjustment laid down in the FRBMA. According to the IMF, "despite the apparent consolidation, off-budget activities increased, deadlines to comply with fiscal targets were extended and the fiscal adjustment was not underpinned by expenditure reform."[75] India's fiscal situation is, without doubt, a severely limiting constraint on the country's ability to boost its innovation capacity.

A large public debt constrains the ability of the government to allocate greater resources to education and public health, and to improve the country's dilapidated infrastructure, all areas where India, as noted earlier, is lagging behind. The inability of the government to introduce expenditure reform is, likewise, a major constraint on policies that might seek to direct greater resources to more productivity-enhancing areas. This year, India is spending close to 4 percent of GDP on regressive subsidies on petroleum, diesel, and various other products, a sum roughly equivalent to what it spends on education and health combined. This is a shocking statistic that highlights the significant need to improve the macroeconomic environment.[76] Without doubt, the deficit is a drag on the economy. A much lower deficit would have been associated with higher growth rates and higher levels of revenue, which would have boosted the ability of the government to respond to pressing social needs.

Not doing business

It takes 13 procedures, a total of 30 days at a cost of 70 percent of income per capita to open up a business in India. In the World Bank's *Doing Business Report 2009*, India ranked 121 (among 181 countries) in this indicator, representing a *drop* of seven places with respect to 2008. Among the 131 countries ranked in the ICI, India has a rank of 100 for the cost of registering property, a rank of 116 for the ease of paying taxes, and a rank of 180 for enforcing contracts. The fact is that bu-

reaucratic red tape and excessive regulation remain serious problems in India, a country afflicted with a pervasive culture of government intervention and control, which adds to business costs, discourages the development of small and medium-sized enterprises, and, given the important role played by entrepreneurship in most forms of innovation, is thus a heavy burden on India's innovative capacity.

Russia's unfulfilled potential

Russia is in many ways a unique case, with a relatively mediocre ranking of 49, well below the rank of countries such as Chile (29), Malaysia (34), and Poland (40), which share broadly similar levels of income per capita (see Tables 6 and 8). Russia has a solid human capital endowment, reflecting decades of investment in education in science and technology. If, as noted earlier, Latin America has a grand total of three Nobel Laureates in science, there are at least ten Russian Nobel Laureates in physics alone. And had Alfred Nobel created a category for mathematics, there is little doubt that Russian mathematicians would have been awarded many prizes, perhaps more than any other nation. At the same time, however, it is a country where there is a huge gap between the stock of resources spent in past decades to foster contributions to knowledge, on the one hand, and, on the other, the kind of output that we would normally recognize today as reflecting achievements in scientific innovation, such as, for instance, patent registration or the presence of identifiable Russian brands in manufactured exports. Soviet technology was able to send the first man into space; it made significant advances in nuclear energy technology; but the context of the Cold War and the inefficiencies of central planning misdirected vast resources to the military-industrial complex, at huge cost in terms of living standards. By the time the Soviet Union collapsed in 1991, it was producing large nuclear submarines, MIG aircraft, and other weapons (sold on credit to its allies in the developing world), but not many consumer goods, and few, if any, manufactured goods with even minimal presence in the global economy. The 1990s witnessed a disorderly transition to a sort of market economy which involved redeployment of labor from the military-industrial complex and other heavy and inefficient industries to the private non-defense sector, particularly light manufacturing, services, and other industries long neglected under the state planning system.

---

[75] International Monetary Fund, 2009b, p. 34.

[76] There is yet another dimension to the fiscal deficit problem which will not be addressed here, having to do with the impact of debt financing on the financial system; it is much easier for the banks to lend to the government than to lend to small and medium-sized enterprises, which are so much at the center of the innovation chain in other countries.

A difficult business environment

There are several factors that help explain the persistence of this gap between its relatively solid educational base and Russia's notable absence among international innovators. First and foremost, 18 years into its transition, Russia has still not established a particularly nurturing business environment. In fact, a case can be made that in some areas, such as levels of corruption, the property rights climate, the lack of independence of the courts, the general level of transparency in the public sector, and in the relations between the government and the business community—what the OECD calls "framework conditions" but which fundamentally refer to the stability and efficiency of the institutions that underpin the market economy—Russia is worse off today than it was five years ago.

This is certainly made unambiguously clear from the good governance indicators compiled by the World Bank and used in the institutional environment pillar of the ICI, as well as by Russia's embarrassingly low rankings in Transparency International's *Corruptions Perceptions Index*—147 among 180 countries in 2008, a drop of *61 positions* since 2003.[77] Russia's deteriorating property rights climate, including for intellectual property, is particularly noteworthy—piracy is rampant in Russia—and perhaps more than any other indicator suggests the severe obstacles which at present exist for the creation of an institutional framework that will encourage innovation.

The high incidence of crime and corruption (ranging from "visits" from tax and fire inspectors to politically motivated expropriations by the state) remains a heavy burden on businesses, imposing heavy costs on them, and, therefore, undermining the ability of Russian companies to compete abroad.[78] Accounting and auditing standards are weak, raising yet another set of concerns about the investment climate. Increasing restraints on freedom of the press highlight the risks for the abuse of power, and the difficulties for civil society to emerge as a constructive counterweight to the growing power of the state. The World Bank's *Doing Business Report* (which provides the indicators that go into the regulatory and legal framework pillar of the ICI) paints a rather uncharitable picture of bureaucracy and red tape in Russia: from rigid labor-market laws and mind-numbing obstacles to the obtaining of licenses—it takes 54 procedures and an average of 704 days to obtain one, at a cost of close to 3,800 percent of income per capita—to difficulties in the payment of taxes and to impediments to international trade. Trading across borders is so laden with red tape in Russia that the country ranks 155th among 181 countries in this particular indicator of the *Doing Business Report*. This is a particularly perturbing indicator, given the need to encourage exports other than resource-based commodities, on which the Russian economy is totally dependent. According to the OECD, the share of high-value added goods in manufacturing exports from Russia to OECD countries is less than 1 percent and is even lower (0.2 percent) in the case of ICT goods. (In Taiwan, in contrast, close to 50 percent of manufactured exports are high-tech exports). Figure 5 presents the ICI's top priorities for policy reform in Russia.

Innovation policies

These extremely unfavorable business environment conditions have had a number of undesirable repercussions. The country is a major exporter of talent. Not surprisingly, capable Russian researchers with a modicum of ambition emigrate at the first available opportunity. There is no significant engagement between the scientific community and the business world. The sort of collaboration and interaction between institutions of higher education and the enterprise sector which have been so instrumental in the development of a vibrant ICT industry in Israel and Taiwan is largely absent in Russia. State funding for research and development to institutions of higher education accounts for less than 5 percent of total state funding to such institutions. This, in turn, means that state funding to science does not play the catalytic role that it has played in other countries to spur innovation. Instead, as noted by the OECD, the emphasis on "institution-based financing tends to protect incumbents and creates few incentives to increase efficiency, productivity or innovation. On the contrary, since much funding is 'cost-based' and allocated with reference to employment levels

---

77   In fact, between 2003 and 2008, Russia has been one of the world's worst performing countries in the *Corruption Perceptions Index*, sharing (undistinguished) company with the likes of Belarus, the Islamic Republic of Iran, Sudan, Uzbekistan, Syria, and Gambia. China's rank fell from 66 to 72; India's rank moved from 83 to 85, and Brazil's from 54 to 80, with Russia having, by far, the worst performance among the largest emerging markets.

78   According to Richard Pipes (2009), "One of the major obstacles to conducting business in Russia is the all-pervasive corruption. Because the government plays such an immense role in the country's economy, controlling some of its most important sectors, little can be done without bribing officials. A recent survey by Russia's Ministry of the Interior revealed, without any apparent embarrassment, that the average amount of a bribe this year has nearly tripled compared to the previous year, amounting to more than 27,000 rubles or nearly US$1,000. To make matters worse, business cannot rely on courts to settle their claims and disputes, and in extreme cases resort to arbitration."

and fixed assets, greater efficiency could lead to loss of funding" (Gianella and Tompson, 2009, p. 20).

The government has attempted to steer policies in the direction of better support for R&D, with the aim of encouraging the emergence of a culture of innovation. It is aware that while levels of overall R&D spending are not low by emerging market standards, such spending remains unduly concentrated on a few sectors, and consists overwhelmingly of state funding, in sharp contrast with other countries, where much of R&D spending comes from the private sector. One way in which a better balance could be achieved in this area would be to phase out fiscal disincentives to enterprise R&D spending through accelerated write-offs. A law passed in June of 2005 on Special Economic Zones was intended to contribute to diversification of Russia's industrial structure and to stimulate innovation. Unfortunately, Russia does not have a good history with such special zones, although they have been a staple of Russian structural reforms since the 1990s. In the specific case of the 2005 law, we are skeptical that it will have the desirable effects—particularly in terms of attracting foreign investment, as Taiwan and Israel have been brilliantly successful in doing—given that "disputes concerning the creation and operation of SEZs are to be settled in Russian courts under Russian law" (Gianella and Tompson, p. 27). In the absence of mechanisms of international arbitration, it is unlikely that foreign investors may want to expose themselves to the lack of independence and arbitrariness of Russian judges and courts and, more generally, to the primitive, opaque nature of the Russian legal system.

Low ICT penetration

Finally, Russia does not do as well as might be expected in the ICI because, with the exception of mobile telephony, it does not have particularly impressive penetration rates for the latest technologies. Even in the area of personal computers—where notable progress has been made in recent years in terms of expanding their use in businesses and households—PC use per 100 inhabitants is about 13.3, putting Russia in 56th place in the world, slightly worse than its rank of 52 in 2006, and broadly in the middle among the 131 economies covered in the ICI. Similar results hold for Internet use: improvements with respect to the recent past, but absolute levels that are not high enough to put Russia above its 64th place in the world.

Other weaknesses undermining innovation potential

Other factors are likely to complicate the authorities' attempts at boosting innovation capacity over the medium term: first is the weakening of a culture of meritocracy in the public sector, with many senior positions in government now going to people with links to the security establishment, who increasingly—and presumably without the required qualifications—find themselves running large state enterprises in the energy and other sectors; second, the return to old authoritarian traditions which sit uncomfortably with the openness and willingness to "challenge the system" that are so common in successful cases of innovation; third is the country's long-term demographic trends, which foresee a rapidly aging and declining population, limit-

Figure 5. Russia: Top priorities for policy reform

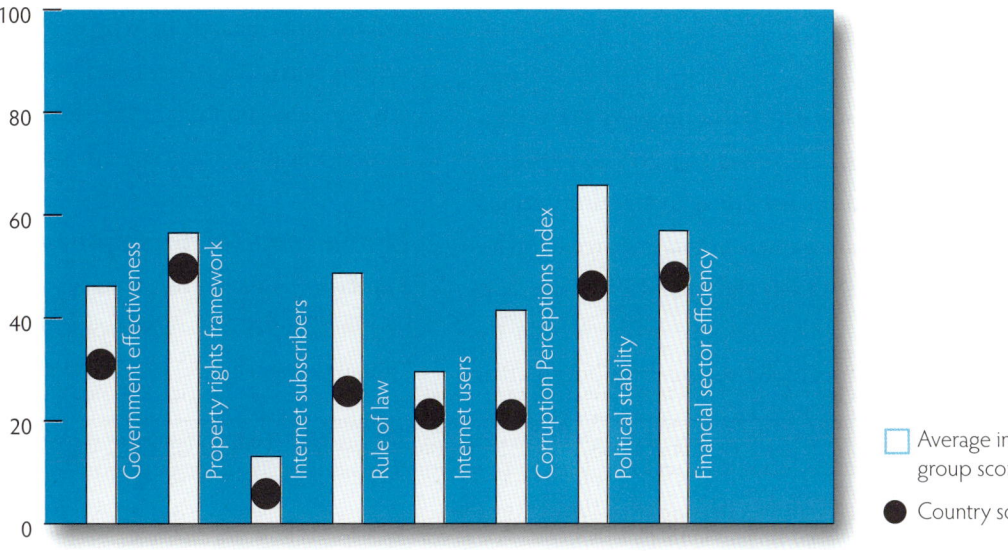

ing the role of the labor force as an engine of economic growth in coming years; finally, an ambivalent attitude toward foreign direct investment, which is welcomed one day, but quickly followed by "renegotiations" of previously agreed contracts with foreign partners, all of this accompanied by the return of old-fashioned ideas about "strategic sectors" which should remain under state control. This has led to a marked increase in the presence of the state in the energy and raw materials sectors. Furthermore, the 2008–2009 financial crisis is projected to result in something close to a 10 percent drop in GDP growth in 2009, and a massive widening of the budget deficit, creating a likely setback for the government's efforts to do more in this critically important area. The sum total of the above suggests that Russia is a classic case of unfulfilled potential—a giant still playing in the little leagues.

Taiwan: Green Silicon Island

A recent and insightful contribution to the debate on the policies that contribute to economic growth is the 2008 study published by the Commission on Growth and Development,[79] which examined the experiences of 13 countries which, beginning in 1960, grew at an annual average rate of at least 7 percent over a period of 25 years or more, and identified those factors which contributed to such remarkable economic performance. The 13 economies examined include Taiwan. And since Taiwan's real growth rate over the 30-year period beginning in 1960 was 9.2 percent, one can assume that it must have been very near the top in this high-growth league. Indeed, between 1952 and 2007, income per capita rose from US$197 to US$16,800, arguably the most remarkable case of catching up seen in the post-World War II period.

Sound policies

A closer look at the Taiwan experience suggests that a combination of sound policies, the strong engagement of the private sector, effective governance, imaginative institutional arrangements, and good macroeconomic management has lifted its population from poverty and helped it join the ranks of the most prosperous and innovative economies in the world.[80] Major investments have been made in both human

resources and infrastructure by both government and the business community, and the benefits of economic growth have been widely shared by all segments of society. Targeted and well thought out government intervention, aimed at facilitating the emergence of a strong private sector role in ICT has worked in Taiwan, because the government has kept active consultative mechanisms in place to attract the input and technical expertise of the private sector, to agree on common approaches, and to bring into its institutions the best technical experts to support both government and business.

A global leader in ICT

Taiwan ranks among the world's top producers of notebook personal computers, flat panel displays, modems, motherboards, and other electronic components and products. In 2007, it ranked fourth globally in the production value of its semiconductor industry (US$44.4 billion) and was first in the world in the production of image display hardware (US$54.5 billion). Taiwan has an impressive capacity for innovation, firm-level technology absorption, collaboration between institutions of higher education and the business community in research, and a pre-eminent position in the use of the latest technologies, from mobile telephones to personal computers and the Internet. Its rank of 13 in the Innovation Capacity Index (Table 6) reflects exceptionally high performance in a number of indicators including patent registration (per capita), in which Taiwan is number 1, schools connected to the Internet (1), R&D worker density (4), tertiary enrolment rate (4), fixed telephone lines (4), students enrolled in science and engineering (5), among others. In fact, Taiwan is ranked 1 in the world in the ICI's Research and Development pillar (Table 7). In research productivity, Taiwan ranked 7th in papers indexed in the 2007 *Science Citation Index*, 7th in papers indexed in *Engineering Index*,[81] and 4th among all countries in US patents granted in 2008. Figure 6 shows some of Taiwan's key strengths.

Human capital development

Although seemingly a disadvantage at the time, the brain drain of the 1960s and 1970s—when some 50,000 of the brightest young Taiwanese went overseas (principally to the United States) for

---

79 See The Growth Commission, 2008, available at: www.growthcommission.org The Growth Report was funded by the World Bank, several industrial country aid agencies (Australia, Canada, the United Kingdom), and some private foundations. The Commission was chaired by Nobel Laureate in Economics Michael Spence.

80 For further details see Dahl and López-Claros, 2006. This section on Taiwan also draws from a visit to Taipei made by López-Claros in February, 2009.

81 National Science Council, 2008, available at: http//www.nsc.gov.tw/tech/

university and advanced studies—allowed Taiwan to build a large pool of qualified and experienced people before its economy was ready to absorb them. From 1985 onwards, incentives drew them back to Taiwan as entrepreneurs, to create start-ups in the science parks, or to take up research, academic, and management positions, bringing not only their knowledge and experience, but also their networks of contacts and working relationships with leading international companies, and enabling today's Taiwanese universities to educate its own manpower for continuing expansion at home. These informal networks, supplemented by overseas offices of various institutes and research centers, facilitate technology transfer, innovation, and strong entrepreneurial relationships.

Launched in 2000, the government's Department of Industrial Technology has vigorously promoted e-business, following four strategic elements: policy, environment, applications, and promotion, with the goal of establishing a global logistics operation system based on a highly efficient e-supply chain framework, linking leading international IT companies (IBM, HP, and Compaq) with 42 Taiwan contract manufacturers, and 15 domestic e-supply chains among domestic IT manufacturers.

Deploying the information society

At the heart of Taiwan's ICT revolution is the Institute for Information Industry (III), a joint government-private sector think tank and management consultancy, promoting the development of the ICT industry and deploying the information society. The III provides a neutral source of expertise independent of both partisan politics and individual corporate agendas, helping Taiwan to increase productivity, raise efficiency, and develop international collaborative projects with key industrial and academic partners and global offices in various important ICT centers. The government contracts a wide range of functions to the III, making use of its human resources in a flexible manner, from proposing policy, providing market analyses, incubating start-ups, developing such concepts as the integrated service model and the digital home, to generating consumer, communications, and computer technologies, and generating over 100 patent applications annually. The III provides professional IT training in both the public and private sectors, develops programs to address the digital divide, creates digital opportunity centers in remote areas and internationally for developing countries, and provides services to small and medium enterprises, as well as disadvantaged and handicapped groups. Over the years, III has provided training to some 400,000 professionals. It also designs and manages projects to strengthen ICT infrastructure, including the planning of e-Taiwan, to extend broadband access to all households, and M-Taiwan, to provide mobile access through a combination of cellular telephone and WLAN networks. As manufacturing moves offshore, it moves the industry forward from tangible to intangible products, and aims to establish best practices in Taiwan as a model for the rest of the world.

With a million or more Taiwanese working in mainland China, trade with that country involves well over US$100 billion in investment—the logical place for Taiwanese businesses to locate

Figure 6. Taiwan: Significant indicators above income group average

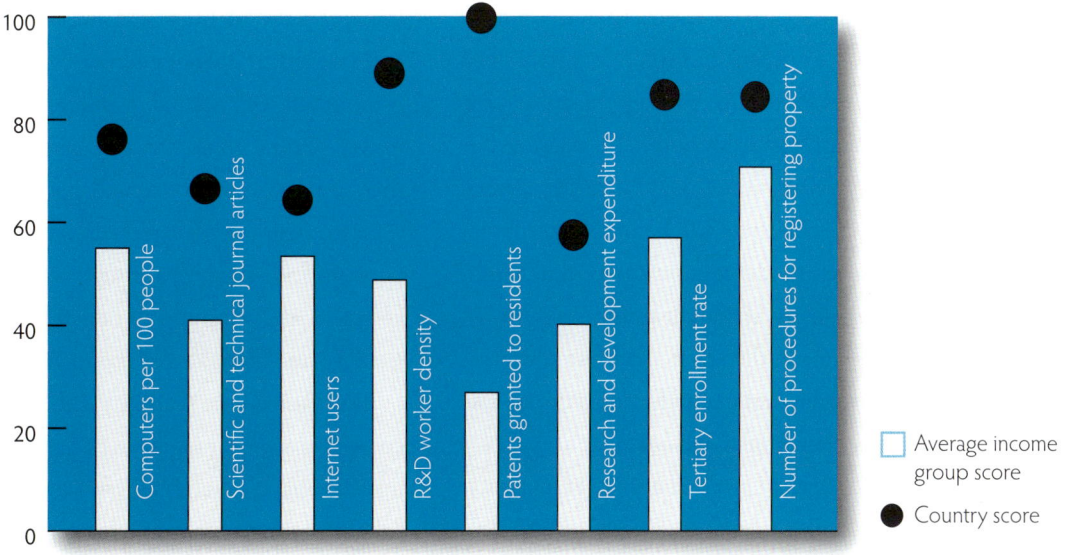

production and take advantage of low land prices and cheap labor. Competition with the mainland is now forcing Taiwan to search for new areas of comparative advantage as a center for research and for corporate headquarters, maintain its engineering and management talent, and invest more in research and collaboration between industry, the research institutes, and universities.

In addition to manufacturing them, Taiwan is quickly adopting ICTs. The III estimated that already in 2007, there were over 10 million Internet users in Taiwan, with a penetration rate of 44 percent, and showing signs of reaching saturation. There were also about 12.5 million mobile Internet subscribers and 4.7 million broadband subscribers.

Taiwan's network of 10 science parks helps incubate start-ups and offers an environment in which companies can take several years to grow before they decide to build their own building on leased government land, shielding them from high property costs. They screen applicants in relation to national priorities and for synergies with park activities. Each science park focuses on a different area, such as semiconductors, integrated circuits, computers and peripherals, telecommunications, precision machinery, biotechnology, and recycling technology, among others. They account for some 40 percent of total exports and imports, or close to US$190 billion and a significant share of government revenue.

In Taiwan, ICTs are not only a matter for business, but play a role in promoting its own social and economic development. The interaction of the two sheds important light on Taiwan's success in this area. However, despite the high value placed on education in Chinese culture, and the efforts made by the government to build human resources, there is still a gap between academia and industry, because the traditional Chinese educational approach at the primary and secondary level does not encourage the kind of innovative thinking necessary for success in scientific research and development, making the student transition to university more difficult. With a work force mostly under 30, the challenge will be to retrain maturing workers as technologies, production processes, and whole industries evolve, to emphasize lifelong learning, using ICTs as well as ongoing programs in the science centers and elsewhere, to sustain momentum and make the educational system as flexible and entrepreneurial as industry.

With the ICT industry having reached a stage of maturity, it will have to support new ventures that leverage Taiwan's comparative advantages, all of which are typical of Porter's innovation-driven stage of development. This may involve venturing into such areas as using ICT to boost alternative energy sources, helping to create digital homes and deliver new services in such burgeoning fields as long-distance patient care and other forms of biomedical research, services aimed at enhancing the quality of life for the elderly, and keeping abreast of developments in the world's leading technology centers to maintain a competitive edge. In this respect, it will be essential to improve the regulatory framework for services, which at times suffer from excessive regulation. This observation is borne out by Taiwan's relatively poor ranking in pillar 3 of the ICI which captures various dimensions of the regulatory environment and where Taiwan's rank of 39 out of 131 countries shows much scope for improvement.

Conclusions

Richard Cooper (2004a, p. 151) makes a compelling case that at the outset of the 21st century technical change and innovation have become "the dominant characteristic" of our time. "New technological ideas," he adds, "combined with social order and the trained human beings who generate and apply them, are the basis for modern economic prosperity." The traditional sources of power and influence—territory, resources, raw manpower, and military might—for centuries the chief determinants of nations' prosperity, are far less important today than they used to be and have given way to a new world in which successful development is increasingly linked to sound policies, to good governance, to effective management of scarce financial resources, and, most important, to the extent to which societies are able to harness the latent capacities of their populations. Successful countries today are not necessarily large geographically or richly endowed with natural resources, nor able to project military power beyond their borders. Increasingly, they are countries that have managed to expand opportunities for their populations through the full exploitation of the opportunities afforded by the world economy through international trade, foreign investment, the adoption of new technologies, macroeconomic stability, and high rates of saving.

The Innovation Capacity Index featured in this chapter correlates a wide-ranging set of relevant factors, policies, and institutional characteristics which are seen as playing a central role in boosting a nation's capacity for innovation. How can countries transform knowledge into value in ways that will result in new products and services, processes and systems? What are the priority policy areas that merit particular attention if countries are to be able to participate successfully in an increasingly complex global economy, requiring growing levels of sophistication? How do these priorities, in turn, depend on a nation's particular stage of development—the quality of its institutions, the human capital endowment of its labor force—and the nature of the political regime against which policies are framed? In building the ICI's theoretical framework, we have established a firm linkage between the stage of development of a given country and the relative importance attached to the many factors boosting innovation capacity. But we have also taken the view, firmly anchored in empirical observation, that democracies tend to be better than authoritarian regimes at encouraging the creation of friendly environments for innovation.

The Innovation Capacity Index is intended to be a policy tool to better examine the broad range of policies and institutions which underpin the creation of an environment conducive to innovation. The methodologies developed allow the identification of country-specific factors which demand priority attention. The reader's attention is directed to the innovation profiles in part 3 of the *Report* which identify, for each country, the top priorities for policy reform. Although this is the first edition of the ICI, the Index will be estimated annually and it is expected that, over time, it will also provide a historical perspective on individual country performance. Above all, by identifying individual country strengths and weaknesses, the Index is intended to stimulate policy dialogue. And the rich body of data used for the calculation of the Index rankings should also provide ample opportunities for the sort of high-minded international comparisons of best-practices which are an essential component of better policy formulation.

To highlight the uses to which the ICI can be deployed, in this chapter we have examined in some depth the innovation capacity of five countries: Sweden, Chile, India, Russia, and Taiwan. Sweden is the ICI's top performing country in 2009, serving as an impressive benchmark for other countries. Yes, Sweden is a rich industrial country with an important presence in the global economy, but there is much in the Swedish approach to innovation that is of particular relevance not only to other industrialized countries, but to many middle-income countries

with aspirations to join the league of top innovators. We are particularly impressed by Sweden's ability to combine open and transparent government, universal social protections, and high levels of competitiveness and productivity to create one of the most innovative economies in the world. Equally impressive is the extent to which an excellent policy framework has turned the private sector into the main engine of innovation.

Chile is an interesting case because it proves that sound policies and good institutions are not the *result* of wealth and prosperity but rather engines for its creation. Chile's performance is far ahead of any other country in Latin America and in many critical areas it is already ahead of the European Union average. A mix of sound macroeconomic management—including arguably one of the most virtuous fiscal policies in the world—institutional reforms, and opening up of the economy to the benefits of free trade, foreign investment, and international competition have combined to create a reliable engine of high growth and poverty reduction. But the authorities have also sought to implement micro policies aimed at enhancing the efficiency of public services through various electronic platforms and at facilitating the use of ICTs more generally. Chile is well poised to catch up with the richer members of the EU, even if some poor performers in the region may occasionally complicate the context for policy implementation.

India is one of the world's most rapidly growing economies and has aspirations to be a global player in the field of technological innovation. Its economic performance over the past two decades has been impressive and has turned India into the world's fourth largest economy. India has favorable demographics, with a growing working age population which, if properly educated, could spur rising productivity and growth. In coming years, however, much more will have to be done to deal with India's disadvantages, including high illiteracy rates, a poorly developed infrastructure, a festering fiscal deficit problem which has pushed the public finances to unhealthy levels of indebtedness, and a regulatory framework characterized by mind-boggling bureaucracy and red tape, which go far to discouraging entrepreneurship and innovation. Still, beyond the benefits of good demographics, India has many features in its favor, including a long political tradition of democracy and rule of law. While its ranking in the ICI (85) is not high, there is enormous scope for the implementation of better policies, including institutional reforms, which might allow India to scale up in the rankings.

Russia's innovation performance lags far behind its true potential. It is a country with a well-established tradition of solid contributions to basic science. In previous decades it was a leader in space exploration, nuclear technology, and aviation. Its transition from the inefficiencies of central planning to the challenges of a market economy has not been easy. During the past five years, the country has lost some steam as a result of the commodity boom which has increased its economic dependence on energy and other raw materials exports. Furthermore, the country does not have a friendly business environment capable of spurring entrepreneurship and allowing the incubation of new ideas and approaches to new products or process creation. Corruption has become an endemic problem of a magnitude most often seen in low-income countries with broken institutions. Its judges and courts lack the independence that might encourage more non-energy investments and its gradual return to authoritarian forms of governance does not bode well for the creation of an environment conducive to various forms of innovation. And yet, there is no intrinsic reason why a country with such a rich complement of human and natural resources and a long and distinguished history of scientific innovation should not catch up with the Swedens of this world.

Taiwan is arguably the most impressive example during the post-World War II period of both the consequences of high growth and the policies that underpin it. That a country should be able to increase its income per capita from under US$200 in 1952 to close to US$17,000 in 2007 is nothing short of astounding. Taiwan's success is attributable to two factors: first, it succeeded in accomplishing many of the good things that have been critical for high growth elsewhere in the world—while taking full advantage of the benefits of international trade and investment and the acquisition of new technologies—and it avoided making the errors that have been such a drag on development in so many other countries. In less than half a century Taiwan transformed itself from a simple agrarian society in the earliest stage of development into a remarkable global technology powerhouse, a world leader in the production of ICT equipment with a supporting infrastructure of science parks, public-private research institutions, and think tanks that have turned Taiwan into one of the world's most prolific innovators. Taiwan's challenge in coming years will be to find creative ways to cooperate with China—an emerging technology power in her own right, with a much lower cost structure—and to move closer to the best performers in the ICI.

Future editions of the *Innovation for Development Report* will provide in-depth analysis of innovation capacity in a growing number of countries. The Innovation Capacity Index will be estimated annually and the results published and analyzed in successive Reports. For obvious reasons, this chapter has covered methodological issues in some detail, as it was thought appropriate to lay out in reasonably explicit form the basic building blocks of the ICI and its underlying assumptions. It is expected, however, that in coming years, the emphasis will shift to analysis of innovation issues as they emerge among the countries covered by the *Report*. Country coverage is also expected to gradually rise over time. Readers are invited to visit a dedicated website at:

www.innovationfordevelopmentreport.org

to find innovation profiles for 63 countries not included in this year's published edition, as well as abstracts and short biographical sketches by the authors who contributed the other papers to this year's *Report*. It is hoped that the framework provided by the *Report* for examining factors, policies, and institutions which contribute to creating an environment that boosts nations' capacity for innovation will prove useful for analysis and policy dialogue in coming years. We expect that these questions will move to center stage in the debate over how best to safeguard human prosperity.

References

Acemoglu, Daron, Simon Johnson, and James Robinson. 2004. "Institutions as the Fundamental Cause of Long-Run Growth." National Bureau of Economic Research Working Paper 10481.

Advanced e-Commerce Institute, III. *Internet in Taiwan*. Available at: http://www.find.org.tw/eng/

———. 2008. *Indicators of Science and Technology 2008*. National Science Council.

———. Institute for Information Industry. Available at: http://www.iii.org.tw/english/

Alvarez Voullième, Carlos, Constanza Capdevila de la Cerda, Fernando Flores Labra, Alejandro Foxley Rioseco, and Andrés Navarro Haeussler. 2006. "Information and Communication

Technologies in Chile: Past Efforts, Future Challenges." *Global Information Technology Report 2006*, Hampshire: Palgrave Macmillan. pp. 71–87.

Balazs, Etienne. 1964. *Chinese Civilization and Bureaucracy: Variations on a Theme*. New Haven and London: Yale University Press.

Boulanger, P. M. 2007. "Political uses of social indicators: overview and application to sustainable development indicators." *International Journal of Sustainable Development* 10(1–2):14–32.

Box, Sarah. 2009. "OECD Work on Innovation—A Stocktaking of Existing Work." STI Working Paper 2009/2. Science and Technology Policy. Paris: OECD.

Chen, Shaohua and Martin Ravallion. 2008. "How have the world's poorest fared since the early 1980s?" (World Bank Development Economics Research Group). In Paul Collier and Jan Willem Gunning (eds.) *Globalization and Poverty*. Cheltenham: Edward Elgar Publishing.

Collier, Paul. 2007. *The Bottom Billion: Why the Poorest Countries Are Failing and What Can Be Done about It*. Oxford University Press.

Cooper, Richard. 2004a. "Half a Century of Development." Weatherhead Center for International Affairs. Harvard University.

———. 2004b. "A Glimpse of 2020." In The Global Competitiveness Report 2004–2005. Hampshire: Palgrave Macmillan. pp. 149–158.

Council for Economic Planning and Development (CEPD). 2002. *Challenge 2008 National Development Plan*. CEPD Series No. (91)048.904.

———. 2008a. *Economic Development, Taiwan, R.O.C. 2008*.

———. 2008b. *Taiwan Statistical Data Book*. Available at: http://www.cepd.gov.tw

———. 2009. Third-Term Plan for National Development in the New Century: Briefing, Taiwan. January.

Dahl, A. L. and A. Lopez-Claros. 2006. "The impact of information and communications technologies on the economic competitiveness and social development of Taiwan." In *The Global Information and Technology Report*. World Economic Forum. Hampshire: Palgrave Macmillan. pp. 107–18.

Diamond, Jared. 1999. *Guns, Germs, and Steel: The Fates of Human Societies*. New York and London: W. W. Norton and Company.

Easterly, William. 2002. *The Elusive Quest for Growth*. Cambridge, MA: MIT Press.

The Economist. 2007. "Chile: Destitute No More." 18 August.

———. 2008. "Hocus-pocus: The real world consequences of producing unreal inflation figures." 14 June.

———. 2009. "A chance to change course." 20 June.

European Innovation Scoreboard. 2007. "Comparative Analysis of Innovation Performance." PRO INNO Working Paper Nº6.

Ferguson, Niall. 2008. *The Ascent of Money: A Financial History of the World*. Allen Lane.

Financial Times. 2008. "Transparency Group Fears for Staff in Bosnia." 22 July.

———. 2008. "Number of poor rises in developing countries—China bucks trend, UN figures show; targets for 2015 still within reach." 12 September.

Friedman, Thomas. 1999. *The Lexus and the Olive Tree: Understanding Globalization*. New York: Farrar Strauss Giroux.

Gianella, Christian and William Tompson. 2007. "Stimulating Innovation in Russia: The Role of Institutions and Policies." Economics Department Working Paper. Paris: OECD.

Goldstone, Jack. 1996. "Gender, Work and Culture: Why the Industrial Revolution Came Early to England But Late to China." *Sociological Perspectives* 39(1 Spring):1–21.

The Growth Commission. 2008. *The Growth Report: Strategies for Sustained Growth and Inclusive Development*. May. Available at: www.growthcommission.org

INSEAD. 2009. *Global Innovation Index 2008–2009*. Available at: http://elab.insead.edu

International Data Corporation of Chile. 2004. "Estudio de Banda Ancha en Chile 2002–2010." Santiago, Chile. March.

International Monetary Fund. 2005. "Review of the IMF's Trade Restrictiveness Index: Background Paper to the Review of Fund Work on Trade." 14 February. Washington, D.C.: International Monetary Fund. Available at: www.imf.org

———. 2009a. World Economic Outlook: Crisis and Recovery. Washington, D.C. April.

———. 2009b. India: Selected Issues. International Monetary Fund Country Report No. 09/186. June.

Kaufmann, D. 2003. "Governance Redux: The Empirical Challenge." *The Global Competitiveness Report 2003–2004*. Hampshire: Palgrave Macmillan.

Kovalev, S. 2007. "Why Putin Wins." *New York Review of Books* 54(18). 22 November.

Landes, David. 1998. *The Wealth and Poverty of Nations*. Little, Brown and Company.

López-Claros, A. 2003. "The Risks and Rewards of BP's Russia Gamble." *The Wall Street Journal*. 17 February.

———. 2004. "Chile: The Next Stage of Development", *Global Competitiveness Report 2004-2005*. Hampshire: Palgrave Macmillan. pp. 111–24.

———. 2005. "Russia: Competitiveness, Growth, and the Next Stage of Development." In *Global Competitiveness Report 2005-2006*, Hampshire, Palgrave Macmillan.

López-Claros, A. and S. Alexashenko. 1998. "Fiscal Policy Issues During the Transition in Russia." *International Monetary Fund, Occasional Paper 155*. Washington, DC

López-Claros, A. and Irene Mia. 2006. "Israel: Factors in the Emergence of an ICT Powerhouse." *Global Information Technology Report 2005–2006*. Hampshire: Palgrave Macmillan. pp. 89–105.

Massachussetts Technology Collaborative. 2008. *Index of the Massachussetts Innovation Economy*. John Adams Innovation Institute of the Massachusetts Technology Collaborative. Available at: http://www.masstech.org/

Mississippi Technology Alliance. Available at: http://www.innovationindex.ms/

Morse, E. and J. Richard. 2002. "The Battle for Energy Dominance." *Foreign Affairs* 81:16–31.

National Science Council. 2008. Indicators of Science and Technology: Taiwan 2008. National Science Council of Taiwan. Available at: http//www.nsc.gov.tw/tech/

Organisation for Economic Co-operation and Development (OECD). 2000. *A New Economy? The Changing role of innovation and information technology in growth*. Paris: OECD.

———. 2008. *Handbook on Constructing Composite Indicators: Methodology and User Guide*. Paris: OECD.

OECD and European Communities. 2005. *Oslo Manual: Guidelines for Collecting and Interpreting Innovation data*, Joint Publication of the OECD and the Statistical Office of the European Communities.

OECD and European Community Joint Research Centre. 2008. *Handbook on constructing composite indicators: Methodology and user guide*. Geneva.

Oregon Innovation Council. 2007. *Innovation Index Oregon 2007*. Oregon Innovation Council. Available at: http://www.oregoninc.org

Owen, D. and D. Robinson. 2003. *Russia Rebounds*. International Monetary Fund.

Pavlidis, P. and W. S. Noble. 2001. "Analysis of strain and regional variation in gene expression in mouse brain." *Genome Biology* 2: Research 0042.1-0042.15.

Pipes, Richard. 2009. "Russia's Pride Could Diminish Its Power." *The Wall Street Journal*. 24 August.

Porter, Michael. 1990. *The Competitive Advantage of Nations*. The Free Press.

Porter, Michael and Scott Stern. 2002. "National Innovative Capacity." *Global Competitiveness Report 2001–2002*. New York: Oxford University Press.

———. 2003. "Ranking National Innovative Capacity: Findings from the National Innovative Capacity Index." *Global Competitiveness Report 2003-2004*. Oxford University Press. pp. 91–115.

———. 1999. *The new challenge to America's prosperity: Findings from the Innovation Index*. Washington, D.C.: Council on Competitiveness.

Presidential Commission for New Information Technologies. 1999. "Chile: Moving Toward the Information Society." Santiago, Chile.

Republic of Chile. 2004. "Agenda Digital 2004/2006." Ministry of the Economy. Santiago, Chile.

Rostow, W. W. 1960. *The Stages of Economic Growth*. Cambridge, MA: Cambridge University Press.

Saeed, A. I., V. Sharov, J. White, J. Li, W. Liang, N. Bhagabati, J. Braisted, M. Klapa, T. Currier, M. Thiagarajan, A. Sturn, M. Snuffin, A. Rezantsev, D. Popov, A. Ryltsov, E. Kostukovich, I. Borisovsky, Z. Liu, A. Vinsavich, V. Trush, and J. Quackenbush. 2003. "TM4: A free, open-source system for microarray data management and analysis." *Biotechniques* 34(2):374–8. Available at: http://www.tm4.org/mev.html

Sala-i-Martin, Xavier and Elsa Artadi. 2004. "The Global Competitiveness Index." *The Global Competitiveness Report 2004–2005*. Hampshire: Palgrave Macmillan. pp. 51–80.

Sells, Michael. 1999. *Approaching the Qur'an*. Ashland, OR: White Cloud Press.

Sen, Amartya. 1999. *Development as Freedom*. Oxford University Press.

———. "Adam Smith's market never stood alone." *Financial Times*. 11 March, 2009.

Shiller, Robert. 2009. "A failure to control the animal spirits." *Fi-

nancial Times. 9 March.

Siegle, Joseph T., Michael W. Weinstein, and Morton Halperin. 2004. "Why Democracies Excel." *Foreign Affairs* 83(5 September/October):57–71.

Smith, Adam. 1994. *An Inquiry Into the Nature and Causes of the Wealth of Nations.* New York: The Modern Library Edition.

Stiglitz, Joseph. 2006. *Making Globalization Work.* London: Allen Lane.

———. 2009. "Progress: What Progress?" *The OECD Observer* 272. Paris. April.

Tanzi, V. 1993. "The Changing Role of the State in the Economy: A Historical Perspective." *IMF Working Paper* 97/114. Washington: International Monetary Fund.

Transparency International. 2008. *Corruption Perceptions Index 2008.* Annual Report. Berlin.

United Nations. "Charter of the United Nations and Statute of the International Court of Justice." Department of Public Information. Available at: http://www.un.org/aboutun/charter/

———. 2004. *Global E-Government Readiness Report 2004: Towards Access for Opportunity.* New York.

United Nations Development Programme (UNDP). Various years. *The Human Development Report.* Available at: www.undp.org

———. *Gender Empowerment Measure* (GEM). Available at: www.undp.org

———. *Gender-related Development Index* (GDI). Available at: www.undp.org

United States Patent and Trademark Office. Available at: http://www.uspto.gov/

World Bank. 2005. "Country Policy and Institutional Assessments: 2005 Assessment Questionnaire." Operations Policy and Country Services. 20 December. Available at: www.worldbank.org

———. 2008a. *Doing Business Report.* World Bank and International Finance Corporation, Washington DC.

———. 2008b. *World Development Indicators,* Washington DC. Available at: http://go.worldbank.org/U0FSM7AQ40

Appendix. Innovation Capacity Index: Variable definitions

Variable	Source	Definition (as described by source) [1]
Pillar 1: Institutional environment		
Good governance		
Voice and accountability	World Governance Institute (WGI)—World Bank	Aggregate indicator. Measures the extent to which country's citizens are able to participate in selecting their government, as well as freedom of expression, freedom of association, and a free media.
Political stability	WGI	Aggregate indicator. Measures the perceptions of the likelihood that the government will be destabilized or overthrown by unconstitutional or violent means, including domestic violence and terrorism.
Government effectiveness	WGI	Aggregate indicator. Measures the quality of public services, the quality of the civil service and the degree of its independence from political pressures, the quality of policy formulation and implementation, and the credibility of the government's commitment to such policies.
Rule of law	WGI	Aggregate indicator. Measures the extent to which agents have confidence in and abide by the rules of society, in particular the quality of contract enforcement, the police, and the courts, as well as the likelihood of crime and violence.
Property rights framework	Aggregate indicator	It is the average of the following aggregate indicators: "Property rights" and "Enforcing contracts."
Property rights	World Bank and WEF	The value of this indicator is given preferentially by the World Bank "Country Policy and Institutional Assessment (CPIA) property rights and rule-based governance" ratings. This criterion assesses the extent to which private economic activity is facilitated by an effective legal system and rule-based governance structure in which property and contract rights are reliably respected and enforced. Each of three dimensions is rated separately: (a) legal basis for secure property and contract rights; (b) predictability, transparency, and impartiality of laws and regulations affecting economic activity, and their enforcement by the legal and judicial system; and (c) crime and violence as an impediment to economic activity. For those countries without this rating, an estimate was made using the World Economic Forum's (WEF) Executive Opinion Survey (EOS) data on property rights and intellectual property protection.
Enforcing contracts	DBR	Average of the three scores corresponding to the World Bank's *Doing Business Report* (DBR) enforcing contracts variables: "number of procedures," "time," and "cost." Indicators on enforcing contracts measure the efficiency of the judicial system in resolving a commercial dispute. The data are collected by studying the codes of civil procedure and other court regulations as well as surveys completed by local litigation lawyers (and, in a quarter of the countries, by judges as well). A procedure is defined as any interaction between the parties, or between them and the judge or court officer. This includes steps to file the case, steps for trial and judgment and steps necessary to enforce the judgment. Time is recorded in calendar days, counted from the moment the plaintiff files the lawsuit in court until payment. This includes both the days when actions take place and the waiting periods between. The respondents make separate estimates of the average duration of different stages of dispute resolution: the completion of service of process (time to file the case), the issuance of judgment (time for the trial and obtaining the judgment) and the moment of payment (time for enforcement). Cost is recorded as a percentage of the claim, assumed to be equivalent to 200 percent of income per capita. Only official costs required by law are recorded, including court and enforcement costs and average attorney fees where the use of attorneys is mandatory or common.
Transparency and judicial independence	World Bank and WEF	The value of this indicator is given preferentially by the World Bank CPIA "transparency, accountability, and corruption in the public sector" ratings. This criterion assesses the extent to which the executive can be held accountable for its use of funds and the results of its actions by the electorate and by the legislature and judiciary, and the extent to which public employees within the executive are required to account for the use of resources, administrative decisions, and results obtained. Each of these three dimensions was rated separately with equal weighting: (a) the accountability of the executive to oversight institutions and of public employees for their performance; (b) access of civil society to information on public affairs; and (c) state capture by narrow vested interests. For those countries without this rating, an estimate was made using the WEF's EOS ratings on "transparency of government policy making," "judicial independence," and "diversion of public funds."

[1] The variable definitions provided here reflect, for the most part, those provided by the compiling organizations themselves.

Variable	Source	Definition (as described by source) [1]
Pillar 1: Institutional environment		
Corruption Perceptions Index	Transparency International (TI)	A country or territory's corruptions perception index score indicates the degree of public sector corruption as perceived by business people and country analysts, and ranges between 10 (highly clean) and 0 (highly corrupt).
Country policy assessment		
Public sector management		
Quality of budgetary and financial management	World Bank, WEF and *Institutional Investor* magazine Country Credit Survey	This indicator is the average of two components: a quality of budgetary and financial management score, as described below, and a credit rating score. The value of the first part of this indicator is given preferentially by the World Bank CPIA "quality of budgetary and financial management" ratings. This criterion assesses the extent to which there is: (a) a comprehensive and credible budget, linked to policy priorities; (b) effective financial management systems to ensure that the budget is implemented as intended in a controlled and predictable way; and (c) timely and accurate accounting and fiscal reporting, including timely and audited public accounts and effective arrangements for follow up. Each of these three dimensions was rated separately. For those countries without this rating, an estimate was made using the WEF's EOS "wastefulness of government spending" ratings. For the credit rating score the country-by-country credit ratings developed by the *Institutional Investor* magazine were used. These are based on information provided by senior economists and sovereign-risk analysts at leading global banks and money management and securities firms. They have graded each country on a scale of 0 to 100, with 100 representing those countries that have the least chance of default. Participants are not permitted to rate their home countries. The individual credit responses are weighted using an institutional investor formula that gives more importance to responses from institutions with greater worldwide exposure and more-sophisticated country analysis systems.
Quality of public administration	World Bank and WEF	The value of this indicator is given preferentially by the World Bank CPIA "quality of public administration" ratings. This criterion assesses the extent to which civilian central government staffs (including teachers, health workers, and police) are structured to design and implement government policy and deliver services effectively. Civilian central government staffs include the central executive together with all other ministries and administrative departments, including autonomous agencies. It excludes the armed forces, state-owned enterprises, and sub-national government. The key dimensions for assessment are: policy coordination and responsiveness; service delivery and operational efficiency; merit and ethics; pay adequacy and management of the wage bill. For those countries without this rating, an estimate was made using the "favoritism in decisions of government officials" and "public trust of politicians" ratings of the WEF's EOS.
Structural policies		
Financial sector efficiency	World Bank and WEF	The value of this indicator is given preferentially by the World Bank CPIA "financial sector" ratings. This criterion assesses the structure of the financial sector and the policies and regulations that affect it. Three dimensions are covered: (a) financial stability; (b) the sector's efficiency, depth, and resource mobilization strength; and (c) access to financial services. These are areas that are fundamental to support successful and sustainable reforms and development. The first dimension assesses the sector's vulnerability to shocks, the banking system's soundness, and the adequacy of relevant institutional elements, such as the degree of adherence to the base core principles and the quality of risk management and supervision. The second dimension assesses efficiency, the degree of competition, and the ownership structure of the financial system, as well as its depth and resource mobilization strength. The third dimension covers institutional factors (such as the adequacy of payment and credit reporting systems), the regulatory framework affecting financial transactions (including collateral and bankruptcy laws and their enforcement), and the extent to which consumers and firms have access to financial services. For those countries without this rating, an estimate was made using the "financial market sophistication," "venture capital availability" and "ease of access to loans" ratings from the WEF's EOS.

Variable	Source	Definition (as described by source) [1]

Pillar 1: Institutional environment

Variable	Source	Definition (as described by source) [1]
Trade openness	World Bank World Trade Indicators (WTI)	TTRI, *Trade Tariff Restrictiveness Index*, (MFN applied tariff) – all goods. This Index summarizes the impact of each country's non-discriminatory trade policies on its aggregate imports. It is the uniform equivalent tariff that would maintain the country's aggregate import volume at its current level (given heterogeneous tariffs). It captures the trade distortions that each country's MFN (most favored nation) tariffs impose on its import bundle using estimated elasticities to calculate the impact of a tariff schedule on a country's imports. These measures are based on actual or current trade patterns and thus do not capture restrictions facing new or potential trade. They also do not take into account domestic subsidies or export taxes. Expressed as a tariff rate.
Foreign direct investment gross inflows	UN Conference on Trade and Development (UNCTAD)	Definitions of foreign direct investment (FDI) used by the UNCTAD WIR are contained in the *Balance of Payments Manual: Fifth Edition* (BPM5) (Washington, D.C., International Monetary Fund, 1993) and the *Detailed Benchmark Definition of Foreign Direct Investment: Third Edition* (BD3) (Paris, Organisation for Economic Co-operation and Development, 1996). According to the BPM5, FDI refers to an investment made to acquire lasting interest in enterprises operating outside of the economy of the investor. Further, in cases of FDI, the investor's purpose is to gain an effective voice in the management of the enterprise. Expressed as percent of GDP.

Macroeconomy

Debt levels	IMF World Economic Outlook (WEO), IMF Country Reports, CIA and World Bank World Development Indicators (WDI)	Gross debt comprises the stock (at year-end) of all government gross liabilities (both to residents and non-residents), in percent of GDP. To avoid double counting, the data are based on a consolidated account (eliminating liabilities and assets between components of the government, such as budgetary units and social security funds). General government reflects a consolidated account of central government plus state, provincial, or local governments.
Fiscal balance	World Bank WDI, IMF Country Reports	Cash deficit/surplus, defined as revenue (including grants) minus expenditures, minus net acquisition of non-financial assets, in percent of GDP.
Macrostability	International Financial Statistics (IFS), IMF WEO and Country Reports	This value is the weighted average of these three scores: "inflation," "interest rate spread," and "national savings rate." The average interest rate spread measures the difference between market short-term lending and deposit rates as published in the IMF's International Financial Statistics. The national savings rate is the share of GDP saved by households within the year. Consumer prices are annual percentage changes in the CPI; we use averages for the year, not end-of-period data.

Variable	Source	Definition (as described by source) [1]

Pillar 2: Human capital, training and social inclusion

Good governance

Education

Adult literacy rate	World Bank WDI	The proportion of the adult population aged 15 years and older which is literate, expressed as a percentage of the corresponding population in a given country, territory, or geographic area, at a specific point in time, usually mid-year.
Secondary gross enrolment ratio	World Bank WDI	Number of pupils enrolled in a given level of education, regardless of age, expressed as a percentage of the population in the theoretical age group for the same level of education.
Tertiary gross enrolment ratio	World Bank WDI	Number of pupils enrolled in a given level of education, regardless of age, expressed as a percentage of the population in the theoretical age group for the same level of education. For the tertiary level, the population used is the five-year age group following on from the secondary school-leaving age.

Variable	Source	Definition (as described by source) [1]

Pillar 2: Human capital, training and social inclusion

Variable	Source	Definition (as described by source) [1]
Expenditure in education	World Bank WDI	Public spending in education includes both capital expenditures (spending on construction, renovation, major repairs and purchases of heavy equipment or vehicles) and current expenditures (spending on goods and services that are consumed within the current year and which must be renewed the following year, including such expenditures as staff salaries and benefits, contracted or purchased services, books and teaching materials, welfare services, furniture and equipment, minor repairs, fuel, insurance, rents, telecommunications, and travel). Expressed in percent of GDP.

Social inclusion and equity policies

Variable	Source	Definition (as described by source) [1]
Gender equity	UN HDR	The value of this indicator is given preferentially by the United Nations (UN) Human Development Report (HDR) "Gender Empowerment Measure" (GEM), a composite index measuring gender inequality in three basic dimensions of empowerment: economic participation and decision-making, political participation, and decision making and power over economic resources. For those countries without this value, an estimate was made using the UNHDR "Gender-Related Development Index" (GDI), measuring average achievement in the three basic dimensions captured in the human development index: a long and healthy life, knowledge, and a decent standard of living, adjusted to account for inequalities between men and women.
Environmental sustainability	2008 Environmental Perfomance Index	The 2008 Environmental Performance Index (EPI) ranks 149 countries on 25 indicators tracked across six established policy categories: environmental health, air pollution, water resources, biodiversity and habitat, productive natural resources, and climate change. The EPI identifies broadly accepted targets for environmental performance and measures how close each country comes to these goals. As a quantitative gauge of pollution control and natural resource management results, the Index provides a powerful tool for improving policymaking and shifting environmental decision making onto firmer analytic foundations.
Health worker density	World Bank WDI	It is calculated as a weighted average of the number of physicians, nurses, and midwives per 1000 people. Physicians are defined as graduates of any facility or school of medicine who are working in the country in any medical field (practice, teaching, research), including generalists and specialists. Nurses include professional, auxiliary, and enrolled nurses and others, such as those in dental and primary care. Midwives include professional, auxiliary, and enrolled midwives.
Inequality measure	UN HDR, World Bank WDI	The ratio of the income or expenditure share of the richest 20 percent group to that of the poorest 20 percent.

Variable	Source	Definition (as described by source) [1]

Pillar 3: Regulatory and legal framework

Doing business

Starting a business

Variable	Source	Definition (as described by source) [1]
Number of procedures	DBR (Doing Business Report)	A procedure is defined as any interaction of the company founder with external parties (for example, government agencies, lawyers, auditors, or notaries). Includes procedures to legally start and operate a company, preregistration (name verification, notarization), registration in the economy's most populous city, and post-registration (social security registration, company seal).
Time	DBR	Time in days required to complete each procedure. It does not include time spent gathering information. Each procedure starts on a separate day. It is considered completed once final document is received. No prior contact with officials is needed. If a procedure can be accelerated for an additional cost, the fastest procedure is chosen.

Variable	Source	Definition (as described by source) [1]
Pillar 3: Regulatory and legal framework		
Cost	DBR	Cost as percent of income per capita required to complete each procedure: official costs only, no bribes, and no professional fees, unless these services are required by law.
Ease of employing workers		
Ease of employing workers	DBR	This value is the average of these three DBR employing worker scores: "difficulty of hiring index," "rigidity of hours index," and "difficulty of firing index." The difficulty of hiring index measures whether fixed-term contracts are prohibited for permanent tasks, the maximum cumulative duration of fixed-term contracts, and the ratio of the minimum wage for a trainee or first-time employee to the average value added per worker. The rigidity of hours index has five components: whether night or weekend work is unrestricted, whether the workweek can consist of 5.5 days; whether the workweek can extend to 50 hours or more (including overtime) for two months a year to respond to a seasonal increase in production; and whether paid annual vacation is 21 working days or fewer. The difficulty of firing index has eight components: whether redundancy is disallowed as a basis for terminating workers, whether the employer needs to notify a third party (such as a government agency) to terminate 1 redundant worker, whether the employer needs to notify a third party to terminate a group of 25 redundant workers, whether the employer needs approval from a third party to terminate 1 redundant worker, whether the employer needs approval from a third party to terminate a group of 25 redundant workers, whether the law requires the employer to consider reassignment or retraining options before redundancy termination, whether priority rules apply for redundancies, and whether priority rules apply for reemployment.
Paying taxes		
Paying taxes	Aggregate indicator	This value is the average of these three DBR paying taxes scores: "number of payments per year," "hours per year," and "total tax rate." The tax payments indicator reflects the total number of taxes and contributions paid per year, the method of payment, the frequency of payment, and the number of agencies involved for this standardized case during the second year of operation. Time is recorded in hours per year. The indicator measures the time to prepare, file, and pay (or withhold) three major types of taxes and contributions: the corporate income tax, value added or sales tax and labor taxes, including payroll taxes and social contributions. Includes collecting information to compute tax payable, completing tax forms, filing with proper agencies, arranging payment or withholding, and preparing separate tax accounting books. The total tax rate measures the amount of taxes and mandatory contributions payable by the business in the second year of operation, expressed as a share of commercial profits. Includes: profit or corporate income tax, social contributions and labor taxes paid by the employer, property and property transfer taxes, dividend, capital gains, and financial transactions taxes, waste collection, vehicle, road, and other taxes.
Protecting investors		
Strength of investor protection	DBR	Strength of investor protection index: The average of the extent of the "disclosure," "extent of director liability," and "ease of shareholder suits" indexes.
Registering property		
Number of procedures	DBR	Procedures to legally transfer title on real property, including: preregistration (checking for liens, notarizing sales agreement), registration in the economy's most populous city, and post-registration (paying taxes, filing title with municipality).
Time	DBR	Time in days required to complete each procedure for registering property. Does not include time spent gathering information. Each procedure starts on a separate day. A procedure is considered completed once final document is received. No prior contact with officials is needed.
Cost	DBR	Cost is recorded as a percentage of the property value, assumed to be equivalent to 50 times income per capita. Only official costs required by law are recorded, including fees, transfer taxes, stamp duties, and any other payment to the property registry, notaries, public agencies, or lawyers.

Variable	Source	Definition (as described by source) [1]

Pillar 4: Research and development

R&D infrastructure

Research and development expenditure	World Bank WDI	Current and capital expenditures (including overhead) on creative, systematic activity intended to increase the stock of knowledge. Included are fundamental and applied research and experimental development work leading to new devices, products, or processes. Expressed as percent of GDP.
Information and communication technology expenditure	World Bank WDI	Includes external spending on information technology ("tangible" spending on information technology products purchased by businesses, households, governments, and education institutions from vendors or organizations outside the purchasing entity), internal spending on information technology ("intangible" spending on internally customized software, capital depreciation, and the like), and spending on telecommunications and other office equipment. Expressed as percent of GDP.
R&D worker density	World Bank WDI	It is calculated as a weighted average of the number of researchers and technicians in R&D per million people. Researchers are people trained to work in any field of science who are engaged in professional research and development activity, usually requiring the completion of tertiary education. Technicians in R&D are people engaged in professional R&D activity, who have received vocational or technical training (usually three years beyond the first stage of secondary education) in any branch of knowledge or technology of a specified standard.
Students in science and engineering	UN HDR	Students in science, engineering, manufacturing, and construction: The share (percent) of tertiary students enrolled in natural sciences; engineering, mathematics, and computer sciences; architecture and town planning; transport and communications; trade, craft, and industrial programmes; and agriculture, forestry, and fisheries.
Scientific and technical journal articles	World Bank WDI	Scientific and engineering technical journal articles per million people published in the following fields: physics, biology, chemistry, mathematics, clinical medicine, biomedical research, engineering and technology, and earth and space sciences.
Schools connected to the Internet	World Bank WDI	Schools connected to the Internet are the share (percent) of primary and secondary schools in the country that have access to the Internet.

Patents and trademarks

Patents granted to residents	*Trilateral Cooperation Statistical Report* (TCSR)	Patents are documents issued by a government office that grant a set of exclusive rights for exploitation (made, used, sold, and imported) of an invention to an inventor or his assignee for a fixed period of time, in exchange for the disclosure and description of the invention. The data correspond to patents granted by the US Patent and Trademark Office (USPTO), European Patent Office (EPO), or Japan Patent Office (JPO). Data for each country represent the highest number of patents granted from either office, according to the 2007 TCSR. Data are per million people.
Trademark applications filed by residents	World Bank WDI	A trademark is any distinctive word, sign, indicator, or a combination of these used by an individual, business organization, or other legal entity to identify that the products and/or services with this trademark have the same origin, and to distinguish them from others in the marketplace or trade. An application for registration of a trademark must be filed with the appropriate national or regional trademark office. Data are per million people.
Receipts of royalty and license fees	World Bank WDI	Receipts between residents and non-residents for the authorized use of intangible, non-produced, non-financial assets and proprietary rights (such as patents, trademarks, copyrights, franchises, and industrial processes) and for the use, through licensing agreements, of produced originals of prototypes (such as films and manuscripts). Data are based on the balance of payments and are on a current US$ per person basis.
Payment of royalty and license fees	World Bank WDI	Payments between residents and non-residents for the authorized use of intangible, non-produced, non-financial assets and proprietary rights (such as patents, copyrights, trademarks, industrial processes, and franchises) and for the use, through licensing agreements, of produced originals of prototypes (such as manuscripts and films). Data are in current US$ per person and are derived from the balance of payments.

Variable	Source	Definition (as described by source) [1]
Pillar 5: Adoption and use of information and communication technologies		
Telephone Communications		
Main (fixed) telephone lines	*International Telecommunication Union (ITU)*	A main line is a (fixed) telephone line connecting the subscriber's terminal equipment to the public switched network, and having a dedicated port in the telephone exchange equipment. This term is synonymous with the terms main station or Direct Exchange Line (DEL) commonly used in telecommunication documents. It may not be the same as an access line or a subscriber. The number of ISDN channels and fixed wireless subscribers should be included. Data are expressed per 100 inhabitants.
Waiting list for main (fixed) lines	ITU	Un-met applications for connection to the Public Switched Telephone Network (PSTN) due to a lack of technical facilities (equipment, lines, etc.). The waiting list should reflect the total number reported by all PSTN service providers in the country. Data are expressed per 1000 inhabitants.
Business connection charge	ITU	Installation (or connection) refers to the one-off charge involved in applying for business basic telephone service. Where there are different charges for different exchange areas, the charge for the largest urban area should be used and specified in a note. Data are expressed as percent of GDP/capita.
Business monthly subscription	ITU	Monthly subscription refers to the recurring fixed charge for a business subscription to the PSTN. The charge should cover the rental of the line but not the rental of the terminal (e.g., telephone set) where the terminal equipment market is liberalized. Separate charges for first and subsequent lines should be stated where appropriate. If the rental charge includes any allowance for free or reduced rate call units, this should be indicated. If there are different charges for different exchange areas, the largest urban area should be used and specified in a note. Data are expressed as percent of GDP/capita.
Residential connection charge	ITU	Installation (or connection) refers to the one-off charge involved in applying for residential basic telephone service. Where there are different charges for different exchange areas, the charge for the largest urban area should be used and specified in a note. Data are expressed as percent of GDP/capita.
Residential monthly subscription	ITU	Monthly subscription refers to the recurring fixed charge for a residential subscription to the PSTN. The charge should cover the rental of the line, but not the rental of the terminal (e.g., telephone set) where the terminal equipment market is liberalized. Separate charges for first and subsequent lines should be stated where appropriate. If the rental charge includes any allowance for free or reduced rate call units, this should be indicated. If there are different charges for different exchange areas, the largest urban area should be used and specified in a note. Data are expressed as percent of GDP/capita.
Mobile cellular communications		
Subscribers	ITU	Refers to the use of portable telephones subscribing to a public mobile telephone service and provides access to Public Switched Telephone Network (PSTN) using cellular technology. This can include analog and digital cellular systems. This should also include subscribers to IMT-2000 (Third Generation, 3G). Subscribers to public mobile data services or radio paging services should not be included. Data are per 100 inhabitants.
Prepaid subscribers	ITU	Number of mobile cellular subscribers using prepaid cards. These are subscribers who, rather than paying a fixed monthly subscription fee, choose to purchase blocks of usage time. Only active prepaid subscribers who have used the system within a reasonable period of time should be included. This period (e.g., 3 months) should be indicated in a note. Data are per 100 inhabitants.
Population coverage	ITU	Mobile cellular coverage of population in percent. This indicator measures the percentage of inhabitants who are within range of a mobile cellular signal, irrespective of whether or not they are subscribers. This is calculated by dividing the number of inhabitants within range of a mobile cellular signal by the total population. Note that this is not the same as the mobile subscription density or penetration.
Connection charge	ITU	The initial, one-time charge for a new subscription. Refundable deposits should not be counted. Although some operators waive the connection charge, this does not include the cost of the Subscriber Identity Module (SIM) card. The price of the SIM card should be included in the connection charge. A note should indicate whether taxes are included (preferred) or not. It should also be noted if free minutes are included in the plan. Data are expressed as percent of GDP/capita.

Variable	Source	Definition (as described by source) [1]
Pillar 5: Adoption and use of information and communication technologies		
Internet, computers, and TV		
Total fixed internet subscribers	*ITU*	The number of total Internet subscribers with fixed access, including dial-up, total fixed broadband, cable modem, DSL Internet, other broadband, and leased line Internet subscribers. Only active subscribers who have used the system within a reasonable period of time should be included. This period (e.g., 3 months) should be indicated in a note. Data are per 100 inhabitants.
Total fixed broadband subscribers	ITU	Total Internet subscribers excluding dial-up Internet: cable-modem (cable tv), DSL, leased line, and others (satellite, fibre, LAN, wireless, wimax...). Total broadband Internet subscribers refers to a subscriber who pays for high-speed access to the public Internet (a TCP/IP connection), at speeds equal to, or greater than, 256 kbit/s, in one or both directions. If countries use a different definition of broadband, this should be indicated in a note. This total is measured irrespective of the method of payment. It excludes subscribers with access to data communications (including the Internet) via mobile cellular networks. Data are per 100 inhabitants.
Internet users	ITU	The estimated number of Internet users per 100 inhabitants. A growing number of countries are measuring this through regular surveys. Surveys usually indicate a percentage of the population for a certain age group (e.g., 15–74 years old). The number of Internet users in this age group should be supplied and not the percentage of Internet users in this age group multiplied by the entire population. In situations where surveys are not available, an estimate can be derived based on the number of subscribers. The methodology used should be supplied, including reference to the frequency of use (e.g., in the last month).
Personal computers	ITU	The number of Personal Computers (PC) measures the number of computers installed in a country per 100 inhabitants. The statistic includes PCs, laptops, notebooks etc., but excludes terminals connected to mainframe and mini-computers that are primarily intended for shared use, and devices such as smart-phones that have only some, but not all, of the functions of a PC (e.g., they may lack a full-sized keyboard, a large screen, an Internet connection, drives, etc).
Television receivers	ITU	The total number of television sets per 100 inhabitants. A television set is a device capable of receiving broadcast television signals, using popular access means such as over-the-air, cable, and satellite. A television set may be a stand-alone device, or it may be integrated into another device, such as a computer or a mobile phone. It may be useful to distinguish between digital and analog signal delivery and between TV sets receiving only a limited number of signals (usually over-the-air) and those that have multiple channels available (e.g., by satellite or cable).
Government ICT usage		
E-government readiness index	UN Global E-*Government Readiness Report*	E-government readiness is a composite index comprising the Web measure index, the telecommunication infrastructure index and the human capital index. E-government is defined as the use of ICT and its application by the government for the provision of information and public services to the people. The aim of e-government therefore is to provide efficient government management of information to the citizen, better service delivery to citizens, and empowerment of the people through access to information and participation in public policy decision making.
Quality of the infrastructure		
Electrification rate	UN HDR	The number of people with electricity access as a percentage of the total population.
Electric power transmission and distribution losses	World Bank WDI	Electric power transmission and distribution losses include losses in transmission between sources of supply and points of distribution and in the distribution to consumers, including pilferage. It is expressed as percent of output.
Roads paved	World Bank WDI	Paved roads are those surfaced with crushed stone (macadam) and hydrocarbon binder or bituminized agents, with concrete, or with cobblestones, as a percentage of all the country's roads, measured in length.

Part 2 Dimensions of Innovation

Chapter 2.1

Enabling Innovative Entrepreneurship through Business Incubation

Mohsen A. Khalil,
Global Information and Communication
Technologies Department,
World Bank Group

Ellen Olafsen,
infoDev, Global Information and
Communication Technologies Department,
World Bank Group

Introduction

A country's primary socioeconomic goal is to improve the quality of life of its citizens. The competitiveness of the economy must be raised, opportunities that empower people to earn sustainable incomes must be created, and problems affecting the population, such as disease and environmental degradation must be alleviated. In pursuing this mission, a country's ability to reformulate the traditional model of economic growth is essential, so that knowledge, technology, entrepreneurship, and innovation are positioned at the center of its development agenda. Innovation, in particular, triggers a virtuous development circle that unleashes human ingenuity to develop and deliver products and services that are needed by the population and increase enterprise competitiveness, while simultaneously creating sustainable incomes and tax revenues that can be reinvested for social and economic gains.

In the development context, innovation should be viewed as changes in thinking, products, processes, organizations, or new ideas which are successfully applied. Innovation in business is thus defined by assessing the novelty of products, services, and processes relative to customers' current perception of value and their experience of alternative offerings. It is linked to performance and growth, through improvements in efficiency, productivity, quality, competitive positioning, and market share.

Inventions with potentially high social and economic value can be found in numerous sources, including the grassroots, academia, small and large enterprises, R&D centers, and government agencies. In today's global knowledge economy, people and institutions also have immediate access to inventions that have already been introduced in other countries and settings. However, the environment often discourages entrepreneurs from bringing inventions to market, regardless of the source. Many are not utilized because they are not adequately tailored to local needs. Thus, countries are faced with the challenge not only of spurring invention domestically or identifying existing inventions abroad that can be adapted to the local environment, but also of *creating the conditions that allow the invention to be coupled with entrepreneurship*, so that the economic and social wealth creation potential of the invention can be realized.

69

Globally, policymakers and their development partners have invested in a range of initiatives to create these favorable conditions, including policy and regulatory incentives, mechanisms to expand access to capital, and education reform. Within this landscape of interventions that link innovation and entrepreneurship is the process of business incubation, characterized by a focus on strengthening dynamic, growth-oriented, early-stage enterprises.

This chapter focuses on the use of business incubation as a tool to help developing countries bring new ideas to the market, and thereby create social and economic wealth. As shown in Box 1, the paper draws on infoDev's extensive experience with supporting business incubation across 80 developing countries, including a comprehensive monitoring and evaluation impact assessment (MEIA), concluded in 2007, which surveyed 49 business incubators in 49 developing countries.

Business incubators within the innovation and entrepreneurship ecosystem

Business incubation is a process aimed at supporting the development and scaling of growth-oriented, early-stage enterprises.[1] The process provides entrepreneurs with an enabling environment at the start-up stage of enterprise development, to help reduce the cost of launching the enterprise, increase the confidence and capacity of the entrepreneur, and link the entrepreneur to the resources required to start and scale a competitive enterprise. Entrepreneurs accepted into the business incubator stay until an agreed upon milestone is reached, often measured in terms of sales revenue or profitability.

Business incubation is one of many tools aimed at fostering innovative enterprise creation and growth. There can be other complementary vehicles, such as business development centers and technology parks. Table 1 illustrates how business incubation is positioned vis-à-vis these two complementary vehicles.[2]

What infoDev refers to as the "innovation and entrepreneurship ecosystem" is an expansion of the so-called "triple helix" framework, also known as the "innovation system." While the innovation system framework is evolving, it emphasizes that there must be sufficient linkages between universities, industry, and government in order to spur innovation and to bring innovation to market (Lundvall, 1992; Nelson, 1993, Fagerberg and Nelson, 2004). Expanding upon this school of thought, infoDev's experience indicates that effective coupling of innovation and entrepreneurship requires what can be described as an ecosystem with active linkages between financiers, academia, policymakers, and the business community (Figure 1). If any one of these linkages is weak or non-existent, the entire system suffers and the ecosystem is not as effective at enabling innovative entrepreneurship as it could be.

Business incubators have a unique position in this ecosystem. They interact with all the actors in the ecosystem, either directly or indirectly, through the enterprises they serve, and feel first-hand the challenges that their clients face when seeking to set up and grow their enterprises, whether the difficulties have to do with regulations, finance, labor, or infrastructure. If these challenges are effectively communicated to the relevant actors in the ecosystem, a valuable feedback loop can be established which benefits not only the incubated enterprises, but innovative entrepreneurs across the economy.

Business incubators offer not only important feedback on

Box 1. infoDev's business incubation experience

infoDev's Business Incubation Network

This Network now spans 189 business incubators in 80 developing countries, 66 of which received direct support from infoDev in the form of technical assistance and grant funding used as co-financing in the start-up phase of the business incubator. The other 123 business incubators have joined infoDev's Business Incubator Network to benefit from the peer-to-peer networking and knowledge-sharing opportunities that the Network offers. More information about infoDev's work in business incubation and about the network can be found at: http://www.idisc.net/en/index.html

infoDev's Monitoring and Evaluation Impact Assessment (MEIA)

The MEIA was completed by an independent consulting company (OTF GROUP), with the aim of assessing the impact of business incubators that had received grant financing from infoDev. Business incubators in 49 developing countries participated in the survey, for which the assessment team carried out on-site observation and interviews. Most of the business incubators included in the assessment were less than five years old. The MEIA findings can be seen at: http://www.infodev.org/en/Project.77.html

---

[1] Although the tool can be used specifically to foster innovative enterprise development, it can also be successfully focused on creating competitive enterprises with high job-creation potential, regardless of whether or not the business concept is innovative.

[2] It should be noted that, in many instances, the distinction between these different tools is blurred; for example, a business incubator may offer business development services to non-incubatees to supplement its revenues; a technology park may include a business incubator to test new ideas, etc.

Table 1. Intermediary vehicles for innovative enterprise development

	Business development centers	Business incubators	Technology parks
Target enterprises	Any small and medium enterprise (SME)	Early-stage enterprises with high growth potential	Emerging and established technology businesses
Key features	• Ad hoc, demand-driven assistance • Focused on a particular issue for which the entrepreneur asks for assistance • Usually broad business support, including training and advisory services	• Emphasis on co-location and "cluster" effect between enterprises • Ongoing , supply and demand-driven assistance until an agreed upon performance milestone has been reached • Integrated mix of intensive strategic and operational support focused on the enterprise in its entirety	• Emphasis on co-location and "cluster" effect between enterprises • Demand-driven assistance • Emphasis on provision of state-of-the-art real estate, office space, and research facilities
Revenue sources	Government /donor subsidies, fee-for-service	Government/donor subsidies, fee-for-service, rent, royalties, equity	Government/donor subsidies, fee-for-service, rent, royalties, equity
Business model	Nonprofit or profit-making		

Figure 1. Innovation and entrepreneurship ecosystem

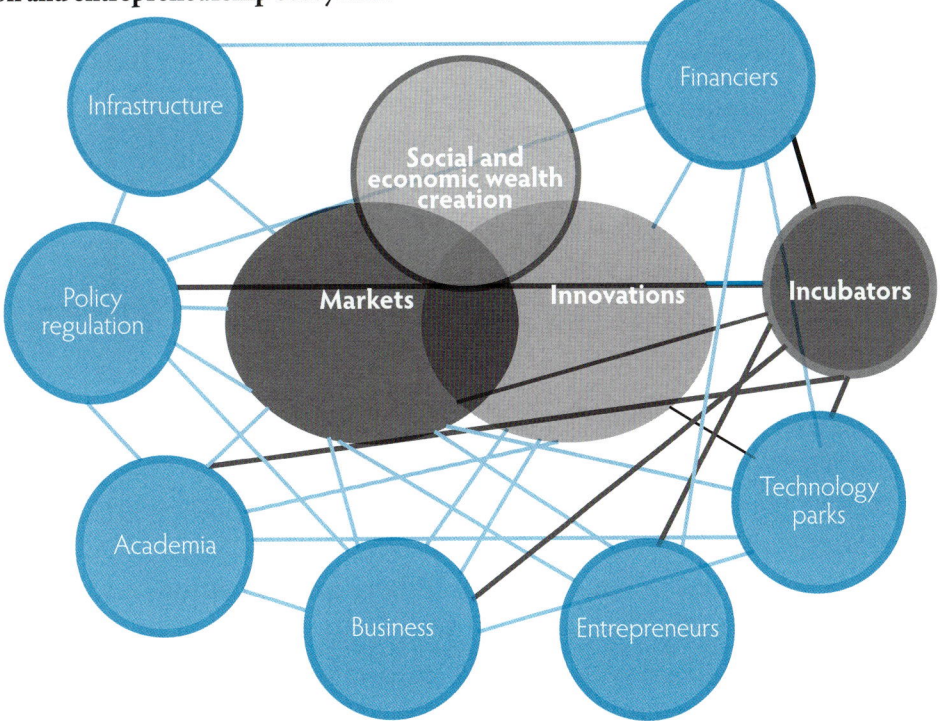

Source: infoDev.

challenges and needs, but also *opportunities* for other actors in the system. For instance, they can offer financiers a pool of high-growth potential investment and lending prospects at reduced risk, given the ongoing assistance that these entrepreneurs continue to receive in dealing with business challenges and opportunities. They can offer academic institutions a vehicle to commercialize research and/or assist graduates with setting up a new business venture, and also provide corporations access to innovative ideas that could potentially strengthen their supply chain, delivery mechanisms, or operations.

In this context, business incubators that manage to create effective relationships with the other actors in the ecosystem can serve as important levers to forge positive change that creates a more enabling environment for innovative entrepreneurs across the economy.

Understanding business incubation

Figure 2 from the Global Entrepreneurship Monitor (GEM) illustrates the stage of enterprise development at which business incubation is targeted. Business incubation is used at the "early-stage entrepreneurial activity" stage in the figure. At this stage, the new venture is more than an idea. The enterprise may already have made its first sales, and the innovator is ready to invest substantial time and resources in pursuing the new venture.[3]

Identifying an early-stage enterprise that will grow significantly if provided with a nurturing environment is not easy. In this respect, the role of the business incubation team is similar to that of a venture capitalist. It "invests" in management, not just in ideas. Good practice business incubators assess both the entrepreneur and the market potential of the business venture in order to determine the potential of the business. A variety of tools and methods are used to assess entrepreneurs and the market potential of business ideas. The process gener-

Figure 2. **Enterprise development stages**

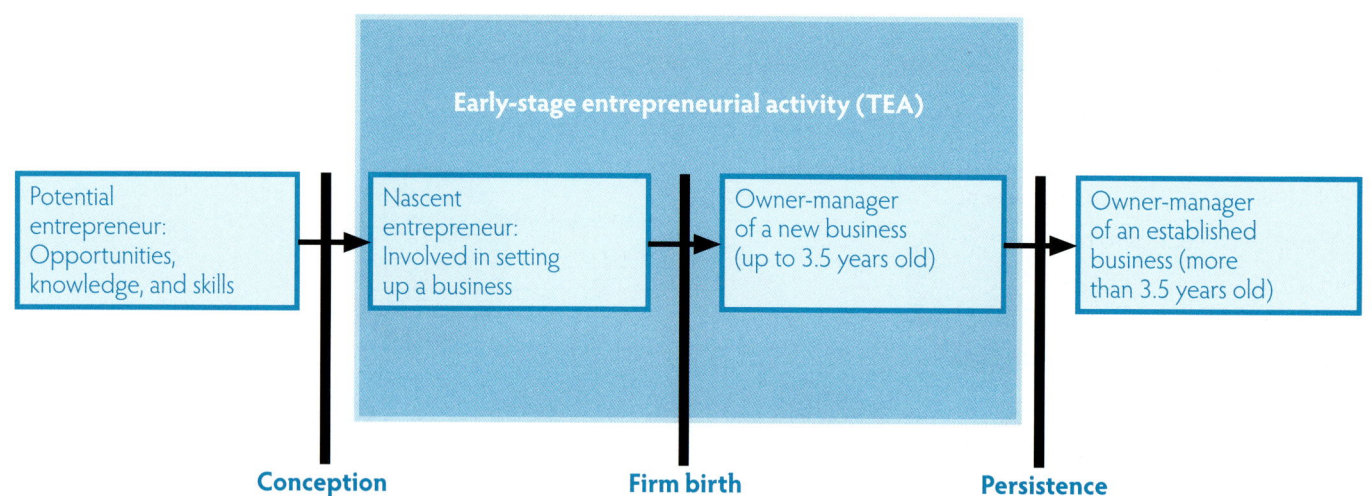

Business incubation is selective. The business incubation process is aimed at assisting growth-oriented entrepreneurs in their quest to grow and become more competitive.[4] Only a subset of entrepreneurs are growth-oriented and pursuing an innovative venture. The GEM index across 38 surveyed countries shows that only 5 to 35 percent of the early-stage entrepreneurs had novel product-market combinations. A critical mass of potential business incubation applicants is, therefore, necessary for business incubation to be an efficient tool for fostering innovative entrepreneurship.

ally involves the engagement of experienced business development professionals, financiers, and industry experts.[5]

Where are innovators and entrepreneurs found?

The human capacity to innovate and become an entrepreneur is everywhere in the world. It is the environment that unleashes human ingenuity or quells it. The GEM measures the percentage of early-stage entrepreneurs with novel product-market combinations. The GEM index shows that there is no specific geographic trend in the location of innovative entrepreneurs.

---

[3] Assistance provided at the "idea" stage is commonly referred to as "pre-incubation," and involves less intensive assistance to allow the entrepreneur to investigate his/her idea further before starting the enterprise.

[4] For the business owner, growth is defined in terms of revenue. Depending on the objectives of the business incubator, job creation potential may be the most important measure of growth, although business incubators and their sponsors should keep in mind that the most innovative firms are not necessarily the biggest creators of jobs; a textile mill is not innovative, but can create a large number of jobs immediately. The production and sale of a technology that enables doctors and nurses to do their jobs more effectively can save lives, but it does not necessarily create a large number of jobs. A business incubator can be an effective tool to promote both types of enterprises, but the incubator stakeholders must be careful about the criteria they set for business incubator applicants, and how the success of the business incubator is measured, so that the set objectives for the incubator can be reached.

[5] For tools used by infoDev's business incubation community, see: http://www.idisc.net/en/index.html

Apart from the lack of a geographic pattern, infoDev's business incubation network does not show a pattern of successful entrepreneurs coming from a specific professional background. Successful entrepreneurs and viable business ideas have come from all sources: from corporations, universities, and the grassroots. There are also excellent examples in the network of individuals who have gone abroad to study or pursue a career, and who then returned to their home country to set up a business venture, thus leveraging their experience and exposure abroad.

Defining the assistance given to entrepreneurs

Once entrepreneurs are accepted into the business incubator, the business incubator analyzes their needs and designs a program to strengthen and accelerate the business. The business incubator is proactive in assisting the clients, and will offer assistance in areas that the entrepreneurs may not be prepared to deal with on their own. The business incubators may also require the incubatees to take training courses to ensure a certain level of management knowledge.

While the exact mix of services depends on what is needed in the local market, business incubators usually provide the following four types of service:

- Shared infrastructure (thus reducing start-up costs), such as office space, meeting rooms, telecommunications, reliable electricity, and in some environments, security services;[6]
- Business advisory services to assist the entrepreneurs with management issues, such as business planning, financial management, marketing, and regulatory compliance on formal matters, such as applications for registration and licensing;
- Financial services, ranging from brokering services to providing seed loans, or taking equity in the enterprise;
- "People connectivity," including mentoring by experienced business professionals, knowledge-sharing with like-minded entrepreneurs, and links to business relationships and opportunities.

The value of a psychologically supportive environment cannot be overemphasized. Most of infoDev's business incubators identified the contrast between entrepreneurship and local values as a key challenge for their clients, and many cited culture as their clients' most significant barrier. Therefore, it is not surprising that entrepreneurs cite the psychological support provided by incubation staff and fellow entrepreneurs in the incubator, who "believe in you and your ideas" as having especially high value. One grateful entrepreneur referred to his incubator as an "oasis of cultural safety."

Business models and sustainability of business incubators

There are four main types of incubator business models: university-based, government owned, non-governmental/not-for-profit entities, or private sector companies. Table 2 shows the high proportion of NGO and non-profit organizations in infoDev's network.

Table 2. Distribution of business incubator models in infoDev's network

University	21%
Government	20%
NGO and non-profit	42%
Private company	17%

Source: http://www.idisc.net

Regardless of ownership structure, one of the most significant challenges of business incubators is to achieve financial sustainability. Financial sustainability, defined narrowly as "earned revenue covering all business incubation expenses," is very rare in both the developed and developing world. Most often, business incubators rely on a mix of revenue sources, including earned revenues based on rents, fees-for-service, and, less commonly, royalties and equity payments, as well as non-reimbursable funding from government and the private sector. For example, according to the MEIA, where the mean age of the incubators surveyed was five years old, 27 percent of incubators indicated that more than 75 percent of their revenue is earned, while over 15 percent indicated that their earned revenue amounted to less than 10 percent. This variability can be explained, in part, by the range of the organizational maturity of the incubators.

---

6   Some business incubators, particularly those targeting a specific industry, provide shared production equipment, often the most expensive part of starting a manufacturing business. Providing entrepreneurs with access to such shared equipment can substantially reduce the start-up costs for entrepreneurs, until they reach a production scale at which investment in equipment carries a lower risk, and until they are able to secure capital from potential investors.

Strategic partnerships and alliances complement earned and non-earned revenues, and are key to the effectiveness and sustainability of business incubators around the world. These partnerships are based on win-win opportunities between institutions, and include a variety of forms of cooperative provision of infrastructure, administration, or services. For example, a bank or business angel may provide non-reimbursable funding to a business incubator, knowing that it in essence is building up future lower-risk, high-growth potential clients. Other examples of such partnerships found across infoDev's network include local companies providing funding, expertise and links to markets—in some cases, through corporate social responsibility programs, but also with an interest in strengthening their own supply chain; others are universities providing brand (and therefore credibility), expertise and space, while benefiting from a vehicle to provide self-employment opportunities for its graduating students and research commercialization opportunities for its professors.

The question of whether or not business incubators should be pushed towards financial sustainability—narrowly defined—is controversial. Relying upon earned revenue as an income source forces a certain discipline on the business incubator, ensuring that it stays market-oriented and provides services that are truly needed by its client companies. At the same time, the very purpose of business incubators is to assist entrepreneurs at the stage in their business life-cycle when they are most volatile and cash-strapped. Most business incubator managers have heard their clients ask for deferred rent payment schedules. Many experience the difficulty their clients have when they are required to pay the full price for assistance with aspects of business, when the incubator can clearly see that the enterprise is cash-strapped and struggling and unable to allocate resources for that purpose.

Another issue often raised has to do with the trust that must be established between the incubator staff and the entrepreneur. If the business incubator becomes overly driven by the need to meet revenue targets, its client entrepreneurs may develop mistrust in the incubator staff, and be uncertain whether the service suggested is actually critical, or whether it has been proposed so that the business incubator can meet its revenue targets.

For these reasons, many business incubators provide their clients with subsidized rents and fee-for-service plans, in

which the subsidies decrease over time, as the enterprise gets on its feet. Royalty schemes—where the incubatee pays an agreed percentage of the additional income earned while under incubation, or possibly for a few years after graduation—can also help to overcome this problem. However, in both cases, the business incubator bears the risk that the enterprise may fail before the cost of incubating the client can be recuperated. Over time, successful enterprises may compensate for those that are not successful, but, in the mean time, incubators struggle with having insufficient cash flow to cover their operating costs.

In order to deliver services and lower operating costs, some incubators are experimenting with providing "virtual services," defined as off-site business incubation, including the use of information and communication services. However, it is not clear whether or not virtual incubation is as effective as the traditional form of incubation, or whether it actually saves money.

The flip side of the sustainability argument is that business incubators who push hard to cover their costs by earned revenues must then change their objectives and work as accelerators of more advanced enterprises, which are less risky and less cash-strapped. This defeats the primary objective of business incubators, viz. to support early-stage entrepreneurs, who have fewer resources and capabilities.

There is a plethora of revenue models among business incubators around the world, and the right revenue model for any given environment can only be arrived at by means of a thorough understanding of the business environment, along with experimentation with various revenue models and prices. As a result of the challenges and opportunities described above, infoDev's experience appears to indicate that the most realistic—and perhaps the most effective—model is one that combines both earned and non-earned revenues. This perspective is echoed by a report requested by the European Commission which assessed the performance of business incubators across Europe (Centre for Strategy and Evaluation Services, 2002).

From a policy perspective, public investment to co-finance the start-up phase of business incubators is justified for at least two reasons: first, as we outline below, because effective business incubation yields economic development returns; second, because from a government budgetary perspective, several assessments have found that government contributions towards business incubation quickly pay for themselves, by generating tax revenues through the enterprises and jobs they generate. For example, over the last 20 years, the government of Brazil has invested 150 million reals in business incubators and technology parks. It is now estimated that graduated enterprises generate 400 million reals annually in tax revenues.[7]

Notwithstanding the challenges of reaching sustainability, long-term sustainability—defined to include strategic alliances and partnerships that offset costs—should be viewed as a success measure. In this respect, as in any viable partnership, there should be various risk-sharing schemes so that all stakeholders share in both costs and revenues, in a way that enhances the success of an incubated entrepreneur, while also supporting the long-term sustainability of the incubator.

Effectiveness of business incubation: infoDev's experience

Appropriate indicators for measuring the *effectiveness* of business incubation as a tool for stimulating the creation and growth of innovative enterprises, include:

- The number of innovative enterprises created
- The viability,[8] revenue size, and growth rate of those enterprises
- The investment size attracted (as a proxy for perceived market value of the enterprise)

Ideally, one would measure these indicators for entrepreneurs that received business incubation assistance, in contrast to a control group in the same industry which did not receive any incubation.

In addition to directly affecting the success of the early-stage enterprises they incubate, business incubators have a broader positive influence on society by helping to commercialize innovations with potentially high impact on human welfare and/or business productivity and competitiveness, and which affect the broader enabling environment for innovative entrepreneurs, and creating new jobs. When assessing the impact of business incubation, it is also important to take into account this broader impact.

Challenges in measuring the effectiveness of business incubation

Limited systematic data are available to measure the effectiveness of business incubation. The lack of data is explained by a number of factors. First, it takes time to see the true results of successful incubation. On average it takes about three to four years to incubate a promising enterprise, and if one would like to measure the viability and growth rate of the incubated firms, one would have to wait at least another three to four years. Empirical evidence in New Zealand suggests that the real growth in revenue (and jobs) often does not occur until between four and seven years after graduation. Among developing countries, apart from a few veterans such as India and Brazil, business incubation is a relatively new concept. In infoDev's network of 189 business incubators in 80 developing countries the mean age of the incubators is only six years. Only now are we starting to see the results of incubation as a tool in developing countries.

A second complicating issue in the assessment of business incubation is that the term "business incubation" is being used loosely to describe a variety of different initiatives that aim to support the development of small and medium enterprises (SMEs), ranging from office parks and business development services to incubation, each with varying objectives, such as, empowerment of disadvantaged groups, job creation, innovation commercialization, and the generation of high-growth enterprises. If one aggregates the results of these so-called "business incubators," one risks trying to evaluate the results of different remedies.

Third, as with other social science interventions, it is difficult to identify a control group against which one can test how the incubated entrepreneurs fared, in comparison with those that did not receive incubation assistance. No entrepreneurs are the same. Moreover, when the business ideas accepted by the incubator are by definition innovative, there are not many other cases against which to compare the outcomes. Finally, it is surprising that many business incubators do not track their results, beyond the simple count of how many enterprises they graduate.

---

[7] For an assessment of the effects of business incubation in Germany, see also Schricke and Liefner, 2006.

[8] One must be careful with using viability as a success measure. For example, a company that has been bought out by a larger firm would no longer exist. However, from an economic perspective, value has still been added to the economy, and the "exit" of the small venture cannot be described as a failure. For example, an evaluation of New Zealand's business incubation program showed that approximately 29 percent of the sample of graduated enterprises experienced a change in ownership as a result of increased shareholders, company mergers, the formation of a new company, or IP buyout/licensing.

Creating innovative enterprises and enhancing viability

Incubation seems an enabling tool for innovative enterprise creation and viability. 150 business incubators in infoDev's business incubation network report that they are currently assisting 12,500 early-stage enterprises, and 92 business incubators report that they have graduated 4200 enterprises. According to the MEIA – which assessed 49 business incubators, over one third

ranging from biogas stoves in Rwanda, to beeswax production technology in the Ukraine, to organic crop boosters in India, a GPS-based bus fleet management system in Brazil, mobile-based electricity vouchers in South Africa, and software solutions that enhance business processes in Romania. Table 3 provides illustrations of innovations in products, services, business models, and production processes, which have been brought to market.

Figure 3. Number of new businesses set up by an incubator

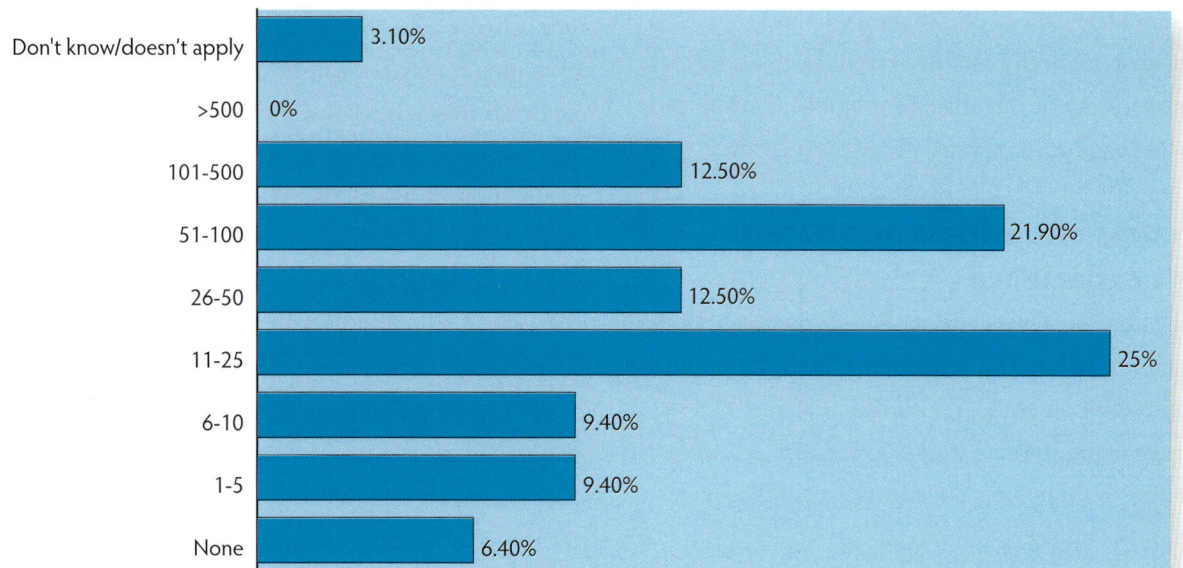

Note: Most of the business incubators surveyed were less than five years old and were not yet at a "mature" enough stage to measure impact.

Source: MEIA.

of the business incubators helped to start more than 50 new businesses (see Figure 3).

A collection of infoDev success stories, described in details in the Appendix to this chapter, profiles 14 innovative enterprises that have graduated from developing country business incubators, and reached the break-even point. These enterprises are mostly five to six years old. In all instances, the enterprises were true start-ups when they entered the business incubator. They had not yet, or barely, made their first sales. Today, these enterprises have reached annual revenues ranging from US$70,000 to US$2.8 million, and employ between six and 32 employees. These ranges appear to be fairly typical for successful enterprises across the infoDev network.

These stories illustrate the range of innovations that can be effectively commercialized with the help of business incubation,

Enhancing enterprise viability

Few countries keep systematic data on the survival rate of graduated incubatees. As shown in Table 4, current statistics indicate that business incubators generally have a very high success rate in generating viable new enterprises, ranging from 80 to 90 percent. By way of comparison, business incubators in infoDev's network—comprised only of developing countries—report that 75 percent of graduated enterprises are still in operation three years after graduation, a much higher number than the general enterprise survival rate. In Brazil, for instance, the survival rate of incubatees is about 80 percent, while 50 percent of all start-up companies do not survive the first year. These numbers are encouraging, given the very difficult business environments in which many of these business incubators operate.

Table 3. Innovations brought to market

Innovation brought to market	Entrepreneur/Innovator
Product	Energy, Environment and Sanitation Company Limited (ESSCO Ltd), in Kigali, Rwanda, produces and sells institutional cooking stoves fueled by biogas. The low-carbon stoves were developed by the Kigali Institute of Science and Technology (KIST), and were brought to market by two Rwandan entrepreneurs. In Rwanda, wood accounts for about 90 percent of household energy consumption. Burning wood emits carbon dioxide, and growing trees absorb it. However, once a tree is cut down for fire wood, it takes several years for a new one to grow to the same carbon absorption level. The collection of fire wood also causes soil erosion, leading to a further adverse impact on the environment and on the livelihoods of the poor. To preserve the environment, the government of Rwanda has, therefore, placed strict limits on deforestation. Thus, the biogas cooking stove is of immediate value for consumers, the protection of the environment, and future generations. As of December 2008, ESSCO's stoves had been on the market for a year and a half, and the company was already earning a small profit. ESSCO was incubated by the KIST Business Incubator. The founders were an engineering professor and a graduate student.
Service	PV Inova, in Porto Alegre, Brazil, is improving the bus transportation experience for both passengers and operators through telephony and tracking products. TELO provides passengers, who cannot afford mobile phones with an inexpensive way to make phone calls during their commute. TELOTrack uses GPS to provide bus fleet managers with the capability to quickly identify and react to problems in public transportation by performing automated identification, diagnosis, and resolution of any deviation or abnormality and to immediately provide critical information to the fleet management team. The products were successfully launched in Porto Alegre and Rio de Janeiro in 2007 and 2008. Three pilots are in operation in the United States, and business negotiations are proceeding in Angola, Argentina, Chile, and South Africa. PV Inova was incubated by the Genesis Institute. The founder was a Brazilian graduate student, who had worked for development banks for several years.
Business model	Expertron, in Pretoria, South Africa, facilitates easy access to prepaid electricity for lower income consumers. When prepaid electricity was introduced in South Africa, municipal cashiers handled sales of prepaid vouchers. Due to the low volume of purchase points, many customers had to travel long distances and stand in long queues to purchase their electricity. As the customers could only afford small amounts of electricity at a time, many had to make the trip several times per month, resulting in a very high cost of electricity for the consumer, and an increase in theft of electricity. Expertron developed a cell-phone vending system to improve service delivery and simultaneously create jobs, by involving people from the community in the process of selling electricity. The system uses standard GSM mobile telephones as affordable point-of-sale (POS) devices to sell and distribute prepaid electricity tokens/vouchers. Any individual having a mobile phone and sufficient funds to purchase prepaid "electricity stock" may become a vendor, and earns a commission on his sales. Today, Expertron has an annual turnover of US$630,000. Expertron was incubated by the Innovation Hub. The founders were three professors in electronic engineering.
Production	Kharpchelo, in Kharkov, in the Ukraine, sells honey and beeswax production equipment to beekeepers. Leveraging his background in engineering, the founder of Kharpchelo has found a way to improve on production technology imported from Russia to develop production equipment that consistently produces high quality honey. Kharpchelo reached US$400,000 in annual revenues in 2008, and employed 20 permanent and 40 seasonal employees. Kharpchelo was incubated by the Kharkov Business Incubator. The founder was an aircraft engineer.

Table 4. Innovations brought to market

Country/Region	Incubated enterprise survival rate	General enterprise survival rate
New Zealand[a]	87% continue to operate after 2 years	69% continue to operate after 2 years
United States[b]	85% continue to operate after 3 years	50% continue to operate after 4 years
Europe[c]	89% continue to operate after 3 years	...
OECD	...	60% continue to operate after 3 years
Germany[d]	90% continue to operate after 3 years	...
Brazil[e]	80% continue to operate after 3 years	50% continue to operate after 1 year
South Africa[f]	80% continue to operate after 3 years	...
infoDev's incubation network	75% continue to operate after 3 years	...

Sources: [a] New Zealand, 2008; [b] National Business Incubator Association; [c] European BIC Network, 2008; [d] Schricke and Liefner, 2006; [e] ANPROTEC Brazil; [f] Small Enterprise Development Agency, 2008.

Accelerating enterprise growth

The hypothesis is that business incubation accelerates enterprise growth, thus saving valuable time and money, and generating social and economic benefits at a faster pace than would otherwise be the case. Ideally, the revenue growth rate of incubated enterprises should be measured against industry benchmarks.

There do not appear to be many systematic studies assessing the revenue growth rates of incubatees, as compared to industry benchmarks in developing countries. However, to illustrate this point, three incubators in Panama, Uruguay, and Costa Rica, all focused specifically on supporting innovative, early-stage, high-growth enterprises have together graduated 63 enterprises with an average annual turnover of US$90,000 at graduation. These enterprises had no, or less than US$15,000, annual turnover at the start of the incubation process, and, on average, were incubated over a period of three years.

New Zealand is one of the few countries that systematically and reliably tracks the impact of business incubation on enterprise growth rates. According to an assessment in 2008, incubator graduates had better revenue growth outcomes than industry benchmarks. Of the graduates reporting turnover, 59 percent surveyed had achieved an average growth rate of 20 percent over the last five years, and 40 percent reported an overall growth rate of at least 150 percent. In contrast, a control group recorded 11 percent of firms achieving a minimum of 150 percent turnover over five years.

Creating jobs

In infoDev's Business Incubation Network, 92 business incubators report that they have graduated 4,230 enterprises employing 62,000 people. This translates into an average of 14 jobs per enterprise. As these business incubators and graduates are still young, it is not yet clear how many additional jobs will be created over time. In Brazil, where business incubation has a longer history, ANPROTEC, its business incubator association, estimates that over the last 20 years, Brazilian incubators have graduated 1,500 enterprises and generated 33,000 jobs, representing an average of 22 jobs per enterprise.

If the broader objective of the business incubator is to facilitate the creation of innovative enterprises, job creation in itself is not a sufficient result indicator, since many start-ups do not necessarily create the most immediate jobs. Nevertheless, the creation of sustainable jobs is an important outcome of support to innovative, early-stage enterprises.

Improving the enabling environment

There are several frameworks for characterizing environments that are conducive to linking innovation and entrepreneurship. Most include an emphasis on legal and regulatory incentives, such as the ease of registering a business, intellectual property protection; incentives for R&D; education reforms stimulating research and more intensive relationships between universities and industry; initiatives to expand access to capital, such as private and public investments in small and medium enterprises (SMEs); and infrastructure improvements to improve electrification, roads, ports and airports, and information and communication services. These are all, of course, critical factors in creating an enabling environment.

What is often overlooked in these frameworks, however, is the importance of "cultural capital."[9] Cultural capital does *not* refer to a nation's ability to innovate. Rather, it refers to the level of tolerance for risk and the interpersonal trust that exists in that society, which affects entrepreneurs' decisions to start a business, and the decisions of others to invest in it. Even in the best of business environments, developing a new innovative venture is risky and requires confidence and patience. Thus, an encouraging environment is key to enable entrepreneurs willing and able to take this risk.

Perhaps the most salient finding of the MEIA was the significant impact of business incubators on the enabling environment achieved through their effective linkages with other actors in the innovation and entrepreneurship ecosystem. Following are a number of examples:

Policy and Regulation. Most of the policies and regulations in the developing countries we analyzed are not optimized for technology entrepreneurs (see Figure 4). Trade laws supporting SMEs exist in fewer than half the surveyed countries; legal incentives to start new businesses and the assurance of intellectual property (IP) protection are weak in three-quarters of the surveyed countries. And over 80 percent of the business incubators reported that regulatory, legal, or policy incentives for new businesses do not exist at all in their countries.

The majority of incubators actively advocate for policy and

---

9   See Fairbanks, "Changing the Mind of a Nation: Elements in a Process for Creating Prosperity."
    Available at: http://www.sevenfund.org/pdf/Changing%20The%20Mind%20of%20a%20Nation.pdf

Figure 4. Regulatory and policy incentives for SMEs

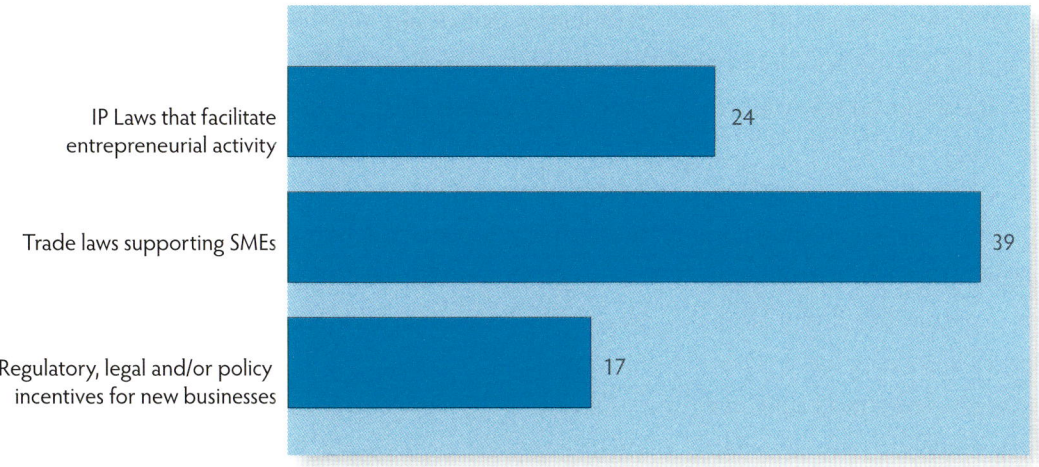

Source: MEIA.

legal reforms that will benefit entrepreneurs and small business owners across the economy. ARC, an incubator in Bulgaria, for example, actively participates in committees and in working groups to advise policymakers, and is currently involved in drafting the National Innovation Strategy and Regional Innovation Strategy for the South Central Region. Another example is CID in Peru, which has formed an association of 11 businesses to draft and guide public policy on SMEs. Several other incubators, including two in Yangling and Tianjin in China, ANPROTEC and RMI in Brazil, TRECSTEP in India, ISTT in Iran, UBICA in the Ukraine, and Ingenio in Uruguay all report that they are regularly consulted by their governments on issues affecting the local business environment for SMEs.

Financing. Eighty percent of the business incubators reported that their incubatees are limited because they do not have access to risk capital and appropriate financial offerings in their local business environments. There was a general feeling that entrepreneurs were "stuck in the middle" between the micro-enterprises served by microfinance institutions and the deals that banks and private investors find attractive. A later study completed by infoDev, entitled "Financing Technology Entrepreneurs and SMEs in Developing Countries," found that the financing gap is particularly pressing, in the range of US$50,000 to US$1 million, and that for SMEs competing in the information and communication technology (ICT) in-

dustry and ICT-related activities, the challenge of accessing growth capital is particularly acute, because these enterprises possess few tangible assets that can be leveraged as collateral for loans.

The MEIA revealed that business incubators have derived a variety of approaches to overcome the financing challenge. A few business incubators have opted to take equity in their client enterprises (e.g., Raizcorp in South Africa and the Panama Technology Business Accelerator). But incubators seek more often to facilitate access to financing for their client, and more broadly to improve the macro-environment for SME financing. For example, CIE-TEC in Costa Rica is collaborating with a local bank to establish a financing fund for SMEs. Octantis in Chile created the country's first network of angel investors. To date, this angel network has invested US$4.3 million in the creation of more than 60 new companies and 12 international patents. The companies have, in turn, achieved aggregate sales of US$30 million.

Culture. Research has shown that environments that embrace risk, diversity of thought and action, and interpersonal trust are correlated with high levels of innovation. However, 85 percent of the business incubators felt that there was very low tolerance for risk or failure in their business environments. Most business incubators identified the contrast between innovative entrepreneurship and risk-averse local values as a key challenge for their clients, and many cited culture as their clients' most significant barrier. Incubator managers in Sri Lanka cite the discouragement entrepreneurs face, as manifested most explicitly by parents who urge their children to get "real" jobs. Incubators in Uzbekistan and Kazakhstan are confronting the additional challenge of cultural legacies where failure (by entrepreneurs) was once seen as criminal.

Thus, business incubators can serve as "an oasis of cultural safety" for entrepreneurs in business environments where entrepreneurship is not encouraged. They also contribute to stimulating more entrepreneurs to pursue their dream of creating their own company, simply by illustrating that it is possible. More than 70 percent of the business incubators said they are working to promote role models to raise awareness of, and confidence in, pursuing entrepreneurship as a career option. An Armenian incubator was described as bringing "a new climate of democratic entrepreneurship" to an economy that he perceived to be dominated by large enterprises and monopolies.

As discussed earlier in this chapter, business incubation is still relatively new in the developing world. As these business incubators mature, more systematic assessments of their effectiveness are needed. That being said, the evidence above illustrates the positive effects that business incubators have already had both in creating viable, innovative, high-growth enterprises, and of positively affecting the broader innovation and entrepreneurship ecosystem.

Key challenges and success factors

While business incubation can be an effective tool to stimulate the creation and growth of innovative enterprises, business incubation is by no means easy. Many business incubators are confronting operational and strategic challenges.

The number one challenge of most business incubators is reaching financial sustainability. It takes time to experiment and arrive at the right revenue model in a given context. While partnerships and strategic alliances are key to both effectiveness and sustainability, it is difficult to manage the expectations and demands of a variety of stakeholders, while at the same time remaining focused on the core objective of the business incubator.

A second important challenge for business incubators is finding and retaining management teams with the right mentality and skill sets. The MEIA established that the effective-

ness of business incubators can be linked directly to the skills, vision, commitment, and entrepreneurial leadership talent of their management teams. Yet 93 percent of the incubators surveyed reported significant difficulties in finding and keeping staff. Thirty percent cited scarcity as an issue, and more than 40 percent said that limited resources kept them from making needed investments in staff. Due to the level of trust necessary to effectively incubate and the many relationships required to run an incubator effectively, frequent management turnover can be very costly to the incubator and hinder the progress of incubatees.

Finally, many developing country business incubators struggle with getting adequate "deal flow." In some instances, the reason for these problems may be that they have not got their "selling proposition" and/or pricing right. But often the culture of entrepreneurship is also a factor, as well as the number of potential candidates in the business incubator's vicinity. As discussed above, only a fraction of entrepreneurs are growth-oriented.

infoDev has identified a number of success factors critical to overcoming these challenges:

- A thorough feasibility assessment is key to success. Business incubators that fail have often not done a thorough job at the feasibility stage.
- There is no one-size-fits-all business incubator model that will work in all contexts. As in any business, the services provided and the business model of the business incubator must be designed in response to local market conditions.
- Founders of business incubators must ensure that managers have the right skills and mentality for the job, and that there is sufficient incentive for that person to stay. They should also be prepared to relinquish control, if the business requires a different set of management skills than those of the founders.
- Business incubators must be set up in such a way that they can operate in a business-minded fashion, even if an academic institution or a government agency is a co-owner or the sole owner.
- Business incubators must ensure that their selection criteria for incubatees are in line with the core objectives of the business incubators and with local market needs, and that there is explicit agreement between all stakeholders on the core objectives.

- A certain size is required for effective business incubation in order to maximize the likelihood that the incubator will achieve financial sustainability. infoDev's experience indicates that 20 to 30 incubatees at any given time and a space of at least 2000m² is ideal in most situations, but this may differ from environment to environment.
- Private sector partnership in business incubation is critical. In the MEIA report, incubators were asked how they would structure their organization differently if they were to start again, and the most frequent answer was greater investor and private sector involvement.

Governments considering investment in business incubation should think of business incubators as an integral part of a broader innovation and economic development program that strengthens the over-all innovation and entrepreneurship ecosystem. As discussed above, the most effective business models for business incubators comprise a mix of earned and non-earned income and, in this context, funding support from well-recognized organizations lends credibility to local business incubators and enables them to get the local buy-in required for long-term sustainability. Furthermore, the willingness of risk capital to invest in the incubated ventures as they graduate is the ultimate market test of the sustainability of the incubator's model.

Conclusion

Innovation is key to social and economic progress. The capability to innovate is innate in human beings everywhere. Policymakers and donor agencies can help facilitate the creation of an innovation and entrepreneurship ecosystem that encourages entrepreneurs, who are willing and eager to take the risk of bringing new ideas to the market, to turn the potential of their idea and ambition into real social and economic value. Business incubation is one vehicle for providing such assistance.

As shown in this chapter, promising results are emerging from the early efforts of business incubation in developing economies. It could be that business incubation will serve an even more important function in these economies, given the absence of an enabling environment and business networks that are more prevalent in the developed world. Business incubation may also play an even more important role, in light of the demographic make-up of developing economies, which have large populations of youth, who, though open-minded and ambitious, do not have the networks, capabilities, or credibility to get a business off the ground. Business incubation may provide the type of support that these young entrepreneurs need to start their own businesses.

References

ANPROTEC Brazil. Available at: http://www.anprotec.org.br/publicacao.php?idpublicacao=538

Bosma, Niels, Zoltan J.Acs, Erkko Autio, Alicia Coduras, and Jonathan Levie. 2008. *Global Entrepreneurship Monitor—2008 Executive Report.* Babson College and Universidad del Desarrollo. Available at: http://www.gemconsortium.org/download/1253659064034/GEM_Global_08.pdf

Center for Strategy and Evaluation Services. 2002. *Final Report: Benchmarking of Business Incubators.* February.

European BIC Network. 2008. *BIC Network in 2007 Facts and Figures.* Brussels. Available at: http://www.ebn.be/

Fagerberg J., D. Mowrey, and R. Nelson. 2004. *The Oxford Handbook of Innovation.* Oxford: Oxford University Press.

Fagerberg, Jan, Martin Srholec, and Bart Verspagen. 2009. *Innovation and Economic Development.* Maastricht: United Nations University and Maastricht Economic and Social Research and Training Centre on Innovation and Technology.

Fairbanks, Michael. "Changing the Mind of a Nation: Elements in a Process for Creating Prosperity." Available at: http://www.sevenfund.org/pdf/Changing%20The%20Mind%20of%20a%20Nation.pdf

infoDev. 2007. "Innovation and Entrepreneurship in Developing Countries: Impact Assessment and Lessons Learned from infoDev's Global Network of Business Incubators." Available at: http://www.infodev.org/en/Project.77.html

———. 2008. *Financing Technology Entrepreneurs & SMEs in Developing Countries: Challenges and Opportunities.* World Bank.

———. 2009. *A Model for Sustainable and Replicable ICT Incubators in sub-Saharan Africa.* World Bank. Forthcoming.

Lundvall, B. A. 1992. *National Systems of Innovation: Towards a Theory of Innovation and Interactive Learning,* London: Pinter Publishers.

Monitoring and Evaluation Impact Assessment (MEIA). Available at: http://www.infodev.org

National Business Incubator Association. Available at: http://www.nbia.org/

Nelson, R. 1993. *National Innovation Systems: A Comparative Analysis*. New York: Oxford University Press.

New Zealand Ministry of Economic Development. 2008. *Incubator Support Programme Evaluation Report*. May.

Schricke, Ester and Ingo Liefner. 2006. "20 Jahre Technologie- und Gründerzentren in Niedersachsen—Eine Untersuchung der regionalokonomischen Effekte," University of Hannover. February.

Small Enterprise Development Agency (SEDA). 2007–2008. *Annual Report 2008*. Pretoria. Available at: http://www.seda.org.za/content.asp?subID=922

Smith, Keith and Jonathan West. 2007. *Innovation Policy, Productivity and the Reform Agenda in Australia: A Framework for Analysis*. Australian Innovation Research Centre. University of Tasmania.

Appendix. More successful infoDev business incubations

Enterprise and incubator	Country	Innovation	Founder
Recycla Chile, incubated by Octantis	Chile	Recycla is the first e-waste recycling company in Chile. Recycla extracts and separates the raw materials of computers, printers, cell phones, and scanners, etc., so that they can later be transformed and re-used. Materials that cannot be re-used, such as batteries and computer screens are treated in a certified hazardous waste treatment center. In 2003, Recycla entered a joint venture with Maxus Technology Inc. of Canada and the U.S., becoming the first Chilean SME to sign an alliance with a NASDAQ-traded company. In August 2003, less than two months after the initiation of operations of the plant, Recycla shipped its first export of recycled copper to China in a 20-foot container. Today, the company is profitable and has an annual turnover of US$2.8 million.	Three Chilean brothers: two were owners of a construction company in the US, and the third was an accountant in Chile with expertise in metal trade and recycling.
GloTech Organics, incubated by TREC-STEP	India	NEMATE GRO is an ecologically friendly and efficient way to improve fertilizer efficiency, which facilitates disease and pest management. SEAMIC is an organic liquid containing six essential nutrients that strengthen the plant's immune system and boost its growth. The effectiveness of the product was demonstrated in 15 field trials for grapes, tomatoes, cluster beans, and carnation flowers, among others. The cluster bean yield doubled and increased in quality; tomato plant root-wilt and stem-rot were eradicated; grape yields increased; and carnations grew faster. The products have reached 2000 farmers in Tamilnadu, Nasik, and Pune, and have substantially increased farmers' incomes. Business links have already been established in Sri Lanka, Malaysia, Mauritius and in several African countries. Today, GloTech Organics is profitable and has an annual turnover of US$71,000.	Engineer working in the R&D department of a large fertilizer company for nine years. The fertilizer company closed its R&D Department, giving him the final push to start his own company.
Focus Solutions	Jordan	Focus Solutions identified a market gap among remedial and legal departments of banks in the Middle East and North Africa, which seek to reduce bad debt, non-performing account receivables, revenue risks, and cost of collections and debt recovery, and to improve their operational risk management compliance for Basel II, the management of write-offs, and performance tracking and reporting. Current customers include CitiBank, Standard Chartered Bank, ABC Group, Arab Bank, and Jordan Bank.	Two university graduates.
Cochlear Implants, incubated by the City of Knowledge	Panama	Cochlear Implants is the first provider of cochlear implant operations in Panama. It is a good example of an entrepreneur commercializing a foreign innovation for the benefit of Panamanians. It is estimated that about 25,000 people in Panama suffer severe or profound hearing loss. Their only option for being able to hear again is to receive a cochlear implant. Cochlear Implants has been serving the hearing-impaired in Panama for a year and already has a waiting list of over 100 people. To date, 60 speech therapists and five local doctors have been trained to perform the procedure.	Dr. Cynthia Guy, an accomplished Panamanian doctor, who studied and worked in the United States.
SoftTechnica	Romania	SoftTechnica provides IT solutions for businesses in Romania. The company specializes in consulting, from analysis and production software, to installation and network configuration, integration of systems and turn-key solutions, including digital signature applications, management information systems for bars, restaurants, beauty salons, and hotels, and for real estate developers, who generate all documents needed for real estate development in Romania. SoftTechnica is becoming known as a high quality provider that is less expensive than foreign service offerings. The company has served more than 100 clients in Romania and has an annual turnover of US$280,000. They are currently exploring expansion into Croatia, Moldova, and Albania where the hospitality industry is expanding and where the products could be easily adapted.	A team of three: two were employed IT professionals in larger international companies, and one was a university graduate.

Enterprise and incubator	Country	Innovation	Founder
TechnoCAD	Romania	TechnoCAD offers computer-aided design (CAD) services in the field of mechanical engineering, focusing on the automotive industry. CAD has been associated with lower product development costs and a shortened design cycle. TechnoCAD provides 3D modeling design for machine building, finite element analysis simulating and indicating stresses and displacements that must be planned for in the engineering, and consultancy to companies that wish to transition from paper to modern CAD systems. TechnoCAD prides itself in the quality of services provided, and currently serves companies in Romania, Germany, France, Austria, Spain, and the U.S. The company currently has an annual turnover of US$800,000 and employs 32 people.	An electronics and telecommunications engineer with 12 years of experience in industrial electronic equipment maintenance.
Naledi3d Factory Ltd., incubated by Maxum	South Africa	In Africa, poor literacy skills and language barriers often pose huge challenges to learning and skills development. The Naledi3d Factory creates visually interactive content based on Virtual Reality (VR). Its intensely visual nature transcends literacy and linguistic barriers by *showing* as opposed to *telling*. The Naledi3d Factory uses the technology to create learning material that is both content- as well as context-rich, in semi-realistic and visual three-dimensional environments. The "***interactive3d learning object***" (or ***i3dlo***) is a self contained piece of visual learning that can be reused in different ways and, most importantly, easily translated into any African language. Even advanced technical subjects can be packaged into re-usable *i3dlo's* and made available to a broader learning community. The benefits of this innovative approach have now been successfully demonstrated in areas as diverse as health, technical, life skills, and agriculture. Naledi3d Factory has reached an annual turnover of US$160,000.	A transport engineer, with a strong ICT component, who had worked for nineteen years at the Council for Scientific and Industrial Research (CSIR). In his last two years at CSIR, he developed a major VR center on behalf of the Council.
Rotasoft, incubated by METU Tech	Turkey	Rotasoft develops "edu-tainment" children's books in Turkish that interact with a computer to bring the book to life with 3D animations. Rotasoft has now partnered with one of Turkey's largest book publishers, which sold more than 5000 books after three months on the market.	Graduates.

Source: infoDev, outreach to business incubators.

Chapter 2.2

THE US PATENT SYSTEM: Does It Strengthen or Weaken Innovation and Progress?

Josh Lerner, Harvard University
Adam B. Jaffe, Brandeis University[1]

Why have patents?

The development and commercialization of new technologies—technological innovation—creates broad social benefits. Over time, it allows us collectively to live longer, healthier lives, to increase our incomes, and to consume a broader array of goods and services. Innovation enables us to live differently from our grandparents. Thus, it is in our collective interests to create social, cultural and legal institutions that foster technological innovation.

The argument for having patents as part of that legal environment is the recognition that the process of technological innovation is expensive. New products and services do not just spring full-grown from the creative mind of an inventor. Though an instantaneous spark of creative genius may start an innovative flame, it typically takes years of research and development to nurture that fire into a commercially viable blaze, with a lot of false alarms along the way. And that nurturing process costs money—often a lot of money.

If technological innovation is socially desirable but expensive, society needs to have institutions that direct time and money into the processes of research and development. One approach to this, at least hypothetically, might be to have the government use money raised through taxes to research and develop new technologies. In fact, the United States government does a lot of R&D, particularly in such areas as defense, space, and the environment, that are themselves important areas of government responsibility. But in our free-enterprise system, we don't think it is a good idea to give the government the job of developing new products and processes for industry. Government is good at many things, but taking entrepreneurial initiative is not one of them.

So if society wants technological innovation, we need institutions that create incentives for private individuals and firms to invest money in the process. And the incentive to invest in R&D must come, ultimately, from an expectation of making a great deal of money if the thing pans out.

To make innovation rewarding, the government must give or grant something valuable to people or firms that produce important innovations. A patent system creates a zone of economic exclusivity for the innovator. In the 17th and 18th centuries, the early British patent system coexisted with "prizes" for people who produced innovative solutions to particular technological problems. Some scholars argue that govern-

---

[1] This work draws from the authors' *Innovation and Its Discontents: How Our Broken Patent System is Endangering Innovation and Progress, and What To Do About It* (2004).

ments or private foundations should offer a major prize as an inducement for drug companies to develop vaccines for tropical diseases, because the people and governments that need these vaccines are too poor to make this research profitable even with patent protection.[2]

While prizes may be effective for drawing forth a specific, desired technology, they may not be effective for stimulating innovation in general. It would be expensive to hand out enough prizes to reward the gamut of industrial innovation; raising the tax money to do this would be unpopular and burdensome; and it would be impossible to decide how big a prize to give to each innovation. Besides, given that the importance of a discovery is usually initially uncertain, some prizes would be too large in some cases and not large enough in others. For instance, while a full £50,000 (several millions of today's dollars) was paid by the British government to John Palmer, the inventor of a new way of organizing the mail, this sum far exceeded that paid to Edward Jenner for his smallpox vaccine, which was responsible for saving millions of lives. Patents, on the other hand, are by their nature proportional to the size of the discovery: the exclusive right to a modest discovery is unlikely to be worth very much, while the exclusive right to an important new technology is usually very valuable. Thus, at least in principle, patents provide an appropriately calibrated reward for different innovations.

The process of patent "examination"

To get a U.S. patent, the inventor or inventors file an application with the Patent and Trademarks Office (PTO).[3] A PTO *examiner* then determines whether the invention meets the standards for patentability. U.S. patent law permits the granting of a patent for

- new processes (e.g., a new approach to brewing beer);
- new machines (e.g., a new tool or automobile carburetor);
- manufactured articles (e.g., a kit to identify an infectious disease);
- new compositions of matter (e.g., a novel type of concrete);
- new and useful improvements of the above;
- distinct and new varieties of plants that are asexually re-

produced; or

- any new, original, and ornamental design for an article of manufacture.

The patent application must pass three other critical tests:

1. **Utility.** Does the invention really do anything, and if so, does it solve the problem it set out to address?
2. **Novelty.** Is the claimed invention really original?
3. **Non-Obviousness.** Even if new, would the claimed invention have been obvious to one skilled in the art at the time of the invention? Even if the invention is new, the law does not allow patenting of inventions that involve only trivial or "obvious" improvements on what has come before, known as "prior art" or anything a skilled practitioner could have modified. In order to get a patent, the invention must pass the tests of "novelty" and "non-obviousness" in relation to the prior art.

The patent *claims* are the legal characterization of what is and is not covered by the patent, and are worded in specific and legalistic language intended to cover as many different devices as possible, giving the holder of the patent maximal power to restrict competitors. Claims also contain language that limits their coverage, so as to distinguish the invention from earlier ones, and establish novelty and non-obviousness.

If the examiner concludes that the applicant's claims are not novel, or are obvious, the applicant may re-draft them, making them more restrictive, and distinguishing the invention from prior art. Thus, the process of patent examination is, to a large extent, one of negotiation between the applicant (typically the applicant's lawyers) and the examiner. The applicant wants claim language that is as broad as possible, while the examiner may insist on restrictions to distinguish the invention from the prior art. Once these parties agree on language that distinguishes the invention from the prior art, the patent is granted. If the claims do not satisfy the examiner (or if the version that satisfies the examiner is so narrow that the applicant judges it not to be worthwhile pursuing), then the patent application is denied. But because the applicant has unlimited opportunities to amend the application to satisfy the exam-

---

[2] Kremer, 2000.

[3] A patent is a legal document issued by a government agency, in the U.S. issued by the PTO, which is part of the Department of Commerce. The patent describes an *invention*, and lists one or more *claims* that specify what the invention does that has never been done before. The patent gives the holder the legal right to prevent anyone else from making, using, selling, or importing an object or device that incorporates any feature covered by the specified claims. This right operates within the U.S., and also blocks importation into the U.S. of goods whose manufacture violate the patent. Though differing in detail, other countries have similar systems. Patents convert the intangible creation of an inventor into "property" that can be bought and sold, or upon which a business can be founded.

iner, a large fraction of all original patent applications in the U.S. are now ultimately granted.

This process involves *only* the applicant (and his or her legal representatives) and the PTO. Other firms or other parties—who might in fact have information regarding the state of the prior art against which the application should be considered—are not allowed to participate. Until recently, the entire application process was secret. No matter how long it took the PTO to resolve a given application, its very existence was kept secret until a patent was granted, and was kept secret forever if a patent was not granted. Under a change made in 1999, most patent applications in the U.S. are now published 18 months after filing. It is still the case, however, that if you learn of an application for a patent on an invention that you think other people had already made years ago, it is difficult to prevent an invalid patent from being granted by the PTO without jeopardizing your ability to later defend yourself.

Once you have a patent

The recipient of a patent (the "patentee") has a legal monopoly that lasts for 20 years from the date the application was filed. While the inventor—hence the patent applicant—must be a human being, the inventor can assign the patent to a company, which then has all of the legal rights associated with the patent. Firms commonly require their employees to assign their patents to the employer, whose lawyers typically handle the application process.

A patent constitutes *intellectual property*. It can be bought and sold, left in a will, given as a wedding present, or left in the attic and forgotten, just like any other property. And a patentee who does not want to sell the patent can rent it to someone else, by granting him a *license* to use the patented technology. The license agreement may or may not require money payments, called *royalties*, from the licensee to the patentee. The big difference between intellectual property and tangible property, however, is that it can be "rented" to multiple people at the same time. Licenses can be granted to multiple parties, with or without restrictions as to how each of the parties may use the technology. In many cases, the patentee neither sells nor licenses the patent, but rather uses it for its own business, and relies on the patent to prevent competitors from using its patented technology.

Infringement

If someone who has not been licensed undertakes activities that are covered by the claims of a patent, they are said to *infringe* the patent. When infringement is suspected, the patentee will typically write a letter demanding that the alleged activity stop, sometimes offering to license the patent, in return for a royalty or other consideration. The recipient of such a letter then has three choices: agree to take a license and pay a royalty, stop the infringement, or continue as before and wait for the patentee's next move.

If negotiations for licensing do not resolve the matter, the patentee can initiate litigation in federal court to enforce the patent. If infringement is proven, the patentee is entitled to a court injunction ordering the activity to cease and to receive *damages* in compensation. In response, the defendant may deny the infringement, and may countersue, claiming that the patent itself is *invalid*. The grounds for invalidity may include fraud or some other malfeasance on the part of the patentee in getting the patent, but the most straightforward basis for invalidity is the charge that the examiner made a mistake and that the invention is not novel or is obvious.

Controlling the risks inherent in innovation

The risk of imitation hangs over all investments in R&D. And the nature of the innovation game seriously skews the profits or returns to innovation. Most investments in new products and processes fail, meaning that their investors lose money. A very small fraction of investments in new products or processes succeed. For the overall "game" of investing in new technology to be worthwhile, the successes must earn enough profit to cover not only their own costs and reasonable return, but also the costs and a reasonable return on those costs for all of the failures. Otherwise, the overall investment strategy will end in loss.

For example, the pharmaceutical industry typically spends more than half a billion dollars per drug.[4] If we combine the risks of failure at the clinical trial and market stages, 80 percent of these expensive testing efforts are a complete loss, never leading to an FDA-approved product. About 14 percent do lead to an FDA-approved product, but do not earn sufficient profits to recoup their own development costs. Only 6 percent earn sufficient profits to recoup investment costs, and this 6 percent must also earn sufficient profits to pay for all the

---

[4] DiMasi et al., 2003.

2.2 THE US PATENT SYSTEM: Does It Strengthen or Weaken Innovation and Progress?

87

losses on the other 94 percent of compounds tested![5]

Neither the managers of firms nor investors like risk. The high risk associated with R&D tends to discourage firms from undertaking it, even if the rewards are reasonably high. Investment in new technology is therefore handicapped by its riskiness, when compared with other forms of spending (for instance, heavier marketing of an existing brand). Furthermore, when a business builds a new factory or buys some new equipment, it doesn't normally worry that its competitors will come and steal the equipment. When a business invests in R&D, it is "building" an asset that it hopes to profit from, just as it does when it builds a factory. But assets from research are intangible, and therefore much easier to steal.

To prevent such theft, a patent allows you to build a security fence around an idea, like the security fence around your factory. Like any fence, it won't necessarily prevent all theft, but if will make it harder and hence make the property more secure, making you more willing to take the risk of building it to begin with.

Three categories of patent risk are left out of this simple description:

1. *Patents are but one of many tools that firms have available to protect their profits from innovation.* Other strategies include secrecy, first-mover advantage, and brand loyalty. Numerous important innovations are never patented, or when they are, the patent protection is not really important to their commercial success. This does not mean that patents are unimportant, but it does mean that their importance in maintaining the flow of new technologies varies across different industries and different kinds of firms.

2. *Inventions don't occur in isolation, but overlap and build on each other.* Thus, if one firm gets a patent, it can retard or stifle inventions that other firms might otherwise undertake, thereby clouding the overall effect of patents on technological progress.

Given that inventing and developing improvements is time-consuming and costly, how do we create *broad* incentives for people to invest in improvements? The obvious answer is to grant patents on improvements, and such patents are indeed allowed, *if the improvement embodies some idea not covered by the patent on the underlying technology.* This seems fair, but is tricky to implement. To make the patent on the original invention useful, its owner must be given some latitude to mod-

ify the invention and still have it be covered by the original patent. If such latitude is too wide, many improvements are likely to fall under the original patent. So there is a tradeoff: granting broad patent protection gives the maximum incentive for "original" inventions, but it may actually discourage improvements.

In principle, subsequent inventors with good ideas about improving an important invention ought to be able to negotiate an agreement with the owner of the original patent that allows the improvement to be implemented. This could be done by granting the improver a license to use the original patent, or by selling or licensing the improvement back to the holder of the original invention. If the improvement is a good one, both the original inventor and the improver have an incentive to see it implemented. In practice, however, such agreements are often difficult to work out.

A related problem is created by the reality that firms are often working more or less in parallel on related research, with multiple firms applying for patents for different versions of the same idea. In principle, although each should be entitled to a patent only on those aspects (if any) of their creation that are unique and truly new, in practice, this is difficult to do. What is more likely is that each will be granted a patent that describes its invention in a way that leaves considerable ambiguity as to whether or not the inventions of the other firms are or are not covered. As a result, all are uncertain about what products they can or cannot legally sell. In the face of potentially overlapping patent grants, the risks associated with bringing new products to market are augmented rather than reduced, because expensive litigation with uncertain outcomes is added to other worries. In such cases, the patent system may well inhibit rather than encourage innovation.

Parallel development sometimes leads to a problem not so much of uncertainty about patent rights, but of patent rights that interfere with each other. A sector characterized by cumulative and overlapping innovation is biotechnology, where there are many patents for extracting or manipulating genetic material and where the development and commercialization of a new genetically engineered product may utilize a number of patented techniques. In principle, licenses could be negotiated to secure all of the needed rights, but as the number of needed licenses gets large this becomes problematic, both in terms of time and money.[6] In some cases, the owners of

---

[5] Pharmaceutical Research and Manufacturers of America, at: http://www.phrma.org/publications/publications/profile02/index.cfm
[6] See Eisenberg (2001) and Blanton (2002).

patents on tools for genetic manipulation insist on "reach through" licensing, whereby someone using the tool would not only need to pay a fee, but also pay royalties derived from the revenues of any product developed using the tool. The royalties on such licenses are typically small—1 percent or less of product revenues—but if a product development effort uses several such tools, the overall profitability of such a development effort can be seriously undermined. In effect, the need to secure multiple reach-through licenses can impose a large tax on the innovation process, and thereby significantly retard innovation.

3. *Some firms and individuals use patents more like grenades, threatening others' property*

Ambiguity about exactly what is covered by related patents held by different firms is not uncommon. Often, the firms holding these patents manage to get along peaceably, like two neighbors who don't know exactly where to divide their lawns but just enjoy their back yards. But some firms seem increasingly to want to do battle over patent rights. Even more worrisome, some firms no longer see patents as defensive weapons to be used to protect their innovations from imitation by oth-

ers, but, rather, wield them offensively to threaten and disrupt the ongoing and future business plans of competitors.

Because of the inherent riskiness of the innovation process, patents are crucial in many cases to providing enough protection that investors are willing to put up the money to develop new technology. But patents are blunt instruments. Because of the complexity of the evolution of technology, the monopoly that they create will sometimes retard rather than encourage competition. In the best of worlds, a patent system is a compromise among competing objectives.

Historical perspective

These issues are not new. Disputes over the way that the patent system works have been the subject of passionate debate for centuries. The same questions are on the agenda today: the ability to reward inventors in a timely manner, the quality of the review provided by patent office procedures, and the risks and burdens created by litigation to enforce patents.

Historical patent controversies are cyclical. The abuses of the Elizabethan era—where the ruler could make grants with the stroke of pen—led to reforms that by the 18th century

Figure 1. Patent policy changes by decade in 60 largest countries

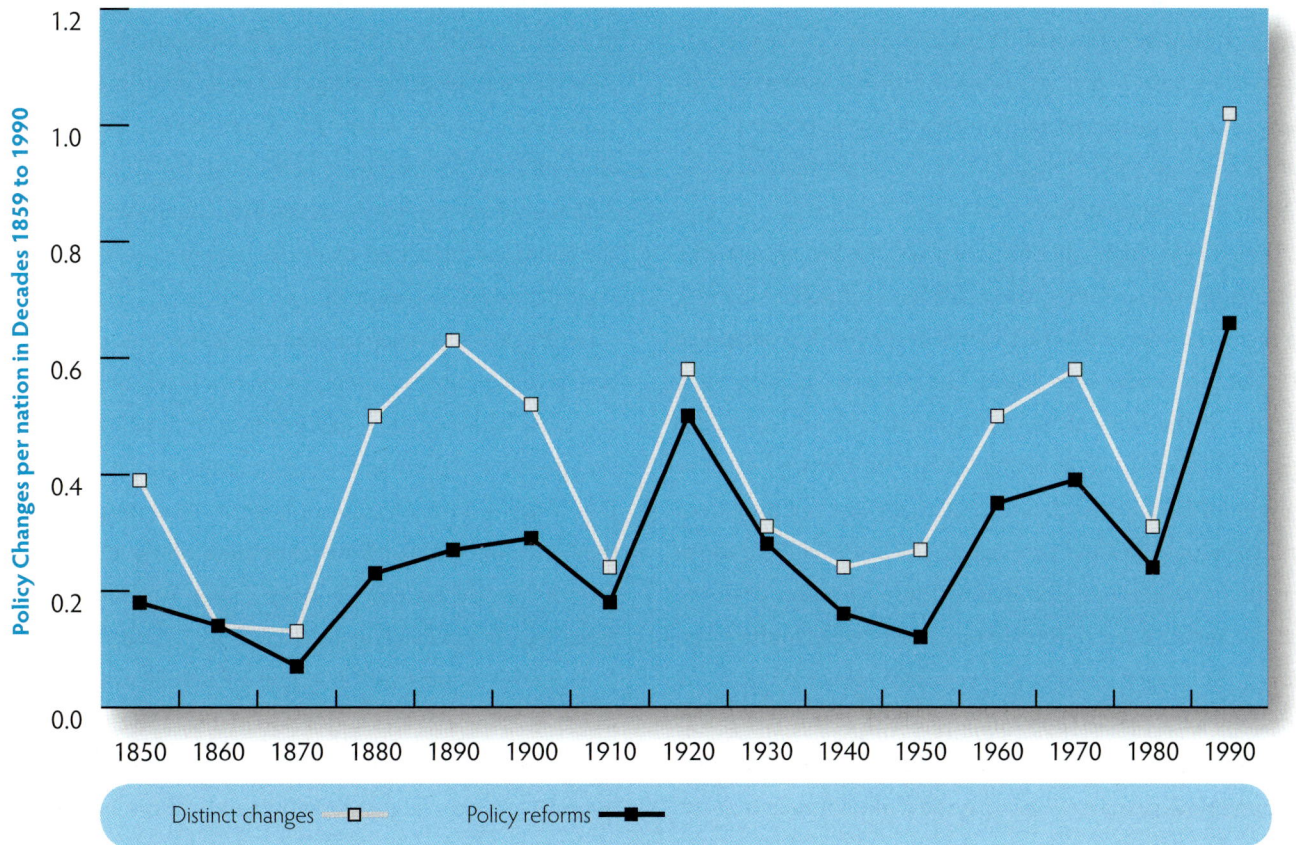

2.2 THE US PATENT SYSTEM: Does It Strengthen or Weaken Innovation and Progress?

90

were judged to have made it too expensive and too difficult to get patents in Great Britain. The 1817 Dutch law made it so easy to get patents, and made them so powerful for their owners, that rising resentment led to the outright abolition of patents in the Netherlands, although later complaints by inventors brought the law back. Far too often, fixes to patent law have created as many problems as they have solved and many unintended consequences!

These examples are not just historical oddities. Focusing on the 60 largest countries (by total economic activity) in 2000—including historical winners that have experienced very rapid growth, such as the United States and Japan, and also those who have encountered economic reversals, such as Argentina and South Africa—we can see the changes in their patent systems from 1859 to 1990. Figure 1 shows the number of policy changes each decade in a number of substantive areas of patent policy.[7]

Because the number of countries in the sample changes over time—countries such as Algeria and India did not exist as independent entities before World War II—we present in each year the ratio of the number of patent reforms undertaken to the number of independent nations in the sample at the beginning of the relevant decade. The figure shows clearly the five visible waves of patent policy change, including the "Patent Controversy" of the 1850s and 1860s discussed earlier, and the most recent wave (in the 1990s), associated with changes implemented to harmonize worldwide patent policy agreed to in the 1993 "Uruguay Round" of multilateral trade liberalization negotiations.

The changes highlighted above were neither unique nor peculiar episodes. Rather, these debates and policy changes are representative of the ways in which the patent system has changed as we have moved into the modern age. The main concern at the time of the Parliamentary rebellion against Queen Elizabeth's patent grants (early 17th century) was the degree of the ruler's discretion. Essentially, she had total freedom to grant rights and how many. The evolution of a modern patent system required the institutionalization of limits on this executive discretion. The same process of limiting the freedom of the ruler—whether monarch, prime minister or president—has played itself out around the globe.

If we do an analysis of the same group of 60 countries over the same time period, we see that the ability of monarchs or

government officials to extend the life of a patent beyond the stated grant period without obtaining any special permission from the legislature has fallen sharply, from nearly eight years in 1850 to less than a year in 2000. We could look at many other ways in which government officials have exercised discretion, such as the right to terminate or license the patent without the patentee's permission, the ability to vary the fee charged patentees, and the prerogative to review patent applications in different ways. These analyses all tell a similar story: the discretion of government officials has, gradually but systematically, been increasingly constrained by specific patent rules.

A prominent feature of both the early British patent controversy and the Dutch controversy was a desire to see that patents are granted only after some kind of systematic test to ensure that they go only to true inventors. Even after arbitrary royal favors were reined in, many countries evolved "registration" systems like that of the Dutch statute of 1817, in which the work of determining whether patents were really valid was devolved to the courts. The cost and disruption that suits over questionable patents entailed was a major concern.

A close look at the history of the patent system shows that there has been a considerable movement from registration to examination systems. Over time, the percentage of nations that examine patents prior to issuing them—rather than automatically granting patents to those who meet the basic requirements—has increased substantially. The share has climbed from 35 percent in 1850 to nearly 70 percent today. And the 70 percent with examination systems includes all of the major industrialized countries, where the bulk of technological innovations originate.

As we shall see, patent protection has become stronger, and this has led to more attention to the way in which these awards are made. Although the quality of the way in which patents are examined hasn't changed—another subject of concern in the U.S. today—these awards are now far more important than they once were, making mistakes more costly. While we emphasize here the experience in the United States, the phenomenon is truly a global one.

Figure 2 depicts one measure of patent protection: the length of time of the longest patent regularly granted, once again averaged across all active nations in our sample. While there have been some ebbs and flows in this measure—note

---

[7] Based on the data-set described in Lerner 2000 and 2002. The changes in patent policy examined are the presence of patent protection for important classes of technologies, the length of patent protection, the amount of time individuals have to put their patent into practice, and major changes in the cost of patent protection.

Figure 2. **Average length of patent awards in 60 largest nations**

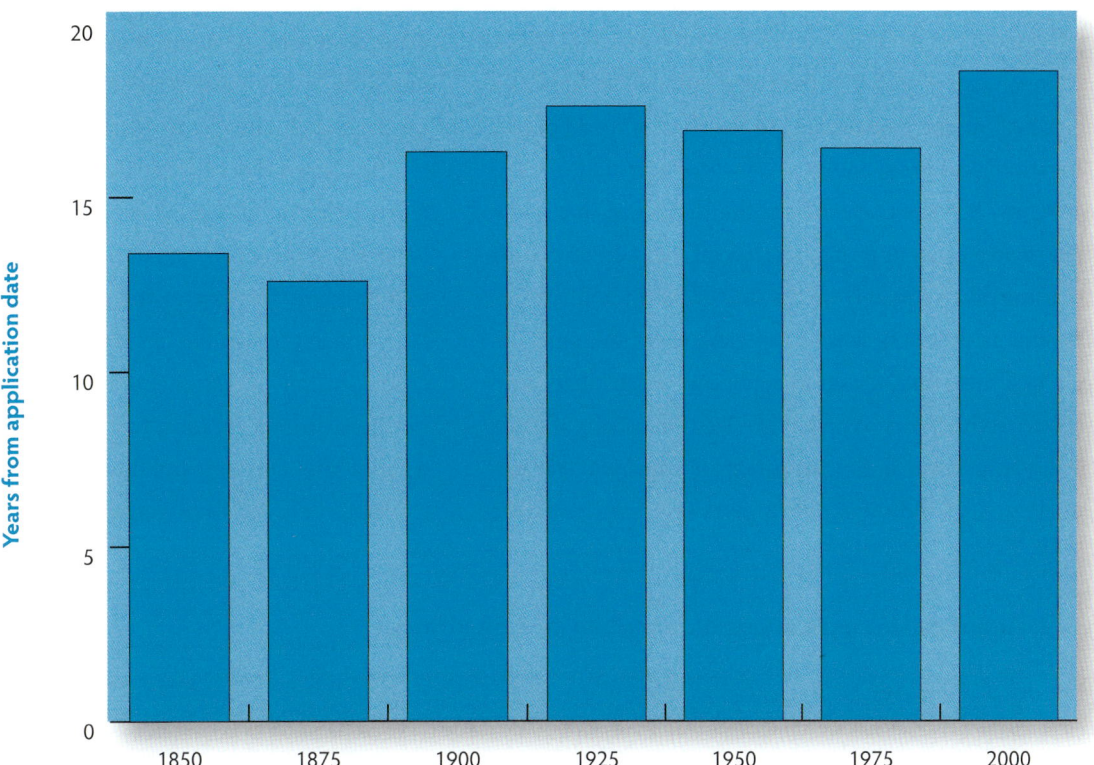

the shortening of the average patent life during the anti-patent movement of the 1860s—the basic trend has been toward longer patent protection. The average patent today is nearly 40 percent longer than that in 1850. This increase in the degree of protection afforded patent-holders implies that assuring the validity of patents—an issue, which has long sparked passionate debate—is particularly critical today.

These episodes in patent history highlight the fact that no one right solution exists to many of these dilemmas. To be sure, we've seen some trends over time: patent office officials have been given less discretion in how they make grants; patent awards have been scrutinized more intensively; and awards have become longer. Different approaches to organizing patent systems have been tried, and many differences persist around the world. However the system is organized, there are likely to be people who think that patents are too easy or too hard to get, and to easy or too hard to enforce.

The silent revolution

An apparently benign change in judicial procedure triggered a whole new era in U.S. patent policy. Almost all formal disputes involving patents are tried in the federal judicial system, following initial litigation in a district court. Prior to 1982,

appeals of patent cases were heard in the various appellate courts, some of which were more than twice as likely to uphold patent claims as others. These differences persisted because the Supreme Court considered these commercial disputes too "banal" to hear patent-related cases.

The result was widespread "forum shopping" in patent cases. Every Tuesday, at noon, patent applicants would crowd the hallway when the list of patent awards was distributed. If their patent was issued, they would immediately instruct their lawyers to file suit—in a patent-friendly district court, such as Kansas City—against some alleged infringer of the newly minted patent. Meanwhile, representatives of firms who might be accused of infringing the issued patent would simultaneously race to the phones, ordering their lawyers to file a lawsuit—but in a district known to be skeptical of patents—e.g., San Francisco—seeking to have the new patent declared invalid. Such dueling lawsuits would usually be combined into a single action, heard in the district court in which the earliest filing was made. Often the fate of the case—and many millions of dollars in damages—would depend on which lawyer had the earlier date stamp on his documents.

In 1982, the U.S. Congress decided to address this problem, perceived to be undermining the effectiveness of patent protection, and threatening U.S. technological and economic strength. It established a centralized appellate court for patent cases, the Court of Appeals for the Federal Circuit (CAFC). The change was presented in the congressional hearings as a benign one, bringing consistency to the chaotic world of patent litigation, and predictability to the enforcement of valid patent rights. But it was clear from the beginning that advocates of stronger patent protection hoped that the new court would come down squarely on the side of patent holders.

And this is precisely what happened. Over the next decade, in case after case, the court significantly broadened and strengthened the rights of patent holders. The share of cases where a district court finding of patent infringement was upheld increased, as did the share of cases reversing an earlier

finding that a patent was not entitled to damages. The CAFC greatly expanded patent-holders' rights along a number of other dimensions, making it easier to shut down a rival's business even before a patent is proven valid, and to extract significantly greater damages from infringers.

The consequences of the CAFC's strengthening of the system of patent enforcement were exacerbated by changes in the behavior of inventors and of the U.S. patent office, which led to a dramatic increase in the number of patent applications filed, and in the number of successful patent applications. Decisions of the CAFC encouraged more patent applications, for three distinct reasons: first, the CAFC included software technologies, business methods, and certain biotechnologies (hitherto believed to be unpatentable) as patentable subject

Figure 3. **Annual patent applications and awards in the United States**

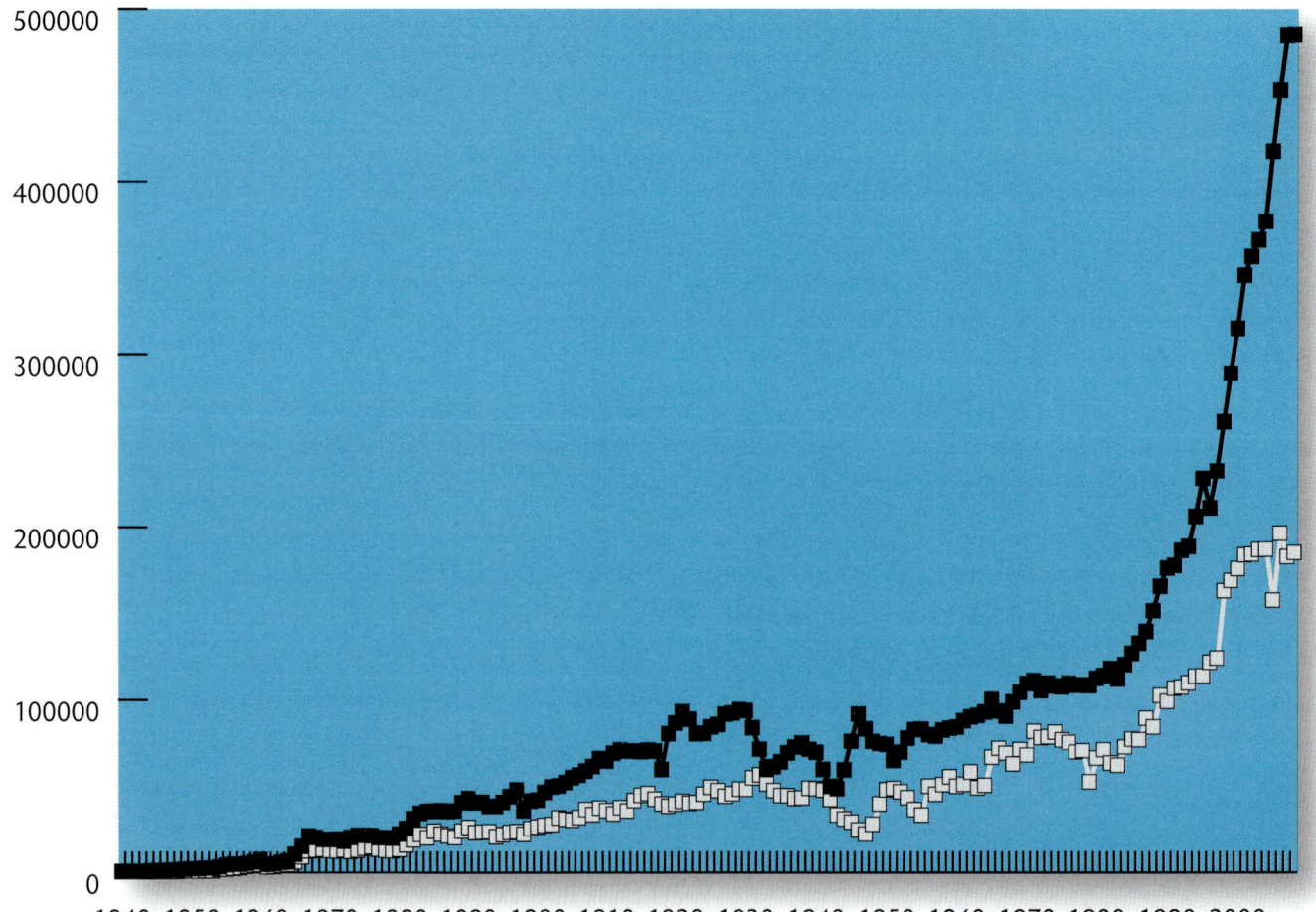

Patent Awards ☐   Patent Applications ■

matter; second, new rulings on the standards of "novelty" and "non-obviousness" made it easier for applicants to qualify for a patent; finally, improved enforceability of granted patents encouraged applications by making the patent right more economically valuable.

As the tide of patent applications began to rise, Congress intervened once again to modify the patent system. In the early 1990s, it converted the U.S. patent office from an agency funded by tax revenues (which collected nominal fees for patent applications), into one funded by the fees it collects. The patent office thus became a "profit center" for the government, collecting more in application fees than it costs to run the agency!

This apparent administrative change had important consequences. Increasingly, the Patent and Trademark Office (PTO) came to view itself as an organization whose mission is to serve patent applicants, an orientation which created strong incentives for it to process applications as quickly as possible, at the lowest possible cost. The result has been a widely perceived decline in the rigor with which the standards of novelty and non-obviousness are applied in reviewing patent applications. This, in turn, encourages more people to apply for dubious patents.

The patent explosion

With so much to gain from the swelling tide, many in the patent community—office officials, the patent bar, and corporate staff—have welcomed these profound shifts in the U.S. patent system.

The weakening of examination standards and the increase in patent applications led to a dramatic increase in the number of patents granted in the U.S. Figure 3 shows that the number of patents granted in the U.S., which had increased at less than 1 percent per year from 1930 until 1982 (the year the CAFC was created), roughly tripled between 1983 and 2002, from 62 thousand per year to 177 thousand per year, an annual rate of increase of about 5.7 percent. Applications ballooned as well to about 350,000 per year.

If this increase in patenting reflected an explosion in U.S. inventiveness, it would be cause for celebration. Unfortunately, the rapid increase in the rate of patenting has been accompanied by a proliferation of patent awards of dubious merit. This disturbing trend is confirmed by international compari-

sons, which show that the number of inventions originating in the U.S. of confirmed worldwide significance grew in the 1990s at a rate less than half that of domestic U.S. patent office grants. It is also confirmed by reference to particular patents granted by the PTO for "inventions" that are not new or are trivially obvious.

Despite the formal process of examination, the system seems more akin to one of registration, in which a determined patentee can get almost any award he seeks, although the granted claims may not be as broad as desired. This is the predictable result when underpaid, inexperienced, and overworked examiners are pushed to resolve cases as quickly as possible, and are given flawed and obsolete tools for finding and searching the prior art. Figure 4 shows how the number of applications that each examiner must handle has grown steadily. Between 1958 and 1975 there were never more than 100 applications received for each examiner. In nine out of eleven years since 1992, the applications per examiner have exceeded this threshold.

While the rejection rates for U.S. patents appear impressive at first glance, these numbers are illusive.[8] The false impression arises from the fact that when patent applicants refile their proposals in response to an initial rejection by the PTO, they are often counted as fresh applications. Fully one-quarter of the seemingly new applications are actually refiled rejected filings. Thus, the success rate is considerably higher.

International comparisons provide further evidence of the decline in U.S. patent quality. The Organisation for Economic Co-operation and Development (OECD) in Paris has been integrating data on patents granted by the U.S. PTO, the European Patent Office, and the Japanese Patent Office. By tracing the links between foreign patent applications and those in the inventor's home country, OECD researchers[9] have identified what they call "families" of patents that correspond to the same underlying invention. An invention successfully patented in the three major patent-granting jurisdictions is relatively important, both because its owner valued it enough to seek protection in all three, and because the examination systems in all three judged it sufficiently novel to merit patenting. The number of such patent families originating in a given country in a given year provides a measure of the number of relatively important inventions produced in that country that year.

The OECD calculations indicate that between 1987 and

2.2 THE US PATENT SYSTEM: Does It Strengthen or Weaken Innovation and Progress?

93

---

8 See Quillen and Webster (2001) and Quillen et al. (2002).
9 The authors have used data provided by Dominique Guellec, Division of Science, Technology and Industry, Organisation for Economic Cooperation and Development, Paris, April 2003.

Figure 4. **Patent applications per patent examiner in the United States**

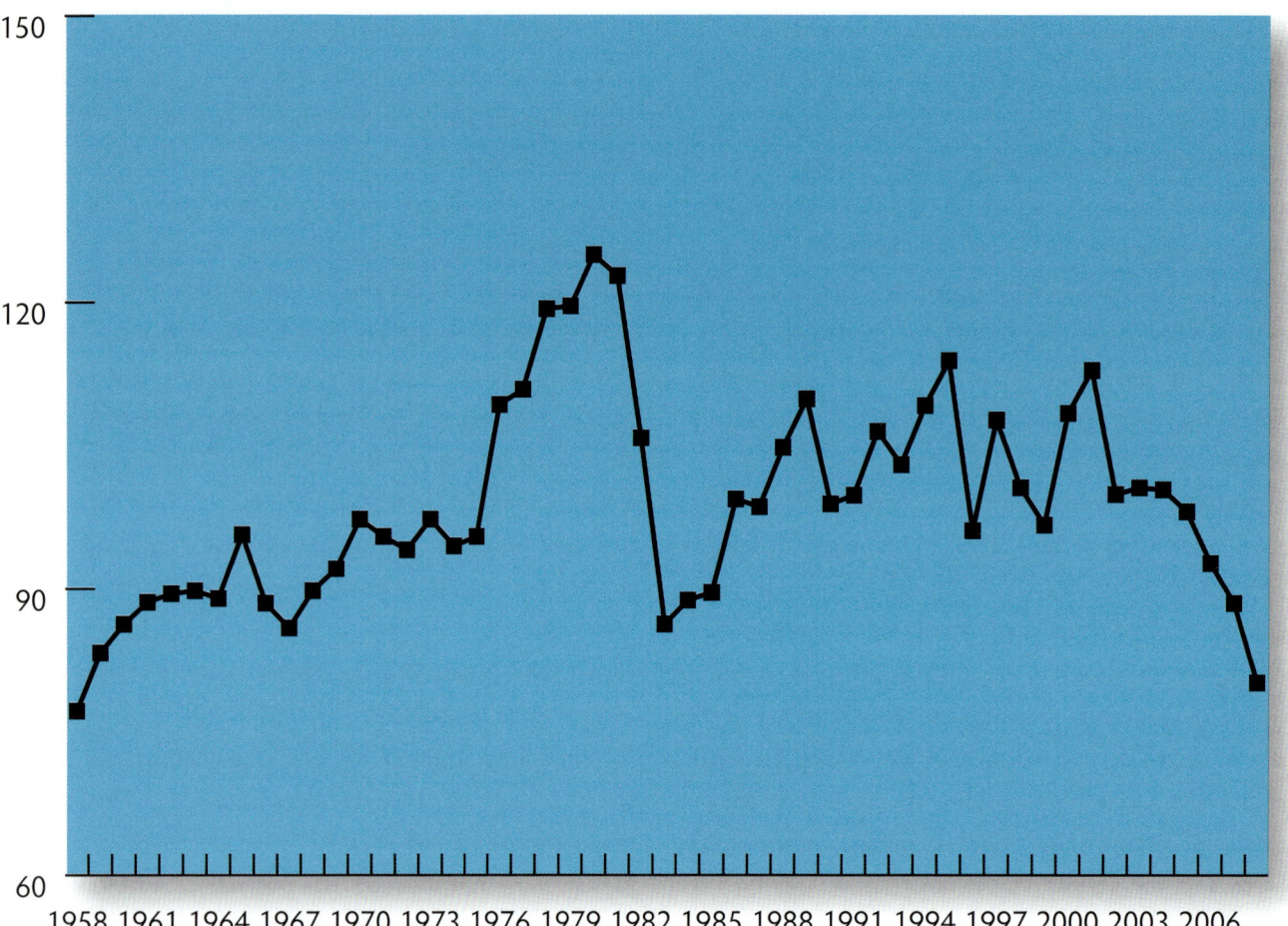

1998, the number of important inventions originating in the U.S. increased by 51 percent. By comparison, the number of successful applications to the U.S. PTO by American inventors increased 105 percent over the same period. If the examination standards in the U.S. were not changing, we might expect successful applications in the U.S. by American inventors to grow at about the same rate as our measure of internationally important inventions originating in the U.S. Actually, we would probably expect families to grow somewhat *faster* than successful U.S. applications, as the process of globalization gradually induces more and more successful U.S. applicants to seek protection around the globe. The fact that the growth in successful PTO applications was, instead, *twice* as large as the growth of international families is hard to explain in any manner other than declining standards in the U.S. PTO, producing an ever-growing proportion of U.S. patents the patent-holders themselves did not think merited patenting elsewhere.

Much of the problem stems from the organization of the

PTO itself. Chronically strained for resources, its officials have struggled to find qualified examiners, particularly in the "new" areas of software, financial methods, and biotechnology, where it had not previously had much expertise. As the CAFC opened the door to new kinds of patents, the few examiners in these new areas were overwhelmed with applications. Examiners of financial patents, for example, often had as little as a dozen hours to assess whether a patent application was truly novel.

Moreover, retaining the few examiners skilled in the new technologies has been difficult. Companies have been eager to hire these examiners, valuable not only for their knowledge of the PTO examination procedure in the new technology, but also for their understanding of what other patent applications are in process but not yet awarded. While corporations and law firms can offer huge salaries, the approximately $40,000 starting federal salary is far below market rates—especially for the examiners of business method patent applications, who are typically required to have not only engineering, but also MBA and law degrees.

The patent litigation explosion

The proliferation of patents on previously existing technologies would sow confusion and legal uncertainty under the best of circumstances, but it has occurred just as the CAFC has been making it easier to *enforce* the rights they convey. The result has been a parallel, and predictable, increase in the number of patent lawsuits. The number of such lawsuits was roughly constant over the 1960s and 1970s, began to rise with the increase in patent awards in the 1980s, and mushroomed in the 1990s, making lawyers, rather than entrepreneurs and researchers, the key players in competitive struggles and rendering the patent system a source of distraction from rather than an incentive to innovation.

The pernicious consequences of the situation can be seen in two kinds of competitive and legal interactions. In the first scenario, an established firm—frequently one whose competitive position and innovative activity are declining—realizes it has a valuable stockpile of issued patents. It then approaches rivals—smaller firms, who do not have extensive financial resources to engage in protracted patent litigation—demanding that they take out licenses to its patents.

Even if the target firm believes that it does not infringe, it may choose to settle rather than fight, as it may simply be unable to finance a protracted court battle or be unwilling to sacrifice investments in R&D and new facilities to finance the fight, not to mention the substantial costs of pre-trial proceedings, extensive documentation, wasted employee time, and unfavorable publicity. Its officers and directors may find themselves held individually liable, or targeted in shareholder lawsuits if the stock price drops.

For some large companies—most notoriously, Digital Equipment, IBM, Texas Instruments, and Wang Laboratories—such patent enforcement activities have become a line of business in their own right. Texas Instruments has in recent years netted close to one billion dollars annually from patent licenses and settlements resulting from its general counsel's aggressive enforcement policy, netting revenue in some years in excess of net income from the sale of products.

In addition to being forced to pay royalties, small firms may reduce or alter their investment in R&D. Evidence from surveys and practitioner accounts suggest that the time and expense of intellectual property litigation is a major consideration when deciding whether to pursue an innovation, especially among smaller firms. Smaller firms tend to shy away from pursuing innovations in areas where large firms have established patent portfolios. The net effect is to suppress innovation by younger, more vibrant concerns.

In the second worrisome scenario, individual inventors seek to "hold up" established firms in their industries. In many cases, these individuals have received a patent of dubious validity, often with overly broad claims. Yet established players have often chosen to settle such disputes, not wishing to risk the uncertainty associated with submitting a complex piece of intellectual property to trial.[10]

Although the escalation in patent litigation over the last two decades may be attributable, in part, to a general trend toward a more litigious society, it is also partly due to the escalation in patenting, since the more patents there are, the more there are to fight over. But there is also a less natural set of forces at work. With the creation of the CAFC and its rulings making it easier for a patentee to prevail, the incentive to sue has been ratcheted upward. And the disastrous deterioration in the examination standards of the over-worked PTO has planted the seeds for thousands of noxious patent weeds, which are now fighting with each other—as well as with the valuable flowers and vegetables—to take over the garden.

As a result of legal and administrative changes made between 1982 and 1990, the PTO has become so overtaxed, and its incentives so skewed towards granting patents, that the tests for novelty and non-obviousness that are supposed to ensure that the patent monopoly is granted only to true inventors have become largely inoperative. Simultaneously, changes in the court system have made patents much more powerful legal weapons than they used to be, with patentees more likely to win infringement suits than was the case before. In other words, in less than a decade, we have converted the patent from a weapon resembling a handgun or a pocket knife

---

[10] Individual inventors will employ various strategems to make the battle more one-sided and drive the large firm to settle the suit, in many cases, demanding a jury trial, presenting themselves as engaged in a "David vs. Goliath" dispute, or choosing a legal jurisdiction where residents are unsympathetic to the defendant. Similarly, individual inventors frequently threaten corporations with the promise that they will obtain a preliminary injunction, which will stop the defendant from using the patented technology even before the trial begins. While an established business might be reluctant to ask for such a drastic measure, individual inventors often feel no such compunction. Given the uncertainty of the trial process, the defendant firm frequently decides to settle with an individual inventor rather than fight.

2.2 THE US PATENT SYSTEM: Does It Strengthen or Weaken Innovation and Progress?

95

2.2 THE US PATENT SYSTEM: Does It Strengthen or Weaken Innovation and Progress?

96

into a bazooka, and then started handing out the bazookas to pretty much anyone who asked for one, despite the legal tests of novelty and non-obviousness. The result has been a dangerous and expensive arms race, which now undermines rather than fosters the crucial process of technological innovation. The so-called "reforms" of the patent system have created a substantial "innovation tax" on some of America's most important and creative firms.

Why is reform so difficult?

The failure of federal efforts to reform the problems of the patent system discussed above is due to several factors:

1. The issues are complex, difficult to understand and clouded by simplistic claims. For instance, because firms use patents to protect innovations, it is frequently argued that "stronger" patents are beneficial for innovation, and virtually any change to the status quo is characterized as "weakening" or "threatening" the patent system. Economists and lawyers are not elevating the dialogue to maximize the system's effectiveness in encouraging innovation.

2. Those with the greatest economic stake in retaining a litigious and complex patent system—the patent bar—have become a powerful lobby against reform.

3. Top executives of technology-intensive firms have not mounted an effective campaign around these issues, perhaps because many of the companies most adversely affected are small, capital-constrained firms without the funds for lobbying. Because the adverse consequences of a malfunctioning patent system are diffuse and indirect even for large firms, there is no consistent voice on this subject from the business world.

4. The ultimate harm of a malfunctioning innovation system is borne by consumers themselves, for whom the adverse consequences are even more indirect and hard to detect. They cannot see that products are more expensive because of litigation expenses and patent royalties, and no one knows about products whose introductions were delayed or cancelled because of patent woes.

The patent system may not be the only policy in which Washington-based discussion has been muddled and the resulting decisions poor. But where so much is at stake, we must still try to preserve what is best for the greater good as part of the discussion.

Goals and objectives for reform

What should reform of the patent system accomplish? While different analysts of the patent landscape have emphasized different aspects of the patent policy problems, there is general agreement on broad goals for reform of the system:

Improve patent quality. "Patent quality" is, to some extent, in the eye of the beholder. Certainly, people are getting patents for inventions that are not new and/or are obvious. We could make it much harder to get a patent on anything. If we did that, the few patents that were issued would be of very high quality, in the sense of being very deserved by the applicant. But the objective of patent quality has to be more than just making sure bad patents *don't* issue. It also has to include making sure that inventors *do* get patents when they have a truly novel, non-obvious invention, that such patents are processed relatively quickly and reliably, and that, once granted, they provide an adequate property right to protect subsequent investment in the invention.

Reduce uncertainty. The primary objective of reform should be to reduce the uncertainty that now pervades the patent system. The sand in the gears of the innovation machine is that companies and individuals must constantly fear that their research and product development may come to naught, because someone is going to assert an as-yet unknown or untested patent against them. Further, when such an assertion of patent infringement is made, the uncertainty about the ability to defend against that assertion often leads either to abandonment of the allegedly infringing technology, or to an agreement to pay possibly unnecessary royalties.

Keep costs under control. The PTO currently spends roughly $1 billion per year for its operations. Patent applicants spend several times that amount and patent litigants billions more. We might think these resources well spent if they achieved a reasonably smoothly functioning system. But the system is not working well, and it is reasonable to wonder whether we need to invest more of society's resources in the patent process. Ideally, the PTO's finances should be decoupled from the amount that it raises in the form of fees, and it should instead spend whatever it takes to ensure high-quality applications. However, dramatically increased resources are not likely to be available, particularly for the operation of the PTO itself. So we need to look for solutions that go beyond throwing money at the problem.

Next steps in reform

Examiners being human, there is an irreducible aspect of judgment in determining if an invention is truly new. Therefore, we cannot hope to have a system in which no "bad" patents ever issue. What is important is to have a system with fewer bad patents. And, since there will always be mistakes, it is important to have a system that functions reasonably well despite the issuance of some bad patents.

Better examination will require more resources. At current application rates, it would be very expensive to give all patent applications a sufficiently thorough examination to reduce the number of bad patents being issued. But a dramatic increase in PTO resources does not seem realistic in the current fiscal environment. Fortunately, it is not necessary to expend vast resources in order to provide reliable examination for *all* patent applications.

Most patents are worthless and unimportant. It is a given feature of innovation to think that our ideas are better than other people think they are. But it is also true of all technological development that the significance of a new idea cannot usually be known when first developed, because that significance depends on subsequent technological and economic developments. Many "good" ideas are patented that never turn out to be worth anything. But the fact that almost all patents are ultimately worthless has an important implication for the "patent quality" problem: if most patents are doomed to be consigned to the dustbin of technological history, what sense is there in spending vast sums to ensure that they all receive high quality examination? The legions of inventors and patent attorneys may not like to think about this, but for the vast majority of patent applications, it will simply never matter—either to the inventor, employers, or competitors—whether the patent is allowed to issue or not.

Most patent examiners go through the motions of making rulings, because rulings have to be made, but they don't matter to the outcome of the game. It is as if no one—officials, players, and coaches—have any idea of the score, or even if the game matters. But they all take it seriously because there is a slim chance that the particular "game" they are playing will turn out (months or years later) to be important. For the ones that do turn out to be important, it will matter a lot if patents are granted that should have been. But for the others, there will never be important technological or economic con-

sequences. And these "others" are the vast majority of all applications in the system.

"Rational ignorance"

If the above is true, then we can think of the poor quality of patent examination as representing what Mark Lemley calls "rational ignorance,"[11] by which is meant that society is rationally choosing to remain ignorant about which patents should be granted by the PTO. In fact, it is reasonably efficient to accept that PTO examination will be of poor quality, and that the cases that really matter will have to be sorted out in the courts. But because only the small fraction of patents that matter will ever get litigated, Lemley argues that the cost of litigation is, overall, efficient.

While we agree that it would be inefficient to provide thorough examination for all applications at the current rate, we disagree that the current situation is acceptably efficient. First, while the out-of-pocket costs of litigation may be tolerable, the intangible costs of a system with pervasive low-quality patents are much higher than just the lawyers' fees for filing and defending patent cases. The uncertainty created by the system for all parties regarding who can legally use what technologies is hard to quantify in dollars, but undermines everyone's incentives to invest in new technology. The loss of new products and processes that never make it to market—or that gain a toehold and are then abandoned after a threatened patent fight—is much larger than the visible costs of patent litigation. Fortunately, changes could be made to improve patent quality without requiring dramatic increases in the resources used in the examination process.

Inventors respond to how the Patent Office behaves. The key to more efficient patent examination is to consider how the examination process affects inventors and firms. Allowing bad patents to issue surely encourages people with bad applications. Applications that would never have been submitted before now look like they are worth a try. Conversely, if the PTO consistently rejected applications for bad patents, people would understand that bad applications are a waste of time and money, and the number of applications would decline in a hurry.

Consider the following experiment: suppose that the PTO could dramatically reduce the issuance of patents on obvious or non-novel inventions by doubling the amount of time that

2.2 THE US PATENT SYSTEM: Does It Strengthen or Weaken Innovation and Progress?

97

---

[11] See Lemley, 2001.

the examiner spent on the average application. If the rate of application were unaffected by this change, it would require an approximate doubling of the PTO budget, since twice as many examiners would be needed to handle the flow of applications in a reasonable period of time. But it is unlikely that the rate of application would be unaffected by a dramatic change in examination standards and hard to know how much the flow of applications would be affected. But if the number of applications made each year were cut in half, then this doubling of examiner effort per patent could be brought about with *no* increase in the overall PTO budget.

This hypothesis is not intended to suggest that the problem is that easy, but to illustrate how the incentives faced by inventors and firms affect the efficiency of the system. As the quality of patent examination has deteriorated, the incentive for submitting marginal patent applications increased. A vicious cycle has emerged in which bad examination increases the application rate, in turn overwhelming the examiners, and reducing examination quality further. If tools could be found to improve patent quality, this feedback would operate in the other direction, reducing the application rate, and freeing up resources to further improve quality.

Potential litigants respond to how the courts behave. When the CAFC issues rulings that increase the chance of the patentee prevailing in an infringement suit, the consequences of this change are not limited to changes in the outcome of specific cases. Such a change in perceived success probabilities changes which disputes are, in fact, litigated. Conversations with attorneys involved in patent disputes indicate that the CAFC's strengthening of the offensive and defensive weapons of the patentee has significantly increased patentees' willingness to bring suit. The change has also significantly decreased the willingness of accused infringers to fight, even when they believe that the patents being used to threaten them are not valid. Constraining litigation, and the uncertainty created for all innovators by the risk of suit, will require a change in these incentives.

Get information to flow into the PTO. Another important aspect of incentives has to do with information: who has it, and what do they do with it? Much of the information needed to decide if a given patent should be issued—particularly information about what related technologies already exist—is in the hands of competitors of the applicant, rather than in the hands of the PTO. There are strong incentives for firms to

share this information. If a competitor of mine has filed a patent application, the last thing I want is to see them get a patent on an application that would have been rejected if the PTO had known about my technology. I would thus have a strong incentive to provide this information—if the PTO gave me an opportunity for input, and if taking advantage of such an opportunity did not create strategic disadvantages for me down the road. So creating opportunities of this sort is another way that the system could exploit the incentives of private parties in order to increase efficiency.

Lest we get overly excited about the beauty of incentives, it is important to recognize that private parties' reactions to the incentives they face can also gum up the works. In particular, any opportunity that we create for outsiders to provide the PTO with information that is unfavorable to their competitors' patent applications will be exploited opportunistically. Even in the case of "good" applications, inventors will probably be only too happy to throw some kind of speed bump in a competitor's path. Thus, any change in procedures that makes it easier for competitors to intervene will probably increase the cost, uncertainty, and delay for valid patent applications.

Ultimately, incentives can mitigate, but not eliminate, the tradeoffs that must be made among the costs of the system, its reliability in screening out bad applications, and the speed and certainty with which good applications lead to issued patents. We could have a system that made very few mistakes, and issued valid patents quickly, but it would be a very expensive system to run, because it would require a lot of time by very experienced examiners. Or, we could have a system that put so many hurdles in the path of an application that bad patents almost never issued, but without a lot of resources such a system would inevitably slow down or deny many valid applications. Or, we can have the existing system, in which we make it so easy to get a patent that a lot of stuff gets through that shouldn't.

We cannot weed out the trash without killing any good stuff, and accomplish this greatly improved sorting without expending more resources. But perfection need not be the enemy of the good. If we pay attention to the incentives that different reforms create for desirable and undesirable behavior, we can recalibrate the system to get a better balance between rapid approval of good applications and reliable rejection of bad ones, and do it without dramatically increasing expenditures.

Building blocks of reform

The essential elements of our plan for patent policy reform and reorganization:

1. Greater resources devoted by the Patent Office to the process of examination, and the efficient use of these resources to bring the day-to-day operations of the PTO into the 21st century;

2. Facilitate pre-grant opposition by creating incentives and opportunities for parties to bring information about the novelty of inventions to the PTO when it is considering a patent grant;

3. The institution of effective re-examinations of granted patents, with a true opportunity to prove invalidity before an open-minded re-examiner, combined with appropriate incentives to discourage frivolous requests for re-examination;

4. To prevent waste on unimportant patents, but ensure sufficient care to avoid mistakes where the stakes are high, provide for multiple levels of review of patent applications, with time and effort escalating as an application proceeds to higher levels;

5. Replace juries with judges and special masters in ruling on claims of patent invalidity based on prior art, so that parties threatened by invalid patents have a reasonable opportunity to make their case.

The first two of these proposals would make the PTO more effective at reasonable cost. The third proposal addresses the reality that the best of all possible PTOs will make mistakes, and the need for a court system capable of rectifying those mistakes.

A less kind, less gentle patent system

Let us not repeat the Dutch patent crisis. We cannot abolish patents—or even weaken the fundamental presumption of validity for appropriately issued patents. We do need to ensure that patents receive appropriate scrutiny to guarantee their validity before they are used to restrict the commercial activities of competitors. While accepting that the PTO will still make mistakes, there must be a judicial system that deals with those mistakes in a balanced way. To achieve this without an infeasible increase in resources for the Patent Office will require significant modifications, carefully tuned to create incentives so that private parties are motivated and have

the opportunity to bring information to bear, but are limited in their opportunities to gum up the works.

Taken as a package, these reforms harness the incentives of private parties to bring information to the table in an efficient way. And they respect the "rational ignorance" principle, by bringing to bear a sequence of more rigorous (and hence more expensive) investigation, as the stakes get higher. Most patents will continue to get a relatively cursory review and then be forgotten. More important ones will get a more rigorous review, and one can presume that fewer mistakes will be made in important cases as a result. For the few cases that really matter and the PTO still got it wrong, the courts will provide a balanced and reasonably reliable final determination as to patent validity. As a result, the uncertainty and patent blackmail that increasingly threaten the whole innovation system would be reduced.

Economists have often been perceived as hostile to the patent system. We do not consider ourselves anti-patent. We are just anti-bad patents, and anti-blackmail made feasible by a court system stacked against those who challenge the bad patents. We want a patent system that can be presumed valid, because it is vital to the continued health of innovation and hence economic growth and prosperity.

Because it has been left to the special interests, patent policy seems arcane and obscure. To be sure, many details of patent law are mind-numbingly complex. But at its heart, the patent system is about three things. It is about technology, the endlessly curious and fascinating process by which new ideas for machines or drugs or computer programs are conceived and developed. It is about people, inventors who create and develop new ways of doing things, business people and lawyers who make decisions about investing in innovation, suing each other, and defending such suits, and government employees who must evaluate the competing claims of different parties and make decisions. And it is about how the rules and procedures established by Congress and the courts affect the way people interact with the underlying process of technological progress. That interaction ultimately affects us all, and so should concern us all.

2.2 THE US PATENT SYSTEM: Does It Strengthen or Weaken Innovation and Progress?

99

References

Blanton, Kimberly. 2002. "Patent 5,693,473" *Boston Globe Magazine*, 24 February.

DiMasi, Joseph A., Ronald W. Hansen, and Henry G. Grabowski. 2003. "The Price of Innovation: New Estimates of Drug Development Costs." *Journal of Health Economics* 22:151–185.

Eisenberg, Rebecca. 2001. "Bargaining Over the Transfer of Proprietary Research Tools: Is This Market Failing or Emerging?" in Dreyfus, Rochelle, Diane L. Zimmerman, and Harry First (eds.) *Expanding the Boundaries of Intellectual Property: Innovation Policy for the Knowledge Society*. Oxford: Oxford University Press, 2001, pp. 223–250.

Kremer, Michael. 2000. "Creating Markets for New Vaccines: Part I, Rationale" and "Creating Markets for New Vaccines: Part II, Design Issues," *Innovation Policy and the Economy* 1:35–118.

Lemley, Mark. 2001. "Rational Ignorance at the Patent Office." Public Law and Legal Theory Working Paper No. 46, School of Law, University of California at Berkeley.

Lerner, Josh. 2002. "150 Years of Patent Protection," *American Economic Review Papers and Proceedings* 92:221–25 (May).

———. 2005. "150 Years of Patent Office Practice." *American Law and Economics Review* 7(1):112–43.

Lerner, Josh and Adam B. Jaffe. 2004. *Innovation and Its Discontents: How Our Broken Patent System is Endangering Innovation and Progress, and What To Do About It*. Princeton, NJ: Princeton University Press.

Pharmaceutical Research and Manufacturers of America, "PhRMA Industry Profile." At: http://www.phrma.org/publications/publications/profile02/index.cfm

Quillen, Cecil D., Jr. and Ogden H. Webster. 2001. "Continuing Patent Applications and Performance of the U.S. Patent Office." *Federal Circuit Bar Journal* 11:1–21 (August).

Quillen, Cecil D. Jr., Ogden Webster, and Richard Eichmann. 2002. "Continuing Patent Applications and Performance of the U.S. Patent and Trademark Office – Extended." *Federal Circuit Bar Journal* 12(1):35–55.

Chapter 2.3

Innovation Policy as Cargo Cult: Myth and Reality in Knowledge-Led Productivity Growth[1]

Alan Hughes[2]

In the immediate post-Second World War years a series of millenarian movements known as "cargo cults"[3] swept through Melanesia. They emerged in the aftermath of intensive US contact in the course of the Second World War. These contacts led to a substantial increase in the material goods available to Melanesian islanders, but the end of the war meant that such material goods became less available as military withdrawal occurred. In these circumstances cargo cults emerged in which prophets would promise the return of cargoes of material goods by their ancestors (often expected to take the form of the Americans) with cargo typically shipped in the airplanes that had been such a common feature of the war experience. The means by which the return of the cargo was to be encouraged varied between different cults in different islands, but frequently involved the ritual preparation and construction of a variety of structures such as airfields, storage facilities, landing strips and associated paraphernalia. Cult members were encouraged to abandon previous cultural practices and often mimicked the behavioural characteristics of Americans (Worsley, 1957; Jarvie, 1964). The emergence of these cults did not lead to the return of material cargo.

There is in my view a danger today that the evolution of innovation policy structures based on copying perceived cultural characteristics and structures of the US innovation system will also fail to deliver the goods. In the case of innovation policy, the cargo is improved economic welfare through improved productivity growth based on enhanced innovation performance. The key 'ritual' structures are increased R&D expenditures; an emphasis upon the commercialization of science through university-based spin-outs and licensing routes in high-technology producing sectors; the promotion of entrepreneurship and new business entry; and a supposed US entrepreneurial culture based on the subsidization of risk taking in venture capital investment and of the development of the SME sector more generally.

[1] Published previously as Chapter 4: "Innovation policy as cargo cult: Myth and reality in knowledge-led productivity growth," in Bessant, J. and T. Venables, (eds), *Creating Wealth from Knowledge: Meeting the innovation challenge*, Edward Elgar, Cheltenham, 2007.

[2] The author is grateful to Richard Lester, Andy Cosh and Michael Scott Morton for many stimulating discussions in this area, to Anna Bullock for help in data preparation, to the Cambridge MIT Institute for financial support for the survey research into US and UK innovation on which this chapter draws, and to the EPSRC for financial support under grant EP/EO23614/1 IKC in Advanced Manufacturing Technologies for Photonics and Electronics—Exploiting Molecular and Macromolecular Materials (which is a 'public space' experiment in fostering commercialization activities).

[3] The study of cargo cults has long engaged anthropologists and their physical manifestations are well established (Worsley, 1957). There is a long and continuing controversy as to their interpretation and meaning in the cultures in which they occur (Jarvie, 1964; Lindstrom, 1993; Jebens, 2004), and the term cargo cult is now more used outside than inside the discipline of anthropology. This is principally a result of the adoption of the term by the scientist Richard Feynman to describe as 'cargo cult science' scientific investigations that fail to deliver the scientific cargo because, while apparently following all the correct forms and structures of scientific investigation, they omit a key ingredient. That key ingredient is due consideration of all the evidence against, as well as for, a hypothesis (Feynman, 1985). The argument in this chapter is in a similar spirit.

These perceived key elements feature centrally in policy debates. For example, in March 2000 the EU adopted the 'Lisbon' strategy to make, within the next decade, the EU the most dynamic and competitive knowledge-based economy in the world. The strategy was explicitly positioned as a response to the observed superior performance of the US economy which had in the previous decade substantially outperformed the European economies. It also explicitly accepted the view that this superior US economic performance was based on the emergence of high-technology sectors such as ICT and biotechnology as key totems of the new knowledge-based economy of the US (European Commission, 2004). Despite the subsequent bursting of the dot.com bubble and an increased awareness of the emerging threat to Europe from India and China rather than the US, these key elements of the innovation and technology strategy connected with Lisbon continue to be emphasized. Thus, in 2004, it was asserted that 'There is overwhelming evidence of the vital importance of boosting R&D as a prerequisite for Europe to become more competitive. To fail to act on that evidence would be a fundamental strategic error ...' (European Commission, 2004: 21). Similarly, it was asserted that entrepreneurship is required to take advantage of technological developments: 'Increasingly, new firms and SMEs are the major sources of growth and new jobs. Entrepreneurship is thus a vocation of fundamental importance, but Europe is not "entrepreneur-minded" enough' (ibid., 28).

Both of these arguments were followed by calls for greater tax subsidization of high-technology investment, R&D expenditures and enhanced policies aimed at boosting entrepreneurship and new entry and reducing risk aversion and the 'stigma of failure' (ibid.).

In relation to enhancing the role of universities, the policy emphasis on spin-offs and licensing 'US style' is often noted:

In recent years, spurred by the experience of the US in particular, policy makers, enterprises, investors and academics throughout the industrialized world have paid increasing attention to the role of universities as drivers of innovation. Many universities have established formal offices and processes for identifying promising discoveries made within their walls and turning them into revenue streams through licensing or spin-outs. (Apax, 2005:4)[4]

The belief in the centrality of university–business links to economic progress and in the commercialization of science through licensing and spin-offs is also explicit in the innovation strategies of many individual countries (OECD, 2001; Yusuf and Nabeshima, 2007).

In this chapter I wish to question these emphases on R&D-intensive high-technology spin-offs from the science base and entrepreneurial science. In doing so it is not my intention to argue that R&D or new entry or the growth of venture capital or university spin-offs do not matter. My contention is rather that they have been greatly exaggerated to the neglect of other key factors when one considers the innovation system as a whole. One of these factors is the importance of the diffusion and use of ICT as a general-purpose technology beyond the ICT and other R&D-intensive high-tech producing sectors. This has enabled 'unexpected' user sectors with negligible conventional R&D spending, such as retailing, to dominate movements in US aggregate productivity growth. A second factor is the dominant role that performance transformation in existing firms plays in driving industry-level productivity compared with the direct role of new entrants. A third is the diversified role played by universities in knowledge exchange that extends beyond a narrow focus on spin-offs and licensing to encompass the creation of human capital and a wide range of formal and informal business interactions. A further factor related to this is the predominant role of customer–supplier interactions in open innovation systems (Chesbrough, 2003) rather than direct university–business interactions. Finally, there is the major role that public procurement policy has played in the USA in the effective provision of public rather than private sector venture capital and the high value placed by US firms on public sector sources of knowledge for innovation. The chapter attempts in the space available to provide an overview of evidence on each of these factors and to consider some broad implications for innovation policy which might be drawn on the basis of that review. In particular it concludes by arguing that the crafting of innovation policy in the context of any specific national innovation system requires a careful consideration of the structural features of that context and the particular opportunities and challenges facing policy practitioners in it. An imperfect interpretation of the experience of one country's system is unlikely to be an appropriate guide to innovation system failure or success elsewhere.

---

[4] While noting the influence of this interpretation of the US model, the Apax report contains a good discussion of the wide range of interactions between universities and the business sector beyond licensing and spin-offs which are necessary to effect knowledge exchange. Hughes (2007) discusses these arguments in the more specific context of UK science and innovation policy.

Interpreting US economic performance

Since so much policy is linked to references to US economic performance, it is useful to begin with a brief overview of it in the recent past. Table 1 shows that the most dramatic feature of US performance since the Second World War is that its recent improvement is heavily concentrated at the end of the last century and at the beginning of this one, when it returned to its long-run trend performance after two decades of relatively low growth performance. The dramatic improvements in pro-

estate, and miscellaneous professional and scientific services. None of these, with the exception of computers and electronics, are in any sense conventionally R&D intensive (Farrell et al., 2005). It's a Wal-Mart- not a Microsoft-led turnaround. The traditionally identified R&D-intensive sectors have not carried most weight.

Wal-Mart, on the back of a major IT-based business structure, has transformed—some people would argue much for the worse—a whole variety of social and economic structures in the

Table 1. **US productivity growth 1947–2003 (real GDP per hour)**

Period	Rate (%)
1947–1972	2.9
1972–1995	1.4
1995–2000	2.5

Sources: McKinsey Global Institute (2001); Farrell et al. (2005).

ductivity growth after 1995 are not, however, due to the direct performance of R&D-intensive high technology industries.

This can be seen if we decompose the aggregate performance into its components. An industry's contribution to the aggregate depends on its own change in productivity growth, and on its size, because the economy is a weighted average of the different sectors.[5]

Decomposing productivity growth in the first period from 1995 to 2000 reveals that six of 59 sectors accounted for the whole of the acceleration in productivity growth. The top three key sectors in the US economy on this basis were wholesaling, retailing, and security and commodity broking. Their joint contribution was twice as great as that of the next three: electronic and electric equipment (semiconductors), industrial machinery and equipment (computers), and telecoms (McKinsey Global Institute, 2001).

None of the top three are technology-intensive sectors in any conventional sense. In the second period, the most recent years for which decomposition data are available, seven sectors accounted for 85 per cent of all the productivity growth. These were retailing, finance and insurance, computer and electronic products, wholesaling, administrative and support services, real

USA and delivered enormous productivity growth in the retailing sector (McGuckin et al., 2005; Foster et al., 2002). Much of this has been linked, as in other service sectors such as transport and financial services, to the implementation of new business models based on ICT and related technologies (Hughes and Scott Morton, 2005, 2006). Wal-Mart's performance is thus an example of the impact of ICT as a general-purpose technology (OECD, 2003a; Helpmann, 1998) in a 'user' rather than a high-tech 'producer' sector (Pilat and Lee, 2001). Microsoft, on the other hand, is a high-tech producer that contributes to the capacity for many of these changes to occur in the 'user' sectors. So in that sense Sam Walton and Bill Gates are complementary; Sam Walton and Wal-Mart are more important to productivity turnaround than Bill Gates and Microsoft, however, because of the scale of the activity that is transformed by the activities of a company such as Wal-Mart when it implements IT-linked business transformations. Differences in services productivity growth account for most of the difference in national productivity performance between the USA, the UK and Europe in the past decade, rather than differences in high-tech producing sectors (Oxford Institute of Retail Management, 2004; Griffith and Harmgart, 2005; Basu et al., 2003; van Ark et al., 2002).

---

[5] More formally, the contribution of sector i to aggregate productivity growth C_i can be expressed as $C_i = \frac{L_0}{L_1} \left(\frac{Y_i}{Y} \dot{Y}_i - \frac{L_i}{L} \dot{L}_i \right)$ where Y_i and L_i are sectoral output and employment growth rates over the period 0 to 1, $\frac{Y_i}{Y}$, and $\frac{L_i}{L}$ are the sectors' shares in output and employment in period 0, and L_0 and L_1 are levels of national employment in time periods 0 and 1 (McKinsey Global Institute, 2001).

High-technology 'producing' sectors are a small part of the economy, especially compared to the technology-using sectors and the services sector more generally. This points to the need to think extremely carefully about the mechanisms by which high-technology activity is diffused through the rest of the economy and not just the scale or productivity performances of high-technology output per se. A focus on high-technology production without a parallel consideration of diffusion or use throughout the innovation system, and the factors affecting that, runs a clear risk of failing to deliver the goods.

Spin-offs and new entry

Now I want to turn to the issue of new spin-offs and their role in productivity performance; I have called this the *golden oldies* versus *the new kids on the block* debate. The new kids on the block are new high-tech spin-off firms that are often attributed such an important role in the science and innovation process. I want to present some facts about spin-offs, especially in the USA, and put them in the context of what is known about the way in which the golden oldies contribute to changes in industry structure and productivity growth.

The first thing is to get a sense of proportion. The US economy has some 500,000 firms starting up each year. That, of course, includes firms of all kinds, from small restaurants to boutique high-tech businesses, not just businesses based on the exploitation of intellectual property (IP) or new products derived from advances in scientific research. In the USA as a whole, in 2004 there were 462 IP-based start-ups where the IP was from a US university. That may be an impressive performance internationally, but its scale has to be borne in mind in interpreting claims of what might be gained in other economies from such spin-offs.

Second, although IP produced by US universities produces results in considerable patenting and licensing activity, it is insignificant numerically compared to the total amount of such research-related activity in the USA. IBM in the year 2005 alone registered 2941 patents with the US Patent office, Canon 1829, and HP 1790. The whole of the University of California (UC) state system, which is one of the most dynamic, productive and innovative university systems in the world, produced 388, MIT 136 and Stanford 90 (US Patent Office, 2005). This is an impressive university performance. It is important, however, to keep it in perspective relative to corporate activity and to think of universities as a quantitatively small but qualitatively important part of a wider system.[6]

Finally, the returns from start-ups and licensing activity are enormously skewed. The following statistics illustrate just how skewed. Only 167 out of 27,322 patents held by 193 US university institutions in 2004 made over $1 million (AUTM, 2005). In the case of Columbia University, Stanford and the UC system, the top five patents accounted for 65 per cent of gross licensing revenues. The chances of winning this lottery are small. That doesn't mean to say you shouldn't try to do it; you can't win unless you buy a ticket, but you have to be realistic about what the odds are. First-mover new start-ups based on radical innovations capable of transforming markets very rarely come to dominate those new markets. In the terminology of Markides and Geroski, such pioneering 'colonisers' of radical new markets rarely survive early market expansion. Fast second movers with rather different 'consolidations' skills come to scale up, dominate and capture maximum value (Markides and Geroski, 2005).[7] Universities also have to be clear about the costs. The vast majority of US university technology licensing offices barely break even or don't make a profit. The gross average annual licensing revenues of the UC system in 2001–2004 of $75 million cost almost $60 million per annum to maintain and manage. Thus in the period 2001–2004 the net contribution of the University of California system's licensing income was $15 million annually, compared to around $235 million of commercial funding of university research (Mowery, 2007).

We can now look at this in a slightly broader way. Instead of just looking at the spin-off activity by US-based universities, we can examine the impact of start-ups as a whole. A substantial amount of work has been done that attempts to decompose the change in productivity in particular industries across the OECD economies in terms of entry, exit and survivor growth (e.g., OECD, 2003b; Bartelsmann et al., 2004). This work breaks down productivity growth between the gains in productivity that are made by the surviving firms which are there throughout the period studied and the transfer of ac-

---

[6] Patent statistics are subject to a number of problems in assessing performance. Companies may patent for strategic reasons, and this strategic significance varies across sectors (see for example Hall, 2004). The broad university–industry picture is, however, clear enough. It is less clear whether the quality of university patents has risen or fallen as their numbers have risen (Sampat et al., 2003; Henderson et al., 1998).

[7] From an innovation system point of view this points to the importance of understanding the interactions between types of firms and the complementarity between spin-offs as a seed-bed of new ideas and the role of subsequent acquisition or replacement by fast second movers.

tivity from lower- to higher-productivity surviving firms. This is the golden oldie effect. The firms are there at the beginning and they are there at the end. Then there's the impact on productivity of firms that leave. If the worst firms drop out, there's a batting average effect and average productivity rises. Finally, there is the effect of new entries, the spin-offs and new start-ups. This is the new kids on the block effect. They enter the system and either die or survive and grow over the period analysed. What is clear from this work is that the vast majority of the productivity growth that is experienced

for the period 1980–92 (Disney et al., 2003). The data relate to establishments that may operate a single plant and multi-plant establishments. Table 2 shows that net entry by singleton establishments accounted for only 15.9 per cent of overall productivity growth, whilst net entry due to the closure and opening of establishments by multi-plant surviving firms accounted for over twice as much (33.2 per cent). Productivity growth within surviving establishments owned by multi-plant businesses accounted for over 44 per cent. Golden oldies, surviving firms, clearly dominate this process.[8] A policy stance

Table 2. Net entry, surviving firm and reallocation components of UK manufacturing establishment productivity growth 1980–1992

Contributors to overall productivity growth	Singleton establishments	Group owned establishments
Surviving establishments' productivity growth	0.6	44.6
Market reallocation between survivors with high and low productivity levels	-0.4	3.9
Market reallocation between survivors with high and low productivity growth	0.4	-2.8
Net entry productivity effect	15.9	33.2

Source: Calculated from Disney et al., 2003.

in any economy and any industry in any time period is driven by the transformation in productivity of the golden oldies; that is, it's the improvement in the performance of the firms that are there all the time. The contribution of survivors (often referred to as the 'within firms' effect) varies between 55 per cent and 95 per cent. The net effect of exits and entry accounts for 20–40 per cent, but most of this is due to the batting average effect of exits. Entry effects are small because of low entry sizes at lower average productivity than incumbents and low survival rates. Only 30–50 per cent of new entrants survive for over five years. Exit and entry rates rise and fall together across countries and over time, with high entry associated with high exit. In the case of the USA the new entry component is typically large and negative, and survival rates are low but survivors on average grow faster. Finally, it is important to note that these studies do not suggest that the USA is characterized by high net entry. Instead it appears that it is characterized by relatively rapid growth of survivors, so it is post entry growth not entry *per se* that matters. To illustrate the effects, we can look at some data from UK manufacturing

that concentrates on driving innovation and productivity by looking only at new independent firms will therefore miss a very important part of the story.

There are some industries and some conditions that are relatively favourable to the success of innovative new entry (Baldwin and Gellatly, 2003; Baldwin, 1993; Gambardello and Malerba, 1999; Audretsch, 1995). The first is where the nature of the technology is constantly changing the basis on which competitiveness can be built. If there is turbulence in the technological regime and entry is relatively low cost, experimentation in new entry may be accompanied by some home runs. Also, if the incumbents—the golden oldies—in an industry are heavily committed to an existing technology, then there's a better chance of a new entity succeeding because the conservatism that goes with very heavy investment in a standard technology makes the incumbents relatively slow to react (Christensen, 1997). Finally, the chances of success are higher if the resources to exploit new business ideas —complementary assets—are not owned by others. If these complementary assets, which are necessary to extract value,

---

[8] It should be noted that the interpretation in the text is rather different from that drawn by the authors who emphasize new entry effects. They choose to regard as 'new entry' new plants introduced by existing multi-establishment businesses. This is clearly not new entry in the sense of new independent firms. Most new plants which open and survive are built by surviving multi-plant firms (the golden oldies).

are owned by somebody else, it is unlikely that they can be appropriated by new independent firms going it alone (Teece, 1996).

The role to be expected for new innovative entry and survival to enhance productivity performance is thus highly context specific. A blanket promotion of new start-ups in support of innovation without careful attention to industry dynamics and the ecology linking new entry and large firm success, and patterns of appropriating value, should be avoided.[9]

Universities and the innovation system

In discussing the role of universities in innovation systems, I shall illustrate my argument with data from a recent survey-based comparison of the UK and US economies. The Centre for Business Research/Industrial Performance Centre (CBR/IPC) US/UK Innovation Benchmarking Survey (Cosh et al., 2006) was carried out in the period March–November 2004. The primary telephone survey covered firms of all sizes from

the survey instrument that related to the interactions between universities and the firms in the survey, as well as drawing on some material on the wider range of interactions that survey firms claimed were relevant to their innovation activities.

Table 3 shows the size distribution of the overall achieved samples in the UK and the US surveys. Approximately two-thirds of the firms in both surveys employ between 10 and 99 people, around one-quarter employ between 100 and 999 people, with the remainder employing over 1000. In order to provide UK–US comparisons that are not contaminated by possible variations between countries in the distribution of responses by sector or by size of firm, I shall focus on the results that are obtained when we form a matched sample. This matched sample consists of 1149 US companies and 1149 UK companies matched by employment size and by sector, where the sectoral matching is at least at the three-digit level. Table 4 shows the sectoral composition of this matched sample, distinguishing between manufacturing and business

Table 3. **Size distribution of UK and US respondent firms in the CBR/IPC survey**

Employment Size	US	UK
10-99	62%	66%
100-999	24%	25%
1000+	14%	9%
N	1540	2129

Source: A. Cosh et al., 2006.

Table 4. **The sectoral composition of a matched sample of UK and US firms**

Sector	High-Tech	Conventional
Manufacturing	28%	38%
Business Services	15%	19%

Source: A. Cosh et al., 2006.

ten employees upwards in the manufacturing and business services sectors. It achieved response rates of 18.7 per cent in the USA and 17.5 per cent in the UK. There was in addition a postal follow-up survey in both countries for firms employing more than 1000 employees. In all, the survey instrument included 200 questions, which generated over 300 variables per firm. In this chapter I shall draw only on those sections of

services and high-tech and conventional sectors within those broad industrial groupings. The distinction between high-technology and conventional sectors is based on the R&D intensity of their activity and the technical composition of their labour force. The survey contains a representative proportion of high-technology businesses in both countries.

One way of looking at the role of university–industry re-

[9] It has been argued that focus on independent growth by new start-ups rather than their acquisition and integration by established firms is also questionable, given the relative strengths of large firms in exploiting or scaling up radical innovations pioneered by new firms (Markides and Gersoki, 2005).

lationships is to locate universities as a source of knowledge for innovation in the wider context of the overall sources of knowledge used by innovation-active firms. The results of an analysis of this kind for firms in the UK–US matched sample are shown in Figure 1.[10]

The picture that emerges is very clear. Customers, suppliers, competitors and the firms' own internal knowledge are the dominant knowledge sources. In both the USA and the UK, universities are relatively low in frequency of use as direct sources of knowledge for innovation. Interestingly, in terms of the proportion of firms reporting universities as a source of knowledge, the UK outstrips the USA. In both countries, use is made of a very wide range of other sources. There is clearly a distributed innovation knowledge system, and in terms of frequency of use, universities are only a small direct part of it.[11] This does not mean that they are not important, but it does

2.3 Innovation Policy as Cargo Cult: Myth and Reality in Knowledge-Led Productivity Growth

Figure 1. Use of sources of knowledge for innovation (% companies)

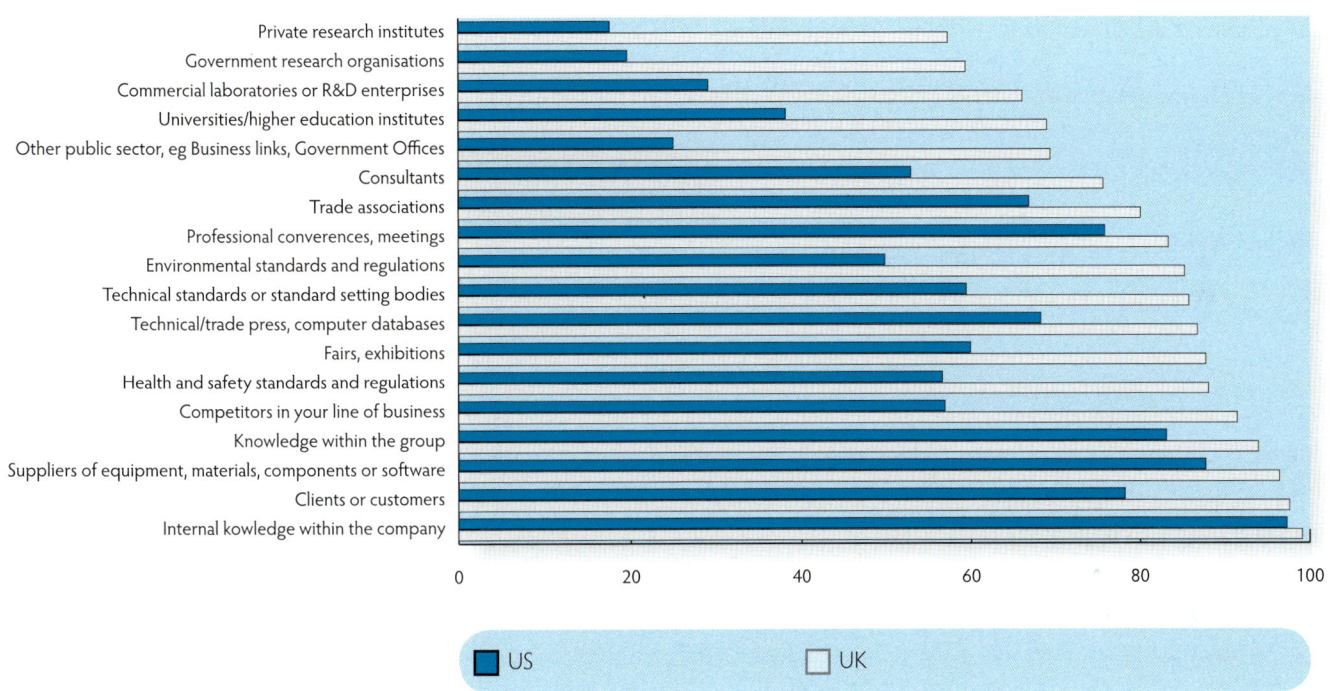

Source: Cosh et al., 2006.

Figure 2. Key sources of ideas or information for innovation in Australian innovating business 2001–2003 (% companies)

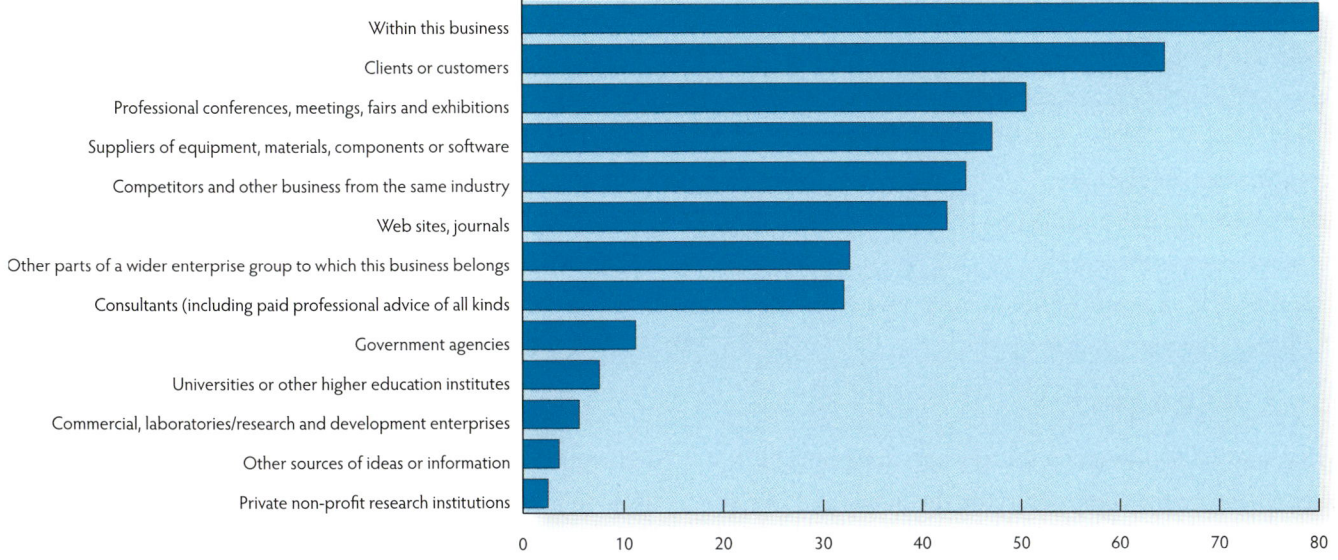

Source: Calculated from ABS, 2006.

[10] The 18 sources identified are consistent with a number of previous innovation surveys including the European Community Harmonised Innovation Survey and the periodic survey of the Small Business Sector in the UK carried out by the CBR since 1991.

[11] These results are similar to those obtained for the USA in the well-known 1994 Carnegie Mellon survey (see for example Cohen et al., 2002).

mean that their contribution has to be seen in the context of a much wider and complex system of innovation information flows. This pattern is not unique to the USA and the UK. The same is true for Australia, for instance, as is apparent from Figure 2, and for the EU more generally.

It is of course possible that frequency of use may not be correlated with the importance placed upon the information obtained. The survey firms were also asked to indicate the value they placed upon the sources of knowledge as well as their use.

The responses are summarized in Table 5, where, follow-

company sources dominate. Internal sources of knowledge plus knowledge obtained from suppliers and customers were ranked most highly as knowledge sources for innovation. In both countries they were followed by technical standards and health and safety regulations as important sources of knowledge from the intermediating and regulatory group. The need to contextualize innovation policy in the circumstances of particular countries, however, is highlighted by the fact that there are significant differences between the UK and the USA in the value placed upon knowledge from the science base, and from the intermediating organizations other than standard settings

Table 5. **High importance of sources of knowledge (% of users of that source)**

	UK %	US %	Ratio(UK/US) x 100
Company sector			
Suppliers of equipment, materials, components, or software	41.5	49.2	84.4
Internal knowledge within the company	79.9	84.5	94.6
Clients or customers	60.9	53.5	113.7
Knowledge within the group	59.4	50.7	117.1
Competitors in your line of business	27.7	20.8	132.9
Intermediating and regulatory organisations			
Consultants	12.5	26.2	47.7
Professional conferences, meetings	14.6	23.9	61.2
Trade associations	15.1	23.5	64.4
Technical/trade press, computer databases	21.5	26.5	80.8
Fairs, exhibitions	17.4	18.0	96.8
Environmental standards and regulations	31.8	46.1	69.0
Technical standards or standard setting bodies	34.6	40.2	86.1
Health and safety standards and regulations	41.3	47.2	87.5
Other public sector e.g. Business links, Government Offices	10.5	38.7	27.1
Scientific knowledge base			
Government research organisations	6.6	24.7	26.6
Private research institutes	7.2	22.9	31.5
Commercial laboratories or R&D enterprises	12.2	28.4	43.0
Universities/ higher education institutes	13.8	27.0	51.3

Source: A. Cosh et al., 2006.

ing Swann (2006), we group sources into three broad categories. These are the company sector, the public and private scientific knowledge base, and a group of intermediating and regulatory organizations. Once again, in both countries the

and regulators. For instance, US firms were almost twice as likely to place a high importance on knowledge gained from consultancies, government research laboratories and other public research laboratories, professional conferences and

Figure 3. Combined use of sources of knowledge for innovation

% Firms: **UK** **US**

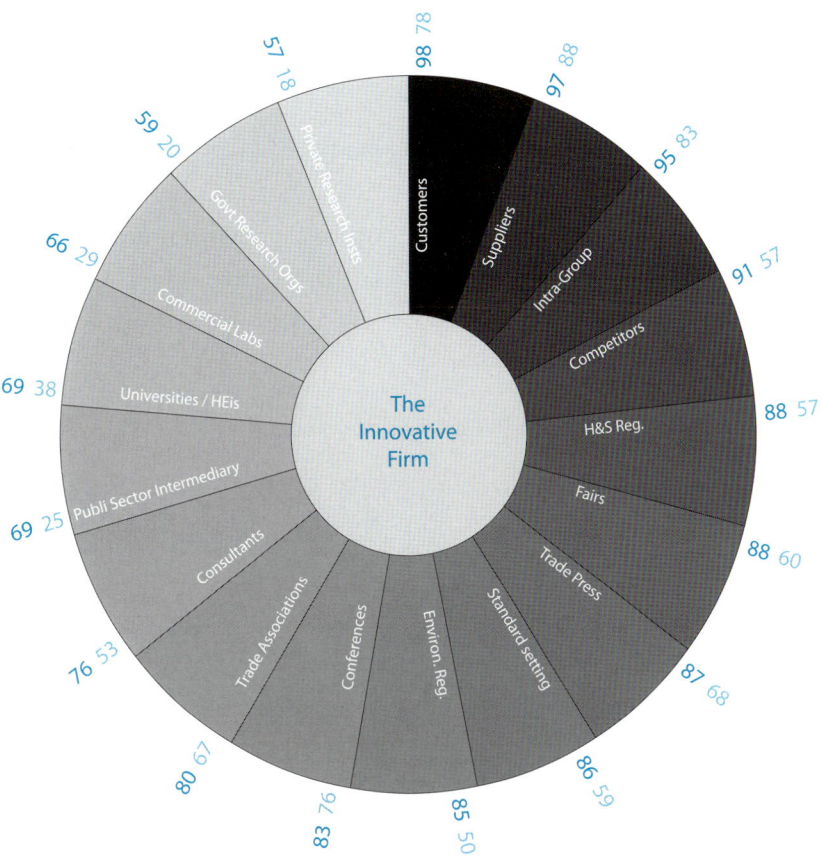

(a) **Use of sources of knowledge for innovation: % companies using each source**

UK	US
0.9	2.7

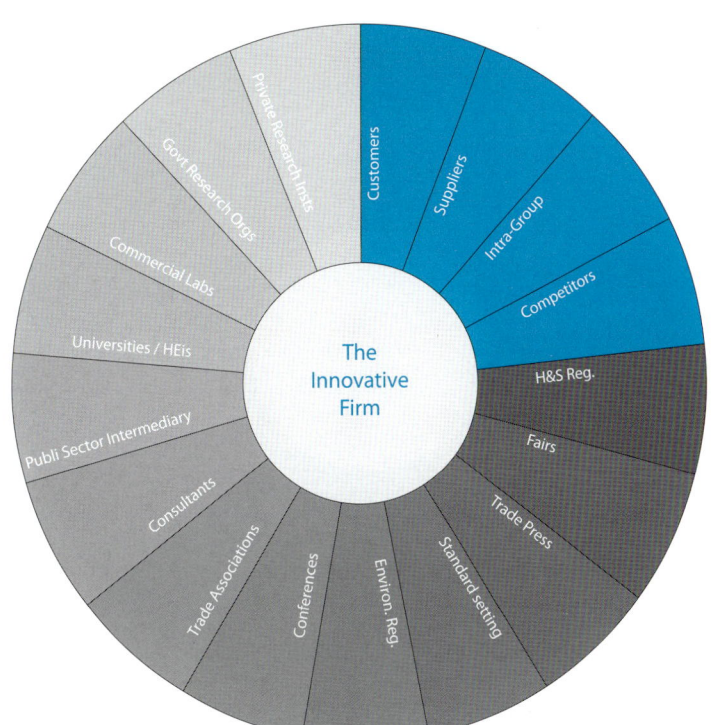

(b) **Use of at least one company source and no other source: % companies**

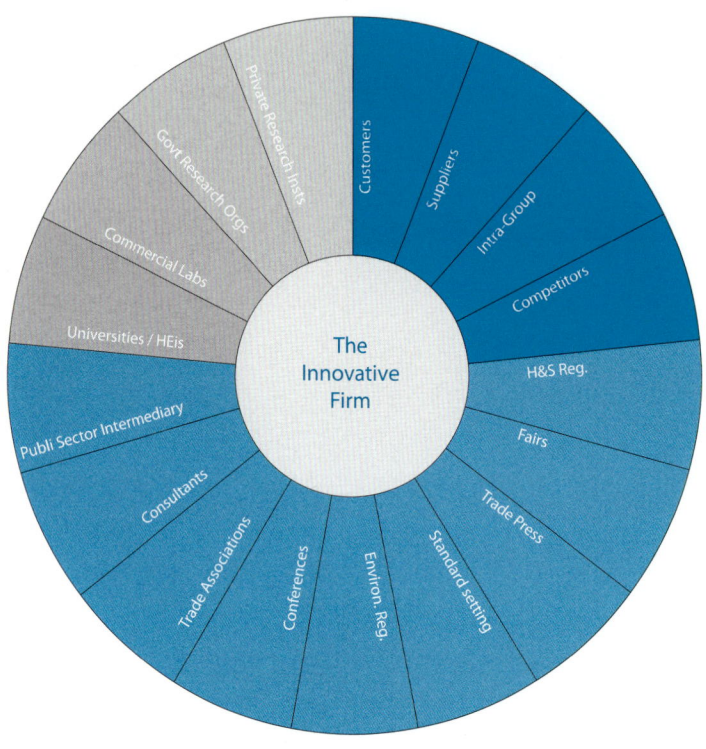

UK	US
0.9	2.7
17.9	40.8

(c) Use of at least one company source and one intermediary source and no others: % companies

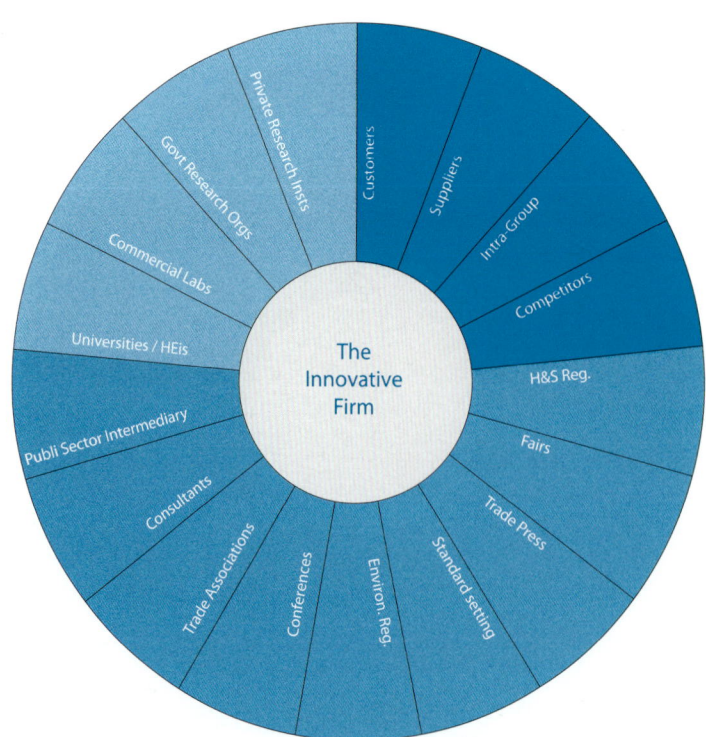

UK	US
0.9	2.7
17.9	40.8
80.2	52.8

(d) Use of at least one source in each group: % companies

Source: Calculated from Cambridge IPC Innovation Benchmarking Survey Database.

trade associations than were UK firms. Moreover, despite being more likely to cite universities as a source of knowledge, UK firms more frequently placed a lower value on it than did US firms. Another difference between the UK and the USA emerges if we probe a little more deeply into the patterns of combined use of sources of knowledge.

Figure 3 (following Swann, 2006) shows in successive quadrants the extent to which firms in the UK and the USA are specialized in their use of sources of knowledge. The first upper left quadrant simply repeats in a different form the contents of Table 5 with the thickness of the bands reflecting the frequency of use of each source of knowledge. The top right-hand quadrant shows the proportion of companies in each country which used at least one source from the company sector, and no other sources. This reveals immediately that although customers, suppliers and competitors and the internal knowledge base of the firm are the most frequently used (and, as we have seen, the most highly valued source), they are almost never used in isolation. When we switch to the bottom left-hand quadrant we identify those firms which used at least one company source and at least one source from the intermediating and regulatory group and no others. Here a significant difference emerges between the UK and the USA.

Over 40 per cent of the US firms used a company source and an intermediary source and no others, while only 17 per cent used this particular combination in the UK. When we turn finally to those companies that used at least one source in each group, we find that the UK firms are far more likely to report using a research base source in combination with the other sources of knowledge in the company and intermediating sectors. It appears therefore that US firms are much more likely to combine company and intermediating sources, whilst UK firms have a much more diffuse use of knowledge sources. Equally, US firms are less likely to use all three knowledge sources and have a more compact knowledge source pattern. Paradoxically, as we have already seen, when they do interact with institutions in the science base, they place a significantly higher value on the outcomes. This raises important questions about the extent to which the value placed upon the science base is enhanced by the use of intermediating institutions between the science base and companies themselves. It also raises the question of whether in the UK the use of so many sources raises difficulties of effective management and reduces their usefulness.[12] In terms of innovation policy this emphasizes the importance of paying attention to the particular structure of the innovation system in which the policy is to

Figure 4. The university role is multifaceted

Educating people
- Training skilled undergraduates, graduates & postdocs

Increasing the stock of 'codified' useful knowledge
- Publications
- Patents
- Prototypes

Providing public space
- Forming/accessing networks and stimulating social interaction
- Influencing the direction of search processes among users and suppliers of technology and fundamental researchers
  - Meetings and conferences
  - Hosting standard-setting forums
  - Entrepreneurship centers
  - Alumni networks
  - Personnel exchanges (internships, faculty exchanges, etc.)
  - Visiting committees
  - Curriculum development committees

Problem-solving
- Contract research
- Cooperative research with industry
- Technology licensing
- Faculty consulting
- Providing access to specialized instrumentation and equipment
- Incubation services

Source: Cosh et al., 2006.

[12] It is interesting to note that an analysis of European Community Harmonised Innovation data shows an inverted U-shaped relationship between innovation performances and the number of knowledge sources used (Laursen and Salter, 2006).

be introduced and an analysis of whether the particular patterns observed, for instance in the USA, are linked to a superior pattern of innovation and productivity performance. It also raises issues of depth as opposed to breadth of interactions.[13]

Once we have looked at the structural position of universities in knowledge flows in the innovation system, it is important to discuss the nature of the interactions between universities and firms. As a precursor to looking at some of the university data arising from the US–UK survey that bear on this issue, it is worthwhile setting out a typology of interactions.

First, universities educate and produce skilled graduates. Second, through their research and dissemination activities, universities increase codified knowledge. University staff publish books and scientific papers, they patent, and in engi-

ties are captured in Figure 4 under the headings of educating people, increasing the stock of codified knowledge and problem solving.

What tends to be less discussed is what Richard Lester and Michael Piore have called 'the public space function of universities' (Lester and Piore, 2004), which is captured in the largest box in Figure 4. This function captures the distinctive role of universities in society and in the innovation system as public spaces in which other interested parties can 'play', if that public space is appropriately structured. This includes a range of 'soft', but nonetheless extremely important activities, to do with network forming, stimulating social interaction, influencing the direction of research processes by identifying commonly experienced problems, setting standards of a technical kind, setting up entrepreneurial centres and so on.

Figure 5. Types of university–industry interaction contributing to innovation (% companies)

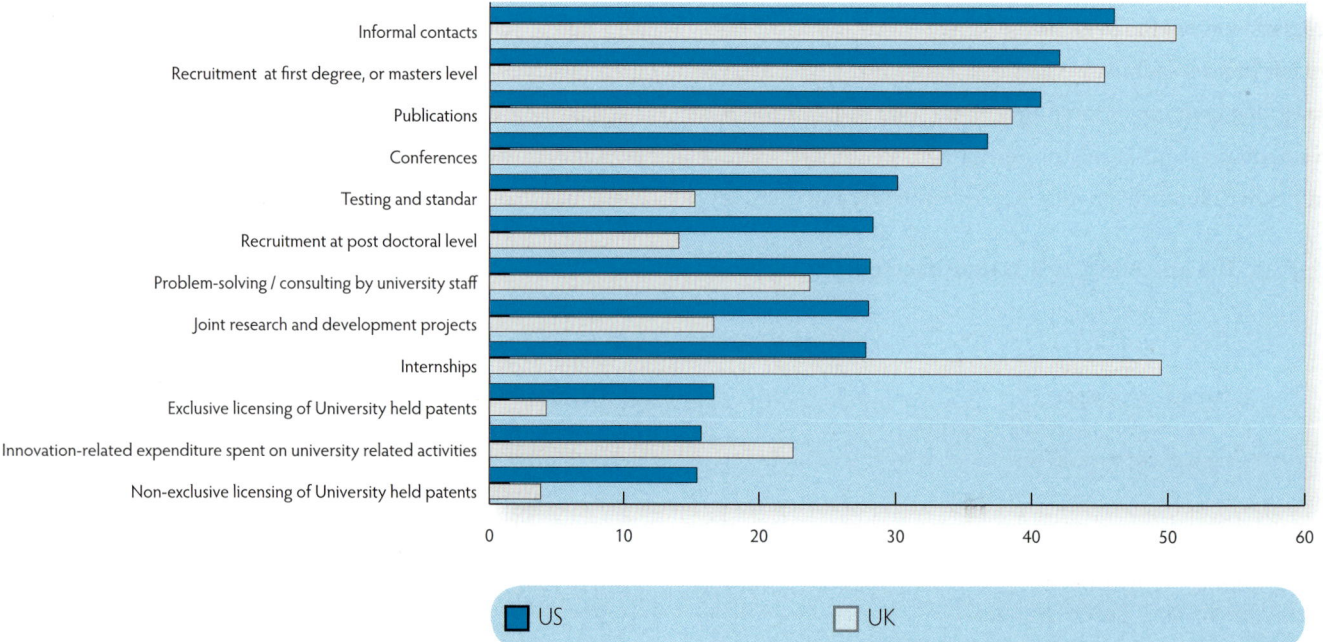

Source: Cosh et al., 2006.

neering faculties may develop prototypes. A very wide range of problem-solving activities is also carried out—often on a regional or local basis, but sometimes on an international basis—directly addressing problems that are brought to the attention of the universities through contract research, cooperative research and faculty consulting. University laboratories may have equipment that can be used for testing various kinds of commercial equipment. These three kinds of activi-

These public space activities permit the discovery of potential complementary interests and the crafting of potential ways to develop them to mutual advantage. They also foster the role of universities as translators and providers of insights into 'new' science. For instance, in the context of the US Advanced Technology Program industrial research participants perceived that 'the university could provide research insight that is anticipatory of further research problems and that it

---

[13] The CBR/IPC Survey also reveals that US firms support these university interactions with a greater commitment of resources than is the case in the UK (Cosh et al., 2006).

could be an ombudsman anticipating and communicating to all parties the complex nature of the research being undertaken' (Hall et al., 2003: 491). It is interesting to explore how these diverse public spaces and other roles are perceived by businesses, and the relative significance of licensing and spin-out formation compared to other interactions. The CBR/IPC Survey sheds some useful light here since respondent firms were asked how they interacted with the universities in their innovation activities and what kind of emphasis they placed on different interactions. Figure 5 reports the results.

It shows that businesses interact across the full spectrum of those elements set out in Figure 5. The most frequent form of interaction is via informal contacts, and it's not only the most frequent—a separate analysis (not shown here) reveals that it is also among the most highly valued (Cosh et al., 2006). All the conventional modes of university output (undergraduates and graduates, and publications and conferences) are frequently cited modes of interaction. In that sense there is no necessary conflict between how the business community says it interacts most with university activities and what academics themselves typically say they want to do.

From the point of view of differences between systems of innovation it is worth noting that US firms appear to use internships more than their UK counterparts,[14] and that they more frequently have an interaction involving innovation-related expenditure with universities. This suggests a greater depth and intensity of interaction in the USA than in the UK, even if US interaction is less frequent. US firms are, however, less, not more, likely to interact via licensing, whether exclusive or non-exclusive. However, when they interact via licensing they value it more highly (Cosh et al., 2006).

It is important to note that these are aggregate figures across manufacturing and business services. In some industries, in particular biomedical sciences, patenting and licensing are significant in terms of frequency of use and qualitative importance (Cohen et al., 2002).

From the point of view of innovation policies outside the USA, it is instructive to note that the intensification of patenting and licensing regimes in US universities has provoked a reaction. This reaction emphasizes the threats posed to the cost and timeliness of effective knowledge exchange and exploitation. In non-biomedical sciences in particular it has been argued that the time and costs involved in negotiating IP

have begun to threaten industrially funded research (Mowery, 2007). Recent research suggests that major US universities are shifting knowledge exchange management beyond patent and licensing to avoid possible adverse reactions on the wider range of interactions. This includes managing wider industrial liaison activities alongside patenting. It also includes negotiating royalty-free licences in some areas as part of industrial funding of research contracts in, for instance, electrical engineering and computer science (Mowery, 2007). If the 'US model' is to guide innovation policy elsewhere, it is as well that the current evolving model rather than the 'old' one is a reference point and the full range of interactions is recognized.

Public policy and venture capital

I now want to turn to the issue of venture capital in the USA, and the view that what is required outside the USA is subsidization of private sector venture capital to promote a more risk-tolerant investment climate. My first point here is that in practice in the USA, one of the most powerful, proactive venture capital supporting activities is public R&D procurement through the Small Business Innovation Research (SBIR) programme (Connell, 2006). The SBIR was established in the 1980s in the middle of the period of very low US productivity growth, when the USA experimented with a range of industrial policy mechanisms to counteract what it perceived, correctly, as its failure to deal with the commercial threat of Germany and Japan.

The SBIR was one of a number of initiatives taken in the course of the 1980s to address this challenge. Many were designed to encourage collaborative and cooperative strategies in relation to innovation policy and productivity performance (Dertouzos et al., 1989; Branscomb et al., 1999; Wessner, 2003). Thus, for example, the 1984 National Cooperative Research Act relaxed anti-trust regulations to facilitate research-based joint venture collaborations. In relation to university–industry links the 1988 Omnibus Trade and Competitiveness Act established, *inter alia*, the Advanced Technology Programme to promote university–industry collaboration. In 1980 the passage of the Bayh–Dole Act was designed to enhance university patenting and licensing based on federally funded research. In the course of the 1980s, several hundred university/industry research centres were also established. By

---

1990 such centres accounted for over $2.5 billion in academic R&D spending (see for example Branscomb et al., 1999; Mowery, 2007).[15]

The SBIR as part of these policy initiatives was specifically established to support businesses with fewer than 500 employees, and provides 100 per cent funded contracts to carry out technologically intensive R&D contractual obligations for US federal agencies. The US federal agencies advertise technical or research-related problems and an open competition results in the award of a contract with potential follow-on contracts. The US government currently mandates 2.5 percent of the total federal agency R&D spending to SBIR, and that, in absolute terms, is a significant sum. It amounts to $2 billion annually, covering 4000 contracts (Connell, 2006). The private venture capital sector in the USA, for comparison, was investing around $1 billion annually in around 200 deals per annum at the seed stage in the period 2005 to 2006 and around $4 billion annually in around 800 larger early-stage deals. This was out of a total annual amount invested in all stages of around $24 billion in those years (Money Tree, 2007). The private venture capital sector in the USA is thus similar in its risk profile to private equity elsewhere, with a focus on later-stage investments and large-scale company buyouts. Only a small proportion of funding goes into seed and early-stage finance. However, the SBIR produces a situation in which much of the very risky early-stage and seed investments are supported by a public sector mandated activity. Some extremely big and successful companies have been assisted in this way. Amgen, Qualcomm and Genzyme, for instance, all have SBIR connections in their origins (Connell, 2006). The SBIR effectively derisks subsequent investment by providing certification and proof of performance capacity in the earlier stages of development for small firms that win these contracts. The balance of evaluation evidence also suggests that SBIR contract winners are more likely to commercialize on the basis of their research and to grow faster than similar firms not funded through SBIR contracts (Lerner, 1999; Audretsch, 2002, 2003; Audretsch, et al. 2002; Wessner, 2001; Wallsten, 2000).

Conclusions

So, what are the overall lessons to draw from this broad overview? The first is that US productivity and growth performance is not based solely on high-tech production per se; it is based on the diffusion of innovations throughout the system, and frequently on the transformation of what people would regard as 'low-tech' sectors (in R&D terms) by general-purpose high technologies based on ICT advances. Second, productivity gains are in general driven by firms that are in existence. Thinking about existing firms and their innovation performance is critically important in the innovation process. Innovation policy should not focus on start-ups alone. Moreover, the role that start-ups may play is conditioned by the nature of particular technological regimes and patterns of appropriability. It is better to think in terms of typologies of commercialization and knowledge exchange, in which new firms' entry and independent growth is one of several potential routes. New firms and spin-offs have an important seedbed role to play but should be understood as part of a more open innovation system in which the interplay between large and small firms and the transformations in large business processes drive innovation and productivity. Third, public sector procurement has potentially a very powerful part to play in supporting private venture capital and bridging the highest-risk gap for early-stage development of research-intensive firms. Fourth, universities have to be seen as part of a complex system. Their direct contribution as a knowledge source is perceived in general by business as relatively small, as compared to other components in the innovation system. Their multifaceted role must be understood within this wider context. The mechanisms for university interaction with business are diverse and may be sector specific. Licensing and spin-offs are only one part of the story. They are significant in only some sectors, and if aggressively pursued may lead to loss of other forms of research funding from business, and high rates of spin-off failure, respectively. A one-size-fits-all economic development or innovation strategy for any country or any university that focuses on licensing and spin-offs alone is not appropriate. An innovation policy that promotes 'public space' interactions is likely to lead— through informal and other interactions—to the discovery and

---

[15] In addition to specific policy initiatives there is also abundant evidence that points to the important role played by federal expenditures, foreign policy related military expenditures generally and the (Defense) Advanced Research Project Agency (DARPA) in particular. This includes, for example, their role in emergence of the Internet, computing and IT as a general-purpose technology (Flamm, 1987; Segaller, 1998; Mowery and Rosenberg, 1998); the development of Silicon Valley (Lécuyer, 2006) and the impact of defence expenditure more generally on the structure and funding of basic applied science (Stokes, 1997).

development of appropriate interaction modes for particular sectors and purposes.

University research is of value and interest to the business sector because it is different. Creating institutional mechanisms that promote access to the space within which this different activity is pursued creates, in turn, the opportunity for the practical utilization of scientific advances, focused problem solving, and the recognition and potential exploitation of commercial opportunities. If the innovation cargo is to be delivered, this space must be adequately designed. If the existing innovation systems are failing to provide it, then discussion of policy for more openness should be high on the agenda.

References

Australian Bureau of Statistics (ABS). 2006. *Innovation in Australian Business 2003* (Reissue). Canberra: Australian Bureau of Statistics.

Apax. 2005. *Understanding Technology Transfer.* London: Apax Partners Ltd.

Audretsch, D. B. (1995. *Innovation and Industry Evolution.* Boston, MA: The MIT Press.

———. 2002. "Public/private technology partnerships: evaluating SBIR-supported research." *Research Policy* 31(1):145–58.

———. 2003. "Standing on the shoulders of midgets: the US Small Business Innovation Research Program. Small Business Economics 20(20):129–35.

Audretsch, David B., Link, Albert N. and Scott, John T. 2002. "Public/private technology partnerships: evaluating SBIR-supported research." Research Policy 31(1):145–58.

AUTM. 2005. *US Licensing Survey FY 2004.* Northbrook, IL: AUTM. US Patent Office. "Patenting by organizations." Washington, DC: US Patent Office.

Baldwin, J. R. 1993. *The Dynamics of Industrial Competition: A North American Perspective.* Cambridge: Cambridge University Press.

Baldwin, J. R. and Gellatly, G. 2003. *Innovation Strategies and Performance in Small Firms.* Cheltenham, UK and Northampton, MA, USA: Edward Elgar.

Bartelsman, E., Haltiwanger, J. and Scarpetta, S. 2004. "Microeconomic evidence of creative destruction in industrial and developing countries." Policy Research Working Paper Series 3464, The World Bank.

Basu, S., Fernald, J. G., Oulton, N. and Srinivasan, S. 2003. "The case of missing productivity growth: or does information technology explain why productivity accelerated in the United States but not in the United Kingdom?" Federal Reserve Bank of Chicago WP8, June.

Branscomb, L. M., Kodama, F. and Florida, R. (eds.). 1999. *Industrializing Knowledge: University Industry Linkage in Japan and the United States.* Boston, MA: MIT Press.

Cambridge Innovation Performance Centre (CBR/IPC) Innovation Benchmarking Survey. at: http://www.cbr.cam.ac.uk/

Chesbrough, H. 2003. *Open Innovation: The New Imperative for Creating and Profiting from Technology.* Boston, MA: Harvard Business School Press.

Christensen, C. M. 1997. *The Innovator's Dilemma. When New Technologies Cause Great Firms to Fail.* Boston, MA: Harvard Business School Press.

Cohen,W. M., Nelson, R. R. and Walsh, J. P. 2002. "Links and impacts: The impact of public research on R&D." *Management Science* 48(1):1–23.

Connell, D. 2006. *Secrets of the World's Largest Seed Capital Fund.* Cambridge, UK: Centre for Business Research, University of Cambridge.

Cosh, A. D., Hughes, A. and Lester, R. 2006. UK PLC: *Just How Innovative Are We?* Cambridge, UK: Cambridge MIT Institute: University of Cambridge. Available at: http://www.cbr.cam.ac.uk/news/160206_Report_only.htm

Dertouzos, M. L., Lester, R. K. and Solow, R. M. 1989. Made in America: *Regaining the Productive Edge.* Boston, MA: MIT Press.

Disney, R., Haskel, J. and Heden,Y. 2003. "Restructuring and productivity growth in UK manufacturing." *The Economic Journal* 113:666–94.

European Commission. 2004. *Facing the Challenge: The Lisbon Strategy for Growth and Employment. Report from the High Level Group Chaired by Wim Kok.* Luxembourg: Office of Official Publications of the European Communities, November.

Farrell, D., Baily, M. N. and Remes, J. 2005. "US Productivity after the Dot Com Bust." McKinsey and Company.

Feynman, R. P. 1985. *Surely You're Joking, Mr Feynman! Adventures of a Curious Character.* New York: W.W. Norton.

Flamm, K. S. 1987. *Targeting the Computer: Government Support and International Competition*. Washington, DC: The Brookings Institution.

Foster, L., Haltiwanger, J. and Krizan, C.J. 2002. "The link between aggregate and microproductivity growth: Evidence from the retail trade." National Bureau of Economic Research NBER Working Paper 9120. August.

Gambardello, A. and Malerba, F. (eds). 1999. *The Organization of Economic Innovation in Europe*. Cambridge: Cambridge University Press.

Griffith, R. and Harmgart, H. 2005. "Retail productivity." The Institute for Fiscal Studies Working Paper WP05/07. London: IFS, December.

Hall, B. H. 2004. "Exploring the patent explosion." CBR-Working Paper,WP 291. Cambridge: Centre for Business Research. University of Cambridge. September.

Hall, B. H., Link, A. N. and Scott, J. T. 2003. "Universities as research partners." *The Review of Economics and Statistics* 85(2):485–91.

Helpmann, E. (ed.). 1998. *General Purpose Technologies and Economic Growth*. Cambridge, MA: MIT Press.

Henderson, R., Jaffe, A. B. and Trajtenberg, M. 1998. "Universities as a source of commercial technology: a detailed analysis of university patenting 1965–1988." *Review of Economics and Statistics* 80(10):119–27.

Hughes, A. 2007. "University industry links and UK science and innovation policy." In S. Yusuf and K. Nabeshima (eds). *How Universities Promote Economic Growth*. Washington, DC: World Bank. pp. 71–90.

———. 2007. "Innovation policy as cargo cult: Myth and reality in knowledge-led productivity growth." In Bessant, J. and T. Venables, (eds). **Creating Wealth from Knowledge. Meeting the innovation challenge**. Cheltenham: Edward Elgar.

Hughes, A. and Scott Morton, M. S. 2005. "ICT and productivity growth—The paradox resolved." CBR Working Paper, WP 316. Cambridge: Centre for Business Research. Cambridge University. December.

———. 2006. "The transforming power of complementary assets." *MIT Sloan Management Review* 47(4):50–58.

Jarvie, I. C. 1964. *The Revolution in Anthropology*. London: Routledge and Kegan Paul.

Jebens, H. (ed.). 2004. *Cargo, Cult and Culture Critique*. Honolulu, HI: University of Hawaii Press.

Laursen, K. and Salter, A. 2006. "Open for innovation: the role of openness in explaining innovations performance among UK manufacturing firms." *Strategic Management Journal* 27(2):131–50.

Lécuyer, C. 2006. *Making Silicon Valley: Innovation and the Growth of High Tech, 1930–70*. Cambridge, MA: MIT Press.

Lerner, J. 1999. "The government as venture capitalist: the long-run impact of the SBIR program." *Journal of Business* 72(3):285–318.

Lester, R. K. and Piore, M. J. 2004. *Innovation: The Missing Dimension*. Cambridge, MA: Harvard University Press.

Lindstrom, L. 1993. *Cargo Cult: Strange Stories of Desire from Melanesia and Beyond*. Honolulu, HI: University of Hawaii Press.

Markides, C. C. and Geroski, P. A. 2005. *Fast Second: How Smart Companies Bypass Radical Innovation to Enter or Dominate New Markets*. San Francisco, CA: Jossey Bass/ Wiley.

McGuckin, R. H., Spiegelman, M. and van Ark, B. 2005. "The US advantage in retail and wholesale trade performance: how can Europe catch up?" The Conference Board Working Paper 1358, New York, March.

McKinsey Global Institute, in association with Solow, R. M., Bosworth, B., Hall, T. and Triplett, J. 2001. *US Productivity Growth 1995–2000: Understanding the Contribution of Information Technology Relative to Other Factors*. McKinsey Global Institute.

Money Tree. 2007. "Money tree report." PricewaterhouseCoopers. Available at: www.pwcmoneytree.com/moneytree/

Mowery, D. 2007. "University–industry research collaboration and technology transfer in the United States since 1980." In S. Yusuf and K. Nabeshima (eds.). *How Universities Promote Economic Growth*. Washington, DC: World Bank. pp. 164–81.

Mowery, D. and Rosenberg, N. 1998. *Paths of Innovation: Technological Change in 20th-Century America*. Cambridge: Cambridge University Press.

OECD. 2001. "Fostering hi-tech spin offs: a public strategy for innovation." *OECD Science Technology Industry Review*. Special Issue 26. Paris.

———. 2003a. *ICT and Economic Growth: Evidence from OECD Countries, Industries and Firms*. Paris: OECD.

———. 2003b. *The Sources of Economic Growth in OECD Countries*. Paris: OECD.

Oxford Institute of Retail Management. 2004. *Assessing the Productivity of the UK Retail Sector*. Templeton College. Oxford. April.

Pilat, D. and Lee, F. C. 2001. "Productivity growth in ICT and ICT using industries: A course of growth differentials in the OECD?" STI Working Papers 2001/4. OECD. June.

Sampat, B. N., Mowery, D. C. and Ziedonis, A. A. 2003. "Changes in university patent quality after the Bayh–Dole Act: a re-examination." International Journal of Industrial Organization 21(9):1371–90.

Segaller, S. 1998. Nerds 2.0.1: *A Brief History of the Internet*. New York: TV Books.

Stokes, D. E. 1997. *Pasteur's Quadrant: Basic Science and Technological Innovation*. Washington, D.C.: The Brookings Institution.

Swann, G. M. P. 2006. "Innovators and the research base: an exploration using CIS4." In *Report for the Department of Trade and Industry/Office for Science and Innovation*. London.

Teece, D. J. 1996. "Competition, cooperation and innovation: organizational arrangements for regimes of rapid technological progress." *Journal of Economic Behaviour and Organization* 8:1–26.

van Ark, B., Inklaar, R. and McGuckin, R. H. 2002. "Changing gear: Productivity, ICT and service industries: Europe and the United States." Research Memorandum GD-60. Gröningen Growth and Development Centre. University of Gröningen.

Wallsten, S. J. 2000. "The effects of government–industry R&D programs on private R&D: The case of the Small Business Innovation Research program." *RAND Journal of Economics* 31(1):82–100.

Wessner, C. W. (ed.) 2001. *The Small Business Innovation Programme SBIR: Challenges and Opportunities*. Washington, DC: National Research Council and National Academy Press.

———. 2003. *Government–Industry Partnerships for the Development of New Technologies: Summary Report*. Washington, DC: National Research Council and National Academy Press.

Worsley, P. 1957. *The Trumpet Shall Sound: A Study of 'Cargo' Cults in Melanesia*. London: MacGibbon and Kee.

Yusuf, S. and Nabeshima, K. (eds). 2007. *How Universities Promote Economic Growth*. Washington, DC: World Bank.

Chapter 2.4

International Mobility of the Highly Skilled: Impact and Policy Approaches

Ester Basri and Sarah Box,[1]
Science and Technology Policy Division,
OECD Directorate for Science, Technology and Industry

Introduction

In seeking to improve productivity performance and underpin long-term growth and development, many countries are placing more emphasis on innovation. This emphasis can only become stronger in the future, as serious issues in such areas as climate, energy, water, and poverty challenge the global community to find new approaches. Putting in place policies that stimulate the creation of new products and services, new processes and operations, and new ways of working, has become a pressing task for many governments.

With a focus on innovation comes a focus on people. Human resources play an essential role in knowledge production and utilization and thus are crucially important to a country's technological and economic development. Skilled people are needed not only in high-technology sectors and research establishments, but increasingly in all sectors of the economy, as the benefits of innovation are sought in a wider range of economic and social activities. The scope and capacity for innovation depends a great deal on the quality of human resources a country can draw on and the way in which these resources are used.

The sources of talented people to support innovative activity are not limited by a country's borders. Migration in search of better economic opportunities, to escape conflict, or to be with family members, has made the movement of people a fairly constant phenomenon over the course of history. More recently, however, the movement of people has intensified, as economic activity has become more globalized and the opportunities for work and study abroad have increased. Against a backdrop of policy change and global market entry by a number of emerging economies, the number of migrants has continued to rise, with the number moving to more developed regions reaching a peak of 3.3 million persons per annum in the period 2000–2005 (IOM 2008, p. 36). Within this group, the migration of highly skilled individuals forms a small but important component. From 1990 to 2000, in net terms, 5 million tertiary-educated adults moved from less developed to more developed countries, while 2 million moved between more developed countries (OECD, 2007a).

This chapter will discuss the international mobility of highly skilled people, with a particular focus on "human resources for science and technology" (HRST).[2] This group of skilled

---

[1] This essay is based on the authors' work in OECD (2008). The views expressed here do not necessarily represent those of the OECD or its member governments.

individuals includes scientists, engineers, and researchers, and plays an important role in stimulating innovative activity. The chapter is in four parts: first, selected data is presented to describe broad patterns of mobility of highly skilled people; second, the importance of mobility for the transfer of knowledge is discussed; third, the effects on receiving and sending countries are presented, with particular focus on the potential impacts of "brain gain," "brain circulation," and the diaspora; and fourth, policy responses of selected OECD countries are presented. The central message is that mobility of the highly skilled has the potential to benefit the migrant, the receiving country and the sending country, but that the policy environment will play an important role in whether this mobility can lift innovative performance.

Data

Although recent years have seen major efforts to improve information on international stocks and flows of highly skilled people, data on cross-border flows of skilled individuals—in particular HRST and certain categories within this classification—remain problematic. Internationally comparable data are difficult to collect because of the heterogeneity of immigration data across countries, especially in definitions and counting methodologies. Drilling down to look at certain categories such as "researchers" is also complicated, due to the way statistical classifications are formed. More broadly, increasingly complex mobility patterns and an expanding geographic range of migration opportunities for the highly skilled increase the challenge of collecting and analyzing rele-

Figure 1. Immigrant and emigrant population 15 years and over with a tertiary education in OECD countries, 2001[1] (in thousands)

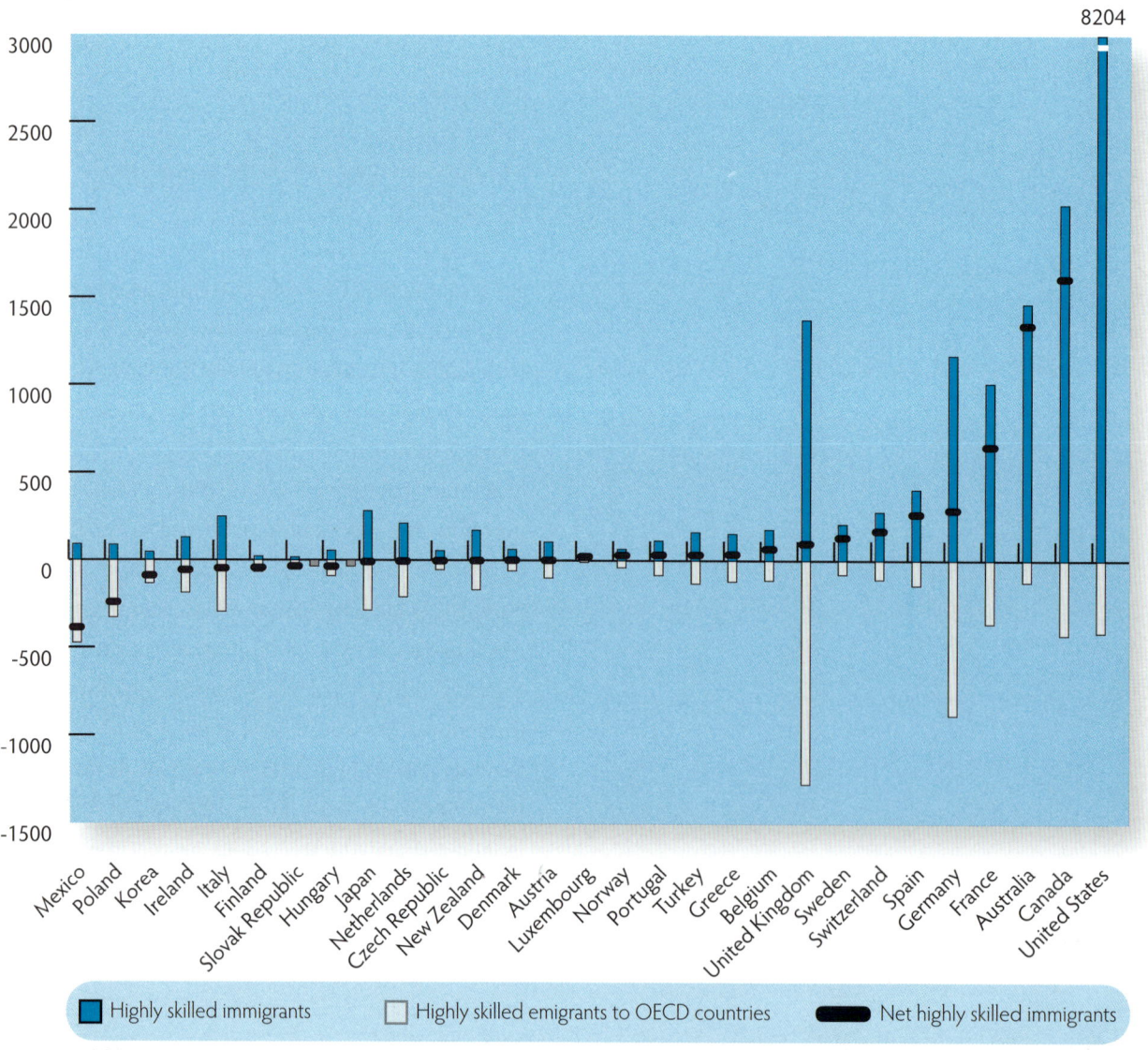

Highly skilled immigrants · Highly skilled emigrants to OECD countries · Net highly skilled immigrants

[1] 2001 or nearest available year.

Source: OECD Database on Immigrants in OECD Countries.

[2] The mobility of some other categories of highly skilled people, such as medical personnel, may pose issues different from those discussed here. For an analysis of mobility and health workforce issues, see OECD (2008a) and the special chapter of OECD (2007b).

vant data. Developing internationally comparable indicators, covering both OECD and non-OECD economies, is clearly an area ripe for further work.

The best internationally comparable data on foreign-born populations and their educational attainment comes mainly from national censuses. Based on the latest round of censuses (undertaken in 2000–2001), the OECD's Database on Immigrants in OECD Countries (DIOC) presents detailed information on the foreign-born population for almost all OECD member countries, as well as some demographic and labor market characteristics of these immigrants, and allows for the calculation of emigration "rates" of tertiary-qualified people to the OECD area for approximately 100 countries. Drawing on this database, Figure 1 shows that most OECD coun-

tries have been the net beneficiaries of highly skilled migrants, with the immigration of those who are highly skilled toward OECD countries from other OECD countries and the rest of the world having exceeded the emigration of the highly skilled from OECD countries to other OECD countries as at 2001. The figures do not include expatriation of the OECD-born highly skilled to non-OECD economies; this is assumed to be relatively uncommon, although, as these economies develop, they will likely exert a greater "pull."

In general, most OECD countries had an "expatriation rate" of less than 10 percent in 2001; in other words, the stock of their tertiary-educated natives living in another OECD country represented less than 10 percent of the total stock of tertiary-educated native-born. However, Ireland, New Zealand,

Figure 2. Share of foreign-born in HRST aged 25–64, in EU-27 and selected countries, 2006[1]

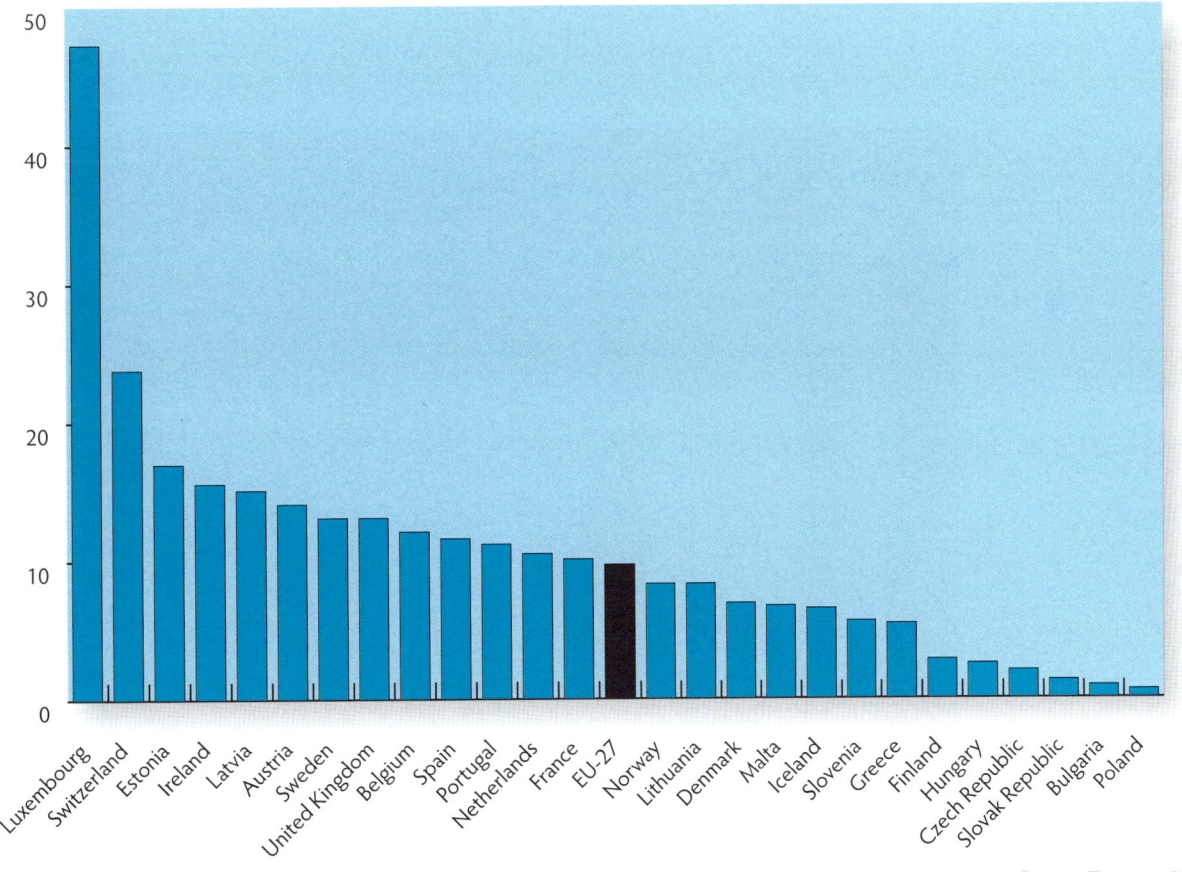

Source: Eurostat, 2007.

[1] For Iceland, Ireland, Luxembourg, Switzerland and Latvia, data from 2005; for Lithuania, data from 2003. EU aggregate does not include Bulgaria, Estonia, Germany, Italy, Malta, Romania, or the Slovak Republic.

[3] Lemaitre (2005) proposes some measures that could improve comparability of migration flow data.

[4] "Emigration rates" are calculated by dividing the number of foreign-born residing in OECD countries and originating in a particular country by the total number of natives of that country, including those no longer living in the country. This does not correspond to the usual definition of an emigration rate, which relates flows of migrants over a certain period of time to the initial stock of persons in the country of origin.

Portugal, and the United Kingdom each had an expatriation rate of more than 10 percent in 2001; in fact, that of Ireland was over 25 percent. The DIOC also allows for the calculation of emigration rates to the OECD area for many non-OECD economies. Among the non-OECD economies with low expatriation rates to OECD countries are most of the large ones, such as Indonesia, Bangladesh, Brazil, India, and China. Smaller countries, particularly islands, such as Jamaica, Haiti, Trinidad and Tobago, and Fiji, tend to have much higher expatriation rates, of more than 60 percent, and in some cases more than 80 percent. African countries also have particularly high expatriation rates of the highly skilled to OECD countries. However, flows are not just South-North; non-OECD economies are also important hosts for migrants. World Bank data suggests that South-South migration accounts for 24 percent of total emigration, although unfortunately the data do not allow for the identification of skilled migrants within this group (Parsons et al., 2007).

As a percentage of highly skilled natives, inflows of highly skilled migrants can be significant for some OECD countries. In Luxembourg, Switzerland, Australia, New Zealand, and Ireland, highly skilled migrants from other OECD countries were equivalent to more than 15 percent of the native-born highly skilled in the country in 2001. Several OECD countries also attract a large number of non-OECD born highly skilled migrants relative to their native-born highly skilled; the ratio was more than 10 percent for Canada, Portugal, the United Kingdom, and the United States in 2001. More recent figures from the European Union also highlight the differences in the significance of foreign skills across countries. Figure 2 shows that while the average share of foreign-born in HRST in the EU-27 was around 10 percent in 2006, this ranged from over 45 percent in Luxembourg, to negligible shares in Poland and Bulgaria.

Another facet of the significance of highly-skilled mobil-

Figure 3. Foreign-born doctorate holders as a percentage of total doctorate holders, 2001,[1] by OECD country of residence

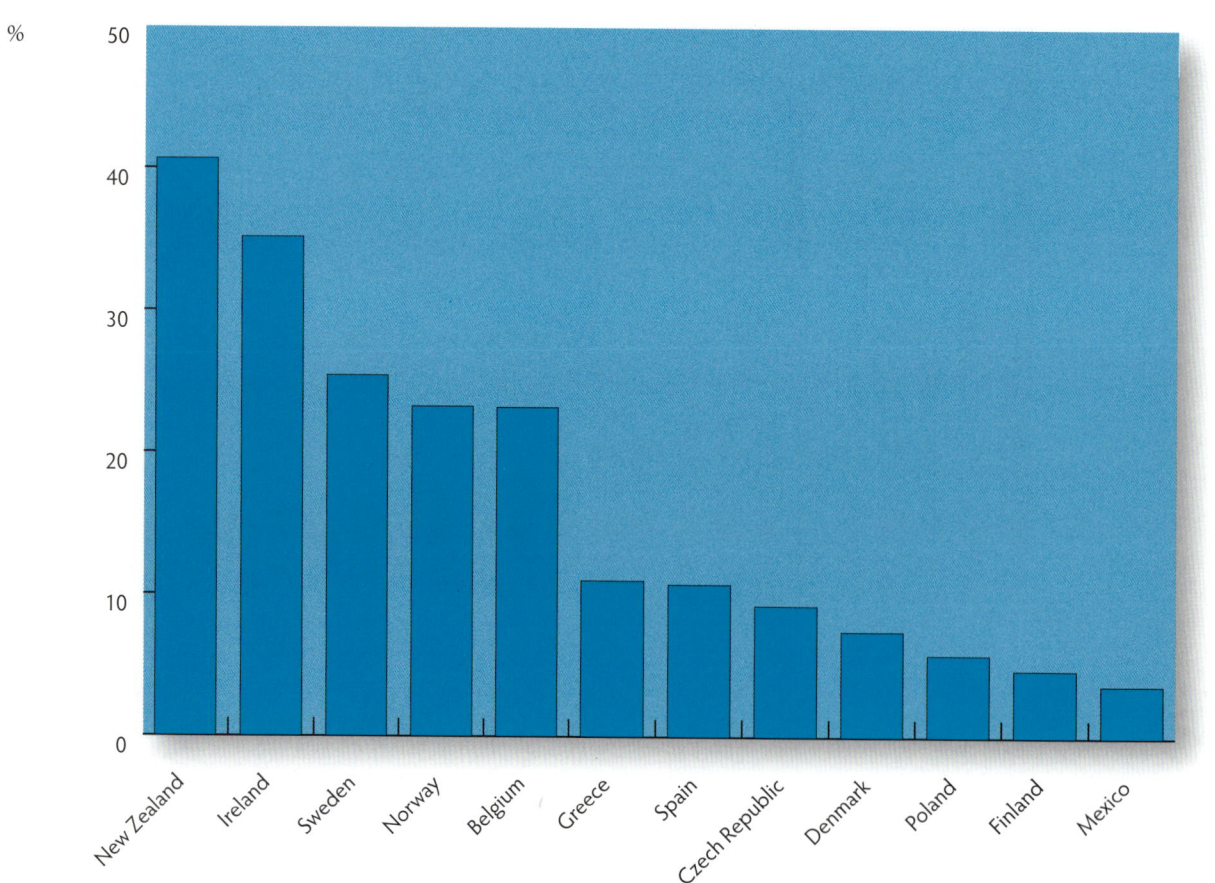

[1] 2001 or nearest available year.

Source: OECD Database on Immigrants in OECD Countries (excluding countries that had provided data for the Careers of Doctorate Holders project).

ity is shown by their contribution to particular skilled workforces. Figure 3 highlights that for some OECD countries, foreign-born doctorate holders form a large proportion of total doctorate holders, who, as a a group, are a particularly mobile population, whose mobility has increased in recent years. Data collected for the OECD/UNESCO/Eurostat Careers of Doctorate Holders project shows that, in European countries for which data are available, 15–30 percent of doctorate holders who are citizens of the reporting country declare another country of residence or stay during the past 10 years (Auriol 2009, forthcoming). Limiting the dataset to recent doctoral graduates (from 1990–2006) lifts this percentage, suggesting that mobility is more prominent among younger or more recent graduates. These figures underestimate the total mobility of doctorate holders, since they are based on the declarations of returnees, and a non-neglible number of citizens may still be abroad.

students who had received science and engineering doctorates from universities in the United States in 2000 found that more than two-thirds were in the United States in 2005, five years after graduating (Finn, 2007). Stay rates differed according to subject field and source country, with the highest stay rates recorded for doctorates in computer/electrical and electronic engineering (76 percent) and for students from China (92 percent) and India (85 percent).

The financial crisis and economic recession that hit the global economy in 2008 will have an impact on the flows of highly skilled workers and students, but the exact pattern of change is hard to predict. With impacts on labor markets lagging behind changes in the real economy, the effects of the economic slowdown will continue for some time to come. Forecasts predict that joblessness in all OECD countries will rise sharply, with the rate of unemployment reaching double digit levels in some countries for the first time since the early

Table 1. Share of science professionals in tertiary-educated workers, circa 2000 (percentages)

	Canada	United States	United Kingdom	Australia	France	Sweden
Among Asian migrants	12.8	20.1	10.9	12.4	14.5	8.2
Among other migrants	9.5	10.3	8.6	8.7	10.6	7.1
Among natives	5.8	7.7	9.6	6.7	8.9	8.5

Note: Science professionals defined as ISCO Group 21 (Physical, mathematical and engineering science professionals).

Source: Database on Immigrants in OECD Countries; OECD (2008d).

Similarly, Table 1 shows that migrants make a significant contribution to the science workforce in some countries; the share of science professionals in the tertiary-educated workforce is frequently higher for migrants, especially those of Asian origin, than for the native population.

The mobility of tertiary-qualified people is also increasingly being accompanied by the international flows of students. The number of students enrolled outside their country of citizenship has risen steadily since 1975, with a particularly sharp increase from 1995 to 2005. In 2005, foreign students made up 7.6 percent of total tertiary enrolment in OECD countries, and 17.5 percent of enrolment in advanced research programs[5] (OECD 2008b, p. 52). The majority of foreign/international students are from non-OECD economies, with China and India the main countries of origin. Students may stay on for further research or employment; a study of foreign

1990s (OECD, 2009a). Migration changes will depend on the situations of individuals and the relative economic performance of the home and host country—or, indeed, potential third-country destinations. Evidence suggests that return migration flows correlate more with economic, social, and political developments in countries of origin, and with the ease of circulation between home and host countries, than with economic conditions in receiving countries (Papademetriou et al., 2009). For students, fluctuating exchange rates also have an influence, with self-financed students needing to reconsider the costs of studying at educational institutions abroad relative to those closer to home. Immigration policies will also play a role; as part of their response to the recession, several countries have tightened their immigration intakes and visa requirements.

Importantly for HRST, the fiscal packages introduced by

---

[5] Advanced research programs are second-stage tertiary studies that lead to the award of an advanced research qualification. The programs are devoted to advanced study and original research and are not based on coursework alone. The programs equate to Level 6 of the International Standard Classification of Education (ISCED).

many OECD countries in response to the economic downturn include expenditures on research, science, and innovation. Given the procyclicality of private investment in innovation, governments are seeking to counter a decrease in private activity in R&D and innovation through such measures as increased R&D funding, investment in R&D infrastructure, and provision of enhanced R&D credits to firms (OECD, 2009b). Many countries are also directing resources towards "green technologies," through support for research, science, and pilot projects. These measures are aimed at promoting sustainable long-term growth, but may also help cushion the impact of the crisis for workers in certain fields and create new opportunities for mobile HRST. Ongoing structural demand for labor in certain sectors of the economy may also mute changes in migration through the business cycle (IOM, 2009).

Mobility and knowledge transfer

Innovation requires learning and the creation of new knowledge through the use, adaptation, and absorption of prior knowledge. In this process, both codified and tacit knowledge are vital, particularly since the latter often provides the spark that leads to advances in science and technology, by providing the combination of information and contextual understanding needed to create something new. Tacit knowledge can be defined as knowledge that cannot be codified and transmitted through documentation, or more broadly as the idea that people can be perceptually or intellectually aware of certain things that help them to interpret and make use of information, even when they cannot easily communicate this awareness to others. Gertler (2003) suggested that this communication difficulty might be because people are not fully conscious of all the "secrets" of their successful performance, or because the codes of language are not well enough developed to permit clear explication.

The importance of mobility of HRST stems from its contribution to the creation of new knowledge, learning, and knowledge diffusion. A great deal of HRST mobility takes the form of movement to places where codified knowledge is produced and used: examples are the movement of full-time students into institutions of formal education, and the mobility of graduates and faculty into foreign universities or into formal R&D labs. But mobility is also an important method for transmitting tacit knowledge. It is thought that tacit knowledge is shared more effectively when people have a common social context, with shared values, language, and culture that facilitate understanding and the building of trust. It is also thought that tacit knowledge is difficult to exchange over long distances, thus requiring a degree of proximity for people to have an effective exchange of ideas.

The diffusion of knowledge in new workplaces is one result of the international mobility of skilled workers. At the firm level, knowledge spreads to colleagues, especially to those in close contact. Power and Lundmark (2004) argue that knowledge and innovation develop most commonly through interaction in the workplace: "If it is in the firm and its various offices and factories that workers predominantly interact and form ideas and knowledge, then the flow of people in and out of such locations may be the most likely channels for local and extra-local sources of knowledge and ideas." A study of academic inventors from six European countries found that knowledge transfer was one of two key variables explaining the mobility of scientists from academia to industry; since not all knowledge is codified in a patent, hiring the inventor gives the new employer access to the crucial tacit knowledge that cannot be sourced by other methods (Crespi et al., 2006). The potential flows of knowledge at the firm level highlight the importance of internal management and knowledge management systems that create appropriate conditions for knowledge diffusion—given that international mobility brings together people with different cultures, languages, and ways of working.

The mobility of skilled people also spreads knowledge at the local or regional level, adding a geographic perspective. Individuals in close spatial proximity can meet and exchange ideas at lower cost than can those who are geographically separated. They are also more likely to have "chance" encounters during which useful knowledge exchanges may occur, and to develop social relationships that may act as conduits for knowledge flows. Internationally mobile workers may therefore influence a wide range of people beyond their actual workplace, through the effects of "knowledge spillovers." Empirical evidence shows that the propensity of innovative activity to cluster spatially is greatest in industries in which tacit knowledge plays an important role. Zucker and Darby (2006) found that "star" scientists and engineers (as defined by their level of authorship) show a clear tendency towards concentration by area, and interpreted this as reflecting both their motivation to cluster with their peers and greater commercial opportunities. They also found that the presence of "stars" in a particular location had a positive and significant effect on the probability of a new firm entering a science or engineering field in that location.

Importantly, while information and communication technology (ICT) may increase the amount of knowledge that can be codified and reduce the importance of face-to-face interactions by mimicking some of the features of these interactions (such as visual communication cues), it is likely that geographic proximity will remain an important factor in knowledge transfer for some time to come. The use of ICT still cannot completely replicate the factors at play in face-to-face communications and geographically proximate networks—in other words, the costs of transferring tacit knowledge across space are still relevant. Von Hippel (1994) suggested that rather than facilitating "anywhere" problem solving, computerization in a world of sticky information would enable researchers, managers, and designers to transfer their work to and among field sites containing sticky information. Some evidence of the ongoing importance of face-to-face interaction comes from studies of scientific collaboration. For example, Gallié and Guichard's (2005) study of two French teams participating in the International Sun-Earth Explorers (ISEE) project with the National Aeronautics and Space Administration (NASA) found that, in spite of tele-conferences and email, researchers still required face-to-face interaction for discussions about important or specialized issues, and for building trust among team members.

Nevertheless, there are some factors that may lessen the need for geographic proximity for knowledge transfer. A study of the Indian diaspora resident in the United States, for example, suggested that co-location and co-ethnicity, as types of relationship that facilitate knowledge flows between inventors, are substitutes (Agrawal et al., 2007). Thus, among inventors who share the same ethnicity, the marginal benefit of co-location is minimal. Other studies suggest that knowledge spillovers may travel across regional and national boundaries if workers are part of a strong "community of practice" (Gertler, 2003) or "collaboration network" (Sorenson et al., 2006). These linkages are said to establish social proximity and to bind members together through shared experiences, expertise, and commitments. Sorenson et al., for example, use patent data to show that links between groups of inventors are particularly important for accessing knowledge of "moderate interdependence," in which knowledge components interact to produce the desired outcome and small errors in the reproduction of these components cause large problems. These links may be through common membership of research groups, or through common intermediaries who have collaborated with members of both teams. These factors point to the potential benefits for countries of enabling and encouraging networks of diaspora (discussed below) and other collegial groups.

Effects on receiving and sending countries

The mobility of highly skilled people has a wide range of effects on receiving countries but is generally regarded as a positive, particularly when dynamic effects related to knowledge flows are considered. Guellec and Cervantes (2002) pointed to a number of possible effects, including: increased R&D and economic activity due to availability of additional highly skilled workers; increased entrepreneurship and creativity; knowledge flows and collaboration with sending countries; possibilities for exports of technology; and bolstering of higher education systems through increased flows of students and academics.

As alluded to earlier, there are also impacts at a geographic level, with highly skilled migrants contributing to clusters of activity and to knowledge spillovers in their new location. The increasing returns to scale that are stimulated by the agglomeration of skilled people can be large, and can "jump-start and maintain economic growth" (World Bank, 2009, p. 160), and there is evidence that human capital in-migration and the innovativeness of a region are significantly related (Faggian and McCann, 2009). Workers are often more productive when they are around other highly skilled people; indeed, areas of denser economic activity are associated with higher labor productivity (Ciccone and Hall, 1996). There is also evidence that people who move to denser areas experience faster human capital accumulation, due to higher rates of interaction with other skilled individuals, a broader range of experience, a bigger pool of role models, better job matching and greater specialization (Glaeser and Maré, 2001).

Importantly, these effects can be beneficial for sending countries as well. Kuhn and McAusland (2006) suggest that the movement of "brains" to larger, wealthier economies can be in the interest of the source country, as these "brains" produce better knowledge (such as more effective medicines or software) abroad than if they had remained at home. This is particularly relevant for the mobility of researchers who produce "public" goods that can be readily shared across borders. There is also evidence that R&D conducted in a foreign country has a positive effect on domestic multifactor productivity, provided the country has the capacity to absorb technology from abroad (Guellec and van Pottelsberghe de la Potterie, 2001). In this way, achieving higher productivity abroad can increase knowledge creation and opportunities for productivity-enhancing knowledge flows back to the source country. However, for sending countries to reap these benefits, the policy environment must be conducive to economic development. Where emigration is driven by a poor policy framework and instability, it will be difficult for sending countries to make use of creative ideas and knowledge flows from their citizens abroad.

In terms of labor market impacts on receiving countries, Guellec and Cervantes (2002) suggest that highly skilled migrants could stimulate wage moderation in high-growth sectors experiencing labor shortages. In general, however, the impact of immigration on labor markets depends on a number of factors and policy settings and can be fairly country-specific. Looking at immigration as a whole, OECD analysis suggests that pressures on real wages from immigration are limited and vanish within a few years (Jean et al., 2007), and impacts on native unemployment are temporary and vanish within four to nine years (Jean and Jiménez, 2007).

The effect of migration on sending countries is complex, depending on the type of migrant, the duration of migration, and the economic situation in both sending and receiving countries. In a world of perfect competition, free mobility of labor is beneficial: migrants receive higher incomes, natives in receiving countries share the immigration surplus, and residents remaining in the sending countries benefit from a rise in land/labor and capital/labor ratios. However, the reality is more complex and nuanced, and it is challenging to disentangle the various effects.

Much migration literature focusing on sending countries looks at effects associated with South-North migration, in particular, "brain drain" and the economics of migrant remittances. The migration of highly skilled workers is thought to decrease living standards and growth in developing source countries in several ways, such as the fiscal cost of educating workers who move abroad, the loss of skills and potential increases in the price of skilled services, and possible negative influences on institutional development (World Bank, 2006; Kapur, 2001). At the same time, however, the costs of emigration must be evaluated against the benefits, in terms not only of wages and remittances but also of the utilization of human capital. The World Bank noted that the benefits from the experience gained by highly skilled emigrants are limited if they are not able to find productive employment in their country of origin (2006, p. 67). It is also possible that limiting "brain drain" may not significantly increase the share of skilled people in a country's working age population, if absolute numbers of emigrants and skilled people are small (Easterly and Nyarko, 2008). Overall, the work on brain drain suggests that country-specific characteristics, including the economic policy environment, are an important consideration when estimating the impact of mobility on sending countries.

There is now an emerging body of literature suggesting that emigration of the highly skilled can have beneficial effects on sending countries (both developed and developing), particularly through channels relating to the transfer and creation of knowledge. From a policy perspective, increasing such knowledge returns is the key to achieving mutual benefits from migration. In addition to tapping into migrants' accumulated knowledge and information when they return permanently, sending countries might also gain through the intermittent or temporary return of their citizens (so-called "brain circulation"), additional human capital creation ("brain gain") as a result of emigration, and the existence of an engaged diaspora. Understanding more about these channels will help countries to develop migration management policies that support economic growth while maintaining coherence with other policy initiatives, including those on development and aid.

Return migration and brain circulation

Return migration and brain circulation can be important conduits for knowledge diffusion, and have been put forward as one answer to concerns about the possible negative effects of brain drain. Under these scenarios, migrants may be seen as a resource or stock, rather than as a net loss to the sending country. For example, it has been suggested that the advanced technological frontier in the United States allowed migrant Indian technology professionals to upgrade their skills substantially and then diffuse their technological knowledge through imitation when they returned home or circulated between the two countries (Kapur, 2001). In China, international mobility promoted international academic exchanges so that Chinese schools were quickly informed about the scientific and technological frontier (Zhang and Li, 2002). Returning and circulating workers can also create permanent networks that assist the flow of knowledge. Saxenian and Hsu (2001) highlighted the strong links between Silicon Valley in California and the Hsinchu-Taipei region of Chinese Taipei, which were built by a repatriate community of United States-educated engineers, who not only transferred capital, skills, and know-how on their return to Chinese Taipei but also maintained links with Silicon Valley's Chinese network and helped to create a social and economic bridge between Silicon Valley and Hsinchu. The links are now maintained in part by a growing population of skilled individuals who work in both places, acting as go-betweens and coordinating economic linkages between the two regions, contributing to the creation of a "two-way thoroughfare" of technology and skills.

kets and work in an environment conducive to the exercise and nurturing of their skills and knowledge." In other words, skills or knowledge are crucially linked to the environment in which they are used. A different environment, or indeed, the lack of conditions for harnessing the skills of returning workers, will result in quite different outcomes. A recent analysis of the role of "high-end talent" in China, for example, highlighted that political stability, the rule of law, and a competitive but fair environment are just as critical as pure economic opportunities in encouraging people to stay or return. Concerns that career advancement is still based on political affiliation rather than pure merit, worries about top-down interference in science, education, and business, and fears of corruption and lack of protection of property rights act to discourage some Chinese from returning (Simon and Cao, 2009, p. 251). The absorptive capacity of the home country also needs to be sufficient. A certain basic level of innovative capability is needed to connect with global networks of knowledge creation, and countries vary greatly in this respect. Furthermore, capacity is needed to harness new organizational and management techniques brought back by returnees, as these can be just as beneficial as technology-related skills.

Table 2. Percentage of the foreign-born population with a duration of stay of ten or more years (by country of residence and level of education)

	Canada	Secondary-educated	Tertiary-educated
Australia	84.1	76.6	67.8
Canada	74.7	71.3	65.1
New Zealand	79	61.2	57.4
OECD (weighted)	67.4	69.4	64.7

Source: OECD (2008d), p. 99.

The highly skilled may be more likely than other cohorts to engage in return migration and brain circulation. Data show that the percentage of tertiary-educated migrants who stay for ten or more years in their foreign country of residence is less than that for primary- and secondary-educated migrants (OECD, 2008d). Table 2 shows that for some countries the difference in duration of stay can be significant.

However, the conditions must be right for return flows to take place and to lead to an equivalent transfer of knowledge. As Ackers (2005) points out, "To achieve such transfers, returning scientists need to be able to reenter local labor mar-

Brain gain

Recent literature suggests that "brain drain" can encourage human capital formation ("brain gain") in sending countries. In particular, the possibility of emigration may encourage skill creation, potentially increasing human capital and growth in the sending country. Regets (2001) noted that three factors influence the incentive of natives to invest more in their own human capital: a) an increase in the domestic return to skills due to the relative scarcity created by the "brain drain"; b) an increase in the expected value of an individual's human capital investment if migration is an option; and c) a reduction in the risk associated with the return to individual human capital investment if migration serves as a labor market stabilizer. The central proposition of this "beneficial brain drain" theory is that if the possibility of emigration encourages more skill creation than skill loss, sending countries may increase their stocks of skills as opportunities to move or work abroad open up. As well as private gains for those who acquire skills, there may be public gains, for example through enhanced intergenerational transmission of skills or spillovers between workers, as some newly skilled workers remain in the sending country or migrate only temporarily.

Empirical findings on the beneficial brain drain have been mixed and ongoing research will be valuable to help confirm evidence and provide input to policymaking. One recent study using a data set on emigration rates by educational levels covering 127 developing countries in 1990 and 2000 revealed that doubling the emigration rate of the working-age (25 years and over) highly skilled induces a 5 percent increase in gross human capital formation among the native population (i.e., residents plus emigrants) (Beine et al., 2008). Adjusting for skilled emigration to estimate the net effect on human capital formation in the source country, the authors found a positive overall effect in countries with low levels of human capital and low migration rates of skilled workers. The net effect appeared to be negative in countries where migration of the highly educated was above 20 percent and/or the proportion of those enrolled in higher education was above 5 percent. The number of countries with negative net effects was slightly larger than those with positive net effects. At an aggregate level, however, the gains of the "winning" (generally the most populated) countries outweighed the losses of the "losing" (generally relatively small) countries. Thus, brain drain migration not only increased the number of skilled workers worldwide, but also the number of skilled workers living in developing countries. Beine et al. concluded that the traditionally pessimistic view of the brain drain had no empirical justification at an aggregate level, but that there were important distributional effects among developing countries.

Diaspora

Diaspora can play a number of beneficial roles for sending countries. In general terms, diaspora provide a source for building networks and a means for keeping in contact with emigrants. By mentoring and serving as role models, diaspora members can boost confidence in sending countries and in overseas investors, who become more familiar with the country and its culture through interaction with diaspora members. Diaspora can also contribute to the creation and diffusion of knowledge by acting as a conduit for knowledge and information flows back to the sending country. For example, there is evidence that the social links of skilled individuals with their home country may increase the probability of knowledge continuing to flow there, even after the individual has moved away (Agrawal et al., 2006). Personal relationships and common institutional affiliations can maintain broad social networks which then facilitate knowledge flows.

There is some evidence of the impact of ethnic scientific and entrepreneurial communities and their ties to their home countries on innovation processes and outcomes. Lazonick (2007), for example, described how Korea's ICT industry began to draw on links to skilled Koreans offshore in the 1980s. An investment by Samsung to design and produce chips involved two parallel groups: one in Silicon Valley that employed 300 American engineers led by five doctorate-trained Korean-Americans with design experience at major American chip companies; and the other in Korea, led by two Korean-American scientists and Korean engineers. Samsung's Silicon Valley unit also trained the company's Korean engineers as part of the process of transferring technology from the United States to Korea. More recently, Kerr (2008) examined whether a larger ethnic research community in the United States improved technology diffusion to foreign countries of the same ethnicity through the acquisition and transfer of codified and tacit knowledge from the United States to the foreign country. He found that foreign researchers cite American researchers of their own ethnicity 30–50 percent more frequently than researchers of other ethnicities, even after controlling for alternative professional/research orientation. Using patenting and manufacturing output data, Kerr also found that growth in American ethnic scientific communities increased foreign output, with output expansion coming both from employment and labor productivity gains. Agrawal et al. (2008) found that knowledge access conferred by the diaspora was particularly valuable in the production of India's most important inventions, as measured by citations received.

A variety of factors affect the role of diaspora in the transfer of technology and knowledge, including the types of people that form diaspora, their reasons for leaving, the political and economic situation in their home country, the income gap between the sending and receiving country, the institutional structures in the home country, and the demand or willingness of the home country to accept outside influence. Overall, successful diaspora networks combine three main features:

1. They bring together people with strong intrinsic motivation;

2. Members play both direct roles (implementing projects in the home country) and indirect roles (serving as bridges and "antennae" for the development of projects in the home country);

3. Initiatives move from discussion to transactions (i.e., there are tangible outcomes) (Kuznetsov, 2006).

In building a successful diaspora, individual "champions" appear essential.

Effects on innovation

It is not easy to find clear quantitative evidence of the impact of mobility on innovation outputs and outcomes. In many cases, causality cannot be established, and it is challenging to construct a counter case. Many variables and factors influence science and technology outcomes and they are hard to untangle. Nevertheless, data and information can be used to build a picture and to establish some links between mobility and broader science and innovation outcomes.

A clear effect of the mobility of highly skilled workers is the increasing internationalization of the labor market for the highly skilled. Both in private industry and academia, foreign staff are sought for their specific knowledge or abilities, their language skills, and their knowledge of foreign markets. Academic staff recruited from overseas are now a significant element of the university workforce in some countries and there is some evidence that their mobility is associated with higher-quality output (Figure 4). One study showed that almost 50 percent of highly cited researchers based in Switzerland, Australia, Canada, Italy, and Germany had research experience outside their home country, and suggested that mobile

129

Figure 4. Share of highly cited researchers with research experience outside of their home country[1] by country of current institution

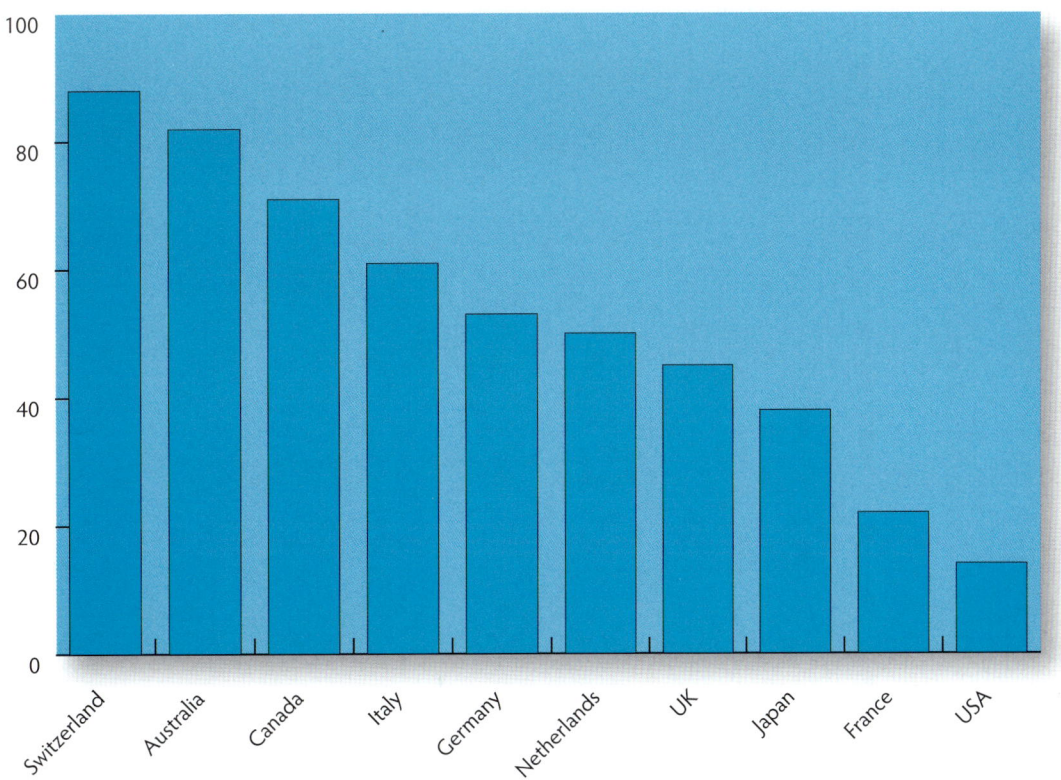

[1] Based on a sample of 494 researchers from the ISI Highly Cited database (1985–2004). Non-home research experience includes post-graduate training, post-doctoral research, and employment.
Source: Evidence Ltd., 2005.

populations can help small research economies to attain a relatively high research performance, through better international linkages and opportunities for collaboration (Evidence Ltd., 2005).

The links between mobility and innovation outputs are less clear, although some evidence suggests that immigrants contribute strongly to patent applications and to the creation of technology firms. Data from the United States, for instance, show that the proportion of patent applications filed with the World Intellectual Property Organization at its US office which name foreign nationals as inventors or co-inventors has increased from 7.6 percent to more than 25 percent between 1998 and 2006 (Wadhwa et al., 2007a). Over 25 percent of the engineering and technology companies started in parts of the United States from 1995 to 2005 also have at least one foreign-born founder (Wadhwa et al., 2007b). More broadly, the world share of patents involving international co-invention increased from 4 percent in 1991–93 to 7 percent in 2001–03, with small and less developed economies particularly active

in international collaboration (OECD, 2007c).

The increased mobility of HRST is paralleled by an increase in collaborative research. Studies from several countries highlight a trend towards international co-authorship of academic articles. For example, from 1995 to 2004, Finland's joint publications with researchers from other EU countries rose by 85 percent (Lehvo and Nuutinen, 2006). Specific patterns of collaboration are likely to be influenced by countries' relative capacity (in terms both of quantity and quality) in various research fields as well as by geographic proximity, a common language, and institutional linkages. For example, evidence from the United States points to links between the number of American doctorates received by foreign students and the percentage of internationally co-authored articles involving the United States and the students' home countries (Figure 5).

Figure 5. Relationship of foreign-born US Science and Engineering doctorate recipients to their country's scientific collaboration with the United States (1994–98 graduates and 1999–2003 articles)

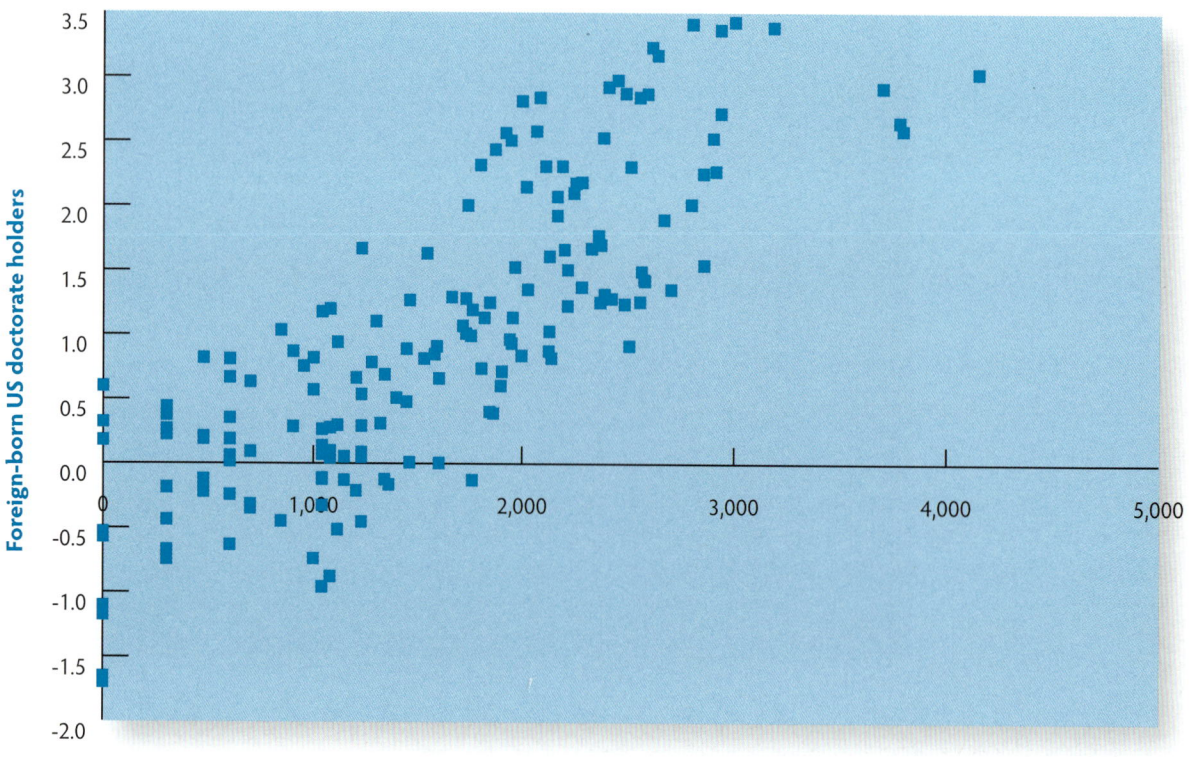

Source: Regets, 2007.

Policy responses

Given the importance of the mobility of highly skilled people for knowledge transfer, many countries have implemented policies to attract and retain HRST talent. An OECD survey of country approaches to HRST mobility found a range of policies in action, such as economic incentives to encourage inflows, immigration-oriented assistance, procedures for recognizing foreign qualifications, social and cultural support, and support for research abroad (see OECD, 2008c, Chapter 4). Each of the policy areas utilized a range of mechanisms, including scholarships, fellowships, grants, facilitated procedures, institutional arrangements, and service centres.

However, there was a wide range of "intensity" with which countries approached the mobility of highly skilled people. Few countries had an explicit mobility strategy, or a strategy to stay in touch with the diaspora, and the range and policy breadth at an operational level varied widely, with some countries focusing on just a few policy mechanisms, while others offered "something for everyone." There was generally more support among countries for inflows of researchers and other HRST than for outflows, perhaps because countries judged outward mobility to be adequate or because they were reluctant to encourage outward mobility, despite arguments about the benefits of brain circulation. The lack of an overall strategy for mobility highlights the risk of incoherence among policies on inflows, outflows, and the diaspora. Ideally, mobility policies should be part of a wider mobility strategy that contributes to the country's objectives for science, technology, and innovation and sets out the rationale for intervention in mobility issues.

National policies appeared generally to target the same population of skilled workers, with little orientation towards particular national scientific and technological interests. In most cases, there were no restrictions on the country of origin (inward mobility) or of destination (outward mobility), so that mobility policies were—at least in theory—globally oriented. Nevertheless, a geographic tendency might emerge as a result of individuals' decisions (e.g., they may be more aware of opportunities close to home or prefer to move shorter distances) or as a result of the efficacy of countries' marketing approaches. The range of support offered for mobility, as opposed to permanent migration, opened the possibility for researchers to use these policies to work in a number of countries.

At an institutional level, information provided by some countries suggested that the most common mobility assistance provided was social support related to language, housing, visas, insurance, and similar issues. Student exchange policies were also popular, as were travel grants for research abroad. The policies and programs offered at an institutional level complemented those offered at the national level, but with a greater focus on information provision and practical social/cultural assistance. Institutional mobility initiatives also tended to provide support for short-term visits, which were less frequently available at the national level.

OECD countries thus have a wide selection of policy tools at their disposal, which they use more or less intensively to promote the mobility of talent.[6] The question then is: what is the role for international mobility policy in the future?

In designing future mobility policies, a key first step is to identify a rationale for intervention and to establish clear objectives. For mobility, the main rationale may be the potential positive externalities from knowledge spillovers and information asymmetry issues. However, countries' economic and social context and governments' overall goals will affect their view of appropriate intervention. The obstacles to mobility commonly cited by policy makers and academics include legal and administrative barriers, lack of funding, personal issues, and language. It is not clear, though, which obstacles stem from a market failure that governments can remedy through policy, and there is no general agreement, even within countries, about the nature of obstacles to mobility. Policy makers also need to consider how obstacles may change in the future and the extent to which issues are specific to HRST and thus warrant a specific approach.

Few mobility policies have been evaluated, so it is difficult to point to best practices. However, some lessons were drawn from evaluation material provided by countries in response to the OECD survey, including the importance of setting appropriate funding levels and program durations for the target population (according to desired skill levels and fields of work). Some interesting points emerged with respect to recipients' personal objectives versus program objectives, and, in particular, whether the long-term goals of programs will be achieved if personal and program objectives differ. The evaluation material showed the importance of good data collection, planned from the outset, to enable an assessment of

---

[6] See www.oecd.org/sti/stpolicy/talent for an inventory of national policy initiatives to encourage the inward and outward mobility of HRST.

the efficiency and effectiveness of a program. More work on evaluating the efficiency, effectiveness, and the impact of mobility policies would be valuable.

Given the differences across countries, it is not possible to identify a "recipe" for what governments should do more or less of, and what should stay the same. One avenue that may hold promise, however, is removing barriers to short-term and circular mobility. Shorter (but potentially repeated) periods abroad may circumvent some of the obstacles that currently hinder HRST mobility and would also support knowledge flows associated with brain circulation and diaspora.

However, successfully reaching policy goals for mobility requires coherence across policy areas. HRST mobility policies and the broader policy environment for innovation need to be complementary. Mobile researchers look for more than simply higher wages when they move across borders; they also want quality research infrastructure, a stimulating research environment and opportunities to explore new areas. In addition, when governments seek to improve innovation outcomes, it is not sufficient to increase the number of the country's skilled HRST. Skilled people must also operate in a system that enables them to use, create, and disseminate knowledge. "Framework" policies (such as those related to education, financial markets, intellectual property, investment, and labor markets) and "innovation-specific" policies (such as those related to public research and financial support for R&D) influence innovation effort and performance, and together can help or hinder a country's efforts to improve their innovative capacity (OECD 2006). Countries must also address shortcomings in national policies that may limit the supply of HRST, since relying extensively on international flows to fill gaps may not be sufficient in a world where many countries are seeking the same pool of workers.

Improving policy coherence also implies considering the consequences of mobility for development in sending countries that are the target of development and aid policies. Linking policy design and implementation in these areas aims to better achieve the goals of both mobility and development and to contribute to more effective management of migration. Three steps to better manage the flows of highly skilled migrants from developing countries and limit the negative effects on these countries might include a) closer monitoring, b) better data collection, and c) partnership arrangements that link recruit-

ment with capacity building in the country of origin (OECD, 2007d). However, it is a two-sided process, with efforts also required by the developing country, particularly in pursuing appropriate policies for stimulating economic and employment growth. The World Bank also recommended that developing countries reap more of the benefits from their diaspora communities in the world's prosperous places by encouraging their economic and political participation at home, and by making it easy for them to retain citizenship, vote, and eventually resettle if they so choose (World Bank, 2009, p. 169).

Conclusions

Alongside sustained growth in foreign direct investment, in trade, and in the internationalization of research and development, the mobility of skilled people, including human resources in science and technology, has become a central aspect of globalization. Migration of talent now plays an important role in shaping the skilled labor forces of many countries. Despite upheavals in the global economy, the mobility of skilled individuals is likely to continue, as people search for better economic opportunities, higher quality research infrastructure, the opportunity to work with leading peers, and freedom to debate cutting-edge issues.

The importance of mobility stems from its contribution to the creation and diffusion of knowledge. Not only does it aid in the production and dissemination of codified knowledge, it is also a key means of transmitting tacit knowledge. This knowledge is spread in the workplace and to geographically proximate individuals and organizations, and can contribute to the emergence of local concentrations of activity. Nevertheless, the gains are not limited to the migrant and the receiving country. The evidence on return migration and brain circulation, beneficial brain drain and diasporas suggests that there are a variety of mechanisms by which migrants can continue to contribute to knowledge creation and innovation in their home countries.

Although specific quantitative evidence of the effect of mobility on innovation outputs and outcomes is limited, there are indications that mobility is leading to greater internationalization of the labor market and that migrants are contributing to firm creation, patents and innovation literature. In this light, many countries have introduced policies to promote mobility of skilled people, particularly scientists, engineers and researchers. Further efforts to remove barriers to short-term mobility and circular migration are likely to support knowledge flows associated with brain circulation and the diaspora. Importantly, however, to reap the gains of these policies, the broader environment for science and innovation must be sound. Innovation is the outcome of a complex process, of which human capital is only one, albeit crucial, part of the equation.

References

Ackers, L. 2005. "Moving People and Knowledge: Scientific mobility in the European Union." *International Migration* 43(5):99–131.

Agrawal, A., I. Cockburn, and J. McHale. 2006. "Gone but not forgotten: knowledge flows, labour mobility and enduring social relationships." *Journal of Economic Geography* 6:571–91.

Agrawal, A., D. Kapur, and J. McHale. 2007. "Birds of a Feather—Better Together? Exploring the Optimal Spatial Distribution of Ethnic Inventors." *NBER Working Paper Series.* Working Paper 12823. Cambridge. MA.

———. 2008. "Brain drain or brain bank? The impact of skilled emigration on poor-country innovation." *NBER Working Paper Series.* Working Paper 14592. Cambridge, MA.

Auriol, L. 2009 (forthcoming). "Careers of Doctorate Holders: Employment and mobility patterns." *STI Working Paper.* OECD, Paris.

Beine, M., F. Docquier, and H. Rapoport. 2008. "Brain drain and human capital formation in developing countries: winners and losers." *Economic Journal* 118 (April):631–52.

Ciccone, A. and R. Hall. 1996. "Productivity and the Density of Economic Activity." *American Economic Review* 86(1 March):54–70.

Crespi, G., A. Geuna, and L. Nesta. 2006. "Labour Mobility of Academic Inventors: Career decision and knowledge transfer." *EUI Working Papers*, RSCAS No. 2006/06. Robert Schuman Centre for Advanced Studies, European University Institute, Florence.

Easterly, W. and Y. Nyarko. 2008. "Is the brain drain good for Africa?" *Brookings Global Economy and Development Working Paper*, No. 19, March.

Eurostat. 2007. "How mobile are highly qualified human resources in science and technology?" *Statistics in Focus: Science and Technology*, 75/2007.

Evidence Ltd. 2005. "Tracking UK and international researchers by an analysis of publication data." Report prepared for the Higher Education Policy Institute. Leeds. June.

Faggian, A. and P. McCann. 2009. "Human capital, graduate migration and innovation in British regions." *Cambridge Journal of Economics* 33:317–33.

Finn, M. 2007. "Stay rates of foreign doctorate recipients from U.S. universities, 2005." Oak Ridge Institute for Science and Engineering. United States.

Gallié, E-P. and R. Guichard. 2005. "Do collaboratories mean the end of face-to-face interactions? An evidence from the ISEE project." *Economics of Innovation and New Technology* 14(6 September):517–32.

Gertler, M. 2003. "Tacit knowledge and the economic geography of context, or The undefinable tacitness of being (there)." *Journal of Economic Geography*, 2003(3):75–99.

Glaeser, E. and D. Maré. 2001. "Cities and Skills." *Journal of Labor Economics* 19(2):316–42.

Guellec, D. and M. Cervantes. 2002. "International Mobility of Highly Skilled Workers: From statistical analysis to policy formulation." in OECD (2002), *International Mobility of the Highly Skilled*. OECD, Paris.

Guellec, D. and B. van Pottelsberghe de la Potterie. 2001. "R&D and Productivity Growth: Panel Data Analysis of 16 OECD Countries." *OECD Economic Studies* 33, 2001/II:103–26.

von Hippel, E. 1994. "Sticky Information and the Locus of Problem Solving: Implications for Innovation." *Management Science* 40(4 April):429–39.

International Organization for Migration (IOM). 2008). *World Migration 2008: Managing Labour Mobility in the Evolving Global Economy*. Switzerland.

———. 2009. "The Impact of the Global Economic Crisis on Migrants and Migration." IOM Policy Brief, March.

Jean, S., O. Causa, M. Jiménez, and I. Wanner. 2007. "Migration in OECD Countries: Labour market Impact and Integration Issues." *OECD Economics Department Working Papers*, No. 562. OECD, Paris.

Jean, S. and M. Jiménez. 2007. "The Unemployment Impact of Immigration in OECD Countries." *OECD Economics Department Working Papers*, No. 563. OECD, Paris.

Kapur, D. 2001. "Diaspora and Technology Transfer." *Journal of Human Development* 2(2):265–86.

Kerr, W. 2008. "Ethnic Scientific Communities and International Technology Diffusion." *Review of Economics and Statistics* 90(3):518–37.

Kuhn, P. and C. McAusland. 2006. "The International Migration of Knowledge Workers: When is brain drain beneficial?" *NBER Working Paper Series*, WP 12761. Cambridge MA.

Kuznetsov, Y. 2006. "Leveraging Diasporas of Talent: Towards a New Policy Agenda." in Y. Kuznetsov (ed.). 2006. *Diaspora Networks and the International Migration of Skills: How countries can draw on their talent abroad, WBI Development Studies*. The World Bank, Washington, DC.

Lazonick, W. 2007. "Foreign Direct Investment, Transnational Migration, and Indigenous Innovation in the Globalization of High-Tech Labor." Revised version of paper presented at the International Forum of Comparative Political Economy of Globalization, 1–3 September 2006, Renmin University of China, Beijing. Available at: *http://faculty.insead.edu/Lazonick/RecentPublications.htm*

Lehvo, A. and A. Nuutinen. 2006. "Finnish Science in International Comparison." Publication of the Academy of Finland 15/06. Helsinki.

Lemaitre, G. 2005. "The Comparability of International Migration Statistics: Problems and Prospects." *OECD Statistics Brief*, No. 9, July.

OECD. 2006. *Economic Policy Reforms: Going for Growth 2006*. OECD, Paris.

———. 2007a. "Trends in International Migration Flows and Stocks 1975–2005." OECD internal working document. 31 May. Paris.

———. 2007b. *International Migration Outlook: Annual Report*. 2007. OECD, Paris.

———. 2007c. *OECD Science, Technology and Industry Scoreboard 2007: Innovation and Performance in the Global Economy*, OECD, Paris.

———. 2007d. *Policy Coherence for Development: Migration and Developing Countries*. OECD, Paris.

———. 2008a. *The Looming Crisis in the Health Workforce: How can OECD countries respond?* OECD, Paris.

———. 2008b. *International Migration Outlook: Annual Report* (*2008* ed.). OECD, Paris.

———. 2008c. *The Global Competition for Talent: Mobility of the Highly Skilled*. OECD, Paris.

———. 2008d. *A Profile of Immigrant Populations in the 21st Century: Data from OECD Countries*. OECD, Paris.

———. 2009a. *OECD Economic Outlook: Interim Report*. March. OECD, Paris.

———. 2009b. "Strategies for aligning stimulus measures with long term growth." Available at: www.oecd.org/dataoecd/12/62/42555546.pdf (accessed 2 June 2009).

Papademetriou, D. and A. Terrazas. 2009. "Immigrants and the Current Economic Crisis: Research evidence, policy challenges, and implications." Migration Policy Institute. January.

Parsons, C., R. Skeldon, T. Walmsley, and A. Winters. 2007. "Quantifying International Migration: A database of bilateral migrant stocks." in Özden, Ç. and M. Schiff (eds.) (2007), *International Migration, Economic Development and Policy*. The World Bank, Washington, DC.

Power, D. and M. Lundmark. 2004. "Working through Knowledge Pools: Labour market dynamics, the transference of knowledge and ideas, and industrial clusters." *Urban Studies* 41(5/6):1025–44, May.

Regets, M. 2001. "Research and Policy Issues in High-skilled International Migration: A perspective with data from the United States." in OECD (2001), *Innovative People: Mobility of Skilled Personnel in National Innovation Systems*. OECD, Paris.

———. 2007. "Research issues in the international migration of highly skilled workers: A perspective with data from the United States." *National Science Foundation Working Paper*, SRS 07-203. June.

Saxenian, A. and J-Y. Hsu. 2001. "The Silicon Valley-Hsinchu Connection: Technical Communities and Industrial Upgrading." *Industrial and Corporate Change* 10(4):893–920.

Simon, D. and C. Cao. 2009. *China's Emerging Technological Edge: Assessing the role of high-end talent*. Cambridge University Press.

Sorenson, O., J. Rivkin, and L. Fleming. 2006. "Complexity, networks and knowledge flow." *Research Policy* 35:994–1017.

Wadhwa, V., G. Jasso, B. Rissing, G. Gereffi, and R. Freeman. 2007a. "Intellectual Property, the Immigration Backlog, and a Reverse Brain-Drain: America's New Immigrant

Entrepreneurs, Part III." Duke University, New York University, Harvard Law School and Ewing Marion Kauffman Foundation. August.

Wadhwa, V., A. Saxenian, B. Rissing, and G. Gereffi. 2007b. "America's New Immigrant Entrepreneurs." Duke University and UC Berkeley. January.

World Bank. 2006. *Global Economic Prospects 2006: Economic Implications of Remittances and Migration*. Washington, DC.

———. 2009. *World Development Report 2009: Reshaping Economic Geography*, The International Bank for Reconstruction and Development/The World Bank, Washington, DC.

Zhang, G. and W. Li. 2002. "International Mobility of China's Resources in Science and Technology and its Impact." in OECD (2002), *International Mobility of the Highly Skilled*. OECD, Paris.

Zucker, L. and M. Darby. 2006. "Movement of Star Scientists and Engineers and High-Tech Firm Entry." *NBER Working Paper Series*, Working Paper 12172. Cambridge, MA.

Chapter 2.5

Grassroots Green Innovations for Inclusive, Sustainable Development

Anil K Gupta,[1]
Indian Institute of Management, Ahmedabad and
National Innovation Foundation

Open innovation[2] or user-driven innovation[3] models have been recognized as important tools even by large, traditional companies which have long relied on internal R&D as a major source of innovations. These companies are not only often unable to meet the needs of their existing clients from within, but they are also constrained in their ability to identify and meet the needs of excluded clients.

Despite the capability of these platforms to generate solutions to many problems, the ability of corporations to influence the lives of common people with a variety of products and services has not increased in the recent past. It is now realized that mere reliance on market forces will not work to fill innovation gaps or for disseminating innovative ideas, products, and services among disadvantaged segments of the population. Thus, there seems to be a crisis in a) the sourcing of ideas which can add value to existing knowledge, b) disseminating innovations in a manner that users can adapt to their local context, and c) co-creating solutions for the future that will ensure ecological integrity and social stability by providing opportunities for an improved life for the most disadvantaged. I am not including in this discussion applications of open-source or free software networks, but am confining myself to a discussion of the hard technologies needed by knowledge-rich but economically poor people.

In this paper, I argue that the Honey Bee Network approach, described below, offers new ways of thinking which can help even the formal, organized sector learn from grassroots innovators and traditional knowledge-holders, and enable them to solve problems in an *affordable, accountable, and accessible* manner. I advance three arguments:

- Lack of material resources spurs knowledge-intensive innovation by common people in the informal sector, and thus provides a basis for sustainability by reducing entropy;

- Many grassroots innovations are important, not only be-

137

---

[1] Much of the content in this paper has evolved from close interaction with professional colleagues and grassroots innovators in the Honey Bee Network (sristi.org) and friends outside. I must thank Ramesh Patel, Puroshottam, Vipin Kumar, Nirmal Sahay, Jayshree, Hiren Prajapati, Riya Sinha, Chinzah Lalmanjuala, Nitin Maurya, Vivek Kumar, Hema Patel, Deepa Tripathi, and Mahesh Patel, for the idea of the distributed supply chain of herbal products; Sivaprasad Chauhan for acknowledging innovators on the package of the products; Mansukh Bhai Jagani for the multi-purpose/functional, motorcycle-powered agricultural machine patented in the U.S.; Amrut Bhai Agrawat for horizontal networking through experimenting farmers workshops, etc. I would like also to acknowledge the support received from many other colleagues in IIMA, SRISTI, GIAN, NIF and SRISTI Innovations.

[2] Chesbrough (2003). Milton Sousa (2009) rightly observes, "It is clear that successful innovation under complexity, uncertainty and change can only be achieved through collaborative approaches that integrate knowledge inside and outside the organization." See also Huston and Sakkab, 2006.

[3] See Allen (1983), quoted by von Hippel (1987). See also numerous papers by von Hippel and his colleagues on user-driven innovations at http://web.mit.edu/evhippel/www/papers/evh-03.htm See in particular von Hippel (2005) and Frank Piller (2008) at: http://www.mass-customization.de/download/piller_2008-pribilla.pdf

Piller reinforces the notion—popular among scholars in this field—that users have incentives other than salaries, monetary compensation or rewards, although many companies do give such awards to the best ideas. The majority do not compensate users, nor do they give them explicit credit on product packaging or brochures. Crowdsourcing, mass sourcing, or distributed strategies for getting ideas

cause they are low in cost and locally sustainable, but also because they offer new problem-solving techniques which can be applied in different contexts;

- Several innovations can be blended by pooling the ideas of different innovators or communities with the scientific and technological knowledge of the formal sector to develop value-added products and services.

If social inclusion is to take place, then many assumptions inherent in the dominant developmental models will have to change, and the role of the state, corporations, and civil society redefined. The classic model of corporate social responsibility will not work in the future, because one cannot first create exclusion and then hope to do something for those who are left out. The strategies for inclusive development will have to build upon the resources in which poor people are especially rich: their knowledge, values, social networks, and institutions. Of course, not all of these factors are equally strong or relevant in every case, but only by recognizing their role will inclusive social development be achieved.

In the first part of this paper, I examine nature as a source of sustainable logic in grassroots innovations. In the second part, I discuss the scope of learning from a variety of green grassroots innovations mobilized by the "Honey Bee Network," an organization developed (on the model of how bees cross-pollinate) for the purpose of learning from the ideas of common people and communities, who are given due credit for their knowledge and ideas.

Learning from nature: A framework for sustainability

In nature, frugality, multi-functionality, simultaneity, and diversity are basic to sustainable resource use. And yet, the institutions that govern the logistics of modern package design, transportation, inventory location, and movement patterns disregard some of these features. In part, this arises because of the way we view markets and their functions in a society in which resources are assumed to be artificially abundant. The recent crisis in the financial markets has highlighted the unrealistic nature of the assumptions underlying commodity markets. This must change.

If 30 percent of goods produced in a rural hinterland are going to be consumed within, say 300 km, or within three months or less, should the storage, packaging, transporta-

tion and the entire supply chain not reflect this reality? In the search for optimal solutions for verticals, are we neglecting the horizontals, and, therefore, the connections among communities? Proximal transactions do not have to be analyzed only in terms of remote supply chains. Today most things are packaged for long-term storage, long distance supply, and multiple points of handling. In a mass consumption society, this was imperative for efficiency. But, with changes in energy and other resource constraints, our assumptions about designing logistical solutions must also undergo basic transformation.

It is here that grassroots knowledge, values, and institutions can perhaps come to our rescue. In every culture, the strengthening of vertical markets has weakened the neighborhood economy. Maintaining individual inventories always consumes more resources than community management—privacy of consumption being celebrated in vertical markets. But, in order for social or community-based/graded/ranked/or influenced consumption to generate more optimal solutions, changes in life style will have to be made. We must realize that autonomy, flexibility, unregulated freedom, and excessive resource consumption carry a high price tag.

Can we moderate our need for autonomy in the short run and gain more autonomy in the long run? Such things as carpooling, the collective purchases of household goods, and shared responsibility for goods and services have already begun to be practised in many cultures in response to the challenge of sustainability. From small scale, scattered, and spontaneous steps, we must now make the transition to large scale, systematic, and organized change in designing our future.

Following are some examples of these changes:

Frugality. When I was a child, shopkeepers would often wrap goods in a piece of old newspaper and tie the package with string. There was no plastic then and even paper bags were costly, to be used only to wrap heavy things. Once the package was brought home, my grandfather would spread the newspaper under the bed and hang the string on a nail, for future reuse. Over the years, natural resources have shrunk, but the scale of our consumption and "footprints" has expanded enormously. The design of packaging material, shelves, transportation systems and consumption patterns will only change when different cycles of production and consumption exchange are reconsidered. Then we will use sturdy packaging material only for long-term and long-distance consumption. This will have an interesting spillover effect: once we modify the logistical chain, community markets will become more competitive. Face-to-face interaction among consumers and producers will take place more often, altering the entire politics of regional and sectoral development.

Multi-functionality. Higher multi-functionality means less waste and better resource utilization. Most cultures in developing countries are multi-functional in their orientation. When goods and services are well designed, multiple functions are considered. This is evident in the grassroots design of user-driven farm machinery and other tools. The skills, resources, and tools for multi-functional design are quite different from those needed for highly specialized and single-function goods and services, such as a motorcycle or two wheeler used only for transportation. Single-function tools have much greater redundancy and waste energy and resources. Multifunction devices and services—such as the same motorcycle or two-wheel scooter used for ploughing land, removing garden weeds, grinding flour, or washing clothes— have much higher feedback loops and thus reduce waste, ensure higher stability, and justify consumption.

Simultaneity. In nature, there is a constant and simultaneous exchange and flow of services and energy in different directions. Such exchanges require different kinds of logistical chains, which in future will have to integrate the anticipated disposal of recyclable, reusable, renewable resources, and also of those which cannot undergo any of these processes, i.e., true waste. In nature, the species which digest biomass co-evolve with biomass-creating species. This co-evolutionary model can also be applied to humans. Logistical systems will have to mimic ecological exchanges in real time, so that environmental load can be reduced and the "sink and the source"[4] can be redesigned. The simultaneity of exchanges at the community level may give rise not only to disposal, but also to innovations in pooling, sourcing, transportation, storage, and consumption. The negative externality will always be higher when all these transactions have to take place at the level of the consumer, whether firm or individual. In other words, coordination of individual choices in the short run to expand

the autonomous choices in the long run has to become the mantra of logistics. For instance, the continuing lack of coordination in purchasing and replacing daily supplies for a household or small enterprise creates excessive individual inventory, waste, more potential energy trapped in immobile resources, and consequently higher cost for internalizing negative externality. This is what is happening today in most "developed," highly urbanized communities worldwide, in which individuals keep much high inventories of such goods to ensure personal comfort and convenience, and do not coordinate their preferences and consumption cycles with their neighbors. Enormous waste of energy and materials results when these goods are not consumed within their safe shelf-life. Although with lower energy prices and higher savings, some societies might manage for the time being, the situation is changing drastically and, some say, irreversibly.

Diversity. On the shelf of any supermarket, small shop, roadside stand, or even home delivery vendor, we invariably find only one or two varieties of a particular vegetable or fruit. Even these few varieties are bred for longer shelf life, more beautiful display, and easier transportation, handling, and storage. If taste and the nutrition suffer, so be it. But that is not necessarily the wish of the majority of people, who would probably appreciate a greater diversity of taste, color, shape, and aroma for both aesthetic and nutritional purposes. But irregularly shaped tomatoes are not considered beautiful. Aesthetics is determined by logistics and the incentive for preserving cultural diversity reduced. Means and ends have become confused, adversely affecting health and nutrition. But the world is beautiful because it is diverse. Agronomy, plant breeding, soil and ecosystem health, human and animal health, and working relationships will undergo complete transformation if diversity in consumption becomes the primary purpose of designing supply chains. There are other implications for logistics: diverse foods or vegetable-dyed textiles would have to be characterized differently and labeled to inform the consumer more accurately about biodiversity and the habitat of raw materials and ingredients. Since vegetable dyes may not always be uniform, consumer preferences will have to accommodate variable coloration. Just as markets created preference for uniformity, the challenge in future will be to do just the opposite. At the same time, packaging and transport logistics

---

[4] Gupta, 2006.

will have to adapt to new needs. For instance, a whole range of technological innovations will be needed to package diverse fruits, vegetables, and other materials. Human needs and preferences for a sustainable world must guide and trigger technological innovations, supply chains and logistical arrangements, not the other way around.

The famous German ethnologist and Nobel Laureate, Konrad Lorenz, gives a fascinating example of this conceptual framework (Reidl, 198[5]). He suggests that when we examine the feathers of birds, or fish fins, or tree branches, we see the limited range of angles at which these feathers, fins, and branches are set. The entire diversity can, in fact, be covered by a range from 15–90 degrees. Reidl draws upon the work of Lorenz when he says that nature has a few designs which she plays with in a parsimonious and frugal manner over and over again.

Until we discover simple principles and incorporate them in the redesign of logistical systems, the world around us will not become more humane, green, compassionate, and collaborative. Creativity and innovation will inevitably follow in the process of opening up the design and implementation process. If such were not the case, large corporations would not turn towards users and other supply chain members for ideas and innovations. Around the world, corporations are recognizing that it is not correct to confuse the system of R&D with innovation and innovations systems with intra-organizational creativity.

The same simplicity, frugality, multi-functionality, and diversity is witnessed in many grassroots innovations. Where do these innovators get their values?

Learning from grassroots innovators

Learning from common people who are not formally trained in a technical institution, who may be illiterate, and who may not know much about "scaling up" is not common today. Sourcing ideas from the ground up is not the first thing that comes to the minds of public policy makers or heads of corporations when they think about social transformation. Even in areas where such technologies as cell phones do penetrate the non-urban "interior," the applications for bridging knowledge and technology gaps often take much longer to come about, if at all. Despite 400 million cell phones sold in India in the last decade, we do not have even 40 applications for empowering knowledge-rich but economically poor people to improve their lives, for creating markets for their cultural or artisanal skills, or for helping them to disseminate successful local solutions. Without major modification in its image-processing capacity, for example, a camera-phone cannot be used to do a microscopic analysis of water or food. There are countless other examples of technology that has not been adapted to serve a larger social good. Hence, either the designers of technologies or other services in the government or private sector do not learn from the people and understand their needs, or they do not have the commitment to be more inclusive. It is possible that some of them do, in fact, wish to be inclusive, but have not found a way of tapping the creative potential of common people, so that social and knowledge gaps can be bridged.

If learning about developing new applications of existing technologies for social inclusion is so difficult, it would appear to be even more difficult to explore and add value to innovations developed by people at the grassroots without outside help. As mentioned earlier, there are concerted efforts being made by some large corporations[6] to use the open-innovation model to involve users in generating solutions for their problems. One excellent example is Lego,[7] where users can create new designs.[8] This began in 1999, when different users hacked the software and posted it on a website. Lego's response was

---

[5] Riedl, 1984.

[6] See the "connect-develop" platform of Procter&Gamble, which aims to develop more than half of its products based on ideas sourced from outside the company in the next few years. The company asks its public: "Do you have a game-changing product, technology, business model, method, trademark, package, or design that can help deliver new products and/or services that improve the lives of the world's consumers? Do you have commercial opportunities for existing P&G products/brands? If so, we'd like to consider a partnership." See: http://www.pgconnectdevelop.com/pg-connection-portal/ctx/noauth/PortalHome.do
See also the bibliography of sources on the open innovation model at: http://www.openinnovation.net/Research/Bibliography.html

[7] Eric von Hippel, Professor at the MIT Sloan School of Management (and close collaborator of the Honey Bee Network) observed, "Lego offers a good example of a smooth transition. . . . Within that firm there were maybe about 20 people who were looking at an open model for new product development. Top management protected and encouraged them, and they are managing to build within an old firm a new way of doing things that is gradually making a transition for the entire Lego company. . . . The transition to open can be done without major disruptions, but it's not easy." See: http://www.deloitte.com/view/en_US/us/Insights/Browse-by-Content-Type/deloitte-review/article/7930c99d77ea2210VgnVCM200000bb42f00aRCRD.html

[8] On 25 August 2009, reviewer Anton Olsen wrote about the new possibilities offered by Lego's potential open-source software and hardware, saying: "However, LEGO has a good policy for groups who wish to take the NXT a little farther. They openly support the hacker community with an open-source version of the NXT firmware, provide detailed hardware information including schematics of the NXT and sensors, and give specifications for interfacing third-party and home-built sensors. They even provide a complete Software Developer Kit (SDK)." See: http://www.wired.com/geekdad/2009/08/hacking-the-nxt-with-legos-blessing

to remain neutral, neither encouraging nor discouraging the hacking. Today, it openly invites users to play with various options and share their results with other users. But it would be a quite different matter if solutions were generated by those grassroots users and taken over by a company such as Lego to develop a commercial product, without sharing the benefits or credit with those who provided the innovative solutions.[9] How can we bridge the gap between the formal sector—private corporations, public/R&D institutions and other international developmental organizations—and the creative and innovative capacities at the grassroots? What can be done to replicate the experience of the Honey Bee Network in India in building bridges with the Council of Scientific and Industrial Research (CSIR) and the Indian Council of Medical Research (ICMR)? How can the private sector not only license the technologies developed by farmers, artisans, mechanics, and other lay people, but also share the benefits and credit, and draw more fully on the creative abilities of common people?

The Honey Bee Network experience[10]

More than 20 years ago, the Honey Bee Network was established, based on four principles:

1. When we learn from common people, they should not remain anonymous, but should get due credit for their knowledge, whether developed by individuals or communities;

2. We should try to connect people to people to cross fertilize their ideas, just as bees carry out cross-pollination; this is possible when we communicate in the local language, compare findings with the knowledge gained from the people, and seek their informed consent if the knowledge is unique;

3. Any commercial benefits accruing from the knowledge provided by people (with or without value addition) should be shared in a fair and just manner with those who have provided their ideas;

4. The process of knowledge exchange is kept transparent, with confidentiality being observed when people so desire, and intellectual property rights respected.

These principles were developed even before the appearance of the Convention on Biological Diversity and other debates on individual knowledge rights. In these two decades, the Honey Bee Network has grown from a few hundred ideas, innovations, and traditional knowledge practices to more than

100,000, pooled in a national database on the subject maintained by the National Innovation Foundation set up in 2000. It goes without saying that not all of these ideas are unique or of significant value.

Five years ago formal agreements were signed with the CSIR and three years ago with the ICMR. Outstanding results have been achieved, proving that grassroots innovations and traditional knowledge can not only generate good solutions to local problems, but in some cases also extend the frontiers of science. This blending of formal and informal science, technology, and innovation systems is likely to increase, as the National Innovation Foundation (NIF)—initially funded with only US$400,000 per year—will probably receive five times this sum in funding when it becomes the Institute of the Department of Science and Technology of the Indian government. Although funding may still not be the only critical factor in bringing about significant change, it will, nonetheless, make a substantial difference to the grassroots innovation movement in the country.

The Society for Research and Initiatives for Sustainable Technologies and Institutions (SRISTI, a voluntary, development organization set up in 1993) soon gave rise, in 1997, to the Grassroots Innovation Augmentation Network (GIAN), a regional incubator for converting innovations to enterprise. The NIF came into being in 2000, with the goal of scaling up the innovation system in the informal sector. With modest, but significant, support from the Sadhbhav Foundation, a private trust, SRISTI also set up a natural product laboratory, to add value to people's knowledge, to take these products to market through SRISTI innovations (a non-profit company), and to pursue innovations in the fields of culture and education. SRISTI has developed many herbal technologies for sustainable resource management. Most of these grassroots innovations are green in nature. There is no alternative but to build on these innovations, while exploring sustainable options in future.

Many of these products are now being manufactured and distributed through an ethical supply chain, such that in sourcing as well as in distribution, local communities are being involved and the benefits shared with the people. In some cases, technologies have been licensed to commercial companies, following a benefit-sharing model developed by SRISTI. The names of the communities and, in a few cases, even the

---

[9] We need to look at the IP that users are sharing with the corporations and for which not all companies share any reward.
[10] See Gupta, 2006, 2007b, 2007d, and 2008a.

photograph of the innovator, have been included on the packaging by the agricultural growth promoter.

The pattern of "celebrity endorsement" is being replaced by endorsements by ordinary innovators and users. If a company markets a small farmer's innovation, he may then endorse the product. For example, a herbal growth promoter for agricultural applications was developed by SRISTI-Sadhbhav Sanshodhan (a natural product lab of SRISTI) based on knowledge provided by a farmer, Popat Bhai. It was then marketed by the Hyderabad-based company Matrix Bioscience with a photograph of the innovator on the bottle. This was a way of announcing to all farmers that one of their own had invented the product and shared his knowledge, thus making the innovation accessible to other farmers. In this way, one individual's creativity can help thousands of others and stimulate other users to participate in the innovation supply chain. Every package carrying such information provides an incentive to users to write back to the NIF if they have an innovation to share.

SRISTI service centers have been set up in different villages by farmer members of the Honey Bee Network to try, demonstrate, and sell herbal crop enhancers with pesticide properties to other farmers. The profit margin given to wholesalers is then distributed among farmers—at a discount of approximately 40 percent of the retail price—who can demonstrate the technology to others. To ensure that people who cannot afford to buy products are included, solutions are also made public by means of booklets, websites, and training sessions.[11] Those who want to create their own product are encouraged to do so. In this way, we are promoting many more models which combine intellectual property and open-source in a creative manner. Recently, in a workshop of innovators who had developed various modifications around a core technology, it was decided to develop a "technology commons," allowing and encouraging people-to-people copying, but requiring a licensing procedure for people-to-firm use. [12]

Many more models of this kind remain to be developed in coming years, and we foresee many changes, such as the following, in the way markets may be organized:

- *Modular design.*[13] More and more modular products will be developed, giving users the option to combine these in the manner they want, and sourcing different modules from different suppliers.[13] For example, a modular cellphone, "smart" component can be attached to an electrical appliance (say, a microwave oven) and remotely switched on or off, so food can be ready at a set time when people come home from work. Just such a device has been developed by Prem Singh, a school dropout. But the switching device is costly. When modular technology is readily available, it will be much cheaper for a user to convert any device into a smart remote-operated one.

- *Collaborative product and service design.* This is likely to become a dominant way of matching the needs of the people with the distributor supply chain. NIF has recently established a FABLAB for digital fabrication (designed by Neil Gershenfeld and his colleagues at MIT) and connected to similar ones in 12 cities across the globe. A grassroots innovator can seek help from any of them to convert her idea into a product with distributed mentoring and design.[15]

- *Low-scale markets offered to large-scale distributor networks.* Today, scale has become the enemy of sustainability. Small-scale needs or demand do not get the attention of leading manufacturers because of the high-cost supply chains they have developed. Niche products can serve niche markets, but they can also serve large modular markets through intermediaries. For example, if people in a particular region enjoy recipes using local, uncultivated but edible, plant ingredients, the foods can be stored, packaged in cheap, biologically safe, materials and made available to local inhabitants by a horizontal supply chain in small scale and at shorter distance. The same products can also be marketed in suitable packaging by global networks, "long tail" fashion,[16] to discerning consumers worldwide.

---

[11] See the innovation and traditional knowledge base at: www.sristi.org These databases represent the single largest source of such knowledge for more than 20 years, and by now should have multiplied many times over. The fact that they have not is an indication that we still seem to be wary of learning from ordinary people. Development foundations and UN organizations would do well to make this a priority goal.

[12] Sinha (2008), based on her doctoral work supported by SRISTI.

[13] See also Sanchez and Collins (2001) and Lau and Yam (2005).

[14] See Ding-Bang Luh and Chia-Ling Chang, 2008.

[15] See a recent book on this subject by Li et al., 2007.

[16] "Long tail" is a type of frequency distribution in which small volumes of hard-to-find items are inventoried and distributed at a significant profit to many customers, in contrast to selling large volumes of a product to a small number of consumers. The group that purchases a large number of "non-hit" items is the demographic called the "long tail." See Anderson, 2004.

- *Differentiation through development and inclusion rather than exclusion.* Many large companies today pride themselves on serving exclusive client groups; however, exclusivity is not only a matter of economic status, but can also be a reward for social and cultural contributions. A teacher distinguished for his or her service to the children in a primary or secondary school is no less important than a client who has sold 50,000 cell phones in a week or a month. The current market system fails to take into account the contributions of people across the spectrum of social, ecological, educational, or cultural groupings, the only models for supporting them being either corporate social responsibility mechanisms or charity—both obsolete.

Simplicity in design will more and more become the rule, such that, in the very near future, an elderly person in India should be able to go to a street corner shop and get a cell phone with only three buttons for contacting three children. If he does not want to call anyone else, he does not need more buttons or a screen. The energy cost will go down, as will the cost of the cell phone. Someone else may want a 6"-screen for educational purposes but not for voice calling. An entire range of applications could emerge, if modular manufacturing were applied.

Grassroots innovation will be taught in every school as a way of becoming an inclusive social person. Developing ideas and solving problems individually and collectively will become an integral part of the education of every child. They will learn to combine the following seven "Es": *ethics, excellence, equity, efficiency, empathy, environment,* and *education,* though in different proportion in different activities.

Innovation as a learning and problem-solving template for social application

Specific innovations may often be less important than the principle behind them. Kanak Das from Assam, North East India, noticed that the condition of the roads in his village was very bad. He was not sure how he could improve the conditions of the road, but asked himself whether it might be possible to make the bumps on the roadwork *for* him by generating en-

ergy. So he developed the first cycle to generate energy from bumps in the road, a concept that can be used in practically all automobiles. Until recently, when a few students at MIT, who had seen the video of Kanak Das' cycle, with its shock absorber-based energy-generating system, such a device had never been developed in the formal sector. [17]

The late Ravjibhai Savalia of Bapunagar developed a frying pan with a ribbed bottom to conserve energy. The Indian Institute of Petroleum tested it and found that, because of the increased surface area, the thermal efficiency of the pan went up by more than 1 percent. If the same technique were applied to heating tubes in a large chemical plant, heat-transfer efficiency and cost savings would increase significantly. Here again, a grassroots innovation could influence the productivity of large corporations.

Peanut crops are grown in rain-fed semi-arid regions with light soils. Yusuf Khan, an innovator in Rajasthan, developed an ingenious groundnut pod-collecting device attached to a tractor. The collector scrapes the soil from the pods, leaving it on the ground, while the pods remain on the sieve. A small-scale entrepreneur in Visakhapatnam saw this innovation and licensed the technology for developing a beach cleaner, adapting a technology developed in a dry region for a wet coastal one.

Following are a few more examples of how ideas from one domain may influence technological development in another:

A farmer used three different plants to develop a herbal pesticide. One was *neem*[18] well known for retarding or stopping the growth of insects and helping in plant protection. When scientist Dr. Dhananjay Tiwari took it up for validation in a joint project of NIF-CSIR, he noticed a phenomenon which had not been reported earlier. When neem was exposed to ultraviolet rays for only 2 to 20 minutes, the effectiveness of the chemical compound *azadirachtin* in the *neem* declined steeply. The longer the exposure, the higher was the degradation. When one of the remaining plants was added to the neem, the degradation stopped, forming a herbal stabilizer. There are many reports of chemical stabilization of reactive potential. If this can be made generic, a unique contribution

---

[17] Although Shakeel Awdhany and others who developed this very sophisticated device acknowledged that they had seen Kanak Das' video, they somehow "forgot" to share the intellectual property rights with the grassroots innovator from India or acknowledge their intellectual debt to him. MIT officials did not respond to my communications on the subject. This is an excellent example of the problem inherent in open-source collaborative product design. The more prominent actors often do not acknowledge the creative contribution of weaker partners, much less share any benefits with them. See http://web.mit.edu/newsoffice/2009/shock-absorbers-0209.html

Kanak Das' innovation was also cited in Gupta (2007c) and was awarded an NIF prize by the by President of India, Honourable Dr. A. P. J. Abdul Kalam in 2002 (see www.nifindia.org/secondaward/press_release.html).

[18] "Azadirachta indica," a tree in the mahogany family, native to India.

will have been made by a grassroots innovator towards the advancement of technology.

The member of a tribe[19] in Orissa had used leaves of a particular plant for ripening bananas. (All fruit ripeners used throughout the world are essentially chemicals known as ethylene inducers.) When the laboratory at the Central Food and Technology Research Institute in Mysore tested this claim, they found that the herbal fruit ripeners not only worked as claimed, but also changed the ratio of reducing to non-reducing sugar. The result was a more nutritious fruit, a botanical development previously unreported in science.[20]

Two brothers in Assam, Mehtar Hussain and Mushtaq Ahmed, developed an inexpensive US$100-windmill to pump water to irrigate a small field. The pump was adapted by the Grassroots Innovation Augmentation Network (GIAN),[21] with the help of NIF for use by salt workers to pump out brine. It was further modified and adapted for use in a desert environment. At a cost of only US$700–US$800, it can be bent 90 degrees to withstand heavy storms, after which it can be raised to its upright position. Recently, an inquiry was received from one of the first nations in the Canadian Arctic, who wanted to use it for generating energy. It is highly unlikely that such an innovation would have emerged in an environment of material abundance.

Such frugal innovations, inspired by "Gandhian engineering," as it is called by Dr. R. A. Mashelkar, Director General of CSIR and President of the Indian National Science Academy, will emerge only in an environment where knowledge is maximized and materials economized.

Lessons for modern organizations and supply chain managers

Following is a selection of concluding recommendations, drawn from the above experience:

Reorganize consumption and production relationships. When technologies are developed by producers who are also users, they better reflect the concerns of both the production and consumption environments. Quality control and frugal design are inevitable consequences.

Invest minimal energy in packaging. A firm will not bother with elaborate packaging if 60 percent of its product is consumed within three months or within 300–500 miles of the point of origin. Products are often over-packaged, as if they were to last a year before consumption and transported more than 1000 miles.

Pay for diversity in taste and appearances. Seldom can one find more than one kind of vegetable or fruit in supermarkets. For supply chain managers, uniform products are easier to pack, transport, and distribute. New ways of sourcing, storing, transporting, and distributing bio-diverse products must be found, as both cultural and bio-diversity are closely linked and mutually reinforcing.

Create frugal design and development processes. A whole range of new platforms will have to be developed to design diverse and affordable products to meet diverse needs. Mass production-based supply chains of uniform design are a thing of the past. Instead, collaborative product design will take place in an environment where the seven "Es" are practiced in an open-sharing platform, and the school curriculum will reflect the importance of building on the ideas of the common people.

Solve unsolved problems. Communities will take ownership of problem solving instead of tolerating such injustices as millions of women carrying water long distances on their heads or manually plucking tea leaves. Time targets and dedicated funds will solve many of these and similar problems which the world now ignores.[22]

Create global markets of grassroots products (g2G).[23] NIF has not only sold products emerging from grassroots innovation on all six continents, but has also received inquiries from all over the world. Grassroots innovation and traditional knowledge-based products must receive adequate attention in global markets. Large-scale changes will have to be implemented in the innovation ecosystem and multimedia, multilanguage, knowledge, and innovation databases much better recognized. Likewise, building on the worldwide emphasis on the positive benefits of microfinance will be an awareness of microventure finance (MVF), a concept currently neglected in the lexicon of social development, to the detriment of innovative potential at the grassroots.

Redesign supply chains. Ethical as well as efficiency criteria will demand that we redesign the supply chains by linking dis-

---

[19] The patent (No: 295/KOL/2007) for the herbal fruit ripener developed by Sahu Budhadeba was filed by NIF on behalf of the innovator on 26 February 2007.

[20] Sinha (2008) based on her doctoral work supported by SRISTI.

[21] See recent papers on linking traditional knowledge and modern science by McGovern et al (2009) and Gupta (2007).

[22] see Gupta, 2009.

[23] See Gupta, 2008.

tributed, decentralized, and diversified sourcing and distribution systems. It is imperative that we redesign production and consumption relationships so as to reduce the gap between producer and consumer. When more and more people contribute their labor, skills, inventiveness, and other resources, we will be able to develop low-cost procurement and distribution systems. For instance, increased numbers of fabrication labs[24] and a network of tool rooms will lead to user-designed products and services. Vertical supply chains will become horizontal. Lifestyles will change to include purchasing from the neighborhood village and community. There will be greater scope for enhanced international trade in value-added products instead of primary commodities. More and more value addition will take place *in situ*, so that its benefits accrue to both producers and workers.

In other words, the entire development paradigm must be rethought, so that the decision-making options of both rich and poor are enhanced and the time frame in which they occur is lengthened.[25] Some people think in terms of surviving the next day, while others have the luxury of being able to plan for the next century. When these time horizons converge, we will have a society in which the skills, knowledge, and resources in which poor people are rich are validated more fully.

References

Allen, R. C. 1983. "Collective Invention." *Journal of Economic Behavior and Organization* 4(1):1–24.

Anderson, Chris. 2004. "The Long Tail." *Wired*. October.

Chesbrough, H.W. 2003. *Open Innovation: The New Imperative for Creating and Profiting from Technology*. Boston: Harvard Business School Press.

Gupta, Anil K. 2006. "From Sink to Source: The Honey Bee Network." In *Innovations*. Summer. Available at: *www.mitpress.mit.edu/innovations*

———. 2007a. "How local knowledge can boost scientific studies." SciDev Net, 15 March. Available at: http://www.scidev.net/en/agriculture-and-environment/opinions/how-local-knowledge-can-boost-scientific-studies.html

———. 2007b. "Breeding, Benefits and Bridges with Modern Science: Giving Innovative farmers their Due." Keynote Paper presented at the Second Session of the Governing Body of the International Treaty on Plant Genetic Resources for Food and Agriculture, Rome, 29 October 2007.

———. 2007c. "The word: Deviant Research." NewScience/Science in Society. 22 September. Available at: http://www.newscientist.com/article/mg19526222.000-the-word-deviant-research.html

———. 2007d. "From farmers' first to laborers' first: Why do we still know so little?" Paper presented at the Workshop on Farmer First Revisited: Farmer Participatory Research and Development Twenty Years On. 12–14 December. London: Institute of Development Studies, University of Sussex.

———. 2008a. "What can we learn from green grassroots innovators: Blending reductionist and holistic perspectives for sustainability science." Lecture delivered at the Centre for International Development, Harvard's Sustainability Science Program, Harvard University, Cambridge, MA. 25 February.

———. 2008b. "G2G—Grassroots to global: The knowledge rights of creative communities." Lecture delivered at Globalization and Justice Conference, Seattle University. 21 February.

Gupta, Anil K. 2009. "The Forgotten Farm Laborer." 14 August. Available at: http://rssww.scidev.net/en/opinions/the-forgotten-farm-laborer.html

Gupta, Anil K., K. K. Patel, A. R. Pastakia, and P. G. Vijaya Sherry Chand. 1995. "Building upon local creativity and entrepreneurship in vulnerable environments." In V. Titi and N. Singh (eds.) Empowerment for Sustainable Development: *Towards Operational Strategies. International Institute for Sustainable Development*. pp.112–37.

von Hippel, E. 1987. "Cooperation between rivals: Informal know-how trading." *Research Policy* 16(6):291–302.

———. 2005. *Democratizing Innovation*. Cambridge, MA: MIT Press.

Huston, L. and N. Sakkab. 2006. "Connect and Develop: Inside Procter & Gamble's New Model for Innovation." Harvard Business Review. March.

Lau, A. K. W. and R. C. M. Yam. 2005. "A case study of product modularization on supply chain design and coordination in Hong Kong and China." *Journal of Manufacturing Technology Management* 16(4):432–46.

Li, W. D., S. K. Ong, A. Y. C. Nee, and C. McMahon (eds.). 2007. *Collaborative Product Design and Manufactur-*

[24] See the pioneering work of Professor Neil Gershenfeld, Director of MIT's Center for Bits and Atoms, which he is trying to link with the needs of grassroots innovators. (Available at: web.mit.edu/spotlight/mobile-fablab)

[25] See Gupta et al., 1995.

ing Methodologies and Applications. Springer Series in Advanced Manufacturing XIV. Available at: http://www.springer.com/engineering/mechanical+eng/book/978-1-84628-801-2

Luh, Ding-Bang and Chia-Ling Chang. 2008. Incorporating users' creativity in new product development via a user successive design strategy." *International Journal of Computer Applications in Technology* 32(4):312–21.

McGovern, Patrick E., A. Mirzoian and G. R. Hall. 2009. "Ancient Egyptian herbal wines." Proceedings of the National Academy of Science (U.S.) 106(18):7361–66. 5 May.

Piller, Frank T. 2008. "Interactive value creation with users and customers." In A.S. Huff (ed.). *Leading Open Innovation.* Munich: Peter Pribilla Foundation. pp. 16–24. Available at: http://www.mass-customization.de/download/piller_2008-pribilla.pdf

Riedl, Rupert. 1984. *Biology of Knowledge: The Evolutionary Basis of Reason.* (Trans., P. Foulkes) New York: Wiley.

Sinha, Riya. 2008. RURAL INDIA: The Silent Innovators, one India one people. SRISTI. 15–17 July.

Sanchez, Ron and R. P. Collins. 2001. Competing and Learning in Modular Markets." *Long Range Planning* 34(6):645–67.

Sousa, Milton. 2009. "Open innovation models and the role of knowledge brokers." Available at: www.ikmagazine.com

Chapter 2.6

Quantifying the Impact of ICTs on Growth in Developing Economies

Markus Haacker,
London School of Hygiene and
Tropical Medicine

Introduction

Innovations in information and communications technologies (ICTs) have had a spectacular economic impact across the developing world. One of the most visible is the deepening and widening of access to information and communication services enabled by modern communication technologies, notably cellular phone services. At the same time, advances in information technologies have transformed and enhanced public and business administration. More generally, ICTs have contributed to changes in global production processes, and enhanced the competitive advantage of developing countries in certain labor-intensive production processes.

While these changes are generally recognized, most of the attention regarding the macroeconomic impact of advances in ICTs has focused on the most developed industrialized countries, where such advances have long been recognized as a major determinant of growth, and where a number of studies have provided estimates of the contribution of ICTs to economic growth. This probably reflects a perception that the impact of ICTs is most pronounced in the leading industrialized countries, the fact that the national accounts for most developing economies are not sufficiently disaggregated for the established approaches to estimating the growth impact of ICTs, and that the markets for ICT equipment in most developing countries are small in absolute terms. Thus, industry data on sales to the respective countries are frequently unavailable or unreliable.

Meanwhile, the impact and potential of ICTs in economic development has long been recognized, as evidenced by ongoing efforts to expand access to ICTs by developing equipment that is affordable and adapted to use in developing countries (e.g., hundred-dollar computers). The "ICT4D" (i.e., ICTs for development) literature documents best practice in utilizing modern ICTs in the context of economic development, which is frequently based on cellular phone technologies.

Below, we argue that the macroeconomic impact of advances in ICTs in developing countries can be analyzed in similar ways as for advanced industrialized countries, and that these technological innovations have significantly affected economic growth in the developing world. Specifically, we adapt the growth accounting approaches to the data available across developing countries, and derive estimates of the mag-

---

[1] Throughout the chapter, the terms "developing countries" or "developing world" broadly refers to "low-income" countries (LICs) and "lower-middle-income" countries (LMICs) as defined by the World Bank. The focus on LICs and LMICs reflects the fact that, in terms of the role of ICTs, many upper-middle-income countries are closer to advanced economies than to developing ones. For our stock-taking exercise, the narrower focus is therefore helpful.

nitude of the growth impact of advances in ICTs for developing economies. In addition to the growth enjoyed by developing countries which succeeded in establishing themselves as producers of ICT equipment, we focus on the growth impact of ICT-related capital deepening. By this, we mean the productivity gains that arise because ICT equipment has become much cheaper (or, equivalently, more powerful for the price). Thus, users of ICT equipment today get much more bang for the buck than they did a few years ago. From a macroeconomic perspective, this means that productivity (and thus growth) increases.

Against this background, we summarize data describing the expanding access to ICTs in developing countries and describe the approaches that have been used to estimate the growth impact of ICTs in industrialized countries. In light of the available data and tools, the next section describes the analytical framework we adopt for our analysis, distinguishing between the production of ICT equipment and capital deepening arising from falling prices of ICT equipment. Finally, we summarize our estimates of the growth impact of ICTs in developing countries, and relate it to estimates available for industrialized countries.

Access to ICTs in developing countries

The economic impact of ICTs is ubiquitous. Modern information technologies have transformed the administration of business and public services, and communication technologies have greatly improved the ease of, and expanded access to, communication and information across the developing world. The immediate benefits of advances in information technologies are seen primarily in the formal sector, as the use of computer hardware requires a substantial capital outlay and a regular power supply (an important constraint in many developing countries). The benefits of modern communication technologies, primarily cellular phone services, are spread more widely.

Measuring the economic impact of advances in information technologies in either developing or industrialized economies poses similar problems. While it is possible to measure the utilization of ICTs, some of the most interesting consequences have to do with the transformation in economic activities enabled by information technologies. In the context of developing countries, examples would include the emergence

of new types of service industries (e.g., call centers) enabled by modern information technologies, and changes in the location in industrial production facilitated by modern ICTs. In both cases, while the economic opportunities are at least partly enabled by advances in ICTs, ICT equipment or related services do not necessarily account for a large share of production costs, and measuring the ICT-related inputs to these economic activities would give a misleading (and understated) measure of the economic impact of ICTs.

Nevertheless, data on access to ICTs provide valuable indicators of the magnitude of their economic impact. First, by comparing the utilization of ICTs across countries, it is possible to draw some inferences about the role these technologies play in the economies of the respective countries. Second, data on expenditures on ICT-related equipment also provide information on its economic value. This aspect has been used by Bayoumi and Haacker (2002) to estimate the economic benefits of declining prices of ICT equipment accruing to the users of such equipment. Alternatively, by integrating data on expenditures and prices of ICT equipment in a growth accounting framework, it is possible to estimate the contribution of advances of ICTs to economic growth. The growth effects occur not only from the *production of ICT equipment*, but also from the *ICT-related capital deepening* as a result of the drop in price of ICT equipment. As the productivity or capabilities of the stock of ICT equipment increase over the years, this arguably has a positive impact on economic growth. Unlike the production of ICT equipment—significant in only a handful of developing countries—all developing countries utilize ICT equipment, so that the growth effects of *ICT-related capital deepening* occur, to differing extents, across the developing world.

This is the approach we are going to follow below. However, in light of the foregoing discussion, it is important to note that the analysis may not capture the full benefits of advances in ICTs in enabling changes in the global production pattern and in transforming the economy. To differentiate from broader economic impact of advances in ICTs, we refer to the direct growth impact associated with the use of ICT equipment as the effects of ICT-related capital deepening.

Access to ICT Equipment

In the absence of sufficiently disaggregated national accounts for most developing countries, and in light of the limited coverage of data covering sales of ICT equipment across countries,[2] the primary source of data on access to ICT equipment is from trade and is available for all countries—at least through the records from trade partners—from the UN Statistics Division (2009). Moreover, as most low- and low-middle-income countries do not produce ICT equipment, net imports of this equipment give a consistent indicator for the absorption of ICT equipment in these countries.

these countries produce IT equipment (we will return to this point momentarily). The picture for communications equipment is similar, with a few small positive outliers and a few large countries with negative net imports, although spending tends to be higher, and fewer countries show net imports virtually equal to zero.

However, net imports can give a misleading picture of the absorption of ICT equipment. Not only do some countries feature negative net imports, but there is also the possibility that other countries produce ICT equipment, even if their net imports are positive overall. To address this problem, we

Figure 1.1. Net imports of ICT equipment in 95 low- and lower-middle-income countries, 2001–2005 (percent of GDP)

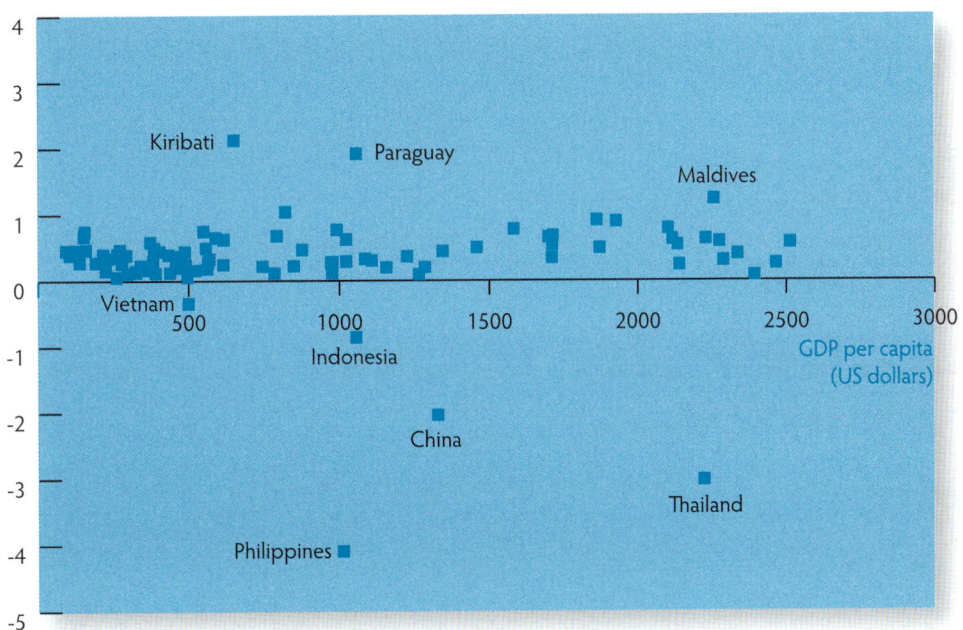

Source: Author's calculations, based on UNIDO (2007), United Nations Statistics Division (2008), Global Insight (2006), and International Monetary Fund (2009).

Figure 1 summarizes available data for net imports of ICT equipment, averaged over the years 2001–05.[3]

Net imports of IT equipment do not vary much across countries with different income levels, with net imports between 0 and 1 percent of GDP. There are three positive outliers, two of which are small island economies where single large trade transactions or unrecorded re-exports may blur the picture. More substantially, five countries (including large ones) feature negative net imports, presumably reflecting that

adopt production figures from UNIDO (2007), which are summarized in Figure 2. According to these data, the number of developing countries where production of ICT equipment plays a significant macroeconomic role is small. The countries covered by UNIDO (2007) include the outliers with negative net imports identified in Figure 1.[4] To cross-check the coverage of the UNIDO (2007) data, we also checked exports and other indicators, and did not find obvious omissions of ICT producers.

---

[2] The most comprehensive international data source is Global Insight's *Global IT Navigator Database* (2006), covering 70 countries; however, these include only 6 low-income countries (out of a total of 50) and 20 low-middle-income countries (out of a total of 54).

[3] For IT equipment, our data include SITC 2 categories 752 (automatic data processing equipment) and 7599 (parts and accessories pertaining equipment in category 752), corresponding to HS 2002 categories 8471 and 847330. The measure of communications equipment we adopt is SITC 2 category 764 (telecommunication equipment, parts, and accessories).

[4] For IT equipment, our data include SITC 2 categories 752 (automatic data processing equipment) and 7599 (parts and accessories pertaining equipment in category 752), corresponding to HS 2002 categories 8471 and 847330. The measure of communications equipment we adopt is SITC 2 category 764 (telecommunication equipment, parts, and accessories).

Figure 1.2. Net imports of communications equipment in 95 low- and lower-middle-income countries, 2001–2005 (percent of GDP)

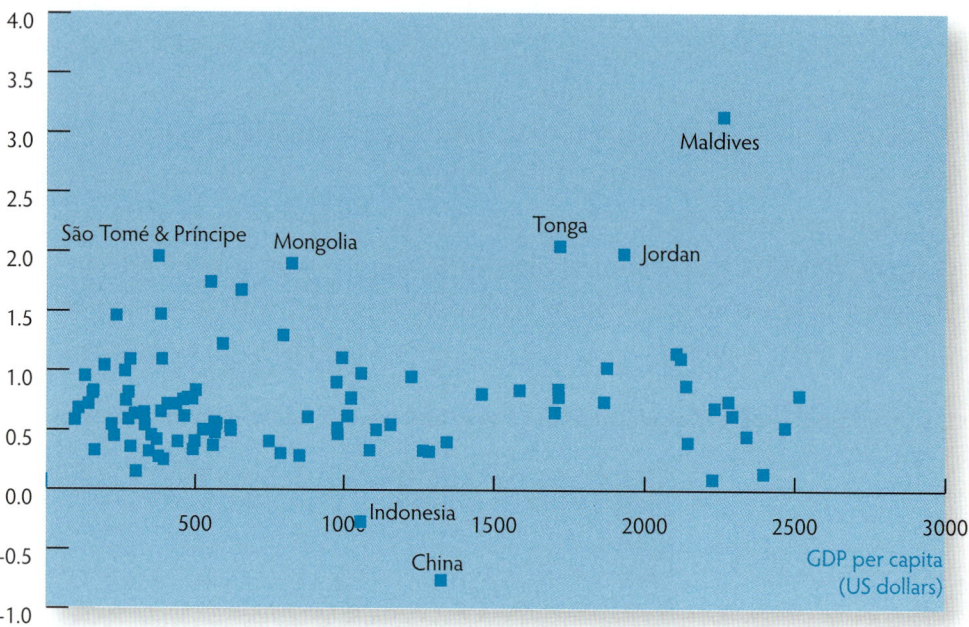

Source: Author's calculations, based on UNIDO (2007), United Nations Statistics Division (2008), Global Insight (2006), and International Monetary Fund (2009).

Figure 2.1. Production of IT equipment in 16 low- and lower-middle-income countries, 2001–2005 (percent of GDP)

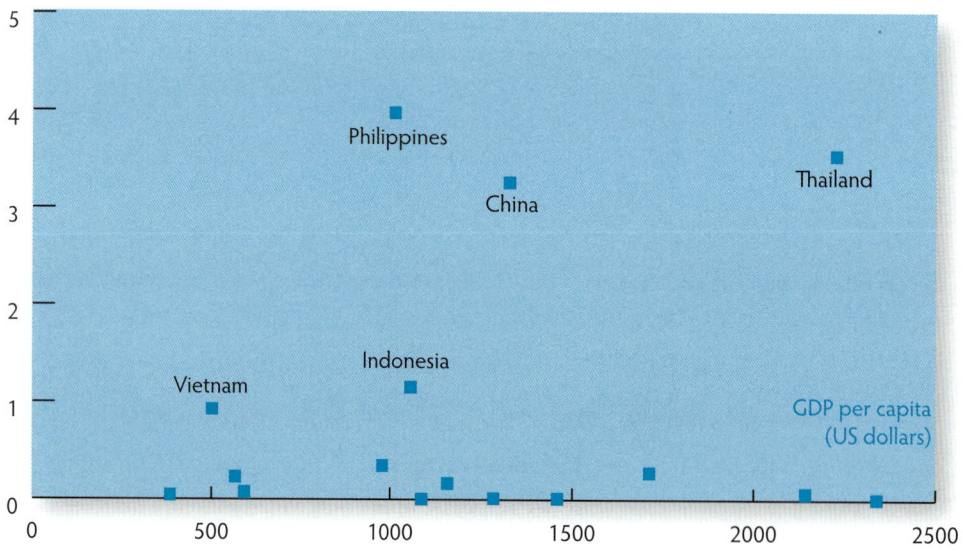

Source: Author's calculations, based on UNIDO (2007), United Nations Statistics Division (2008), Global Insight (2006), and International Monetary Fund (2009).

Thus, overall, we find that net imports obtained from international trade data give a consistent measure of the absorption of ICT equipment for most developing countries. However, for about a dozen countries, it is necessary to take account of domestic production, and for these, domestic absorption is essentially obtained as the sum of domestic production and net imports. [5]

---

[5] To obtain estimates of domestic absorption from production and net import figures, it is necessary to apply certain weights, e.g., to account for certain trade costs.

Figure 2.2. **Production of communications equipment in 16 low- and lower-middle-income countries, 2001–2005 (percent of GDP)**

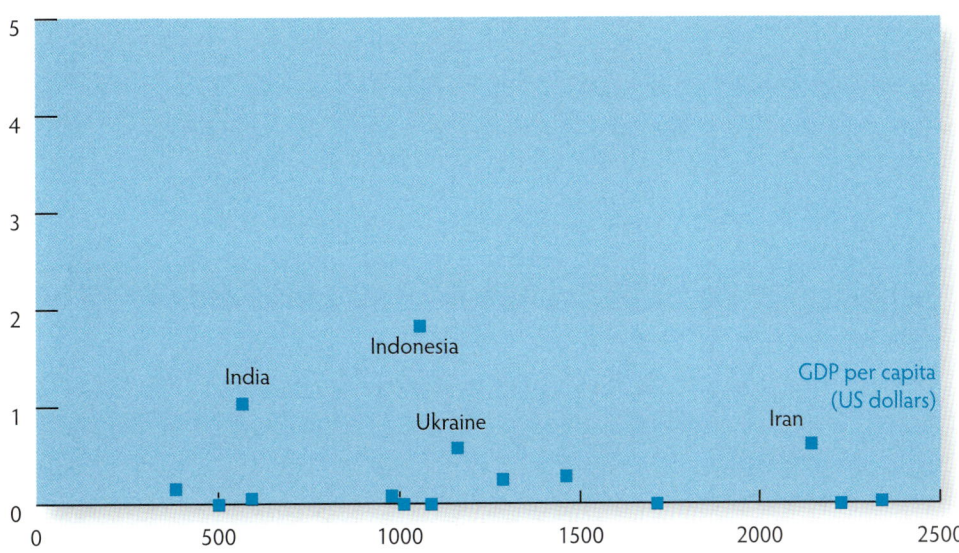

Source: Author's calculations, based on UNIDO (2007), United Nations Statistics Division (2008), Global Insight (2006), and International Monetary Fund (2009). Excludes China as data were unavailable.

Access to communication services

While our analysis focuses on the use of ICT equipment, another important dimension of economic transformations in developing countries enabled by advances in ICTs is the increasing access to communication services. This is captured partly in our analysis below, as ICT equipment does imply investments in both communications infrastructure and end-user equipment. However, as these data only provide very indirect information on the use of communication services,

we complement our analysis with a brief excursion discussing trends in access to communication services in developing countries, and the role played by advances in communication technologies.

Crude data on the use of communication services are available across countries, as the use of such services generally requires some form of subscription (including active SIM cards). Figure 3 provides a snapshot of the latest available data (2007) on the number of phone subscriptions across developing economies.

Figure 3.1. **Mainline and cellular phone subscriptions and GDP per capita, 2007**

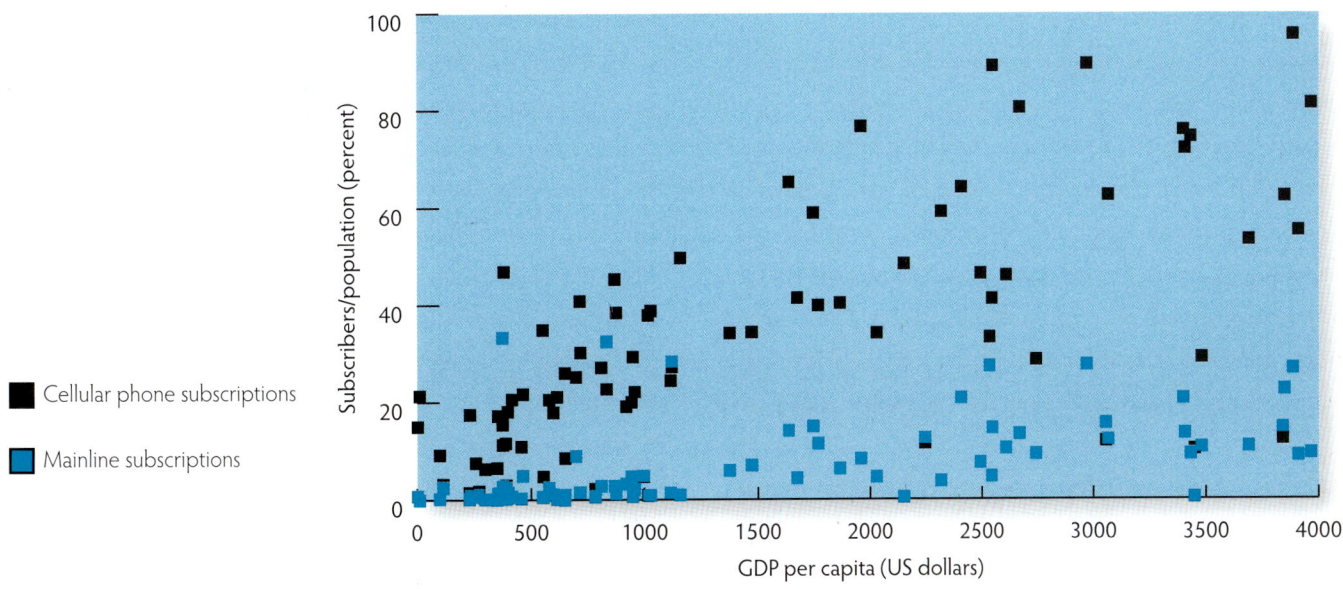

Source: ITU (2009).

Figure 3.2. Cellular phone subscriptions (percent of total)

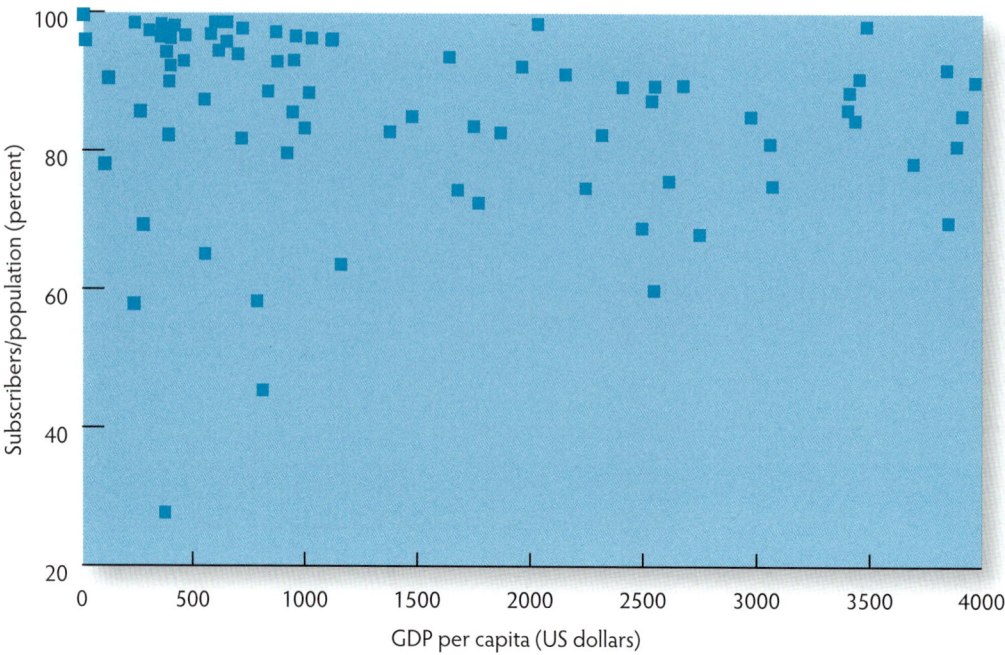

Source: ITU (2009).

Both mainline subscriptions and cellular phone subscriptions increase with GDP per capita. The number of mainline subscriptions is below the equivalent of 10 percent of the population for low-income countries, and below 25 percent of the population for middle-income countries. In most developing countries, the number of cellular phone subscriptions is much higher—frequently several times larger—than the number of mainline subscriptions. In fact, as Figure 3.2 suggests, it can be argued that cellular phone technologies have transformed access to communications in the poorest developing countries, where the share of cellular phone subscriptions is typically above 90 percent.

Figure 3 also accentuates a fundamental difference in the way in which developing and advanced economies expand access to communication services associated with the spread of cellular phone technologies. Whereas in advanced economies cellular phone technologies primarily contribute to a deepening of access to communication (as most users have additional direct access to communication services), in developing countries they primarily contribute to the widening of access, greatly expanding the number of people with access to services.

Cellular phone subscriptions exceed the number of mainline subscriptions in most developing countries. Among low-income countries (broadly, those with a level of GDP per capita below US$1,000 in Figure 3), cellular phone services play a subordinate role in some, but exceed 70 percent of subscriptions in the majority of countries.

Figure 4 complements the snapshot just discussed with a summary of trends in access to phone services.

Among the most striking features is the rise of China, where the total number of subscriptions increased from 0.6 percent of the population in 1990 (about the same as other low-income countries at the time) to 69 percent in 2007, much higher than the average for low-middle-income countries.[6] However, the share of cellular phone subscriptions (60 percent) is unusually low by international standards. For the other countries or regions, Figure 4 illustrates the expansion in access to communication services that took place since the mid-1990s, enabled by cellular phone services, which accounted for 6 percent of total subscriptions in 1995, but 77 percent as of 2007. This shift has been most pronounced in sub-Saharan Africa,[7] where the share of cellular phone subscriptions in low- and low-middle-income countries has grown from 2 percent in 1995 to 95 percent in 2004, resulting in a significant increase in access to communications services.

Our discussion of access to communication services also provides an illustration of the kind of effects of falling prices of ICT equipment that we aim to capture. As we will see later, investment in communications services did accelerate over

---

[6] China and India moved from the low- to the lower-middle-income bracket within the period shown.

[7] Totals for low- and lower-middle-income countries include sub-Saharan African countries. Additionally, the total for low- and lower-middle-income countries in sub-Saharan Africa is shown separately to illustrate the crucial effect of cellular phone technologies in that region.

Figure 4.1. Access to mainline phone services (number of subscriptions in percent of population)

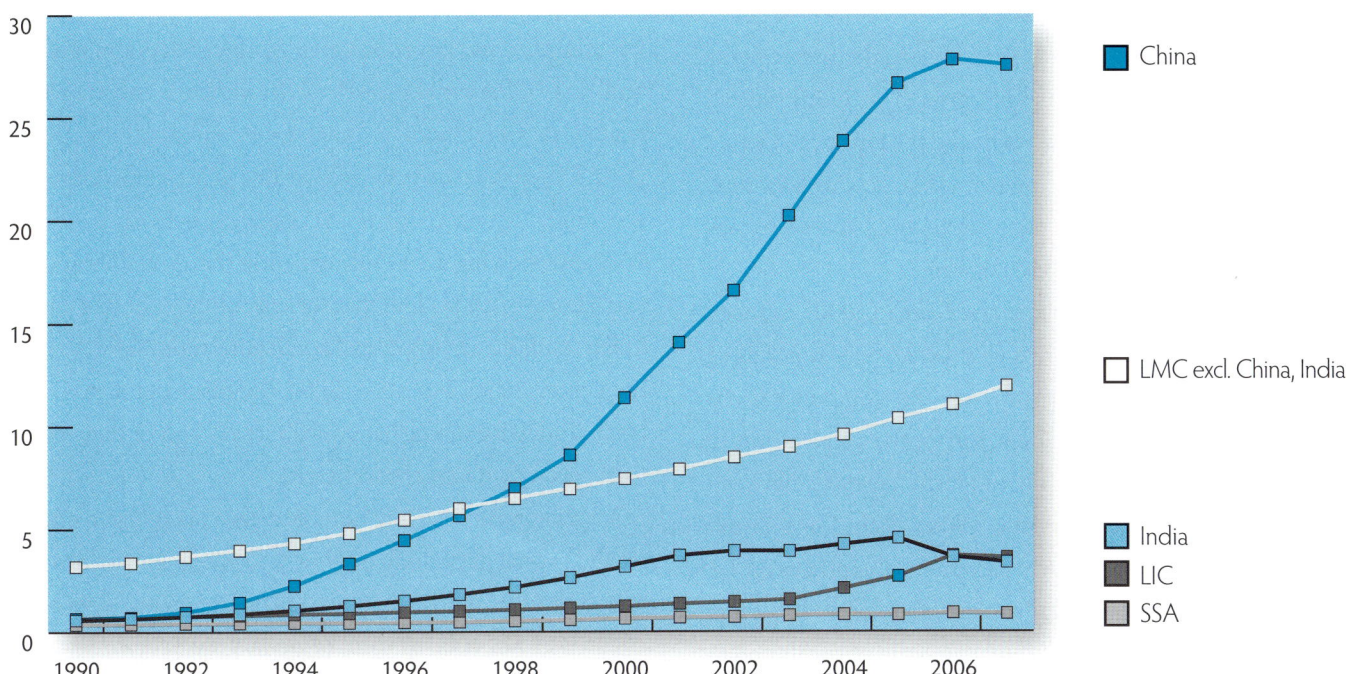

Source: ITU (2009).

the period shown in Figure 4, but only moderately. Thus, it cannot plausibly explain the expansion in access to communication services, and implies that the expansion in access is the result of "more bang for the buck," i.e., falling prices of communications services driven by technological progress,[8] resulting in an expansion in economic activities, hence, economic growth.

Figure 4.2. Access to cellular phone services (number of subscriptions in percent of population)

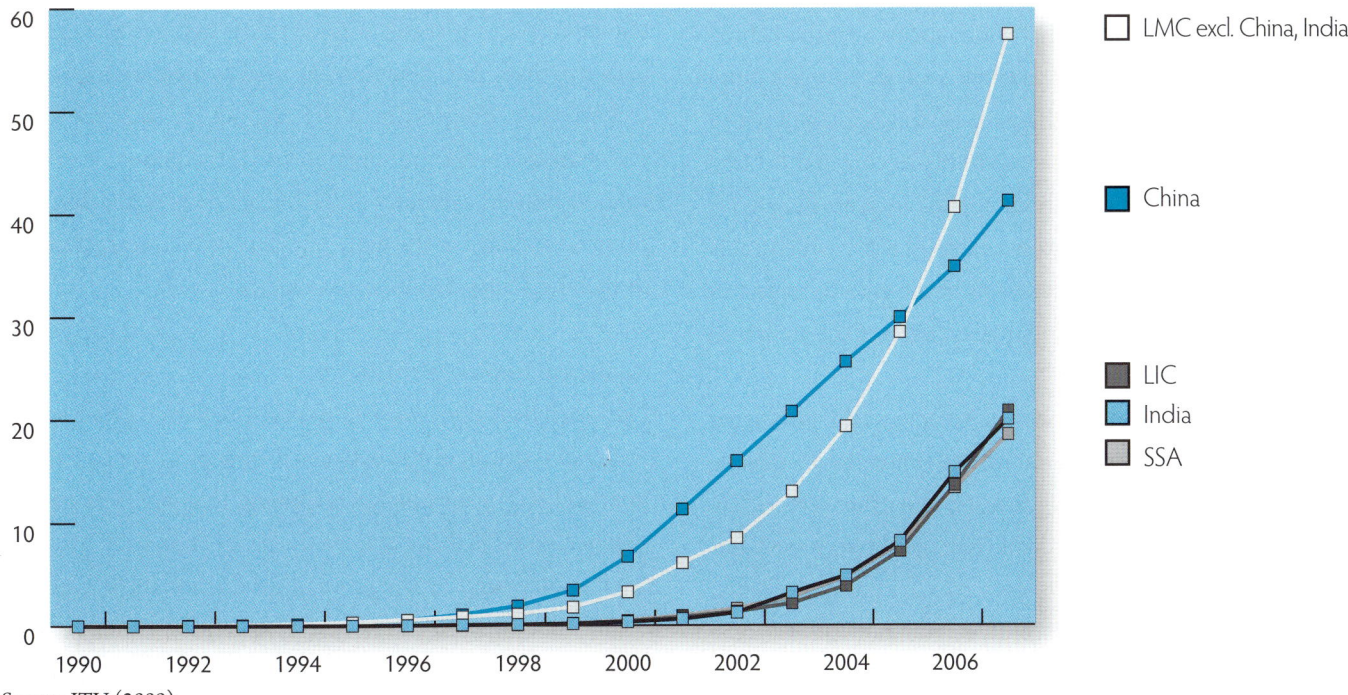

Source: ITU (2009).

[8] In addition to reducing the costs of certain services, technological advances in communication services frequently result in an increase in competition in the communications sector, contributing to the decline in prices.

Review of the growth impact of ICTs in industrialized countries

The spectacular advances in ICTs are reflected in assessments of the determinants of economic growth in industrialized countries, notably the United States and the European Union. Below, we highlight selected contributions to this literature, to provide some context for our analysis and because our approach draws on aspects of this literature.

For the United States, Oliner and Sichel (2000) estimate that IT-related capital deepening (mainly reflecting falling relative prices of hardware, software, and communications equipment) accounted for about half of the acceleration of labor productivity growth in non-farm business between 1974–90 and 1996–99. Similarly, Oliner, Sichel, and Stiroh (2007) find that advances in ICTs directly contributed 1.1 percentage points to the growth in labor productivity in 2000–06, of which 0.6 percentage points can be attributed to ICT-related capital deepening. Adopting a broader concept of output (GDP and consumption of durable goods), Jorgenson, Ho, and Stiroh (2008) obtain estimates of the growth impact of ICTs of similar magnitude (for a more comprehensive measure of output), also estimating the impact of ICT-related capital-deepening at 0.6 percentage points in 2000–06, but finding a smaller impact from production of ICT equipment (0.4 percentage points).

The literature on sources of economic growth in the EU or the OECD differs from that which focuses on the United States in one dimension central to our analysis: it covers more than one country, making the issue of consistency between national data relevant. In this regard, Colecchia and Schreyer (2002) find that the methods applied in generating price indices for ICT products across countries differ widely. This is a crucial point for our assessment of the growth impact of ICTs, as explained below in our presentation of the analytical framework.

Van Ark (2001) distinguishes between ICT-producing manufacturing and service industries, intensive ICT-using manufacturing and service industries, and other sectors. He finds that productivity growth differentials between the United States and most European countries are partly explained by a larger and more productive ICT-producing sector in the United States, but also by larger productivity contributions from ICT-using industries and services in the United States.

The study by Inklaar, Timmer, and van Ark (2007) is based on input and output data on the industry level, distinguishing 26 industries and covering seven advanced economies. They find that "differential growth performance is most strongly related to differences in total factor productivity (TFP) growth," whereas the pattern regarding ICT capital deepening was similar across countries, accelerating between 1995 and 2000, but slowing down subsequently. Van Ark, O'Mahoney, and Timmer (2008) find that the contribution of ICT capital in Europe has been lower and accelerated more slowly than in the United States. For 1995–2004, they estimate that ICT-related capital deepening has contributed 0.5 percentage points to the growth of labor productivity.

Jorgenson and Vu (2007) analyze the contribution of ICT-related capital deepening to growth in the 110 countries covered by the Penn World Tables, including many developing countries.[9] For 2000–04, out of global growth of 3.75 percent annually, Jorgenson and Vu (2007) attribute 0.42 percentage points to ICT-related capital deepening. Across major regions, the contributions of ICT-related capital deepening range from 0.27 percentage points for 28 countries in sub-Saharan Africa to 0.47 percentage points for the G-7 economies.

Growth impact of ICTs in developing economies

Following a brief discussion of our analytical framework, our discussion proceeds in three steps: first, we present the data on prices of ICT equipment used to measure the rate of technological progress in ICTs; second, for those developing countries where production of ICT equipment plays a role, we present estimates of its impact on growth; third, we present estimates of the growth impact of ICT-related capital deepening across developing countries.

Analytical framework

Our analytical framework (discussed in more detail in the Appendix) distinguishes between the *production of ICT equipment* and *ICT-related capital deepening*. The growth impact of advances in ICTs that occurs in the production of ICT equipment is measured by the value of production of ICT equipment, multiplied by the rate of productivity growth in this sector.

The growth impact of advances in ICTs through ICT-re-

---

[9] Their study is based on data on ICT-related spending for 70 countries only, and proxies for the other countries, including most of the developing countries captured by their study.

Figure 5. **Rates of price decline of ICT equipment, 1991–2006 (annual rate of decline, in percent)**

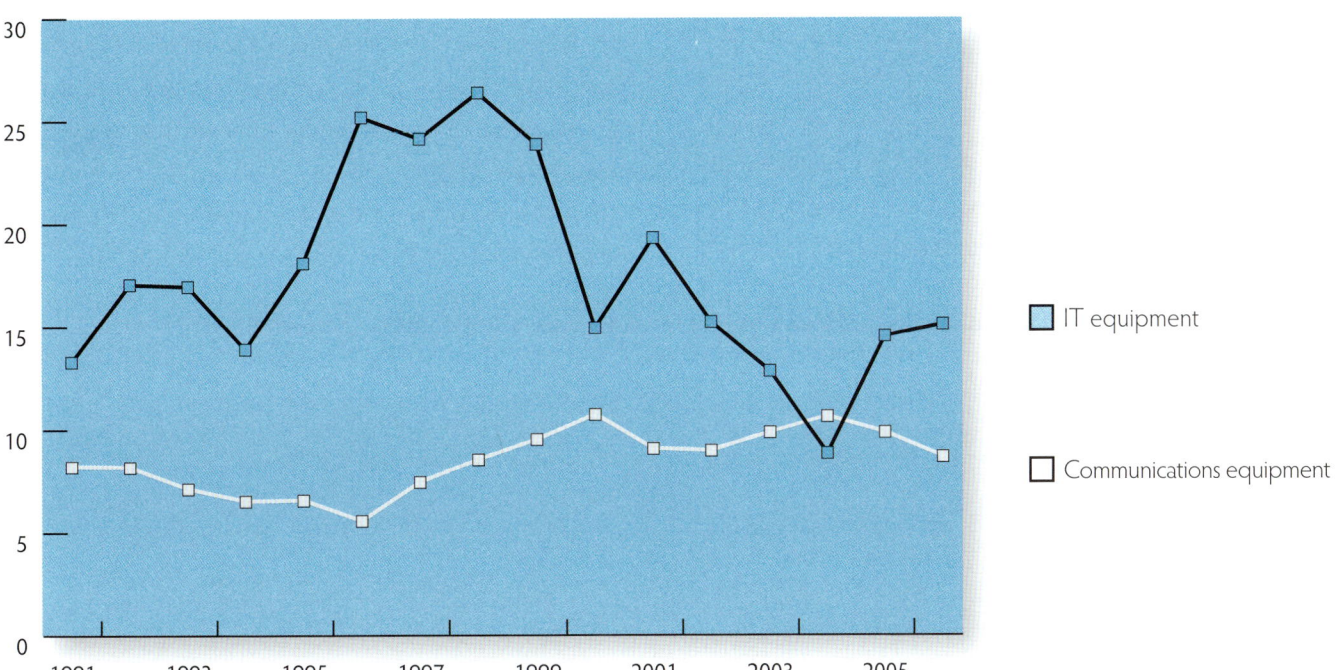

Source: Author's calculations, as described in text, based on data from US Department of Commerce, Bureau of Economic Analysis (2008), US Department of Labor, Bureau of Labor Statistics (2008), and Doms (2005).

lated capital deepening occurs in two ways: first, as the price of ICT equipment drops, the rate of return on investments in ICT equipment increases, and investment is channeled towards ICT equipment to take advantage of this higher rate of return. Second, the drop in the price of ICT equipment and reallocation of investment has an effect similar to a productivity shock—i.e., output increases for a given value of inputs. As higher output translates into higher savings and, in turn, investment, growth also increases over the following years, until the economy reaches a new "steady-state" equilibrium.

Measuring technological progress in ICTs

The factor that drives growth impact is the increasing productivity of ICT equipment and of ICT-related capita deepening mirrored by declining prices of this equipment. However, price series for ICT equipment across developing countries do not exist. Taking the cue from studies of the growth impact of ICTs in developed economies, and reflecting that ICT equipment is highly tradable, our price series is based on US data, which are frequently used in international studies because they are known to capture well the technological advances in ICTs.

The annual rates of price decline for ICT equipment, nota-

bly for IT equipment, have fluctuated considerably between 1991 and 2006 (Figure 5), ranging from 8.8 percent to 26.4 percent for IT equipment, and from 5.6 percent to 10.7 percent for communications equipment. While the rate of technological progress in the production of IT equipment attained a peak in the second half of the 1990s, the price declines of communications equipment were highest in 2000–06.

ICT Production

Regarding the growth impact of advances in ICTs arising from the production of ICT equipment, we match the production data from Figure 2, covering the period 2001–05, with the data on declining prices in Figure 5. Our estimates of the implications of increasing productivity in the production of ICT are summarized in Figure 6. The first finding concerns the geographic distribution of the gains in GDP growth. Mirroring the concentration of the production of ICT equipment in upper-middle- and high-income countries, only a few developing countries show any growth gains from the production of ICT equipment. The second finding is that, where they occur, the growth gains are not very large. In only four countries do we observe growth gains equivalent to 0.4 percentage points or higher: Philippines and Thailand (where most gains arise from the production of IT equipment), Indonesia (where the gains are split roughly evenly between IT and communications equipment), and China (where our data understate the impact, as we do not have production figures for communications equipment). Thus, overall, we find that the production of ICTs is concentrated in a few countries, and that the magnitude of the growth impact of advances in ICTs in the production of ICT equipment is moderate.

ICT-related capital deepening

The growth impact of ICT-related capital deepening depends on three factors: the rate of decline in the prices of ICT equipment (illustrated in Figure 5), the scale of investment in ICT equipment, and features of the structure of the economy, notably the capital intensity of production.

Figure 7 summarizes trends in investment rates in IT equipment and communications equipment (over the 1990–2006 period.

Figure 6.1. **Growth impact of production of IT equipment in 16 countries, 2001–2005 (percent of GDP)**

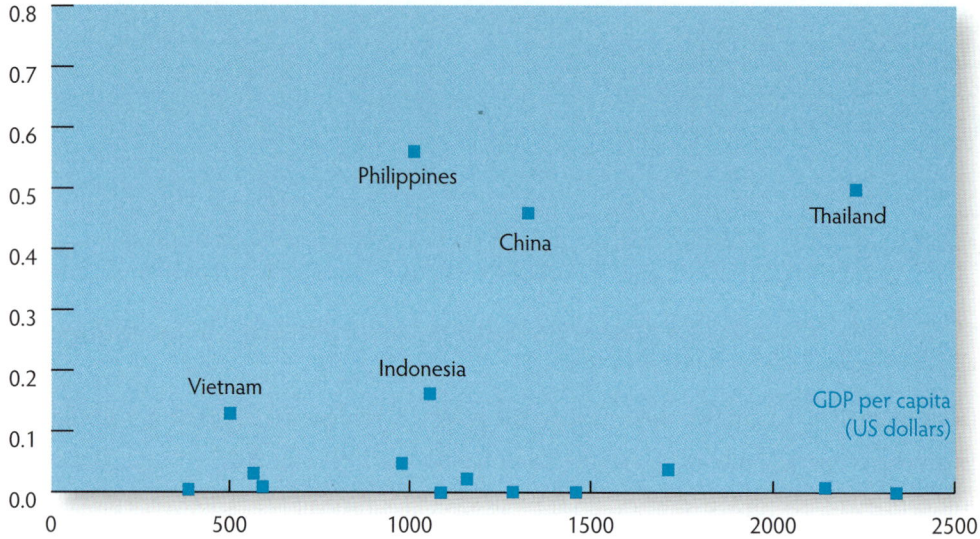

Source: Author's calculations, based on UNIDO (2007), United Nations Statistics Division (2008), Global Insight (2006), and International Monetary Fund (2009).

Figure 6.2. **Growth impact of production of communications equipment in 16 countries, 2001–2005 (percent of GDP)**

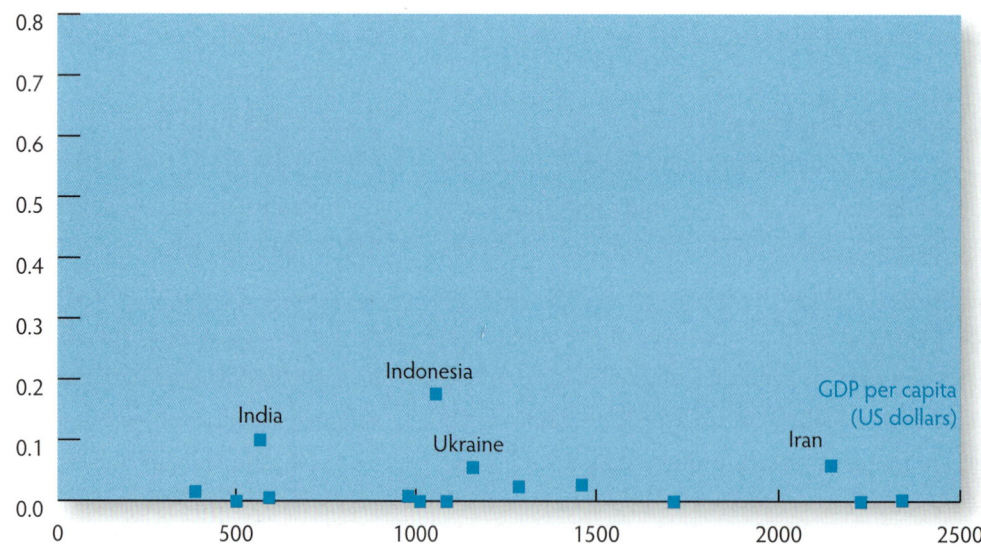

Source: Author's calculations, based on UNIDO (2007), United Nations Statistics Division (2008), Global Insight (2006), and International Monetary Fund (2009). We exclude China, as data were unavailable.

Figure 7.1. IT-related investment in low- and lower-middle-income countries, 1990–2006 (percent of GDP)

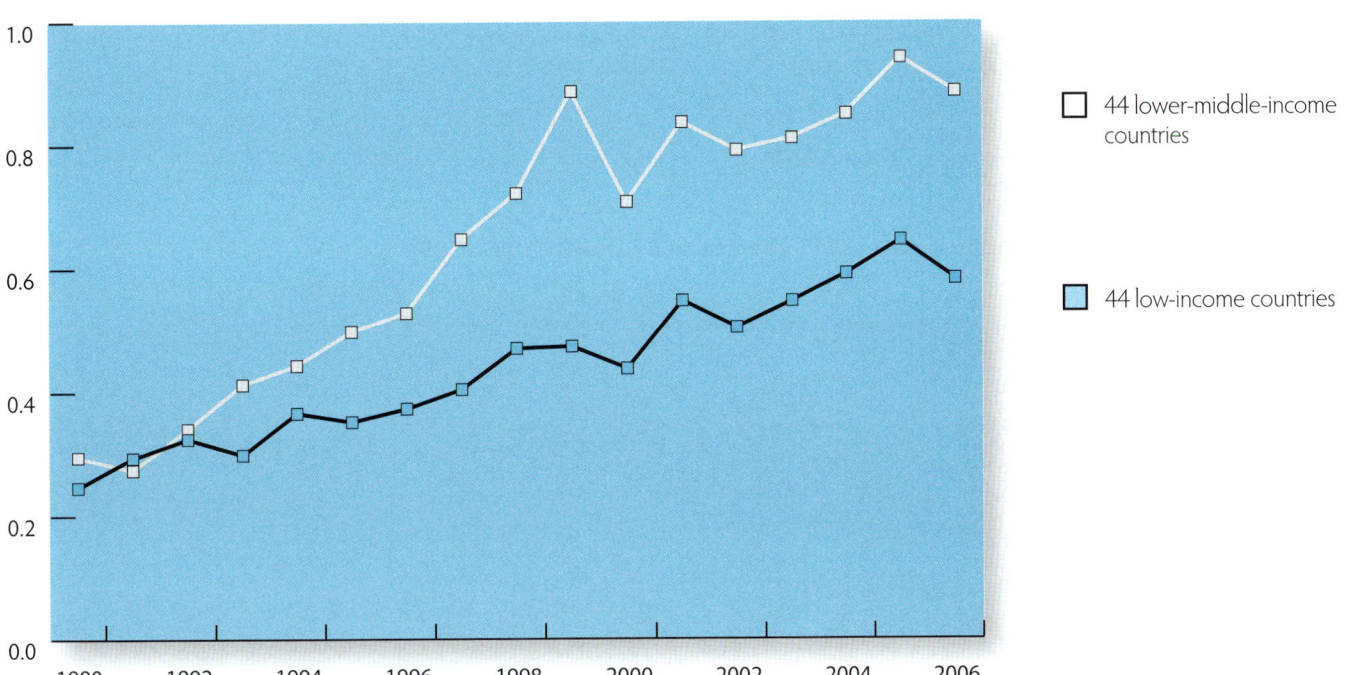

Source: Author's calculations, based on UNIDO (2007), United Nations Statistics Division (2008), Global Insight (2006), and International Monetary Fund (2009).

Figure 7.2. Communications-related investment in low- and lower-middle-income countries, 1990–2006 (percent of GDP)

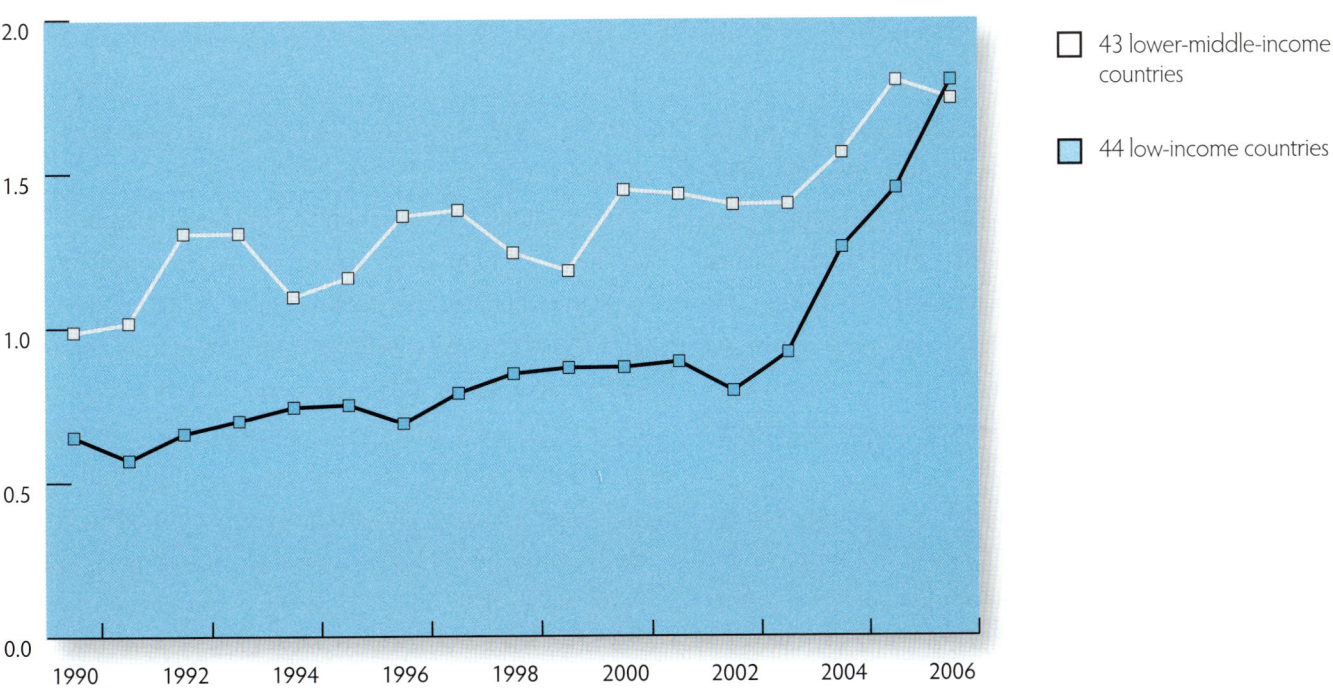

Source: Author's calculations, based on UNIDO (2007), United Nations Statistics Division (2008), Global Insight (2006), and International Monetary Fund (2009).

Overall, investment in ICT equipment, in percent of GDP, roughly doubled between 1990 and 2006. As for the roles of information and communications technologies, respectively, we see that the level of investment in communications equipment was more than twice that in IT equipment. Figure 7 also illustrates some notable differences in the evolution of

ment in communications equipment are considerably higher as a rule than investment rates for IT equipment—according to Figure 7, by a factor of about 2 on average—the rates of price decline for IT equipment are higher. On average, they exceed the rates of price decline for communications equipment by a factor of about 2.2 (over the 1990–2006 period),

Figure 8.1. Contribution of declining prices of IT equipment to growth, 1991–2006

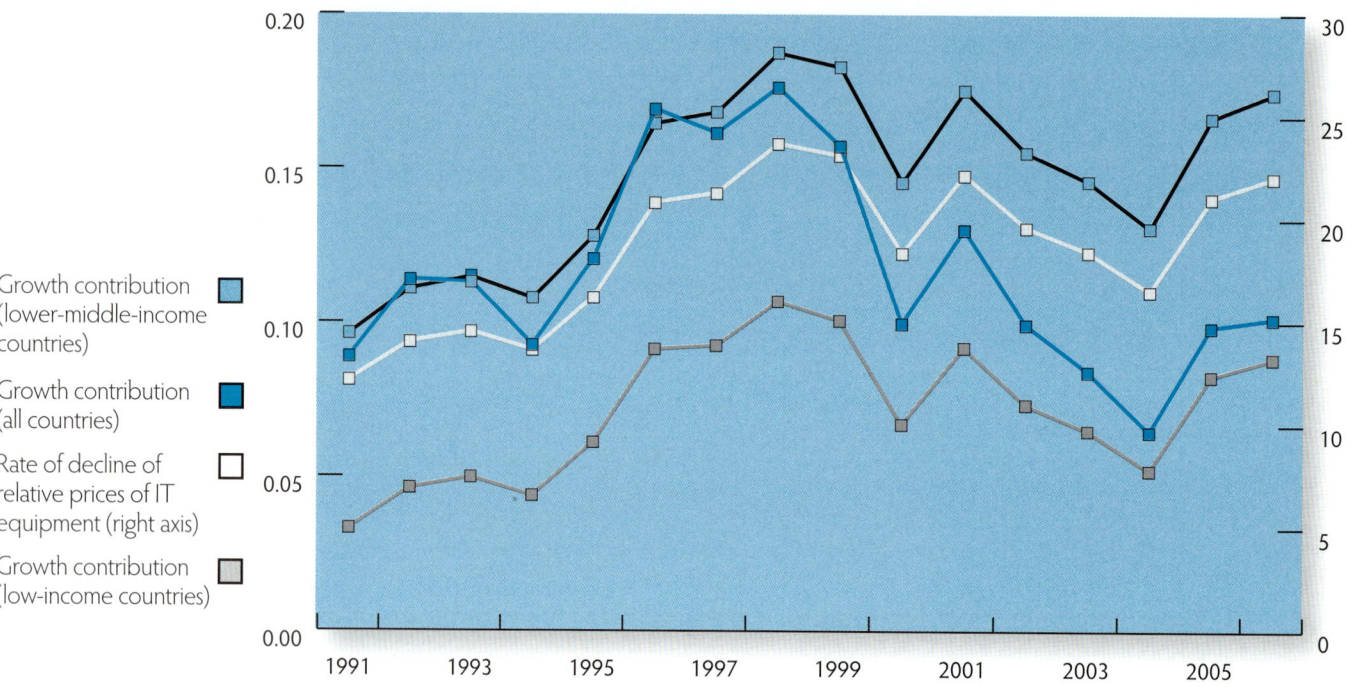

Growth contribution (lower-middle-income countries)

Growth contribution (all countries)

Rate of decline of relative prices of IT equipment (right axis)

Growth contribution (low-income countries)

Source: Author's calculations, as described in text.

the role of IT and communications equipment between low-income countries and lower-middle-income countries. Starting at about the same level, investments in IT-related equipment have accelerated markedly in lower-middle-income countries, as compared to low-income countries. At the same time, investment in communications equipment started out lower in low-income countries, but has accelerated markedly since 2002, catching up with low-middle-income countries by 2006.[10]

Figure 8 and Table 1 summarize our estimates for the impact of declining prices of ICT equipment on economic growth. The magnitude of the contributions of IT equipment and communications equipment to growth, respectively, are similar, rising from about 0.09 percent at the beginning of the period covered to 0.13 towards the end. While rates of invest-

with the result that the magnitude of the growth effects is similar.

One interesting exception to this broad picture is seen in the years 1996–2000, during which the contribution of capital deepening from declining prices of IT equipment to growth peaks, and exceeds the contribution from communications equipment. This is the period which has motivated much of the early work on the economic impact of advances in ICTs in the United States (e.g., Jorgenson (2001), or Oliner and Sichel (2000)). Our estimates are in line with this earlier literature, partly because our international price data are based on US price indices. However, our distinction between the direct effects of shocks to prices of IT equipment and the indirect effects through capital accumulation—which arise as the economy gradually moves towards the steady-state growth

[10] The increase in investment in communications equipment in low-income countries after 2002 is spread across a large number of low-income countries. While investment in communications equipment rose by at least 1 percent of GDP in 20 low-income countries between 2002 and 2006, it fell by at least 1 percent of GDP in only one country.

Figure 8.2. Contribution of declining prices of communications equipment to growth, 1991–2006

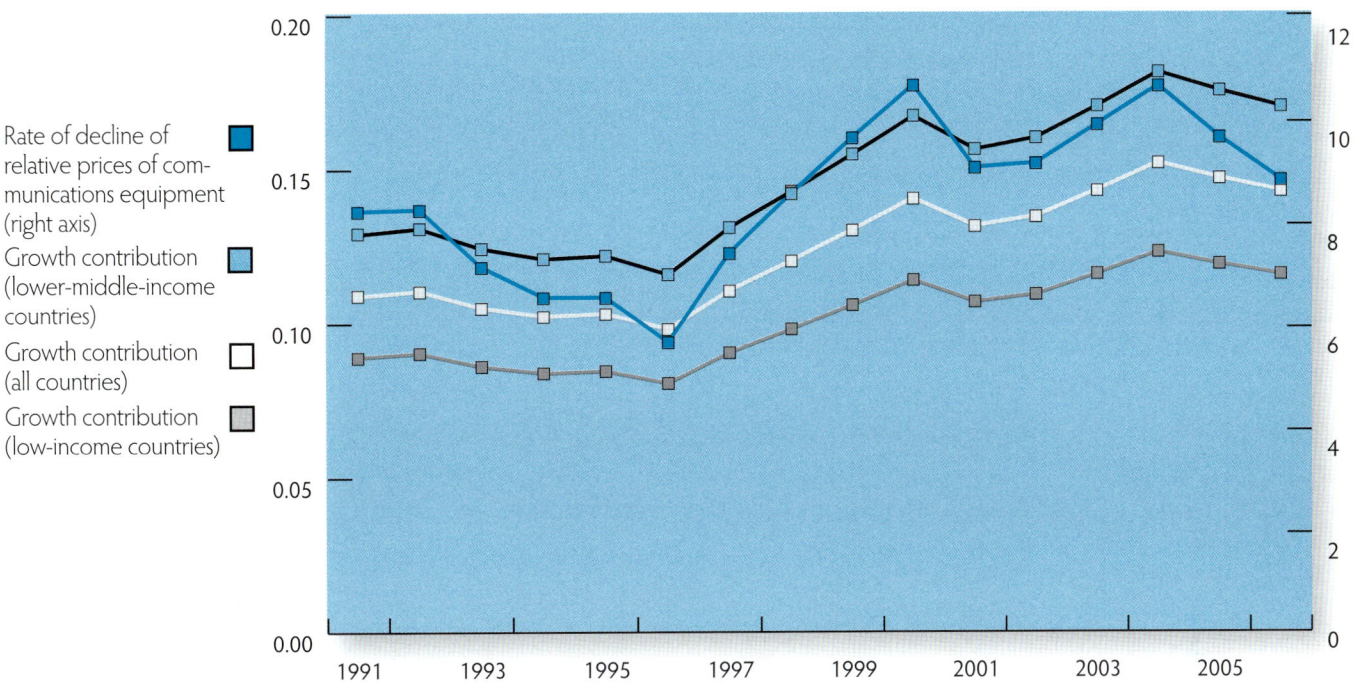

Source: Author's calculations, as described in text.

Figure 8.3. Direct and indirect growth impact of declining prices of IT equipment, 1991–2006

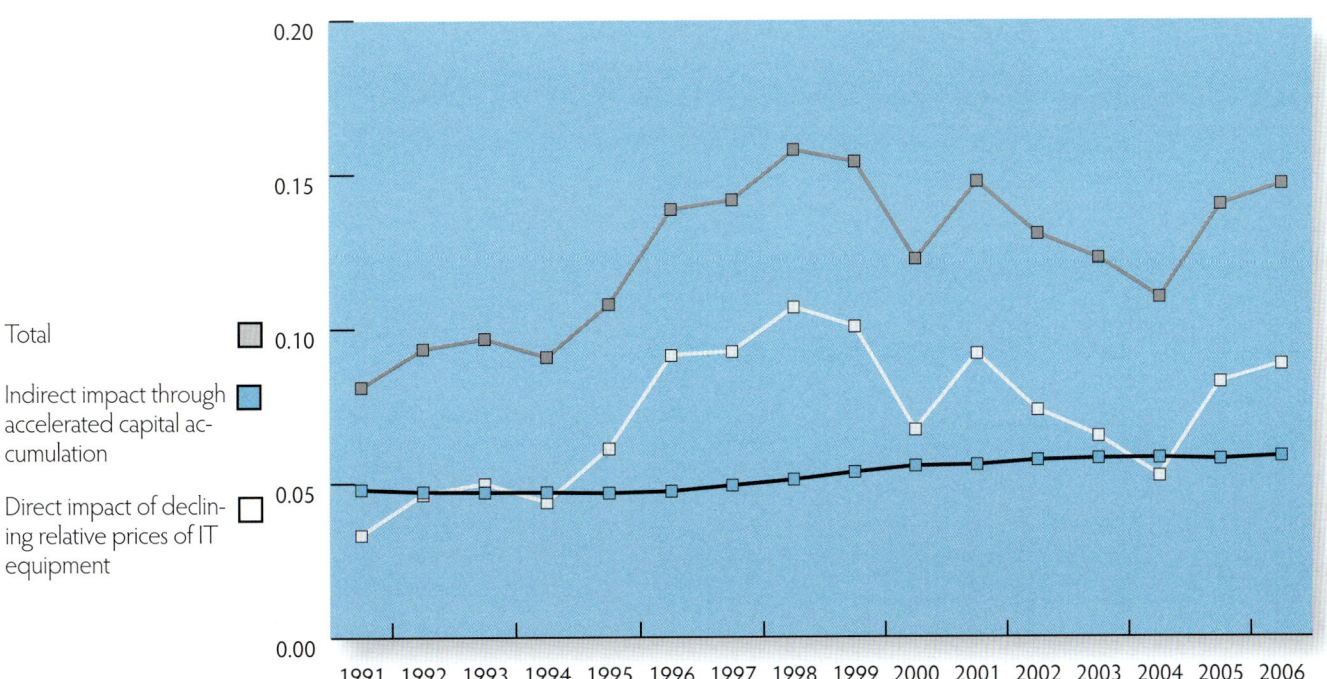

Source: Author's calculations, as described in text.

Figure 8.4. Direct and indirect growth impact of declining prices of communications equipment, 1991–2006

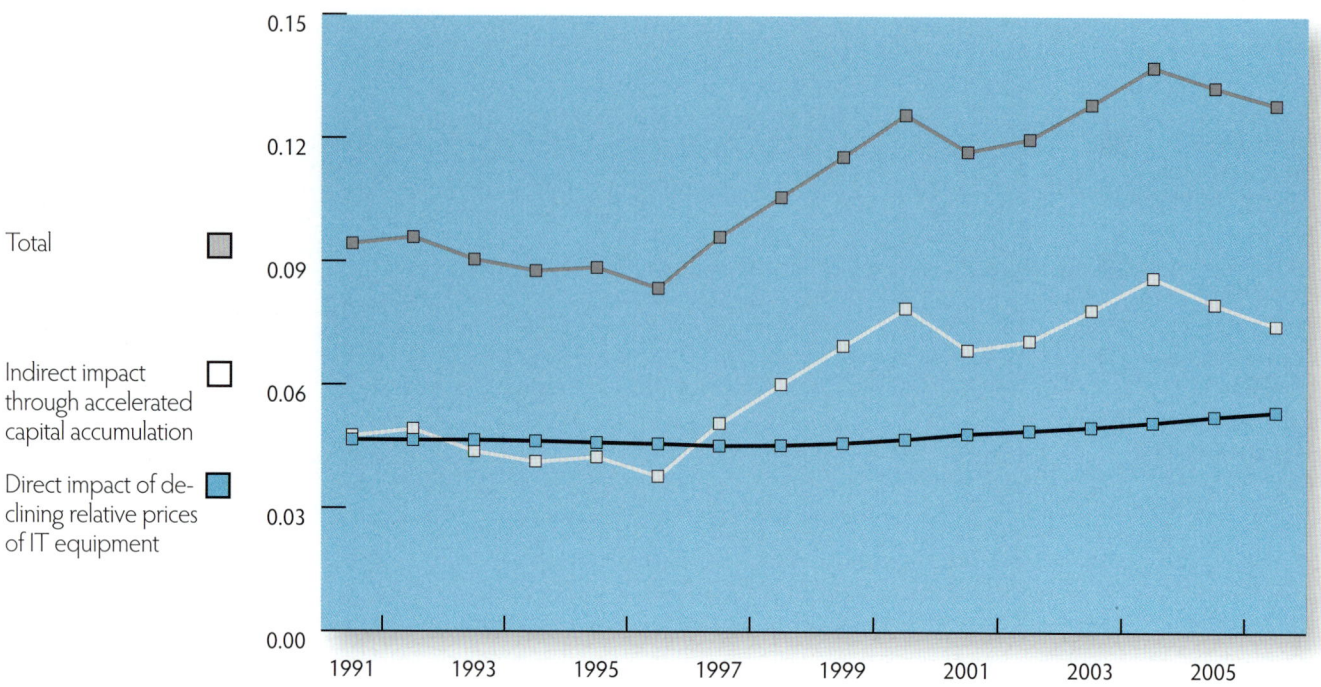

Source: Author's calculations, as described in text.

Figure 8.5. Contribution of declining prices of ICT equipment to growth in low-income countries, 1991–2006

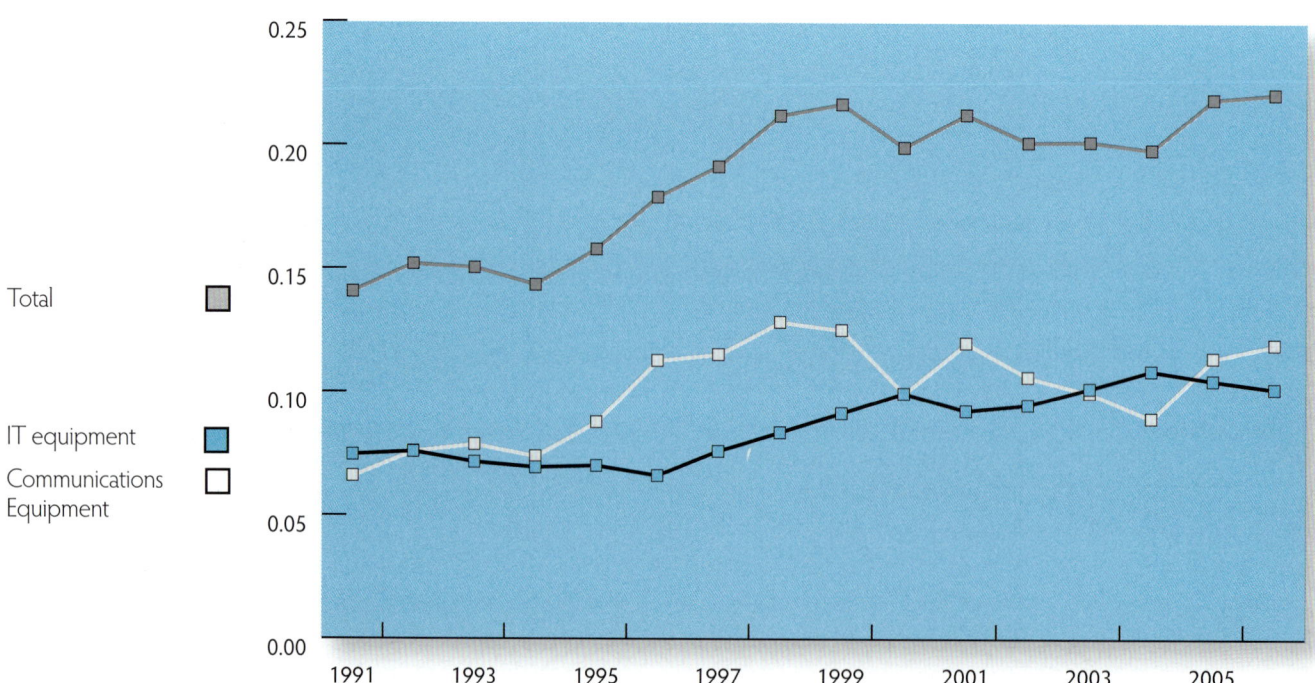

Source: Author's calculations, as described in text.

Figure 8.6. Contribution of declining prices of ICT equipment to growth in lower-middle-income countries, 1991–2006

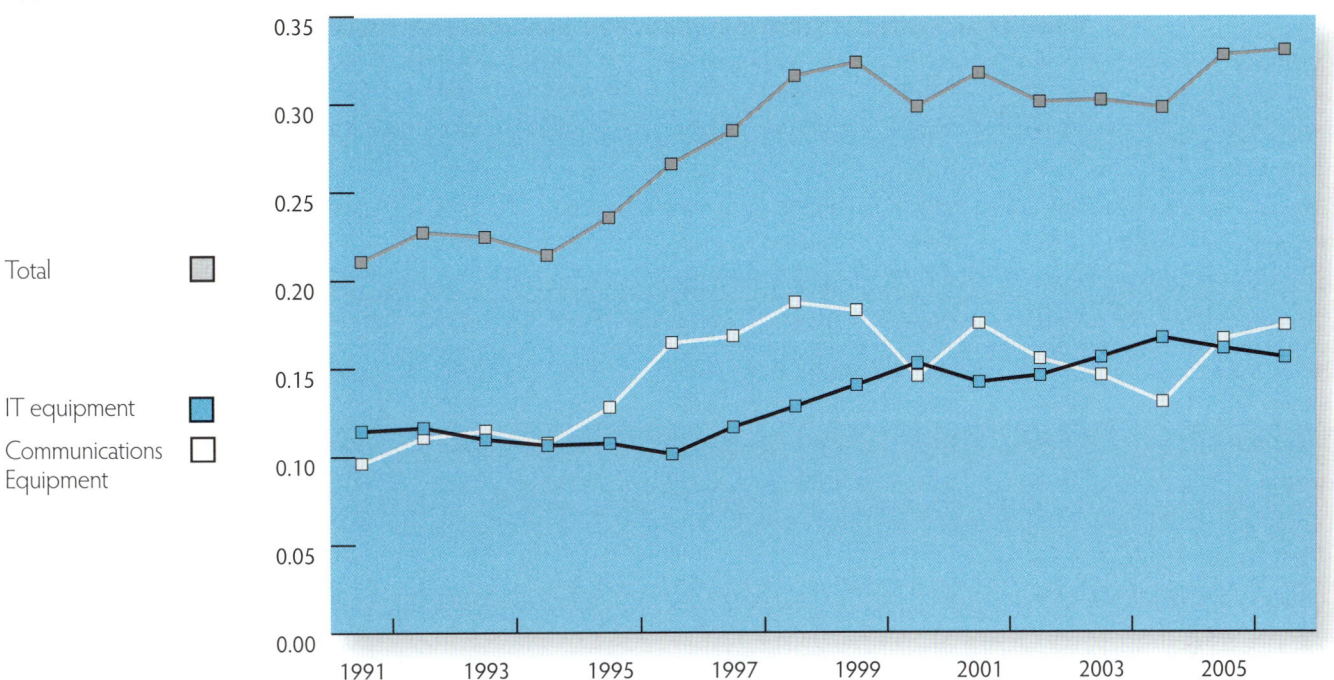

Source: Author's calculations, as described in text.

path following a shock—provides a more differentiated picture, as the dampening impact on growth of the slowdown in the rate of decline of relative prices is partly offset by a gradual increase in the induced effects through capital accumulation.

As regards the role of the direct impact of declining relative prices of equipment and the indirect effects through an induced acceleration in capital accumulation, we find that the magnitude of the direct and indirect effects is similar; i.e., about half of the growth effect of a drop in the prices of ICT equipment is realized immediately, while the other half occurs gradually over subsequent years. Almost all of the variations in the growth impact of falling prices of ICT equipment on a year-to-year basis reflect the direct effects of changing prices. However, changes in the indirect impact play some role over longer time horizons, and contribute about one-sixth to the acceleration of the growth impact of advances in ICTs between 1990 and 2006.

Figures 8.5, 8.6 and Table 1 summarize the overall growth effects of falling relative prices of ICT equipment for low- and lower-middle-income countries. Overall, the impact is about

one-third smaller in low-income countries, compared to lower-middle-income countries. To gain an understanding of the magnitude of the growth impact from ICT-related capital deepening in developing countries, it is also useful to compare it to existing estimates for the impact of advances in ICTs in the most advanced economies. In doing so, it is important to compare our estimates with those of ICT-related capital deepening only—as many studies of the growth impact of ICTs in developed countries also include contributions from the production of ICT equipment—and with further productivity gains attributed to broader economic transformations enabled by ICTs.[11] The most direct match to our data are the estimates quoted by Oliner, Sichel, and Stiroh (2007) and by Inklaar, Timmer, and van Ark (2008). Based on these, the growth impact of ICT-related capital deepening in developing countries is about one-half of the growth impact in the most advanced industrialized countries.

[11] The latter would need to be based on a complete growth accounting framework which we do not use, owing to lack of data on labor inputs.

Table 1. Impact of ICT-related capital deepening on growth in selected countries, 1991–2006

Country	IT equipment			Communications equipment			Total ICT equipment		
	1991–1995	1996–2000	2001–2006	1991–1995	1996–2000	2001–2006	1991–1995	1996–2000	2001–2006
Bangladesh	0.04	0.06	0.06	0.03	0.04	0.05	0.07	0.10	0.10
China, Mainland	0.14	0.22	0.21	n.a.	n.a.	n.a.	n.a.	n.a.	n.a.
Egypt	0.03	0.04	0.04	0.07	0.08	0.09	0.09	0.12	0.13
Ethiopia	0.06	0.09	0.09	0.06	0.07	0.08	0.12	0.16	0.16
India	0.08	0.12	0.11	0.10	0.11	0.14	0.18	0.23	0.25
Indonesia	0.05	0.08	0.08	0.14	0.16	0.19	0.19	0.24	0.27
Nigeria	0.05	0.07	0.07	0.06	0.07	0.09	0.11	0.14	0.15
Pakistan	0.05	0.08	0.07	0.05	0.06	0.07	0.10	0.13	0.14
Philippines	0.09	0.14	0.13	0.21	0.24	0.29	0.30	0.39	0.43
Vietnam	0.09	0.14	0.13	0.14	0.16	0.20	0.23	0.30	0.33
Country groups (unweighted averages)									
All countries covered	0.09	0.14	0.13	0.09	0.11	0.13	0.19	0.25	0.26
Low-income countries	0.08	0.12	0.11	0.07	0.08	0.10	0.15	0.20	0.21
Lower middle-income-countries	0.11	0.17	0.16	0.11	0.13	0.15	0.22	0.30	0.31

Note: Data on the (substantial) production of communications equipment in China were unavailable, so we could not construct estimates of investment in communications equipment.

Source: Author's calculations.

Conclusions

Motivated by the apparent role of ICTs in developing countries, we set out to quantify the growth impact of technological advances in ICTs across the developing world. To this end, we analyzed existing data on the production of ICT equipment and built a dataset covering the absorption of ICT equipment in essentially all developing economies.

Our analysis suggests that the direct growth impact arising from technological advances in the production of ICT equipment plays a subordinate role in the developing world. We were able to identify only four countries where productivity gains in the ICT-producing sector contributed more than 0.4 percentage points to growth in the 2001–05 period. Moreover, as these countries export most of their ICT-related output, they face a negative terms-of-trade shock as the prices of ICT equipment decline, and these growth increments do not buy much in terms of increased consumption possibilities.

However, advances in ICTs do affect economic growth across the developing countries as lower prices of ICT equip-ment result in ICT-related capital deepening. Across the countries covered, the growth impact of ICTs has increased from 0.19 percent annually in 1990–05 to 0.26 percent annually in 2000–06. Reflecting higher rates of ICT-related investment, the growth impact is stronger in low-middle-income countries—rising from 0.22 percent annually in 1990–05 to 0.31 percent annually in 2000–06—than in low-income countries, where it accounted for 0.15 percent annually in 1990–05, rising to 0.21 percent annually in 2000–06.

Regarding the sources of these growth increments, we see that they are split more or less evenly between capital deepening related to IT equipment (computer hardware, etc.) and to communications equipment (switching, transmission, and user equipment). While investment in communications equipment has been roughly twice as high as investment in IT equipment, the rate of technological progress in IT equipment has been higher.

References

van Ark, Bart. 2001. "The Renewal of the Old Economy: An International Comparative Perspective." STI Working Paper No. 2001/5. Paris: OECD.

van Ark, Bart, Mary O'Mahony, and Marcel P. Timmer. 2008. "The Productivity Gap between Europe and the United States: Trends and Causes." *Journal of Economic Perspectives* 22(1):25–44.

Bayoumi, Tamim, and Markus Haacker. 2002. "It's Not What You Make, It's How You Use IT: Measuring the Welfare Benefits of the IT Revolution Across Countries." IMF Working Paper No. 02/117. Washington DC: International Monetary Fund.

Colecchia, Alessandra, and Paul Schreyer. 2002. "The Contribution of Information and Communication Technologies to Economic Growth in Nine OECD Countries." OECD *Economic Studies* 34:153–71.

Doms, Mark. 2005. "Communications Equipment: What Happened to Prices?" in Corrado, Carol, John Haltiwanger, and Daniel Sichel (eds.), *Measuring Capital in the New Economy: Studies in Income and Wealth*, Vol. 65. Chicago: University of Chicago Press.

Global Insight. 2006. *Global IT Navigator Database*. As of June 2006. Boston MA: Global Insight.

Inklaar, Robert, Marcel P. Timmer, and Bart van Ark. 2008. "Mind the Gap! International Comparisons of Productivity in Services and Goods Production." *German Economic Review* 8(2):281–307.

International Monetary Fund (IMF). 2009. *World Economic Outlook Database*. April 2009 edition. (Washington DC: IMF.

Jorgenson, Dale W. 2001. "Information Technology and the US Economy." *American Economic Review* 91(1):1–32.

Jorgenson, Dale W. and Kevin J. Stiroh. 2000. "Raising the Speed Limit: US Economic Growth in the Information Age." *Brookings Papers on Economic Activity*, 2000, No. 1:125–235.

Jorgenson, Dale W. and Khuong Vu. 2007. "Information Technology and the World Growth Resurgence." *German Economic Review* 8(2):125–45.

Jorgenson, Dale W., Mun S. Ho, and Kevin J. Stiroh. 2008. "A Retrospective Look at the US Productivity Growth Resurgence." *Journal of Economic Perspectives* 22(1):3–24.

Oliner, Stephen D., Daniel E. Sichel. 2000. "The Resurgence of Growth in the Late 1990s: Is Information Technology the Story?" *Journal of Economic Perspectives* 14(4):3–22.

Oliner, Stephen D. and Kevin J. Stiroh. 2007. "Explaining a Productivity Decade." *Brookings Papers on Economic Activity*, 2007, No. 1:81–137.

Senhadji, Abdelhak. 2000. "Sources of Economic Growth: An Extensive Growth Accounting Exercise." *IMF Staff Papers* 47(1):129–57.

United Nations Industrial Development Organization (UNIDO). 2007. *Industrial Statistics Database – 4-Digit Level of ISIC Code (Revision 2 and 3)*. Vienna: UNIDO.

United Nations Statistics Division. 2009. Commodity Trade Statistics Database (available online at http://comtrade.un.org/db As obtained 31 July 2009.

United States Department of Commerce, Bureau of Economic Analysis. 2008. Producer Price Indices. At: http://www.bls.gov/ppi/home.htm As obtained 15 June 2008. Washington DC: US Department of Commerce.

United States Department of Labor, Bureau of Labor Statistics. 2008. *National Income and Product Accounts*. At: http://www.bea.gov/national/nipaweb/Index.asp As obtained 15 June 2008. Washington DC: US Department of Labor.

Appendix: Analytical framework

Compared to the available literature on the growth impact of ICTs in major industrialized countries, the principal challenge to estimating the growth impact of ICTs in developing economies is the limited availability of data. As most developing countries do not produce ICT equipment, output data are available for only a few countries and do not clearly identify intermediate inputs. Investment data disaggregated enough for our purposes—at a minimum identifying ICT-related investment—are generally unavailable. Sectoral output data may exist, but the classification of sectors is not consistent across countries, and it is not clear whether the classification of sectors as ICT-intensive adopted in industrialized countries is relevant in the context of developing countries. Moreover, in the absence of sector-specific input data, our data only allow us to follow output growth, rather than productivity growth, across sectors. As the tertiary sector (including many sub-sectors classified as IT-intensive in the "ICT and growth" literature) generally expands in the course of economic development, higher growth rates in this sector cannot be attributed to the impact of ICTs. Another major constraint is labor market data, which are unavailable or unreliable in many developing countries.

As regards the growth impact of ICT production, we have adopted a crude approach, applying the rate of price decline for the respective category of ICT equipment to the production figures reported by UNIDO (2007). These estimates are subject to a wide margin of error, as we do not control for intermediate inputs (including ICT products) which might be imported. In addition to the limitations imposed by the lack of consistent cross-country data, there is another main reason why we focus on ICT-related capital deepening rather than ICT production. Producers of ICT equipment face falling prices for their ICT-related products, and literally cannot buy anything based on the improved productivity in the ICT sector. To be specific, because we use the decline in prices to measure the rate of technological progress, the impact of productivity gains is offset by a negative terms-of-trade effect one-to-one. Thus, there is a zero impact on living standards of increased GDP growth owing to productivity gains in the ICT-producing sector.[1]

As for the growth impact of ICT-related capital deepening, our analysis distinguishes between ICT and non-ICT capital, and focuses on the growth impact of ICT-related capital deepening. The production function of an economy can be summarized by

$$Y = AK_{non\text{-}ICT}^{\alpha_1} K_{ICT}^{\alpha_2} L^{1-\alpha_1-\alpha_2} \qquad (1)$$

and the growth impact of ICT-related capital deepening is equal to $\alpha_2 \cdot \dfrac{dK_{ICT}}{K_{ICT}}$, i.e., the growth rate of the capital stock of ICT equipment $\dfrac{dK_{ICT}}{K_{ICT}}$ weighted by the parameter α_2 (the elasticity of output with respect to ICT capital).

To obtain estimates of the growth rate of ICT capital, we follow the approach adopted by Jorgenson et al. (2008) in constructing the capital stocks from investment data and estimates of depreciation rates for the respective type of equipment. To this end, we exploit the fact that most developing countries do not produce ICT equipment. Our estimates of ICT-related investment are therefore based on the data on net imports of ICT equipment (discussed above), based on trade data from the United Nations Commodity Trade Database (see United Nations Statistics Division, 2008). To obtain data on ICT-related investment, these data need to be adapted in two directions: first, for the small number of developing countries producing ICT equipment (documented in UNIDO, 2007), domestic production is added to net imports to obtain an estimate of domestic absorption of ICT equipment. Second, domestic investment data include certain import costs, margins, and taxes. To obtain estimates of ICT investment compatible with national accounts data, we therefore apply a mark-up, which we obtain based on a comparison between domestic sales data (e.g., where available from Global Insight, 2006) and import data.

The second factor we need to take into account is the rate of depreciation of ICT equipment. In light of the rapid rate of technical progress, prices of ICT equipment (especially of IT hardware like computers) decline rapidly. Existing ICT equipment thus loses value and, correspondingly, the contributions (at lower prices) of new ICT investment in the ICT capital stock are much larger than nominal investment data would suggest at first sight. As ICT equipment is highly tradable, we follow international practice (as the cross-country studies noted above) and use the US price indices for IT and communications equipment, respectively. Moreover, the physical rate of depreciation of IT equipment is higher than

---

[1] This point is also discussed in some more detail in Bayoumi and Haacker (2002).

for other forms of capital. Our assumptions of a depreciation rate of 31 percent for IT equipment and of 11 percent for communications equipment are informed by Jorgenson and Stiroh (2000).

Finally, we need to estimate the parameter α_2. Our estimates are informed by cross-country estimates of the overall share of capital (i.e., $\alpha_1 + \alpha_2$; see Senhadji, 2000), and country level investment data, using the data on ICT investment described above and data on overall investment from IMF (2008). If the rates of both economic and physical depreciation of the different types of capital were the same, the parameters α_1 and α_2 would be proportional to the investment rates for the respective capital goods. In the present case, the parameter α_2 (pertaining to ICT equipment) is somewhat higher than that suggested by the investment share, as ICT-related investment is discouraged by the high rate of depreciation.

In this framework, the growth impact of a drop in the price of ICT equipment occurs through two channels: first, the decline in price implies that the rate of return to ICT capital increases, and investors reallocate investment to ICT equipment until the rates of return to the different types of capital are equalized. This results in an immediate increase in output savings. Subsequently, as savings exceed depreciation, output increases gradually, owing to increased capital accumulation, until the economy achieves a new steady-state equilibrium.

Chapter 2.7

Good Governance for Sustained Growth and Development

Daniel Kaufmann,
The Brookings Institution[1]

Introduction

The importance of governance and institutions for economic growth and development has been debated for many years. Even today the international community is ambivalent about it, less due to the weight of evidence than for geopolitical reasons.

Until fairly recently, core aspects of governance were taboo for international financial institutions (IFIs). In some, even the spelling out of the word corruption was banned from official documents. For about a decade, beginning in the mid-1990s, there was a shift at donor agencies such as the World Bank, and a gradual embrace of the challenge of governance by many other development organizations. In recent years there has been a reversal, as the focus of resistance to governance reforms in powerful countries has grown, causing a shift of priorities among some aid agencies, and resulting in a downgrade in the importance of governance and anti-corruption in international institutions such as the World Bank and the IMF.

Nonetheless, it is possible nowadays to discuss and debate the reality of governance worldwide in a more open and transparent fashion than was possible decades ago. And it is also possible to witness concrete steps being taken in reformist countries, drawing from evidence-driven policymaking.

The topic of governance is complex and multidisciplinary in nature. The initial lessons learned have not been conclusive and continue to be debated and open to a variety of interpretations. Thus, it is vital that we question and advance our understanding of this field by utilizing the results of empirical research, which often reveals evidence at odds with long and popularly held beliefs.

After presenting some initial definitions of governance as a subject of study, this chapter explores the evidence, based in large measure on research carried out since the late 1990s when I was at the World Bank. Some of the findings presented here are therefore not new. Yet they are as relevant now as they were at that time. We then describe the importance of governance to development and relate how it is linked to the performance of various countries over the recent decade 1998 to 2008. This leads to a discussion of the link between good governance, transparency, and the control of corruption and growth, development and security, introducing the notions of legal and illegal corruption and "soft capture."

167

[1] Currently Senior Fellow, The Brookings Institution; previously Director at the World Bank Institute; however, the views here are the author's own and do not reflect institutional positions. The research reported in this chapter draws from longstanding collaboration with various co-authors, particularly Aart Kraay, Massimo Mastruzzi, and Joel Hellman. The drafting of this Chapter benefitted invaluably from the collaboration of Augusto López Claros, who through the years contributed substantively to my own research and writings, and from Yasmina Mata. Without their input and initiative, this chapter would not have been a reality. Responsibility for errors remains mine.

The implications of these governance issues for foreign aid, and the importance of freedom of the expression, transparency, and political participation are explored in the concluding section, where some strategies for reform are suggested.

Governance defined

Governance can be defined as the set of traditions and institutions, formal and informal, which determine how authority is exercised in a country for the common good. This set of institutions encompasses three main areas: political, economic, and institutional. The first area includes the process of selecting, monitoring, and replacing governments; the second the capacity to formulate and implement sound policies and deliver public services; the third concerns the respect of citizens and the state for the institutions that govern economic and social interactions among them.

In order to more closely understand, measure, and analyze governance, Aart Kraay and I embarked over a decade ago at the World Bank on a project to construct Governance Indicators, departing from the simple definition suggested above. We collected and studied scores of datasets from dozens of organizations around the world, and proposed a strategy to organize and make sense from such disparate data to build a set of composite governance indices. They cover well over 200 countries, based on more than 350 variables, gathered from over 30 institutions worldwide.[2]

Specifically, the Worldwide Governance Indicators (WGI) includes six key dimensions of institutional quality or governance, and measure, through two indicators each, the political, economic and institutional dimensions of governance described above. The following six dimensions are measured:

1. *Voice and accountability* – measuring political, civil and human rights
2. *Political instability and violence* – measuring the likelihood of violent threats to, or changes in, government, including terrorism
3. *Government effectiveness* – measuring the competence of the bureaucracy and the quality of public service delivery
4. *Regulatory burden* – measuring the incidence of market-unfriendly policies
5. *Rule of law* – measuring the quality of contract enforcement, the police, and the courts, as well as the likelihood of crime and violence
6. *Control of corruption* – measuring the exercise of public power for private gain, including both petty and grand corruption, and state capture

Figure 1. WGI Control of corruption, 2008: selected countries

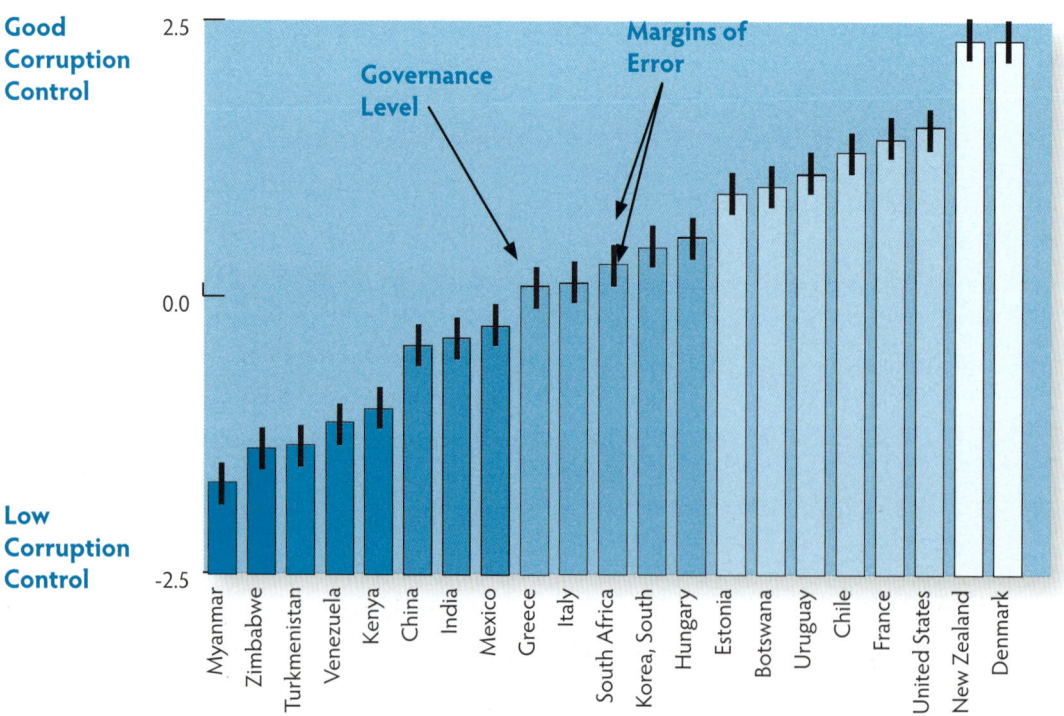

Source: Kaufmann et al., 2008.

---

[2]　See Kaufmann et al., 2008.

In Figure 1 we depict one of the six governance dimensions, namely control of corruption, for the most recent data available, namely 2008. In the Figure we see a subset of the world sample, and both the point estimate as well as the margins of error are shown, in order to emphasize that any measure of governance or investment climate is subject to a non-trivial margin of error.

Such margins of error imply that caution must be exercised in making comparisons across countries (and over time): small differences are unlikely to be significant, and thus the scores of countries close to each other in ranks or scores are likely to be locked in a 'statistical tie.'

At the same time, given that the margins of error are not very large (as seen in Figure 1), it is possible to meaningfully distinguish between different clusters of countries: those exhibiting good governance (eight at right), middling governance (five in center), subpar governance (China, India, and Mexico), and governance crisis (five at left).

Governance matters

It is therefore evident that, in contrast with past practice, there are now substantial efforts being made to measure governance. The WGI we have been constructing for over a dozen years is one such effort. There are also other important efforts in the field of composite indexes, such as the Transparency International (TI) *Corruption Perceptions Index* (CPI), for one dimension of governance, corruption), the Ibrahim Index of African Governance (for one region), among others.

Why are these indexes useful? Does governance really matter sufficiently to justify the significant efforts at measurement, index construction, and analysis?

Reformers in many governments, as well as investors, civil society leaders, and the international aid community increasingly view governance as being key to development, and to improving the investment climate; hence, its vital importance for innovation.

This, in turn, has increased the demand for monitoring the quality of governance in a country over time. Furthermore, some aid donors have come to the view that aid flows have a stronger impact on development in countries with good institutional quality.

The data is very clear about the vast differences in the level of governance performance across countries. In Figure 2, we depict the WGI governance performance levels in 2008, for

Figure 2. Governance around the world: A comparative perspective

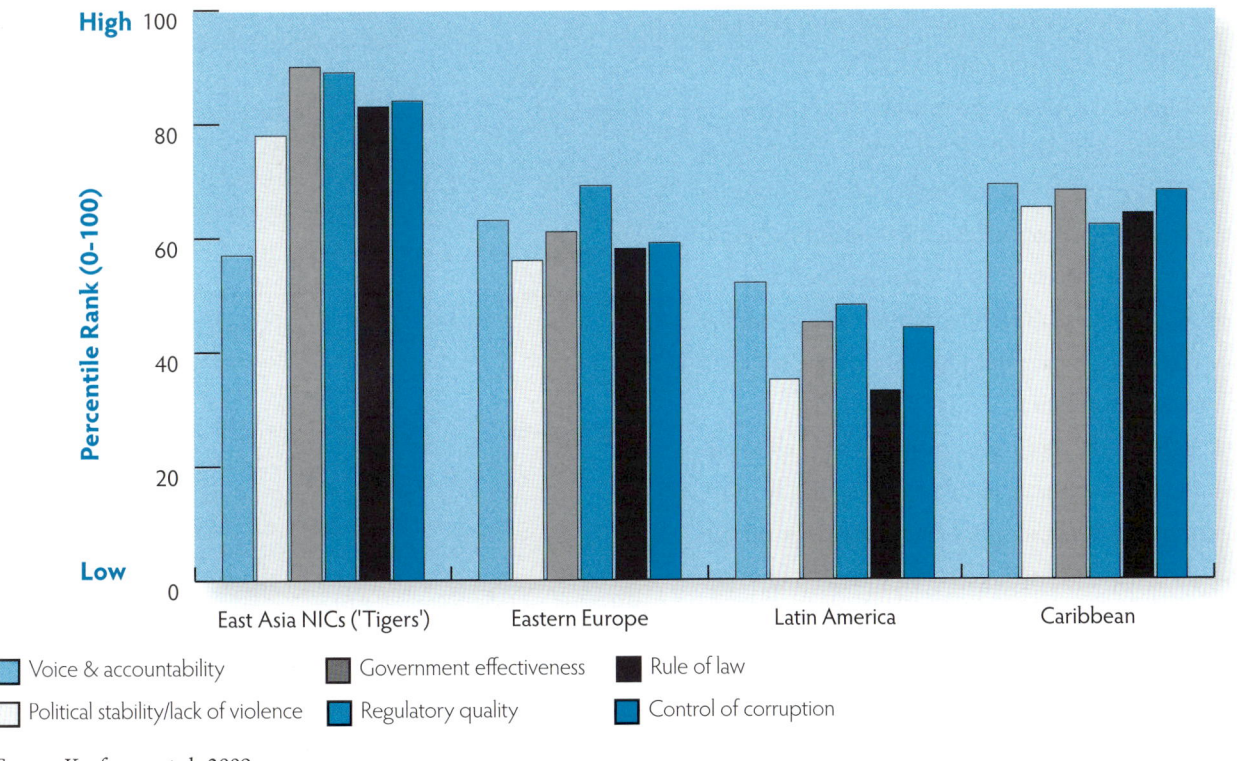

Source: Kaufmann et al., 2009.

each one of the six components, and for selected regions. Even at the general regional level—an average masking significant variation across countries within each region—there are marked differences, both comparing across regions, as well as across the six governance components (Figure 2).

On average, Latin America nowadays lags behind Eastern Europe in governance, and within Latin America, the challenges of violence and rule of law stand out in particular. In contrast, the East Asian "Tigers" are further ahead in general, although, relatively speaking, they still face serious challenges in voice and democratic accountability.

Further, the analysis of the individual data sources for the Governance Indicators shows that there is no convincing evidence of significant improvement in world-wide governance during the last decade.

Since 1998, we find that almost one-third of the countries in the world have experienced a significant change in some important dimension of governance (at a 75 percent confidence level). These changes tend to be evenly distributed between countries that exhibited an improvement and those that deteriorated in their governance. Such a balanced pattern of improvement and deterioration, coupled with the many countries that did not experience a significant change, explains the finding of a world-wide stagnation in governance on average.

As suggested, however, such a worldwide average masks substantial variation in governance performance across many countries. In Table 1, we can see which countries experienced important changes in the various governance dimensions.

Overall, these findings remind us that, while changes in institutional quality are often gradual, there are also countries which have achieved sharp improvements—or suffered rapid deterioration—over a decade or less. They also challenge the common perception that, while deterioration in a particular country *can* take place rather quickly, improvements are of necessity slow and incremental.

Challenging the "institutional pessimists," Table 1 shows an extensive list of countries that have improved markedly in selected dimensions of governance since the late 1990s. As we can see, this also challenges the "Afro-pessimists," since we can see in the same Table that there are a number of countries in Africa which have improved in a rather short period of time, even if it is still the case that other countries have not. As shown in Table 1, roughly as many countries in Africa show declines in these

Table 1. Significant changes in governance worldwide this past decade: 1998–2008*

Voice and accountability

Significantly worsened	Belarus, Côte d'Ivoire, Eritrea, Ethiopia, Fiji, Gabon, Iran, Laos, Morocco, Nepal, Philippines, Singapore, Solomon Islands, Thailand, Tunisia, United Arab Emirates, Venezuela, Yemen, Zimbabwe
Significantly improved	Afghanistan, Albania, Burundi, Chile, Congo, Croatia, Ghana, Hong Kong, Indonesia, Iraq, Kenya, Lesotho, Liberia, Macedonia, Niger, Nigeria, Peru, Serbia, Sierra Leone, Suriname, Tanzania, Turkey, Uganda, Zambia

Regulatory quality

Significantly worsened	Argentina, Bolivia, Côte d'Ivoire, Ecuador, Eritrea, Gabon, Guinea, Maldives, Myanmar, Sri Lanka, Togo, Uruguay, Venezuela, Zimbabwe
Significantly improved	Angola, Armenia, Azerbaijan, Belarus, Bosnia-Herzegovina, Bulgaria, Croatia, Democratic Republic of Congo, Georgia, Iraq, Japan, Liberia, Libya, Mauritius, Rwanda, Serbia, Slovak Republic, Tajikistan, Uzbekistan

Rule of law

Significantly worsened	Argentina, Bolivia, Chad, Côte d'Ivoire, Ecuador, Eritrea, Italy, Kyrgyzstan, Lebanon, Maldives, Mauritania, Nepal, Philippines, Thailand, Trinidad and Tobago, Venezuela, Zimbabwe
Significantly improved	Albania, Algeria, Estonia, Georgia, Hong Kong, Kiribati, Latvia, Liberia, Qatar, Rwanda, Serbia, St. Kitts and Nevis, St. Lucia, St. Vincent and the Grenadines, Tajikistan, Vanuatu

Control of corruption

Significantly worsened	Cote d' Ivoire, Egypt, Eritrea, Greece, Italy, Kuwait, Kyrgyzstan, Laos, Lebanon, Malaysia, Maldives, Mauritania, Morocco, Philippines, Sudan, Thailand, Zimbabwe
Significantly improved	Albania, Cape Verde, Colombia, Croatia, Democratic Republic of Congo, Estonia, Georgia, Hong Kong, Indonesia, Liberia, Rwanda, Serbia, Slovak Republic, St. Lucia, Tanzania, Ukraine, United Arab Emirates, Zambia

*Note: Based on the WGI aggregate indicators for 212 countries. Significant changes in this table do include those for which the statistical confidence level exceeds 75 percent. In the majority of cases, however, the confidence level exceeds 90 percent.

Source: Kaufmann, Kraay, and Mastruzzi, 2009.

particular governance dimensions as show improvements.

Specifically, there have been significant improvements since 1998 in *voice and accountability* in countries such as in Chile, Bosnia, Croatia, Serbia, Ghana, Indonesia, Sierra Leone, Slovak Republic, and Peru, while a significant deterioration has taken place in countries such as Côte d'Ivoire, Zimbabwe, Kyrgyzstan, Venezuela, Belarus, and Nepal.

Similarly, a deterioration in *rule of law* during that period has taken place in a number of countries, such as Ethiopia and Argentina, while significant improvements in *government effectiveness* have taken place in Bulgaria, among others.

In this context, it is telling that there are clusters of countries that have been improving, in comparison with others. For instance, there is some evidence of improved governance in a number of dimensions in some Caribbean countries, in contrast with much of Latin America. Particularly telling is the story of the post-socialist transition countries. Those transition countries, which in the mid-1990s were promised potential entry to the European Union—upon fulfilment of appropriate economic and institutional reforms, as embodied in the *acquis communautaire*[3]—have experienced a decade-long improved trend in governance, while many of the post-socialist CIS countries (which were not offered such a window of opportunity) stagnated.

The governance vs. wealth dilemma: Which one comes first?

How fundamental are good governance and the control of corruption for growth, development and security? In spite of the myriad contributions to the field by many authors, there are still serious unresolved questions and debates in the development community. These debates not only refer to the importance of governance and corruption in development. They also center around the willingness, capacity, and effectiveness of the international community to help countries improve in these areas.

The explosion of empirical research over the past decade, coupled with lessons from countries' own experience, have given us a more solid basis for judging many of the effects of governance on development, and the effectiveness—or lack thereof—of strategies to improve it.

At the most basic level, data from the governance indicators at first reveal a very high correlation between good governance and key development outcomes across countries.

Figure 3 shows in very simple terms the close link between governance—in this case utilizing rule of law as a proxy—and national income per capita.[4]

Yet these robust correlations in themselves represent a "weak" finding in terms of policy application. This is because such simple correlations do not shed light either on the direction of causality, or on whether an omitted ("third") correlated variable is the fundamental cause accounting for the effects on developmental outcomes.[5] Thus, we have three possible explanations for the strong positive correlation between incomes and governance: a) better governance exerts a powerful effect on per capita incomes; b) higher incomes lead to improvements in governance; c) there are other factors that make countries both richer and also better governed.

Some claim that the link between governance and income does not mean that better governance *boosts* incomes, but, rather, the reverse, that higher incomes automatically translate into better governance. That is, richer countries are better able to afford the costs associated with providing a competent

Figure 3. The development dividend from good governance

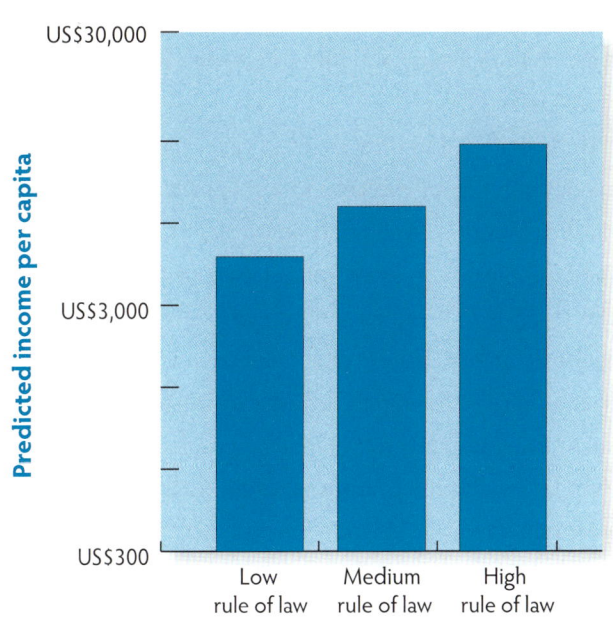

Note:
i)   Vertical axis is in log scale.
ii) Sources for calculations: WGI for the horizontal axis; Kaufmann et al., 2009; the vertical axis measures predicted GDP per capita on the basis of Instrumental Variable (IV) results for each of the three categories; estimations based on various authors' studies, including Kaufmann and Kraay, 2002.

Sources: Kaufmann et al., 2008; income per capita (in PPP terms) is from Heston et al., 2002; and CIA World Factbook, 2001.

---

3   The *acquis communautaire* is the term given to the entire body of European Union law, and the content, principles, and political objectives of the Treaties on which the Union is founded.
4   See Kaufmann, 2004a, p. 144; Kaufmann et al., 2003; Heston, et al., 2002; and CIA World Factbook (2001).

government bureaucracy, sound rule of law, and an environment in which corruption is not condoned.

However, our research does not support this claim. It is misleading to suggest that corruption is due to low income, and thus, to invent a rationale for discounting bad governance in poor countries. We implemented a novel methodology that enabled us to separate out the effects of per capita income on governance, and found evidence that this effect is certainly not positive, and, if anything, negative.

Of course, this does not mean that the simple correlation between governance and per capita income is negative, since this is dominated by the strong positive effects of governance on income. In terms of the specifics of the particular methodology used, we implemented an empirical framework which allowed for the identification of causal effects running in both directions between governance and per capita income.

Indeed we found a significant causal impact of improved governance on per capita income: the effects of improved governance on income in the long run are found to be very large, with an estimated 300 percent improvement in per capita income associated with an improvement in governance by one standard deviation, and similar improvements in reducing child mortality and illiteracy. Other researchers have found similar effects.

To illustrate: an improvement in rule of law by one standard deviation from the current levels in Ukraine to those "middling" levels prevailing in South Africa could lead to a three-fold increase in per capita income in the long run. A larger increase in the quality of rule of law (by two standard deviations) in the Ukraine (or in other countries in the former Soviet Union) to the much higher level in Slovenia or Spain would further multiply this income per capita increase.

Similar results emerge from other governance dimensions: a mere one standard deviation improvement in voice and accountability from the low level of Venezuela to that of South Korea, or in control of corruption from the low level of Indonesia to the middling level of Mexico, or from the level of Mexico to that of Costa Rica would also be associated with an estimated fourfold increase in per capita incomes, as well as similar improvements in reducing child mortality by 75 per-

cent and major gains in literacy.

Therefore, the evidence points to better governance as being the *cause* of higher economic growth and improved development—including, of course, a boosting of a country's capacity to more quickly improve levels of productivity and the ability to make better use of existing technologies and modern management processes. But, as we detail below, not the other way around: higher incomes in themselves do not automatically translate into improved governance.[6]

In fact, while there is a rapidly growing literature that shows the causal effects of better governance on higher per capita income, this is not the case for identifying causation in the opposite direction, from per capita income to governance. Traditionally, identification of the first direction of causality has been done with the aid of instrumental variables, such as the main language or settler mortality patterns, which we have utilized as instruments to arrive at the very large estimates of the effects of governance on income. Yet no good instruments exist for testing the reverse causality direction, namely, from per capita income to improved governance. Thus, we utilized a different technique to test whether there was a major effect of incomes on governance.

This finding of an absence of (or even possibly negative) feedback from per capita income to governance has two implications: first, a strategy of waiting for improvements to come automatically as countries become richer is unlikely to succeed. Second, in the absence of positive feedback from per capita income to governance, we are unlikely to observe virtuous circles in which better governance improves incomes, which, in turn, lead to further automatic improvements in governance.

Together, these two implications point to the fundamental importance of positive and sustained reform efforts to improve governance in countries where it is lacking. While these findings apply across the globe, it is a timely reminder now to countries like the United States in the aftermath of the Wall Street debacle—more on this later—as well as to international financial institutions, where the challenge of enhancing aid effectiveness, particularly in many African and South Asian countries, remains elusive.

The fact that good governance is not a "luxury good" to

---

[5]   The method for untangling the directions of causation underlying the strong correlations is explained in detail in Kaufmann and Kraay, 2002.

[6]   The gathering of a major governance dataset and the construction of the aggregate indicators themselves (through the particular Unobserved Component Model) give us important additional information: the margins of error for each country estimate for the Governance Indicators. These additional data permit us to implement a different, rarely used strategy to estimate the effect of incomes on governance: namely the utilization of non-sample information, known as the "out-of-sample" technique. Using this technique, we find no evidence of positive feedback from higher per capita income to better governance outcomes. See Kaufmann and Kraay, 2002.

which a country automatically graduates when it becomes wealthier means, in practical terms, that leaders, policymakers, and civil society need to work hard and continuously at improving governance within their countries.

Corruption in business

It is important to understand the possible reasons for the absence of positive feedback from per capita income to governance when designing strategies to improve governance and combat corruption. Contrary to conventional wisdom, the public sector is not the sole shaper of the investment climate faced by domestic firms and foreign investors in a country. Similarly, the private sector is not the passive recipient of the investment climate.

In reality, there is a complex interplay between corporate and public sector governance and policy-making, whereby powerful segments of the private sector also play a very important role in shaping key public policy, legislation, and regulations which constitute the rules of the game, and the business environment within which these corporations operate.

The worldwide empirical evidence from the enterprise sector itself is illustrative. Based on the survey of thousands of enterprises, the Executive Opinion Survey (EOS), conducted annually by the World Economic Forum, among others, it is possible to show the prevalence of different corrupt practices, as reported by the firms. As we see in Figure 4, contrary to popular belief, in recent times it is not the traditional forms of administrative bribery which are particularly prevalent, but instead corruption in the judiciary, procurement, and in state capture by powerful enterprises which have become more dominant. In this figure, drawn from survey micro-data, it is important to note a similar pattern which emerges, as compared with the aggregate governance indicators shown in Figure 2: Latin America, and the countries of the former Soviet Union (CIS) rate well below the new European countries (formerly Eastern Europe), and even further below the newly industrialized countries (NICs) of East Asia.

More generally, even the definitions and views as to what constitutes the investment climate tend to underestimate the importance of governance factors. Until very recently, the focus has been on a rather narrow and traditional set of factors comprising the investment climate, emphasizing economic, financial, and legal regulations by fiat, but divorced from the political dimensions of governance. In the literature, the treatment of the concept of the investment climate itself is not in tune with what the enterprises themselves report in surveys of what matters the most for their operations.

Figure 4. Unbundling Corruption: Executive Opinion Survey 2006

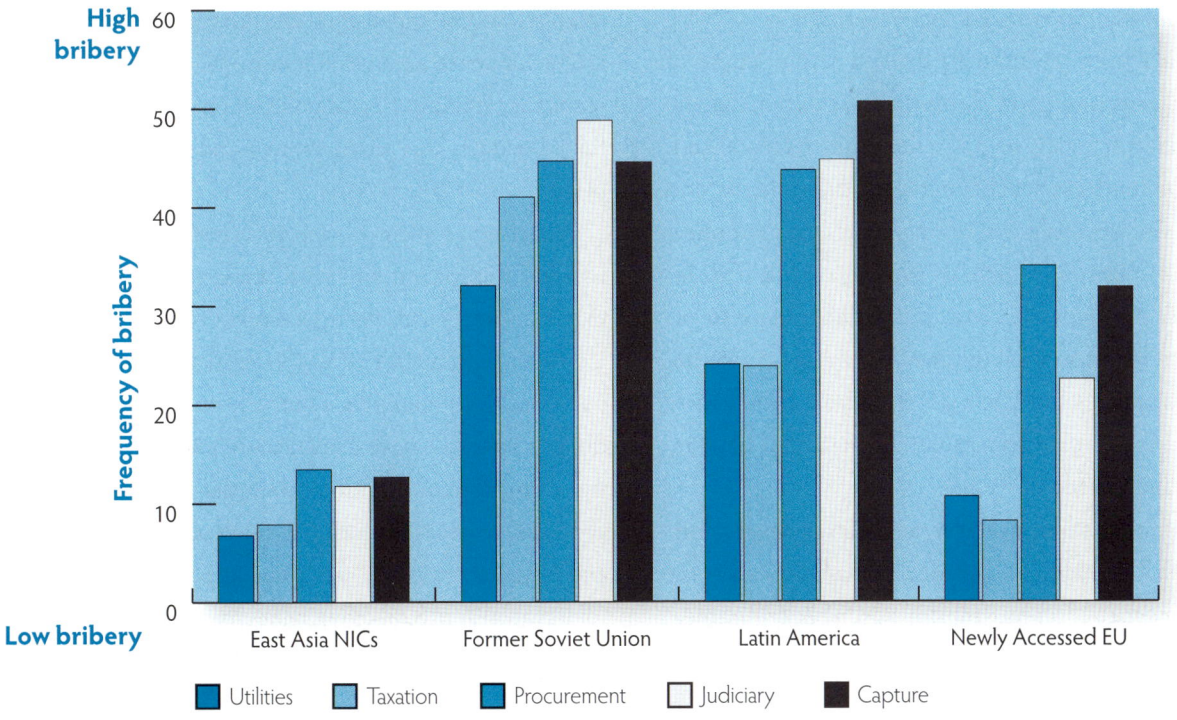

Source: EOS firm survey, 2006. Question: In your industry, how commonly do firms make undocumented extra payments or bribes connected with permits/utilities/taxation/awarding of public contracts/influencing laws and regulations/the judiciary? (Scale of responses from "very common" to "never occurs").

State capture

Based on empirical evidence, we have advanced an explanation for this negative feedback: the phenomenon of *state capture*, whereby powerful companies or elite individuals—"oligarchs" in extreme cases—bend the regulatory, policy, and legal institutions of a nation for their private benefit, thus exerting undue and illicit influence in *shaping* the laws, policies, and regulations of the state.[7] This is typically done through high-level bribery, lobbying, or influence peddling. We noted above, as depicted in Figure 4, its high prevalence in many regions of the world.

When institutions of the state are "captured" by vested interests in this way, or, more subtly, when powerful vested interests exert undue influence in shaping the rules of the game for their own benefit, entrenched elites in a country can benefit from a worsening status quo of misgovernance and can successfully resist demands for change *even as per capita income rises.*

In fact, in researching the impact of state capture on private sector development, with my colleague Joel Hellman we found that those post-socialist transition countries which largely avoided capture enabled their private sectors to grow their output and investment twice as fast as the firms in captured economies (as in other Eastern European countries, e.g., Russia). This is depicted in Figure 5.

Broadening the definition: Legal and illegal acts

Corruption has been traditionally defined as the "abuse of public office for private gain." Behind this definition lies the image of a predatory state, seen as a huge outstretched hand, extorting firms for the benefit of politicians, high officials, and bureaucrats. Conventional wisdom points to low corruption within OECD countries, which, on average, ranked at about the 90th percentile. However, corruption is not unique to developing countries, nor has it declined on average. Some developing countries, such as Chile and Botswana, exhibit *lower* levels of corruption than some fully industrialized nations, such as Italy and Greece.

Yet here again, there are averages, and it is a relative standing. Moreover, the measures captured in these indicators relate to the standard definition of corruption, namely the abuse of public office for private gain, focusing on the illegal nature of acts such as bribery, for the purpose of illicit private gain at the expense of the public.

There has historically been an epistemological and legalistic set of biases in the study of corruption, which, as hinted earlier, relies on a definition which, in light of current realities and problems, warrants an open challenge. Specifically, the traditional notion of "abuse of public office for private gain" has often been interpreted in a legal sense to mean committing an illegal act, and, more broadly, places exclusive focus on the public sector.

However, the reality of corruption in this context is twofold: first, it most often involves collusion between at least two parties, typically from the public *and* the private sector, for a corrupt act to take place; second, where the rules of the game, laws, and institutions have been shaped, at least in part, to benefit certain vested interests, some forms of corruption may be *legal* in some countries.

Even in strong states, such as in rich OECD countries, powerful conglomerates and influential firms can have significant influence in shaping regulatory policy. For instance, soft forms of political funding are legally permitted in some countries, through the creative use of legal loopholes. Such political funding may exert enormous influence in shaping institutions and policies benefiting the contributing private interests, at the expense of the broader public welfare. The undue influence exercised by some Wall Street firms over the regulatory framework and political process in the US powerfully illustrates this phenomenon, and was one of the causes of the global financial crisis.

A similar problem is seen in favoritism in procurement, where a transparent and level playing field for competition may be absent, without necessarily involving illegal bribery. Consider the situation in which legislative votes or executive decisions in sectoral policy-making—e.g., in telecommunications or energy—have been unduly influenced by either private campaign contributions to legislators, or by private favors provided to decision-makers.[8] Such subtle forms of capture and "legal corruption" exist: an expectation of a future job for a regulator in a lobbying firm, or a campaign contribution

---

[7] Hellman et al., 2003. See also more recent evidence in some Latin American countries, emerging from the governance and anticorruption diagnostics (GAC) of the World Bank Institute (WBI) at: http://www.worldbank.org/wbi/governance/capacitybuild/

[8] For example, the cost to society of bribing a bureaucrat to obtain a permit to operate a small firm pales in comparison with, say, a telecommunications conglomerate that corrupts a politician to shape the rules of the game so as to obtain a grant for monopolistic rights, or investment banks influencing the regulatory and oversight regime governing them. In such cases, corruption would be considered to have taken place, even if the act was not strictly illegal. As countries become industrialized, governance and corruption challenges do not disappear. They simply morph and become more sophisticated, whereas the transfer of briefcases stuffed with cash may be less frequent.

with strings attached. In many countries this may be legal, even if unethical. In industrialized nations undue influence is often legally exercised by powerful private interests, which in turn influence the nation's regulations, policies and laws. This has dire consequences, as we see in the various forms of corruption underlying the current global financial crisis that began in the U.S.

Moreover, the private/public sector governance challenge is not confined to the domestic players in a country. In spite of the fact that the OECD Anti-Bribery Convention came into force over five years ago, many multinational corporations headquartered in OECD countries still bribe abroad, at times affecting public policy, and more generally undermining public governance in emerging economies. While the OECD Convention deserves credit for some concrete progress—there is an increase in the number of investigations in a few OECD countries—such progress has been slow, uneven, and altogether absent in some countries.

Consequently, sharper focus on less traditional definitions of corruption is called for, akin to the simple notion of "privatization of public policy."[9] Such an alternative definition would focus on the key mediating "institution" committing the abuse of power, namely undue influence, by vested interests. It would also provide for neutrality in terms of the legality, or lack thereof, of the "corrupt" act itself.

Furthermore, such a definition would be neutral as regards the private or public nature of the sector players, thereby implicitly recognizing the important and activist role that some in the private sector also play. In particular, it would force us to also scrutinize the role of corporate ethics, in their legal and illegal dimensions, alongside the often-cited role of public sector ethics. Finally, within such a broad definition, responsibility resides with *both those who exert undue influence, and those who are unduly influenced.*

The governance challenge in the United States

As is clear from the above, the study of corruption must include acts that may be legal in the strict narrow sense, but where the rules of the game have been bent. Would this broader view of corruption result in different corruption ratings? Absolutely. For example, over the past few years, traditional measures of corruption, such as the *Corruption Perceptions Index* by Trans-

parency International, have placed the U.S. among the least corrupt nations in the world, ranking 18th out of 180 rated countries. In stark contrast, when, in 2004, I calculated an index of "legally corrupt" manifestations—measured through the extent of undue influence through political finance and powerful firms influencing politicians and policy making—the U.S. rated in the bottom half among the 104 countries surveyed. Countries such as the Netherlands, Norway, Denmark, and Finland exhibited low levels of "legal corruption" (ranking 1st through 4th, respectively). Yet the U.S. was rated 53rd, a few ranks below Italy, while Chile was ranked 18th. Also rated better than the U.S. were countries like Botswana, Colombia, and South Africa. Illustrating this, Figure 6, also based on the Executive Opinion Survey of the World Economic Forum, shows quantitatively how firms themselves see the level of corruption in their respective regions and countries. Of particular interest is the gap between the United Sates—where overt forms of bribery are not prevalent, but where "legal corruption" and soft capture is frequent—and the Nordic countries as well as the East Asian NICs. The political antecedents to the financial crisis experienced in the US were, in fact, suggested by this type of evidence, consistent with regulatory capture and, linked to it, the role of money in politics.

Figure 5. **State capture by powerful firms halves private sector growth**

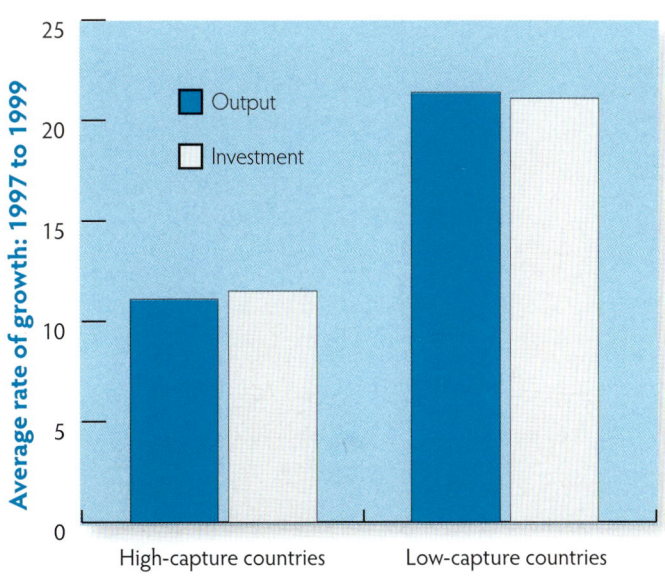

Source: Hellman et al., 2003.

---

[9] In more detail, meaning: unduly influencing public policy or the provision of a public good to the benefit of the 'influencer' agent, at the expense of public welfare.

Corruption and capture are important causes of the current global financial crisis. But it is also urgent to face up to the consequences of what is being dubbed the "new world order." There is a rapid—unprecedented in peacetime—expansion in the role and scope of government in "market economies." This new, overarching role of government, taking place in the U.S. and other large economies, is occurring at five levels:

- the public sector is reshaping regulation;
- governments are becoming owners of financial institutions;
- governments are bailing out selected private concerns through quick and massive infusions of funds;
- governments are providing a huge fiscal stimulus into infrastructure;
- governments are extending the social (and housing) safety net for millions of vulnerable citizens.

There are governance and corruption risks in each of these areas. Lobbyists are already at the door. These new risks are not exclusive to the U.S., but apply to other G-7 countries, including Russia and China, among others. With the U.S. in the lead, current global estimates of disbursed and planned bailout funds approach US$3 trillion, while cumulative global plans for fiscal stimulus approach US$2 trillion.

The bribery industry around the world

In order to give an approximation of the importance of corruption, one might pose the question: How large is the corruption "industry" worldwide? It is very difficult to obtain even a rough estimate of the size of the corruption industry, given its hidden nature, for corruption and bribery typically operate in the dark. This obscurity of corruption makes official estimates virtually impossible to obtain, not to mention unreliable. Nonetheless, thanks to the increasing availability of particular questions in enterprise and household surveys, which ask for quantitative estimates of bribery, it is possible, under certain conditions, to make calculations, and to extrapolate for the whole population.

In interpreting the results of this exercise, significant caution applies, given the margin of error in the data, the assumptions in the extrapolation exercise itself, and the fact that some forms of corruption are not quantified through this approach—e.g., budgetary leakages, or asset theft within the public sector. Bearing such serious caveats in mind, an estimate of the extent of annual worldwide transactions that are tainted by corruption comes close to US$1 trillion.[10]

Transparency

Partly because there is a higher comfort level with techno-

Figure 6. Bribery vs. "legal corruption" (and soft capture), 2004

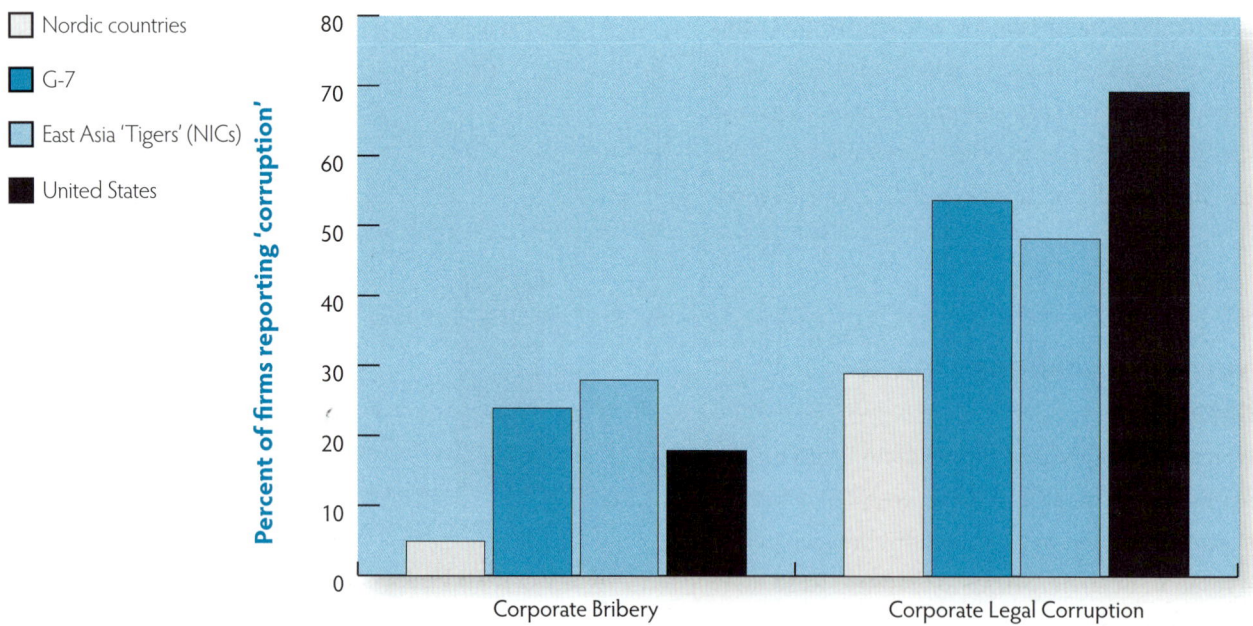

Source: Kaufmann, 2004b; author's calculations based on EOS Survey 2004.

[10] The margin of error of this estimate being obviously large, it may well be as low as US$600 billion; or, at the other end of the spectrum, it could well exceed US$1.5 trillion, or about 2.5 percent of world GDP.

cratic "fixes," traditional themes such as public sector management (including civil service reforms, codes of conduct, etc.) continue to be given significant prominence in the aid community. By contrast, transparency has been an underemphasized pillar of institutional reforms.

That there has been relatively little progress on the ground in this area is regrettable, in view of the influential conceptual contributions of a number of Nobel laureates, who have developed a framework linking the citizen's right to know and access to information with development outcomes.[11] Even popular lore alludes to the importance of transparency, as illustrated by the old adage "sunlight is the best disinfectant."

Yet not only does the implementation of transparency-related reforms remain chequered on the ground virtually everywhere, but, in contrast with other dimensions of governance, such as the rule of law, corruption, and the regulatory burden, there is a large gap between the extent of the conceptual contributions and the progress on its measurement and empirical analysis. There has been a particular paucity of literature on transparency which breaks down or unbundles transparency into its specific components, such that it becomes usable as policy advice and intervention.[12]

Thus, we made a modest contribution to the empirical understanding of various dimensions of transparency by undertaking construction of a transparency index for 194 countries, based on over 20 independent sources. Country ratings and their margins of error were generated, for an aggregate transparency index with two sub-components: economic/institutional transparency and political transparency. The results suggest enormous variation across countries in the extent of their transparency. In fact, a high level of transparency is not the exclusive domain of a particular region, or of rich countries, and there are transparency-related challenges in countries in each region, as illustrated in Figure 7.[13]

We find that transparency is associated with better socioeconomic and human development indicators, as well as with higher competitiveness and lower corruption. In presenting concrete policy initiatives, we suggest that much progress can be achieved without inordinate resources. In fact, transparency reforms are substantial net savers of public resources, and can obviate the necessity for excessive regulations or rules.

And transparency reforms need not remain abstractions at the level of rhetoric any longer.

In the next section, we provide some examples of concrete reforms, which some countries have taken selectively, and which many more could consider undertaking comprehensively.

Concrete transparency reforms

Since research shows clearly that transparency helps improve governance and reduce corruption—essential ingredients for better development and faster economic growth—the international community and individual countries must pay closer attention to this issue. Within a concerted, practical, and comprehensive pro-transparency strategy, a basic checklist of concrete reforms, which countries may use for self-assessment, a report-card of sorts, might include the following items:

- public disclosure of assets and incomes of candidates running for public office, public officials, politicians, legislators, judges, and their dependents;
- public disclosure of political campaign contributions by individuals and firms, and of campaign expenditures;
- public disclosure of all parliamentary votes, draft legislation, and parliamentary debates;
- effective implementation of conflict-of-interest laws, separating business, politics, legislation, and public service, and adoption of a law governing lobbying; publicly blacklisting of firms that have been shown to bribe in public procurement (as done by the World Bank); and a requirement to "publish what you pay" by multinationals working in extractive industries;
- effective implementation of freedom-of-information laws, with easy access for all to government information;
- freedom of the media (including the Internet);
- fiscal and public financial transparency of central and local budgets, adoption of the IMF's Reports on Standards and Codes framework of fiscal transparency, detailed government reporting of payments from multinationals in extractive industries, and open meetings involving the country's citizens;
- disclosure of actual ownership structure and financial status of domestic banks, and enhanced standards of financial

---

[11] See Stiglitz (1999) and Sen (1981).

[12] Our research attempted to partly fill these empirical and policy-related gaps. In Bellver and Kaufmann (2005), we reviewed the existing literature, and presented various definitions of transparency, with a view to providing an empirical framework of worldwide indicators on various dimensions of transparency. These initial empirical results are intended to help bring about concrete policy and institutional innovations related to transparency reforms.

[13] There is even significant variation in transparency within countries, such as differences in performance between the economic/institutional and political dimensions of transparency, or, related to this, differences in the way institutions within a country operate as regards transparency.

Figure 7. Transparenting transparency: Toward an index of overall country transparency, selected countries, 2004

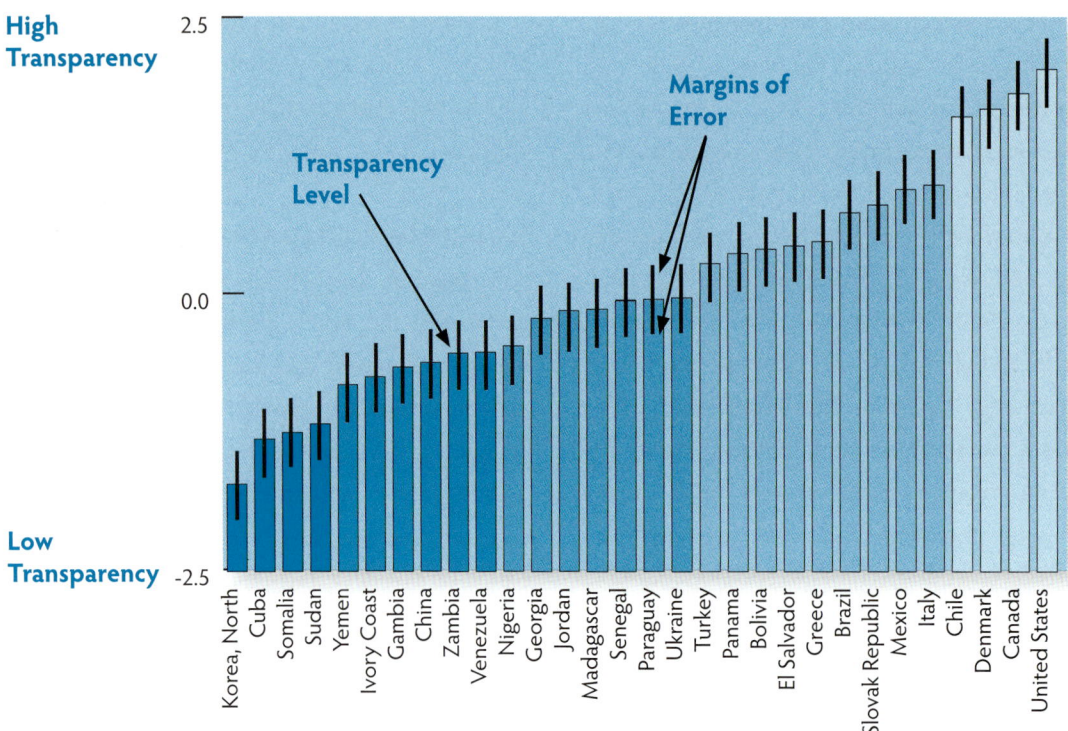

Note: Selected countries are presented for illustration, and due to margins of error, no precise ranking is warranted.

Source: Bellver and Kaufmann, 2005.

disclosure of financial firms;

- transparent (Web-based) competitive procurement;
- periodic implementation and publicizing of country governance, anti-corruption and public expenditure tracking surveys, such as those supported by the World Bank;
- Transparency programs at the city level, including budget disclosure and open meetings.

Of course, transparency reforms are not the only institutional reform priorities. IFIs and donors can complement these reforms by continuing to support traditional core competencies, helping with capacity-building, sharing knowledge, and focused reforms in key institutions in emerging economies, such as in the judiciary, customs, and tax and procurement. Further, at the municipal level, and in the context of decentralization, the donor community can also help to further institutional progress and anti-corruption in emerging economies.

These targeted reforms supporting highly vulnerable institutions would, however, have to be adapted to the specific country realities, and thus might vary considerably from country to country in their priority and in specific design.

In some countries, the first priority identified might be to support procurement reforms, strengthening accountability institutions in parliament, and freedom of the press; in others, it may be reforms in the judiciary, women's rights, and the revamping of customs. As we see in Figure 8, on average there is a strong link between freedom of the press, gender equality, and transparency, on the one hand, and corruption control, on the other. In-depth governance diagnostics at the country level are thus required first, to be empirically driven to inform policy making and empower reformists in and outside of government. Second, these diagnostics require working closely with experts and institutions within the country, which ought, itself, take the lead in such reforms, allowing donors to play an important, but supportive, role.

The role of foreign aid

Governance matters significantly for aid effectiveness. While some have challenged their findings, the widely known work by Burnside and Dollar[14] on assessing aid effectiveness shows, on the basis of cross-country aggregate data, that the quality

---

14 Burnside and Dollar, 1999.

of policies and institutions of the aid recipient country is critical. It is at least as revealing, however, to explore these links at the micro-economic level, focusing, for instance, on the effectiveness of investment projects, which show that institutions matter for project effectiveness.[15] Also, our calculations of World Bank-funded projects suggests that if there is high corruption in an aid-recipient country, the probability of project success, of institutional development impact, and of long-term sustainability of the investment, is much lower than in countries with better governance.

There is a notion that donor agencies can "ringfence" projects in highly corrupt countries and sectors, and, by such insulation or inoculation against corruption, somehow guarantee that it is efficiently implemented, and that the development objectives are attained, even where other projects fail. This is unrealistic. With the possible exception of some humanitarian aid projects, the notion that the aid community can fully insulate projects from a country's overall corrupt environment is not borne out by the evidence. The data suggest that when a systemic approach to governance, civil liberties, rule of law, and control of corruption is absent, the likelihood of an aid-funded project being successful is greatly reduced.

Some argue that there is not much that the international financial institutions (IFIs) can do to help countries improve governance and controlling corruption, even if a country is not viewed as facing a historical or culturally deterministic fate to stay with poor governance for many generations to come. Some development experts are sceptical about the ability of IFIs and donors to help countries improve their governance, either because of a conviction that the "macro" matters more—a mistaken belief in historical determinism—or, the more nuanced view, because the interventions needed to improve governance are politically sensitive and, therefore, very difficult for outsiders to encourage.

Indeed, countries themselves must shoulder responsibility and take the lead in implementing often difficult political and institutional reforms. There are areas that fall outside the mandate of IFIs, such as promotion of fair multi-party elections. But it is within the ability of IFIs and donors to do something about initiatives to encourage transparency, freedom of information, and an independent media, participatory anti-corruption programs led by the country, and gender equality—all of which have been underemphasized so far in the fight against corruption and in the quest for enhanced aid effectiveness.

Pathways for reform: Implications for innovation

Clearly, additional income flows alone will not improve governance. The findings described in this chapter emphasize the need to revisit conventional advice on strategies to improve public governance. In fact, having ignored the private-public governance nexus for a very long time, the international community has often erred in its emphasis on conventional public sector interventions as a key instrument to help countries improve governance, often drawing from standard templates from industrialized countries.

Simply put, traditional public-sector management interventions have not worked, because they have focused on technocratic organizational "fixes," often supported through technical assistance, the importation of hardware, organizational templates, and visits by "experts" from rich countries. Instead, further focus is needed on other aspects that have a political dimension.

Figure 8. Freedom of the press, women's rights, and transparency is associated with corruption control (emerging economy sample, 135 countries)

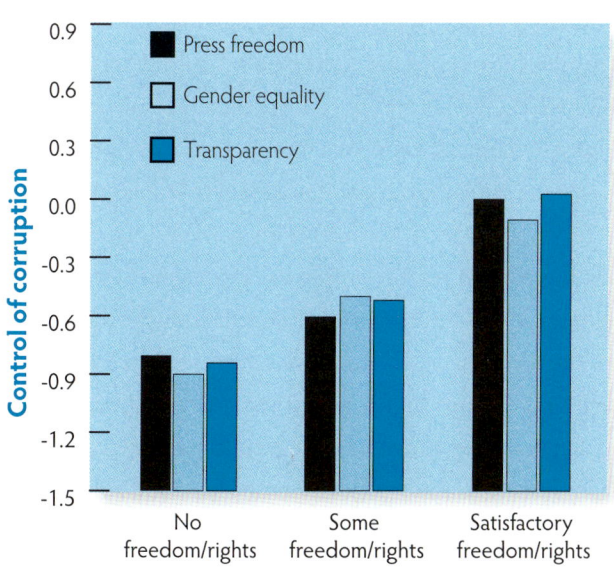

Sources: For press freedom: Freedom House; for gender equality: Country Policy and Institutional Assessment (CPIA), 2004; for control of corruption: Kaufmann et al., 2005, available at: http://www.worldbank.org/wbi/governance/govdata

---

[15] See Isham et al. (1997) and Dollar and Levin (2005).

In particular, addressing the interaction between corporate strategies and public governance (mediated by the "institution of influence") is of particular interest. Specifically, the findings on undue influence and state capture point to the limits of traditional public-sector measures (such as incessant drafting of new laws and ethics manuals, creating new anti-corruption agencies, or launching anti-corruption campaigns).

Moreover, in some settings, an instinctive tendency to over-regulate, which may take place in the throes of a corruption scandal, is not infrequent, and can be counter-productive. Excessive regulations not only do not address the more fundamental causes of corruption, but often create further opportunities for bribery. There is ample evidence suggesting that the burdens of bureaucracy and red tape strongly discourage the creation of small businesses and, hence, retard the emergence of an entrepreneurial class in the country. Although anti-corruption commissions, revised laws and awareness-raising campaigns have been put in place, they have had limited success. The focus on petty or administrative bribery has been misplaced at the expense of high-level political corruption.

To the extent that regulations are often self-imposed, a move to a system that considerably reduces the scope of mindless rules, aimed mainly at providing opportunities for corruption, may be one important element in moving the country to a higher growth plateau that will also begin to nurture a culture of innovation. Overall, these anti-corruption initiatives-by-fiat appear to have little impact, and often serve as politically expedient ways to react to the pressure to "do something" about corruption. Often, this results in neglect of more fundamental and systemic governance reforms.

At the same time, there is no doubt that an improved regulatory framework is needed for financial sectors, as illustrated by the current reform proposals, and incipient implementation in Washington, London, and Brussels in the aftermath of the financial crisis. This does not mean, however, that very tight regulations by fiat across the board will be optimal. A balance should be struck, in which transparency and disclosure feature prominently, and innovation and some measure of risk taking are not unduly discouraged.

Given the long list of interventions that have not worked, as well as the role often ascribed to historical and cultural factors in explaining governance, it is easy to fall into the pessimist camp. That would be a mistake. First, historical and cultural factors are far from deterministic—witness, for instance, the diverging governance paths of neighbouring countries in the southern cone of Latin America, the Korean peninsula, the transition economies of Eastern Europe, and in southern Africa.

Among these, one notable example refers to the rather divergent development paths adopted by Chile and Argentina over the past couple of decades. Chile has had Latin America's best growth performance, witnessing a tripling of GDP per capita between 1980 and 2008 and, since the return to democracy in 1989, it has exhibited one of the most dramatic reductions in poverty in the world. On the other hand, Argentina at present has no access to capital markets, having defaulted on its obligations to official creditors. A combination of extremely competent macroeconomic management and comprehensive institutional and structural reforms have increased Chile's presence in the global economy and sheltered its population during times of stress. Chile is the only country in 2009 whose sovereign credit rating has been upgraded, amidst a large number of downgrades, affecting sovereigns, banks, and corporations.

Second, there are strategies that offer particular promise. The coupling of progress on improving voice and participation—freedom of expression and gender mainstreaming—with transparency reforms can be particularly effective, as we saw earlier in Figure 8.

Unfortunately, progress in these areas of political and institutional governance, such as freedom of the press, gender equality, and transparency, has been mixed in many countries in the world. This disappointing reality highlights the pitfalls of focusing only on formalistic political changes. For instance, over the past 20 years, there has been a substantial increase in the number of electoral democracies across emerging economies, with dozens more countries joining the ranks of countries holding elections. However, improved formal polity has not always translated into improved freedoms for the press, increased citizen voice, or opportunities for women. For instance, out of the 119 countries which Freedom House classified as electoral democracies in 2008, 47 are in fact classified as *not* having a fully free press.[16]

The data for Africa is also telling. According to Freedom House, there has been significant progress in the area of political rights over the past two decades. Yet press freedoms, which

---

[16] Freedom House, available at: http://www.freedomhouse.org

it has been tracking since 1995, have not improved. There is evidence, in fact, that some deterioration may even have taken place in recent times in a number of African countries.

In sum, while in many countries in the world there has been progress in selected political rights areas, this has not always resulted in enhanced media freedoms, gender equality, or political and institutional transparency. And this matters a great deal, because where there is progress in these areas, progress can also be expected in corruption control. There is nothing deterministic about corruption, yet difficult political and systemic institutional reforms are often needed.

Conclusion

In this chapter, we have shown that broader definitions are needed to explain the evidence presented by research showing the decline or improvement in over 200 countries on a broad array of governance indicators. These indicators include voice and accountability, political instability and violence, government effectiveness, the regulatory burden, the rule of law, and control of corruption.

Empirical research over the past decade, coupled with lessons from countries' own experience, have given us a more solid basis for judging many of the effects of governance on development, and the effectiveness—or lack thereof—of strategies to improve it.

There is an important causal link between good governance and key development outcomes across countries, especially between various governance components (voice and democratic accountability, governance effectiveness, and control of corruption) and national income per capita. Our research disproves the common assumption that becoming rich is a precondition for a country to "afford good governance," to have a competent government bureaucracy, sound rule of law, and an environment in which corruption is not condoned. Contrary to popular belief, corruption is not the direct result of low income, and good governance is not a 'luxury good.' Indeed, the evidence points to better governance as being the *cause* of higher economic growth and improved development, and not the reverse.

Based on empirical evidence, we introduced the notions of "state capture" and "legal corruption" to explain why traditional definitions and views of the investment climate—usually focused on the public sector—have tended to under-

estimate the importance of governance factors and why they do not accurately reflect what enterprises themselves report as being of greatest significance for their operation.

When powerful companies or elite individuals bend the regulatory, policy, and legal institutions of a nation for their private benefit, thus exerting undue influence in shaping the laws, policies, and regulations of the state, they become a force for successfully resisting demands for change, even as per capita income rises. They thus pose a collective challenge for effective world governance, and render ineffective many ongoing efforts to combat corruption. Such a broader view of corruption, one which takes into consideration "legal" acts, would result in a different definition of corruption and would also significantly alter corruption ratings.

Finally, our research shows clearly the relationship between strategies for transparency, gender equality, freedom of expression, and public participation not only with better socio-economic and human development indicators, but with higher competitiveness and lower corruption. Reforms in such areas are not only *not* costly, but have proven to be net *savers* of public resources, obviating the necessity for excessive regulations or rules. The international aid community, which has recently lowered the priority accorded to governance and anti-corruption, and thus further jeopardized aid effectiveness, would be well advised to rethink its strategies and embrace more fully these good governance approaches. This would entail greater selectivity in the types of operations undertaken, and also in the choice of which governments receive large amounts of aid.

There is clear evidence that, contrary to the popular wisdom, additional income flows alone will not improve governance and that improved governance *results* in higher incomes, not the other way around. Countries must shoulder responsibility and take the lead in implementing often difficult political and institutional reforms.

References

Acemoglu, D., S. Johnson, and J.A. Robinson. 2001. "The Colonial Origins of Comparative Development: An Empirical Investigation." *American Economic Review* 91(5):1369–1401.

Bellver, A. and D. Kaufmann. 2005. "Transparenting Transparency: Initial Empirics and Policy Applications." World

Bank Policy Research Working Paper. Washington, D.C.

Burnside, C. and D. Dollar. 1999. "Aid, Policies, and Growth." Policy Research Working Paper Series 1777. World Bank.

CIA World Factbook, 2001.

Country Policy and Institutional Assessment (CPIA), 2004. Available at: http://go.worldbank.org/7NMQ1P0W10

Dollar, D. and V. Levin. 2005. "Sowing and Reaping: Institutional Quality and Project Outcomes in Developing Countries." Policy Research Working Paper Series 3524. World Bank.

Freedom House, 2005. Available at: http://www.freedomhouse.org

Hall, R. and C. Jones. 1999. "Why do Some Countries Produce so Much More Output per Worker than Others?" *Quarterly Journal of Economics* 114:83–116.

Hellman, J., G. Jones, and D. Kaufmann. 2003. "Seize the State, Seize the Day: State Capture, Corruption, and Influence in Transition Economies." *Journal of Comparative Economics* 31(4):751–73.

Heston, A., R. Summers, and B. Aten. 2002. Penn World Table Version 6.1. Philadelphia: Center for International Comparisons at the University of Pennsylvania (CICUP).

Isham, J., D. Kaufmann, and L. Pritchett. 1997, "Civil Liberties, Democracy, and the Performance of Government Projects." *World Bank Economic Review* 11(2):219–2.

Kaufmann, D. 2004a. "Governance Redux: The Empirical Challenge." *The Global Competitiveness Report 2003–2004.* World Economic Forum. New York: Oxford University Press. pp. 137–164.

———. 2004b. "Corruption, Governance and Security: Challenges for the Rich Countries and the World." *The Global Competitiveness Report 2004–2005.* World Economic Forum. Hampshire: Palgrave Macmillan. pp. 83–102.

———. 2005. "Myths and Realities of Governance and Corruption." The Global Competitiveness Report 2005–2006. World Economic Forum. Hampshire: Palgrave Macmillan.

———. 2009. "Corruption and the Global Financial Crisis." Forbes, 27 January. At: http://www.forbes.com/2009/01/27/corruption-financial-crisis-business-corruption09_0127corruption.html

Kaufmann, D. and A. Kraay. 2002. "Growth without Governance." *Economist* 3(1):169–229.

Kaufmann, D., A. Kraay, and M. Mastruzzi. 2003. "Governance Matters III: Governance Indicators 1996–2002." Policy Research Working Paper 3106. World Bank.

———. 2005. "Governance Matters IV: Governance Indicators for 1996–2004." Policy Research Working Paper 3630. World Bank.

———. 2008. "Governance Matters VII: Aggregate and Individual Governance Indicators 1996–2007." Policy Research Working Paper 4654. World Bank.

———. 2009. "Governance Matters VIII: Governance Indicators for 1996–2008." June. Available at: www.govindicators.org

Naím, M. 2005. *Illicit: How Smugglers, Traffickers and Copycats are Hijacking the Global Economy.* Doubleday.

Sen, A. 1981. *Poverty and Famines: An Essay on Entitlement and Deprivation.* Oxford University Press.

Stiglitz, J. 1999. "On Liberty, the Right to Know, and Public Disclosure: The Role of Transparency in Public Life." Oxford Amnesty Lecture.

World Bank. 2004. "Doing business in 2004." Washington, DC. At: http://www.doingbusiness.org/Documents/DB2004-full-report.pdf

World Bank Institute (WBI). "Governance and Anti-Corruption." Available at: http://www.worldbank.org/wbi/governance/capacitybuild/

Chapter 2.8

Dynamics and Challenges of Innovation in Germany

Alexander Ebner,
Johann Wolfgang Goethe-University,
Frankfurt, Germany
Florian A. Täube,
European Business School,
Oestrich-Winkel, Germany

Introduction

The economy of the Federal Republic of Germany, labeled a 'growth miracle' in the 1950s, emerged as the growth engine of Europe throughout the 1960s and 1970s. Despite the slowdown of growth rates during the 1980s, it remained Europe's strongest and most innovative economy (Harhoff, 2008), making its unique brand of a 'social market economy' a role model, combining technological innovativeness, international openness, and industrial competitiveness with an extensive welfare system. Since the 1990s and 2000s, however, the various challenges posed by economic globalization, technological change, locational competition, demographic pressures, persistent mass unemployment, and the fiscal burdens of reunification with East Germany have exercised a pressure for institutional reform. Indeed, current German policy discourse is preoccupied with the need for infusing more entrepreneurial drive into the economy at large, in line with the formation of a knowledge-based economy. This reform orientation also affects the debate on innovation: the primary task is the restructuring of Germany's innovation system in the direction of an entrepreneurial approach that combines institutional flexibility in the research and educational systems with promotion of entrepreneurship in both start-ups and established firms, while providing adequate risk capital and manpower. Thus, innovation in Germany is a reflection of more extensive institutional changes which are transforming Germany's postwar coordinated 'social market economy' into an institutional hybrid whose shape is not yet clear—apart from becoming decidedly more entrepreneurial.

In this chapter, we outline existing conceptual frameworks for assessing innovation dynamism in a country, combining the neo-Schumpeterian notion of a national innovation system with Michael Porter's concept of 'national innovative capacity' (Furman, Porter, and Stern, 2002) and David Audretsch's approach to the 'entrepreneurial society' (2007). Next, we survey the relevant institutional determinants of the German economy, addressing such issues as the trade regime, competition law, labor relations, the financial system, and entrepreneurship policies. We then highlight the basic features of the German innovation system, in particular pointing to factors such as education and training, R&D, and university-industry relations. We emphasize two salient developments in financial markets, namely the advent of venture capital and of high-growth stock markets. Finally, we investigate patenting and the role of regulatory conditions for new ventures and provide an outlook on the future challenges to German innovation performance in the context of globalization.

Assessing innovation: Knowledge, institutions, and entrepreneurship

Innovation may be perceived as an interactive process requiring complex institutional arrangements, specifically, the coordination of the interplay between entrepreneurship and organization in the commercialization of knowledge by introducing novelty into the domains of markets and industries. In other words, technological innovation may be perceived as an outcome of 'embedded entrepreneurship'; that is, a collective process shaped by formal and informal institutional frameworks and knowledge infrastructures (Ebner, 2009). The systems innovation approach offers a useful Schumpeterian perspective for addressing these topics and emphasizes the innovation-driven character of capitalist development. It examines the impact of institutional networks on the generation and assimilation of innovations within a given territorial setting, and highlights the conditions for organizational learning in stimulating innovative capability. Businesses comprise the principal terrain for innovation, but other institutional elements, such as R&D facilities, education and training programs, and financial, legal, and patent systems are taken into consideration as components of those public and private networks which contribute to the introduction of new technologies (Freeman, 2002). Thus, knowledge is viewed as a fundamental resource for innovation, whereas organizational learning is assessed as the most important underlying process. Industrial structures and the institutional set-up of an economy then determine the shape and performance of a national or regional innovation system, whose structural constellations yield a specific entrepreneurial potential. Thus, innovation systems include those structures of governance which handle the innovative contributions of public and private goods in a manner supportive of economic growth (Lundvall et al., 2002).

Institutional configurations indicate the specifics of an innovation system in a given territory or country. Accordingly, as the nation-state provides the most relevant indicator of economic order, innovation-related interactions and related mar-

ket activity, the systems concept has been most extensively applied to compare national innovation systems, augmented by analyses at the regional and sectoral level. Indeed, it makes common sense that the systems approach considers how decisively the policies of particular governments, national laws and a shared culture in an institutional arena affect the intensity and direction of technological innovation (Lundvall, 1992; Nelson and Rosenberg, 1994). Corresponding efforts in comparative institutional analysis have taken a general interest in the diversity of national models of capitalist development and the institutional basis of advantages in their respective innovation patterns (Hall and Soskice, 2001). The description of institutional characteristics even allows for characterizing distinct types of national innovation systems, such as 'myopic' or 'dynamic.' Myopic systems are typical in the United States and the UK with short-term modes of technology investment, as compared with dynamic systems of innovation in late-industrializing countries such as Germany and Japan, which tend to recognize the long-range character of technological investment and complement market processes with specific policies for technological learning (Patel and Pavitt, 1994).

With the onset of globalization, however, these national characteristics are being transformed, depending on country-specific decisions and path dependencies. In particular, the extent to which firms, research centers, and government agencies internationalize their innovation-oriented activities and interactions has become highly relevant, indicating the way in which national innovation systems are moving towards greater international openness and market competition (Galli and Teubal, 1997). However, despite this tendency, the most crucial modes of interaction remain at the national level. There may be supranational parallels, but these do not yet substitute for the institutional competence of nation-states. Besides, even though tendencies toward structural convergence persist, they have the effect of promoting even greater institutional specialization and divergence among national innovation systems (Freeman and Soete, 1997).

In this context, national innovative capacity, as outlined by Porter and others, implies those very institutional and structural features of a national innovation system, which promote the international competitiveness of firms and industries, and thus reflect the competitive advantages of nations in an increasingly globalized economy. National innovative capacity is defined as the capability of a country to produce and commercialize new technologies over the long term. In so doing, a country can reflect institutional and structural features, such as variation in economic geography, firm-level spillovers, and cross-country differences in industry and technology policy, as indicated by public R&D expenditures and the enforcement of intellectual property rights. In the systems framework, the linkages between innovation infrastructures and industrial clusters are critical for maintaining dynamic growth, augmented by appropriate skilled human resources, well-structured research endowments, and adequate venture capital (Furman, Porter, and Stern, 2002).

Entrepreneurship parallels these concerns. The systems framework somewhat neglects certain aspects in the microfoundations of innovation, namely, the role of new business ventures in the ongoing structural change towards a knowledge-based entrepreneurial economy. Indeed, since 2000, major impulses for employment creation and income growth in the OECD have been generated by start-up enterprises of diverse scale and scope well beyond the operations of large corporations operating in global production and service networks. According to Audretsch's 'knowledge spillover theory of entrepreneurship,' cross-national differences in entrepreneurial performance may be explained by institutional contexts. Those which are rich in knowledge and new ideas will tend to generate more entrepreneurial opportunities (Audretsch, 2007). Accordingly, a new policy approach is emerging which focuses on enabling the creation and commercialization of knowledge by promoting the startup and viability of new firms involved in knowledge-based entrepreneurship. This policy approach encompasses multiple levels of activity in a pattern of systemic linkages, ranging from the individual to the enterprise, and to the cluster or network, which might involve an industry or sectoral dimension, or a spatial dimension, such as a district, city, region, or even an entire country. Related policy instruments include distinct policies for taxation, immigration, or education, as well as more direct instruments, such as the public provision of resources for finance or training (Audretsch, Grilo, and Thurik, 2007).

However, given the empirical complexity of entrepreneurship, innovation and economic growth, we must not consider the perspectives and approaches described above as being mutually exclusive. It is more appropriate to view the eco-nomic performance of a nation or region as the outcome of a particular system of innovation. It is the system which constitutes the backbone of an institutional infrastructure that promotes specific innovation capacity in an evolving entrepreneurial society. Next, in this integrative perspective, we examine the institutional determinants of the German economy, in particular its trade regime, competition regime, labor and welfare state issues, and entrepreneurship policies. This will prepare the reader for a critical discussion of the German innovation system, highlighting the key areas of technological advantages, education and training, R&D, and university-industry relations.

Institutional determinants of the German economy

The German economy illustrates outstanding growth over a very long period, beginning with the successful recuperation of the German Empire in the last quarter of the 19th century, accomplished through major efforts in education, training, and innovation. Indeed, both modern university- and the science-based research firms originated in the late 19th century as examples of institutional innovation in Germany (Keck, 1993; Buenstorf and Murmann, 2005). In combination, these formed the backbone of an industrial system of professional engineers that enabled Germany (and the United States) to overtake British technological leadership, in industries such as synthetic dyes (Murmann, 2003; Jones, Wuchty, and Uzzi 2008). This was followed by the German 'economic miracle' of the 1950s which turned Western Germany into an export-oriented economic power in the world economy, the home of major multinational enterprises, and the home market of globalizing industries such as electronics, pharmaceuticals, and automobiles. Nonetheless, the various challenges posed by globalization, technological change, intensified geographic competition, demographic pressures, persistent mass unemployment, and the fiscal burdens of reunification have hampered German economic performance since the 1990s. These have made it difficult for Germany to put in place those institutional reforms which are necessary to establish a more vibrant growth trajectory in line with the demands of a knowledge-based economy. The underlying drive to maintain international competitiveness—despite all the internal rigidities and frictions that hamper Germany's political and economic system—is evident in the openness of the German economy.

Actually, Germany is not only the production base for almost one quarter of European GDP—making Germany the growth engine of the Common Market—it also excels as the most decisive export-oriented economy in Europe, with a competitive industrial base that produces almost one third of its GDP exclusively for export. In fact, since 2003 Germany is the "world champion of exports" (Wu, 2008), with around 10 percent of world exports. Figure 1a illustrates total exports and Figure 1b the even higher share of exports of manufactured goods. The latter reflects the relevance of the manufacturing sector in Germany and its competitive advantage in terms of innovation, not cost. In Germany, manufacturing as a share of GDP has been stable since 1995, relative to the UK, U.S., France, and Japan, in all of which it has been declining.

Figure 1a. **Total exports (2003–2007)**

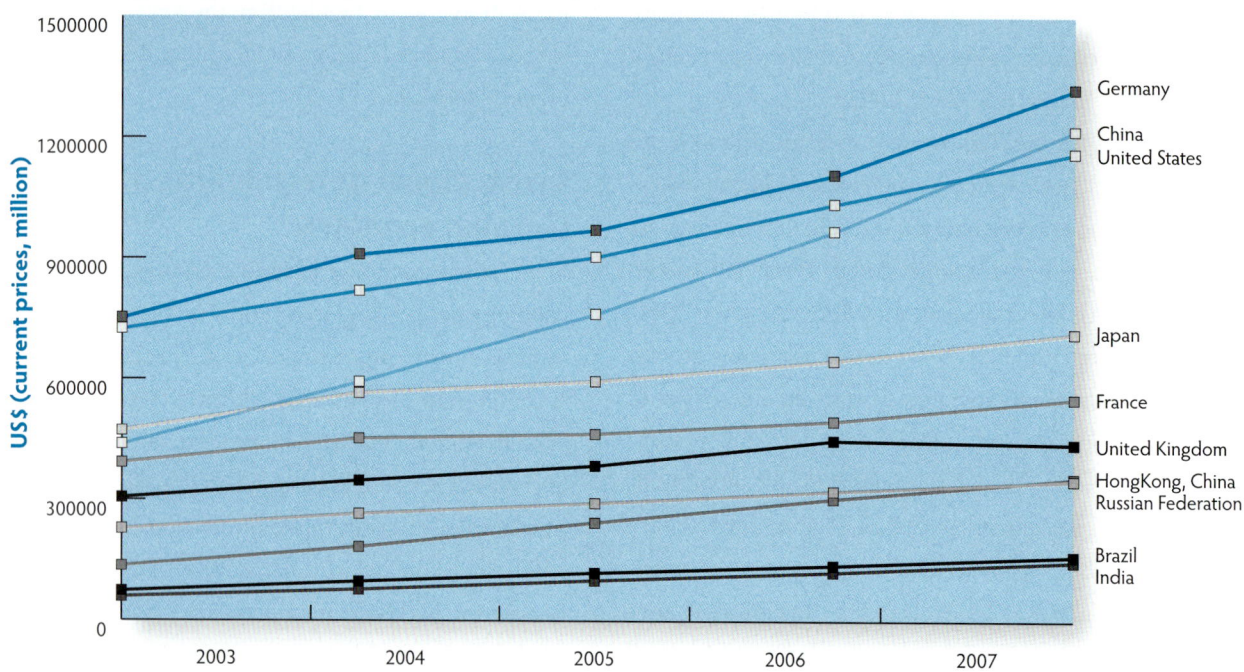

Figure 1b. **Manufacturing exports (2003–2007)**

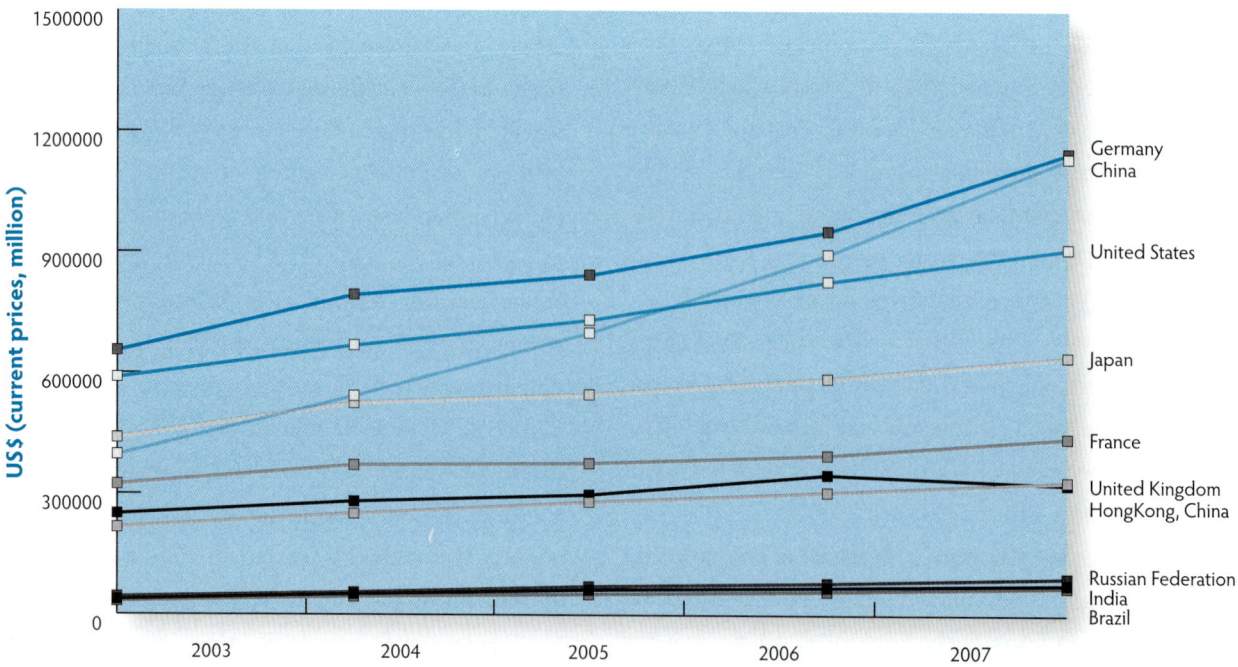

Source: WTO Statistics database. Available at: http://stat.wto.org/StatisticalProgram/WSDBStatProgramHome.aspx?Language=E

In terms of exports, China is catching up, and, having already overtaken Japan in 2004 and the U.S. in 2007, is now a close second behind Germany (Wu, 2008).

Structurally, the German trade regime is subject to the various rules and regulations of the European Union, which has exercised an impact on trade issues ever since the Rome Treaties of 1957 installed the Customs Union in Western Europe. However, this relatively high degree of openness to foreign trade and investment is subject to restriction when it comes to areas such as agriculture and some industries, in which non-tariff protectionism is bolstered by subsidies and related instruments of selective industrial policy, in particular those having strategic significance. Moreover, the German economy displays institutional peculiarities—viz. the historic and only very recently changing role of the banking sector and labor unions in non-market decision making—which may be responsible for the relative underperformance in capital inflow (Siebert, 2005).

The openness of the German economy is also promoted by a distinct competition regime, governed by European Union regulations. Yet post-war efforts to boost Germany's competitiveness to reflect market principles have persistently influenced the European policy agenda, resulting in adoption of policy approaches to antitrust legislation, mergers, cartels, and to liberalization, internationalization, and state aid which parallel the concepts found in traditional German competition policy.

Institutional reform to activate entrepreneurial initiative is characteristic of labor and welfare state policy. Some one-third of German GDP is allocated to social policy issues, financing an extensive welfare state which provides budgets for unemployment benefits, health care, and old-age pensions. The linking of social and labor market policy has been a fundamental concern since the 1990s, confronting the pressures of reunification, globalization, and demographic change. Following the failure of neo-corporatist labor market initiatives, the Schröder government implemented the so-called 'Hartz Commission' proposals for reform of administration and the social security system. One key focus of these reforms, as bundled in the Agenda 2010 program, is the drive to motivate the unemployed to undertake entrepreneurial initiatives, using material incentives that would lower the reservation wage and facilitate an earlier re-entry into the labor market (Czada, 2005).

Corresponding efforts to promote greater flexibility in wage setting and labor regulations indicate that the heyday of German neo-corporatism is definitely over. The role of both unions and employer associations in labor relations is decreasing. As a result, not only are firm-level wages more easily adapted to potentially more volatile local conditions, but there is greater fragmentation of organized interest groups. This latter consequence may lead to intense distributional conflicts. Still, the concept of 'co-determination,' which gives the unions a strong standing in the supervisory boards and worker's councils of large firms, is still in place, even though its abolition had been repeatedly predicted.

This pattern of change also applies to entrepreneurship policy, which combines basic concerns for competition and labor market policy with a drive for innovation. While traditional types of industrial policy have been largely focused on large firms in both sunrise and sunset sectors of the German economy, the most pressing task is the promotion of entrepreneurship on all organizational levels, regardless of the scale and scope of the firms involved. Indeed, traditional industrial policy has basically neglected the dense networks of small and medium-sized enterprises (SMEs) that are the foundation for employment, training, and income generation in Germany's industrial system of flexible specialization, with its regional focus in south-west Germany. During the 1990s, however, the special role of business start-ups in promising new industries was recognized, leading to initiatives such as the BioRegio contest, aimed at the formation of innovative networks in biotechnology among public and private sector in 17 regions all over Germany, which competed for public funding. In this manner, industrial policy was reshaped to dovetail with a greater regional and sectoral differentiation in the German innovation system. This, in turn, provided the institutional underpinnings for technological change and competitive performance (Heidenreich, 2005; Annesley, 2004).

Dynamics of the German innovation system

The German innovation system is characterized by a pattern of predominantly incremental and process innovation in its key manufacturing industries, based on a skilled and relatively autonomous workforce. It combines strengths in high-quality competitive innovations in the chemical, pharmaceutical, mechanical engineering, automotive and electronics industries with close links between those industries and public research institutes such as the Max Planck Society in basic research and the Fraunhofer Society in applied research. The corresponding technological advantages are found predominantly in the electronics, chemical, automotive and machine tool industries, indicating a bias towards innovation in high-skill, high-quality products, which are, however, different from the R&D-intensive segments of high-technology industries. This pattern of innovation reflects specialization advantages that are deeply rooted in the configuration of firms, industries, and institutional frameworks alike. While the underlying long-term approach to industrial coordination has been comparatively successful in established technological paradigms, it may lead to rigidities and a loss of efficiency in times of rapid technological change, when more short-term and market-oriented approaches have greater competitive impact (Harding and Soskice, 2000; Hall and Soskice, 2001). The institutional setup of German innovation may be illustrated as follows, showing the specific public, private and intermediate sectors of Germany's coordinated market economy:

knowledge-intensive services. Moreover, Germany's share of international R&D investment was only about 20 percent, far below the OECD average; the same holds for the 20 percent share of basic research in the overall R&D portfolio (Legler, Krawczyk, and Leidmann, 2009; Rammer et al., 2004; Prange, 2005). These statistics already point to a kind of 'entrepreneurship gap,' and indicate the need for change in the institutional setup of the German innovation system, away from a focus on innovation support that benefits research consortia among established large firms with an international reach. A telling illustration of this problem is provided by empirical assessments of attitudes towards innovation problems in national populations, using such sources as the World Values Survey or other primary data (Mitchell et al., 2000; Witt and Redding, 2009). Theoretical models integrate *cognitive* and *normative* with more formal *regulatory* dimensions (Busenitz, Gomez, and Spencer, 2000). In the case of Germany, it is primarily science and technology, which are appreciated, whereas the view of start-up activities and entrepreneurial risk-taking remains well below the OECD average (Belitz and Kirn, 2008).

The reconfiguration of the German innovation system to-

Major components of the German innovation system

Public sector	Intermediate sector	Intermediate sector
• Federal Government and Administration • State Governments and Administration • Universities and technology colleges • Max Planck Institutes	• Fraunhofer Society • Vocational training institutions	• Electronics industry • Chemical industry • Automotive industry • Machine tool industry

Based on this institutional framework, Germany's gross expenditures on R&D equaled about 2.5 percent throughout the decade since 2000, well above OECD average. This effort was shaped by an expansion of the private sector and a gradual decline in public spending, a tendency that differs markedly from expanding public expenditures in the United States and Japan.

The private sector provides almost two-thirds of R&D funding in Germany, although the number of sectors receiving those funds has narrowed, reducing the industrial base of gross domestic expenditure on research and development (GERD). The automotive, chemical and machine tool industries are outstanding in this regard, in contrast to international trends which show a focus on high-tech industries and

wards greater entrepreneurship is well represented by the case of German biotechnology, which exercises a technology-driven pressure for change (Casper, Lehrer, and Soskice, 1999). In science-based industries such as biotechnology, firms are intensively engaged in cooperation with universities and non-university research institutes. They are pro-active in the exchange of knowledge with both domestic and international science bases (Jones, Wuchty, and Uzzi, 2008). This catalytic role of scientific knowledge in forming innovation networks is due to the increasing costs of innovation, the drive for closer interaction between basic and applied research, as well as between users and producers (Powell, Koput, and Smith-Doerr, 1996). Taking the biotechnology industry as an example, the direction for institutional change in the German innovation system becomes clear.

Although the European Union has emerged as a key actor in providing governance and regulation, regional authorities are also increasingly important. For example, in the area of the financing of innovation, Germany's traditional bank-centered financial system has undergone reforms which will improve access to local as well as international venture capital. Greater flexibility is also the target of ongoing reforms in both the research and educational systems, as Germany faces a shortage of qualified personnel for the commercialization of scientific knowledge. Innovation activities are moving beyond the confines of established large firms, with research SMEs engaging in partnerships and alliances with local knowledge providers and national as well as international companies and becoming crucial players in global knowledge networks. On the model of Baumol's "David-Goliath symbiosis" (2002), established firms have increasingly taken ideas from SMEs, e.g., the pharmaceutical industry from biotech firms, or through new Internet platforms such as Innocentive, that is, from wherever solutions to a problem are to be found. This tendency towards greater diversity in entrepreneurship and innovation is being supported by innovation policies in the multi-level setting of national as well as regional and European initiatives (Kaiser and Prange, 2004).

Most recently, the Merkel government has implemented a so-called 'High-tech strategy for research and development' that should boost R&D spending by €6 billion, although even this sum is not sufficient to reach the goal of 3 percent GERD in 2010, as was announced in the ambitious earlier plan. But the need for financial resources is not the only problem. Even more pressing is the need for adequate manpower to maintain adequate innovation performance in an entrepreneurial and science-based context. Some 100,000 additional research scientists and engineers—representing a substantial increase in human capital—would be needed to achieve a GERD of 3 percent. This gap in human resources becomes even more problematic when we take into account the current decline in the number of engineering graduates. To address this gap, the government initiated a program in 2000 (the greencard for foreign IT experts) to attract international manpower in knowledge-intensive industries. However, neither this effort, nor the reforms of the German system of higher education in 1998 and 2002, which were aimed at increasing the autonomy

of universities to make them more compatible with international models, could meet the needs. Indeed, it would appear that what is required to meet the need for qualified personnel, especially in science-based industries, is a concerted effort at the level of Europe as a whole (Grupp, Schmoch, and Breitschopf, 2008).

Financing innovation

The mode of strategic non-market decision making in the business sector mentioned above (e.g., in labor relations) also applies to the German financial system, which relies on bank-based governance procedures for capital allocation. While the system has been undergoing a major shift to a stronger role for capital markets, it still differs markedly from those of Britain or the United States, which are market-based. This is reflected in Germany's underdeveloped corporate governance system. As a consequence, the post-war model of a coordinated market economy with close linkages between banks and industry, based in a neo-corporatist setting of intermediate institutions, the so-called *Deutschland AG*, is in the process of partial dismantling and restructuring. The German economy has become an institutional hybrid, subject to path-dependent institutional change (Streeck, 2009). The most significant changes to the financial landscape have been the advent of venture capital (VC) and, more recently, stock market segments focused on entrepreneurial firms and their need for growth capital, such as the NASDAQ or the German Neuer Markt, which failed spectacularly after the burst of the dot-com bubble (Vitols and Engelhardt, 2005).

Venture capital (VC)—sometimes called risk capital—arose first in the United States and its more risk-friendly economy. In Germany, however, VC is still relatively marginal, given the size of the economy and the strong tradition of bank financing. However, together with the UK, Germany accounts for 50 percent of VC investment in Europe. Despite the fundamental differences in the financial systems of the UK and Germany, the former being market-based and the latter bank-oriented, the German VC landscape is closer to that of the UK than of Japan, another economy with a relationship-oriented financial system dominated by banks. Banks are a major source of VC finance in most countries, and they are particularly important in Germany and Japan, while in Israel corporations are most prominent and in the UK pension funds. Interestingly, however, government funding plays a more important role in the UK than in Germany (Mayer, Schoors, and Yafeh, 2005).

In contrast to independent and corporate private equity providers, banks and government funds tend to have a less pronounced role in corporate governance and in monitoring the companies they finance, and often serve only as bridge investors (Tykvová, 2006). There is evidence that European venture capitalists engage in less monitoring and thus adopt a more hands-off approach to their portfolio companies, as compared to those in the United States. The use of convertible securities is markedly lower in Europe than in the U.S. (Schwienbacher, 2008). Furthermore, investors have less control, fewer veto rights, and use common equity in coun-

tries of German legal origin, relative to those of socialist, Scandinavian, or French legal origin (Mayer, Schoors, and Yafeh, 2005; Cumming and Johan 2008). For the bank-based systems of Germany and Japan, one potential explanation lies in the finding that prior relationships with a company in the venture capital market increase the likelihood of banks granting a loan (Hellmann, Lindsey, and Puri, 2008).

Given the more risk-averse nature of German investors, VC is more easily available—especially from institutional sources—after a new venture has passed the early growth stage and begun to prove itself. In other words, very early seed-funding, from what some call 'business angels,' is rarely available from financial investors in Germany. Therefore, entrepreneurs are restricted in their risk-taking due to the limited availability of

Table 1. Sources of funds and characteristics of VC investments

	Funds	Banks	Insurance companies	Pension funds	Corporate investors	Individual investors	Individual investors	Other institutions
Germany	187	0.59**	0.22**	0**	0.16**	0.36*	0.09**	0.21**
Israel	119	0.51	0.11**	0.02**	0.60**	0.36	0.01**	0.54
Japan	62	0.56	0.43	0**	0.27	0.21**	0.03**	0.80**
United Kingdom	140	0.44	0.36	0.49	0.26	0.45	0.24	0.55

Panel A. Sources of external funds for the VC industry

	Funds	Early	Middle	Late
Germany	187	0.68**	0.89*	0.74
Israel	98	0.93**	0.49**	0.28**
Japan	57	0.15**	0.19**	0.65**
United Kingdom	140	0.48	0.84	0.80

Panel B. VC Investments by stage

	Funds	Life sciences	IT and software	Electronics and semiconductors	Manufacturing	Other industries
Germany	183	0.84	0.81**	0.73	0.68	0.73
Israel	95	0.57**	0.89	0.49**	0.24**	0.09**
Japan	56	0.55**	0.86**	0.11**	0.14**	0.34**
United Kingdom	140	0.81	0.96	0.79	0.78	0.75

Panel C. VC investments by industry

Notes: Panel A refers to sources based on binary and not mutually exclusive responses to a question of whether or not a particular fund uses a certain source; Panel B is based on discrete and not mutually exclusive responses by funds and reports the proportion of funds investing in different investment stages: "early" refers to seed and start-up, "middle" to expansion and growth, and "late" to later stages; Panel C refers to the sectors of investment using five groups of industries: life sciences, IT and software, electrical and semiconductors, manufacturing and chemicals, and other industries.

* Mean values which are statistically different from those of the UK at the 10 percent level.
** Mean values which are statistically different from those of the UK at the 5 percent level.
Source: Meyer, Schoors, and Yafeh, 2005.

funds, even in new high-technology industries, such as bio-technology. German VCs do not favor any industry in partic-ular, as compared with their strong inclination to IT and soft-ware in Israel and Japan; however, there is a slight tendency towards chemicals and manufacturing. In contrast, in Israel, known for its thriving high-tech industry, VC funds focus on the type of early-stage investment with which VC is common-ly associated (Mayer, Schoors, and Yafeh, 2005). Table 1 illus-trates the sources of funds and characteristics of investments by VC firms in Germany, Israel, Japan, and the UK.

The failure of the Neuer Markt had serious implications for VC funding, because venture capitalists see stock markets as their exit route to reap the benefits of their risk investment through initial public offerings (IPO). Originally, the Neuer Markt was designed to mimic a number of aspects of the US NASDAQ, enabling young firms an easier route to capital markets than the regular market segment, thereby increasing the match between supply and demand of risk capital. In par-ticular, it was supposed to bring

- greater transparency for investors, particularly for smaller "outsider" investors who did not have intimate access to company management;
- liberal listing requirements, which allowed relatively new, as well as loss-making, companies to get a listing;
- increased protection for small shareholders, e.g., in defin-ing a minimum period of time after the IPO during which inside investors could not sell their shares;
- greater liquidity, that is, the ability to buy or sell shares near the current market price, provided though a system of designated sponsors obligated to provide bid-ask mar-ket quotes.

While the Neuer Markt started slowly—there were only 12 listings in 1997—by 2000, it had become the most impor-tant market for growth stocks in Europe, overtaking the UK's AIM and France's Nouveau Marché by a large margin (with 50 percent market capitalization) and attracting IPOs from other countries, including the U.S. However, this was largely attributed to the dot-com bubble. After it burst, the Neuer Markt never picked up again in activity and eventually had to be discontinued. Competing explanations for the breakdown of the Neuer Markt abound and most scholars ascribe its fail-ure to agency problems related to lax regulation and moral hazard of financial market actors, as illustrated by a number

of scams during the high noon of the New Economy frenzy. Institutional theories, on the other hand, take a more nuanced approach and dig deeper into the intricate web of relation-ships in German society. For instance, Vitols and Engelhardt (2005), use the "varieties of capitalism" perspective to inves-tigate the Neuer Markt. They find that there was a mismatch between the liberal market institution Neuer Markt and the generally coordinated market economy of Germany. In par-ticular, most firms in the German economy are rather con-ventional, as compared to the entrepreneurial ones that are striving for growth markets in the Neuer Markt. The arche-typical, conventional German automotive or manufacturing firms emphasize incremental innovation and low risk, while entrepreneurial firms are less risk-averse. Moreover, labor markets are an important determinant of the risk profiles of firms; Germany never developed an entrepreneurial culture of risk takers like those in the America's high-tech community of Silicon Valley, where experienced managers, scientists, and personnel were always in short supply for the growth compa-nies of the New Economy.

Using the VC approach, Vitols and Engelhardt explain not only the failure of the Neuer Markt, but also why it never re-covered to earlier levels, while IPOs in the U.S. and UK did. Hence, there seem to be some fundamental differences in the institutional structure of the German innovation *system* itself, which increases the likelihood that partial attempts toward in-novation and change are going to fail. In other words, more entrepreneurial-oriented financial markets would need to be complemented by similar measures in other realms.

Patents and other legal issues

We outlined above Germany's strong bureaucratic tendency and its adherence to a sometimes stifling, regulatory system. A number of factors relating to the business environment are regularly observed by the World Bank under its Doing Busi-ness initiative. Table 2 illustrates such measures as "starting a business," "employing workers," "registering property," "pro-tecting investors," or "enforcing contracts" and aggregates them in an "Ease of Doing Business Rank."

It should be noted that most of the leading economies are small, such as those of Singapore, New Zealand, Hong Kong, and Denmark. As the table shows, Germany has a disappoint-ing rank of 25, lagging behind many industrialized countries,

Table 2. Top 25 countries for doing business

Ease of doing business rank	Economy	Starting a business	Dealing with construction permits	Employing workers	Registering property	Getting credit	Protecting investors	Paying taxes	Trading across borders	Enforcing contracts	Closing a business
1	Singapore	10	2	1	16	5	2	5	1	14	2
2	New Zealand	1	2	14	3	5	1	12	23	11	17
3	United States	6	26	1	12	5	5	46	15	6	15
4	Hong Kong, China	15	20	20	74	2	3	3	2	1	13
5	Denmark	16	7	10	43	12	24	13	3	29	7
6	United Kingdom	8	61	28	22	2	9	16	28	24	9
7	Ireland	5	30	38	82	12	5	6	18	39	8
8	Canada	2	29	18	32	28	5	28	44	58	4
9	Australia	3	57	8	33	5	53	48	45	20	14
10	Norway	33	66	99	8	43	18	18	7	7	3
11	Iceland	17	28	62	15	28	70	32	34	3	16
12	Japan	64	39	17	51	12	15	112	17	21	1
13	Thailand	44	12	56	5	68	11	82	10	25	46
14	Finland	18	43	129	21	28	53	97	4	5	5
15	Georgia	4	10	5	2	28	38	110	81	43	92
16	Saudi Arabia	28	50	45	1	59	24	7	16	137	57
17	Sweden	30	17	114	10	68	53	42	6	55	18
18	Bahrain	49	14	26	18	84	53	15	21	113	25
19	Belgium	20	44	37	168	43	15	64	43	22	8
20	Malaysia	75	104	48	81	1	4	21	29	59	54
21	Switzerland	52	32	19	13	12	164	19	39	32	36
22	Estonia	23	19	163	24	43	53	34	5	30	58
23	Korea	126	23	152	67	12	70	43	12	8	12
24	Mauritius	7	36	64	127	84	11	11	20	76	70
25	Germany	102	15	142	52	12	88	80	11	9	33

Source: World Bank, 2009.

such as fellow OECD members U.S., UK, or Japan. Germany also trails behind many emerging economies, such as Thailand, and Malaysia. Most surprisingly, among the countries ahead of Germany are the likes of Georgia, Saudi Arabia, and Bahrain. While Germany scores better than the average in dealing with construction permits, getting credit, trading across borders, and enforcing contracts, it scores dramatically low in employing workers, starting a business, and protecting investors. Probably the most telling statistic with regard to entrepreneurial policies, however, is the relationship between "Starting a Business" and "Closing a Business." Apart from Korea, no other country in the top 25 scores as poorly as Germany. In fact, with the exception of Japan, the difference

in ranking between "Starting a Business" and "Closing a Business" for all other countries is at least 39 points.

The one area in which Germany has a clear advantage is the legal system covering property rights, in particular for intellectual property (IPR). Germany is arguably one of the countries with the strongest IPR protection and contract enforceability. As a result, Germany is one of the leading countries in the world in terms of triadic patenting (Figure 2). Most importantly, it is the leading patentee in Europe, accounting for more than 40 percent of triadic patents, and making it the third most active economy after the U.S. and Japan. Moreover, according to reports by the OECD, Germany also ranks fourth with respect to triad patents relative to GDP and third

Figure 2. Triadic Patent Families

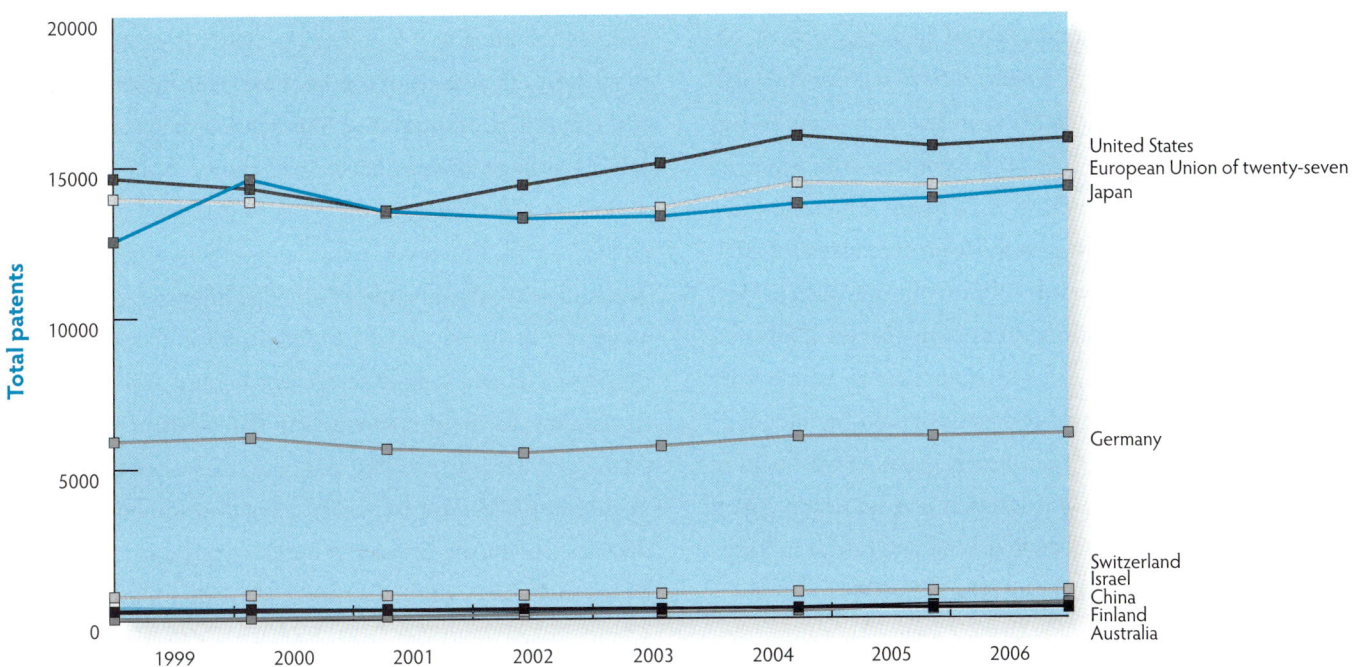

Source: OECD, Database for Structural Analysis, 2008.

relative to population (OECD, 2007). In particular, Germany is a world leader in technological niches such as nanotechnology, fuel cells, and wind energy, as well as other environmentally friendly technologies. More broadly, however, German industry is not focusing its R&D efforts in high-technology sectors such as pharmaceuticals, IT, and aerospace, but rather in mature sectors ones such as automobiles, chemicals and machine tools. In other words, Germany's presence in dynamically growing high-tech and services sectors is low both in terms of output as well as R&D (Harhoff, 2008).

However, the ability to patent and enforce IPR is not always a necessary, certainly not a sufficient, condition to actually commercialize inventions, as in the case of the digital MP3 format, invented by the Fraunhofer Society in Germany and commercialized by US and Japanese firms. Referring back to Schumpeter's (1939) distinction between invention and innovation, having an idea or even developing a new product is not enough to successfully put it on the market and benefit from it economically.

One reason behind Germany's weakness in high-technology is that innovation processes in Germany are mainly oriented toward incremental innovation (Soskice, 1997). Across Europe, this is done predominantly in the large R&D labs of established and mature firms, whereas other countries, notably the U.S., have a larger share of young firms which account, for instance, for more than 50 percent of new pharmaceuticals (Sapir et al., 2003). The low rate of entrepreneurship is often ascribed to bureaucratic red tape that impedes new venture formation. These include the cost of establishing a new venture, in particular one of limited liability, in order to protect the founders' private assets. For instance, the capital requirements for establishing a limited-liability company in Germany is €25,000, as compared to the low 4-digit figure in other EU countries. This stems from the traditionally creditor-oriented values in Germany, which impose a serious ex-ante obstacle to firm formation. The reason that this is so serious is that limited liability companies are particularly appropriate for knowledge-intensive businesses (Harhoff, 2008).

Another component of regulatory impediments is the tax system, which offers weak incentives for innovation and entrepreneurship. The uncertain and intangible nature of innovation makes it less suitable for credit financing, as high risk cannot be collateralized by assets such as machinery. However, the German tax system and other arrangements have led to a high dependence of SMEs upon credit, as outlined in the previous section. This is best exemplified by the differential

193

taxation of profits from innovation financed through debt vs. equity, the latter being almost twice as high. This kind of incentive leads to higher amounts of debt financing, with the associated risk of bankruptcy, which, in turn, reduces the inclination of firms to take further risks in the form of innovative activities. This imbalance is lower in countries with generally lower taxation levels than those of Germany. Similarly, there is a comparative disadvantage in the form of reduced possibilities for tax loss carry-forwards. While countries such as Britain, Sweden, and even France do not impose restrictions on this, Germany limits it in order to minimize the cross-border transactions of multinational companies. As an (unintended) consequence, this reduces the attractiveness of investment in innovative projects of longer duration and uncertain payoff patterns. With lower capital endowments and turnover, SMEs are most affected by this measure (Harhoff, 2008).

Yet another key player in the German innovation system is the competition regime of the Federal Cartel Office (*Bundeskartellamt*), which governs competition regulations in harmony with the European Commission and its Directorate for related issues. Next to the matter of business mergers and monopolistic market settings, the infrastructure conditions of competition are subject to its operations. This emphasizes the primacy of the concept of the 'dominant position' of a firm, with all of its various forward and backward linkages in the market. The reports and recommendations of the German Monopolies Commission (*Monopolkommission*), consisting of selected academic and business representatives, parallel these policy concerns. Yet, despite its seemingly clear-cut mandate, the actual practice of German competition policy is typically filled with exemptions from the more rigid rules. This means, for instance, that mergers leading to market dominance may be legal when they are in line with industrial policy concerns for R&D synergies, illustrating a problem of rules, exemptions and efficiency trade-offs in the domain of competition and innovation that is also prevalent on the European policy level (Motta, 2004; Kühn, 1997).

Conclusion

In conclusion, innovation in Germany presents challenges that require an urgent institutional response. First, venture capital is in short supply. The federal government implemented a law on the modernization of the framework for capital equity in 2008, aimed at setting incentives for providing equity for startup enterprises that were not yet subject to procedures for initial public offerings. Because of its inflexible regulations, the effects of this law have been limited. Second, while the share of innovative SMEs in Germany is above the OECD average, their relatively low equity shares constitute a hindrance for R&D, as the latter is primarily financed by equity capital. Moreover, public support for R&D in SMEs is insufficient. Third, while knowledge-intensive services are an employment-creating expanding sector in the world economy, the German service sector lags behind. This is an area demanding focus in reconfiguring the German innovation system. Fourth, the research and educational system must be made more attractive to human capital in the high-tech industries, enhancing university-industry relations, in particular when it comes to the commercialization of knowledge in science-based industries (EFI, 2009).

To summarize, the current challenges of the German innovation system may be analyzed in terms of the following SWOT analysis:

Strengths	Weaknesses
• Human resources in manufacturing and engineering R&D • Strong performance in private sector R&D • Sustained output of triadic patents • Combination of cost competiveness and industrial innovativeness • Strong base of large production-oriented firms	• Lack of human capital in high-tech industries • Insufficient supply of venture capital • Lack of innovative small enterprises • Slow growth of knowledge-intensive services
Opportunities	**Threats**
• Knowledge flows in an open economy • Combining technological advantages with global production networks • Institutional experimentation in a multi-level system of innovation • Promoting entrepreneurial culture with a balanced welfare system	• Fragmentation of innovation systems may result in industrial dualism • Selective technology policy may obstruct promising paradigms • Cluster support may lead to the closure of industry networks • Ill-conceived political-economic reforms may result in stagnation

194

While the German economy is still one of the strongest and most innovative in the world, its orientation is very backward-looking. The structure of industrial sectors is centered on mature industries of the early 20th (and even 19th) century, indicating that most innovation in these areas is incremental, rather than radical. This might strengthen its resilience in economic crises such as the current one with its steady growth of new product development. However, it might have more adverse effects, if other countries use the crises for restructuring to reposition themselves in the new, high-tech and service industries of the future. One of the areas in which Germany is a world leader is green technologies. However, thus far, most technological innovations in this realm have come from the more mature automotive and chemical industries. In order to broaden the industrial basis for sustainable innovation, Germany has to adapt the framework for entrepreneurship, in particular in the field of high-technology industries. As a note of caution, this is not meant to suggest that extant practices and existing industries should be abandoned; rather, that Germany must complement its strengths with new institutional rules, thereby enabling and providing incentives for innovation through a more flexible regulatory and financial environment.

References

Annesley, C. 2004. *Postindustrial Germany: Services, Technological Transformation and Knowledge in Unified Germany.* Manchester: Manchester University Press.

Audretsch, D. B. 2007. *The Entrepreneurial Society.* New York: Oxford University Press.

Audretsch, D. B., I. Grilo, and A. R. Thurik. 2007. "Explaining Entrepreneurship and the Role of Policy: A Framework." In Audretsch, D. B., I. Grilo, and A. R. Thurik (eds.), Handbook of Research on Entrepreneurship Policy, Cheltenham: Elgar, pp. 1–16.

Baumol, W. 2002. "Entrepreneurhips, innovation and growth: The David-Goliath symbiosis." *The Journal of Entrepreneurial Finance and Business Ventures* 7:1–10.

Belitz, H. and T. Kirn. 2008. „Deutlicher Zusammenhang zwischen Innovationsfähigkeit und Einstellungen zu Wissenschaft und Technik im internationalen Vergleich." In DIW Vierteljahreshefte zur Wirtschaftsforschung, Vol.2: Nationale Innovationssysteme im Vergleich, Berlin: Duncker und Humblot, pp. 49–64.

Buenstorf, G. and J. P. Murmann. 2005. "Ernst Abbe's scientific management: theoretical insights from a nineteenth-century dynamic capabilities approach." Industrial and Corporate Change 14(4):543–78.

Busenitz, L., C. Gomez, and J. Spencer. 2000. "Country Institutional Profiles: Unlocking Entrepreneurial Phenomena." *Academy of Management Journal* 43(5):994–1003.

Casper, S., M. Lehrer, and D. Soskice. 1999. "Can High-Technology Industries Prosper in Germany? Institutional Frameworks and the Evolution of the German Software and Biotechnology Industries." *Industry and Innovation* 6(1):5–24.

Czada, R. 2005. „Die neue deutsche Wohlfahrtswelt: Sozialpolitik und Arbeitswelt im Wandel." In S. Lütz und R. Czada (eds.). *Wohlfahrtsstaat – Transformation und Perspektiven*, Wiesbaden: VS. pp.127–54.

Cumming, D., and S. A. Johan. 2008. „Preplanned exit strategies in venture capital." *European Economic Review* 52(7):1209–41.

Ebner, A. 2009. *Embedded Entrepreneurship: The Institutional Dynamics of Innovation.* London and New York: Routledge.

Expertenkommission Forschung und Innovation (EFI). 2009. "Gutachten zu Forschung" Innovation und technologischer Leistungsfähigkeit 2009, Berlin: EFI.

Federal Statistical Office (Statistisches Bundesamt). 2007. *National accounts: Germany's interrelations with the global economy—An analysis of imports and exports 2007.* Wiesbaden: Destatis.

Freeman, C. 2002. "Continental, national and sub-national innovation systems—Complementarity and economic growth." *Research Policy* 31(2):191–211.

Freeman, C. and L. Soete. 1997. *The Economics of Industrial Innovation*, (3rd ed.), London: Pinter.

Furman, J. F., M. Porter, and S. Stern. 2002. "The Determinants of National Innovative Capacity." *Research Policy* 31:899–933.

Galli, R. and M. Teubal. 1997. "Paradigmatic Shifts in National Innovation Systems." In C. Edquist (ed.). *Systems of Innovation: Technologies, Institutions, and Organizations.* London: Frances Pinter. pp. 342–70.

Grupp, H., U. Schmoch, and B. Breitschopf. 2008. „Perspektiven des deutschen Innovationssystems: Technologische

Wettbewerbsfähigkeit und wirtschaftlicher Wandel." In B. Blättel-Mink und A. Ebner (eds.). *Innovationssysteme: Technologie, Institutionen und die Dynamik der Wettbewerbsfähigkeit.* Wiesbaden: VS. pp.249–66.

Hall, P. A. and D. Soskice. 2001. „An Introduction to Varieties of Capitalism." in P. A. Hall and D. Soskice (eds.), *Varieties of Capitalism: The Institutional Foundations of Comparative Advantage.* Oxford: Oxford University Press. pp.1–68.

Harding, R. and D. Soskice. 2000. "The End of the Innovation Economy?" In Harding, R. and W. E. Paterson (eds.). *The Future of the German Economy: An End to the Miracle?* Manchester: Manchester University Press. pp. 83–99.

Harhoff, D. 2008. „Innovation, Entrepreneurship und Demographie." *Perspektiven der Wirtschaftspolitik* 9 (Special Issue):46–72.

Heidenreich, M. 2005. "The Renewal of Regional Capabilities: Experimental Regionalism in Germany." *Research Policy.*34:739–57.

Hellmann, T., L. Lindsey, and M. Puri. 2008. "Building Relationships Early: Banks in Venture Capital." *Review of Financial Studies* 21(2):513–41.

Jones, B. F., S. Wuchty, and B. Uzzi. 2008. "Multi-University Research Teams: Shifting Impact, Geography, and Stratification in Science." *Science* 322:1259–62.

Kaiser, R. and H. Prange. 2004. "The Reconfiguration of National Innovation Systems: The Example of German Biotechnology." *Research Policy* 33:395–408.

Keck, O. 1993. *The National System for Technical Innovation in Germany, in R. Nelson (ed.), National Innovation Systems: A Comparative Analysis.* New York: Oxford University Press, pp. 115–57.

Kühn, K-U. 1997. "Germany." In Graham, E.M. and J. D. Richardson (eds.), *Global Competition Policy.* Washington, DC: Institute for International Economics, pp.115–49.

Legler, H., O. Krawczyk, and M. Leidmann. 2009. *FuE-Aktivitäten von Wirtschaft und Staat im internationalen Vergleich.* Hannover: NIW.

Lundvall, B.-Å. (ed.). 1992. *National Systems of Innovation: Towards a Theory of Innovation and Interactive Learning.* London: Frances Pinter.

Mayer, C., K. Schoors, and Y. Yafeh. 2005. "Sources of funds and investment activities of venture capital funds: Evidence from Germany, Israel, Japan and the United King-dom." *Journal of Corporate Finance* 11(3):586–608.

Lundvall, B.-Å., B. Johnson, E. S. Andersen, and B. Dalum. 2002. "National systems of production, innovation and competence building." *Research Policy* 31:213–31.

Mitchell, R., B. Smith, K. Seawright, and E. Morse. 2000. "Cross-Cultural Cognition and the Venture Creation Decision." *Academy of Management Journal* 43(5):974–93.

Motta, M. 2004. *Competition Policy: Theory and Practice.* Cambridge: Cambridge University Press.

Murmann, J. P. 2003. *Knowledge and Competitive Advantage: The Coevolution of Firms, Technology, and National Institutions.* Cambridge: Cambridge University Press.

Nelson, R. (ed.) 1993. *National Innovation Systems: A Comparative Analysis.* New York: Oxford University Press.

Nelson R. R. and R. Rosenberg. 1994. "American Universities and Technical Advance in Industry." *Research Policy* 23(3):323–48.

OECD. 2007. *Compendium of Patent Statistics.* Paris: OECD.

———. 2008. *Database for Structural Analysis.* Paris: OECD.

Patel, P. and K. Pavitt. 1994. "National Innovation Systems: Why They Are Important, and How They Might Be Measured and Compared." *Economics of Innovation and New Technology* 3(1):77–95.

Powell, W.W., K.W. Koput, L. Smith-Doerr. 1996. "Interorganizational Collaboration and the Locus of Innovation: Networks of Learning in Biotechnology." *Administrative Science Quarterly* 41(1):116–45.

Prange, H. 2005. *Wege zum Innovationsstaat: Globalisierung und der Wandel nationaler Forschungs- und Technologiepolitiken.* Baden-Baden: Nomos.

Rammer, C., W. Polt, J. Egeln, G. Licht, and A. Schibany. 2004. *Internationale Trends der Forschungs- und Innovationspolitik: Fällt Deutschland zurück?* Baden-Baden: Nomos.

Sapir, A., P. Aghion, G. Bertola, M. Hellwig, J. Pisani-Ferry, D. Rosati, J. Viñals, and H. Wallace. 2003. "An Agenda for a Growing Europe: Making the EU Economic System Deliver." Report of an Independent High-Level Study Group established on the initiative of the President of the European Commission. Available at: http://www.euractiv.com/ndbtext/innovation/sapirreport.pdf

Schumpeter, J. A. 1939. *Business Cycles.* New York: McGraw-Hill.

Schwienbacher, A. 2008. "Venture capital investment practices in Europe and the United States." *Financial Markets and Portfolio Management* 22(3):195–217.

Siebert, H. 2005. *The German Economy: Beyond the Social Market*. Princeton: Princeton University Press.

Soskice, D. 1997. "German Technology Policy, Innovation, and National Institutional Frameworks." *Industry and Innovation* 4(1):75–96.

Streeck, W. 2009. *Re-Forming Capitalism: Institutional Change in the German Political Economy*. Oxford: Oxford University Press.

Tykvová, T. 2006. "How do investment patterns of independent and captive private equity funds differ? Evidence from Germany." *Financial Markets and Portfolio Management* 20(4):399–418.

Vitols, S. and L. Engelhardt. 2005. "National Institutions and High Tech Industries: A Varieties of Capitalism Perspective on the Failure of Germany's 'Neuer Markt.'" Discussion Paper SP II 2005–03. Wissenschaftszentrum Berlin.

Witt, M. A. and G. Redding. 2009. "Culture, meaning, and institutions: Executive rationale in Germany and Japan." *Journal of International Business Studies* 40(5):859–85.

World Bank. 2009. *Doing Business 2009 Report*. Palgrave Macmillan, Available at: http://www.doingbusiness.org

Wu, Liming. 2008. "German export to remain world's biggest in 2008: Difficult year ahead." *China View*. Available at: http://news.xinhuanet.com/english/2008-12/19/content_10528088.htm

Chapter 2.9

From Enlightenment to Enablement: Opening up Choices for Innovation

Andrew Stirling, University of Sussex

Introduction

Since the Enlightenment, we have tended to think of scientific and technological progress as linear and cumulative. In the high-level debate over "the knowledge society," this is still the way these crucial issues are treated in worldwide governance. Technology policy is routinely described as indiscriminately "pro-innovation" and its critics labelled generally "anti-technology." Scope may be conceded for debates over risk, or the distribution of costs and benefits. But the main challenge is seen as a competitive race along a pre-ordained track. As a key feature of the "knowledge society" and a founding theme of the Enlightenment, it is a remarkable fact that this linear understanding is just plain wrong. This paper will explore some implications.

The truth is that, in any given area, science and innovation may actually advance in many alternative directions. As in biological evolution, while many pathways for progress are possible, not every path that is feasible and viable will actually be realized. At each stage of development, societies pursue only a restricted subset of the diverse potentialities. As processes of evolution unfold, certain pathways encountered earlier are "closed down," while other possibilities are "opened up." Whether deliberately, blindly, or unconsciously, societies choose certain possible orientations rather than others for change in science and technology.

These choices are driven by multiple factors in complex decision making processes. Directions of change are particularly susceptible to the exercise of power. Many questions arise. To what extent are choices deliberate and democratic? Is public policy open, inclusive and accountable in dealing with links between technological risk, scientific uncertainty, social values, political priorities and economic interests? What are the relationships between social and technological progress, on the one hand, and public participation and responsible precaution, on the other? What are the most appropriate and practical ways, under different conditions, to get the best out of specialist expertise, while engaging stakeholders, learning from different experiences, and empowering the least privileged groups in society?

In beginning to pose such questions, worldwide "knowledge societies" are facing a new transition—potentially comparable in significance to the Enlightenment itself. Beyond simply recognising the possibility of progress in knowledge and innovation, we are beginning to engage with the realities of the multiple, contending directions for advance. This paper argues that by becoming more clear-eyed and empowered about the possibilities for a more deliberate steering of progress, we face the opportunity to move from Enlightenment to what we might call "Enablement." As with other such transitory opportunities, it remains unclear whether we will make this choice or pass it by.

Two faces of technological vulnerability

This dynamic of continuously branching choices is characterized by two important kinds of vulnerability. The first is society's vulnerability to technology; that is, people, their environments, and fellow creatures are perpetually vulnerable to the unforeseen, unintended or contested consequences of our evolving technological commitments. Some examples might include offensive weaponry, nuclear materials, toxic chemicals, urban congestion, alienating architecture, commodity crops, intensive husbandry, zoonotic diseases, processed foodstuffs, and fossil fuels. In each case, it is the possibilities of alternatives that make these exposures "vulnerabilities" rather than immutable conditions of existence.

The second vulnerability is the converse of the first: that of technology to society. Entirely feasible and viable technological pathways are themselves vulnerable to being foreclosed, especially at their incipient stages, by circumstance or contrary societal forces. Examples include renewable energy, sustainable agriculture, preventive health care, green chemistry, public media, socialized transport, open source software, community architecture, etc. No matter how much more favorable a particular path may seem, it can rapidly become impossible to shift course once certain formative moments have passed.

Complex historic forces determine how societies selectively commit to certain technological pathways as opposed to others. Some systematic mechanisms result in channelling a restricted subset of possible directions. For instance, though they may originate in essentially random patterns, the simple positive feedback dynamics of market "lock-in" may direct the course of change, as with the ubiquitous, but dysfunctional QWERTY keyboard—the result of 19th century mechanical typewriter design requirements—which persists in today's highly competitive computer products. Similar mechanisms of path-dependency and lock-in characterize such artefacts as bicycles, automobiles, road systems, prisons, nuclear power, computer software, chemical production, civil engineering, and weapons systems, all of which reflect the needs, preferences, values, and interests of rather restricted social groups. This is also true of the routines, practices, and thought paradigms of even the most successful and influential innovating organizations, which become imprinted in resulting technologies and the trajectories which they promote.

Cultural expectations may also assert the sensibilities of relatively privileged social actors, such as entrepreneurs, investors, regulators and opinion makers. Once established, these socio-technical interests can become institutionalized and acquire their own momentum at the expense of less-privileged alternatives, and may, in turn, become virtually autonomous, "capturing" ostensibly neutral (or even supposedly contending) social actors. This phenomenon is often observed in such areas as nuclear infrastructure, the fossil fuel and automotive industries, industrial chemicals, genetic modification, cigarette manufacture, food additives, pharmaceuticals, and military systems. In this way, early assertive expectations over which pathway will be followed can be self-fulfilling. Investors, suppliers, regulators, and customers will often pick winners on the grounds of perceived inevitability, rather than judgements of superiority. Expectations can thus be self-reinforcing, foreclosing even what all agree to be preferable long-run options.

In recent times, this foreclosing of contemporary technological choice is in many ways intensified by increasingly transnational capital flows, regulatory standardization, trade harmonization, market concentration and globalising governance, all of which may exert a homogenizing effect on what might otherwise be more varied selection environments. Such developments may reduce global diversity in areas of technology choice, such as food production, energy services, public health, materials management, urban mobility, information, and communication. Although real world complexities do allow for some degree of persistent technological diversity, it remains impossible to realize fully all physically feasible—or even functionally viable—technological configurations, with no guarantee that even the most favorable long term pathways will be utilized.

Who pays the highest price for this foreclosure of choice? One pervasive consequence of the indeterminacy of technological vulnerability is that, despite the diversity and complexity of these branching choices, adverse repercussions tend to fall most heavily on those people with the least resources, privilege, or power. This is true for three main reasons: first, because technological evolution implies change and uncertainty. As shown with tragic frequency in earthquakes, floods, droughts, hurricanes and epidemics, even where there exists general parity of exposure across rich and poor, the impact tends to fall most damagingly on the least affluent and most excluded of people. Poverty impairs adaptive capacity and resilience; subsistence farmers are unable to follow recommended practices; the lowest paid workers operate outside health and safety law; product usage regulations fail to account for the way children play; toxic waste management excludes export to poorer countries. And, as with natural disasters, pre-existing social conditions of marginality exacerbate vulnerability to even the most general of the unforeseen, unintended and contested consequences of technological commitments.

Second, technological vulnerabilities, as distinct from natural ones, bear even more disproportionately on the least powerful because of the systematic tendency for preference to be given to those technological pathways which favored existing privileged interests. The adverse effects associated with modern processed food, for example, fall disproportionately on those who are most marginalized, even within affluent populations. The third reason follows distinctly from this. Not only are the poor vulnerable to the technological choices of the rich, but the technological choices that might most favor the interests of the poor are also disproportionately liable to being foreclosed.

Governance of technological risk

In recognising that vulnerability to the consequences of technological choices fall most heavily on the least powerful, it is interesting to note the desire on the part of contemporary institutions of technology governance to express socially progressive aspirations. "Sustainability," "equity," and "poverty-reduction" feature prominently as the declared motivations behind international policy making.[1] Taken at face value, these suggest serious commitments to reducing the adverse effects of technological choices. In order to deliver on such

socially progressive claims, one might expect that technology policies and strategies would address the underlying challenges of technology choice. Of course, even given that such lofty ambitions to remedy all the deeply-entrenched uncertainties are unlikely to be successful, it seems reasonable to judge the general efficacy of technology governance by how seriously leaders actually engage with these fundamental realities.

It is, therefore, quite striking that much high-level discourse in technology governance does not reflect these issues. Rather than highlighting the pros and cons of alternative pathways, the reality of choice itself is denied, exacerbating the associated vulnerabilities. This was eloquently illustrated by the President of the UK Royal Academy of Engineering in the globally-broadcast BBC Reith Lectures, who portrayed history quite explicitly as a one-track "*race to advance technology*," with the challenge being simply "*to strive to stay in the race.*" He asserted that technology "*will determine the future of the human race*," rather than the other way around. Existing patterns of technology are seen as self-evidently good, with the role of the public being simply to "*recognise ... and give* [technology] *the profile and status it deserves.*" [2]

According to these elite representations of technology change, Prime Ministers and European Commissioners, for instance, routinely defer to the supposedly determining role in decision making of unspecified notions of "sound science," while public misgivings over particular technologies are either misrepresented or stigmatized as being "anti-science" or "anti-technology." Senior politicians treat dissent over specific aspects of unfolding directions of technological change not as legitimate evaluative positions, but as prejudiced and unreasonable. Indeed, to a former deputy director of the United Nations, criticisms of particular technologies were interpreted as indiscriminate anti-technology fears, reflecting a "flat earth society, opposed to modern economics, modern technology, modern science, modern life itself."

Political rhetoric will typically advocate the "way forward," without specifying a direction. Yet it seems only in the field of technological progress that this polemic has been elevated to hegemonic status. There is no doubt that the political implications are expedient. But they may also reflect more emergent forms of dissonance. The multiple possible vectors for progress are reduced to a single scale. Technology is invested with its own agency. Attention is fixated on actuality rather

---

[1] See, for example, UNEP, 1997; Millennium Development Declaration, 2000; and Obama, 2009.
[2] Broers, 2005.

than potentiality. Value is held to be self-evident. Progress is teleologically defined by whatever unfolds. In all these ways, conventional elite discourses on technological progress appear reminiscent of "pre-operational" thought in child development. In other words, they are akin to "baby talk."

Whether intentional or not, the effect of this language renders it more than just rhetoric. In effect, it denies even a vocabulary for dissenting interests in technology choice. It undermines the very discussion of the adverse effects of technological choice and risks alienating those dedicated political resources, practices, and institutions which have been so hard-won in other areas and which have facilitated healthy, critical, democratic politics of social choice. The least powerful are further disempowered.

One possible exception to this picture serves to underscore the general pattern: in the area of climate change, problems of vulnerability to ill-advised technology choice are undeniable. Emerging climate change policies are unprecedented in the scale of deliberate societal aims to remedy the obvious adverse effects of existing technologies. Even here, though, there is scepticism over the mismatch between targets and established market trends. And, despite the explicit values proclaimed, specific technologies continue to be seen in remarkably unitary terms. The transition to a low-carbon economy is often treated simply as a matter of "management," with associated choices and values implicitly self-evident and devoid of political content. Despite the many low-carbon possibilities, it is routinely claimed on behalf of options favored by incumbent interests (such as nuclear power), that there exists "no alternative." When such misleading assertions are challenged, the back-up arguments are that we should do everything. Each position, like the mainstream discourse described above, excludes the real challenges of prioritization and commitment in technology choice.

To recognize that in this area, as in others, societies face real technological choices in no way trivializes the monumental scale and urgency of dealing with the effects of climate change. With options including carbon capture and storage, various forms of geo-engineering, a multitude of frameworks for demand efficiency and energy service innovations, alternative varieties of nuclear power, centralized continent-scale renewable energy infrastructures or shifts towards a diversity of new distributed small-scale sustainable energy resources,

there is a plethora of feasible low-carbon pathways. With contrasting pros and cons and enormous potential for scale economies and learning-by-doing in every case, each of these could plausibly be considered as a potentially central element in dedicated climate technology policies. Granted, any strategy must inevitably involve some diversity and not all of these options can be fully realized together. However, climate change policy remains a critical arena within which technology governance *amplifies*, rather than reduces, the risk of "closing down" technology choice.

This alarming dearth of attention to choice should not be taken to imply that all aspects of technological vulnerability are entirely neglected in mainstream governance. For example, the prominent field of risk regulation involves a worldwide framework of institutions and practices of formidable scale and complexity. In areas such as occupational health, consumer safety and environmental pollution—if not yet in the field of climate change—there is no doubt that this infrastructure has been responsible for significant reductions in the adverse effects that might otherwise have been presented by unfettered market-based processes. The point is, however, that existing provisions for risk regulation address only a limited subset of the complex issues raised by multiple technological potentialities. Far from highlighting choices between radically contrasting orientations for technology, mainstream risk management tends to focus on modifying the details of existing paths. The resulting effects of regulation can thus serve to further enhance "lock-in" to those already existing pathways and the erection of barriers to more radical change. In other words, by concentrating political attention on circumscribed notions of risk, conventional regulation—despite its incidental benefits—serves at times to reinforce vulnerabilities to narrow or restrictive technological pathways.

In order to substantiate this serious claim, it is important to consider more carefully how conventional risk regulation routinely *excludes* scrutiny of alternative options or claimed benefits, as tends to be the case across almost every sector and virtually every jurisdiction. In such areas as food safety, chemical pollution, or genetically modified organisms, risk assessment typically focuses in a narrow fashion on highly codified notions of "evidence" concerning the probabilities of restricted kinds of hazard in particular favored technologies. In addition to the evidence presented in the voluminous analytic literature, the author can also testify from personal experience on a number of regulatory advisory bodies that the criteria for regulatory intervention are typically very demanding. Rarely is any consideration given to complex or additive effects or cumulative trajectories. Even where a focus of concern lies in the functioning of relevant laws, it is typically assumed that social actors comply with the applicable regulations. Only if a given adverse property of a new product can be found to be absent from any other technology on the market, is this considered as grounds for regulatory action. Thus, the baseline for acceptable risk is effectively taken as the performance associated with the *most harmful existing product*, no matter how negative this is acknowledged to be. The unfolding of associated technological pathways is thereby addressed as a succession of single incremental cases, each taken in isolation, and subject only to testing at the level of the lowest common denominator in contemporary practice.

In order to be credible in this highly restricted, asymmetrical discourse, those who advocate for the values or social benefits of alternative technologies must articulate their position in more covert ways. They must substitute legitimate social evaluation with what will be accepted as "science-based" grounds for concern over the physical harms to human health or environment which are threatened by the mainstream technology. These critics must demonstrate these harms rigorously in advance, often requiring expensive forms of audit and analysis and according to highly demanding standards of proof. Even in the field of medicine and human health, where risk-regulatory intervention tends to be most stringent, the demands are systematically stacked against those who are sceptical of existing directions of technology change. Here, as elsewhere, it is asserted as a principle of "sound science" that concerns must be evidence-based in highly circumscribed ways, even if the salient features in question are—as is by definition the case with much innovation—substantively novel in their details. Yet, in a highly unscientific corollary of this, absence of evidence of harm is routinely treated in risk regulation as if it constituted *evidence of absence* of harm. These are some of the ways in which risk regulation routinely helps to promote mainstream industrial interests over potentially viable alternatives in such areas as chemical production, nano-technologies, pharmaceuticals, genetically modified organisms, civil engineering, transport infrastructures, information,

communication and military systems. Clearly, this invocation of science in risk regulation prompts further consideration of the roles of knowledge in technology choice.

Power and knowledge in the social appraisal of risk

Political, economic and institutional forms of power are not just *implicated* in the tangible business of technology choice. They are also routinely entangled in the substance, limits and interactions of the contending knowledges that help to inform and condition these choices. Indeed, deeply engrained conceptions of the nature of knowledge can serve to compound technological vulnerabilities in a number of ways.

The first unfounded assumption is that the mere marketability of a particular innovation is sufficient authority for presuming that it is socially acceptable. In other words, if, in the view of market actors, a particular next step in an innovation can be shown to work, then this is *prima facie* evidence that it signifies "progress" and that it is somehow inevitable. As we have seen, established regulatory structures qualify this picture only in cases where exceptional risks are identified. Attention to wider political interests, ethical issues or cultural values is typically given only where an innovation may offend the strongest sensibilities of established religions: for example, in the ethical preoccupations with reproduction or control over one's body, rather than equity in pharmaceutical priorities, infringements of commons in release of genetic modifications, or the risk of organized violence from military weapons. But even for those risks that are subject to such explicit ethical considerations, the effect (as in risk regulation) is more often to modulate—and even reinforce—established practice than to open up alternatives. Little room is left for scrutiny of the purposes or motivations that drive the favoured directions for science and technology. In this way, technical *feasibility* is effectively treated as a proxy for social *acceptability*. In this way, technology governance systematically excludes crucial issues explored by such thinkers as Aristotle, Kant, and Habermas, who have shown that knowledge is an insufficient moral basis for action.[3] Just because we know *how* to do some possible thing, does not mean that we *should* do it.

A second false assumption is that if knowledge is adequate to enable an innovation, then it will give us a complete understanding of the consequences. Clearly, our knowledge of

---

[3] Habermas, 1984.

the consequences of technology choice—both positive and negative—is, in any given area, seriously incomplete. Humanity's experience with unexpected carcinogenic, mutagenic, neuro- and repro-toxic and endocrine-disrupting effects of synthetic chemicals show repeatedly that we are as vulnerable to the harmful risks from entirely novel mechanisms of otherwise well-functioning innovations, as from known hazards. Therefore, even when we have the best available information, it is difficult to determine unequivocally which of a range of alternatives may prove most favorable. Yet the absence of *documented* risk continues to be asserted as sound scientific grounds for presuming that existing strategies in such new areas as nano-materials are acceptable. Since contemplating the unknown necessarily requires imagining beyond the available evidence, it is treated as "unscientific" in conventional risk regulation. What might truly be thought unscientific, however, is this effective denial of the unknown, which obscures the fact that knowledge of the familiar and of the novel are not necessarily one and the same, and further privileges favored pathways of innovation over existing alternatives. Once again, to assume completeness of knowledge and address seriously only those potential risks for which there is already evidence excludes the ancient wisdom of Lao Tzu and Socrates, as well as of modern economists such as Knight, Keynes, and Loasby, that what we don't know is as important as what we do know.

A third wrong, but less acknowledged, assumption about the nature of knowledge compounds this dilemma. Even the most apparently complete knowledge may nonetheless be indeterminate in its implications. In other words, no matter how much we think we know, we will always be subject to surprise. This may be because pertinent knowledge is unevenly distributed in society: different aspects of the consequences of contrasting choices will be known to varying degrees in different communities. We are especially vulnerable to this where—as with Rumsfeld's famous "known knowns"—we are complacent about what is supposedly "known," without paying due attention either to the meta-criteria by which this itself can be known—by what means and to whom—or to the crucial social origin and context of the knowledge in question. For example, approval for the halo-hydrocarbon refrigerants and aerosols which later caused stratospheric ozone decline was initially driven by complacent acceptance that these substances were benign. Yet the knowledge of specialists concerning

the vulnerability of the atmosphere to these compounds was effectively ignored for many years until later awarded a Nobel Prize.[4] Even in the rigorously codified and exhaustively explored field of mathematics, Gödel showed axiomatically that apparently complete domains of knowledge may always conceal indeterminacies.

A fourth problem area arises because the relationship between knowledge and ignorance is the inverse of what is conventionally presumed. Even if the fallacious assumptions cited above are avoided, it might still seem reasonable to expect that *increasing* knowledge will at least *decrease* our ignorance. This is why risk assessment frequently suspends judgement pending further research, presuming that the resulting increased knowledge will dispel our ignorance. Unfortunately, hard-won but oft-forgotten experience shows this also to be a precarious assumption. For example, advances in knowledge resulting from complex systems research and enhanced computing capabilities reveal chaotic nonlinear dynamics—and thus imminent surprises—in even the most determinate of systems. Knowledge of specific outcomes in such fields as climatology, oceanography, and ecology may thus come to be recognized as less well-founded after such advances than before. Increased knowledge has actually led to increased ignorance.

A fifth incorrect assumption holds that knowledge is additive. If it is conceded that knowledge is distributed across different groups and that pooling this may increase ignorance, as discussed above, surely we can at least be confident that adding such knowledges together at least means increasing the total stock of knowledge? Unfortunately, this is also not necessarily true. In regulating GM crops, for example, the varying understandings of geneticists, virologists, cell biologists, soil scientists, ecologists, agronomists, economists, and sociologists are fundamentally in tension, and so inimical to simple aggregation. The same is true of the "knowledge" of biotechnology entrepreneurs, chemical producers, plant breeders, industrial agriculturalists, and subsistence farmers, since such knowledge is strongly conditioned by their diverse social contexts and disparate cultural values. Yet risk assessment proceeds as if it were possible simply to add together different knowledge "inputs" and arrive at a single more comprehensive—or even more "objective"—picture.

Sixth, risk regulation assumes that "facts" and "values" are

---

[4] Farman, 2001.

effectively independent. At heart, knowledge is assumed to be constituted by facts, irrespective of any conditioning values or interests. This is the essence of underlying, rigid institutional distinctions between "risk assessment" and "risk management." Again, this is manifestly incorrect. Our understandings do not exist in innocent isolation, but are unconsciously intertwined with contingent experience and interests. In sectors such as nuclear power, genetically-modified crops, proprietary pharmaceuticals and nanotechnology, for instance, knowledge is actively shaped by our wider social, economic, and technological commitments. Vast infrastructures are constructed not only on the basis of what we think we know, but also of what we wish for—and act as if it were so. This does not simply increase exposure to associated ignorance. It also forms powerful pressures to exaggerate convenient knowledge and suppress inexpedient ignorance. The more we are committed to what we think we want and know, the greater is the pressure to exclude that we might be wrong.

This latter powerful political dynamic may help to shed light on the inexplicable persistence in risk regulation of these unsafe assumptions about the nature of knowledge. Whatever the reasons, each assumption tends to compound our technological vulnerabilities. Together, they exacerbate exposure to risk and detract from real understanding of the consequences of technological alternatives. They provide a screen behind which those powerful interests which determine how knowledge is represented can adopt the most expedient interpretations, thus compounding specific risks. Further, they provide a pretext for the active dismissal of inconvenient knowledge possessed by marginal groups. And such dismissal can apply as much to progressive social interventions intended to reduce or forestall unforeseen, unintended, or contested effects as they do to the technological pathways themselves.

Precaution in the regulation of technological risk

Recent years have seen the growth of an important institutional response to these neglected entanglements of power and knowledge in technology governance: the precautionary principle.[5] Associated controversies play out in academic literatures on risk, in environmental science, social science, international law, and feature prominently in mainstream political discourse. Nurtured in the earliest multilateral initiatives for environmental protection in the 1970s, precaution (*Vorsorge*) first came to legal maturity in German environmental policy in the 1980's. Since then, it has been championed by environmentalists and strongly resisted by some of the industries they challenge. Diverse formulations of the principle proliferate in international instruments, national jurisdictions, and policy areas. From a guiding theme in European Community (EC) environmental policy, it has become a general principle of EC law, and a repeated focus of attention in high-stakes trade disputes.

Applying especially to technological risks in areas such as food safety, chemicals, genetic modification, telecoms, nanotechnology, climate change, and public health, precaution has until recently been particularly controversial in the United States. Elsewhere, however, its influence has extended from environmental regulation, to wider policy making on issues of risk, science, innovation, and world trade. As it has expanded in scope, so precaution has grown in profile and authority and in its general implications for the governance of technology.

Sometimes generally characterized as an injunction that "*it is better to be safe than sorry*," precaution has been subject to a storm of strongly-asserted criticisms—specifically, that it is ill-defined, intrinsically "irrational," inherently favors discriminatory measures and implies a blanket rejection of technology. It is striking, even in academic debate, how much of this criticism avoids engaging with the real form taken by precaution, let alone the wider implications. Although there exists a variety of variously permissive or stringent forms, this can be illustrated by focusing the canonical version of precaution, expressed in Principle 15 of the 1992 Rio Declaration:

"In order to protect the environment, the precautionary approach shall be widely applied by States according to their capabilities. Where there are threats of serious or irreversible damage, lack of full scientific certainty shall not be used as a reason for postponing cost-effective measures to prevent environmental degradation."

By considering even this relatively early and straightforward expression of the precautionary principle, we can see how misguided are many of the most prominent criticisms.

First, far from being ill-defined, the precautionary principle actually hinges on the presence in decision making of two particular properties: a potential for irreversible harm and a lack of scientific certainty. Thus, precaution is not a detailed

---

[5] Stirling, 2009.

decision rule in its own right, but, as its name conveys, a general principle. Just as principles like proportionality or cost effectiveness are partly defined by their methods—such as risk assessment and cost-benefit analysis—so, too, precaution is as much about methods and processes of appraisal, as about rules and instruments in risk management. It makes no more sense to say the principle on its own is "ill-defined" than to say this of any other such general principle.

Second, rather than being intrinsically "irrational," precaution simply involves being transparent over the evaluative presumptions under which rationality is to be applied. As we have seen, though they play an essential role in conventional risk regulation, values are often concealed by the language of "sound science." Precaution, on the other hand, explicitly articulates a position under which qualities of environmental integrity and human health are on balance favored over the more restricted sectoral or strategic institutional interests asserted by incumbent market actors. Real irrationality lies in the denial that risk science is devoid of values.

Third, there is the undifferentiated "anti-technology" rhetoric that we have already examined. Far from necessarily implying the blanket rejection of even single technological pathways, precaution actually refers to the *reasons for action*, not to the substance or stringency of the consequent actions themselves. It may, thus, as readily lead to strengthened standards, containment strategies, licensing arrangements, labelling requirements, liability provisions, compensation schemes, substitution measures, and research strategies as the much-feared bans or phase-outs. Precaution is about being more deliberate in our technology choices.

Fourth—and contrary to concerns that it inherently favors discriminatory measures—precaution applies in principle symmetrically to all technological or policy alternatives in any given context. There is no reason why it should be felt to favor one pathway over another. Precaution thus constitutes a general discipline in technology choice, under which environmental and human values are rendered more explicit and transparent and the intensity and orientations of commitments become a matter for deliberate political engagement.

In short, regulatory innovations prompted by the precautionary principle are responding to each of the flawed assumptions found in the last section to underlie conventional risk-based approaches to the governance of technological vulnerabilities. The explicit value of precaution addresses both the insufficiency of knowledge as a moral basis for action and the reflexive intertwining of knowledge and interests. The focus on scientific uncertainty addresses properties of incompleteness and uncertainty in knowledge that are otherwise neglected in risk regulation. Finally, and more indirectly, by prompting further reflection and more sophisticated practices in response to uncertainty, precaution helps focus greater attention on the sometimes inverse relationship between knowledge and ignorance and the lack of coherence between and among different knowledges.

The essential contribution made by the precautionary principle is, therefore, to provide a framework under which to broaden out the processes through which societies come to understand the implications of our possible technological choices. By focusing policy attention on uncertainties of a kind that are otherwise neglected or denied, precaution acts to help extend and enrich the ranges of issues, the arrays of options, the varieties of scenarios, the palettes of methods, and the pluralities of perspectives that are engaged in the social appraisal of alternative technological pathways.

Risk, power and public engagement

It was shown earlier how various problematic assumptions and political processes serve to "close down" social commitments around technological pathways favored by existing interests. Part of this phenomenon is the way that power also operates in the institutions and practices of appraisal, to condition not only the concrete choices themselves, but even the form of our knowledges concerning possible alternatives. Liability law, for instance, often allows private decision makers effectively to ignore those possible forms of harm which may reasonably be claimed to be unknown. Even if damages actually transpire, circumscribed definitions of harm, time constraints, procedural rules, compensation limits, fault restrictions, and channelling of responsibility may all serve to protect the beneficiaries from the repercussions of their optimistic assumptions. Likewise, the practice of insurance—for those protected by the terms of contract—apparently translates intractable conditions of uncertainty into a more comfortable state of actuarial risk.

The effect of all these institutions and procedures is to close down not only technology choices, but also what counts as legitimate or plausible representations of knowledge on the associated implications and meanings. This, in turn, provides the vital political resource of justification, thus allowing "decisions" to be conceived, asserted and defended, and "trust" and "blame" to be effectively managed. As a result, powerful incumbent interests manage to further externalize the consequences of the uncertainty and inevitability of technology choice. The inconvenient limitations of knowledge do not disappear, of course, but are simply rendered invisible. It is then only a matter of time before they bite back with the tragic inevitability of Bhopal, Chernobyl, or the global "credit crunch." In this self-reinforcing dance of imperatives, restricted, risk-based methods for addressing technology choice are both produced by, and actively help reproduce, the wider political dynamics. This is the predicament neatly described by Beck as "organized irresponsibility."

It is against this background, that we may come to better understand the real significance of increasing moves towards public engagement on questions of technology choice. Across all parts of government, business and civil society, diverse forms of this discourse are now burgeoning. Champions arise well beyond practitioners and social scientists and emerge in places as diverse as the European Commission, Greenpeace, the House of Lords, the Royal Commission on Environmental Pollution, government departments such as the Department of Innovation, Universities and Skills (DIUS) and large corporations such as Unilever, as well as within established institutions of science, engineering and medicine from the Royal Society and the Wellcome Trust to the Research Councils. Yet attention typically focuses more on *how* engagement takes place rather than *why*. This is especially true with political choices over the directions taken by science, technology, and innovation.[6]

Public engagement here has many faces. Variously pursued as "citizen participation," "inclusive deliberation," or "stakeholder dialogue," it takes place both in and with contrasting publics. Specific approaches include citizen juries, focus groups, consensus conferences, interactive websites, strategic commissions, and stakeholder panels. Yet amidst the clamor, this basic question of "why?" has no single answer. It prompts a variety of equally reasonable but contending responses. Is public engagement about enriching and invigorating our democracy? Is it about fostering trust and acceptance? Or does

it try to build better, more robust pathways for science and technology? Under different circumstances and from different perspectives, different points are emphasized. The question gets more complex—and more intrinsically political.

Central here are the neglected realities of scientific and technological progress discussed earlier. As we have seen, whether in agriculture, energy, ICT, materials or public health, technical and institutional innovations may unfold in a variety of directions. Low-carbon energy strategies may focus on efficient use, smart grids, carbon capture, nuclear fission, or centralized and distributed renewables. The path to sustainable agriculture is variously claimed by organic farming, advanced cultivation, GM crops, and non-GM biotechnologies. Responses to the shortage of human organs are promised by embryonic or adult stem cells, xenotransplantation, various medical technologies, or preventive public health. Innovation for public health might more generally prioritize proprietory pharmaceuticals for treating relatively innocuous diseases of the rich, or "open source" responses to some of the most devastating afflictions of the poor. It is against the background of the pressing realities of choice that we can consider the fundamental political dynamics underlying discussions of public engagement.

In short, the answer to the question "Why engage?" receives different, equally reasonable responses, depending on how public engagement is perceived, designed, implemented, and evaluated. First, a dominant view among many academics, commentators, and practitioners is that public engagement is about enhancing the *democracy* of scientific and technological choices. In this view, engagement is justified, even if the choices that arise are agreed to be less effective, efficient, or timely. As long as the *process* itself is more enriching, empowering, or fair, then democratic aims are satisfied. The design (and evaluation) of engagement is geared to counter undue influence from vested interests and ensure qualities like accessibility, transparency, equity, and legitimacy in the course of decision making.

In contrast, the linear, Enlightenment view of progress taken in the world of policy making focuses more on outcomes and less on process. Here, public engagement is a means to an end, fostering commodities like acceptance, credibility, and blame management (for the directions of change favored by incumbent interests) or trust and strategic intelligence (sup-

---

[6] Jasanoff, 2007.

porting associated institutions and policies). This more *instrumental* rationale hinges on relatively narrow institutional aims, concerning political *justification*, rather than on qualities of the engagement process or supporting vigorous political debate to enable more legitimate choice.

Of course, there is a spectrum of such instrumental positions. There may often be flexibility over which precise outcome is favored, as long as it is effectively justified. Like conventional consultation, expert committees or risk assessments, public engagement can help here in the vital political tasks of maintaining consent and managing conflict concerning whatever recommendations should arise. But in other cases, there will be a clear idea of the particular outcome to be justified. Even without overt manipulation, there are many ways in which engagement—like expert analysis—can be framed so as to favor the "right" answer. By subtle (possibly inadvertent) shifts in process design, particular sites can be selected, specific products approved, or individual policies legitimated. Again, this is not a partisan point. It applies as much to an environmental NGO looking for radical changes in energy or transport behaviors as it does to powerful industrial interests defending the *status quo* in present technologies and policies. Whether such an instrumental motivation is judged good or bad depends on the point of view. Either way, the design (and evaluation) of engagement is focused not on process, but on privately favored outcomes (such as trust, acceptance, or blame avoidance).

The third general motivation for public engagement in technology choice also hinges more on outcomes than process. Here, though, the merits are not judged in terms of narrow sectional interests. Instead, they appeal to widely-recognized substantive qualities, such as reducing impact, protecting health, enhancing precaution, or promoting social well-being. Though details differ, all agree as to the overall desirability. For instance, a corporation may be genuinely open-minded about which products to develop, but simply wish to understand the needs and values of potential customers and the wider society. Similarly, organizations such as government departments, regulatory agencies, scientific academies, and intergovernmental bodies all agree that broad public engagement at the earliest stages in the development of a technology can help gather relevant experience and knowledge, and so provide early warning of possible problems.

When it is realized that technological progress occurs as much by intrinsically political choices as by the inevitable unfolding of our knowledge of Nature, then this argument for public engagement is not romantic. Bearing in mind the complexities of knowledge discussed earlier, this simply recognises that public engagement can draw on the relevant knowledge of users, consumers, or local communities to help test more rigorously the assumptions underlying expert perspectives and so confer more robust and plural results. Specialist expertise is essential, but it is not sufficient in order to definitively compare, prioritize, or distribute different forms of benefit or harm. This is not just about validating subjective judgements over issues like the prioritization of avoiding injuries or disease, harm to workers or children, or the impact on biodiversity or jobs. Nor is it primarily about fairness or democratic legitimacy in political processes. A substantive rationale for public engagement aims rather at ensuring deeper, broader, and richer consideration of relevant options, issues, uncertainties, and values. It is in this way that we might hope to enable more socially robust choices; and so in this very real sense, "better" technologies.

Opening up directions for choice

It is for these reasons that there can be no single final or definitive answer to the question "why engage the public on scientific and technology choices?" Responses will inevitably vary by circumstance, perspective, and timing. We may wish simultaneously to nurture democratic process and promote more specific and private instrumental ends on the lines outlined above. But these motives have different implications for the ways in which we view and carry out public engagement in science and technology.

There are particularly serious implications for the *evaluation* of engagement. Since they vary with motivation, evaluation criteria may display odd contradictions and circularities. In the British government's 2003 dialogue exercise about genetic modification (GM) of foods,[7] one of the evaluation criteria was the impact on decision making. Since the outcome was rather sceptical about GM, it failed to justify more positive government policy. As a result, it was not particularly influential. This contributed to under-performance in the official evaluation, which was cited, in turn, as a (circular) reason for government caution over the exercise in the first

---

7 DEFRA, 2003.

place. To include "policy influence" as an evaluative criterion for well-conducted public engagement (rather than for wider governance) is a sure sign that there are unrelated underlying motivations.

Taking account of all these complexities—and the backdrop of branching technology choices discussed earlier—we can draw a distinction between initiatives that try to open up decisions on science and technology and those that close down.[8] Conventional approaches to public engagement tend to assume that the most desirable general outcome is the achievement of closure (a verdict in a citizen jury or consensus in a consensus conference). This appears simultaneously to fulfil the functions of democratic process, practical justification, and the identification of substantively "best" options. Yet it is just this kind of closing down that presents some of the most acute problems. If closure takes place invisibly within a specific engagement process, then questions arise as to what the role of established democratic institutions should be. How representative, legitimate, or accountable are the included participants or procedures? Might a similar exercise have arrived at different conclusions if it were structured or informed in a different way? What was the opaque (possibly accidental) influence of power *within* the engagement process?

Instead, we may use a range of different approaches to achieve a complementary role for public engagement exercises on science and technology. Rather than aiming at closing down around a single recommendation to policy making, approaches such as open space, deliberative mapping, interactive modelling, multi-criteria mapping, scenario workshops, and dissensus groups instead transparently open up implications of different possible choices. They explore in detail—and open to external view—the ways in which alternative viable directions for science and technology appear favorable under contrasting assumptions, conditions or perspectives. They offer richly detailed information concerning interactions between options, values, and knowledges. The resulting "plural and conditional" recommendations provide a more authentic reflection of the irreducible political complexities. Such recommendations are plural because, while they may rule out some, they outline a range of potentially justifiable actions. They are conditional because each recommendation is qualified by associated values, assumptions, or contexts.

Although possibly inconvenient to officials or managers attempting to prescribe decisions, responsible politicians or chief executives may actually welcome this deeper information. For every senior civil servant insisting that practical advice must take the form of a single sentence in a one-page briefing, there is a beleaguered Minister wondering how much their latitude for choice has been constrained (and *vice versa*). Despite the apparently greater humility and caution of this opening-up approach, it can also—by clearly identifying pathways that appear unfavorable under *all* viewpoints—add to the robustness of decisions. Where engagement highlights alternatives, the resulting justification is also more credible. Choices are still made, but decisions are better informed, more transparent, and at the right level.

An opening-up approach to public engagement can help nurture a richer, more vibrant, and mature politics of technology choice. It recognizes that different knowledges, values and interests favor different, equally feasible, directions for innovation. This is not postmodern "anti-science." Just because a number of directions are viable does not mean that "anything goes." In fact, this approach is more realistic about science and technology and celebrates its many possibilities. Just as what Robert Merton called "organized scepticism" is recognized as a fundamental quality in science, so pluralism and dissensus in public engagement can help build more rational social discourse about science and technology. And by making processes of closure more transparent, systematic opening up is also more consistent with existing procedures for democratic political accountability. Thus, public engagement helps to enable, rather than suppress, a healthier politics of choice.

It is through the progressive institutional and methodological innovations discussed in this paper that we may hope to meet the challenges of more honest, open, and deliberate steering of the continuing processes of scientific and technological choices with which we began. In particular, precaution offers a framework for increasing the breadth, diversity, and humility of our use of knowledge in the face of uncertainty over innovation. It reminds us that choices in science and technology are often conditioned by quite proximate political, economic, and institutional interests, and that we might therefore wish to balance this with more explicit attention to general values of human well-being and environmental integrity. Likewise, rather than simply fostering understanding, trust, or acceptance, public engagement offers ways to be

---

[8] Stirling, 2008.

more mature, explicit, and accountable when dealing with the implications of a plurality of possible choices. Where engagement yields divergent outcomes in different contexts, it opens the door to pursuit of a greater diversity of pathways, under different social and political conditions.

Only by acknowledging the limitations of current mainstream Enlightenment notions of progress, can we come to appreciate the real value of these new developments. They take us away from impoverished fixations with "how fast?" "how far?" and "who leads?" in a race along some preordained track. In their place, we engage with more open questions that do greater justice to the real multivalent genius of science and technology: "which way?" "who says?" and "why?" When mainstream policy debates on innovation in knowledge societies begin openly to empower this more challenging and overtly political kind of question, then we will know that we are truly moving from Enlightenment to Enablement.

References

Beck, U. 1992. *The Risk Society.* London: Sage.

Broers, A. 2005. *The Triumph of Technology: Lecture 1 of the 2005 Reith Lectures. BBC*, London. Transcript (16/7/6) at: http://www.bbc.co.uk/radio4/reith2005/lecture1.shtml

Department for Environment, Food and Rural Affairs (DEFRA). 2003. *GM Nation: Findings of a Public Debate.* London: Department for Environment, Food and Rural Affairs. At: [16/12/2008] at: http://www.gmnation.org.uk

Farman, J. 2001. "Halocarbons, the Ozone Layer and the Precautionary Principle." European Environment Agency, at: http://www.eea.europa.eu/publications/environmental_issue_report_2001_22/issue-22-part-07.pdf

Habermas, J. 1984. *The Philosophical Discourse of Modernity.* Cambridge: Polity.

Jasanoff, S. 2007. *Designs on Nature: Science and Democracy in Europe and the United States.* Princeton, NJ: Princeton University Press.

Millenium Development Declaration. 2000. *United Nations General Assembly Resolution A/RES/55/2, 18th September 2000.* At [30/3/6]: http://www.un.org/millennium/declaration/ares552e.pdf

Ministerial Declaration. 1995. Fourth International Conference on the Protection of the North Sea. Esbjerg: Denmark, 8–9 June.

Modan, B. and S. Billharz (eds.). 1997. "Sustainability Indicators: Report of the Project on Indicators of Sustainable Development." United Nations Environment Programme. Chichester: John Wiley.

Obama, B. 2009. Technology. Section of "*the Agenda*" posted on Whitehouse Website. At [4/9]: http://www.whitehouse.gov/agenda/technology/

OSPAR. 1992. Convention for the Protection of the Marine Environment of the North-East Atlantic. At: http://www.ospar.org/eng/html/welcome.html

Stirling, A. 2008. "Opening Up and Closing Down: Power, Participation and Pluralism in the Social Appraisal of Technology." *Science Technology and Human Values* 33(2):262–294, March. At: http://sth.sagepub.com/cgi/content/abstract/33/2/262

———. 2009. "The Precautionary Principle." In J-K. Olsen, S. Pedersen, V. Hendricks (eds.), *Blackwell Companion to the Philosophy of Technology*. Oxford: Blackwell.

Chapter 2.10

How do Emerging Markets Innovate? Evidence from Brazil and India

Simon Commander,
European Bank for Reconstruction and
Development (EBRD) and Altura Advisers

Introduction

Accounts of growth in the developed or OECD economies over the past quarter century are replete with reference to the importance of new technology. Indeed, the advent of the computer and its widespread adoption across firms has been likened to the impact of earlier key innovations, such as the steam engine and/or electricity. More generally, information and communications technology (ICT) has been grouped as a new general purpose technology that has, in turn, been associated with new waves of innovation that have driven productivity growth and ultimately acceleration in the growth rates of countries.

Interestingly, however, both the productivity and growth consequences of ICT adoption have varied widely across countries and regions. This, in turn, has spawned considerable enquiry into the factors behind such variation.[1] For example, one such puzzle has been the fact that the productivity effects of ICT adoption and production have been notably smaller in Europe than in North America. Part of the reason for this appears to be the different ways in which ICT has been adopted and managed. As such, the associated organizational dimensions of the new technology appear to play an important role in explaining differences in outcomes.[2] For instance, Bloom, Sadun, and Van Reenen (2006) found that the productivity of ICT capital has been significantly higher in US-owned establishments than in other firms operating in the UK. This difference appears to be related to differences in the effectiveness of the co-investments that US-owned firms have made.[3]

Most of the research on the impact of new technology has, quite naturally, been concerned with the developed or OECD economies. The extent of adoption and the consequences for firm and economy-wide performance in emerging markets and developing countries remains much less well understood.[4] Yet, a priori, it could be expected that the gains from adoption in these locations should be large. Further, it is generally argued that innovation in developing countries is characterized mainly by adoption of technologies off the existing shelf, rather than by the outright creation or development of new technologies themselves. As such, the main feature of innovation in developing countries is likely to be the application of technologies that have been developed elsewhere. The chal-

---

[1] See, for example, Jorgenson (2001), Oliner and Sichel (2000), and Stiroh (2004).
[2] See Brynjolfsson and Hitt (2000 and 2004).
[3] See Basu et al. (2003).
[4] World Bank (2006) is an exception.

lenge is then to understand what factors explain the pace of adoption in developing countries—such factors could readily include policy and financing constraints—and the consequences of that adoption.

This paper looks at precisely these issues in the context of ICT adoption in two major emerging market economies: Brazil and India, drawing on some unique evidence for some 1000 manufacturing firms. The paper summarizes the main conclusions of Commander, Harrison, and Menezes-Filho (2009).[5] The dataset that is used was specifically collected to look at adoption, its timing and consequences, as well as the constraints that face developing country firms. This latter aspect is important, given the widespread evidence that suggests that it is the presence of constraints in the business environment and their intensity that often differentiates developing from advanced market economies.[6]

The results of the analysis show that not only are there differences in the timing of adoption and the patterns of ICT use across the two countries, but that there are also differences within the countries themselves. When ICT is adopted and applied, there tends to be a large and positive impact on performance that cuts across sectors and countries. This effect may be enhanced by complementary organizational and managerial changes, although such changes are particularly hard to measure. Interestingly, using information only for India, it

of weak institutions and infrastructure appears to have been important in holding back the adoption of new technology, as well as limiting the scale of any beneficial effects. This has some important implications for policy.

Describing the data

The information used in this paper consists of a unique firm-level survey of nearly one thousand firms in Brazil and India The survey was implemented in both countries through a series of face-to-face interviews. Data was mostly collected for several points in time, namely 2003, 2002, and 2001. In each country, we selected a target of 500 firms in six, 3-digit manufacturing branches: auto-components, soaps and detergents, electrical components, machine tools, wearing apparel, and plastic products. Stratification was by industry, region, and size (employment), with quota sampling. In India, firms were sampled in nine states. In Brazil, firms were sampled in seven regions.

Table 1 provides some basic descriptive statistics for the sample for each country, broken down for mean and median values of size (employment), sales, materials and wage shares, and capital intensity, as well as the mean rate of growth in sales and employment over the period 2001–2003. There is a good deal of variation in all variables. With respect to employment size, however, the median values are actually very

Table 1. Descriptive statistics for full sample

| | Brazil | | | | | India | | | |
	Mean	Median	s.d	Obs.		Mean	Median	s.d	Obs.
Employment	207	70	431	387		367	70	1074	476
% Change in employment	22	7.8	63.2	368		19.7	14.3	37.7	471
% Change in sales	57.8	25	128	294		31.8	23.1	56.5	447
Materials share	0.44	0.41	0.31	194		0.41	0.4	0.25	433
Wage share	0.22	0.16	0.25	195		0.09	0.05	0.14	446
Capital intensity	0.75	0.32	1.19	156		0.56	0.25	1.03	395
ICT Intensity	0.04	0.01	0.18	278		0.02	0	0.13	430
Changes in ICT intensity	0.02	0	0.13	273		0.02	0	0.13	430

Note: Levels are for 2003 and changes are for the 2-year period 2001-2003.

emerges that infrastructure constraints—themselves closely correlated with other constraints—not only vary widely across the states or regions of India, but in those states where infrastructure is weaker, there is evidence of both lower *adoption* of ICT and lower *returns* to ICT adoption. A combination

similar across countries, as is the ranking by branch. Average employment growth has been quite similar in both countries, although median growth in India was double that in Brazil. In terms of shares, the major difference between the countries is with respect to labor. In India the mean and median wage

---

[5] See Commander, Harrison and Menezes-Filho (2009) for a full description of the data and analysis.
[6] For an interpretation of that literature, see Commander and Svejnar (2009).

shares were only 30 to 40 percent of those in Brazil. Capital intensity was also higher in Brazil but by a far smaller margin.

The six branches of manufacturing that have been sampled were picked not only because they provide significant variation in their production processes—and hence in their likely adoption of ICT—but also because they comprise a significant component of output and employment in manufacturing in both countries. In India, these six branches account for nearly 17 percent of total manufacturing employment and over 20 percent of value added. In Brazil these shares were around 30 percent and 32 percent respectively.

In terms of policy, both countries have seen clear changes over our reference period, particularly in India. Telecommunications have been liberalized, with significant entry of new providers, particularly for mobile services and internet service providers in India. With respect to the trade regime in India, tariff rates for the six branches fell on average by over

riod—with the exception of auto-components that received protection of 46 percent. In the case of labor legislation, in neither country was there significant change over our reference period.

ICT adoption

Table 2 provides some descriptive statistics regarding the extent of ICT adoption by 2003. The main adoption indicator was constructed from responses to a question regarding the degree of ICT in a given firm. These ranged from ICT not being used at all to all processes being automated and integrated into a central system.[7] Each response was scored for each firm for 2003. In addition, a usage index is also reported. The index was put together from responses concerning the intensity of use of ICT for four firm functions: accounting services, inventory management, marketing and product design, and the production process. These two sets of indicators give some

Table 2. Measures of ICT adoption, 2003

	Brazil					India			
	Mean	Median	s.d	Obs.		Mean	Median	s.d	Obs.
Summary measures									
Adoption index	3.5	4	1.22	491		2.94	3	1.05	476
Usage index	11.64	12	3.48	461		10.71	10	3.36	473
Hardware									
ICT capital as % of sales	4.18	0.59	17.78	278		3.34	0.44	17.01	379
PCs per employee	0.28	0.2	0.29	379		0.22	0.15	0.25	473
Servers per employee	0.04	0.02	0.07	372		0.02	0	0.05	473
Workforce usage									
% of non-production workers using PCs	69.6	90	37.9	484		53.9	59	34.6	476
% of production workers using ICT-controlled machinery	23.3	10	31.2	468		15.3	6	23.3	473

60 percent between 1999 and 2005. At the start of the period, the average tariff rate was 33.5 percent, falling to 15 percent by 2005, except in electronics, where it was only 1.9 percent. In Brazil, most trade liberalization occurred between 1990 and 1995, so that by 1998, Brazilian tariffs were mostly close to the Indian rates that existed in 2005, and hence were substantially lower than the Indian tariff rates at the start of the pe-

sense of both the depth and breadth of ICT in any given firm. What can be seen from Table 2 is that mean adoption and usage rates in India tend to be slightly lower in India than in Brazil. In addition, it is also true that firms with little or no adoption of ICT are far higher in India. Indeed, over 40 percent of Indian firms were using ICT in a minimal way, as against 25 percent in Brazil. ICT capital expressed as a share of sales, as

[7] The possible responses and their scores were: IT is not used at all=1; IT is used only for some offices along with accessing the Internet, e-mailing =2; IT is used for some advanced applications. Most processes are automated, but there is no integration into a central system=3; Most processes are automated and some of them are integrated into a central system=4; Almost all processes are automated and integrated into a central system=5.

well as other hardware indicators, also showed higher readings in Brazil than in India. In addition, the share of workers using ICT was significantly higher in Brazil.

As regards the characteristics of firms adopting ICT, it was found that in both countries, the size of the firm was generally positively and significantly associated with the indicators of adoption and usage. The age of the firm mostly did not matter,

sample dropped to just below 200, due to the fact that many Brazilian firms only reported ranges rather than levels for key financial variables.

The results reported in Table 3 were robust and reasonably similar for the two countries. All base variables were highly significant. The ICT capital stock was also highly significant in both cases. A main finding of the analysis was that the coeffi-

Table 3. Results from production functions: Signs and significance (Brazil results reported first)

	-1	-2	-3
Employment	+ : + Both***	+ : + Both***	+ : + Both***
Materials	+ : + Both ***	+ : + Both ***	+ : + Both ***
Capital	+ : + Both***	+ : + Both***	+ : + Both***
ICT	+ : + ** ***		+ : + * insig
Adoption = 3		+ : + insig ***	+ : + insig ***
Adoption = 4		+ : + ** ***	+ : + ** ***
Adoption = 5		+ : + ** *	+ : + * insig
Observations	132; 335	132; 335	132; 335
R-squared	0.85; 0.87	0.86; 0.87	0.86; 0.88

Notes: *, ** and *** indicate significance at the 10 percent, 5 percent and 1 percent levels respectively; Insig=insignificant

although in Brazil older firms tended to have higher ICT per worker and per unit of sales. The extent of unionization of the workforce mostly did not matter. In terms of ownership, multinationals in Brazil used ICT more intensively; in India this is true for foreign joint ventures. Firms with foreign ownership or participation tend to have higher ICT use or adoption.

ICT and its consequences for productivity

In a detailed analysis involving the estimation of firm-level production functions for both Brazil and India, the main coefficients of interest were the elasticity of output with respect to ICT capital, and the implied (gross) rate of return to ICT capital for any given firm. Table 2 summarizes the results, where controls for industry, region/state and age have been introduced. For simplicity, only the sign and significance level are indicated. The number of observations in the Brazil

cients on ICT capital corresponded to very high median rates of return in both countries. However, although high returns to ICT investment have also been found in developed countries,[8] this might be partly explained by a high cost of ICT capital due to depreciation and obsolescence. It might also be conjectured that ICT may itself be correlated with other omitted variables, such as skills and other firm characteristics which may explain why returns to ICT are high. However, further careful exploration of these possibilities tended to confirm the validity of these high estimates. None of the tests—including use of instrumental variables—resulted in large changes to the estimated ICT coefficients.

A secondary question of interest is whether the sort of complementary investments and changes that have been found to be important in the OECD context have also been present in these two large emerging markets. To look at this,

[8] See, for example, Brynjolfsson and Hitt (2000); Stiroh (2004).

several measures of organizational change were introduced to see whether the returns to ICT investment were affected. The measures used were, first, whether a firm "removed a level of hierarchy or reduced the number of reporting levels" and whether such change was explicitly related to IT over the past three years. The second measure was whether a firm had "improved monitoring of individual workers or groups of workers" related to IT. The third was whether a firm had "improved management decision-making based on up-to-date information." Once firms that made minimal or no use of ICT were excluded, there was a very strong and positive effect of reducing hierarchies in Brazil; indeed for Brazilian firms that use ICT more intensively, the return to ICT capital stock is only significantly positive if they also undertake organizational change at the same time. But in India this effect was absent. The other measures of organizational change do not appear to have any significant impact in either country. It is, of course, quite possible that these measures do not accurately measure the organizational and managerial changes that have been occurring. And that may matter.

A further consideration of significance from a policy perspective is whether the use of ICT in these two developing countries has been associated with any change in the skill composition of an adopting firm's workforce. There is now copious evidence from OECD economies that the introduction of ICT has shifted the relative demand for skills, with the result that firms rely to a greater extent on more skilled workers.[9] If this were also the case in developing countries, with their relatively large shares of unskilled workers, this might have significant implications for the employment rates of different skill groups, as well as for wage inequality.

The evidence from this dataset is that, in common with OECD economies, there has been a shift in relative labor demand in Brazil and India.[10] For both production and non-production workers, the share of those with either upper-secondary or college education clearly increased between 2001 and 2004, while that for less educated workers declined. For non-production workers, the shift was mainly towards college-educated workers, at the expense of those with only upper-secondary education. Interestingly, Harrison (2008) also finds that ICT adoption itself has been positively associated

with regional differences in the supply of educated workers, with ICT adoption being negatively associated with a region's relative wage for more educated workers. These features suggest a process at work that has much in common with that in the OECD and, in a similar vein, emphasizes the importance of continuing investment in education and skills. This appears particularly important in Brazil, where long-standing inadequacies in the educational system have led to skills shortages, and raised barriers to innovation.

Impact of infrastructure and other constraints on adoption and productivity

As mentioned above, a characteristic of many emerging markets is the presence of significant impediments to doing business and, hence, to limitations in the quality of the overall business environment. This prompts the question as to whether features of the policy and institutional environment affect both the adoption of, and the returns to, ICT. With the dataset used in this study, it was possible to explore these questions further, using only information on India.[11]

In the first instance, the impact of infrastructure quality and labor regulation on ICT adoption and, hence, of ICT capital intensity in different Indian states was analyzed.[12] Identification of the impact of infrastructure quality and labor regulation on the estimated returns to ICT capital was then undertaken. Two measures of the institutional environment in which firms make their decisions about ICT investment were used. The first was the number of days a firm experienced "power-related problems (power cuts or surges, either partial or total) from the public grid." The value was then averaged within states, creating a variable that ranged from 10.3 days in Tamil Nadu to 33.0 in Delhi. This measure can be taken as a good indicator of the quality of general infrastructure. The cross-state correlations between the state mean number of days disrupted by power-related problems and the proportion of firms in a state reporting that a particular factor constrained their ICT adoption were almost all high and significant, particularly for skills, labor laws, the number of customers and suppliers using the internet, and a lack of government support for ICT. In addition, the correlations across states between the different reported constraints were also extremely high,

9   See, for example, Bresnahan et al. (2002).
10   The full analysis is in Harrison (2008).
11   There were too few observations in Brazil to undertake the same analysis, and the measures of infrastructure quality and labor market regulation were also not available for Brazil.
12   Similar analysis has been done by Besley and Burgess (2002) when looking at the impact of labor regulation on manufacturing growth in India.

suggesting a cluster of poor institutions and/or aspects of the economic environment in some states that are not conducive to ICT adoption.

A second state-level measure of the institutional environment that was used was a labor regulation index for 1995 constructed by Besley and Burgess (2004). This variable comes from state-specific text amendments to the Industrial Disputes Act 1947, which were coded as either pro-worker (positive), pro-employer (negative), or neither, in order to construct a cumulative index of labor market regulation. For the states in the sample, the index ranges from 4 in the most pro-worker state (West Bengal) to –2 in the most pro-employer states (Andhra Pradesh and Tamil Nadu).

Both poorer infrastructure and more pro-worker labor regulation were found to be associated with lower levels of ICT investment. The size of the estimated effects was economically significant. Moving from the state with the highest mean number of days disrupted (Delhi with 33.0) to the state with the lowest (Tamil Nadu with 10.3) was associated with, on average, a 400 percent increase in ICT intensity. While moving from the state with the most pro-worker labor regulation index (West Bengal with 4) to the state with the most pro-employer index (Andhra Pradesh or Tamil Nadu with –2) was associated, on average, with more than a 200 percent increase in ICT intensity.

Poorer infrastructure quality and more pro-worker labor regulation were both associated with significantly lower levels of ICT adoption across Indian states. These findings have important implications for policy. Certainly changes to the power sector, both in terms of management and ownership appear desirable in India, while further attempts to limit the impact of restrictive labor laws and practices could be expected to boost not only ICT investment, but also employment more generally.

Conclusion

The last couple of decades have seen extraordinary changes in the way that firms function and use technologies. The great ICT revolution has been increasingly shown to be a major factor behind the acceleration in productivity in North America and, to a lesser extent, in Europe and Japan. In the rich countries, the impact of the ICT revolution has been through two principal channels: first, the creation or production of new technology itself—a process that remains significantly dominated by the rich economies and the U.S. in particular—and second, the use, or application of the new technology by firms in pursuing their respective businesses. In the case of the developing and emerging economies, it is this second channel that is mainly of interest, given their role as adopters of extant technology, rather than the creators of new technology per se. Although, more recently, the growth of a buoyant software industry, particularly in India, has been associated with some shift into the creation of new products and business practices,[13] it is fair to say that the developing market issue of interest is mainly about how, why, and when their firms adopt technologies, and with what consequences for productivity and firm growth. In this sense, the obvious question addressed in this paper has concerned the way in which developing country firms—in this instance manufacturing firms in selected sectors in Brazil and India—have moved, with a predictable lag, to take on new technologies. In particular, the paper has thrown light on the timing and consequences of ICT adoption, using specially raised data from approximately 1,000 firms. The results—taken with the obvious caveats about generalization, given the sample size and composition, and the relatively limited number of data points—are quite striking. With respect to the adoption of new technology, larger sized firms and foreign ownership tend to be associated with higher adoption. In both countries, adoption has also been associated with a higher share of educated workers (particularly in Brazil),[14] and, hence, with a change in the skill mix, a phenomenon that has also been well-documented for the OECD. The evidence also shows that Brazilian firms have, on average, adopted more ICT than their Indian counterparts, and have used their ICT more intensively. However, firms operating in states with good institutional arrangements in India tend to have adoption rates similar to their Brazilian comparators. Certainly, the variation between countries in adoption is far smaller than the aggregate data would suggest.

With respect to the association between ICT and productivity, the analysis has established that, also consistent with evidence from OECD economies, there have been very high returns to ICT adoption. There is some limited evidence in Brazil that high returns have been located in firms that have also undertaken investments in organizational change.

Finally, given the growing concern in policy circles with the

---

[13] See Commander (2005).

[14] See Harrison (2008) for a detailed analysis of these dimensions using the same dataset.

quality of the business environment, it appears that, when using the Indian evidence, poorer infrastructure quality in particular is associated not only with a lower level of ICT adoption, but also with lower returns to ICT capital investment. It is clear that much of the policy challenge in India should focus on addressing the sources at state level of these inefficiencies and institutional weaknesses.

References

Basu, Susanto, John G. Fernald, Nicholas Oulton, and Sylaja Srinivasan. 2003. "The case of the missing productivity growth: Or, does information technology explain why productivity accelerated in the United States but not the United Kingdom?" *NBER Macro-Economics Annual.*

Bloom, Nick, Rafaella Sadan, and John Van Reenen. 2006. "It ain't what you do, it's the way you do I.T.: Investigating the productivity miracle using US multinationals." Centre for Economic Performance. April. Mimeo.

Breshnahan, Tim, Erik Brynjolfsson, and Lorin Hitt. 2002. "Information technology, workplace organization and the demand for skilled labor." *Quarterly Journal of Economics* 117(1):339–76.

Brynjolfsson, Erik and Lorin Hitt. 2000. "Beyond computation: Information technology, organizational transformation and business performance." *Journal of Economic Perspectives* 14(4):23–48.

———. 2004. "Computing productivity: Firm-level evidence." *Review of Economics and Statistics* 85(4):793–808.

Commander, Simon (ed.). 2005. "The Software Industry in Emerging Markets." London: Edward Elgar.

Commander, Simon, Rupert Harrison, and Naercio Menezes-Filho. 2009. "ICT adoption and productivity in developing countries: New firm-level evidence from Brazil and India." *Review of Economics and Statistics,* forthcoming.

Commander, Simon and Jan Svejnar. 2009. "Institutions, ownership and competition: The efficiency of firms in 26 transition economies." *Review of Economics and Statistics.* Forthcoming.

Harrison, Rupert. 2008. "Skill biased technology adoption: Firm-level evidence from Brazil and India." Institute for Fiscal Studies. June. Mimeo.

Jorgenson, Dale. 2001. "Information technology and the US economy." *American Economic Review* 91(1):1–32.

Oliner, Stephen and Daniel Sichel. 2000. "The resurgence of growth in the late 1990s: Is information technology the story?" *Journal of Economic Perspectives* 14(4):3–22.

Stiroh, Kevin. 2004. "Reassessing the role of IT in the production function: A meta analysis." Federal Reserve Bank of New York. Mimeo.

World Bank. 2006. *Information and Communication for Development: Global Trends and Policies.* Washington, D.C.

Chapter 2.11

Technology and Innovation for Addressing Climate Change: Delivering on the Promise

Laura Altinger,
United Nations Economic Commission
for Europe

Introduction

That climate change is occurring and that it is caused by man are the unequivocal conclusions of the Fourth Assessment Report (AR4) of the Intergovernmental Panel on Climate Change (IPCC), reflecting broad scientific consensus in this matter. Many key observed climate indicators are already moving beyond their natural variability patterns, including global mean surface temperature, global ocean temperature, the extent of Arctic sea ice, sea-level rise, precipitation, ocean acidification, and extreme climate events. Since 2000, fossil fuel emissions—the main cause of anthropogenic climate change—have risen at an accelerating 3.4 percent per annum.

Since the publication of AR4, new research has revealed further, discomforting evidence of global climate change. Global ocean temperature—considered to be a better indicator of change in the climate than air temperature variations, given that oceans store the largest amount of heat in the Earth's surface—has risen significantly in recent years. Current estimates show that ocean warming is about 50 percent higher than was reported in AR4. The rate of sea-level rise has been on the increase since 1993, largely explained by the loss of ice from Greenland and Antarctica, with new projections suggesting a rise of around a metre or more by 2100. Further, a dramatic reduction in the area of Arctic sea ice in summer has been discernible since the publication of AR4, with a loss of almost 2 million square kilometres in the minimum area covered in 2007, and a similar decrease occurring in 2008 (University of Copenhagen, 2009).

Previously observed climatic imbalances are considered to have been further exacerbated by "feedback effects." For example, since an ice-free ocean absorbs more heat than an ice-covered ocean, the loss of Arctic sea ice creates feedback in the climate system which results in more warming. Similarly, water vapour, which contributes to the natural greenhouse effect that makes Earth habitable, follows and amplifies changes in global temperatures, including those caused by human activity. Another important feedback mechanism derives from naturally occurring carbon sinks, which act to neutralize CO_2 emissions. Their ability to absorb emissions has been weakened by increasing ocean acidification, ocean circulation changes, as well as water, temperature, and nutrient constraints on land CO_2 uptake. In addition, latent carbon pools are increasingly released into the atmosphere, e.g., through thawing of the Arctic permafrost.

In parallel, the burden on the environment from human activities is increasing. Greater economic and population growth are, and will continue to be, major determinants of global greenhouse gas (GHG) emissions. For example, accelerating energy needs of large developing countries, such as the drive by India and China to boost their standards of living and growing populations, are likely, at least in the short run, to be met by more coal-fired power plants, which are rapidly turning these countries into important GHG emitters.

The effects of climatic warming include increasing temperatures, higher climate variability, increased precipitation in wetter areas and declines in drier areas, as well as a higher incidence of extreme weather events, such as heat waves, storms, and floods.

While all human and natural systems will be adversely affected by even a modest rate of climate change, poor and developing countries and communities are particularly vulnerable, due to their geographic location and limited resources. Many developing countries will be hit hard by climate change and will face a variety of development risks and challenges, especially those related to ecosystems, agriculture and food security, water management, infrastructure, and health (World

Bank, 2009; IPCC, 2007a).

The scale of the climate change challenge suggests that technology and innovation will have to play a pivotal role in global mitigation efforts to reduce GHGs and in adaptation designed to decrease vulnerability to the adverse impact of climate change. Existing technologies can already be deployed to stabilize or reduce global emissions and to limit or manage risk. New technologies will also play an important role in effecting the future long-term transformation of our economies onto green and sustainable growth trajectories.

While addressing climate change through technology and innovation will yield positive environmental effects, it also offers potentially vast co-benefits that contribute to other developmental goals. Economic and social progress during the past centuries has been achieved mainly due to technology. Innovation policy and climate change mitigation and adaptation through technology and innovation should therefore be pursued in a mutually reinforcing way.

In this paper we review the key sectoral and regional trends and necessary technologies for tackling climate change, including the associated cost estimates. We then discuss existing challenges in bringing technologies to bear upon climate change, and the policy and institutional structures which support or impede this process.

Figure 1. GHG emissions by sector in 2004 (in %)

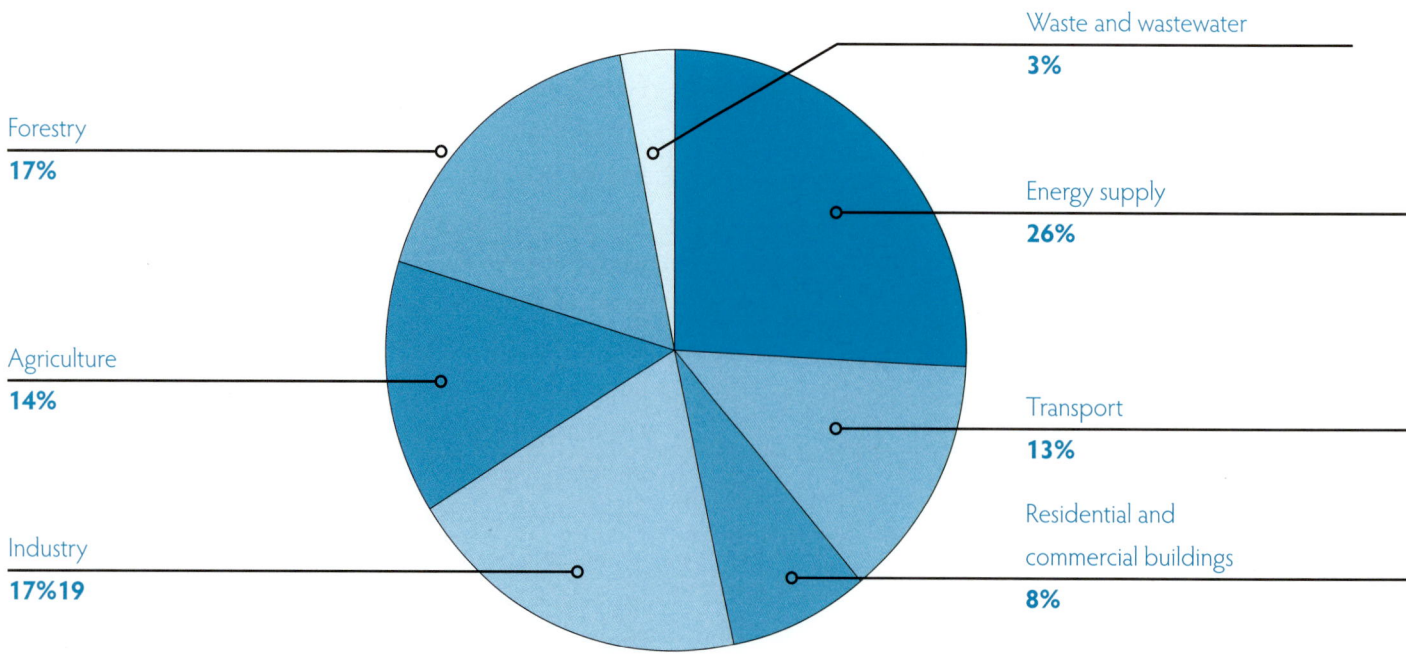

Source: IPCC, 2007d.

Table 1. GHG emissions by country grouping and region

UNFCCC country grouping	Share of global GHG emissions, 2004 (in %)	Share of population (in %)	Share of global GDP (in %)	GHG emissions per unit of GDP (kg CO_2eq / US$)
Developed countries (Annex I)	46.4	19.7	56.6	0.683
Developing countries (non-Annex I)	53.6	80.3	43.4	1.055
Regions:				
Africa	7.8			
Central Eastern Europe	9.7			
Middle East	3.8			
Latin American and Caribbean	10.3			
East Asia	17.3			
South Asia	13.1			
U.S. and Canada	19.4			
Japan, Australia, New Zealand	5.2			
Europe	11.4			

Source: IPCC, 2007d.

The role of technology in addressing climate change

This section reviews the key sectoral and regional trends of GHG emissions which are causing global warming, highlighting those which are a priority for technology and innovation in tackling climate change.

Emission trends by sector

Figure 1 shows the approximate share of global emissions from the different economic sectors. Contributing roughly one-quarter (27 percent) of total anthropogenic CO2-eq emissions in 2004, energy supply (electricity and heat generation) is the most dominant source of GHG emissions, followed by industry (19 percent), land use and land-use change (17 percent), agriculture (14 percent), transport (13 percent), residential, commercial and service sectors (8 percent) and waste (3 percent).[1] Since 1970, the largest growth in GHG emissions has come from power generation and road transport.

Emission trends by region

On a geographic basis, countries are divided into developed and developing, according to the classification used by the United Nations Framework Climate Change Convention (UNFCCC), which distinguishes Annex I (developed country Parties) from non-Annex I (developing country Parties) to the Convention. Table 1 shows that in 2004, Annex I countries accounted for 46.4 percent of global GHG emissions, and 20 percent of world population, while non-Annex I countries account for the remaining 53.6 percent and 80 percent of world population (IPCC, 2007d).

GDP per capita and population growth were the main drivers of the rise in global emissions between 1970 and 2004. To some extent, these income and population effects were offset by declining carbon and energy intensities. Projections show that global GHG emissions will increase between 25 and 90 percent by 2030 relative to 2000, without a change in policy. Up to three-quarters of the increase in emissions is expected to come from developing countries (IPCC, 2007d).

---

[1] Some uncertainty remains with regard to the estimates of CH_4 and N_2O emissions (prevalent in agriculture) and CO_2 from agriculture and forestry.

Emission reduction targets

To prevent dangerous anthropogenic interference with the climatic system, AR4 estimates that the global mean temperate rise would need to be contained to 2°C. This translates into emissions pathways which will have a bearing on the ultimate success of achieving this temperature target. There is widespread agreement that stabilizing GHG concentrations at 450 to 750 parts per million (ppm)—the range thought sufficient to contain serious damage to the planet, will require significant innovation and large-scale adoption of emission-reducing tech-

to be made by 2020, and then 2050, to stay within the range that is thought sufficient to produce a safe outcome. In projection scenarios run by the International Energy Agency (IEA), global savings of CO_2 emissions would amount to as much as 48 Gt (IEA, 2008).

Technology for mitigation

The types of technology identified as playing a central role in mitigation efforts include energy efficiency improvements from production through to end use; solar, wind, nuclear fis-

Figure 2. Major technologies and their GHG-abatement potential up to 2020, total 19 Gt CO₂e

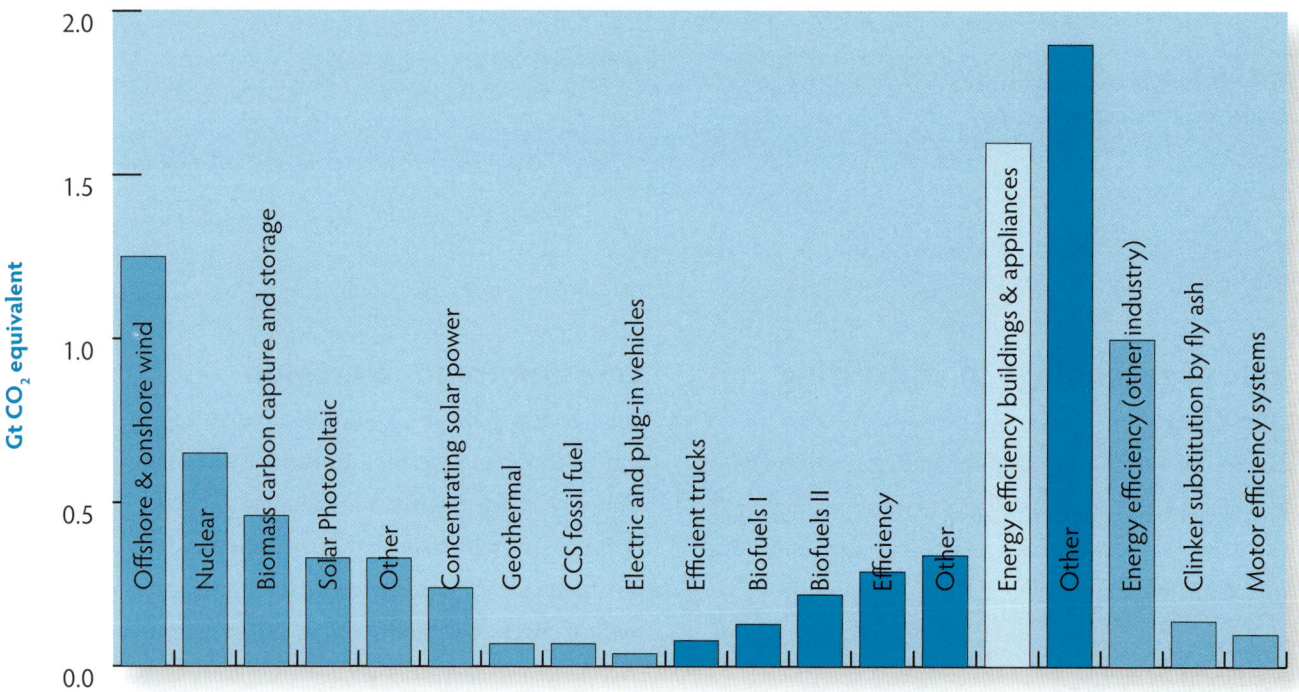

Source: Tomlinson, 2009.

nologies (IPCC, 2007d). The AR4 gives an example of the approximate timeline for the target emission pathways.

The sheer scale of the climate challenge is enormous. To achieve the targets set out in AR4, global emissions would have to be reduced by 19 Gt of CO_2 equivalent (GtCO₂e) by 2020 (McKinsey, 2009). Half of this reduction could be expected to be achieved by land use, land use change, and forestry (LULUCF),[2] while the other half could be tackled by applying existing or near-market technologies and resource efficiency measures.

However, much deeper cuts in GHG emissions would have

sion and fusion, geothermal, biomass, and clean fossil technologies, including carbon capture and storage; energy from waste; hydrogen production from non-fossil energy sources, and fuel cells (IPCC, 2007d).

Four major polluting sectors: energy, transport, buildings, and industry hold the largest potential for applying these technologies with major GHG-abatement potential in the near and long term. Figure 2 shows the scale of emission reductions which could be achieved by developing or deploying existing or near-market technologies and efficiency improvements across these four major sectors (Tomlinson, 2009).

---

2 "Land use, land-use change and forestry (LULUCF)" is defined by the UN Climate Change Secretariat as a greenhouse gas inventory sector that covers emissions and removals of greenhouse gases resulting from direct human-induced land use, land-use change and forestry activities.

Although a substantial part of these emissions savings derive from efficiency savings in the buildings, industrial, or transport sectors, wind and nuclear technologies are highlighted as having major abatement potential in the near-term.

In the longer run-up to 2050, the main technologies that may be relied upon to deliver some four-fifths of the required reductions are in the technological pipeline for full commercialization after 2020. However, in order for this to happen,

Box 1. Energy efficiency in residential and commercial buildings

The built environment is a large contributor to global warming. When including the emissions from electricity use, CO_2 emissions from the building sector amount to almost a quarter of global emissions. In addition, other GHG emissions from cooling and refrigeration contribute to some 15 percent of total GHG emissions of buildings. Therefore, enhancing energy efficiency in buildings is an important measure for reduction of GHG emissions from buildings and is seen as the most cost-effective mitigation opportunity in the sector.

Most technologies for achieving energy efficiency in buildings already exist widely, are cost-effective, and have been successfully applied. These include passive solar design, high-efficiency lighting and appliances, highly efficient ventilation and cooling systems, solar water heaters, insulation materials and techniques, high-reflectivity building materials, and multiple glazing. The largest energy savings can be realized in new buildings through an integrated and environmentally sustainable design process.

Carbon dioxide mitigation in buildings offers an array of concomitant benefits, including improvements in indoor and outdoor air quality, higher energy security, as well as increased comfort, health, and quality of life.

Several market barriers have been identified, including: the high cost of obtaining reliable information on energy efficiency measures, misaligned incentives between landlords and tenants, limited access to finance, energy price subsidies, and fragmentation of the industry and design process into multiple professions and trades.

Governments can promote the energy efficiency savings in buildings through a number of measures. Education, training, and information-sharing initiatives are vital. Appliance standards and building energy codes and labelling should be continuously updated. Not only can governments lead by example in their public procurement policies, but they can also ensure that energy pricing and other financial support are incentive-compatible.

Source: IPCC (2007d), Chapter 6.

Figure 3. Major technologies and their GHG-abatement potential up to 2050, total 48 Gt CO_2e

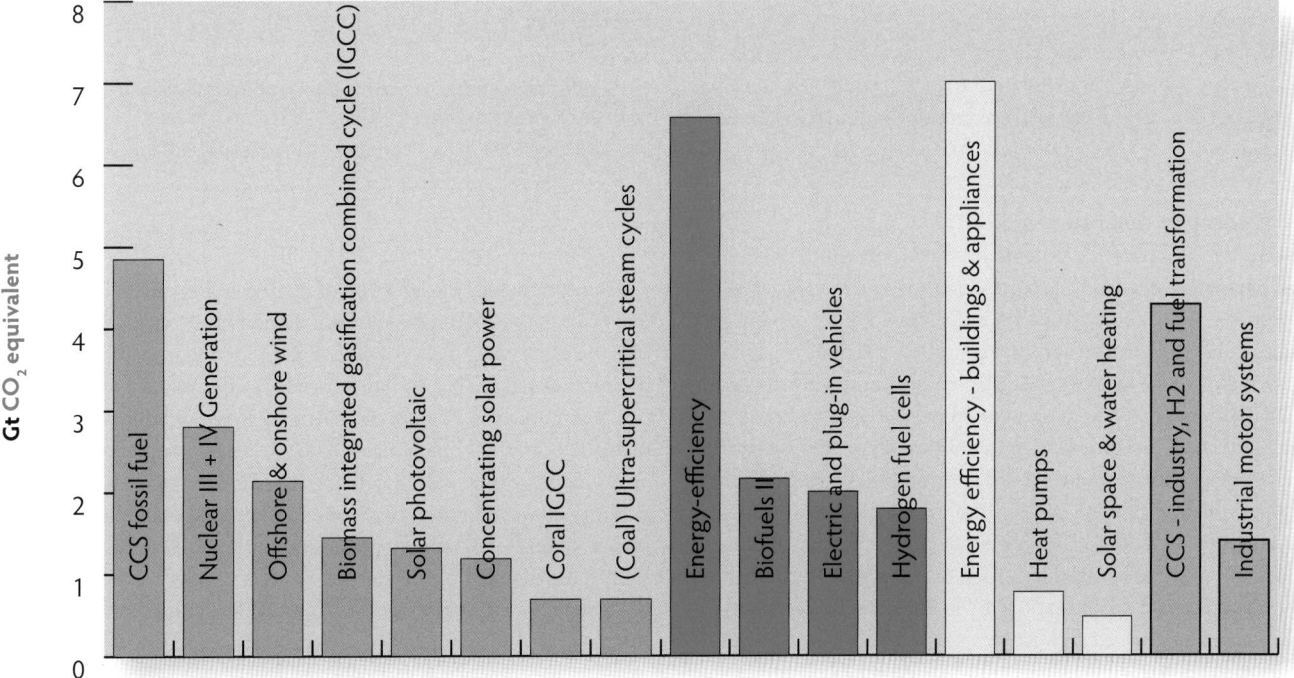

Source: Tomlinson, 2009.

these will still require investment in research, development, and demonstration to become operational. Figure 3 illustrates their theoretical abatement potential up to 2050.

Energy-efficiency measures still dominate the landscape up to 2050. As shown earlier in Table 1, as compared to developed countries, developing countries still have a more emissions-intensive production process, which in the longer term will be reduced as more efficient technologies become accessible to them.

Ongoing research into new technologies to mitigate climate change currently being pursued in the area of geo-engineering may also offer promise. Two major research areas focus on ways of removing carbon dioxide from the atmosphere and reducing the amount of sunlight that comes to earth.

Box 2. Carbon capture and storage (CCS)

Carbon capture and storage (CCS) is the process of capturing carbon dioxide and other greenhouse gases at the point of emission and storing them, either in underground reservoirs (geological sequestration) or by injection into oceans (ocean sequestration), or by converting them into solid materials.

CCS is particularly relevant to fossil fuel-based power plants, which are widely expected to remain the mainstay of energy production well into the next decades, especially in middle-income countries, such as China and India, which will see their energy supply greatly expand during this period of rapid industrialization.

In order to be sequestered from power plants, carbon dioxide must be captured as a relatively pure gas. Although some industrial processes routinely capture CO_2 as a by-product, existing capture technologies are not yet cost-effective for widespread use. Costs to separate out CO_2 account for some three-fourths of total CCS costs. Technologies likely to be exploited in this context include absorption, adsorption, low-temperate distillation, gas separation membranes, and mineralization and bio-mineralization.

Geological carbon dioxide sequestration typically uses sites such as depleted oil and gas reservoirs, shale formations with high organic content, unmineable coal seams, or underground saline formations. Many of these applications yield value-added by-products or other advantages.

The potential for mitigating climate change is huge. The International Energy Agency has estimated that CCS could account for around 20 per cent of the emission reductions required in the energy and industrial sectors by 2050, making it one of the single most effective instruments for combatting global warming.

However, there are also concerns about the safety of carbon storage, especially the risk that it will escape and migrate to the earth's surface or contaminate drinking water. In the light of the possible dangers and the lack of understanding about and long experience with such technology, there is currently a good deal of public reluctance to accept CCS. This public perception is viewed as one of the main barriers to progress.

Another key challenge will be to make it cost-effective and commercially viable. CCS plays a central role in the European Union's efforts to contain global warming. Currently, there is work being done in the EU to set up some dozen demonstration plants that will be up and running by 2015, in order to test a variety of CCS technologies, with the aim of making the best ones commercially viable by 2020. Nevertheless, the added cost of building and operating each CCS demo plant has been estimated to exceed €1billion per plant or about €80-100 per ton of carbon dioxide. This cost would have to be cut in half in order for the technology to be commercially viable. Progress will also be contingent on a higher future price of carbon.

There are also regulatory challenges, such as drafting new regulations to govern the transport and storage of carbon dioxide, which must be met.

Source: Carbon Sequestration Leadership Forum (CSLF), available at: www.cslforum.org and Zero Emissions Platform, available at: http://www.zeroemissionsplatform.eu/

Box 3. Hydrogen fuel cell vehicles

Hydrogen fuel cells have the potential to replace the internal combustion engine in vehicles and to provide power in stationary and mobile power applications. They typically use an electrochemical process to convert hydrogen and oxygen into electricity.

When compared to the conventional combustion-based technologies used in power plants and passenger vehicles, fuel cells offer many benefits. First, they produce smaller quantities of greenhouse gases. Fuel cell vehicles with an on-board fuel reformer will emit two-thirds less pollution than a gasoline combustion engine. In the case where pure hydrogen is used, fuel cell vehicles are carbon neutral, emitting only heat and water as by-products. Moreover, they are expected to achieve efficiencies on the order of 40 to 45 percent vis-à-vis internal combustion engines, which convert only about 15 per cent of the energy from gasoline to turn the wheels of a car or truck.

Many types of fuel cells already exist, such as Polymer Electrolyte Membrane Fuel Cells, Alkaline Fuel Cells, Solid Oxide Fuel Cells, Phosphoric Acid Fuel Cells, and Molten Carbonate Fuel Cells.

The commercial introduction of hydrogen fuel cells to the market faces two major challenges: cost and durability. As with CCS technology, the high cost of fuel cell power systems does not yet allow them to be commercially viable, such that they have an advantage over conventional technologies. A fuel cell system for transportation applications would need to cost approximately US$30/kW to be competitive.

An additional problem is that fuel cell systems are not yet sufficiently durable. In the case of transportation applications, fuel cell power systems will need to achieve roughly the same level of durability and reliability of current automotive engines, i.e., a 5,000-hour lifespan, to be competitive. They must also be able to function over the range of vehicle operating conditions (40°C to 80°C).

Source: International Partnership for the Hydrogen Economy, http://www.iphe.net

Economic costs

The expected economic cost of reorientation to a low- or zero-emissions world is immense. The *Stern Review* has pointed out that the earlier action is taken to mitigate climate change, the more cost-effective such intervention will be. Based on economic modelling, the *Review* estimates that the overall costs of climate change would amount to a loss of at least 5 percent of global GDP for each year in the foreseeable future. Taking into account a wider set of risks and damages, this could rise to 20 percent of GDP or more. By contrast, the costs of action now could still be limited to some 1 percent of global GDP per year.

Nevertheless, required annual average investment up to 2050 is still estimated to be close to $1 trillion, or over 1 percent of world GDP. Total net additional investment for developing countries up to 2030 is estimated to be in the range of 2005 US$200–210 billion only for mitigation, while additional adaptation costs are estimated at around US$ 49–171 billion per year (UNFCCC, 2008).

To rise to the challenge of achieving the technological revolution necessary for climatic stabilization, effective policies and institutions are needed both to support the development of GHG-abatement technology, as well as to finance its global diffusion and adoption. It is vitally important that economic prices of goods and services better reflect their real costs, especially those to the environment.

Delivering on the promise of technology to combat climate change

The process of development, transfer, and diffusion of these climate technologies will require policy intervention at both the domestic and international level, in both developed and developing countries.

Developing environmentally sound technologies

As already discussed above, a very large investment will be needed to generate the required innovation effort. Much of this effort can be expected to be undertaken by the private sector. Although innovation is not limited to research and development (R&D), it is a well-tracked indicator of innovative activity. The empirical literature finds substantial R&D capacity in the principal sectors and companies relevant to GHG emissions (Newell, 2008). But more needs to be done to re-orient this effort toward green technologies.

It is well-known that market failures associated with innovation—that is, deficiencies in markets to allocate resources—exist, and must be corrected for by governments. This is particularly true for innovation in the context of climate change (Stern, 2007).

Cap-and-trade systems

In the context of climate change, the biggest market failure has arguably been the missing market for carbon emissions, a classic environmental externality. As a result, private agents do not face the full social costs of their GHG emissions, resulting in a level of emissions which is much greater than the socially optimal level.

Creating a positive price for GHG emissions based on the "polluter pays" principle—for example, through cap-and-trade systems, such as the European Trading Scheme (ETS) or the Kyoto Protocol (KP) of the UNFCCC—is therefore the first step for potentially internalizing this environmental externality and realigning polluters' incentives both to pollute less and to make full use of existing emissions-abatement technologies. This additional market demand for such technologies attracts higher R&D on the supply side, aimed at producing new or better pollution-abatement technologies.

The incentive to invest in new environmental technology will be influenced by many external factors as well as systemic features of the cap-and-trade systems. For example, if the level of the cap is set too generously, there will be little incentive for making behavioral changes including investments in alternative environmental technologies. Many GHG emission sources may not be covered. Similarly, uncertainty about the continuity or future of the cap-and-trade regime will certainly deter investors in new technologies. The carbon price is another major factor influencing investment. Investment in new technologies is a positive function of the carbon price. The higher the price of carbon, the more incentives there will be to exploit new opportunities offered by environmentally sound technologies (ESTs). Therefore, in a downturn, there will be fewer incentives to invest in clean technologies, unless the government provides support, for example, in the form of production or consumption subsidies, additional demand through public procurement, or resources to boost R&D.

Carbon tax

A second, positive development in this respect has been the introduction of a carbon tax in countries such as Sweden and Denmark. Such schemes bring the cost of production closer to its true cost, by internalizing the environmental externality. However, the challenge lies in establishing a suitable price for emissions that correctly reflects underlying social costs. Carbon taxes give both consumers and producers an incentive to reduce their carbon emissions, through the introduction of carbon-cutting technologies. France is currently in the process of debating the imposition of a new carbon tax for 2010, with current proposals by the Rocard Commission setting the price of carbon emissions at EUR 32 per ton.

The carbon tax that exists in some European countries is superimposed over and above the commitments called for carbon emissions in the European Trading System and the Kyoto Protocol, both of which apply to all European Union (EU) countries. A carbon tax could have a positive impact on technology development, because it has the effect of lowering the relative price of environmentally friendly technologies. How it interacts with the cap-and-trade regime is an empirical matter. A priori, a higher tax could make producers more likely to meet their cap by investing in ESTs, as least those which have come into the price range as a result of the carbon tax.[3]

Other market failures

Other market failures have been described in the extensive R&D literature (de Coninck, 2008; Stern, 2007) on this topic. These include spillover effects, adoption externalities, and incomplete information, all of which will cause private markets to underinvest in new technologies for emissions abatement and may warrant specific public policy intervention to directly target R&D.

For example, firms that invest in R&D for climate technologies may not be fully able to capture all the associated benefits from their investments in innovation—regardless of intellectual property rights—due to spillovers to other technology producers and users, to long asset lives for this type of GHG-reducing technology, and to the fact that the market for such innovations is very dependent on the stability of the regulatory or policy regime. This is likely to result in underinvestment by the private sector.

Second, adoption externalities for new technology, such as learning-by-using, learning-by-doing, or network externalities can also prevent R&D investors from reaping the full reward for their investments. Typically, there is a benefit associated with the overall scale of technology adoption, which may be limited for certain GHG-abatement technologies.

Finally, incomplete information arises from the large uncertainty associated with the returns to investment in innovation, especially where the international policy regime under the UNFCCC has yet to be fully determined and against the background of substantial time lags between initial discovery and profitable market penetration. Equally important, there is a mismatch between the owner of residential or commercial property, who typically makes the decision about the level of investment in energy-efficient buildings, and the tenant who has to pay the utility bills.

In this context, the role for government is clear. It must ensure that there is a positive price attached to carbon that fully internalizes the environmental externality, creating the right incentives for the market to reduce emissions and to pursue investment and R&D for producing and consuming environmentally sound technologies, otherwise known as "market-demand pull." In addition, government must support this innovation process by creating enabling conditions, some of which might include ensuring adequate human capital skills and technological competencies relevant to clean-energy and environmentally sound technologies (especially scientists and engineers); a supporting physical infrastructure for greater technological sophistication; adoption of active measures to promote technology diffusion; and strong linkages between firms and key research and development (R&D) actors (World Bank, 2008). For example, to support private sector developments in the energy sector, governments could usefully supply smart grids or storage technologies.

As usual for public policy, governments should fund that part of the innovation process least likely to be tackled by the private sector, such as pre-commercial research. Some experts have even argued in favor of public support for technology demonstration projects, for example in the area of carbon capture and storage (CCS), due to their high cost and the non-appropriable nature of returns to knowledge generated through such projects (Newell, 2008).

---

[3] A carbon tax means that each ton of CO_2 produced will cost a company that much more to emit than without the tax, making it more worthwhile for them to invest in environment technologies that help to reduce CO_2. The cost of the ESTs is compared to the savings that accrue from each ton of CO_2 emissions avoided throughout the lifetime of the new technologies. Therefore, a carbon tax will reduce the relative price of ESTs, bringing them into the profitable range, especially those that are least costly, or those with the greatest potential of reducing emissions.

Technology-oriented agreements (TOAs) have also been put forward as a means of helping to address global climate change through technology (de Coninck et al., 2008; Newell, 2008). TOAs that cover technology mandates, standards, or incentives, appear to offer particularly large potential to be effective in environmental terms and can operate in conjunction with emissions-reduction policies. Examples include the Multilateral Fund under the Montreal Protocol, the Global Environment Facility (GEF), the European Union Renewables Directive, and proposals for a CCS technology mandate or a Zero-Emission Technology Treaty.

Transferring climate change technology to developing countries

The growth of carbon emissions over the next decades will come mainly from developing countries, especially China and India. Although developing countries do not yet have binding emissions targets to fulfil, they will play an increasingly important role in the international effort to curb GHG emissions through ESTs. This is likely to be a major issue at the 15th session of the UNFCCC Conference of the Parties (COP-15) in Copenhagen in December 2009.

The debate centers around how to construe the "shared vision" for long-term cooperative action, contained in the Bali Action Plan, including a long-term global goal for emission reductions, to achieve the ultimate objective of the UNFCCC, in accordance with the principle of common but differentiated responsibilities and respective capabilities, and taking into account social and economic conditions and other relevant factors. Developing countries emphasize that the responsibility for climate change and its consequences must lie with developed countries as they are responsible for most GHG emissions to date. Therefore, they are generally reluctant to accept binding emissions targets but would like progress on technology transfer and financing from developed countries, while emphasizing sustainable economic development and climate change adaptation.

Due to the urgency of action on climate change, others argue that the failure by developing countries to accept legally binding emissions targets—even if these initially only represent GHG stabilization rather than reduction—would be a serious shortcoming for any internationally negotiated agreement.

According to recent research, the intensity of innovation is closely related to per capita income. Developing countries typically lack the ability to generate cutting-edge innovations. Therefore, they generate technological progress by relying on the adoption and adaptation of pre-existing but new-to-the-market or new-to-the-firm technologies (World Bank, 2008). With few exceptions, most technological improvement across the developing world in years to come will still rely heavily on technology diffusion from more advanced countries, although advanced technologies often have to be adapted to local conditions, which can require further innovation.

For developing countries, therefore, the process of acquiring climate change technology is likely to occur mainly through the transfer of climate technological knowledge and equipment for emissions reductions from developed countries and associated financial support. This is explicitly recognized by the legally binding commitments by developed countries to support technology transfer to developing countries, as well as its financing, under the UNFCCC.

Technology transfer to developing countries relies on two major mechanisms, international trade and foreign direct investment (World Bank, 2008). These interact with domestic factors that determine the technical absorptive capacity of the economy to ensure technological progress. It is expected that technology transfer of environmental goods and services will be provided through the same or similar mechanisms.

However, the transfer of technology to developing countries will be eased as soon as they commit to binding targets to reduce emissions either under the Kyoto Protocol or its successor, develop carbon markets of their own, or at least begin to adopt voluntary standards for cutting GHG emissions. In the absence of such mechanisms, most of their technology demand will typically be focused on adaptation. This means that international efforts to push climate-mitigation technology to reach these countries may not be fully sustainable.

On the other hand, official development assistance (ODA) that conforms to criteria of sustainable development may hold promise in the more immediate future to help transfer ESTs, especially if that assistance can be used to leverage private sector investment in climate-friendly technology.

Imports of high-technology goods and services can help developing countries raise the quality of their own products or improve efficiency of production processes. They also offer possibilities for copying technologically advanced products or processes. Technology diffusion tends to follow regional patterns, with the largest transfers coming from natural trading partners. Overall, increased openness to trade has allowed especially middle-income countries with good absorptive capacity to attain important benefits from high-technology exposure.

Trade can be expected to become increasingly important as a technology transfer mechanism, once there is significant demand from developing countries for mitigation technologies. At present, the most pressing climate change agenda in developing countries tends to emphasize climate change adaptation and the management of climatic risks, which require rather different technologies.

Foreign direct investment (FDI) is a promising channel for the transmission of climate technology to developing countries. Although it is difficult to quantify, FDI is likely to involve a substantial degree of technological transfer, including know-how and business process technology. It can also be directed at the host country's R&D efforts, with clear positive spillover effects, especially if inputs are sourced locally.

Licensing is yet another mechanism by which developing countries can gain access to technology. However, in countries with weak intellectual property regimes, firms will be less willing to license technology for fear that it will be copied.

A current challenge for FDI is its dramatic slowdown in the wake of the financial crisis, the global downturn, and protectionist tendencies, all of which may limit its practical relevance in the immediate context of climate change technology transfer. FDI will be a more important mechanism for larger host countries, as the potential lack of commercial viability has been identified as a key barrier to technology transfer via this mechanism (Schneider et al., 2008). Another barrier is the lack of information about the investment opportunities and the host market.

There is an ongoing debate about barriers to international trade in environmental goods and services, including intellectual property rights (IPRs). IPRs, in the form of trademarks or patents, can promote innovation by allowing innovators to capture the value associated with creating new ESTs by assigning temporary monopoly rights which yield high returns to innovators. Nevertheless, the high prices associated with such innovations can also impede diffusion of these products or processes. Research on this issue is ongoing.

Diffusion of technologies within countries

A very important factor for technological achievement is the rate with which technology spreads within a country. This will depend on many factors, such as the degree to which the workforce is technologically literate, whether the investment climate promotes investment and permits the creation and expansion of firms using higher-technology processes, the extent to which it permits access to capital, and the extent to which its public institutions promote the diffusion of critical technologies where private demand or market forces are inadequate.

To foster the required technology transfer for climate change, developing countries must therefore focus on improving the policies and institutions that are conducive to foreign investment. These include the rule of law, a sound regulatory regime, transparency and good governance, market openness, an adequate human capital skill base, and adequate protection of intellectual property rights.

Technology transfer under the UNFCCC process

The recognition that technology must play a key role in mitigating and adapting to climate change underlies the global process under the UNFCCC and has given rise to a long string of commitments.

To help implement sustainable development outcomes, Agenda21 explicitly called for the "favourable access to and transfer of environmentally sound technologies, in particular to developing countries." It also referred to the necessary "economic, technical, and managerial capabilities for the efficient use and further development of transferred technology" (Article 34.4).

Binding commitments undertaken by UNFCCC Parties under Article 4 include promoting and cooperating in the "development, application and diffusion, including transfer, of technologies, practices and processes that control, reduce or prevent anthropogenic emissions of greenhouse gases" (Article 4.1(c)). In addition, developed countries would "promote, facilitate and finance, as appropriate, the transfer of, or access to, environmentally sound technologies and know-how to … developing country Parties" (Article 4.5).

The Marrakech Accords adopted the Technology Transfer Framework (TTF) to enhance the implementation of Article

4.5 of the UNFCCC through the improved transfer of and access to ESTs and know-how. It identified five key themes to support implementation: technology needs and needs assessment; technology information; enabling environments;[4] capacity building; and mechanisms for technology transfer. These form the basis of the Technology Transfer Framework (TTF) established by Decision 4/CP.7.

Most recently, at the 13th session of the Conference of the Parties (COP-13), the Bali Action Plan (BAP) launched the long-term cooperative action (LCA) process, in preparation for the adoption of a decision at the forthcoming COP-15. This encompasses "the removal of obstacles to, and provision of financial and other incentives for, scaling up of the development and transfer of technology to developing country Parties in order to promote access to affordable environmentally sound technologies" (1(d)(i)), as well as "cooperation on research and development of current, new and innovative technology" (1(d)(iii)) and in specific sectors (1(d)(iv)).

To accelerate the implementation of the TTF, the Expert Group on Technology Transfer (EGTT) was created by Decision 4/CP.7, which put together a strategy for the post-2012 technology development and transfer under the Convention. Country-led technology needs assessments (TNAs) identify the climate change technology priorities, especially of developing country Parties, and the barriers to technology transfer across key sectors. The UNFCCC secretariat also runs the Technology Transfer Clearing House (TTClear).

The EGTT has recently elaborated a strategy paper for the long-term perspective beyond 2012 to facilitate technology transfer through potential mechanisms that could have large-scale mitigation and adaptation impact across the world. Other criteria used to evaluate these mechanisms included flexibility regarding the needs of different countries, effectiveness across sectors, the ability to leverage private-sector capital, cost-effectiveness, ease of implementation, and the potential to be self-sustaining. The strategy distinguishes three stages: R&D, technology demonstration and deployment, and diffusion.

For the period up to 2030, it recommends a greatly expanded public and private R&D effort, including stronger centres of innovation particularly in developing countries, greater private-sector and public-sector investment flows, and more emphasis on strengthening institutional capacity and enabling environments in developing countries. A final element would consist of sectoral planning and cooperation, which would oversee implementation of programmes according to economic sectors at global, regional, or national levels.

The Clean Development Mechanism (CDM)

The CDM is one of the three additional implementation mechanisms established by the Kyoto Protocol to complement the domestic GHG reduction targets encompassing developed nations and transition economies, as implemented by Annex I Parties to the Convention. The CDM aims to provide these Parties with the option of meeting their emissions reduction targets in the most cost-effective manner, frequently outside their national borders.

The CDM is currently the only market mechanism aimed at triggering changes in developing countries to achieve GHG emissions reductions. CDM projects typically involve investments by firms in Annex I countries in developing countries, aimed at reducing current or future emissions. Such projects must satisfy certain demanding criteria, which require that projects be "additional"—that is, that they would not have occurred without the CDM mechanism—and that they support the sustainable development objectives of the host country. For example, a typical project could be using new technologies to capture methane gas from landfills and convert it into electrical power that is then returned to the grid. For this reason, it has also been viewed through the lens of a technology transfer mechanism.

Recent studies (Schneider et al., 2008; Seres, 2008) suggest that technology transfer does occur under the CDM mechanism, both as equipment and know-how.

However, the value of certified emissions reductions (CERs) that derive from CDM projects, raising the profitability of investments, fluctuates substantially, related to the carbon price in other carbon markets, especially the European Trading System which is still the dominant carbon market in terms of size. Such uncertainty surrounding the price will negatively impact technology-transfer decisions, and has also led to the departure from the market of a substantial number of experienced and well-informed project developers because of the collapse of the carbon price.

Other limits of the CDM are that the largest share of projects are directed to a handful of non-Annex I countries, typi-

---

[4] Definition given by FCCC/CP/2001/13/Add.1: The enabling environments component of the framework focuses on government actions, such as fair trade policies, removal of technical, legal and administrative barriers to technology transfer, sound economic policy, regulatory frameworks and transparency, all of which create an environment conducive to private and public sector technology transfer.

cally large middle-income countries, such as China, India, and Brazil, which are also major GHG emitters and, therefore, logically attract investments and technology for GHG mitigation. Scale also appears to be important, suggesting that investment will flow to larger countries. Smaller countries and small GHG emitters are, therefore, not going to be major recipients of CDM investments, at least not in the short to medium term.

Another potential shortcoming of the CDM mechanism is that its strength stems from the compliance shortfall of countries with GHG reduction obligations under the KP. If such shortfall occurs because firms in developed countries are not relying on the introduction of environmentally friendly technologies in their production processes to reduce their emissions, but would, instead, rather purchase offsets or transfer technologies that are already viable under the CDM, this will detract from the technology push that is also needed to bring key new environmental technologies to the market with much more significant GHG emission-reduction potential.

Conclusions

The sheer scale of the climate change challenge means that technology that achieves substantial GHG emissions abatement will play a leading role in mitigating and adapting to climate change at the global level.

According to the UNFCCC, we already possess, or are close to developing, the key technologies to limit GHG emissions across key sectors. In the medium term leading to 2020, technologies with the biggest potential for GHG reduction include those that increase energy efficiency in buildings, appliances, industrial processes, and transport, namely offshore and onshore wind, nuclear energy, biomass carbon capture and storage, and solar energy (photovoltaic and concentrating solar power).

In the long run leading to 2050, key technologies that will be relied upon to make the substantial cuts in GHG emissions required to keep the planet safe will target energy efficiency in all key sectors: carbon capture and sequestration for power generation and industry, nuclear power, biofuels, wind, electric and plug-in vehicles, and hydrogen fuel cells.

In this process of acquiring and applying environmentally sound technologies, we distinguished between developing innovative technologies for climate change, transfer of technol-

ogy to developing countries, and diffusion of technologies.

The private sector is expected to play the lead role in developing technologies. However, important market failures in the context of climate change and, more generally, technology and innovation must be addressed by the public sector. These include creating mechanisms, such as carbon markets, that will fully internalize the cost of GHG emissions, a classic environmental externality. In this context, policymakers should make efforts to ensure that the cap is set at an appropriate level, so that there will be sufficient incentive to invest in alternative environmental technologies and to expand the coverage of the systems to achieve a wider sectoral and geographic scope. All efforts should be made to shape the regulatory regime so that the carbon price is high enough to create the right incentives for this decision-making process. Global policymaking should agree on the parameters of a future regime as soon as possible, in order to avoid prolonging the period of uncertainty which is likely to contribute to deterring investment in environmentally sound technologies.

To overcome the market failures, governments have an important role to play in research and development in groundbreaking technologies, especially at the early stages of technological development. This includes policy to ensure an adequate pool of human capital.

In a downturn, governments can make up some of the lost demand by boosting investment in supportive infrastructure for new technologies or green fiscal stimulus spending aimed at environmental technology development. These efforts can be usefully supplemented by carbon taxes.

Technology transfer to developing countries through international trade and foreign direct investment is likely to be limited, at least in the short to medium term, to large middle-income developing countries which are already generating significant GHG emissions. Without legally binding commitments or other mechanisms for regulating GHG emissions in these countries, there is unlikely to be much market pull for such technologies, with the result that opportunities to exploit new markets, especially in smaller countries, will remain very limited. In addition, the focus of climate change activities in smaller developing countries is typically on adaptation to changing conditions, including good risk management. Other barriers to trade and investment may also be hindering these potential mechanisms for technology transfer to developing countries, including intellectual property rights, or scale and information constraints.

Successfully addressing climate change through technology and innovation also necessitates adequate enabling conditions in developing countries, to ensure that such technology enjoys rapid and pervasive diffusion. This implies the need for good governance, rule of law and other enabling conditions which can offer vast co-benefits for strengthening economic development. Given that much of the economic and social progress during the past centuries can be attributed to technology, the case of environmental technologies need not be different. Other benefits can include improved local air quality and greater energy security.

Finally, technology transfer under the UNFCCC process adequately recognizes the need for developed countries to support the technology transfer of clean technology, as well as to provide substantial financing. The focus on providing information and the proposed sectoral focus are equally promising developments. However, to date, the Clean Development Mechanism under the Kyoto Protocol offers perhaps the most promising technology transfer mechanism. To boost this role, policymakers should again ensure that carbon prices remain at a level that will guarantee the profitability of private sector investments under this mechanism. However, as this is a market-driven mechanism, it is likely to maintain its geographic focus in developing countries that are large, or potentially large, GHG emitters. In order not to stymie the market-pull financing flows that must take place mainly in developed countries, the parameters set by existing cap-and-trade systems must carefully balance the need to limit the scope of offsetting emissions against the benefits of the CDM in transferring technologies to developing countries.

References

Aldy, J. & R. Stavins. 2008. "The Role of Technology Policies in an International Climate Agreement." Belfer Center for Science and International Affairs. 3 September.

Carbon Sequestration Leadership Forum (CSLF). Available at: www.cslforum.org

Committee on Climate Change (CCC). 2008. "Building a Low-Carbon Economy: The UK's Contribution to Tackling Climate Change." London: Committee on Climate Change.

de Coninck, H. et al. 2008. "International technology-oriented agreements to address climate change." *Energy Policy* 36:335–56.

Frondizi, I. 2009. *Guide to the clean development mechanism.* Rio de Janeiro: Imperial Novo Milênio, Ministério da Ciência e Tecnologia, UNDP and UNCTAD.

German Advisory Council on Global Change (WBGU). 2008. *Climate Change as a Security Risk.* London and Sterling, VA: Earthscan.

International Energy Agency (IEA). 2008. *Energy Technology Perspectives: Scenarios and Strategies to 2050.* Paris: OECD/IEA.

Intergovernmental Panel on Climate Change (IPCC). 2000. *Methodological and technological issues in technology transfer.* Report of Working Group III of the Intergovernmental Panel on Climate Change. Cambridge: Cambridge University Press.

———. 2007a. *Climate Change 2007: Synthesis Report.* Contribution of Working Groups I, II and III to the Fourth Assessment Report of the Intergovernmental Panel on Climate Change. Geneva: IPCC.

———. 2007b. *Climate Change 2007: The Physical Science Basis.* Contribution of Working Group I to the Fourth Assessment Report of the IPCC. Geneva: IPCC.

———. 2007c. *Climate Change 2007: Impacts, Adaptation and Vulnerability.* Contribution of Working Group II to the Fourth Assessment Report of the IPCC. Geneva: IPCC.

———. 2007d. *Climate Change 2007: Mitigation of Climate Change.* Contribution of Working Group III to the Fourth Assessment Report of the IPCC. Geneva: IPCC.

International Partnership for the Hydrogen Economy. Available at: http://www.iphe.net

McKinsey & Company. 2009. Pathways to a Low-Carbon

Economy: Version 2 of the Global Greenhouse Gas Abatement Cost Curve: McKinsey & Company.

McKinsey & Company. 2009. Unlocking energy efficiency in the US economy: McKinsey & Company. Available at: http://www.mckinsey.com/clientservice/electricpower-naturalgas/US_energy_efficiency

McKinsey Global Institute. 2008. *The carbon productivity challenge; curbing climate change and sustaining economic growth.* McKinsey Global Institute.

Newell, R. 2008. "International Climate Technology Strategies." Discussion Paper 08-12, Harvard Project on International Climate Agreements. Belfer Center for Science and International Affairs, Kennedy School of Government. October.

OECD. 2009. "The Economics of Climate Change Mitigation: How to build the necessary global action in a cost-effective manner." Economics Department Working Paper 701. Paris: OECD.

Schneider, M., A. Holzer, and V. Hoffmann. 2008. "Understanding the CDM's contribution to technology transfer." *Energy Policy* 35:2930–38.

Seres, S. 2008. "Analysis of Technology Transfer in CDM Projects." Paper prepared for UNFCCC Registration and Issuance Unit CDM/SDM. December. Available at: http://cdm.unfccc.int/Reference/Reports/TTreport/TTrep08.pdf

Stern, N. 2007. *The Economics of Climate Change: The Stern Review.* Cambridge: Cambridge University Press.

Tomlinson, S. 2009. "Breaking the Climate Deadlock, Technology for a Low Carbon Future." Joint project between The Climate Group and The Office of Tony Blair.

United Nations Economic Commission for Europe. 2009. *Annual Report,* UNECE. Geneva.

United Nations Framework Convention on Climate Change (UNFCCC). 2008. "Investment and financial flows to address climate change: An update." Technical paper FCCC/TP/2008/7. 26 November.

University of Copenhagen. 2009. *Synthesis Report from Climate Change—Global risks, challenges and decisions,* Copenhagen 2009, 10–12 March. Available at: www.climate-congress.ku.dk

World Bank. 2008. *Global Economic Prospects: Technology Diffusion in the Developing World,* Washington, D.C.: World Bank.

———. 2009a. *International Trade and Climate Change: Economic, Legal, and Institutional Perspectives,* Washington, D.C.: World Bank.

———. 2009b. *Development and Climate Change: the World Bank Group at Work.* Washington, D.C.: World Bank.

Zero Emissions Platform. Available at: http://www.zeroemissionsplatform.eu/

Chapter 2.12

Innovation and Social Development in Latin America

Hernán Rincón,
President, Microsoft Latin America

How is innovation related to economic development? The question we should really be asking ourselves first is: what does economic development mean for Latin America now?

When we refer to the so-called "underdeveloped" countries, we usually think of development as the capacity to move from a low to a high-income economy, through policies and government commitment which enable a high social impact.

For the "developed" countries, economic development seems to imply going back to the kind of seemingly "stable" high-income economy that prevailed before the current financial crisis. (Indeed, many of the measures taken to deal with the effects of the crisis seem intended to reproduce the conditions that existed before the onset of the crisis).

I cannot agree completely with either conception. First, because there are many gray areas in our understanding of what is "underdeveloped" or "developed." Second, I am not at all convinced that the correct way to measure development is by determining whether a given country either has a low- or high-income per capita. We see many so-called "developed" countries with a long record of wars, where discrimination is rife, unable to provide high quality education or low access to health care plans for their citizens—not exactly characteristics that we have come to recognize as elements of a "developed" civilization. Thus, economic development in its broadest sense is not necessarily perfectly correlated with economic growth—as measured by GDP—any more than GDP is a fully satisfactory measure of economic well-being. Third, there is no going back to where we were prior to the financial crisis, for the developed *and* for the developing countries.

Without delving too much here into an intellectual and philosophic discussion of what economic development means, I wish to expand some of the previous statements, to help us better understand where innovation comes in to the picture.

In a recent editorial in which I analyzed the financial crisis, I wrote that it is highly unlikely that either developed or developing countries are going to return to the conditions that prevailed before the financial crisis. Why? Because the world has changed. In the business and economic sectors, it is as simple as that.

This is the conclusion I came to after reading a post from marketing and business consultant Steve Yastrow on the blog of another distinguished economist, Tom Peters, in which they were debating whether or not the economic crisis was a traditional "recession." The main idea is that we are not in a recession but in a period of "recalibration." Everything is different now: a new economy, a new status quo. And this is where the most important question came up: how are governments, organizations, and companies handling their own recalibration?

Their first challenge has been to accept the new situation. It is not simply a matter of tightening belts and lowering costs, in the hope that the market will recover. Once the financial crisis comes to an end, the economic environment will not be the same. Competitors will be different, and so will clients. And those who have not recalibrated their business or organizations will find themselves in trouble.

During periods of recalibration, changes are more profound than they may seem at first, and we have to know how to read them carefully. Consider, for example, what happened in the photography industry, where industry leaders defined themselves as companies selling chemical products, such as rolls of film, photographic paper, and offered services such as fast processing. They did not foresee the explosion of digital cameras on the consumer market, leaving fast processing in the dust, a dim memory of a bygone century. Another example of how not being ready to recalibrate can undermine development is the Internet, and how it has changed our lives. Wireless systems around the world are being successfully used, especially in developed countries and in some countries in Latin America, where traditional telecommunications are lagging behind demand. The Internet means being connected and having access to global information. Success or social impact is impossible without Internet access nowadays. Countries or communities which are helping their citizens get connected are thriving. A good example is the province of San Luis, Argentina, with its "San Luis Digital" project, through which the entire province has free wireless Internet access. As Governor Adolfo Rodríguez Saá explained, they are faced with the "problem" of truck drivers stopping at the roadside to get connected! Hardly what you would call a problem!

Unified Communications at the Housing for Workers National Fund Institute (Infonavit) in Mexico, gave the institution the tools for instant messaging, VoIP, and video and audio conference capabilities. This not only meant simplified communications, but the new collaboration tools added a productivity of 30 additional minutes per day per employee. This is a perfect example of recalibration and how new technologies and innovation can increase efficiency and economic growth.

The second challenge of the financial crisis and of the recalibration process was to understand how governments, organizations, partners, and clients changed in this globalized and interconnected economy. Think about conversations and dialogues that now take place in language 2.0, basically a language that has been completely transformed to fit the new digital environment. Digital language is about online personal relations, where communications are centered on personal identity and professional interests.

An interesting symptom of the US economy at present is that only three of every four companies have a corporate website. Only 11 percent of companies service online purchasing. While 28 percent accept online orders, only 21 percent use other sites or Web pages to sell their products online.

Think about your country, your government, your companies, your organizations, or your communities. Have they recalibrated? Or are they on the road to recalibration?

The answer to this challenge lies in innovation. Why innovate? Because it allows organizations to find new answers to complex problems, such as how to operate with greater levels of efficiency and efficacy, or even how to create new business models adapted to the new reality.

Think about how Information Technology (IT) can help in this innovation process. IT today forms the backbone of our everyday lives: how we work, learn, and communicate. From using technology, to improving the performance of employees, enhancing collaboration between the workforce and the community, analyzing or serving the needs of clients, and reducing costs.

It is time to recalibrate. Countries and companies that understand this and know how to recognize this opportunity will not only survive this period, they will thrive in the new economic environment.

So returning to our earlier question: what does economic development mean for Latin America now? Latin America is in a better position than ever before. I am optimistic about the region, after seeing clear signs of growth in some countries, even in turbulent times. When the financial crisis hit, Latin America was well prepared, not only because of the continent's innate experience of surviving several economic, political, and social crises over the last century, but also because of the growth experienced over the last five years. In most countries, the much stronger economic policies which were implemented (compared, for instance, with those that prevailed in the region in the 1980s and early 1990s) lowered inflation and interest rates, improved the public finances and reduced external debt, and substantially increased foreign exchange reserves, positioning Latin America to better face the downturn and adapt to the new environment.

This gave the region an advantage, and also a new opportunity. Rapid government action is critical in order to quickly restore confidence, stimulate job creation, as well as consumer and business investment. But in the rush to respond, we should not lose sight of the vital importance of investing for the future.

Investing in innovation is fundamentally about investing in people. In today's knowledge-driven world, innovation depends on people who are technically sophisticated, have critical thinking skills, and possess a working knowledge of math and science. This obviously means that investment in education has to be an essential element of any economic development plan.

To quote Bill Gates, from a recent speech at the Government Leaders Forum–Americas, held in March 2009:

> Because, as important as it is to create jobs and jumpstart economic growth now, we need to keep our sights set on a bigger goal. If we stay focused on expanding access to education, we will increase the number of people who contribute to economic growth and who benefit from it. In the process, we'll unleash the entrepreneurial powers of our citizens, and increase the number of smart minds that are focused on innovation and on solving some of the hard problems we face.

So investing in education means increasing teacher skills, expanding access to technology, and improving math and science education in schools at every level, from elementary to university. Also vitally important is the training of the men and women already in the workforce for jobs that require experience using technology.

Now is also the time to invest in Research and Development (R&D). Great ideas will be born in universities throughout Latin America, hence it is important to boost the capacity of universities and research centers. When these ideas are properly coordinated with the business community they will become the next generation of products and services, leading to economic growth.

In this context, intellectual property protection is more important than it has ever been in the past, as it creates the fundamental incentive for innovation in the academic and business communities, and the clear rules that gives businesses the confidence they need to grow their investment.

Although we have also revised our budgets at Microsoft, we have not cut back in our commitment to R&D, which, of course, plays a critical role in getting out of any crisis. Governments cannot act alone, and they can help enormously in fostering R&D. Governments can benefit even more now from partnerships in this new reality, since the private sector has a significant role to play in the new economy. IT is a direct source of jobs and a major contributor to GDP growth. IT is also responsible for more than 50 percent of the productivity gains in other industries, according to the Organization for Economic Co-Operation and Development (OECD).

All this provides the right context for defining how innovation can have a direct impact on economic development and growth. The kind of economic development needed today is very different from the past. Countries must now focus on creating world-class knowledge economies. And this is particularly good news for developing economies, because the transformation from industrial and agrarian economies to knowledge economies is a great global leveler, one that innovation and technology can drive.

In his book *The World Is Flat*, Thomas L. Friedman identifies ten forces that flattened the world, nearly all of them made possible by IT-based innovation. These IT innovations—the PC-driven information revolution, the Internet, productivity, workflow and collaboration software, increasingly low-cost and ubiquitous digital devices—all represent the building blocks of knowledge economies. IT helps companies increase efficiency and productivity. It contributes to network effects, such as lowering transaction costs and speeding more innovation. The Internet promotes global trade by connecting buyers and sellers, and by cutting market entry costs. IT investments in youth can cause a shift toward higher-skilled workers who can earn higher wages. And, of course, a thriving IT industry itself is a key driver of economic growth.

Realizing the immense potential of information technology as a driver of economic growth and development is even more critical right now. As the current crisis starts to subside, new opportunities will emerge for countries and companies that are leaders in innovation. History shows that new technologies are born out of every downturn. Some countries in Latin America were ready and took advantage of the financial crisis, emerging faster than others. In an earlier period, such opportunities were the exclusive domain of the few who could access and capitalize on great ideas. Now, thanks to technology recovery, recalibration, and innovation, resources are available to many more. Hopefully, someday soon, they will be available to all.

Microsoft's role

Microsoft Unlimited Potential works to enable social and economic opportunity. Thirty years ago, Microsoft began with the dream of a computer on every desk and in every home. Today, for the more than 1 billion people reached so far, life has changed dramatically. Information is available instantaneously. Personal and professional connections are made easily. Products and services are sold conveniently. Progress is achieved more readily. But for more than 5 billion people all over the world, the opportunity to learn, connect, create, and succeed remains elusive.

Several barriers stand in the way of effectively reaching these underserved communities, including environmental or infrastructure obstacles, localization issues, the need for personalized solutions, and the prohibitive cost of technology.

To better serve these people, Microsoft Unlimited Potential combines advanced technologies and strong partnerships with governments, international organizations, NGOs, educational institutions, and technology and service partners. Ultimately our mission is to promote sustained social and economic opportunity for those at the middle and bottom of the world's economic pyramid: the 5 billion people who have yet to fully benefit from the latest technological advances.

In the short term, Unlimited Potential aims to reach the next 1 billion people by 2015, by exploring solutions in three interrelated areas. Each is crucial to developing sustained economic opportunity:

- Transforming education
- Fostering local innovation
- Creating jobs and opportunities

In these three areas, Microsoft Unlimited Potential can create the greatest possible impact by building a virtuous cycle

of sustained social and economic development. This cycle drives communities and helps build connections to form new communities. It is fueled by local and global partnerships and, most important, ultimately becomes locally sustainable. Sustainability is a key indicator of effective programs and activities, and constitutes our long-term measure of success.

Microsoft is honored to be an industry leader, a global corporate citizen, and an active partner to hundreds of governments and thousands of businesses and community organizations around the world, bringing the same spirit of innovation to global citizenship which is already a defining feature of the international business community. Success has meant that we now have the opportunity—and the responsibility—to use our resources and influence to help make the world a better place for all.

IT can also help level the developing world in their relationships with developed countries. Random innovation doesn't necessarily bring concrete results. What is needed is focused innovation to compete. Economies now need to foster innovation in order to compete in a global market. Any company in Latin America can compete with an Asian company, or sell services and products to a European government. Innovation means many things, it includes many components, but the common denominator is the use of technology.

Fostering Innovation in Latin America

Innovation in Latin America requires focus in four main areas:

Training: There is a crucial need to train not only the existing, but the future work force. This will enable to insert countries and people in the global knowledge economy.

Fostering local innovation: Fostering the creation of local development of innovative technologies and systems, instead of the adoption of technologies from abroad, is crucial for innovation. Each country must be in a position to respond to their own needs, in the context of its own laws, taxes, systems, and culture. This implies the creation of a local intellectual property ecosystem. A study by International Data Corporation (IDC) shows that the launch of Windows 7 alone will have significant impact on the technology ecosystem. The IDC calculated that for every dollar of revenue generated by

Microsoft between the launch of Microsoft Windows 7 in October 2009 and end-2010 in Brazil, for example, the ecosystem beyond Microsoft will reap US$22.62. And companies from Brazil, Russia, India and China from the Microsoft ecosystem are expected to invest more than US$13 billion by the end of 2010 in the development, marketing, and support products and services built around Windows 7.

Global competition: For a country to compete, local companies must be on top of technology and innovation adoption, use, training, and skills. Innovation must be included in business, partnership, and employment models. For example, Telmex in Mexico, Codelco in Chile, Polar in Venezuela, Bradesco in Brazil can all compete on the same level with companies around the world. The difficulty arises for small- and medium-sized enterprises (SMEs) that have to compete with similar SMEs in developed countries. At this point, not many in Latin America are able to compete. They need help with investment in innovation, so that they can increase productivity and efficiently use new technologies.

Social responsibility: Innovation must reach government agencies and entities, to help them become transparent, secure, and more responsive to their citizens. Innovation and technology in education and health are vital in meeting basic needs of populations. Development is only possible through a highly educated and skilled community. Education is the change agent, and health makes it possible. For example, 23 of Microsoft's 110 worldwide Innovation Centers are located in Brazil; two of the 17 high-tech research facilities financed by Microsoft in the world are being developed in Brazil, in partnership with the São Paulo Research Foundation (FAPESP). In the second half of 2009, Microsoft will celebrate the opening of a new Internet Security Laboratory and Interoperability Laboratory (the sixth such laboratory in Brazil) in partnership with trustworthy institutions such as Serviço Nacional de Aprendizagem Comercial (SENAC, National Commercial Training Service) and Instituto Tecnológico de Aeronáutica (ITA, Aeronautical Institute of Technology).

The future of Latin America lies in investing in our people, in giving them the tools they need for real digital inclusion. Some people say that being successful is a matter of luck. I believe that being successful is about embracing opportunities. The success of governments, businesses, and individuals is measured by the opening of opportunities for others and the positive impact they have on people's lives. Although markets have changed, and governments and businesses around the world are recalibrating, we must remain committed to enabling people and organizations throughout the world realize their full potential.

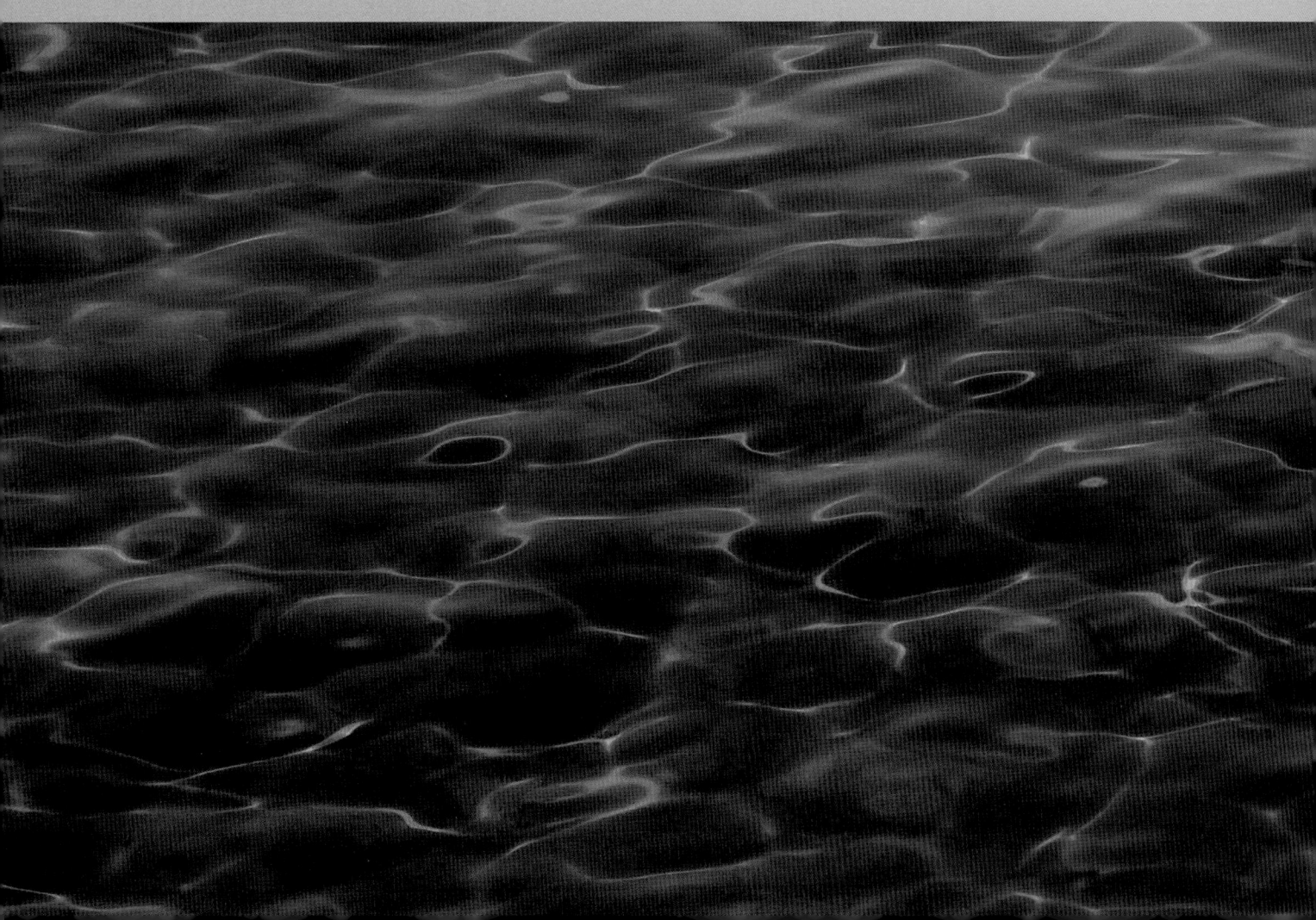

Part 3 Innovation Profiles

Reading the Innovation Capacity Index 2009–2010 Country Profiles

The Innovation Capacity Index 2009–2010 (ICI) ranks a total of 131 countries. This year's published edition of the Innovation for Development Report includes innovation profiles for 68 of the 131 countries covered by the Index, accounting for approximately 96 percent of world GDP. The remaining 63 innovation profiles are available at: www.innovationfordevelopmentreport.org

The profiles provide a bird's eye view of the particular situation in each country and of the different dimensions of its innovation capacity.

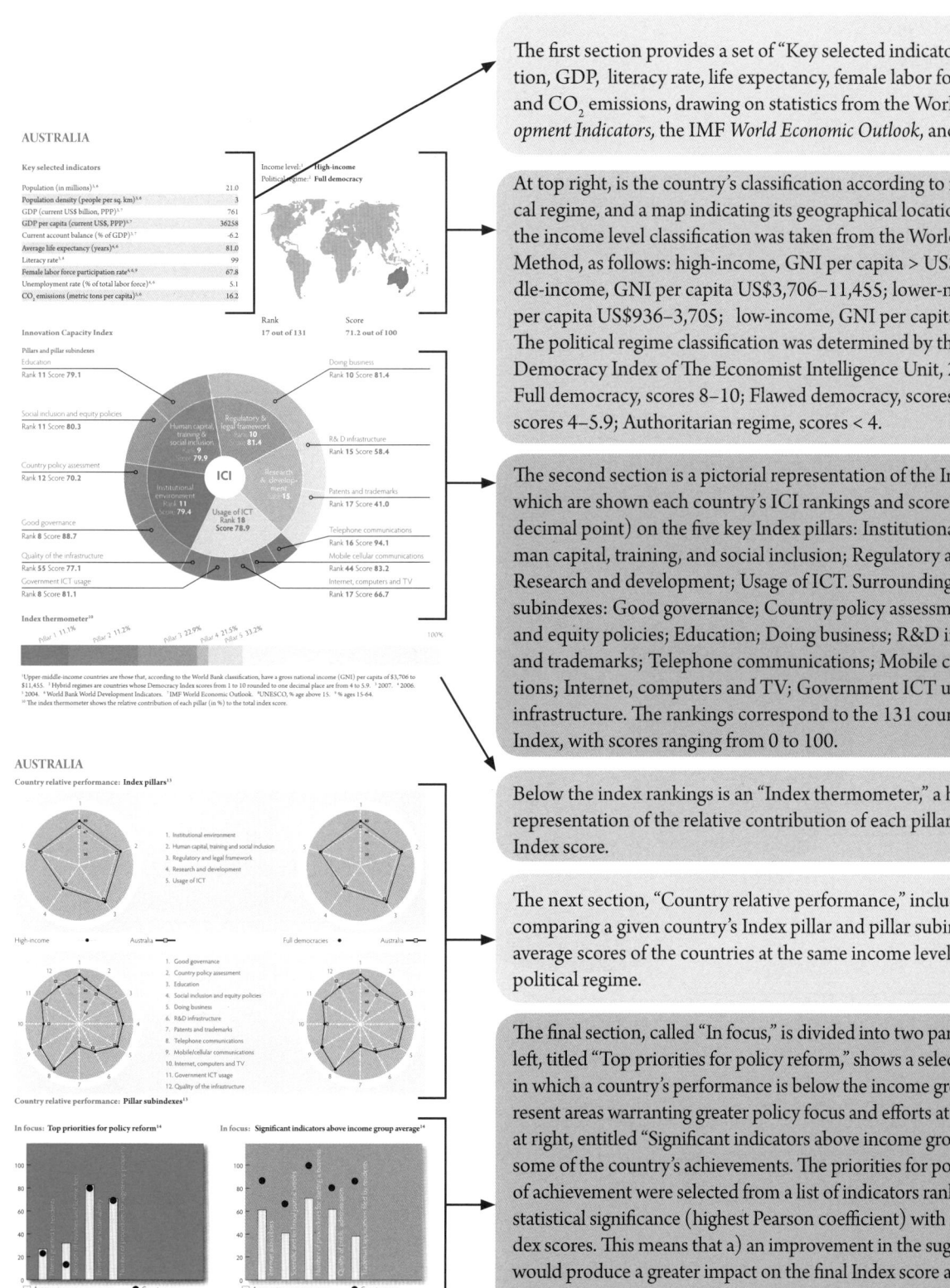

The first section provides a set of "Key selected indicators" related to population, GDP, literacy rate, life expectancy, female labor force participation rate, and CO$_2$ emissions, drawing on statistics from the World Bank's *World Development Indicators,* the IMF *World Economic Outlook,* and UNESCO.

At top right, is the country's classification according to income level and political regime, and a map indicating its geographical location. For this year's Index, the income level classification was taken from the World Bank 2007 Atlas Method, as follows: high-income, GNI per capita > US$11,456; upper-middle-income, GNI per capita US$3,706–11,455; lower-middle-income, GNI per capita US$936–3,705; low-income, GNI per capita < US$935.
The political regime classification was determined by the country's score in the Democracy Index of The Economist Intelligence Unit, 2008, as follows: Full democracy, scores 8–10; Flawed democracy, scores 6–7.9; Hybrid regime, scores 4–5.9; Authoritarian regime, scores < 4.

The second section is a pictorial representation of the Index, in the center of which are shown each country's ICI rankings and scores (rounded to the first decimal point) on the five key Index pillars: Institutional environment; Human capital, training, and social inclusion; Regulatory and legal framework; Research and development; Usage of ICT. Surrounding these are the pillar subindexes: Good governance; Country policy assessment; Social inclusion and equity policies; Education; Doing business; R&D infrastructure; Patents and trademarks; Telephone communications; Mobile cellular communications; Internet, computers and TV; Government ICT usage; Quality of the infrastructure. The rankings correspond to the 131 countries covered by the Index, with scores ranging from 0 to 100.

Below the index rankings is an "Index thermometer," a horizontal graphic representation of the relative contribution of each pillar score to the overall Index score.

The next section, "Country relative performance," includes four radial charts comparing a given country's Index pillar and pillar subindex scores with the average scores of the countries at the same income level and having a similar political regime.

The final section, called "In focus," is divided into two parts. The bar graph at left, titled "Top priorities for policy reform," shows a select group of indicators in which a country's performance is below the income group average. These represent areas warranting greater policy focus and efforts at reform. The bar graph at right, entitled "Significant indicators above income group average," highlights some of the country's achievements. The priorities for policy reform and areas of achievement were selected from a list of indicators ranked according to their statistical significance (highest Pearson coefficient) with respect to the final Index scores. This means that a) an improvement in the suggested areas of reform would produce a greater impact on the final Index score and rankings, and b) that high marks on these significant indicators contributed to raising the country's ranking and score with respect to countries in its income group.

241

ALGERIA

Key selected indicators

Population (in millions)[3,4]	33.9
Population density (people per sq. km)[3,4]	14
GDP (current US$ billion, PPP)[3,5]	225
GDP per capita (current US$, PPP)[3,5]	6533
Current account balance (% of GDP)[3,5]	23.2
Average life expectancy (years)[4,6]	72.0
Literacy rate[3,7]	75
Female labor force participation rate[4,6,8]	38.9
Unemployment rate (% of total labor force)[4,9]	15.3
CO_2 emissions (metric tons per capita)[4,10]	6.0

Income level:[1] **Lower-middle-income**

Political regime:[2] **Authoritarian regime**

Innovation Capacity Index

Rank	Score
76 out of 131	**46.7 out of 100**

Pillars and pillar subindexes

Education
Rank **66** Score **59.9**

Social inclusion and equity policies
Rank **91** Score **46.1**

Country policy assessment
Rank **30** Score **59.9**

Good governance
Rank **102** Score **33.0**

Quality of the infrastructure
Rank **45** Score **82.8**

Government ICT usage
Rank **91** Score **35.2**

Doing business
Rank **99** Score **57.6**

R&D infrastructure
Rank **84** Score **18.0**

Patents and trademarks
Rank **88** Score **0.7**

Telephone communications
Rank **77** Score **77.8**

Mobile cellular communications
Rank **56** Score **76.8**

Internet, computers and TV
Rank **92** Score **6.8**

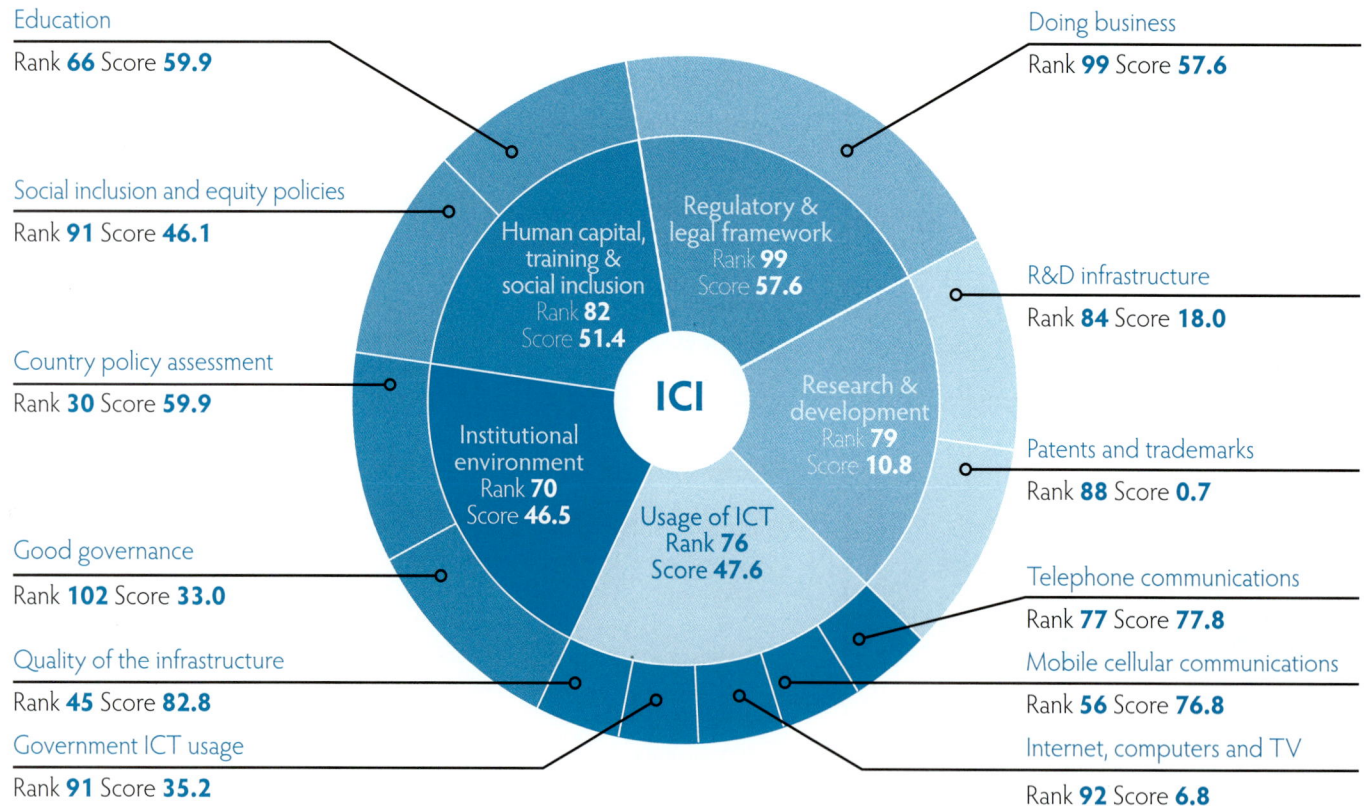

Index thermometer[11]

Pillar 1 29.8% Pillar 2 33.0% Pillar 3 24.7% Pillar 4 2.3% Pillar 5 10.2%   100%

---

[1] Lower-middle-income countries are those that, according to the World Bank classification, have a gross national income (GNI) per capita of $936 to $3,705.   [2] Authoritarian regimes are countries whose Democracy Index scores from 1 to 10 rounded to one decimal place are less than 4.   [3] 2007.   [4] World Bank World Development Indicators.   [5] IMF World Economic Outlook.   [6] 2006.   [7] UNESCO, % age above 15.   [8] % ages 15–64.   [9] 2005.   [10] 2004.   [11] The index thermometer shows the relative contribution of each pillar (in %) to the total index score.

242

ALGERIA

Country relative performance: **Index pillars**[12]

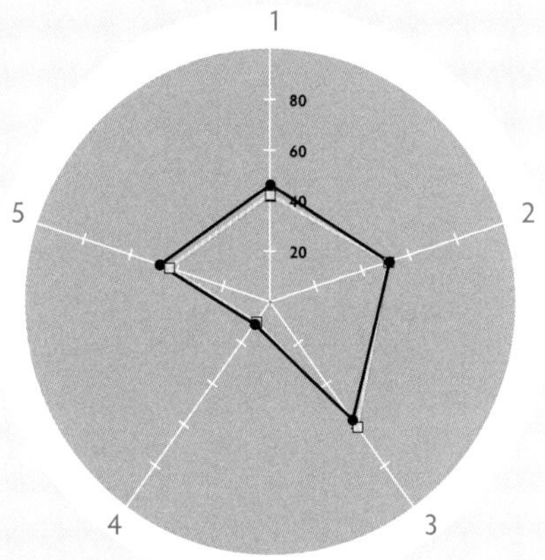

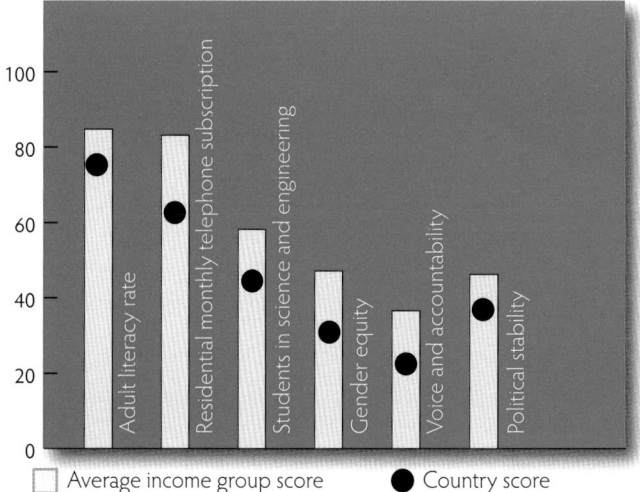

1. Institutional environment
2. Human capital, training and social inclusion
3. Regulatory and legal framework
4. Research and development
5. Usage of ICT

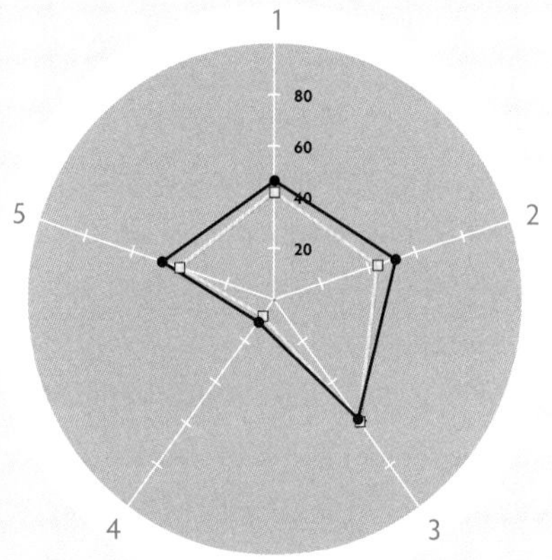

Lower-middle-income ☐   Algeria ●——

Authoritarian regimes ☐   Algeria ●——

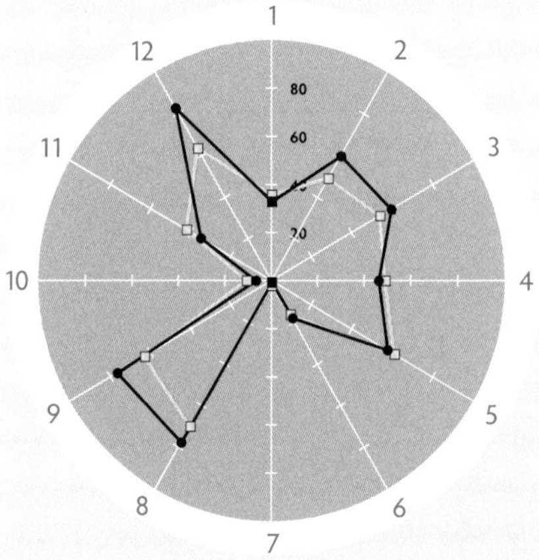

1. Good governance
2. Country policy assessment
3. Education
4. Social inclusion and equity policies
5. Doing business
6. R&D infrastructure
7. Patents and trademarks
8. Telephone communications
9. Mobile/cellular communications
10. Internet, computers and TV
11. Government ICT usage
12. Quality of the infrastructure

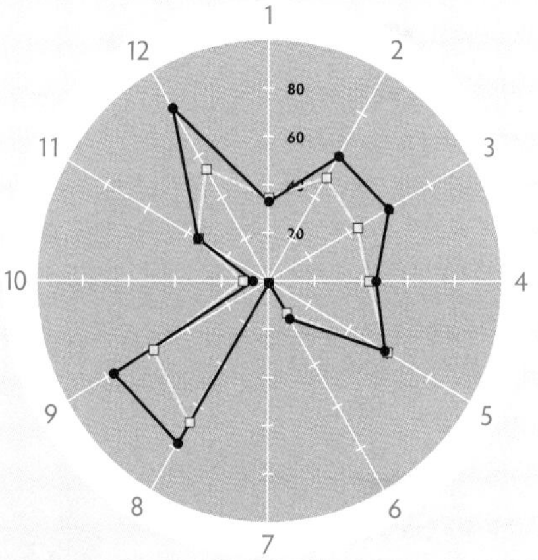

243

Country relative performance: **Pillar subindexes**[12]

In focus: **Top priorities for policy reform**[13]

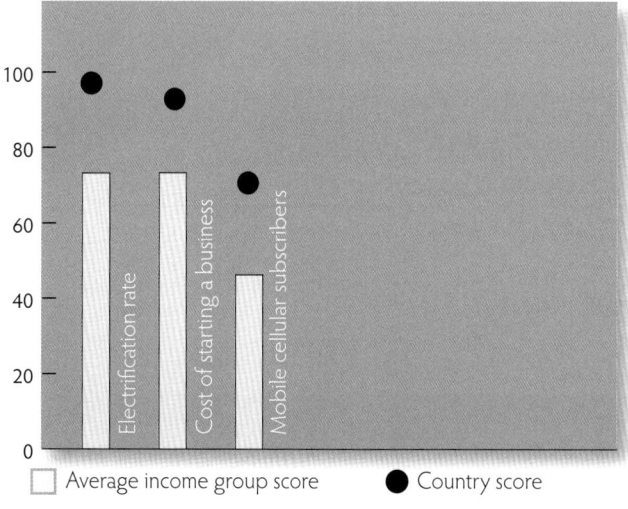

☐ Average income group score   ● Country score

In focus: **Significant indicators above income group average**[13]

☐ Average income group score   ● Country score

[12] This analysis compares pillar and subindex country scores with respect to the average scores of countries grouped according to income level and political regime (see notes 1 and 2).   [13] Significant country indicators with the greatest distances above or below the income group average scores (i.e., achievements and targets for policy reform, respectively) were selected and ordered according to their Pearson correlation coefficients (highest first) with respect to the final index scores, to produce a ranking.

ARGENTINA

Key selected indicators

Population (in millions)[3,4]	39.5
Population density (people per sq. km)[3,4]	14
GDP (current US$ billion, PPP)[3,5]	524
GDP per capita (current US$, PPP)[3,5]	13308
Current account balance (% of GDP)[3,5]	1.1
Average life expectancy (years)[4,6]	75.0
Literacy rate[3,7]	98
Female labor force participation rate[4,6,8]	62.2
Unemployment rate (% of total labor force)[4,6]	10.2
CO_2 emissions (metric tons per capita)[4,9]	3.7

Income level:[1] **Upper-middle-income**

Political regime:[2] **Flawed democracy**

Rank	Score
66 out of 131	**49.2 out of 100**

Innovation Capacity Index

Pillars and pillar subindexes

Education
Rank **30** Score **70.4**

Social inclusion and equity policies
Rank **49** Score **59.0**

Country policy assessment
Rank **125** Score **37.6**

Good governance
Rank **81** Score **38.8**

Quality of the infrastructure
Rank **66** Score **67.2**

Government ICT usage
Rank **39** Score **58.4**

Doing business
Rank **88** Score **61.3**

R&D infrastructure
Rank **70** Score **23.7**

Patents and trademarks
Rank **30** Score **16.3**

Telephone communications
Rank **45** Score **86.4**

Mobile cellular communications
Rank **39** Score **84.7**

Internet, computers and TV
Rank **57** Score **20.0**

Circular chart labels:
- Human capital, training & social inclusion — Rank **41** Score **63.6**
- Regulatory & legal framework — Rank **88** Score **61.3**
- Institutional environment — Rank **104** Score **38.2**
- Research & development — Rank **46** Score **20.0**
- Usage of ICT — Rank **49** Score **56.5**
- ICI

244

Index thermometer[10]

Pillar 1 19.4% Pillar 2 32.3% Pillar 3 24.9% Pillar 4 6.1% Pillar 5 17.2%   100%

[1] Upper-middle-income countries are those that, according to the World Bank classification, have a gross national income (GNI) per capita of $3,706 to $11,455. [2] Flawed democracies are countries whose Democracy Index scores from 1 to 10 rounded to one decimal place are of 6 to 7.9. [3] 2007. [4] World Bank World Development Indicators. [5] IMF World Economic Outlook. [6] 2006. [7] UNESCO, % age above 15. [8] % ages 15–64. [9] 2004. [10] The index thermometer shows the relative contribution of each pillar (in %) to the total index score.

ARGENTINA

Country relative performance: **Index pillars**[11]

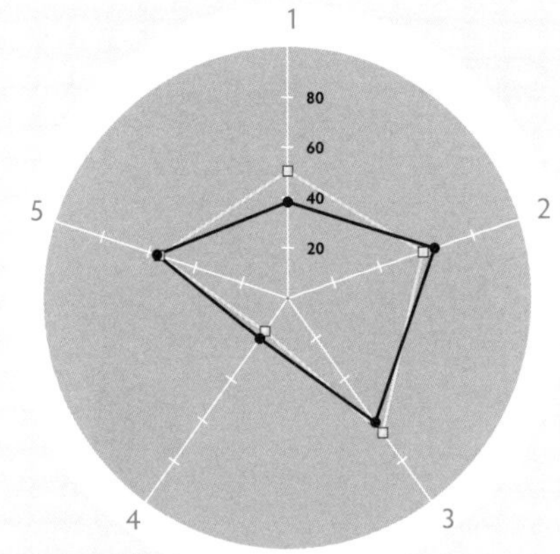

1. Institutional environment
2. Human capital, training and social inclusion
3. Regulatory and legal framework
4. Research and development
5. Usage of ICT

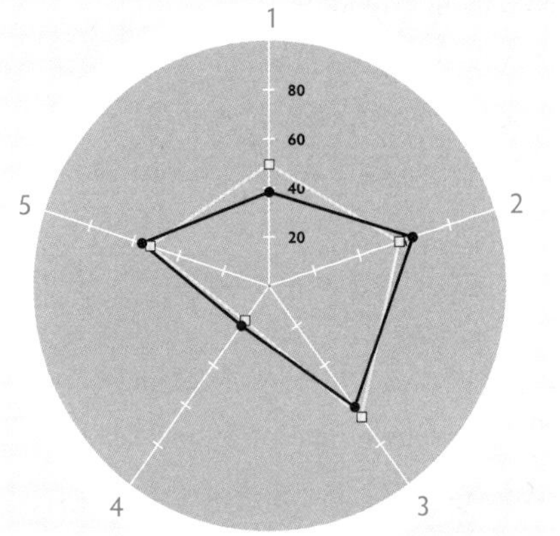

Upper-middle-income ▫   Argentina ●━━

Flawed democracies ▫   Argentina ●━━

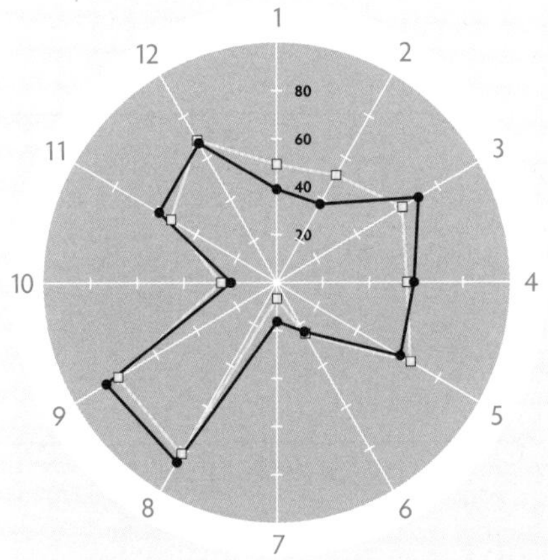

1. Good governance
2. Country policy assessment
3. Education
4. Social inclusion and equity policies
5. Doing business
6. R&D infrastructure
7. Patents and trademarks
8. Telephone communications
9. Mobile/cellular communications
10. Internet, computers and TV
11. Government ICT usage
12. Quality of the infrastructure

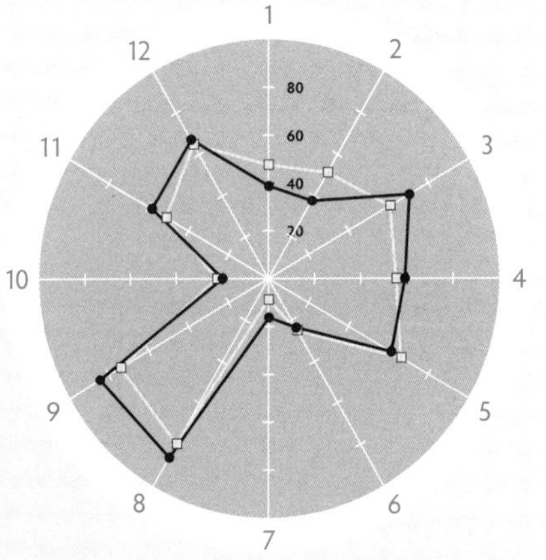

Country relative performance: **Pillar subindexes**[11]

In focus: **Top priorities for policy reform**[12]

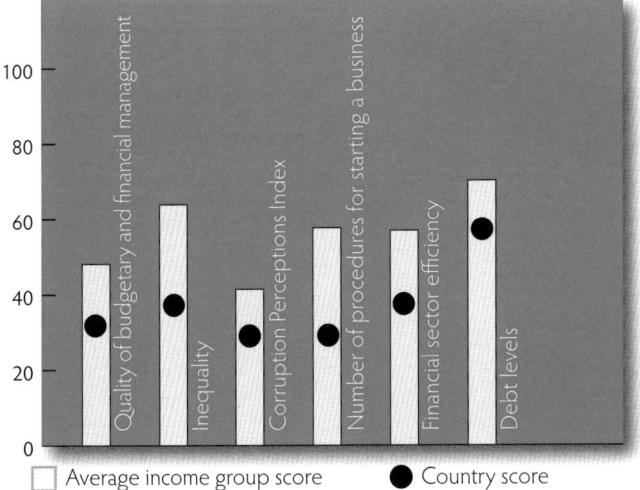

□ Average income group score   ● Country score

In focus: **Significant indicators above income group average**[12]

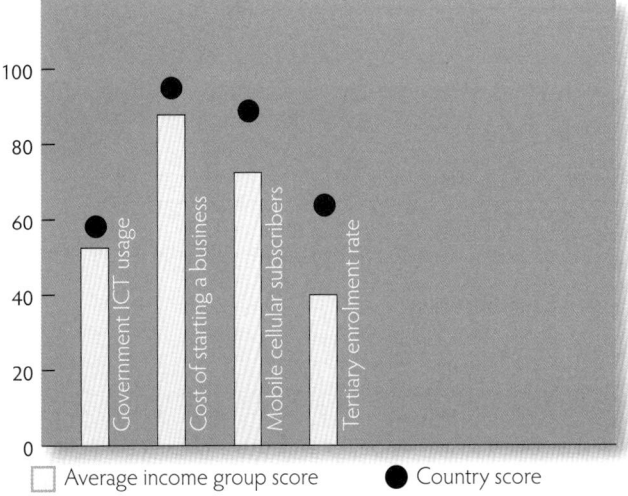

□ Average income group score   ● Country score

[11] This analysis compares pillar and subindex country scores with respect to the average scores of countries grouped according to income level and political regime (see notes 1 and 2).

[12] Significant country indicators with the greatest distances above or below the income group average scores (i.e., achievements and targets for policy reform, respectively) were selected and ordered according to their Pearson correlation coefficients (highest first) with respect to the final index scores, to produce a ranking.

AUSTRALIA

Key selected indicators

Population (in millions)[3, 4]	21.0
Population density (people per sq. km)[3, 4]	3
GDP (current US$ billion, PPP)[3, 5]	761
GDP per capita (current US$, PPP)[3, 5]	36258
Current account balance (% of GDP)[3, 5]	-6.2
Average life expectancy (years)[4, 6]	81.0
Literacy rate[7, 8]	99
Female labor force participation rate[4, 6, 9]	67.8
Unemployment rate (% of total labor force)[4, 10]	5.1
CO_2 emissions (metric tons per capita)[4, 11]	16.2

Income level:[1] **High-income**

Political regime:[2] **Full democracy**

	Rank	Score
Innovation Capacity Index	**17 out of 131**	**71.2 out of 100**

Pillars and pillar subindexes

Education
Rank **11** Score **79.1**

Social inclusion and equity policies
Rank **11** Score **80.3**

Country policy assessment
Rank **12** Score **70.2**

Good governance
Rank **8** Score **88.7**

Quality of the infrastructure
Rank **55** Score **77.1**

Government ICT usage
Rank **8** Score **81.1**

Doing business
Rank **10** Score **81.4**

R&D infrastructure
Rank **15** Score **58.4**

Patents and trademarks
Rank **17** Score **41.0**

Telephone communications
Rank **16** Score **94.1**

Mobile cellular communications
Rank **44** Score **83.2**

Internet, computers and TV
Rank **17** Score **66.7**

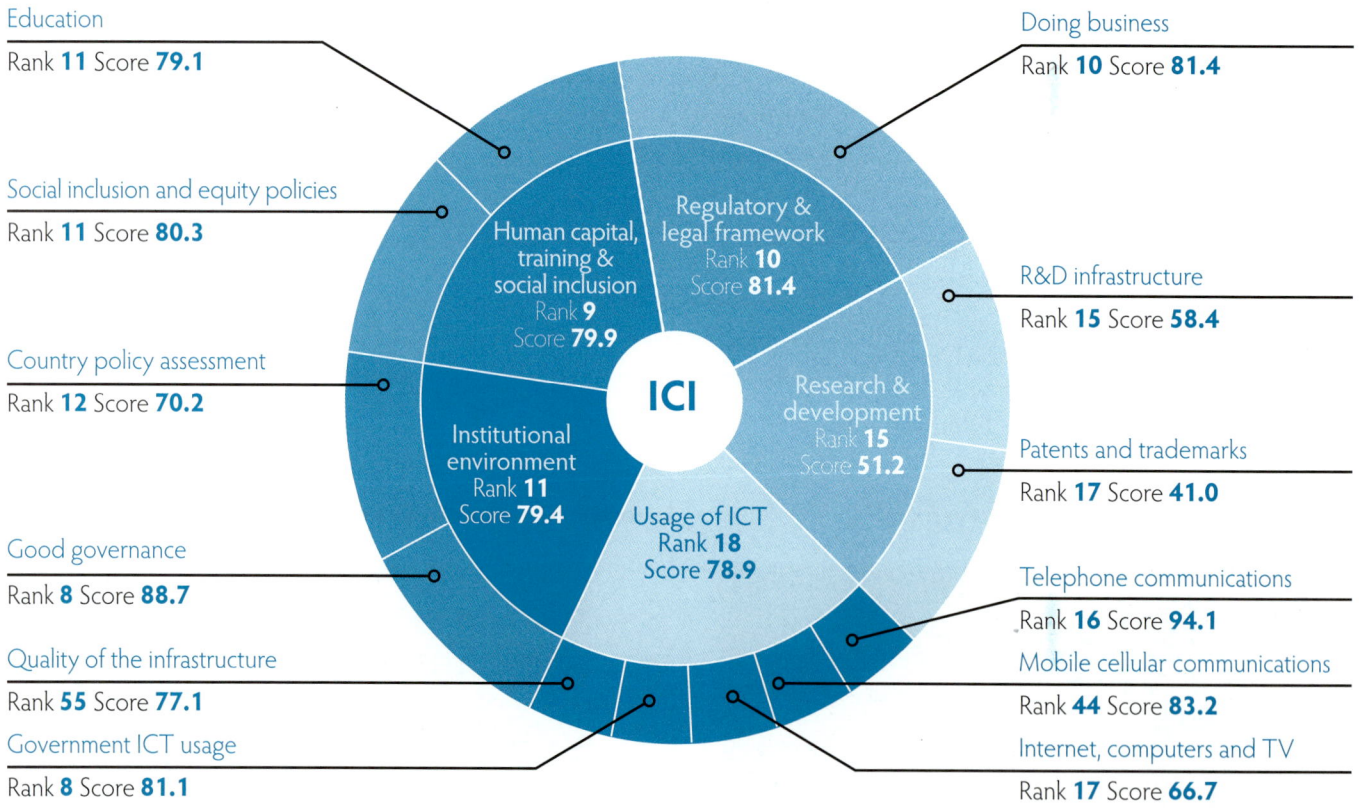

Human capital, training & social inclusion Rank **9** Score **79.9**

Regulatory & legal framework Rank **10** Score **81.4**

Institutional environment Rank **11** Score **79.4**

Research & development Rank **15** Score **51.2**

Usage of ICT Rank **18** Score **78.9**

ICI

Index thermometer[12]

Pillar 1 11.1% Pillar 2 11.2% Pillar 3 22.9% Pillar 4 21.5% Pillar 5 33.2% 100%

246

[1]High-income countries are those that, according to the World Bank classification, have a gross national income (GNI) per capita of $11,456 or more. [2] Full democracies are countries whose Democracy Index scores from 1 to 10 rounded to one decimal place are of 8 to 10. [3] 2007. [4] World Bank World Development Indicators. [5] IMF World Economic Outlook. [6] 2006. [7] Estimate. [8] UNESCO, % age above 15. [9] % ages 15–64. [10] 2005. [11] 2004. [12] The index thermometer shows the relative contribution of each pillar (in %) to the total index score.

AUSTRALIA

Country relative performance: **Index pillars**[13]

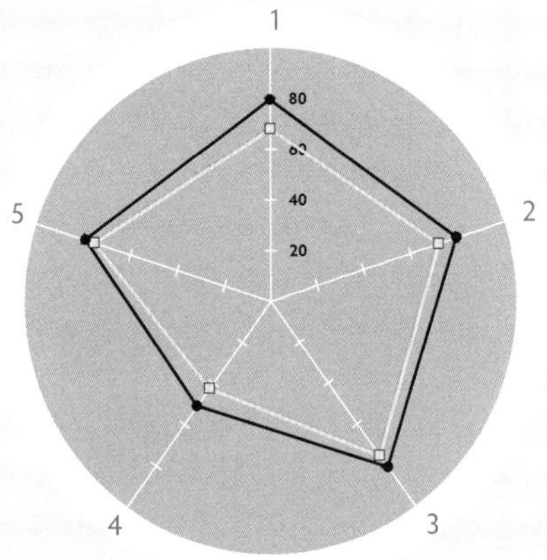

1. Institutional environment
2. Human capital, training and social inclusion
3. Regulatory and legal framework
4. Research and development
5. Usage of ICT

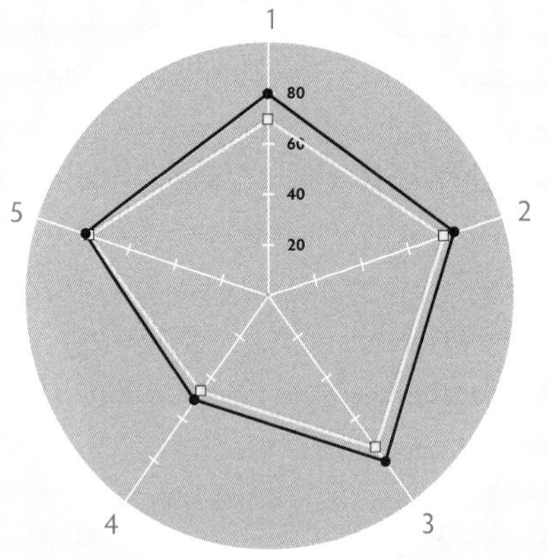

High-income ☐   Australia ●—   Full democracies ☐   Australia ●—

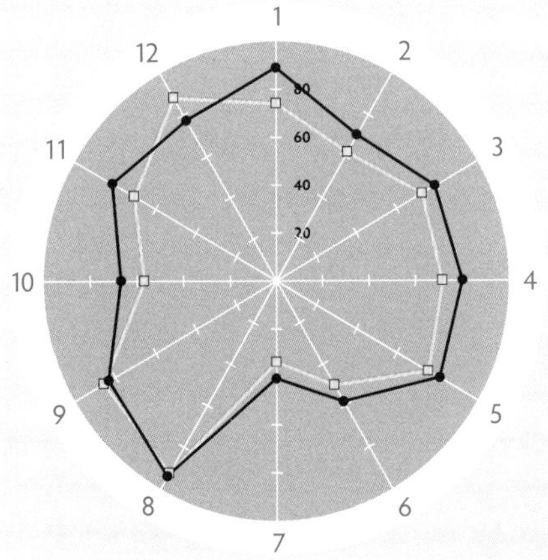

1. Good governance
2. Country policy assessment
3. Education
4. Social inclusion and equity policies
5. Doing business
6. R&D infrastructure
7. Patents and trademarks
8. Telephone communications
9. Mobile/cellular communications
10. Internet, computers and TV
11. Government ICT usage
12. Quality of the infrastructure

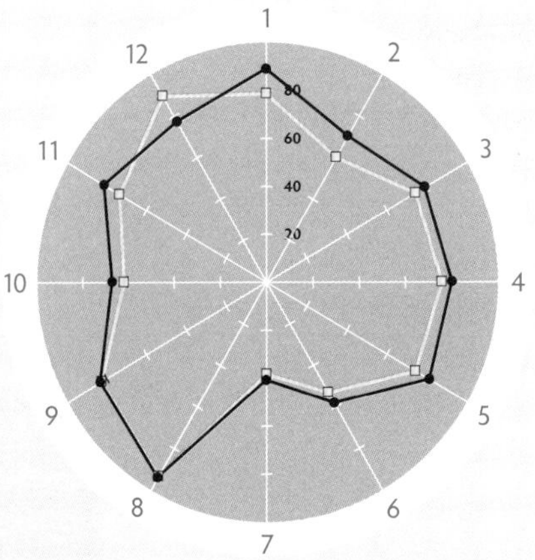

Country relative performance: **Pillar subindexes**[13]

In focus: **Top priorities for policy reform**[14]

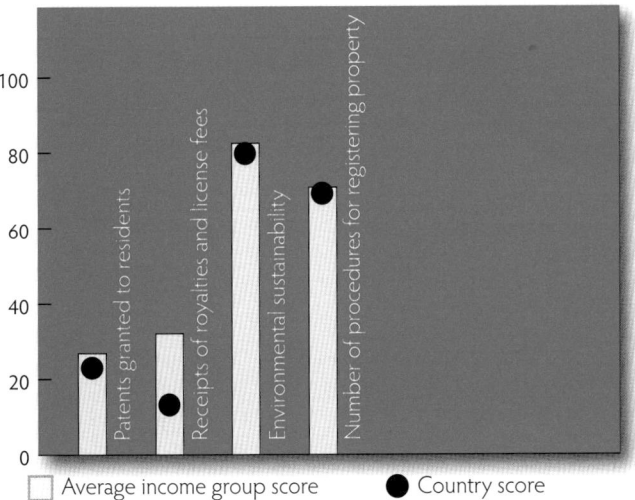

☐ Average income group score   ● Country score

In focus: **Significant indicators above income group average**[14]

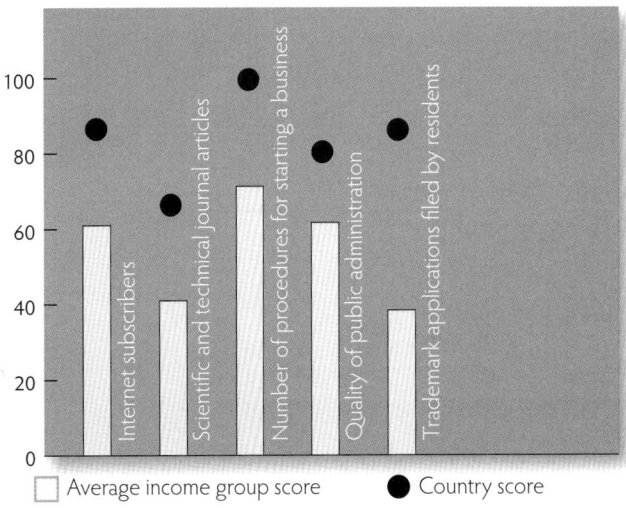

☐ Average income group score   ● Country score

[13] This analysis compares pillar and subindex country scores with respect to the average scores of countries grouped according to income level and political regime (see notes 1 and 2). [14] Significant country indicators with the greatest distances above or below the income group average scores (i.e., achievements and targets for policy reform, respectively) were selected and ordered according to their Pearson correlation coefficients (highest first) with respect to the final index scores, to produce a ranking.

247

AUSTRIA

Key selected indicators

Population (in millions)[3,4]	8.3
Population density (people per sq. km)[3,4]	101
GDP (current US$ billion, PPP)[3,5]	318
GDP per capita (current US$, PPP)[3,5]	38399
Current account balance (% of GDP)[3,5]	2.7
Average life expectancy (years)[4,6]	79.8
Literacy rate[7,8]	99
Female labor force participation rate[4,6,9]	64.5
Unemployment rate (% of total labor force)[4,10]	5.2
CO_2 emissions (metric tons per capita)[4,11]	8.5

Income level:[1] **High-income**

Political regime:[2] **Full democracy**

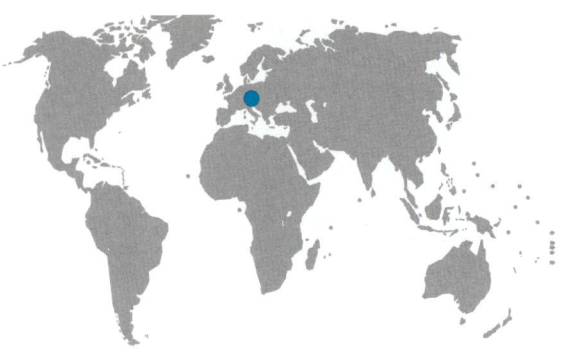

Innovation Capacity Index

Rank	Score
23 out of 131	**66.7 out of 100**

Pillars and pillar subindexes

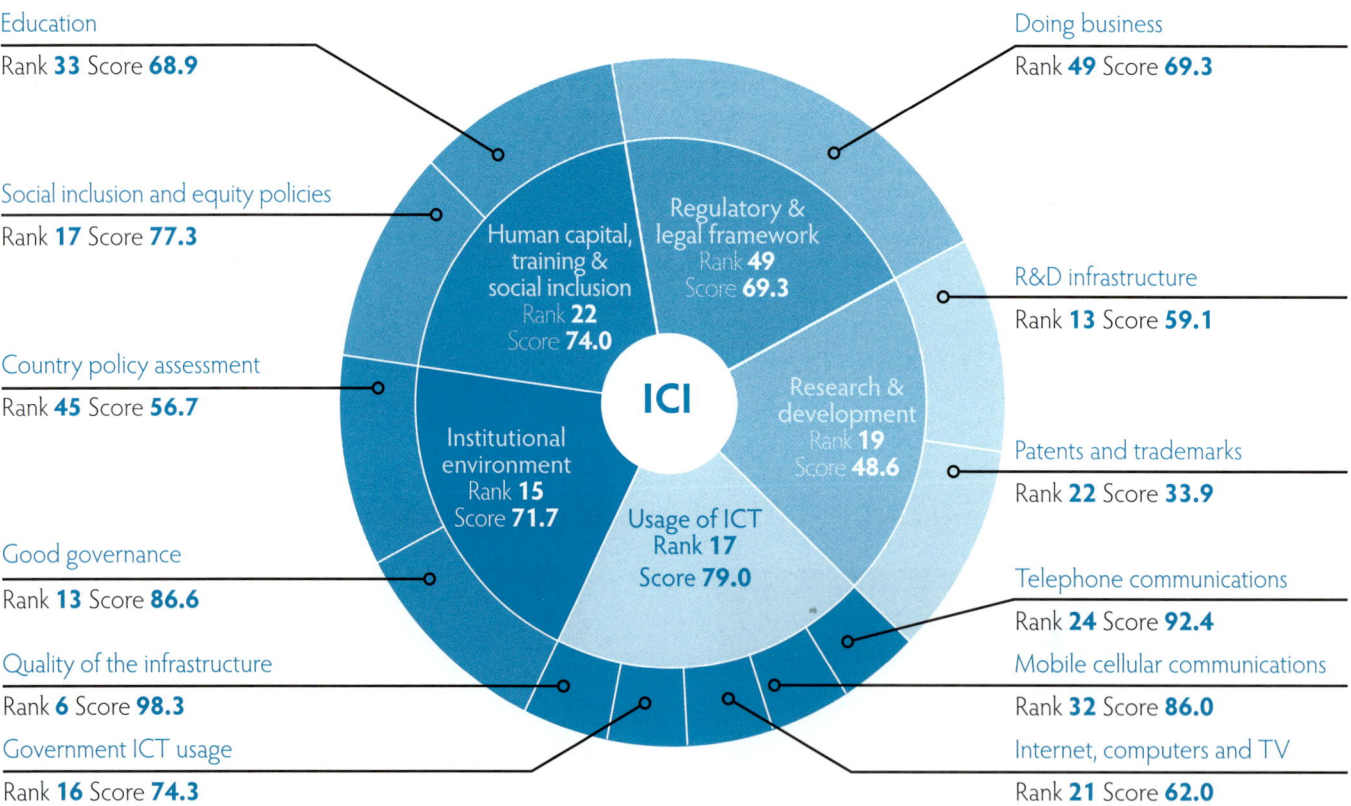

Education
Rank **33** Score **68.9**

Social inclusion and equity policies
Rank **17** Score **77.3**

Country policy assessment
Rank **45** Score **56.7**

Good governance
Rank **13** Score **86.6**

Quality of the infrastructure
Rank **6** Score **98.3**

Government ICT usage
Rank **16** Score **74.3**

Doing business
Rank **49** Score **69.3**

R&D infrastructure
Rank **13** Score **59.1**

Patents and trademarks
Rank **22** Score **33.9**

Telephone communications
Rank **24** Score **92.4**

Mobile cellular communications
Rank **32** Score **86.0**

Internet, computers and TV
Rank **21** Score **62.0**

Human capital, training & social inclusion Rank **22** Score **74.0**

Regulatory & legal framework Rank **49** Score **69.3**

ICI

Institutional environment Rank **15** Score **71.7**

Research & development Rank **19** Score **48.6**

Usage of ICT Rank **17** Score **79.0**

Index thermometer[12]

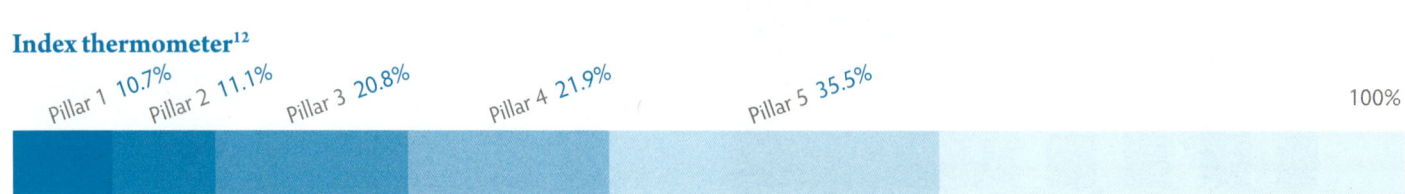

Pillar 1 10.7% Pillar 2 11.1% Pillar 3 20.8% Pillar 4 21.9% Pillar 5 35.5% 100%

[1] High-income countries are those that, according to the World Bank classification, have a gross national income (GNI) per capita of $11,456 or more. [2] Full democracies are countries whose Democracy Index scores from 1 to 10 rounded to one decimal place are of 8 to 10. [3] 2007. [4] World Bank World Development Indicators. [5] IMF World Economic Outlook. [6] 2006. [7] Estimate. [8] UNESCO, % age above 15. [9] % ages 15–64. [10] 2005. [11] 2004. [12] The index thermometer shows the relative contribution of each pillar (in %) to the total index score.

AUSTRIA

Country relative performance: **Index pillars**[13]

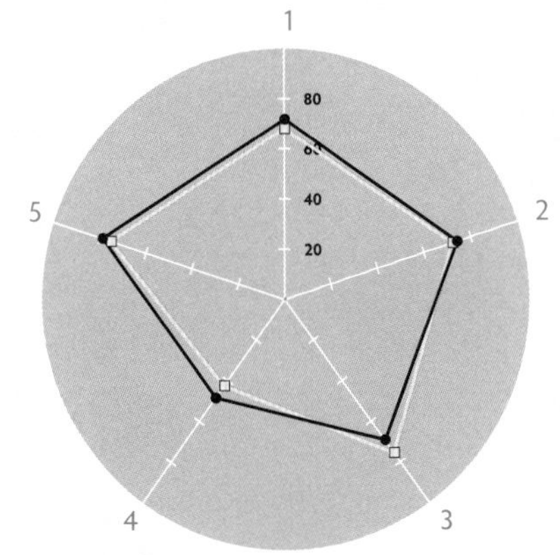

1. Institutional environment
2. Human capital, training and social inclusion
3. Regulatory and legal framework
4. Research and development
5. Usage of ICT

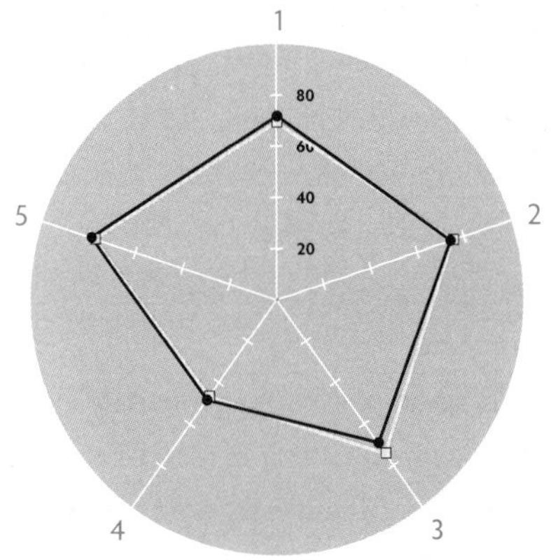

High-income ☐   Austria ●—

Full democracies ☐   Austria ●—

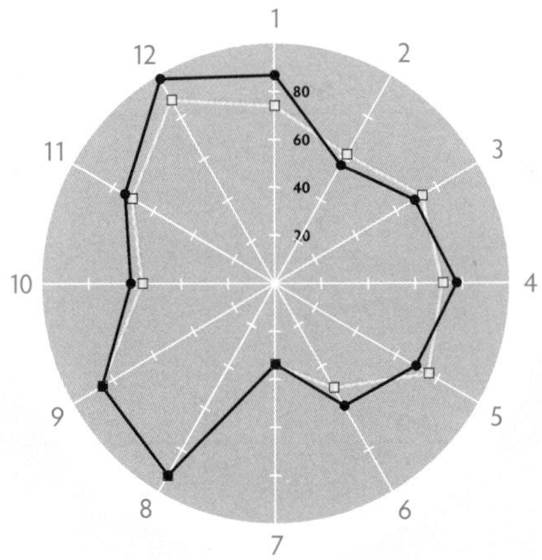

1. Good governance
2. Country policy assessment
3. Education
4. Social inclusion and equity policies
5. Doing business
6. R&D infrastructure
7. Patents and trademarks
8. Telephone communications
9. Mobile/cellular communications
10. Internet, computers and TV
11. Government ICT usage
12. Quality of the infrastructure

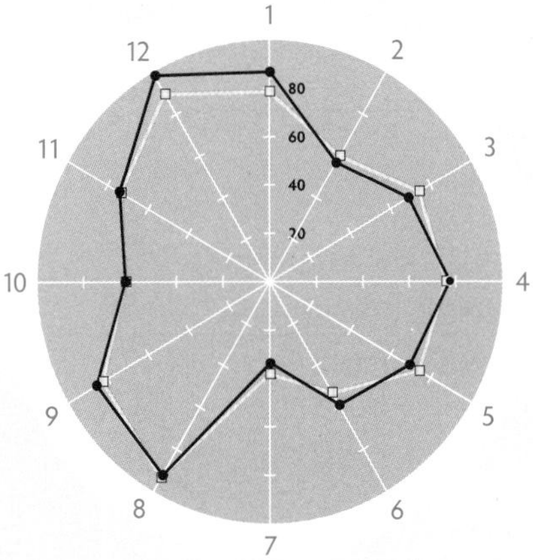

Country relative performance: **Pillar subindexes**[13]

In focus: **Top priorities for policy reform**[14]

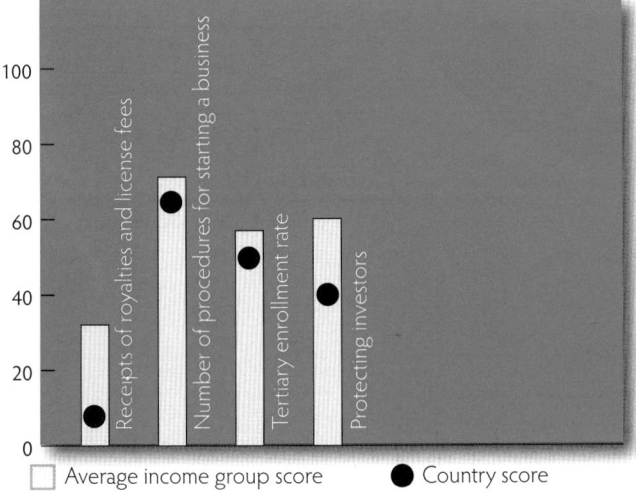

☐ Average income group score   ● Country score

In focus: **Significant indicators above income group average**[14]

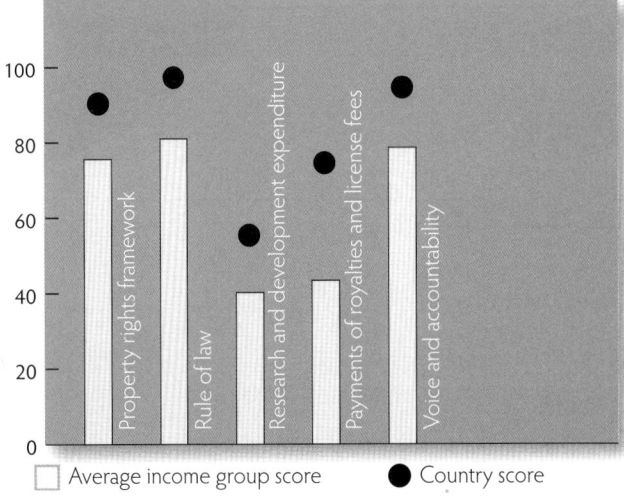

☐ Average income group score   ● Country score

---

[13] This analysis compares pillar and subindex country scores with respect to the average scores of countries grouped according to income level and political regime (see notes 1 and 2).

[14] Significant country indicators with the greatest distances above or below the income group average scores (i.e., achievements and targets for policy reform, respectively) were selected and ordered according to their Pearson correlation coefficients (highest first) with respect to the final index scores, to produce a ranking.

BELGIUM

Key selected indicators

Population (in millions)[3,4]	10.6
Population density (people per sq. km)[3,4]	351
GDP (current US$ billion, PPP)[3,5]	376
GDP per capita (current US$, PPP)[3,5]	35273
Current account balance (% of GDP)[3,5]	3.2
Average life expectancy (years)[4,6]	79.5
Literacy rate[7,8]	99
Female labor force participation rate[4,6,9]	57.8
Unemployment rate (% of total labor force)[4,10]	8.1
CO_2 emissions (metric tons per capita)[4,11]	9.7

Income level:[1] **High-income**

Political regime:[2] **Full democracy**

Innovation Capacity Index

Rank **22 out of 131**

Score **67.6 out of 100**

Pillars and pillar subindexes

Education
Rank **18** Score **75.7**

Social inclusion and equity policies
Rank **6** Score **85.6**

Country policy assessment
Rank **26** Score **60.5**

Good governance
Rank **18** Score **79.2**

Quality of the infrastructure
Rank **24** Score **91.4**

Government ICT usage
Rank **24** Score **67.8**

Doing business
Rank **26** Score **74.4**

R&D infrastructure
Rank **20** Score **45.6**

Patents and trademarks
Rank **15** Score **50.0**

Telephone communications
Rank **20** Score **93.3**

Mobile cellular communications
Rank **41** Score **84.4**

Internet, computers and TV
Rank **24** Score **56.5**

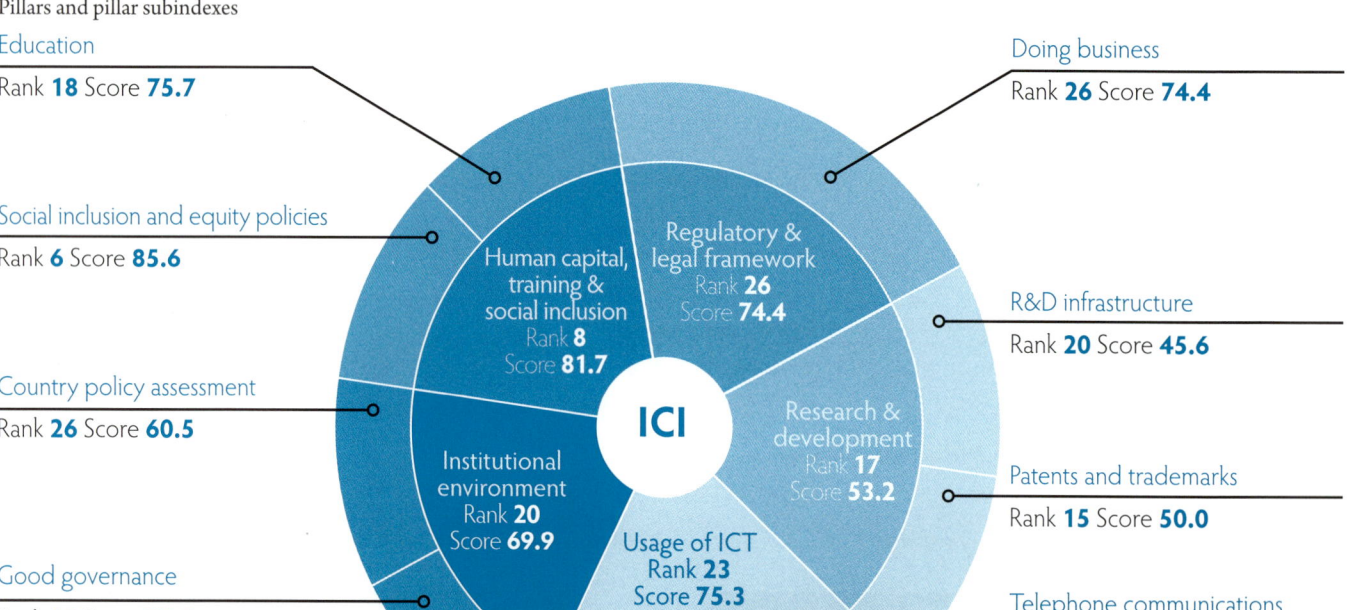

Human capital, training & social inclusion
Rank **8** Score **81.7**

Regulatory & legal framework
Rank **26** Score **74.4**

Institutional environment
Rank **20** Score **69.9**

Research & development
Rank **17** Score **53.2**

ICI

Usage of ICT
Rank **23** Score **75.3**

Index thermometer[12]

Pillar 1 10.3% Pillar 2 12.1% Pillar 3 22.0% Pillar 4 22.2% Pillar 5 33.4% 100%

250

---

[1] High-income countries are those that, according to the World Bank classification, have a gross national income (GNI) per capita of $11,456 or more.
[2] Full democracies are countries whose Democracy Index scores from 1 to 10 rounded to one decimal place are of 8 to 10. [3] 2007. [4] World Bank World Development Indicators. [5] IMF World Economic Outlook. [6] 2006. [7] Estimate. [8] UNESCO, % age above 15. [9] % ages 15–64. [10] 2005. [11] 2004.
[12] The index thermometer shows the relative contribution of each pillar (in %) to the total index score.

BELGIUM

Country relative performance: **Index pillars**[13]

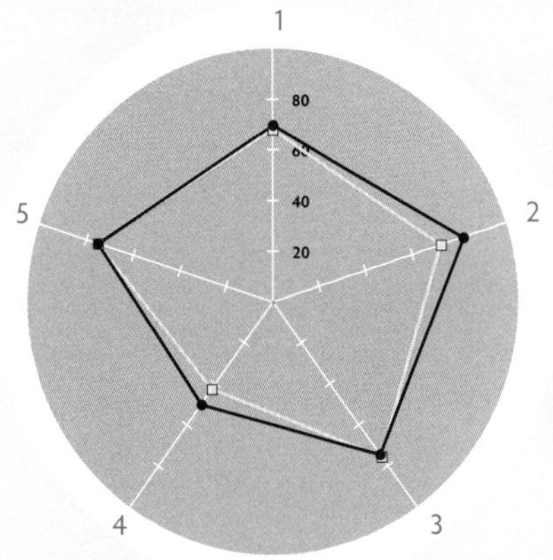

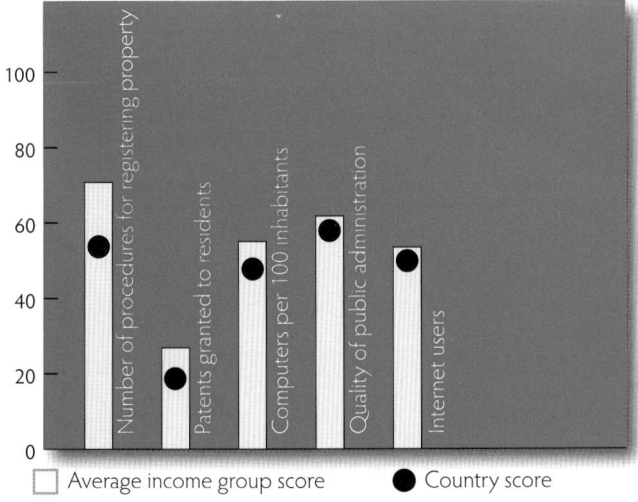

1. Institutional environment
2. Human capital, training and social inclusion
3. Regulatory and legal framework
4. Research and development
5. Usage of ICT

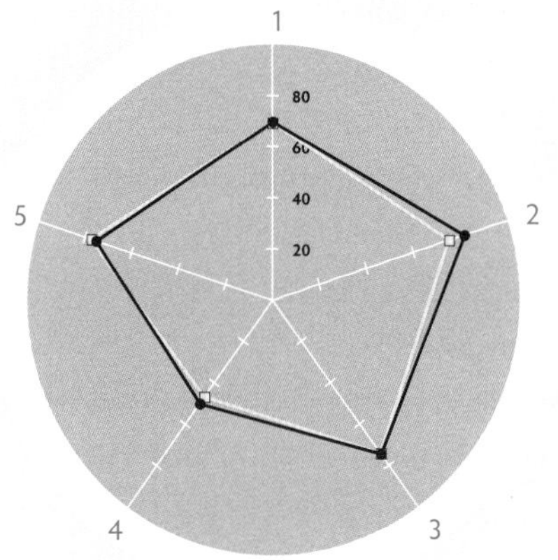

High-income ☐ Belgium ●——

Full democracies ☐ Belgium ●——

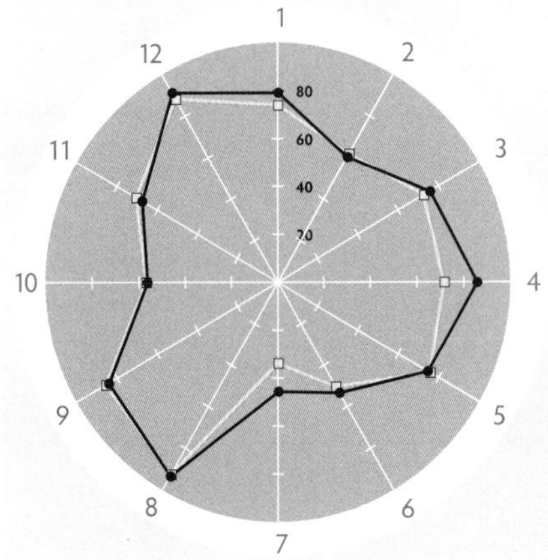

1. Good governance
2. Country policy assessment
3. Education
4. Social inclusion and equity policies
5. Doing business
6. R&D infrastructure
7. Patents and trademarks
8. Telephone communications
9. Mobile/cellular communications
10. Internet, computers and TV
11. Government ICT usage
12. Quality of the infrastructure

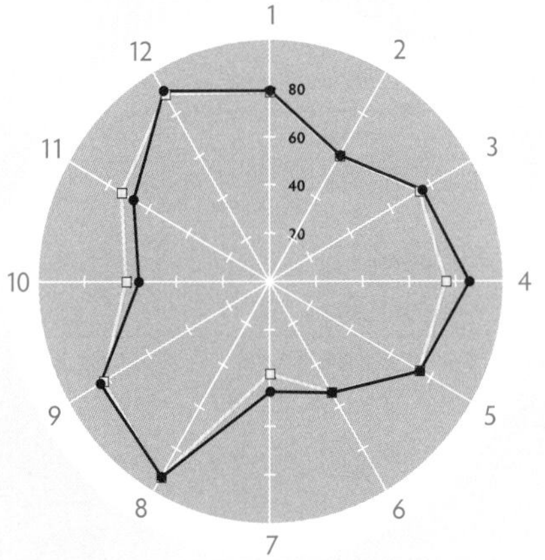

Country relative performance: **Pillar subindexes**[13]

In focus: **Top priorities for policy reform**[14]

In focus: **Significant indicators above income group average**[14]

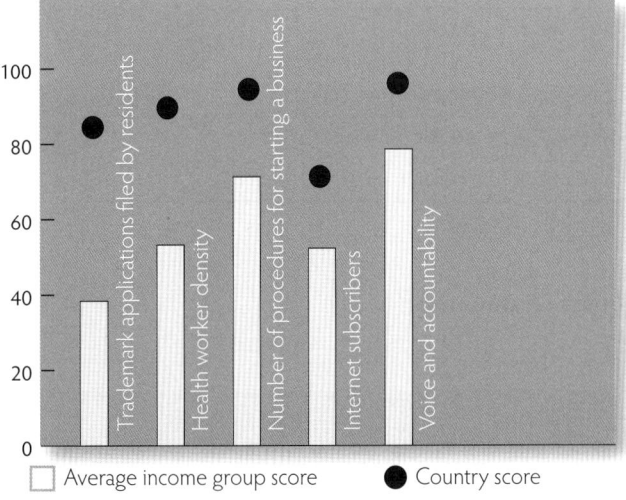

☐ Average income group score ● Country score

[13] This analysis compares pillar and subindex country scores with respect to the average scores of countries grouped according to income level and political regime (see notes 1 and 2).

[14] Significant country indicators with the greatest distances above or below the income group average scores (i.e., achievements and targets for policy reform, respectively) were selected and ordered according to their Pearson correlation coefficients (highest first) with respect to the final index scores, to produce a ranking.

BOTSWANA

Key selected indicators

Population (in millions)[3,4]	1.9
Population density (people per sq. km)[3,4]	3
GDP (current US$ billion, PPP)[3,5]	26
GDP per capita (current US$, PPP)[3,5]	16450
Current account balance (% of GDP)[3,5]	16.8
Average life expectancy (years)[4,6]	49.8
Literacy rate[3,7]	83
Female labor force participation rate[4,6,8]	47.5
Unemployment rate (% of total labor force)[4,9]	23.8
CO_2 emissions (metric tons per capita)[4,10]	2.4

Income level:[1] **Upper-middle-income**
Political regime:[2] **Flawed democracy**

Innovation Capacity Index

Rank **67 out of 131**

Score **49.1 out of 100**

Pillars and pillar subindexes

Education
Rank **91** Score **48.9**

Social inclusion and equity policies
Rank **95** Score **45.5**

Country policy assessment
Rank **18** Score **66.7**

Good governance
Rank **34** Score **63.9**

Quality of the infrastructure
Rank **92** Score **49.8**

Government ICT usage
Rank **89** Score **36.5**

Doing business
Rank **60** Score **66.8**

R&D infrastructure
Rank **94** Score **12.9**

Patents and trademarks
Rank **92** Score **0.6**

Telephone communications
Rank **66** Score **79.7**

Mobile cellular communications
Rank **58** Score **76.6**

Internet, computers and TV
Rank **99** Score **4.3**

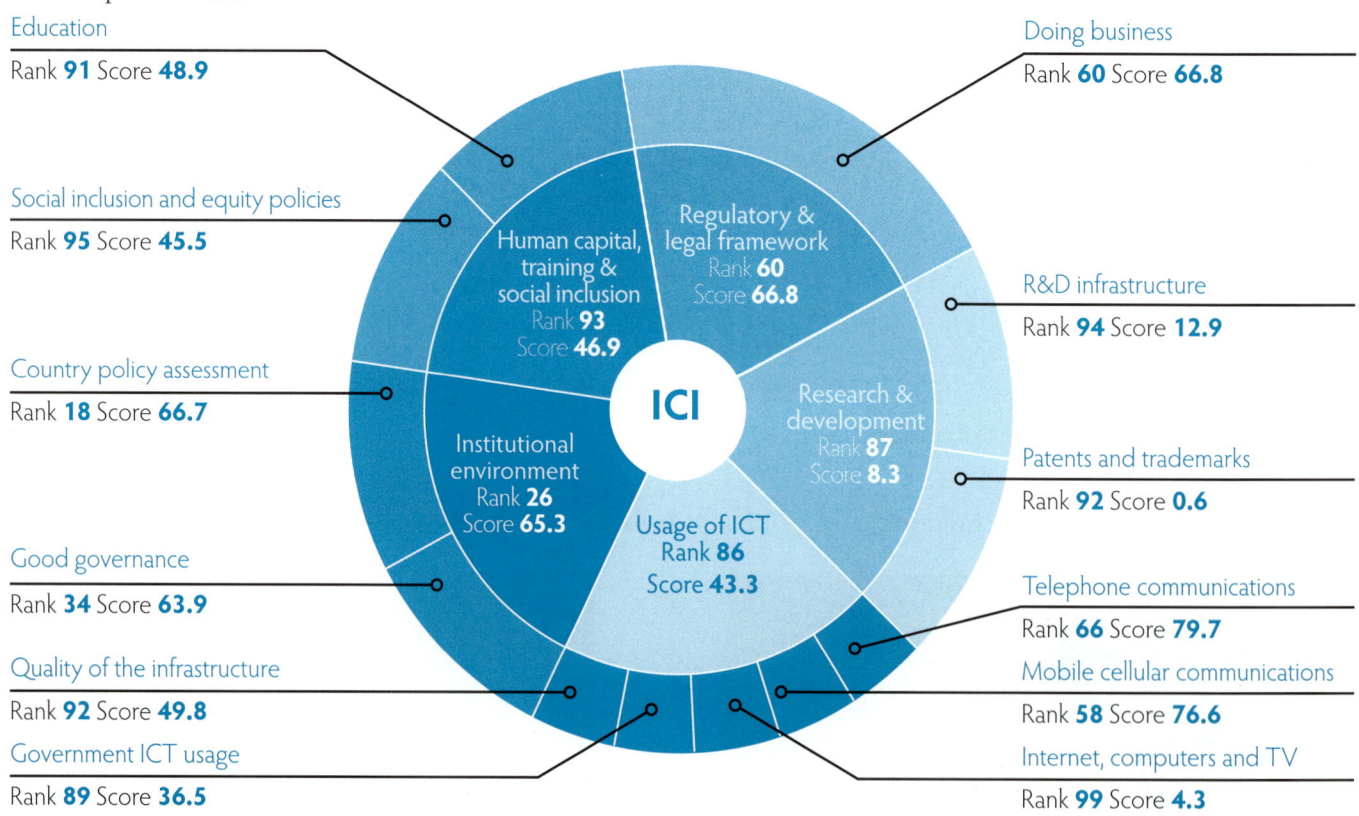

Index thermometer[11]

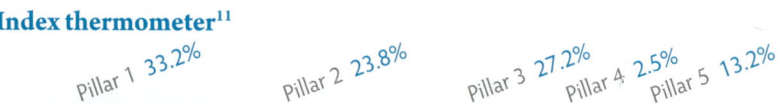

Pillar 1 33.2% Pillar 2 23.8% Pillar 3 27.2% Pillar 4 2.5% Pillar 5 13.2%   100%

[1] Upper-middle-income countries are those that, according to the World Bank classification, have a gross national income (GNI) per capita of $3,706 to $11,455. [2] Flawed democracies are countries whose Democracy Index scores from 1 to 10 rounded to one decimal place are of 6 to 7.9. [3] 2007. [4] World Bank World Development Indicators. [5] IMF World Economic Outlook. [6] 2006. [7] UNESCO, % age above 15. [8] % ages 15–64. [9] 2003. [10] 2004. [11] The index thermometer shows the relative contribution of each pillar (in %) to the total index score.

BOTSWANA

Country relative performance: **Index pillars**[12]

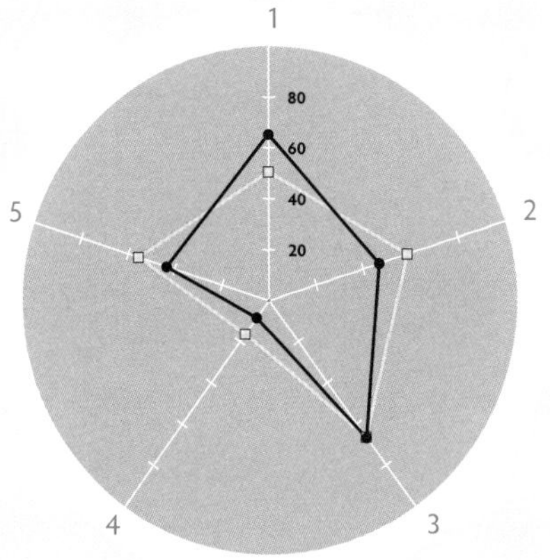

1. Institutional environment
2. Human capital, training and social inclusion
3. Regulatory and legal framework
4. Research and development
5. Usage of ICT

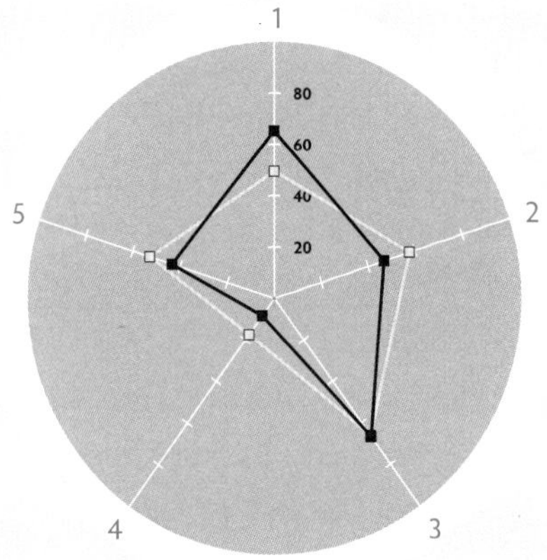

Upper-middle-income ☐   Botswana ●—

Flawed democracies ☐   Botswana ●—

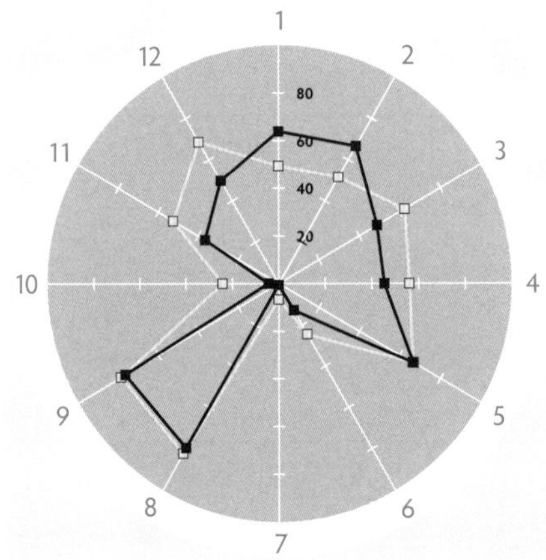

1. Good governance
2. Country policy assessment
3. Education
4. Social inclusion and equity policies
5. Doing business
6. R&D infrastructure
7. Patents and trademarks
8. Telephone communications
9. Mobile/cellular communications
10. Internet, computers and TV
11. Government ICT usage
12. Quality of the infrastructure

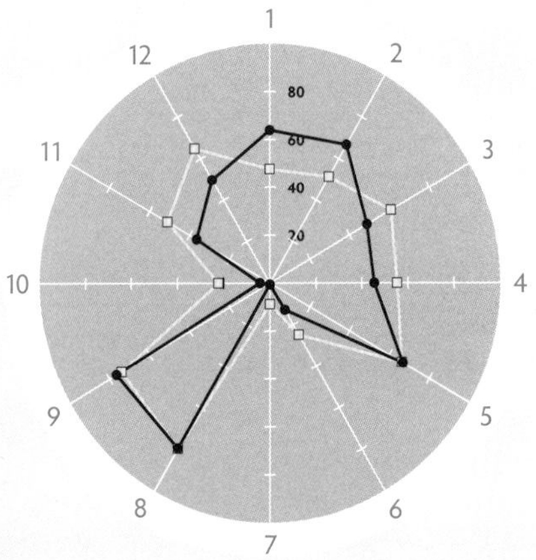

Country relative performance: **Pillar subindexes**[12]

253

In focus: **Top priorities for policy reform**[13]

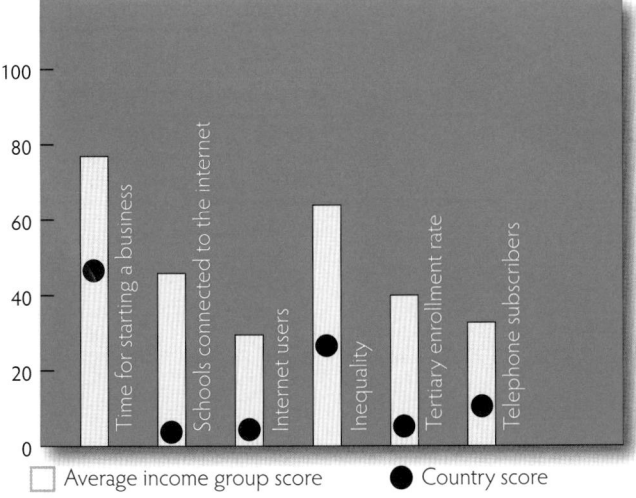

☐ Average income group score   ● Country score

In focus: **Significant indicators above income group average**[13]

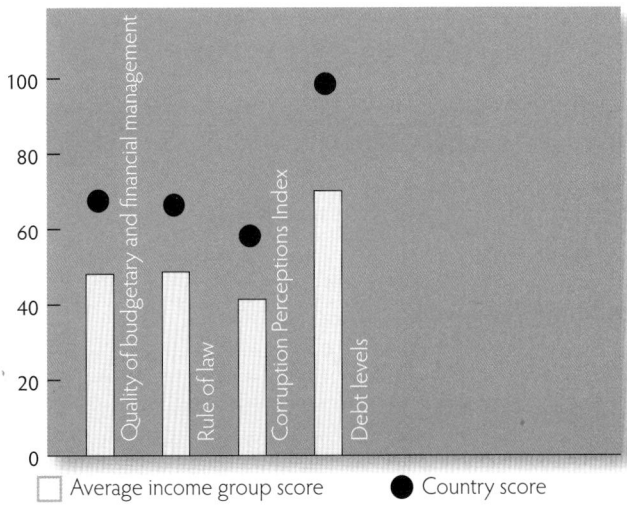

☐ Average income group score   ● Country score

---

[12] This analysis compares pillar and subindex country scores with respect to the average scores of countries grouped according to income level and political regime (see notes 1 and 2).

[13] Significant country indicators with the greatest distances above or below the income group average scores (i.e., achievements and targets for policy reform, respectively) were selected and ordered according to their Pearson correlation coefficients (highest first) with respect to the final index scores, to produce a ranking.

BRAZIL

Key selected indicators

Population (in millions)[3,4]	191.6
Population density (people per sq. km)[3,4]	23
GDP (current US$ billion, PPP)[3,5]	1836
GDP per capita (current US$, PPP)[3,5]	9695
Current account balance (% of GDP)[3,5]	0.3
Average life expectancy (years)[4,6]	72.1
Literacy rate[3,7]	91
Female labor force participation rate[4,6,8]	61.5
Unemployment rate (% of total labor force)[4,9]	8.9
CO_2 emissions (metric tons per capita)[4,9]	1.8

Income level:[1] **Upper-middle-income**

Political regime:[2] **Flawed democracy**

Innovation Capacity Index

Rank
87 out of 131

Score
45.2 out of 100

Pillars and pillar subindexes

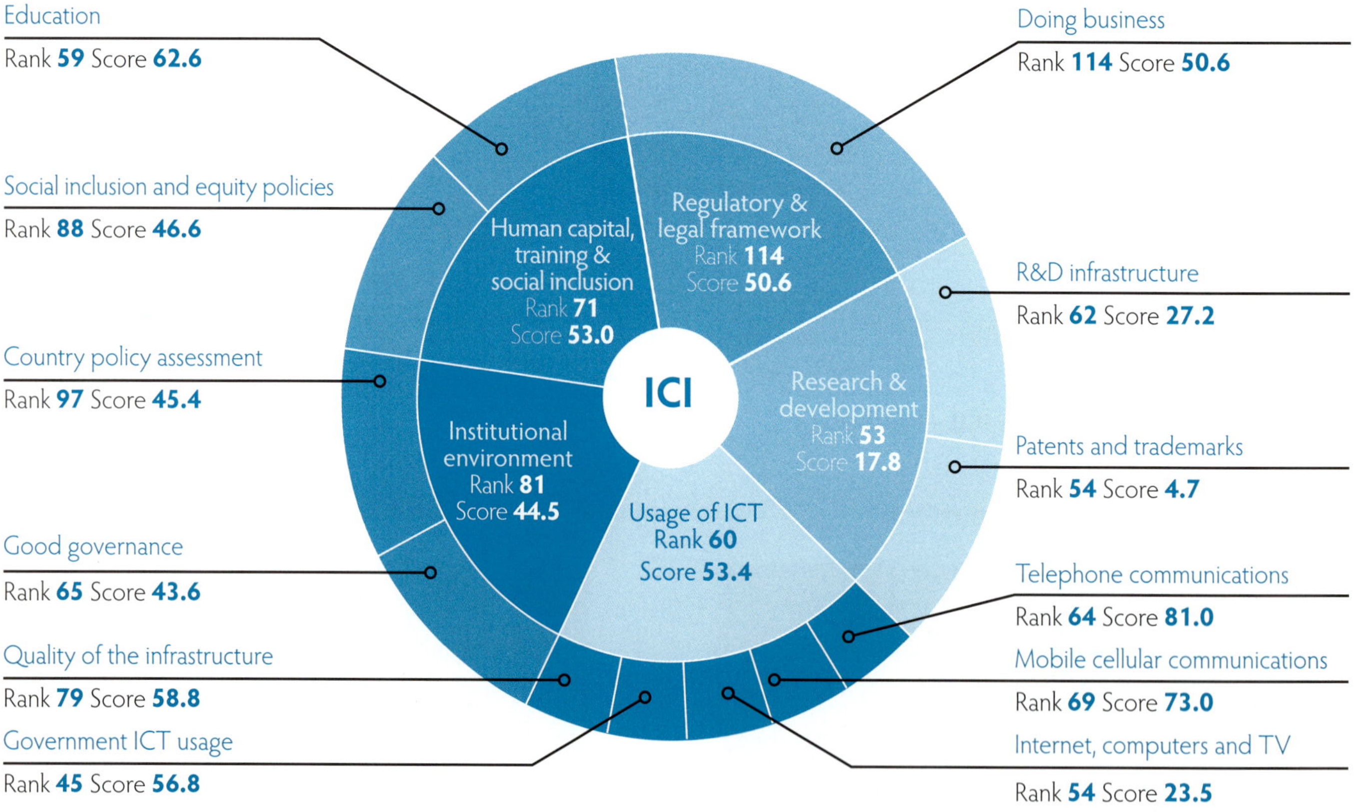

Education
Rank **59** Score **62.6**

Social inclusion and equity policies
Rank **88** Score **46.6**

Country policy assessment
Rank **97** Score **45.4**

Good governance
Rank **65** Score **43.6**

Quality of the infrastructure
Rank **79** Score **58.8**

Government ICT usage
Rank **45** Score **56.8**

Doing business
Rank **114** Score **50.6**

R&D infrastructure
Rank **62** Score **27.2**

Patents and trademarks
Rank **54** Score **4.7**

Telephone communications
Rank **64** Score **81.0**

Mobile cellular communications
Rank **69** Score **73.0**

Internet, computers and TV
Rank **54** Score **23.5**

ICI

Human capital, training & social inclusion
Rank **71** Score **53.0**

Regulatory & legal framework
Rank **114** Score **50.6**

Research & development
Rank **53** Score **17.8**

Institutional environment
Rank **81** Score **44.5**

Usage of ICT
Rank **60** Score **53.4**

Index thermometer[10]

Pillar 1 24.6%   Pillar 2 29.3%   Pillar 3 22.4%   Pillar 4 5.9%   Pillar 5 17.7%   100%

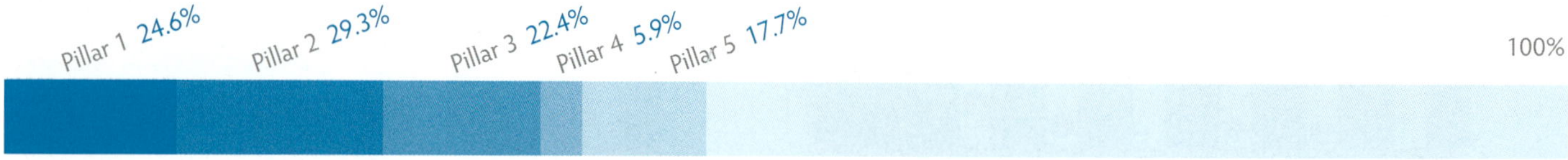

[1] Upper-middle-income countries are those that, according to the World Bank classification, have a gross national income (GNI) per capita of $3,706 to $11,455. [2] Flawed democracies are countries whose Democracy Index scores from 1 to 10 rounded to one decimal place are of 6 to 7.9. [3] 2007. [4] World Bank World Development Indicators. [5] IMF World Economic Outlook. [6] 2006. [7] UNESCO, % age above 15. [8] % ages 15–64. [9] 2004. [10] The index thermometer shows the relative contribution of each pillar (in %) to the total index score.

254

BRAZIL

Country relative performance: Index pillars[11]

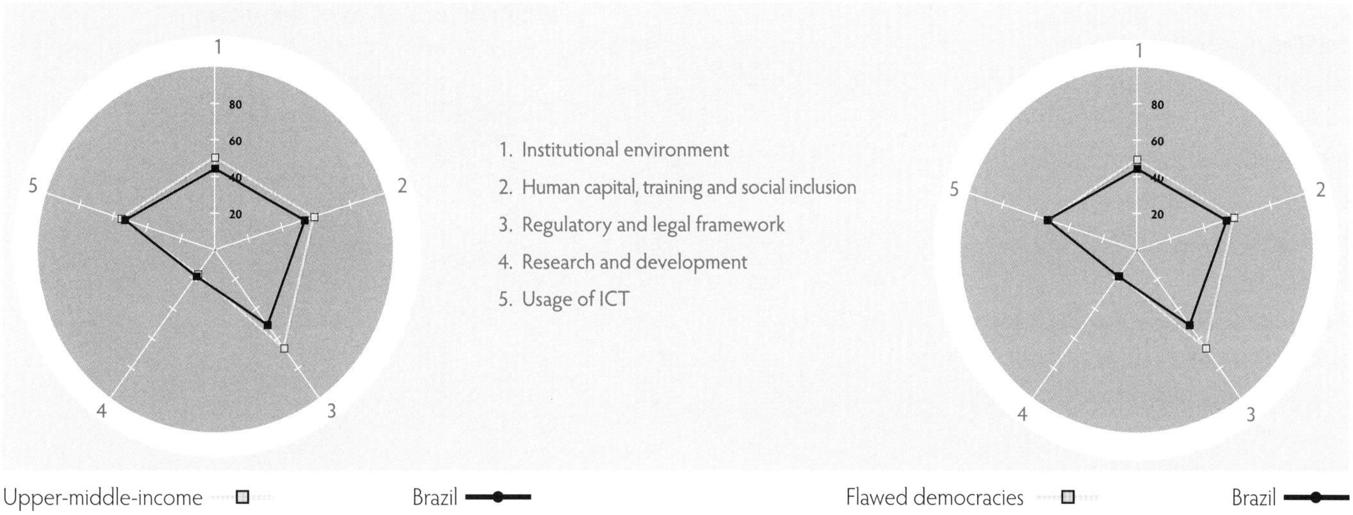

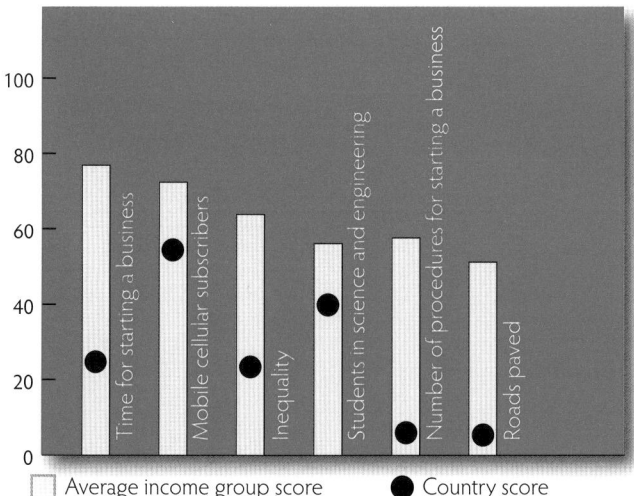

1. Institutional environment
2. Human capital, training and social inclusion
3. Regulatory and legal framework
4. Research and development
5. Usage of ICT
Upper-middle-income
Brazil
Flawed democracies
Brazil

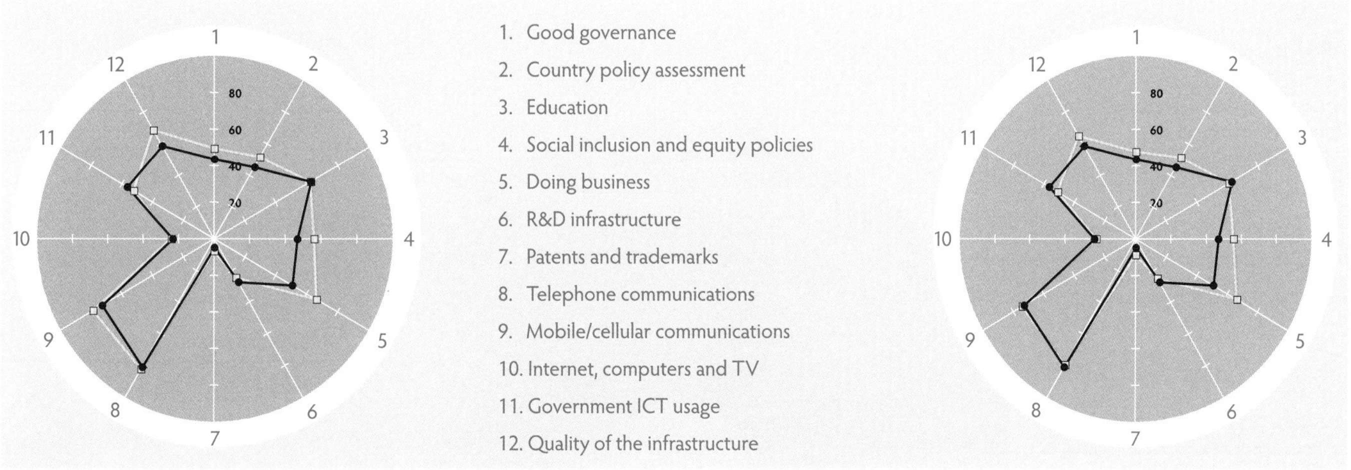

1. Good governance
2. Country policy assessment
3. Education
4. Social inclusion and equity policies
5. Doing business
6. R&D infrastructure
7. Patents and trademarks
8. Telephone communications
9. Mobile/cellular communications
10. Internet, computers and TV
11. Government ICT usage
12. Quality of the infrastructure

Country relative performance: Pillar subindexes[11]

In focus: Top priorities for policy reform[12]

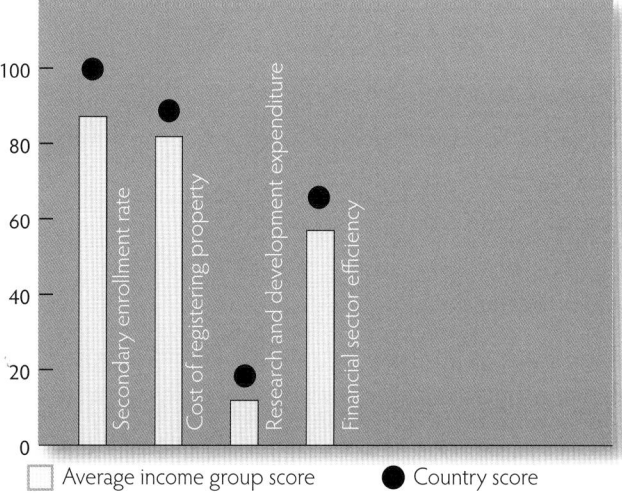

Time for starting a business
Mobile cellular subscribers
Inequality
Students in science and engineering
Number of procedures for starting a business
Roads paved
Average income group score
Country score

In focus: Significant indicators above income group average[12]

Secondary enrollment rate
Cost of registering property
Research and development expenditure
Financial sector efficiency
Average income group score
Country score

[11] This analysis compares pillar and subindex country scores with respect to the average scores of countries grouped according to income level and political regime (see notes 1 and 2).

[12] Significant country indicators with the greatest distances above or below the income group average scores (i.e., achievements and targets for policy reform, respectively) were selected and ordered according to their Pearson correlation coefficients (highest first) with respect to the final index scores, to produce a ranking.

BULGARIA

Key selected indicators

Population (in millions)[3,4]	7.6
Population density (people per sq. km)[3,4]	70
GDP (current US$ billion, PPP)[3,5]	86
GDP per capita (current US$, PPP)[3,5]	11302
Current account balance (% of GDP)[3,5]	-21.4
Average life expectancy (years)[4,6]	72.6
Literacy rate[3,7]	98
Female labor force participation rate[4,6,8]	51.6
Unemployment rate (% of total labor force)[4,9]	10.1
CO_2 emissions (metric tons per capita)[4,10]	5.5

Income level:[1] **Upper-middle-income**

Political regime:[2] **Flawed democracy**

Innovation Capacity Index

Rank	Score
33 out of 131	**57.7 out of 100**

Pillars and pillar subindexes

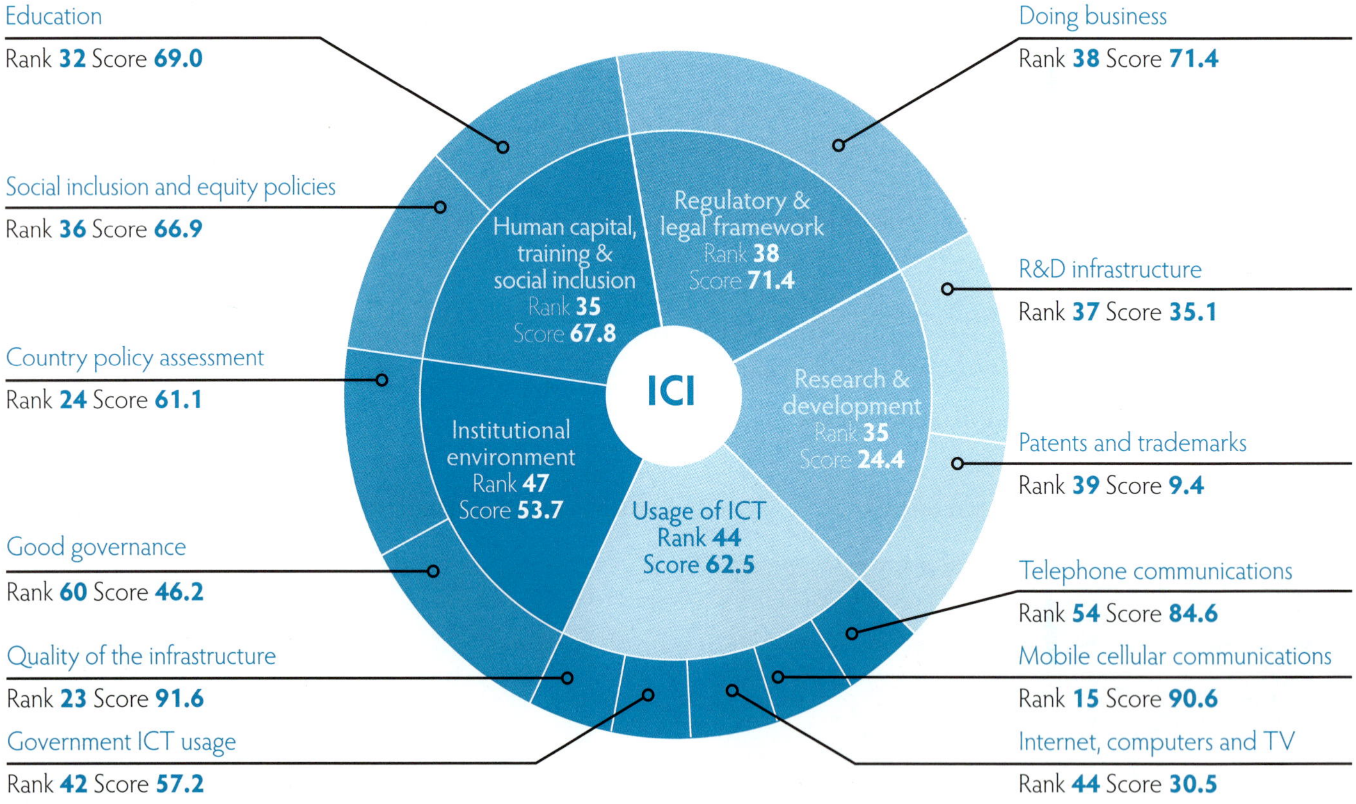

Education
Rank **32** Score **69.0**

Social inclusion and equity policies
Rank **36** Score **66.9**

Country policy assessment
Rank **24** Score **61.1**

Good governance
Rank **60** Score **46.2**

Quality of the infrastructure
Rank **23** Score **91.6**

Government ICT usage
Rank **42** Score **57.2**

Doing business
Rank **38** Score **71.4**

R&D infrastructure
Rank **37** Score **35.1**

Patents and trademarks
Rank **39** Score **9.4**

Telephone communications
Rank **54** Score **84.6**

Mobile cellular communications
Rank **15** Score **90.6**

Internet, computers and TV
Rank **44** Score **30.5**

Center diagram labels:
- ICI
- Human capital, training & social inclusion — Rank **35** Score **67.8**
- Regulatory & legal framework — Rank **38** Score **71.4**
- Institutional environment — Rank **47** Score **53.7**
- Research & development — Rank **35** Score **24.4**
- Usage of ICT — Rank **44** Score **62.5**

Index thermometer[11]

Pillar 1 23.3% Pillar 2 29.4% Pillar 3 24.8% Pillar 4 6.3% Pillar 5 16.2%   100%

256

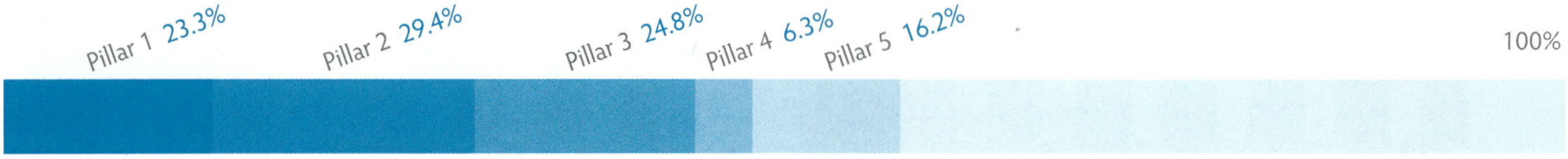

[1] Upper-middle-income countries are those that, according to the World Bank classification, have a gross national income (GNI) per capita of $3,706 to $11,455. [2] Flawed democracies are countries whose Democracy Index scores from 1 to 10 rounded to one decimal place are of 6 to 7.9. [3] 2007. [4] World Bank World Development Indicators. [5] IMF World Economic Outlook. [6] 2006. [7] UNESCO, % age above 15. [8] % ages 15–64. [9] 2005. [10] 2004. [11] The index thermometer shows the relative contribution of each pillar (in %) to the total index score.

BULGARIA

Country relative performance: **Index pillars**[12]

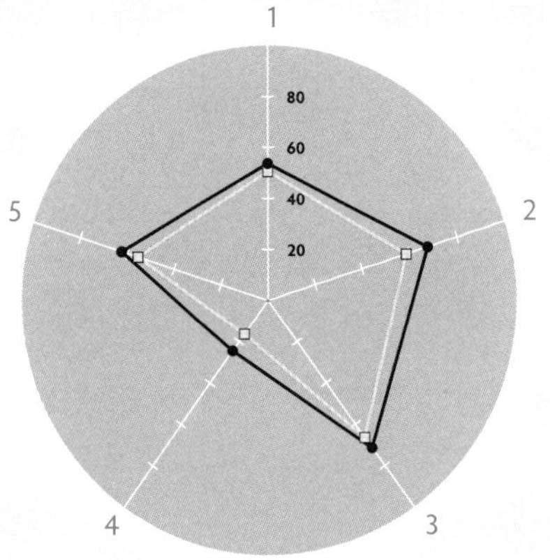

1. Institutional environment
2. Human capital, training and social inclusion
3. Regulatory and legal framework
4. Research and development
5. Usage of ICT

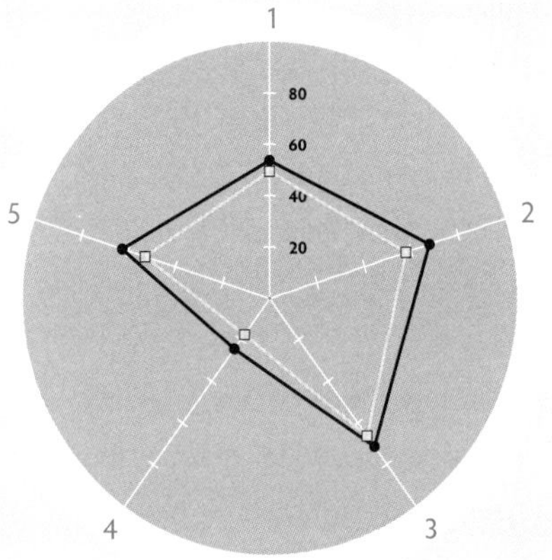

Upper-middle-income ☐   Bulgaria ●

Flawed democracies ☐   Bulgaria ●

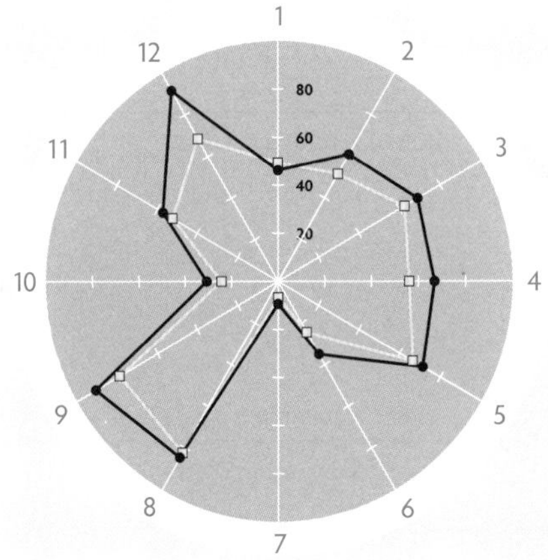

1. Good governance
2. Country policy assessment
3. Education
4. Social inclusion and equity policies
5. Doing business
6. R&D infrastructure
7. Patents and trademarks
8. Telephone communications
9. Mobile/cellular communications
10. Internet, computers and TV
11. Government ICT usage
12. Quality of the infrastructure

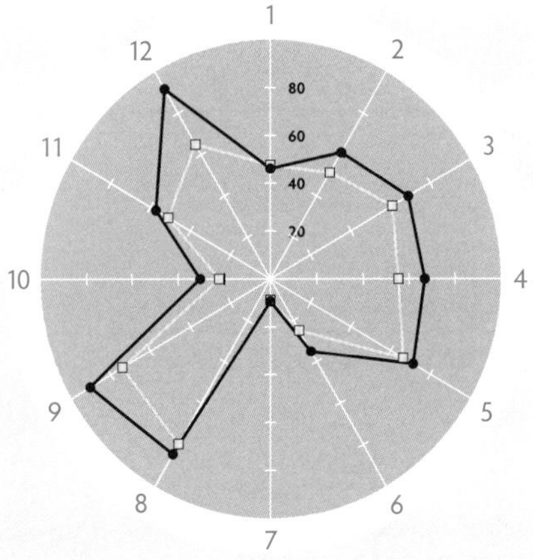

Country relative performance: **Pillar subindexes**[12]

In focus: **Top priorities for policy reform**[13]

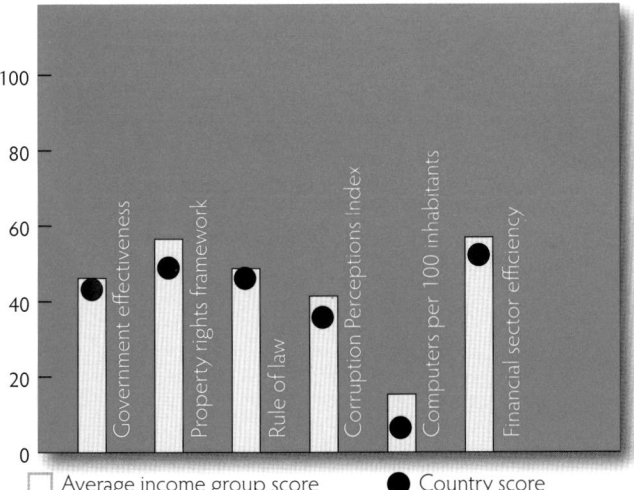

Government effectiveness · Property rights framework · Rule of law · Corruption Perceptions Index · Computers per 100 inhabitants · Financial sector efficiency

☐ Average income group score   ● Country score

In focus: **Significant indicators above income group average**[13]

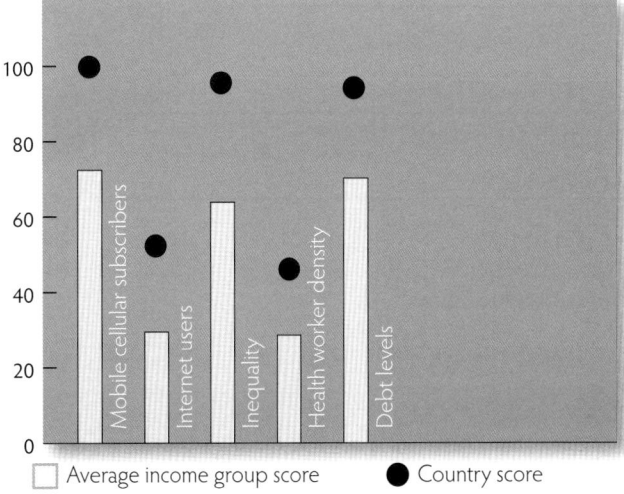

Mobile cellular subscribers · Internet users · Inequality · Health worker density · Debt levels

☐ Average income group score   ● Country score

[12] This analysis compares pillar and subindex country scores with respect to the average scores of countries grouped according to income level and political regime (see notes 1 and 2).

[13] Significant country indicators with the greatest distances above or below the income group average scores (i.e., achievements and targets for policy reform, respectively) were selected and ordered according to their Pearson correlation coefficients (highest first) with respect to the final index scores, to produce a ranking.

CANADA

Key selected indicators

Population (in millions)[3,4]	33.0
Population density (people per sq. km)[3,4]	4
GDP (current US$ billion, PPP)[3,5]	1266
GDP per capita (current US$, PPP)[3,5]	38435
Current account balance (% of GDP)[3,5]	0.9
Average life expectancy (years)[4,6]	80.4
Literacy rate[7,8]	99
Female labor force participation rate[4,6,9]	73.2
Unemployment rate (% of total labor force)[4,10]	6.8
CO_2 emissions (metric tons per capita)[4,11]	20.0

Income level:[1] **High-income**
Political regime:[2] **Full democracy**

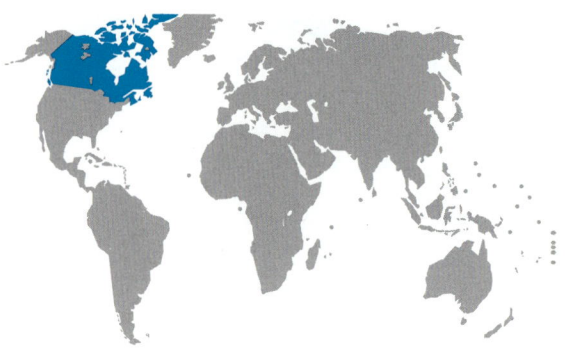

Innovation Capacity Index

Rank **7 out of 131** Score **74.8 out of 100**

Pillars and pillar subindexes

Education
Rank **12** Score **77.5**

Social inclusion and equity policies
Rank **12** Score **80.1**

Country policy assessment
Rank **26** Score **60.5**

Good governance
Rank **11** Score **88.1**

Quality of the infrastructure
Rank **55** Score **77.1**

Government ICT usage
Rank **7** Score **81.7**

Doing business
Rank **3** Score **88.8**

R&D infrastructure
Rank **14** Score **58.8**

Patents and trademarks
Rank **11** Score **48.9**

Telephone communications
Rank **3** Score **98.2**

Mobile cellular communications
Rank **78** Score **64.6**

Internet, computers and TV
Rank **3** Score **87.2**

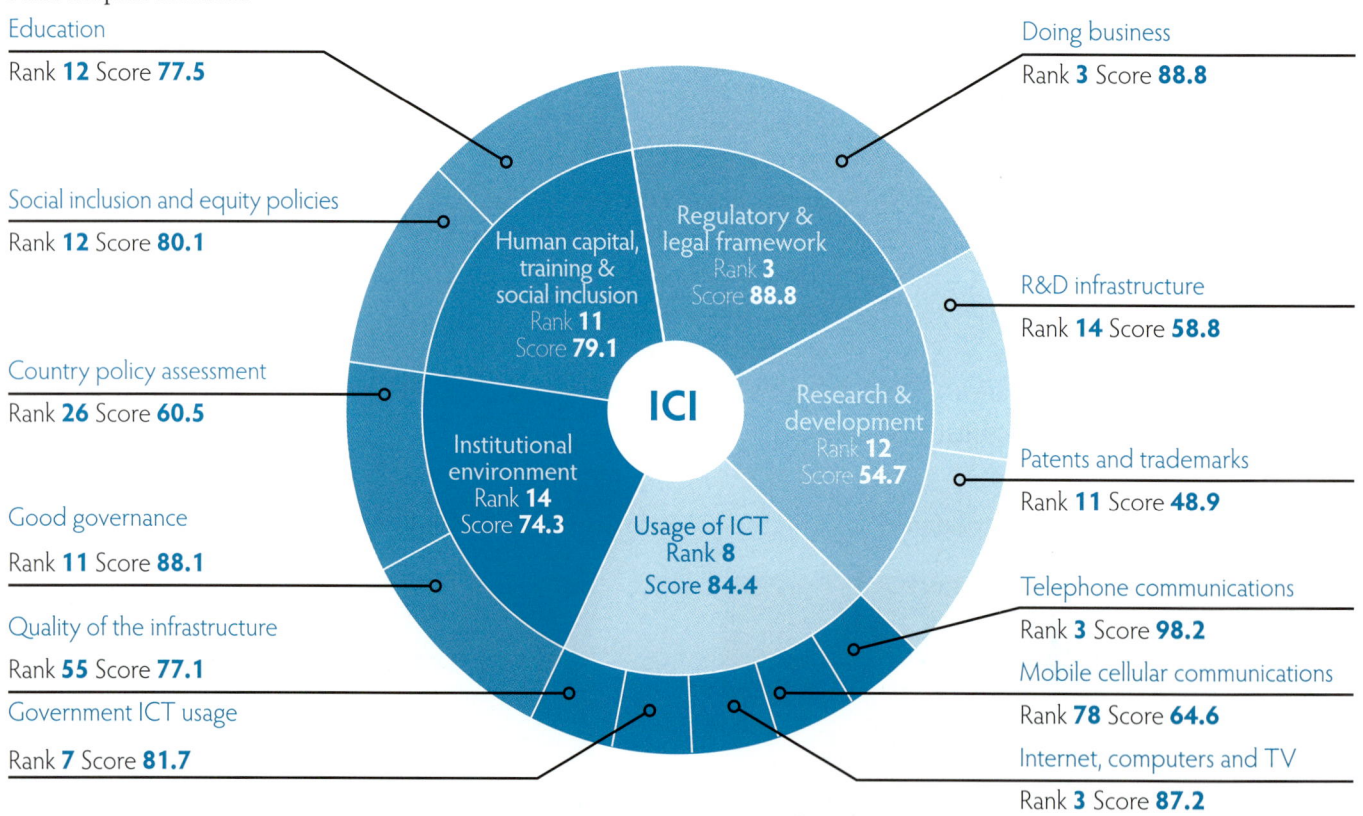

Index thermometer[12]

Pillar 1 9.9% Pillar 2 10.6% Pillar 3 23.7% Pillar 4 21.9% Pillar 5 33.8% 100%

[1] High-income countries are those that, according to the World Bank classification, have a gross national income (GNI) per capita of $11,456 or more. [2] Full democracies are countries whose Democracy Index scores from 1 to 10 rounded to one decimal place are of 8 to 10. [3] 2007. [4] World Bank World Development Indicators. [5] IMF World Economic Outlook. [6] 2006. [7] Estimate. [8] UNESCO, % age above 15. [9] % ages 15–64. [10] 2005. [11] 2004. [12] The index thermometer shows the relative contribution of each pillar (in %) to the total index score.

258

CANADA

Country relative performance: **Index pillars**[13]

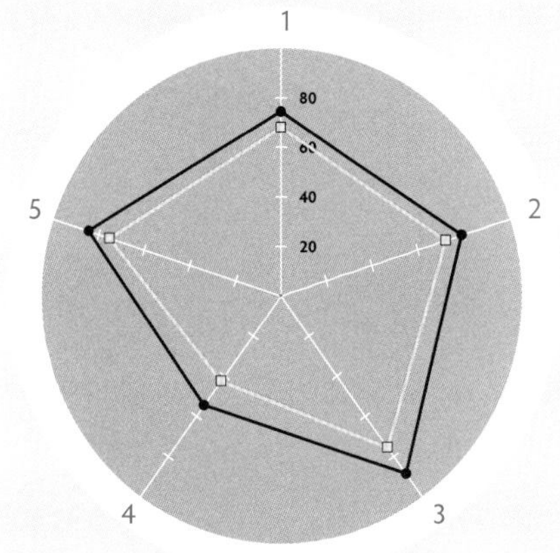

1. Institutional environment
2. Human capital, training and social inclusion
3. Regulatory and legal framework
4. Research and development
5. Usage of ICT

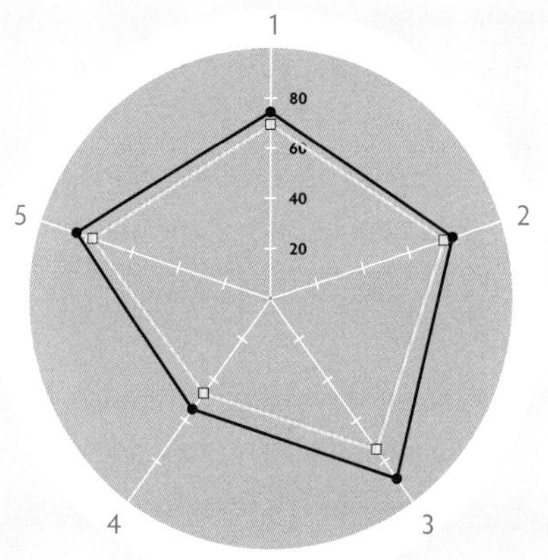

High-income ☐   Canada ●——

Full democracies ☐   Canada ●——

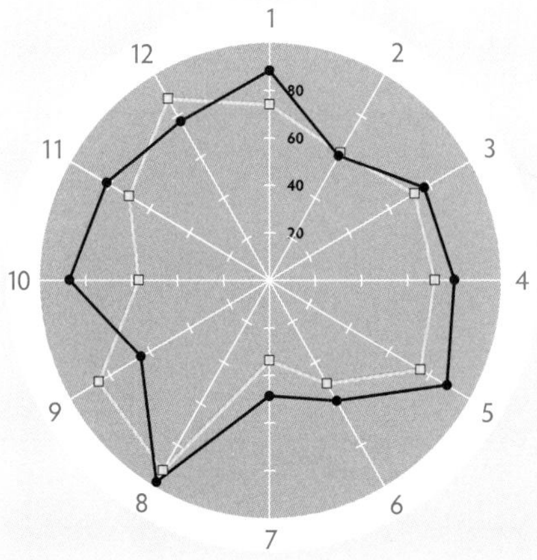

1. Good governance
2. Country policy assessment
3. Education
4. Social inclusion and equity policies
5. Doing business
6. R&D infrastructure
7. Patents and trademarks
8. Telephone communications
9. Mobile/cellular communications
10. Internet, computers and TV
11. Government ICT usage
12. Quality of the infrastructure

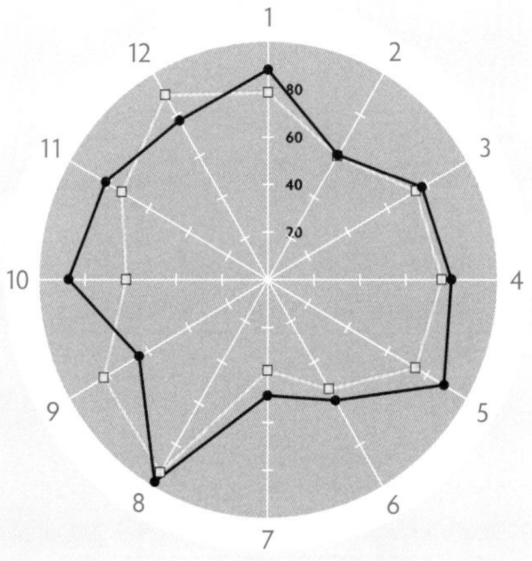

Country relative performance: **Pillar subindexes**[13]

In focus: **Top priorities for policy reform**[14]

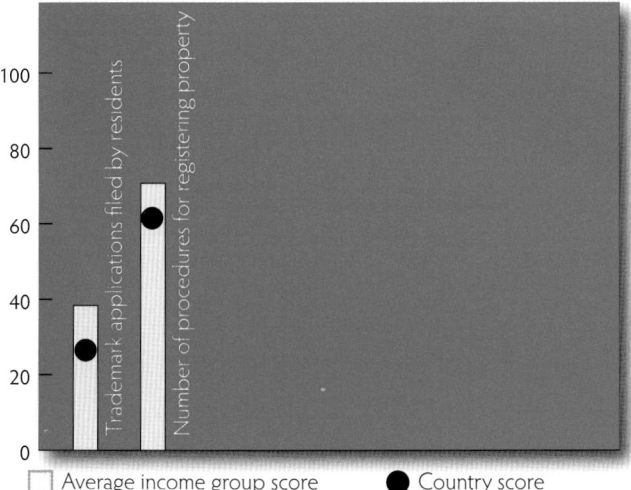

☐ Average income group score   ● Country score

In focus: **Significant indicators above income group average**[14]

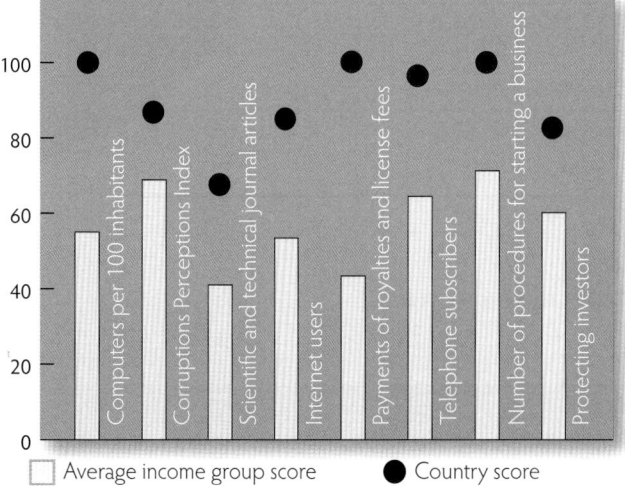

☐ Average income group score   ● Country score

[13] This analysis compares pillar and subindex country scores with respect to the average scores of countries grouped according to income level and political regime (see notes 1 and 2).

[14] Significant country indicators with the greatest distances above or below the income group average scores (i.e., achievements and targets for policy reform, respectively) were selected and ordered according to their Pearson correlation coefficients (highest first) with respect to the final index scores, to produce a ranking.

CHILE

Key selected indicators

Population (in millions)[3, 4]	16.6
Population density (people per sq. km)[3, 4]	22
GDP (current US$ billion, PPP)[3, 5]	231
GDP per capita (current US$, PPP)[3, 5]	13936
Current account balance (% of GDP)[3, 5]	3.7
Average life expectancy (years)[4, 6]	78.3
Literacy rate[3, 7]	97
Female labor force participation rate[4, 6, 8]	41.3
Unemployment rate (% of total labor force)[4, 9]	6.9
CO_2 emissions (metric tons per capita)[4, 10]	3.9

Income level:[1] **Upper-middle-income**

Political regime:[2] **Flawed democracy**

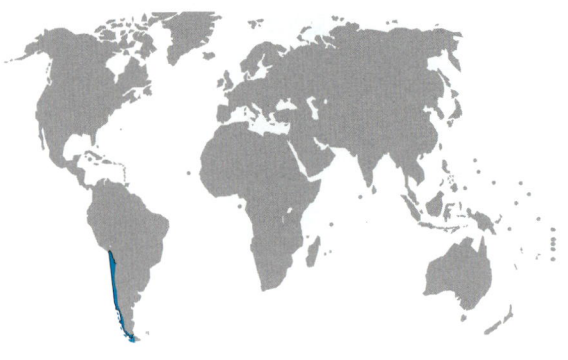

Rank	Score
29 out of 131	**59.4 out of 100**

Innovation Capacity Index

Pillars and pillar subindexes

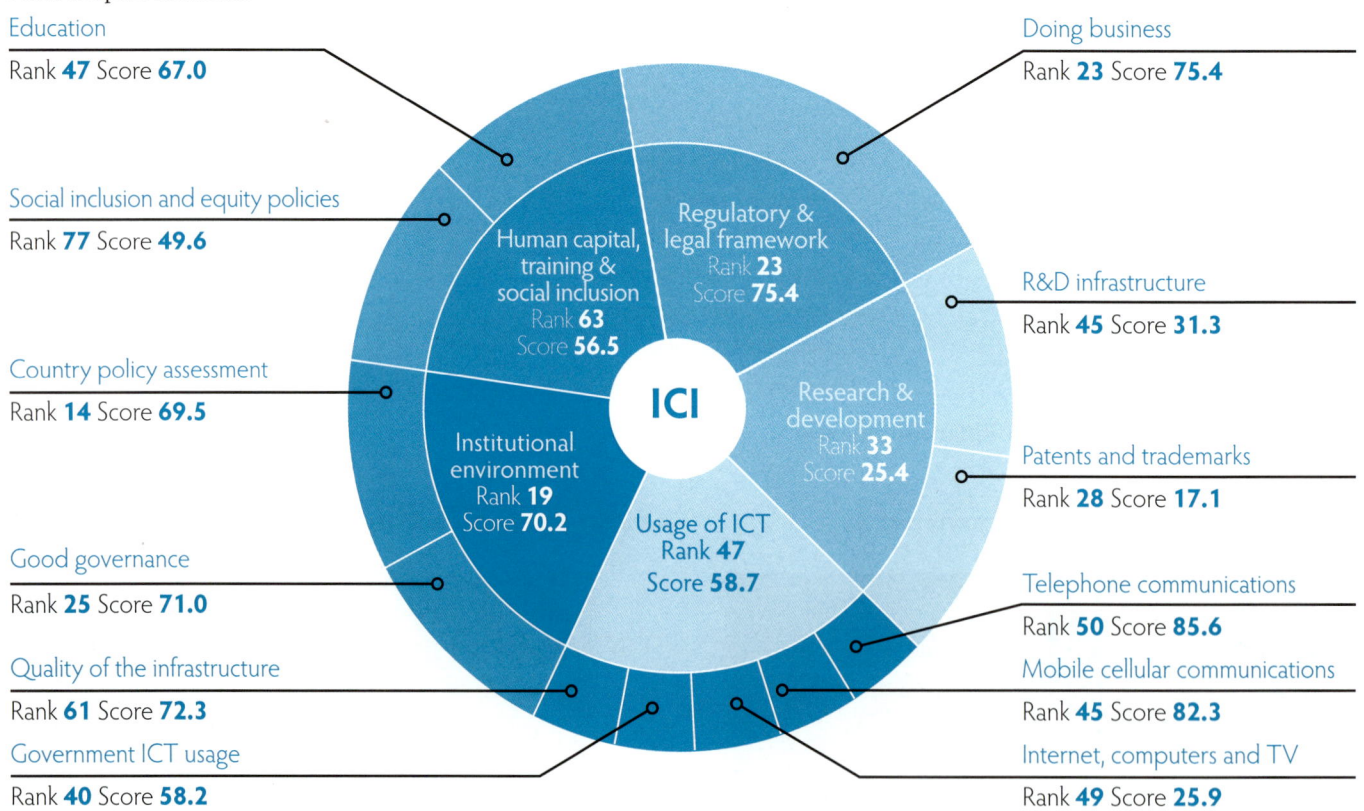

Education
Rank **47** Score **67.0**

Social inclusion and equity policies
Rank **77** Score **49.6**

Country policy assessment
Rank **14** Score **69.5**

Good governance
Rank **25** Score **71.0**

Quality of the infrastructure
Rank **61** Score **72.3**

Government ICT usage
Rank **40** Score **58.2**

Doing business
Rank **23** Score **75.4**

R&D infrastructure
Rank **45** Score **31.3**

Patents and trademarks
Rank **28** Score **17.1**

Telephone communications
Rank **50** Score **85.6**

Mobile cellular communications
Rank **45** Score **82.3**

Internet, computers and TV
Rank **49** Score **25.9**

Human capital, training & social inclusion
Rank **63**
Score **56.5**

Regulatory & legal framework
Rank **23**
Score **75.4**

Institutional environment
Rank **19**
Score **70.2**

Research & development
Rank **33**
Score **25.4**

Usage of ICT
Rank **47**
Score **58.7**

ICI

Index thermometer[11]

Pillar 1 29.6% Pillar 2 23.8% Pillar 3 25.4% Pillar 4 6.4% Pillar 5 14.8% 100%

[1] Upper-middle-income countries are those that, according to the World Bank classification, have a gross national income (GNI) per capita of $3,706 to $11,455. [2] Flawed democracies are countries whose Democracy Index scores from 1 to 10 rounded to one decimal place are of 6 to 7.9. [3] 2007. [4] World Bank World Development Indicators. [5] IMF World Economic Outlook. [6] 2006. [7] UNESCO, % age above 15. [8] % ages 15-64 [9] 2005. [10] 2004. [11] The index thermometer shows the relative contribution of each pillar (in %) to the total index score.

CHILE

Country relative performance: **Index pillars[12]**

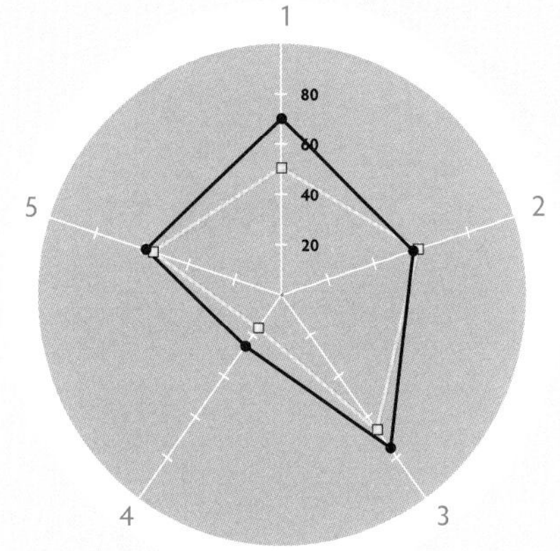

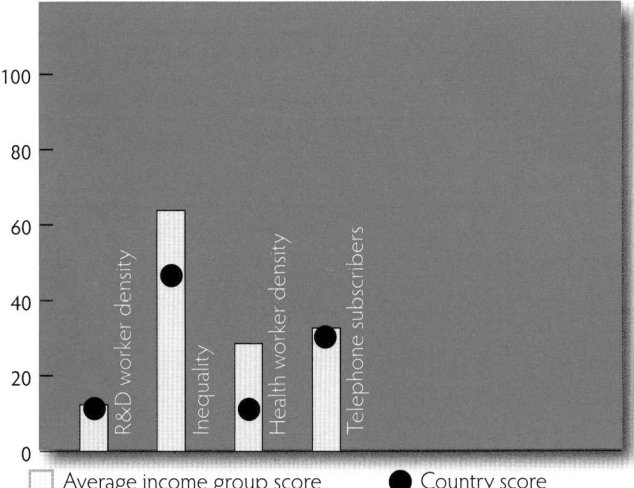

1. Institutional environment
2. Human capital, training and social inclusion
3. Regulatory and legal framework
4. Research and development
5. Usage of ICT

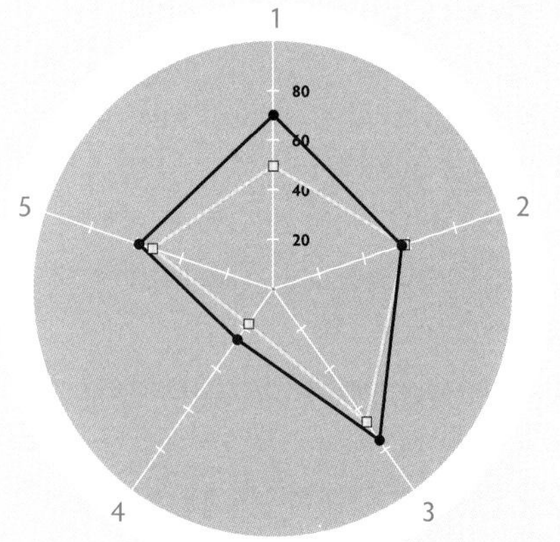

Upper-middle-income ☐   Chile ●━━━

Flawed democracies ☐   Chile ●━━━

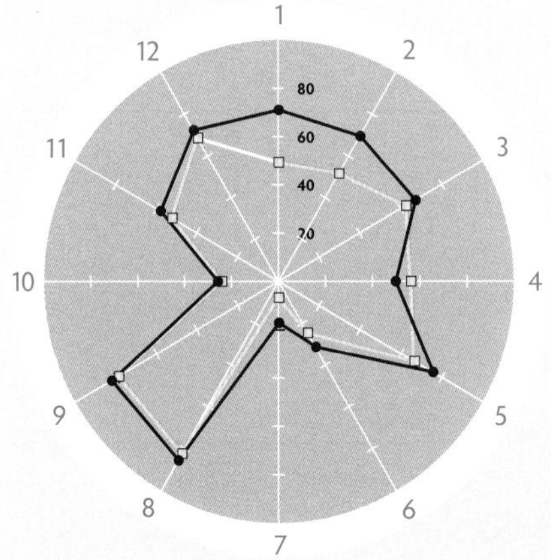

1. Good governance
2. Country policy assessment
3. Education
4. Social inclusion and equity policies
5. Doing business
6. R&D infrastructure
7. Patents and trademarks
8. Telephone communications
9. Mobile/cellular communications
10. Internet, computers and TV
11. Government ICT usage
12. Quality of the infrastructure

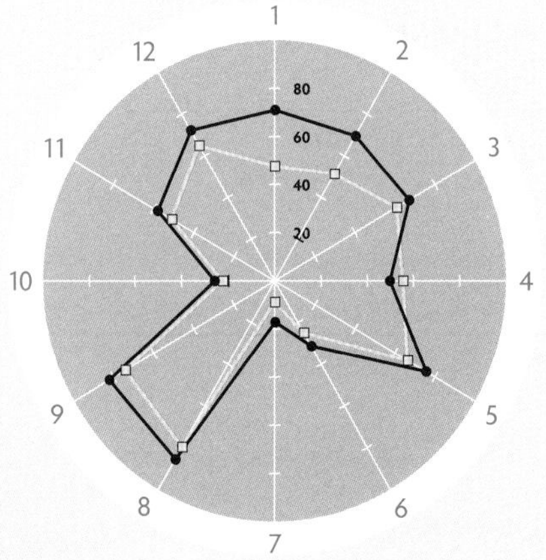

Country relative performance: **Pillar subindexes[12]**

In focus: **Top priorities for policy reform[13]**

In focus: **Significant indicators above income group average[13]**

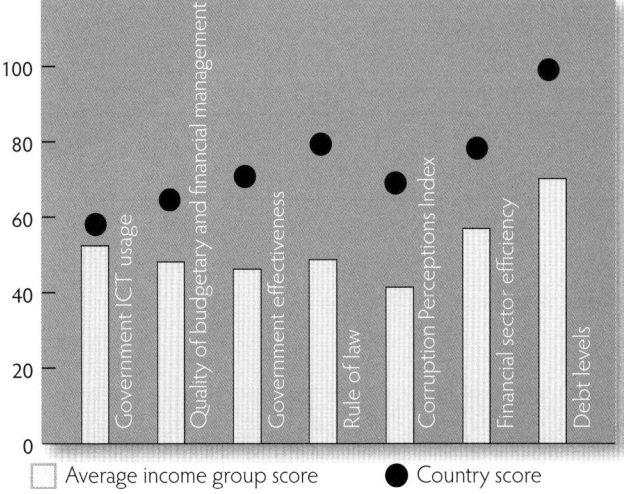

☐ Average income group score   ● Country score

☐ Average income group score   ● Country score

---

[12]   This analysis compares pillar and subindex country scores with respect to the average scores of countries grouped according to income level and political regime (see notes 1 and 2).

[13]   Significant country indicators with the greatest distances above or below the income group average scores (i.e., achievements and targets for policy reform, respectively) were selected and ordered according to their Pearson correlation coefficients (highest first) with respect to the final index scores, to produce a ranking.

CHINA, PEOPLE'S REPUBLIC OF

Key selected indicators

Population (in millions)[3,4]	1320.0
Population density (people per sq. km)[3,4]	142
GDP (current US$ billion, PPP)[3,5]	6991
GDP per capita (current US$, PPP)[3,5]	5292
Current account balance (% of GDP)[3,5]	11.1
Average life expectancy (years)[4,6]	72.0
Literacy rate[3,7]	93
Female labor force participation rate[4,6,8]	75.4
Unemployment rate (% of total labor force)[4,9]	4.2
CO_2 emissions (metric tons per capita)[4,10]	3.9

Income level:[1] **Lower-middle-income**

Political regime:[2] **Authoritarian regime**

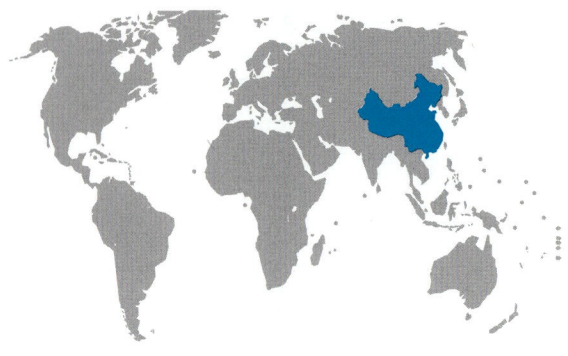

	Rank	Score
Innovation Capacity Index	**65 out of 131**	**49.5 out of 100**

Pillars and pillar subindexes

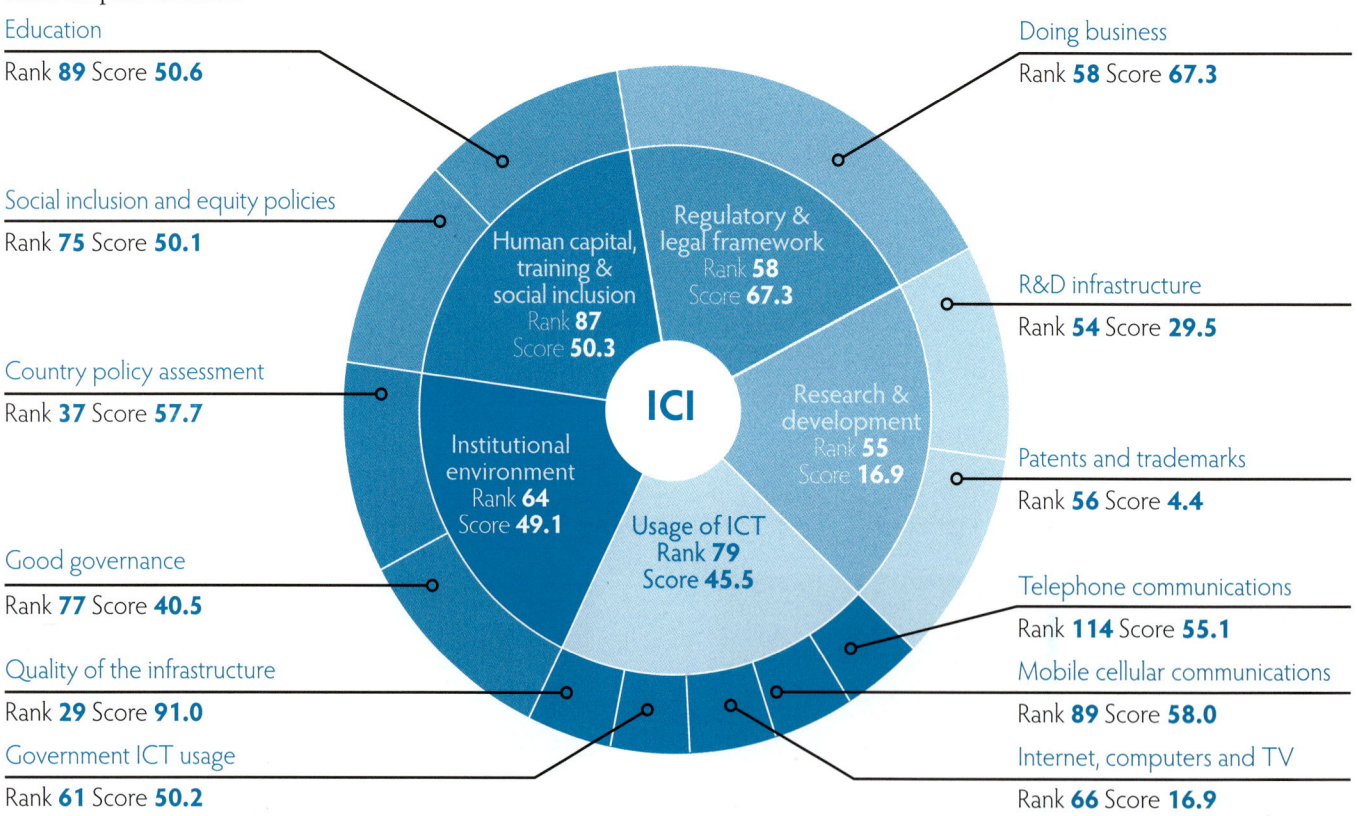

Education
Rank **89** Score **50.6**

Social inclusion and equity policies
Rank **75** Score **50.1**

Country policy assessment
Rank **37** Score **57.7**

Good governance
Rank **77** Score **40.5**

Quality of the infrastructure
Rank **29** Score **91.0**

Government ICT usage
Rank **61** Score **50.2**

Doing business
Rank **58** Score **67.3**

R&D infrastructure
Rank **54** Score **29.5**

Patents and trademarks
Rank **56** Score **4.4**

Telephone communications
Rank **114** Score **55.1**

Mobile cellular communications
Rank **89** Score **58.0**

Internet, computers and TV
Rank **66** Score **16.9**

Human capital, training & social inclusion
Rank **87** Score **50.3**

Regulatory & legal framework
Rank **58** Score **67.3**

Institutional environment
Rank **64** Score **49.1**

Research & development
Rank **55** Score **16.9**

Usage of ICT
Rank **79** Score **45.5**

ICI

Index thermometer[11]

Pillar 1 29.8% Pillar 2 30.5% Pillar 3 27.2% Pillar 4 3.4% Pillar 5 9.2%   100%

[1] Lower-middle-income countries are those that, according to the World Bank classification, have a gross national income (GNI) per capita of $936 to $3,705.   [2] Authoritarian regimes are countries whose Democracy Index scores from 1 to 10 rounded to one decimal place are less than 4.   [3] 2007.   [4] World Bank World Development Indicators.   [5] IMF World Economic Outlook.   [6] 2006.   [7] UNESCO, % age above 15.   [8] % ages 15–64.   [9] 2005.   [10] 2004.   [11] The index thermometer shows the relative contribution of each pillar (in %) to the total index score.

CHINA, PEOPLE'S REPUBLIC OF

Country relative performance: **Index pillars**[12]

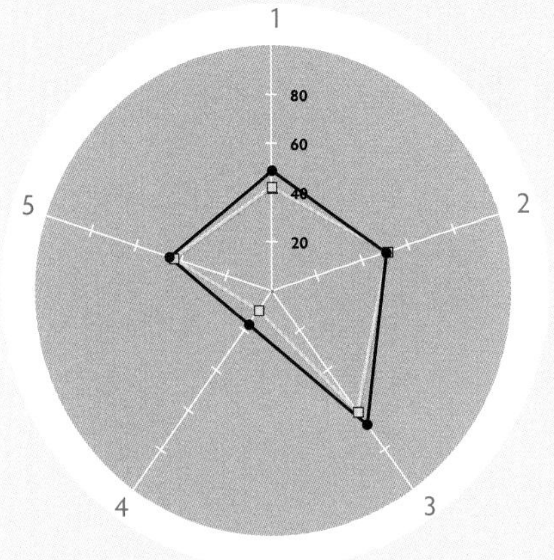

1. Institutional environment
2. Human capital, training and social inclusion
3. Regulatory and legal framework
4. Research and development
5. Usage of ICT

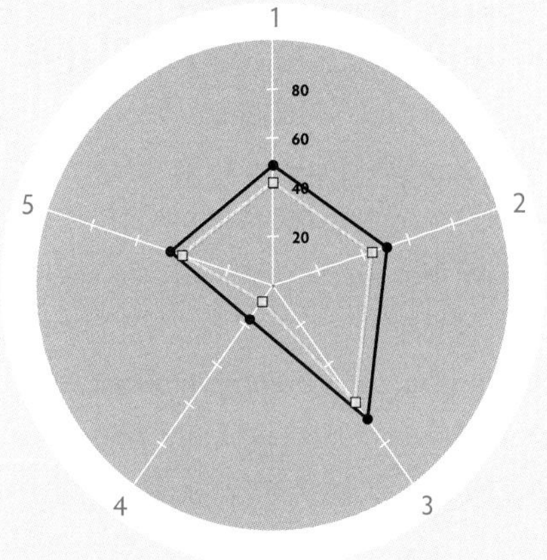

Lower-middle-income ·····□·····   China ●———●

Authoritarian regimes ·····□·····   China ●———●

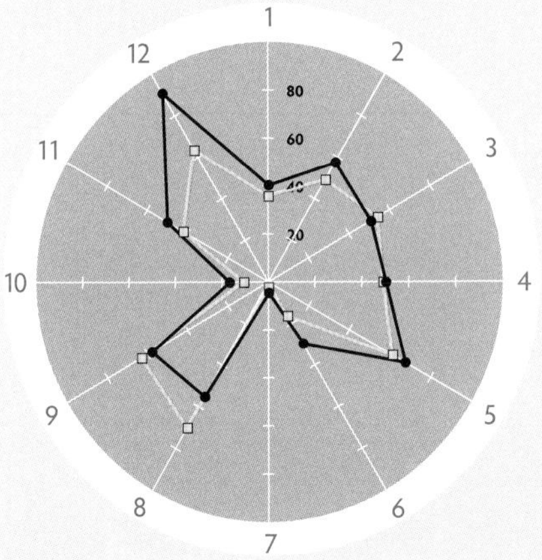

Country relative performance: **Pillar subindexes**[12]

1. Good governance
2. Country policy assessment
3. Education
4. Social inclusion and equity policies
5. Doing business
6. R&D infrastructure
7. Patents and trademarks
8. Telephone communications
9. Mobile/cellular communications
10. Internet, computers and TV
11. Government ICT usage
12. Quality of the infrastructure

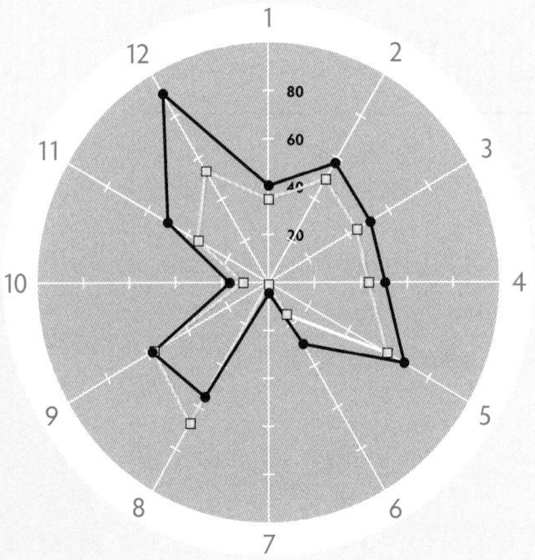

263

In focus: **Top priorities for policy reform**[13]

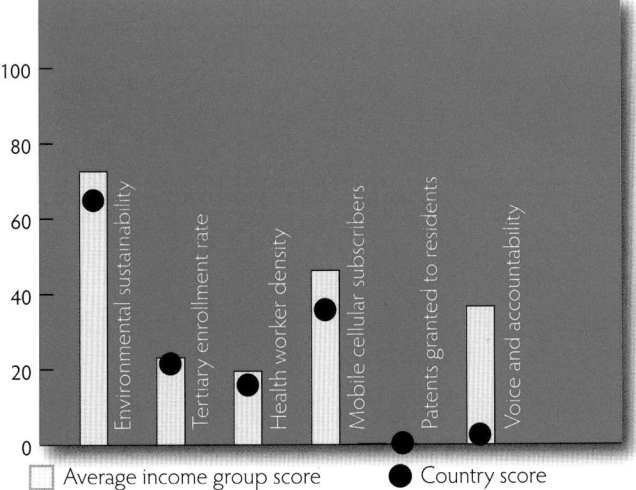

□ Average income group score   ● Country score

In focus: **Significant indicators above income group average**[13]

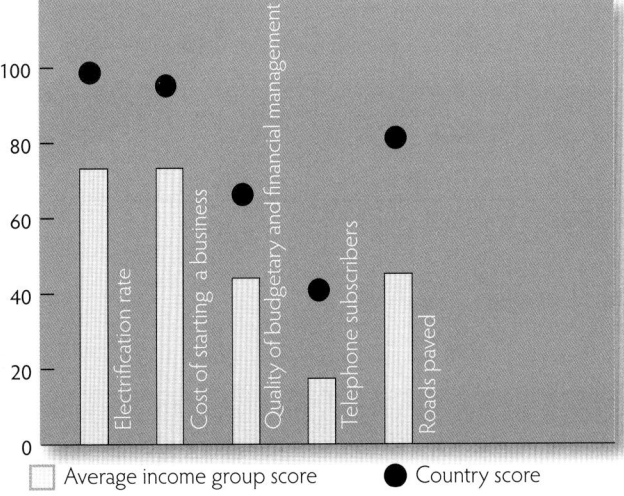

□ Average income group score   ● Country score

---

[12] This analysis compares pillar and subindex country scores with respect to the average scores of countries grouped according to income level and political regime (see notes 1 and 2).

[13] Significant country indicators with the greatest distances above or below the income group average scores (i.e., achievements and targets for policy reform, respectively) were selected and ordered according to their Pearson correlation coefficients (highest first) with respect to the final index scores, to produce a ranking.

COLOMBIA

Key selected indicators

Population (in millions)[3,4]	46.1
Population density (people per sq. km)[3,4]	42
GDP (current US$ billion, PPP)[3,5]	320
GDP per capita (current US$, PPP)[3,5]	6724
Current account balance (% of GDP)[3,5]	-3.8
Average life expectancy (years)[4,6]	72.6
Literacy rate[3,7]	94
Female labor force participation rate[4,6,8]	66.7
Unemployment rate (% of total labor force)[4,9]	9.5
CO_2 emissions (metric tons per capita)[4,10]	1.2

Income level:[1] **Upper-middle-income**

Political regime:[2] **Flawed democracy**

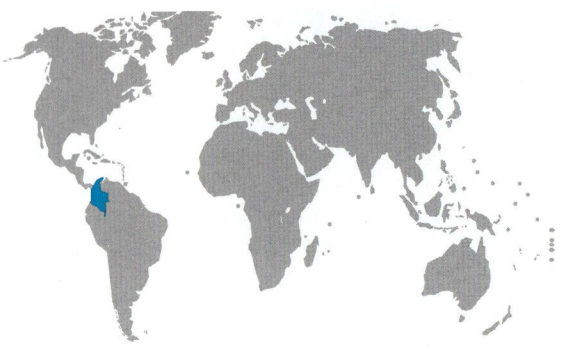

Innovation Capacity Index

Rank	Score
72 out of 131	**48.0 out of 100**

Pillars and pillar subindexes

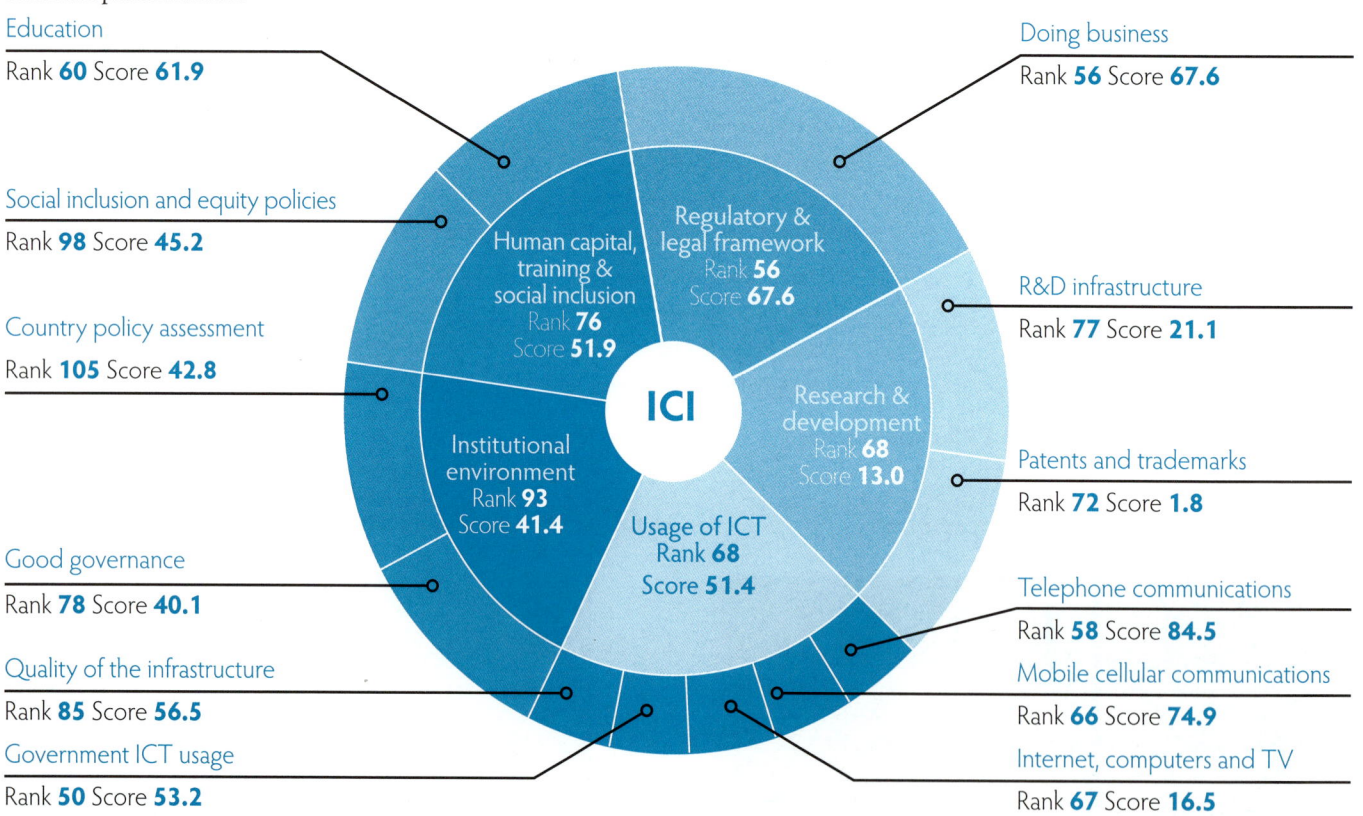

Education
Rank **60** Score **61.9**

Social inclusion and equity policies
Rank **98** Score **45.2**

Country policy assessment
Rank **105** Score **42.8**

Good governance
Rank **78** Score **40.1**

Quality of the infrastructure
Rank **85** Score **56.5**

Government ICT usage
Rank **50** Score **53.2**

Doing business
Rank **56** Score **67.6**

R&D infrastructure
Rank **77** Score **21.1**

Patents and trademarks
Rank **72** Score **1.8**

Telephone communications
Rank **58** Score **84.5**

Mobile cellular communications
Rank **66** Score **74.9**

Internet, computers and TV
Rank **67** Score **16.5**

Human capital, training & social inclusion
Rank **76** Score **51.9**

Regulatory & legal framework
Rank **56** Score **67.6**

Institutional environment
Rank **93** Score **41.4**

Research & development
Rank **68** Score **13.0**

Usage of ICT
Rank **68** Score **51.4**

ICI

Index thermometer[11]

Pillar 1 25.9% Pillar 2 32.4% Pillar 3 28.2% Pillar 4 2.7% Pillar 5 10.7% 100%

[1] Lower-middle-income countries are those that, according to the World Bank classification, have a gross national income (GNI) per capita of $936 to $3,705. [2] Flawed democracies are countries whose Democracy Index scores from 1 to 10 rounded to one decimal place are of 6 to 7.9. [3] 2007. [4] World Bank World Development Indicators. [5] IMF World Economic Outlook. [6] 2006. [7] UNESCO, % age above 15. [8] % ages 15–64. [9] 2005. [10] 2004. [11] The index thermometer shows the relative contribution of each pillar (in %) to the total index score.

COLOMBIA

Country relative performance: **Index pillars**[12]

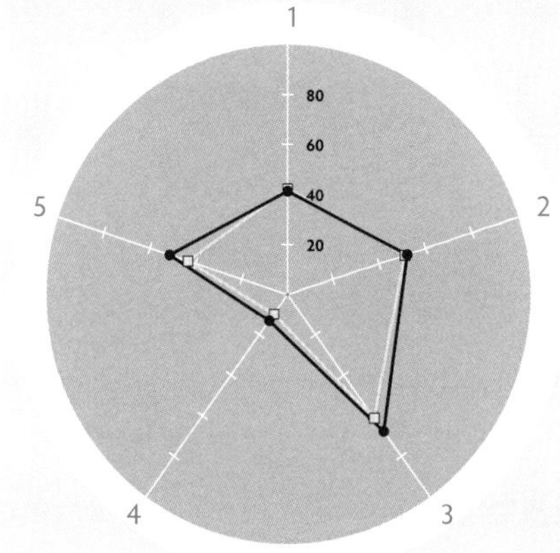

1. Institutional environment
2. Human capital, training and social inclusion
3. Regulatory and legal framework
4. Research and development
5. Usage of ICT

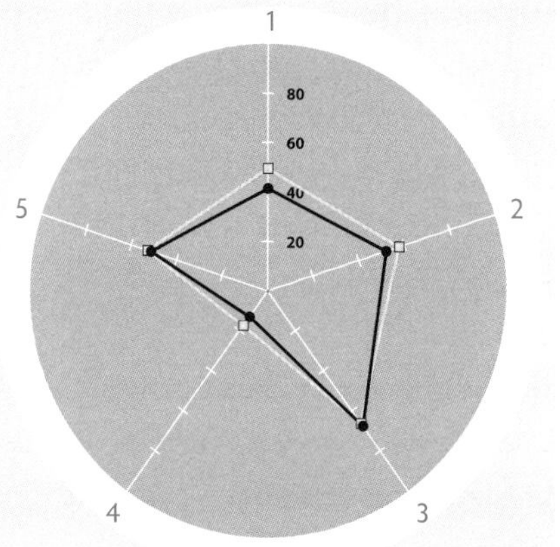

Upper-middle-income □ Colombia ●

Flawed democracies □ Colombia ●

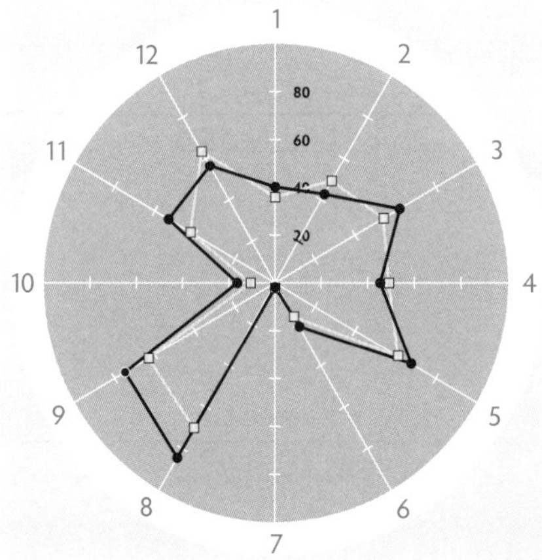

1. Good governance
2. Country policy assessment
3. Education
4. Social inclusion and equity policies
5. Doing business
6. R&D infrastructure
7. Patents and trademarks
8. Telephone communications
9. Mobile/cellular communications
10. Internet, computers and TV
11. Government ICT usage
12. Quality of the infrastructure

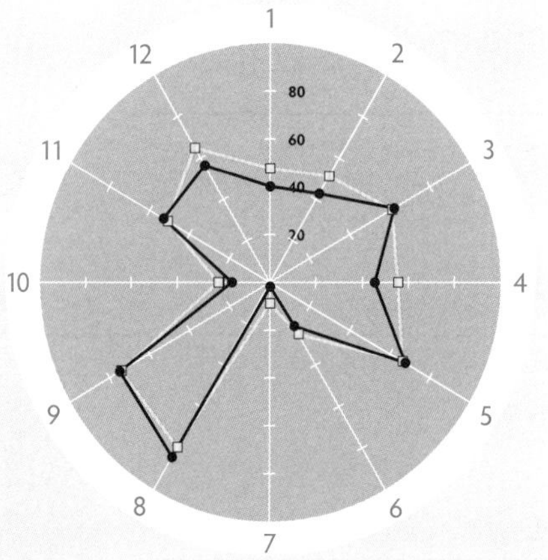

Country relative performance: **Pillar subindexes**[12]

In focus: **Top priorities for policy reform**[13]

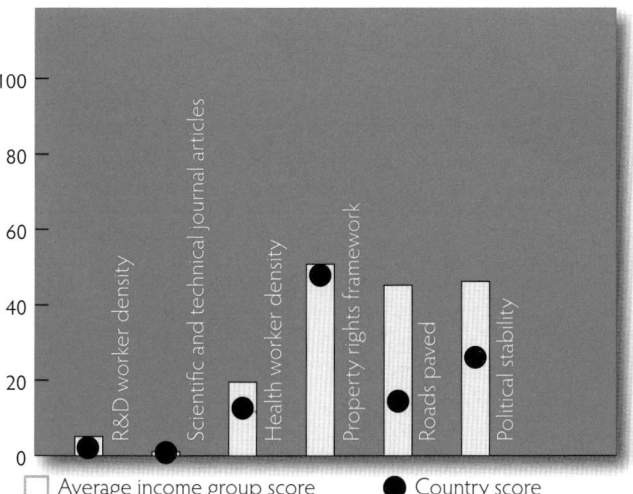

□ Average income group score   ● Country score

In focus: **Significant indicators above income group average**[13]

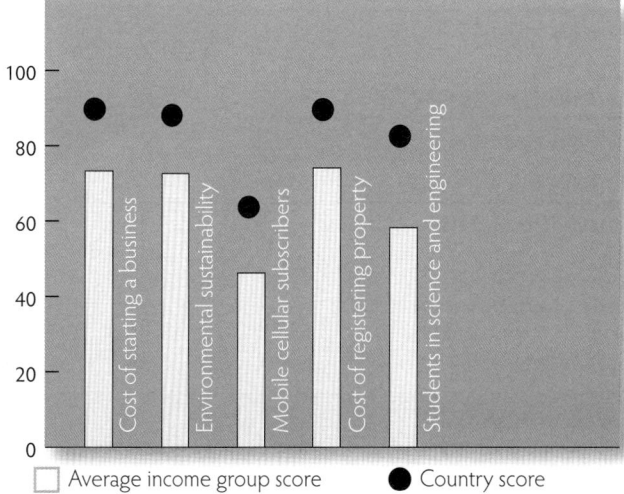

□ Average income group score   ● Country score

[12] This analysis compares pillar and subindex country scores with respect to the average scores of countries grouped according to income level and political regime (see notes 1 and 2).

[13] Significant country indicators with the greatest distances above or below the income group average scores (i.e., achievements and targets for policy reform, respectively) were selected and ordered according to their Pearson correlation coefficients (highest first) with respect to the final index scores, to produce a ranking.

COSTA RICA

Key selected indicators

Population (in millions)[3, 4]	4.5
Population density (people per sq. km)[3, 4]	87
GDP (current US$ billion, PPP)[3, 5]	46
GDP per capita (current US$, PPP)[3, 5]	10300
Current account balance (% of GDP)[3, 5]	-5.8
Average life expectancy (years)[4, 6]	78.7
Literacy rate[3, 7]	96
Female labor force participation rate[4, 6, 8]	50.0
Unemployment rate (% of total labor force)[4, 9]	6.6
CO_2 emissions (metric tons per capita)[4, 10]	1.5

Income level:[1] **Upper-middle-income**

Political regime:[2] **Full democracy**

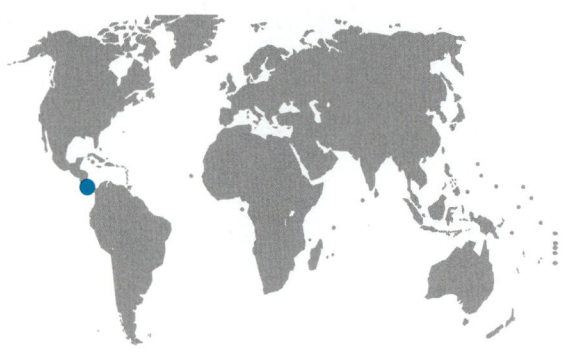

Innovation Capacity Index

Rank	Score
58 out of 131	**51.5 out of 100**

Pillars and pillar subindexes

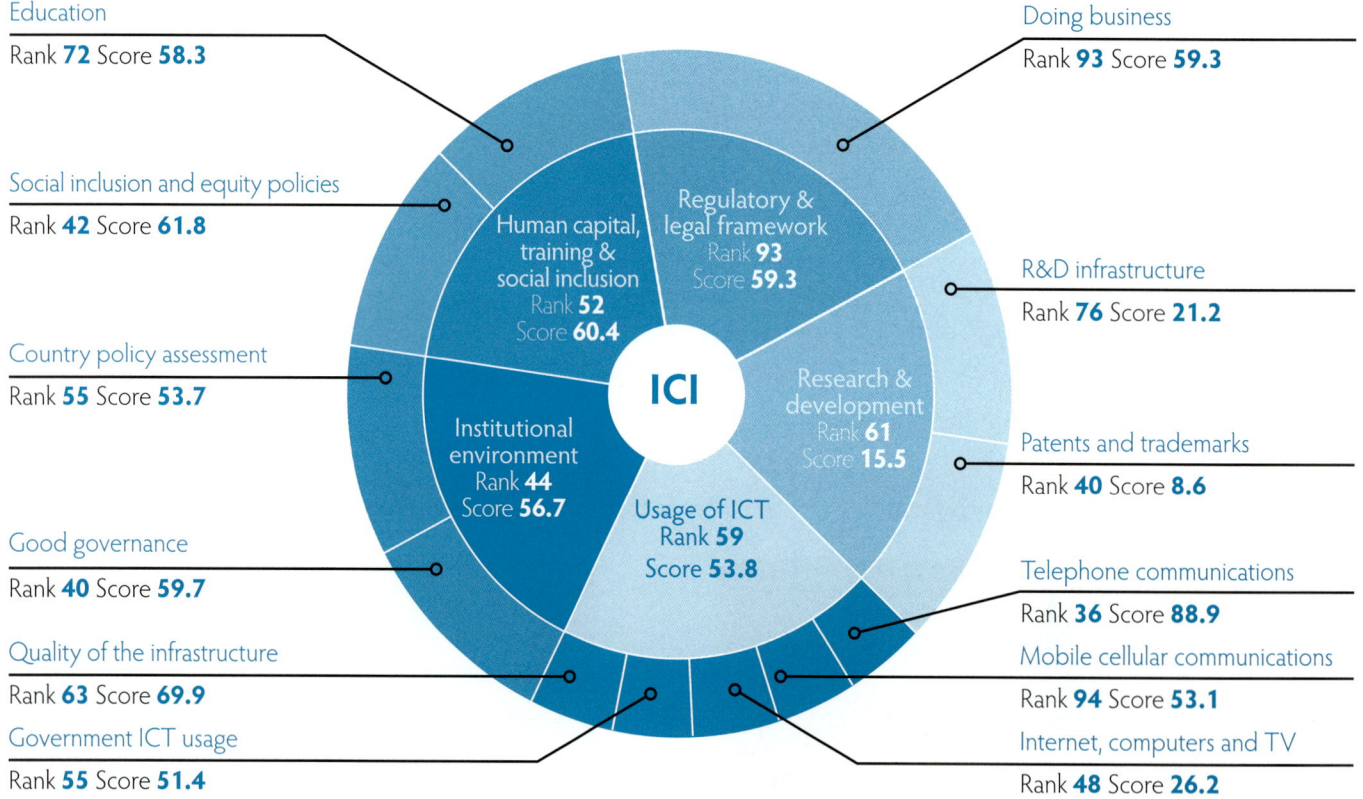

Education
Rank **72** Score **58.3**

Social inclusion and equity policies
Rank **42** Score **61.8**

Country policy assessment
Rank **55** Score **53.7**

Good governance
Rank **40** Score **59.7**

Quality of the infrastructure
Rank **63** Score **69.9**

Government ICT usage
Rank **55** Score **51.4**

Doing business
Rank **93** Score **59.3**

R&D infrastructure
Rank **76** Score **21.2**

Patents and trademarks
Rank **40** Score **8.6**

Telephone communications
Rank **36** Score **88.9**

Mobile cellular communications
Rank **94** Score **53.1**

Internet, computers and TV
Rank **48** Score **26.2**

Human capital, training & social inclusion
Rank **52** Score **60.4**

Regulatory & legal framework
Rank **93** Score **59.3**

Institutional environment
Rank **44** Score **56.7**

Research & development
Rank **61** Score **15.5**

Usage of ICT
Rank **59** Score **53.8**

ICI

266

Index thermometer[11]

Pillar 1 27.5% Pillar 2 29.3% Pillar 3 23.0% Pillar 4 4.5% Pillar 5 15.7%   100%

[1] Upper-middle-income countries are those that, according to the World Bank classification, have a gross national income (GNI) per capita of $3,706 to $11,455.   [2] Full democracies are countries whose Democracy Index scores from 1 to 10 rounded to one decimal place are of 8 to 10.   [3] 2007.   [4] World Bank World Development Indicators.   [5] IMF World Economic Outlook.   [6] 2006.   [7] UNESCO, % age above 15.   [8] % ages 15–64.   [9] 2005.   [10] 2004.   [11] The index thermometer shows the relative contribution of each pillar (in %) to the total index score.

COSTA RICA

Country relative performance: **Index pillars**[12]

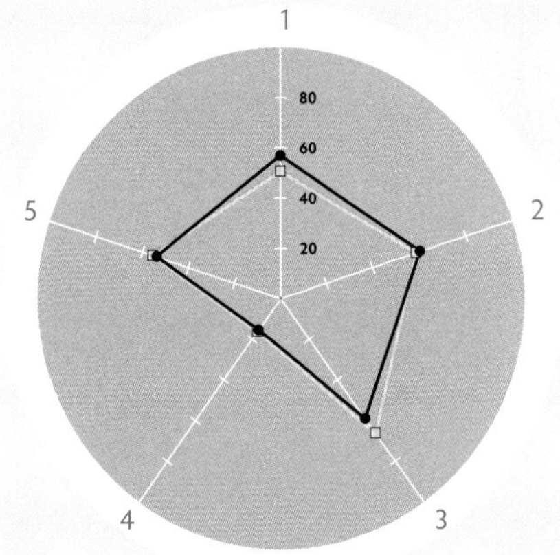

1. Institutional environment
2. Human capital, training and social inclusion
3. Regulatory and legal framework
4. Research and development
5. Usage of ICT

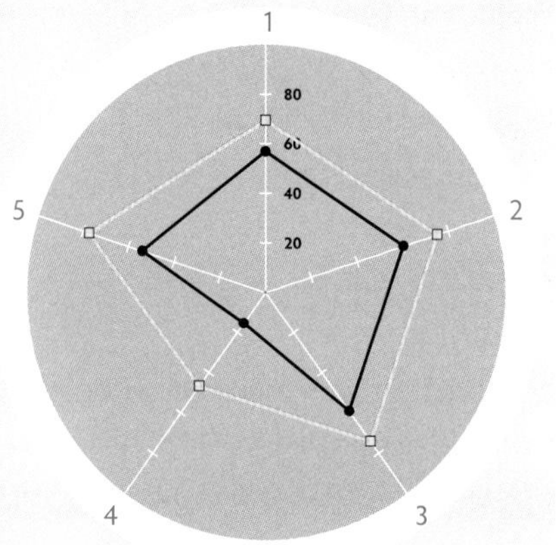

Upper-middle-income ⬜ ⟶ Costa Rica ●━

Full democracies ⬜ ⟶ Costa Rica ●━

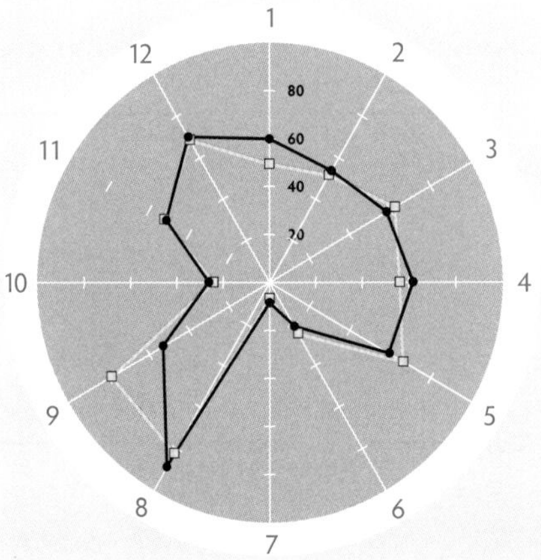

1. Good governance
2. Country policy assessment
3. Education
4. Social inclusion and equity policies
5. Doing business
6. R&D infrastructure
7. Patents and trademarks
8. Telephone communications
9. Mobile/cellular communications
10. Internet, computers and TV
11. Government ICT usage
12. Quality of the infrastructure

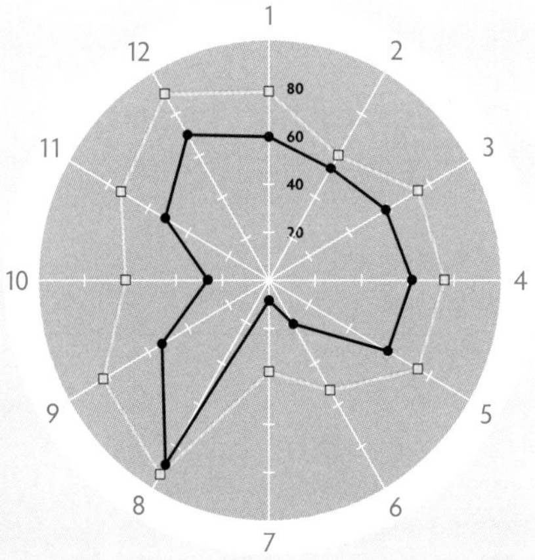

Country relative performance: **Pillar subindexes**[12]

In focus: **Top priorities for policy reform**[13]

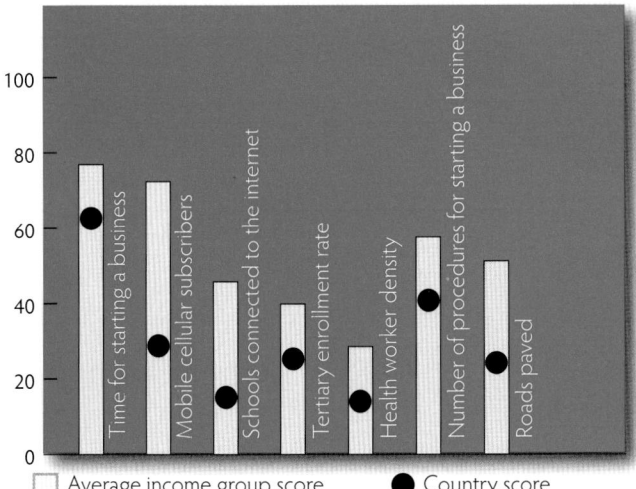

☐ Average income group score   ● Country score

In focus: **Significant indicators above income group average**[13]

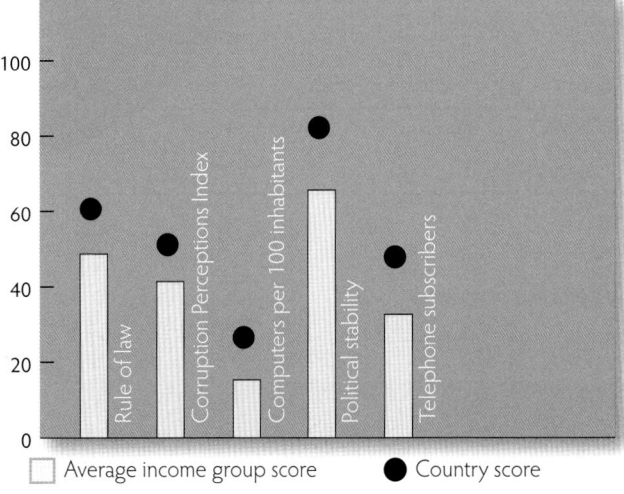

☐ Average income group score   ● Country score

[12] This analysis compares pillar and subindex country scores with respect to the average scores of countries grouped according to income level and political regime (see notes 1 and 2).

[13] Significant country indicators with the greatest distances above or below the income group average scores (i.e., achievements and targets for policy reform, respectively) were selected and ordered according to their Pearson correlation coefficients (highest first) with respect to the final index scores, to produce a ranking.

CZECH REPUBLIC

Key selected indicators

Population (in millions)[3, 4]	10.3
Population density (people per sq. km)[3, 4]	134
GDP (current US$ billion, PPP)[3, 5]	249
GDP per capita (current US$, PPP)[3, 5]	24236
Current account balance (% of GDP)[3, 5]	-2.5
Average life expectancy (years)[4, 6]	76.5
Literacy rate[7, 8]	99
Female labor force participation rate[4, 6, 9]	64.4
Unemployment rate (% of total labor force)[4, 10]	7.9
CO_2 emissions (metric tons per capita)[4, 11]	11.5

Income level:[1] **High-income**

Political regime:[2] **Full democracy**

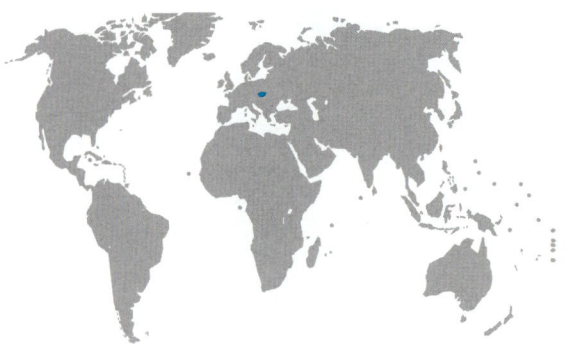

Rank	Score
32 out of 131	**58.0 out of 100**

Innovation Capacity Index

Pillars and pillar subindexes

Education
Rank **28** Score **71.1**

Doing business
Rank **54** Score **68.0**

Social inclusion and equity policies
Rank **21** Score **72.5**

R&D infrastructure
Rank **21** Score **52.6**

Country policy assessment
Rank **66** Score **52.1**

Patents and trademarks
Rank **36** Score **13.0**

Good governance
Rank **37** Score **61.5**

Telephone communications
Rank **40** Score **87.4**

Quality of the infrastructure
Rank **11** Score **96.8**

Mobile cellular communications
Rank **12** Score **91.4**

Government ICT usage
Rank **25** Score **67.0**

Internet, computers and TV
Rank **33** Score **41.1**

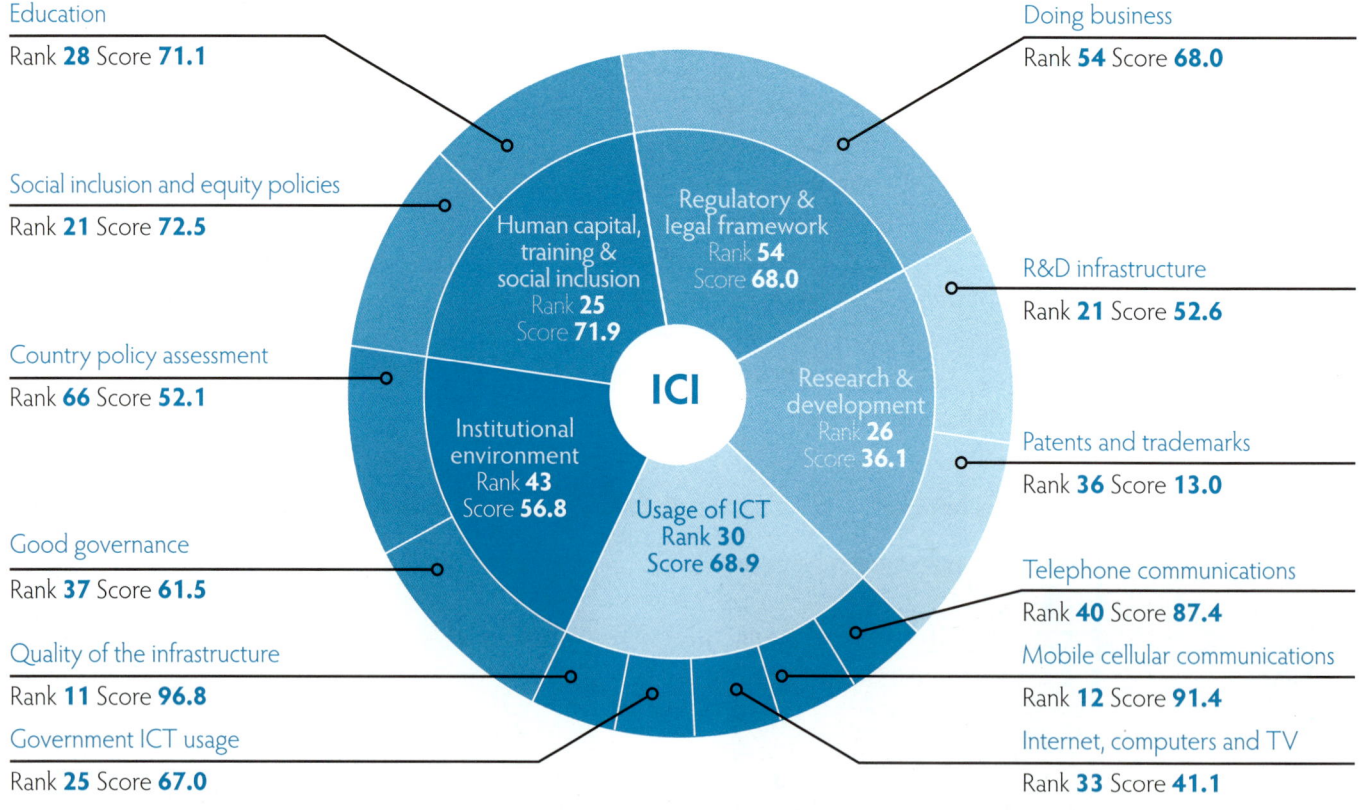

ICI

Human capital, training & social inclusion
Rank **25** Score **71.9**

Regulatory & legal framework
Rank **54** Score **68.0**

Research & development
Rank **26** Score **36.1**

Institutional environment
Rank **43** Score **56.8**

Usage of ICT
Rank **30** Score **68.9**

268

Index thermometer[12]

Pillar 1 9.8% Pillar 2 12.4% Pillar 3 23.5% Pillar 4 18.7% Pillar 5 35.7% 100%

[1] High-income countries are those that, according to the World Bank classification, have a gross national income (GNI) per capita of $11,456 or more. [2] Full democracies are countries whose Democracy Index scores from 1 to 10 rounded to one decimal place are of 8 to 10. [3] 2007. [4] World Bank World Development Indicators. [5] IMF World Economic Outlook. [6] 2006. [7] Estimate. [8] UNESCO, % age above 15. [9] % ages 15–64. [10] 2005. [11] 2004. [12] The index thermometer shows the relative contribution of each pillar (in %) to the total index score.

CZECH REPUBLIC

Country relative performance: **Index pillars**[13]

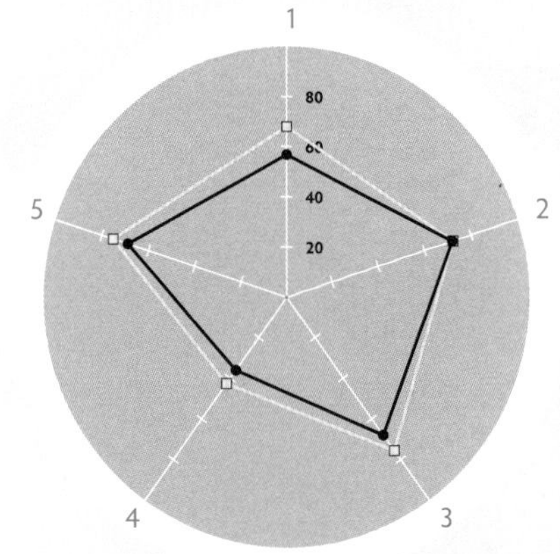

1. Institutional environment
2. Human capital, training and social inclusion
3. Regulatory and legal framework
4. Research and development
5. Usage of ICT

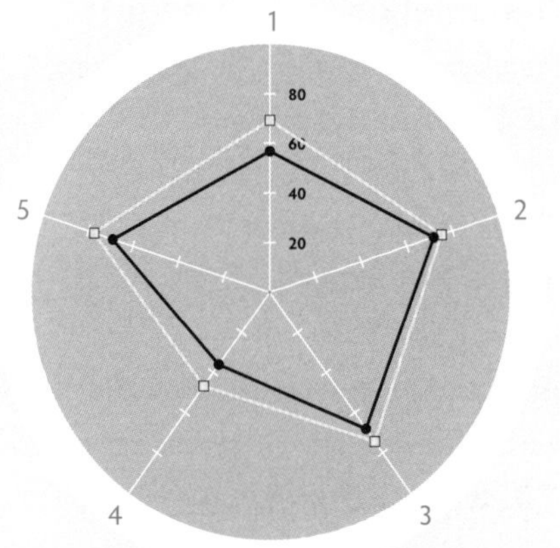

High-income   □ ·····   Czech Republic ●——●

Full democracies   □ ·····   Czech Republic ●——●

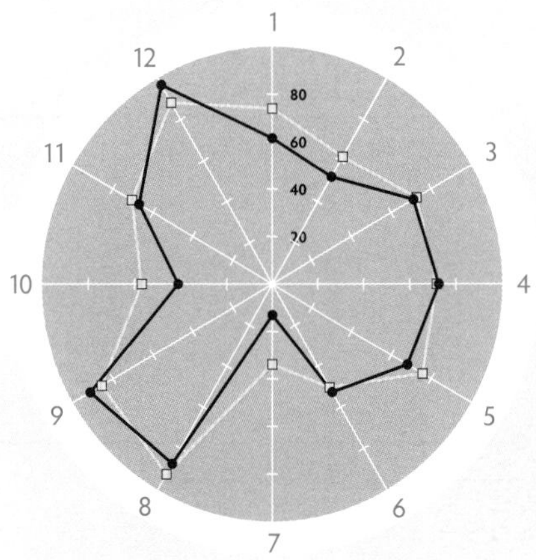

1. Good governance
2. Country policy assessment
3. Education
4. Social inclusion and equity policies
5. Doing business
6. R&D infrastructure
7. Patents and trademarks
8. Telephone communications
9. Mobile/cellular communications
10. Internet, computers and TV
11. Government ICT usage
12. Quality of the infrastructure

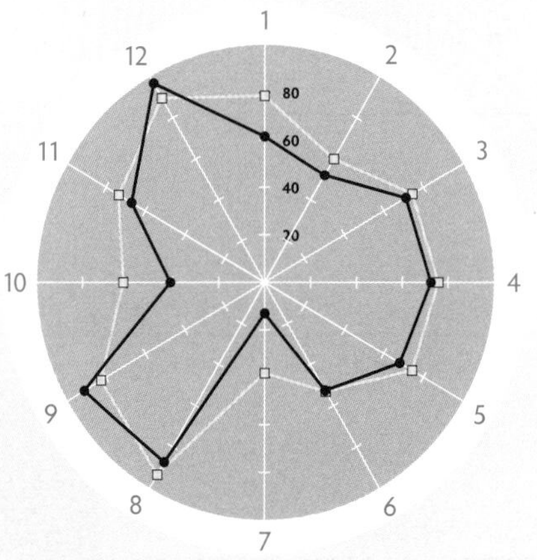

Country relative performance: **Pillar subindexes**[13]

269

In focus: **Top priorities for policy reform**[14]

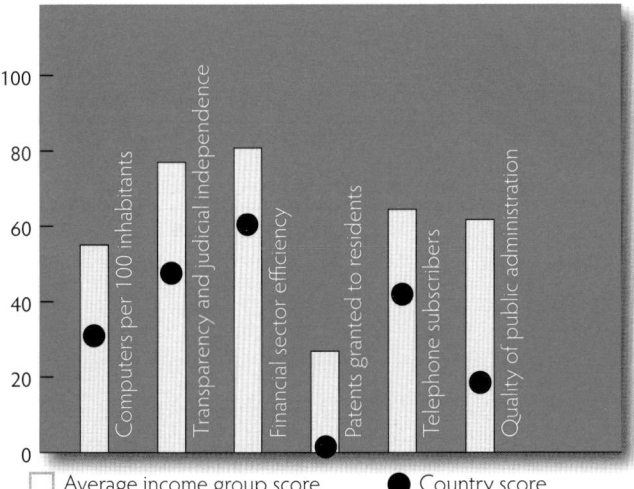

Bars: Computers per 100 inhabitants; Transparency and judicial independence; Financial sector efficiency; Patents granted to residents; Telephone subscribers; Quality of public administration

□ Average income group score   ● Country score

In focus: **Significant indicators above income group average**[14]

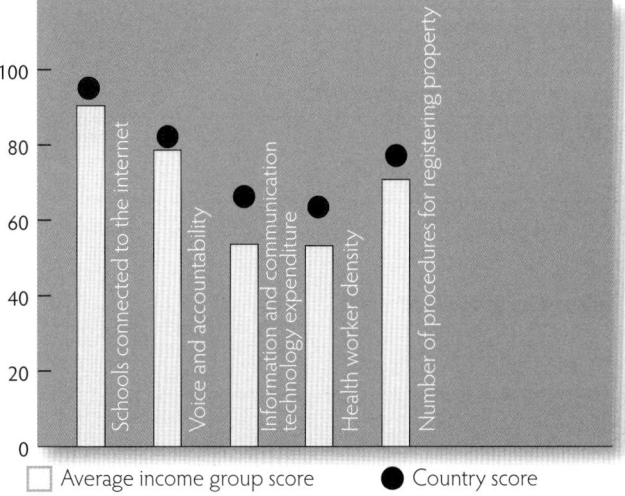

Bars: Schools connected to the internet; Voice and accountability; Information and communication technology expenditure; Health worker density; Number of procedures for registering property

□ Average income group score   ● Country score

---

[13] This analysis compares pillar and subindex country scores with respect to the average scores of countries grouped according to income level and political regime (see notes 1 and 2).

[14] Significant country indicators with the greatest distances above or below the income group average scores (i.e., achievements and targets for policy reform, respectively) were selected and ordered according to their Pearson correlation coefficients (highest first) with respect to the final index scores, to produce a ranking.

DENMARK

Key selected indicators

Population (in millions)[3, 4]	5.5
Population density (people per sq. km)[3, 4]	129
GDP (current US$ billion, PPP)[3, 5]	204
GDP per capita (current US$, PPP)[3, 5]	37392
Current account balance (% of GDP)[3, 5]	1.1
Average life expectancy (years)[4, 6]	78.1
Literacy rate[7, 8]	99
Female labor force participation rate[4, 6, 9]	74.0
Unemployment rate (% of total labor force)[4, 10]	4.8
CO_2 emissions (metric tons per capita)[4, 11]	9.8

Income level:[1] **High-income**

Political regime:[2] **Full democracy**

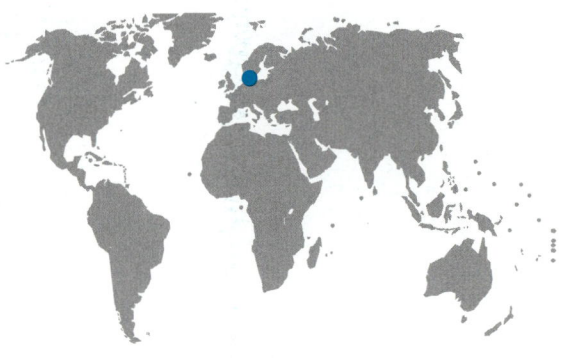

Innovation Capacity Index

Rank	Score
11 out of 131	**73.3 out of 100**

Pillars and pillar subindexes

Education
Rank **9** Score **79.6**

Social inclusion and equity policies
Rank **5** Score **85.7**

Country policy assessment
Rank **5** Score **73.7**

Good governance
Rank **1** Score **93.6**

Quality of the infrastructure
Rank **2** Score **99.0**

Government ICT usage
Rank **2** Score **91.3**

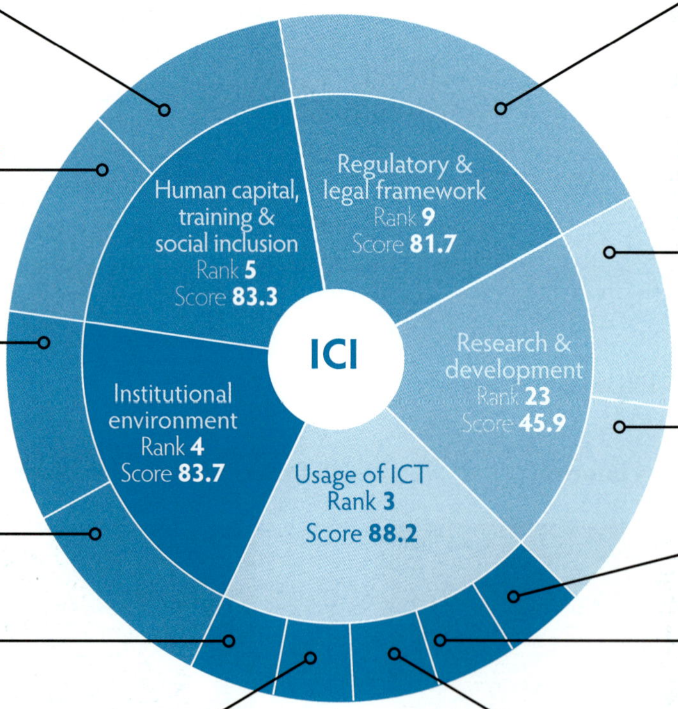

Doing business
Rank **9** Score **81.7**

R&D infrastructure
Rank **11** Score **65.6**

Patents and trademarks
Rank **26** Score **18.4**

Telephone communications
Rank **12** Score **95.2**

Mobile cellular communications
Rank **70** Score **72.6**

Internet, computers and TV
Rank **4** Score **84.1**

ICI

Human capital, training & social inclusion
Rank **5** Score **83.3**

Regulatory & legal framework
Rank **9** Score **81.7**

Institutional environment
Rank **4** Score **83.7**

Research & development
Rank **23** Score **45.9**

Usage of ICT
Rank **3** Score **88.2**

Index thermometer[12]

Pillar 1 11.4% Pillar 2 11.4% Pillar 3 22.3% Pillar 4 18.8% Pillar 5 36.1% 100%

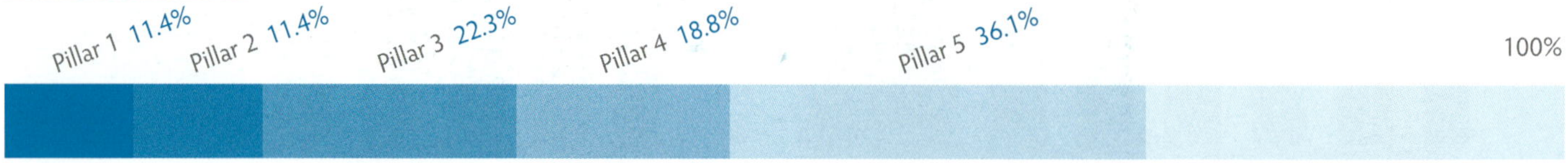

[1] High-income countries are those that, according to the World Bank classification, have a gross national income (GNI) per capita of $11,456 or more.
[2] Full democracies are countries whose Democracy Index scores from 1 to 10 rounded to one decimal place are of 8 to 10. [3] 2007. [4] World Bank World Development Indicators. [5] IMF World Economic Outlook. [6] 2006. [7] Estimate. [8] UNESCO, % age above 15. [9] % ages 15–64. [10] 2005. [11] 2004. [12] The index thermometer shows the relative contribution of each pillar (in %) to the total index score.

270

DENMARK

Country relative performance: **Index pillars**[13]

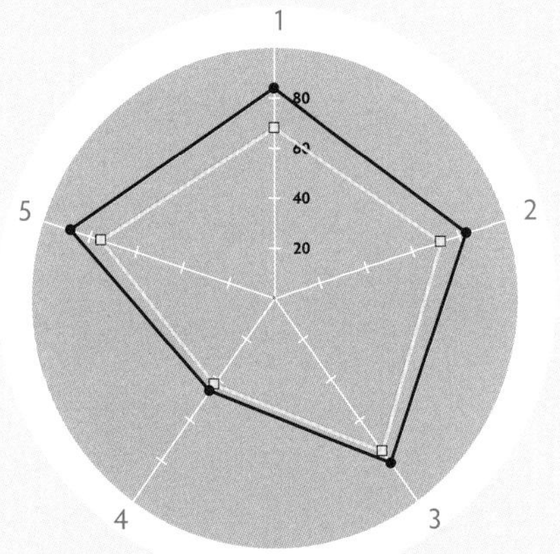

1. Institutional environment
2. Human capital, training and social inclusion
3. Regulatory and legal framework
4. Research and development
5. Usage of ICT

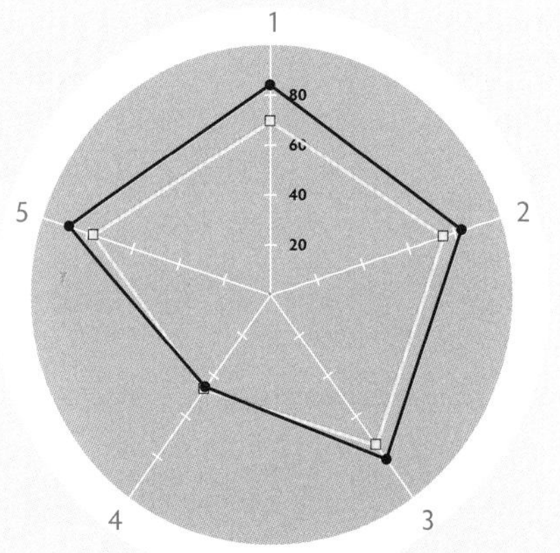

High-income ☐ ⋯⋯⋯ Denmark ●——

Full democracies ☐ Denmark ●——

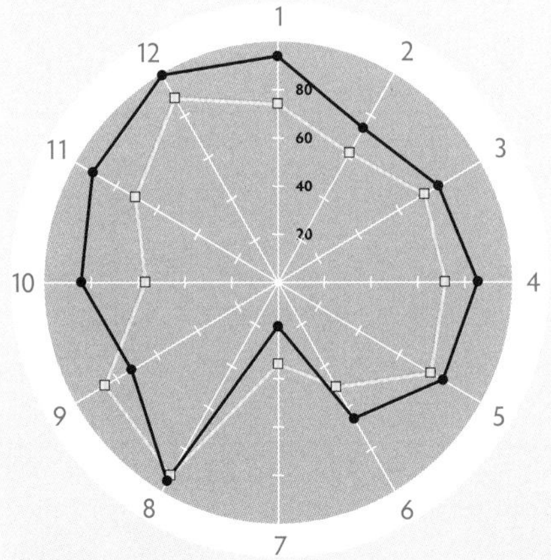

1. Good governance
2. Country policy assessment
3. Education
4. Social inclusion and equity policies
5. Doing business
6. R&D infrastructure
7. Patents and trademarks
8. Telephone communications
9. Mobile/cellular communications
10. Internet, computers and TV
11. Government ICT usage
12. Quality of the infrastructure

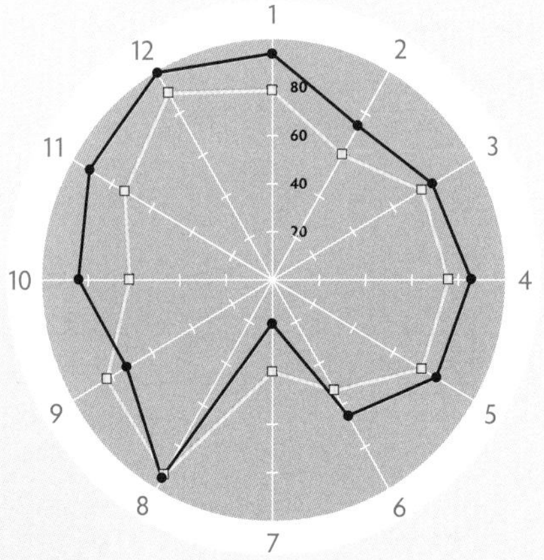

Country relative performance: **Pillar subindexes**[13]

In focus: **Top priorities for policy reform**[14]

In focus: **Significant indicators above income group average**[14]

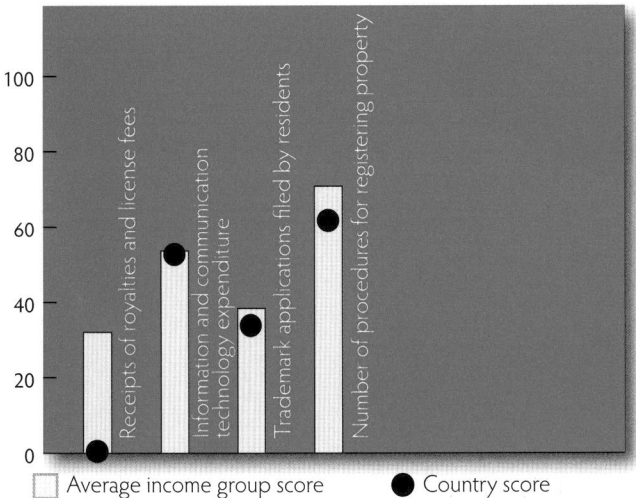

☐ Average income group score ● Country score

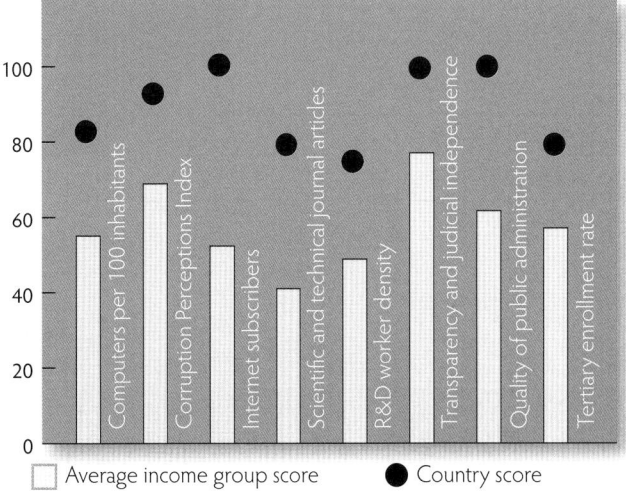

☐ Average income group score ● Country score

[13] This analysis compares pillar and subindex country scores with respect to the average scores of countries grouped according to income level and political regime (see notes 1 and 2).

[14] Significant country indicators with the greatest distances above or below the income group average scores (i.e., achievements and targets for policy reform, respectively) were selected and ordered according to their Pearson correlation coefficients (highest first) with respect to the final index scores, to produce a ranking.

EGYPT, ARAB REPUBLIC OF

Key selected indicators

Population (in millions)[3,4]	75.5
Population density (people per sq. km)[3,4]	76
GDP (current US$ billion, PPP)[3,5]	404
GDP per capita (current US$, PPP)[3,5]	5491
Current account balance (% of GDP)[3,5]	1.5
Average life expectancy (years)[4,6]	71.0
Literacy rate[3,7]	72
Female labor force participation rate[4,6,8]	21.6
Unemployment rate (% of total labor force)[4,9]	10.7
CO_2 emissions (metric tons per capita)[4,9]	2.2

Income level:[1] **Lower-middle-income**

Political regime:[2] **Authoritarian regime**

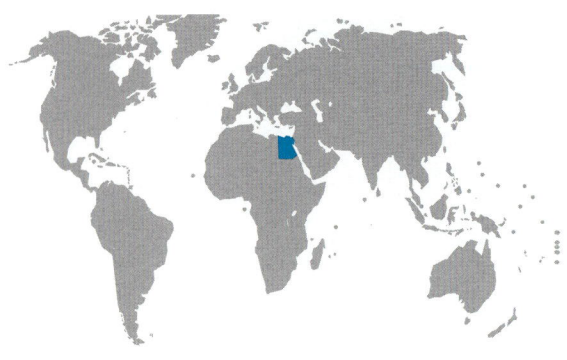

Rank	Score
79 out of 131	**46.3 out of 100**

Innovation Capacity Index

Pillars and pillar subindexes

Education
Rank **68** Score **58.7**

Social inclusion and equity policies
Rank **83** Score **48.0**

Country policy assessment
Rank **118** Score **39.7**

Good governance
Rank **88** Score **36.1**

Quality of the infrastructure
Rank **42** Score **84.8**

Government ICT usage
Rank **71** Score **47.7**

Doing business
Rank **68** Score **65.8**

R&D infrastructure
Rank **72** Score **23.0**

Patents and trademarks
Rank **95** Score **0.4**

Telephone communications
Rank **71** Score **79.0**

Mobile cellular communications
Rank **83** Score **60.6**

Internet, computers and TV
Rank **85** Score **9.3**

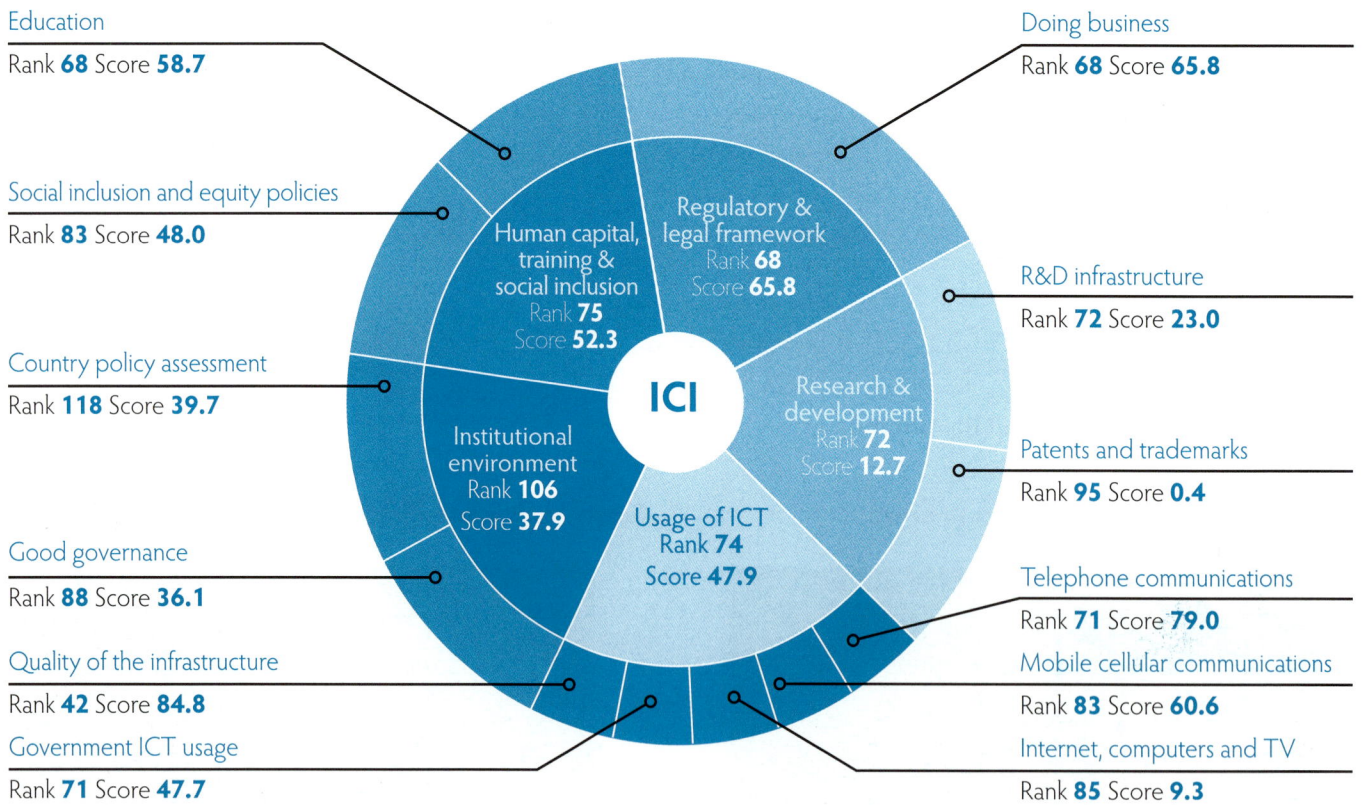

Index thermometer[10]

Pillar 1 24.6% Pillar 2 33.9% Pillar 3 28.4% Pillar 4 2.7% Pillar 5 10.3% 100%

[1] Lower-middle-income countries are those that, according to the World Bank classification, have a gross national income (GNI) per capita of $936 to $3,705. [2] Authoritarian regimes are countries whose Democracy Index scores from 1 to 10 rounded to one decimal place are less than 4. [3] 2007. [4] World Bank World Development Indicators. [5] IMF World Economic Outlook. [6] 2006. [7] UNESCO, % age above 15. [8] % ages 15–64. [9] 2004. [10] The index thermometer shows the relative contribution of each pillar (in %) to the total index score.

EGYPT, ARAB REPUBLIC OF

Country relative performance: **Index pillars**[11]

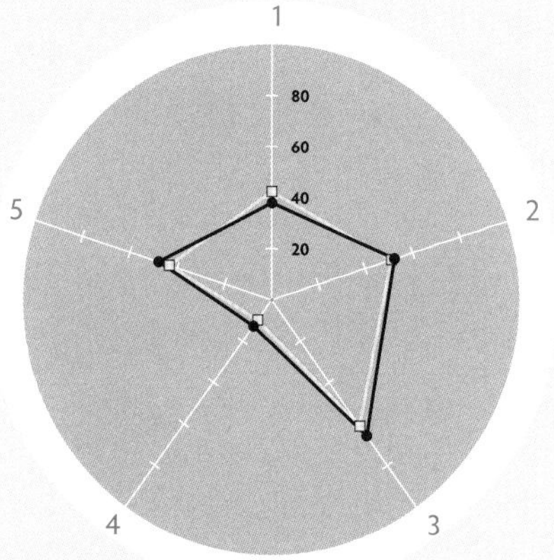

1. Institutional environment
2. Human capital, training and social inclusion
3. Regulatory and legal framework
4. Research and development
5. Usage of ICT

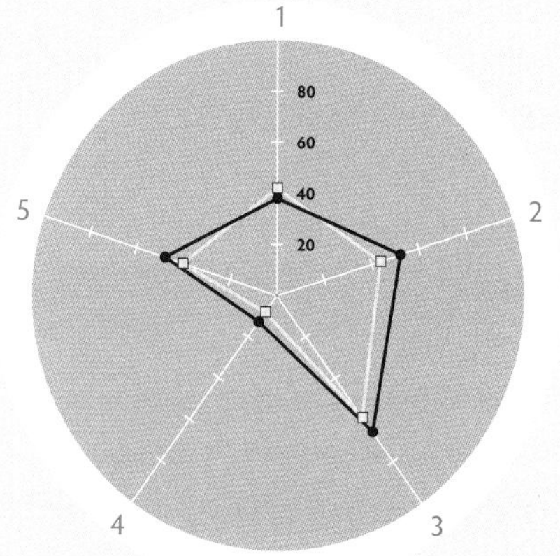

Lower-middle-income ☐   Egypt ●—   Authoritarian regimes ☐   Egypt ●—

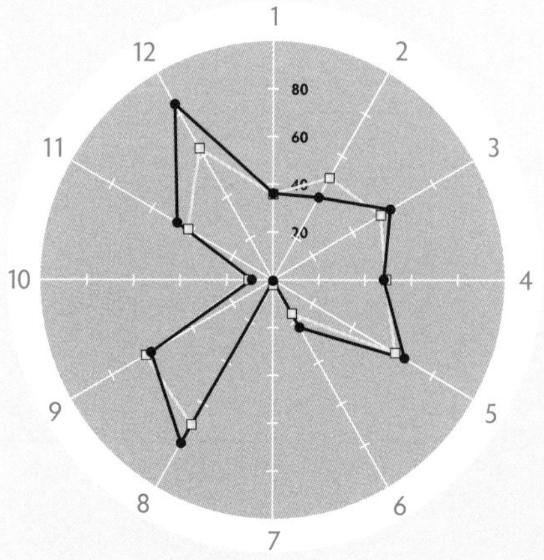

1. Good governance
2. Country policy assessment
3. Education
4. Social inclusion and equity policies
5. Doing business
6. R&D infrastructure
7. Patents and trademarks
8. Telephone communications
9. Mobile/cellular communications
10. Internet, computers and TV
11. Government ICT usage
12. Quality of the infrastructure

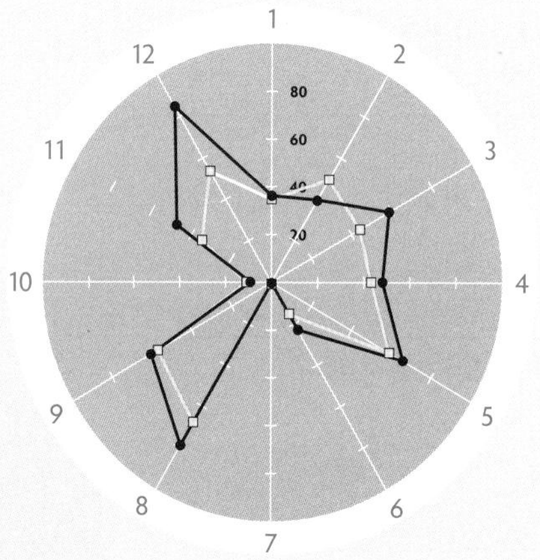

Country relative performance: **Pillar subindexes**[11]

In focus: **Top priorities for policy reform**[12]

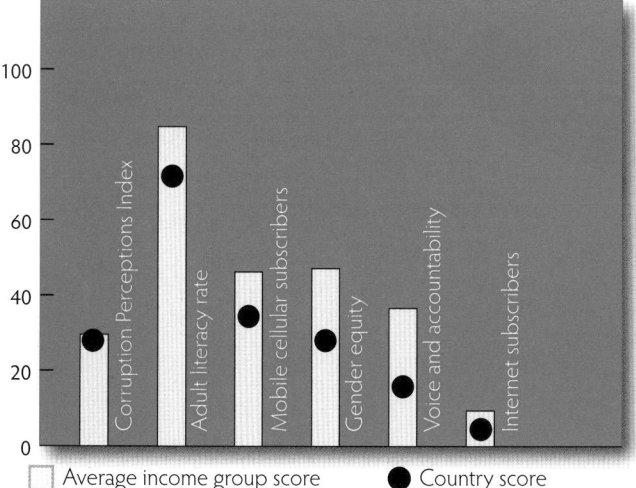

☐ Average income group score   ● Country score

In focus: **Significant indicators above income group average**[12]

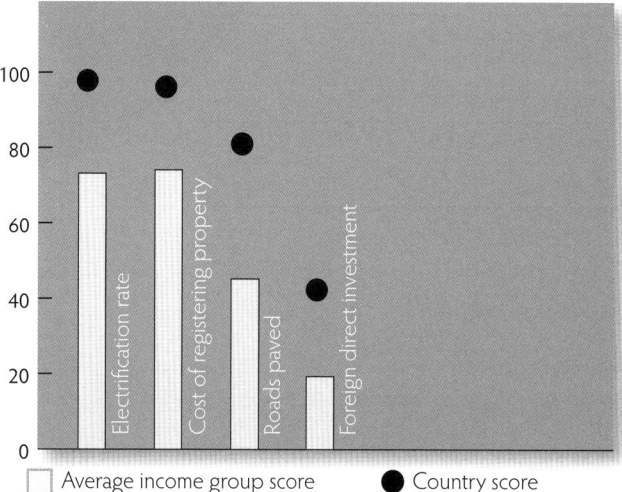

☐ Average income group score   ● Country score

[11] This analysis compares pillar and subindex country scores with respect to the average scores of countries grouped according to income level and political regime (see notes 1 and 2).

[12] Significant country indicators with the greatest distances above or below the income group average scores (i.e., achievements and targets for policy reform, respectively) were selected and ordered according to their Pearson correlation coefficients (highest first) with respect to the final index scores, to produce a ranking.

EL SALVADOR

Key selected indicators

Population (in millions)[3,4]	6.9
Population density (people per sq. km)[3,4]	331
GDP (current US$ billion, PPP)[3,5]	42
GDP per capita (current US$, PPP)[3,5]	5842
Current account balance (% of GDP)[3,5]	-4.8
Average life expectancy (years)[4,6]	71.5
Literacy rate[3,7]	85
Female labor force participation rate[4,6,8]	51.1
Unemployment rate (% of total labor force)[4,6]	6.6
CO_2 emissions (metric tons per capita)[4,9]	0.9

Income level:[1] **Lower-middle-income**
Political regime:[2] **Flawed democracy**

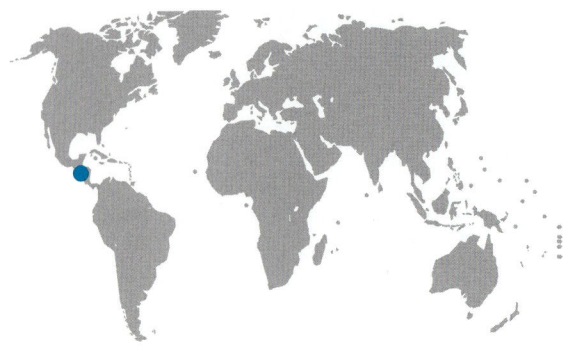

Rank 70 out of 131   **Score** 48.3 out of 100

Innovation Capacity Index

Pillars and pillar subindexes

Education — Rank **62** Score **60.9**

Social inclusion and equity policies — Rank **93** Score **45.7**

Country policy assessment — Rank **92** Score **47.4**

Good governance — Rank **63** Score **44.2**

Quality of the infrastructure — Rank **75** Score **60.2**

Government ICT usage — Rank **63** Score **49.7**

Doing business — Rank **69** Score **65.6**

R&D infrastructure — Rank **95** Score **12.7**

Patents and trademarks — Rank **64** Score **2.8**

Telephone communications — Rank **46** Score **86.3**

Mobile cellular communications — Rank **27** Score **87.8**

Internet, computers and TV — Rank **87** Score **8.9**

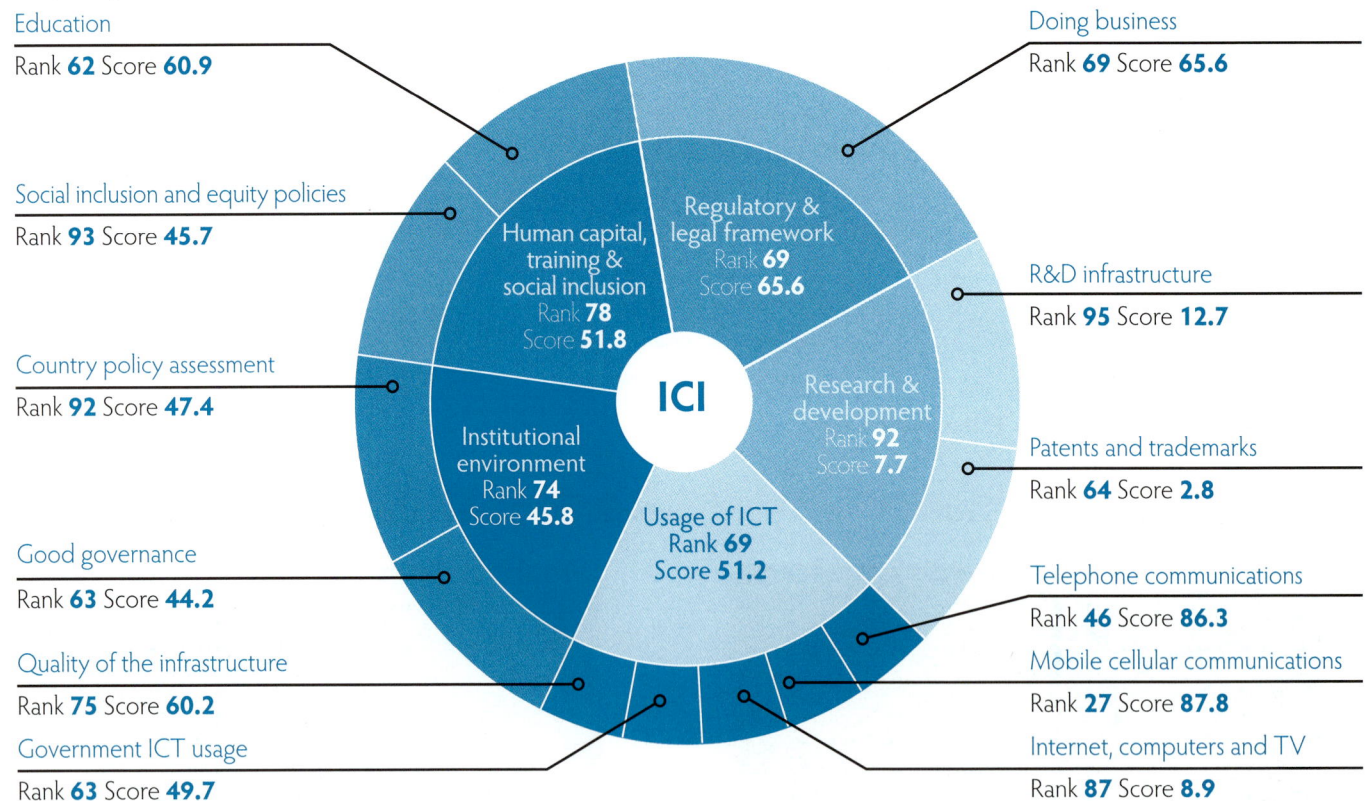

Index thermometer[10]

Pillar 1 28.5% Pillar 2 32.2% Pillar 3 27.2% Pillar 4 1.6% Pillar 5 10.6%   100%

[1] Lower-middle-income countries are those that, according to the World Bank classification, have a gross national income (GNI) per capita of $936 to $3,705. [2] Flawed democracies are countries whose Democracy Index scores from 1 to 10 rounded to one decimal place are of 6 to 7.9. [3] 2007. [4] World Bank World Development Indicators. [5] IMF World Economic Outlook. [6] 2006. [7] UNESCO, % age above 15. [8] % ages 15–64. [9] 2004. [10] The index thermometer shows the relative contribution of each pillar (in %) to the total index score.

274

EL SALVADOR

Country relative performance: **Index pillars**[11]

1
2
3
4
5
80
60
40
20

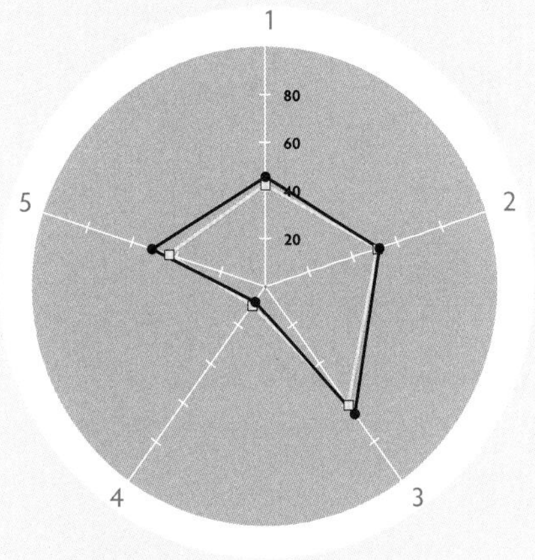

1. Institutional environment
2. Human capital, training and social inclusion
3. Regulatory and legal framework
4. Research and development
5. Usage of ICT

1
2
3
4
5
80
60
40
20

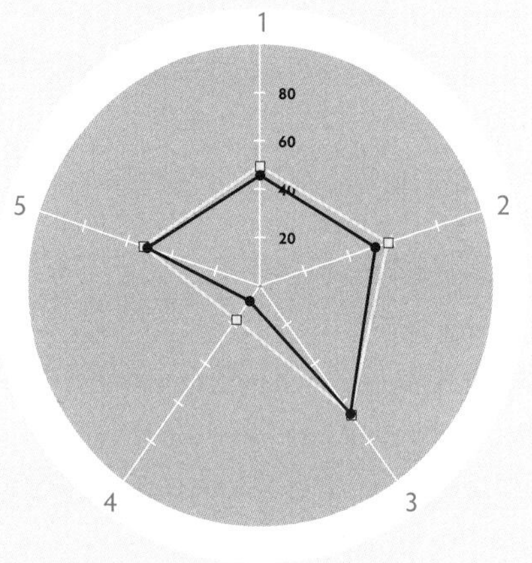

Lower-middle-income □ El Salvador ●

Flawed democracies □ El Salvador ●

1
2
3
4
5
6
7
8
9
10
11
12
80
60
20

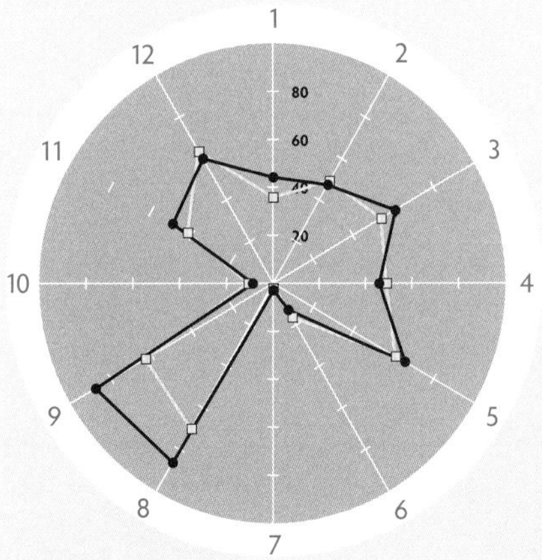

1. Good governance
2. Country policy assessment
3. Education
4. Social inclusion and equity policies
5. Doing business
6. R&D infrastructure
7. Patents and trademarks
8. Telephone communications
9. Mobile/cellular communications
10. Internet, computers and TV
11. Government ICT usage
12. Quality of the infrastructure

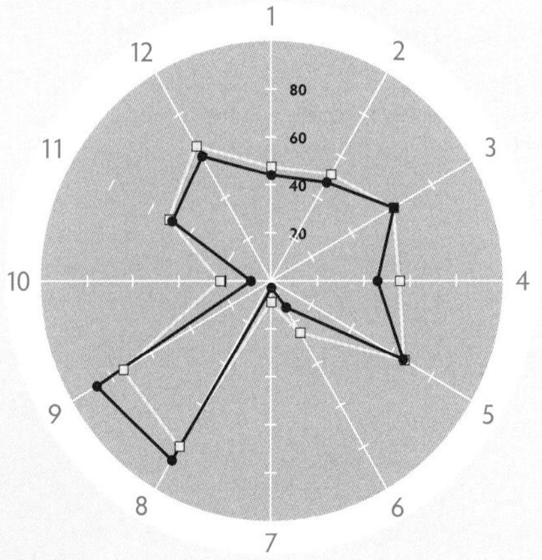

1
2
3
4
5
6
7
8
9
10
11
12
80
60
40
20

Country relative performance: **Pillar subindexes**[11]

In focus: **Top priorities for policy reform**[12]

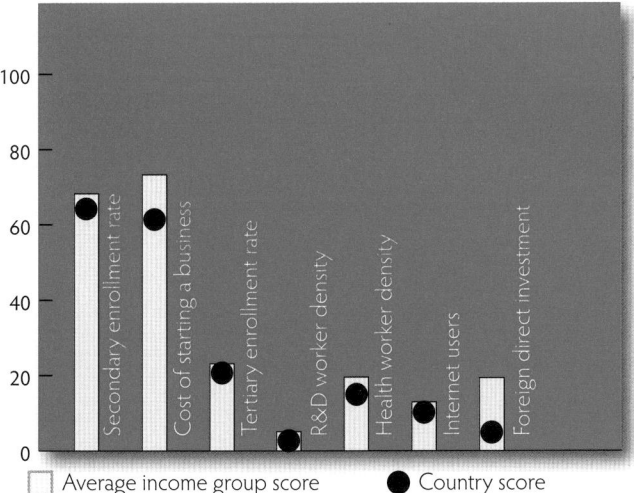

100
80
60
40
20
0
Secondary enrollment rate
Cost of starting a business
Tertiary enrollment rate
R&D worker density
Health worker density
Internet users
Foreign direct investment

□ Average income group score ● Country score

In focus: **Significant indicators above income group average**[12]

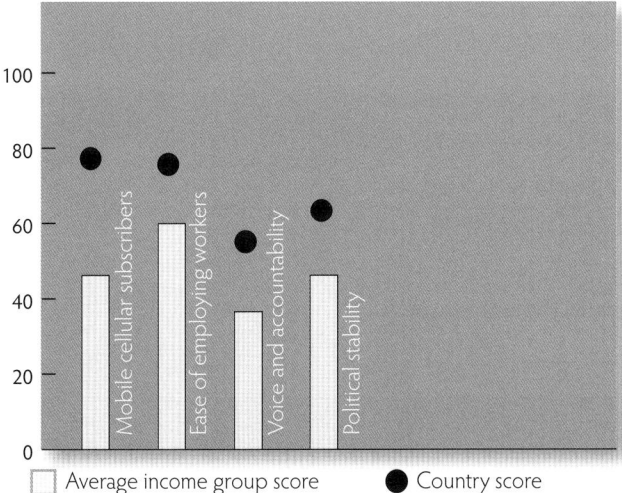

100
80
60
40
20
0
Mobile cellular subscribers
Ease of employing workers
Voice and accountability
Political stability

□ Average income group score ● Country score

[11] This analysis compares pillar and subindex country scores with respect to the average scores of countries grouped according to income level and political regime (see notes 1 and 2).

[12] Significant country indicators with the greatest distances above or below the income group average scores (i.e., achievements and targets for policy reform, respectively) were selected and ordered according to their Pearson correlation coefficients (highest first) with respect to the final index scores, to produce a ranking.

ESTONIA, REPUBLIC OF

Key selected indicators

Population (in millions)[3,4]	1.3
Population density (people per sq. km)[3,4]	32
GDP (current US$ billion, PPP)[3,5]	28
GDP per capita (current US$, PPP)[3,5]	21094
Current account balance (% of GDP)[3,5]	-16.0
Average life expectancy (years)[4,6]	72.6
Literacy rate[3,7]	100
Female labor force participation rate[4,6,8]	64.6
Unemployment rate (% of total labor force)[4,9]	7.9
CO_2 emissions (metric tons per capita)[4,10]	14.0

Income level:[1] **High-income**

Political regime:[2] **Full democracy**

	Rank	Score
Innovation Capacity Index	**25 out of 131**	**62.7 out of 100**

Pillars and pillar subindexes

Education
Rank **3** Score **83.6**

Social inclusion and equity policies
Rank **25** Score **70.4**

Country policy assessment
Rank **14** Score **69.5**

Good governance
Rank **22** Score **71.7**

Quality of the infrastructure
Rank **90** Score **53.7**

Government ICT usage
Rank **13** Score **76.0**

Doing business
Rank **18** Score **77.3**

R&D infrastructure
Rank **31** Score **41.0**

Patents and trademarks
Rank **38** Score **11.9**

Telephone communications
Rank **34** Score **89.9**

Mobile cellular communications
Rank **24** Score **88.1**

Internet, computers and TV
Rank **22** Score **58.8**

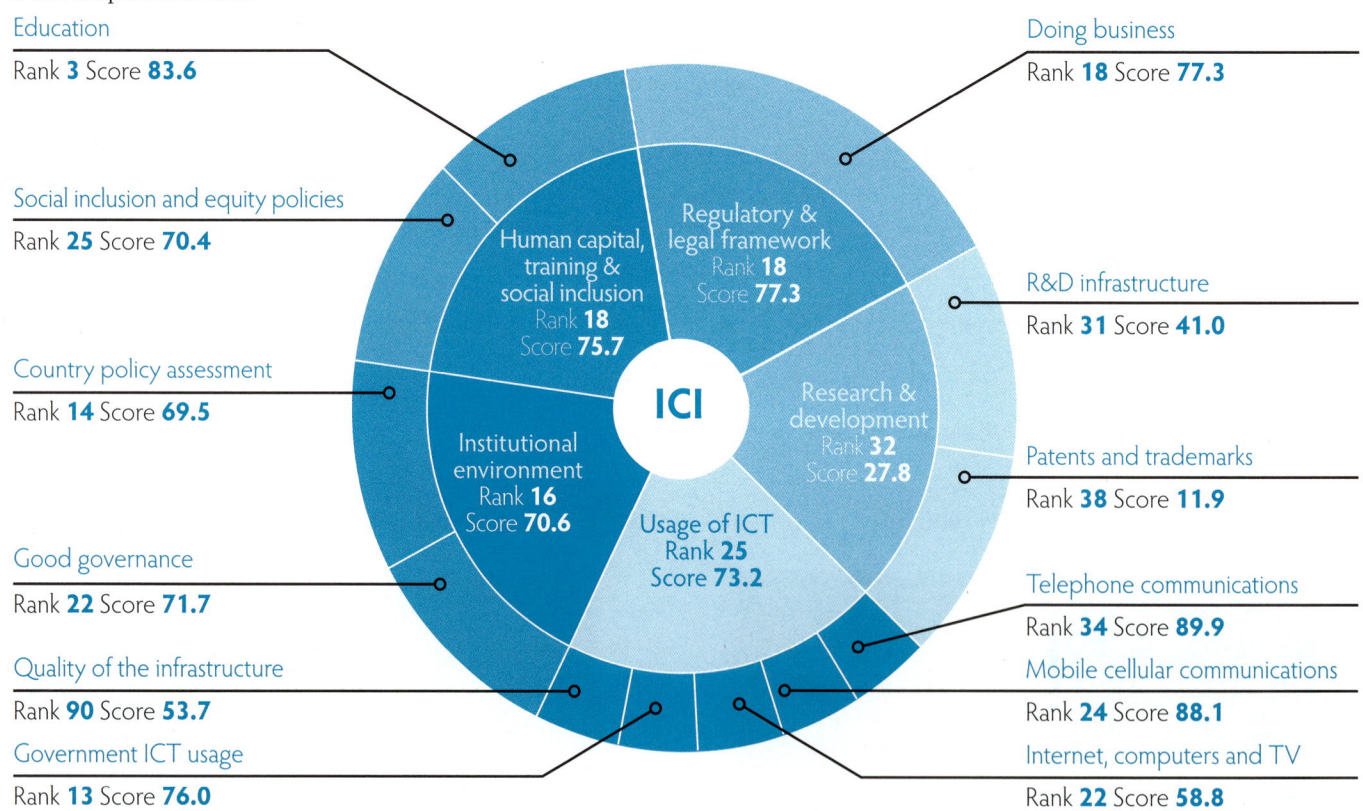

ICI

Human capital, training & social inclusion
Rank **18** Score **75.7**

Regulatory & legal framework
Rank **18** Score **77.3**

Research & development
Rank **32** Score **27.8**

Institutional environment
Rank **16** Score **70.6**

Usage of ICT
Rank **25** Score **73.2**

Index thermometer[11]

Pillar 1 16.9% Pillar 2 18.1% Pillar 3 24.7% Pillar 4 11.1% Pillar 5 29.2% 100%

---

[1] High-income countries are those that, according to the World Bank classification, have a gross national income (GNI) per capita of $11,456 or more.
[2] Full democracies are countries whose Democracy Index scores from 1 to 10 rounded to one decimal place are of 8 to 10. [3] 2007. [4] World Bank World Development Indicators. [5] IMF World Economic Outlook. [6] 2006. [7] UNESCO, % age above 15. [8] % ages 15–64. [9] 2005. [10] 2004. [11] The index thermometer shows the relative contribution of each pillar (in %) to the total index score.

ESTONIA, REPUBLIC OF

Country relative performance: **Index pillars**[12]

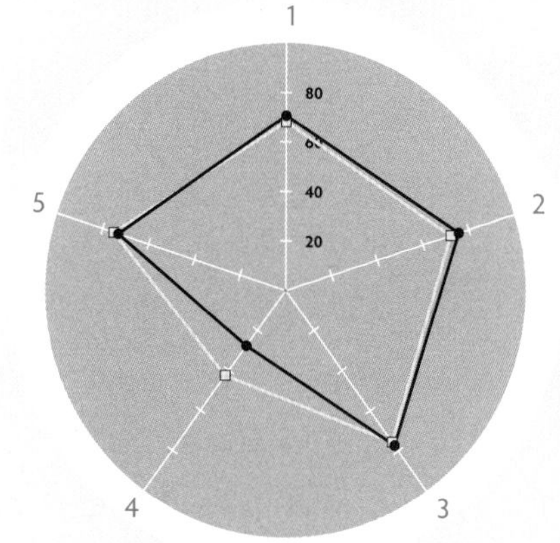

1. Institutional environment
2. Human capital, training and social inclusion
3. Regulatory and legal framework
4. Research and development
5. Usage of ICT

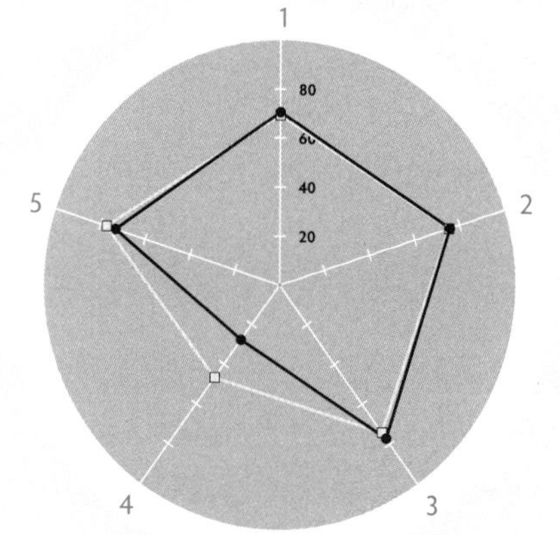

High-income ☐   Estonia ●——

Full democracies ☐   Estonia ●——

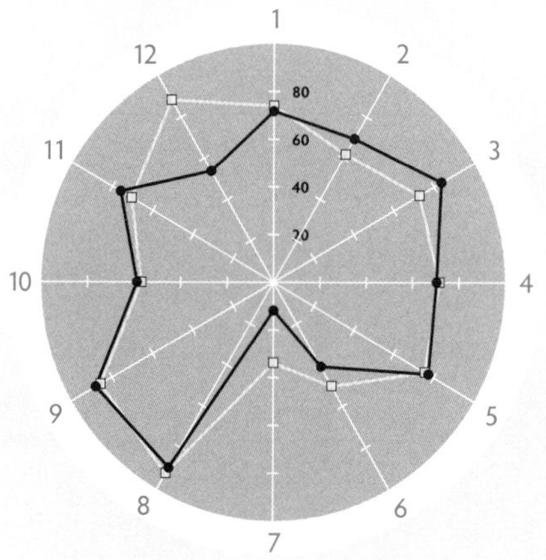

1. Good governance
2. Country policy assessment
3. Education
4. Social inclusion and equity policies
5. Doing business
6. R&D infrastructure
7. Patents and trademarks
8. Telephone communications
9. Mobile/cellular communications
10. Internet, computers and TV
11. Government ICT usage
12. Quality of the infrastructure

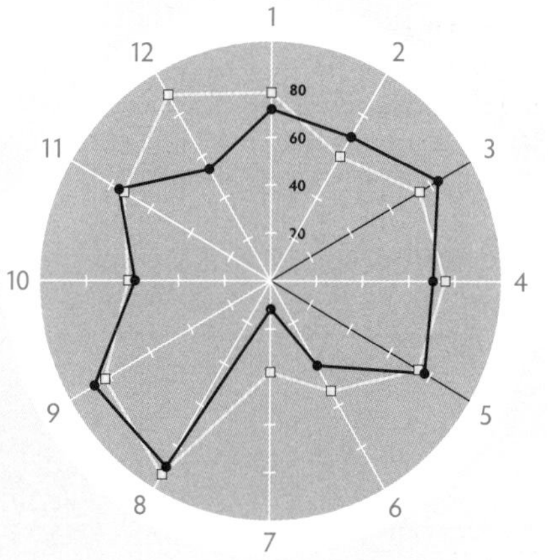

Country relative performance: **Pillar subindexes**[12]

In focus: **Top priorities for policy reform**[13]

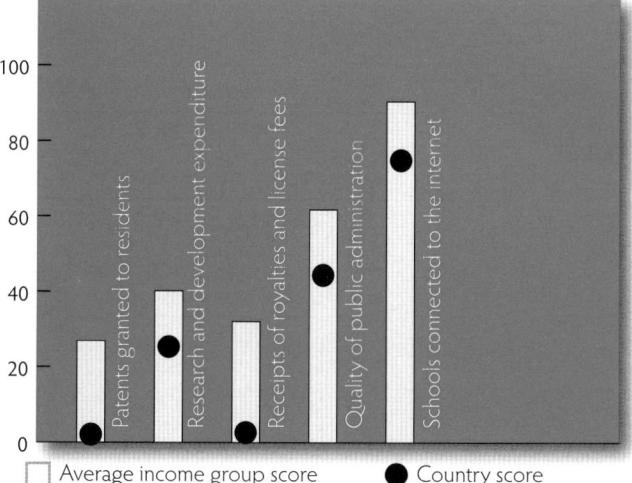

☐ Average income group score   ● Country score

In focus: **Significant indicators above income group average**[13]

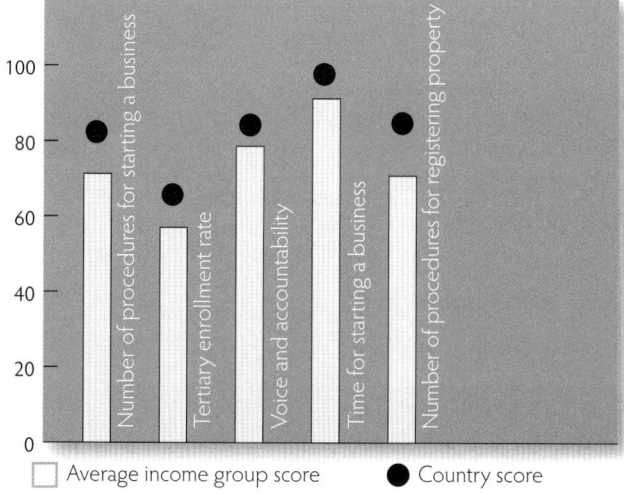

☐ Average income group score   ● Country score

277

[12]   This analysis compares pillar and subindex country scores with respect to the average scores of countries grouped according to income level and political regime (see notes 1 and 2).

[13]   Significant country indicators with the greatest distances above or below the income group average scores (i.e., achievements and targets for policy reform, respectively) were selected and ordered according to their Pearson correlation coefficients (highest first) with respect to the final index scores, to produce a ranking.

FINLAND

Key selected indicators

Population (in millions)[3,4]	5.3
Population density (people per sq. km)[3,4]	17
GDP (current US$ billion, PPP)[3,5]	185
GDP per capita (current US$, PPP)[3,5]	35280
Current account balance (% of GDP)[3,5]	4.6
Average life expectancy (years)[4,6]	79.2
Literacy rate[7,8]	99
Female labor force participation rate[4,6,9]	72.8
Unemployment rate (% of total labor force)[4,10]	8.4
CO_2 emissions (metric tons per capita)[4,11]	12.6

Income level:[1] **High-income**

Political regime:[2] **Full democracy**

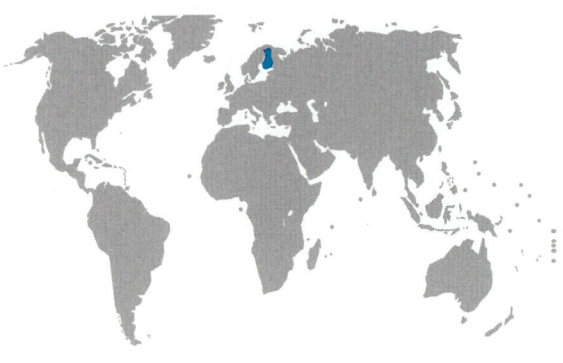

Innovation Capacity Index

Rank	Score
2 out of 131	**77.8 out of 100**

Pillars and pillar subindexes

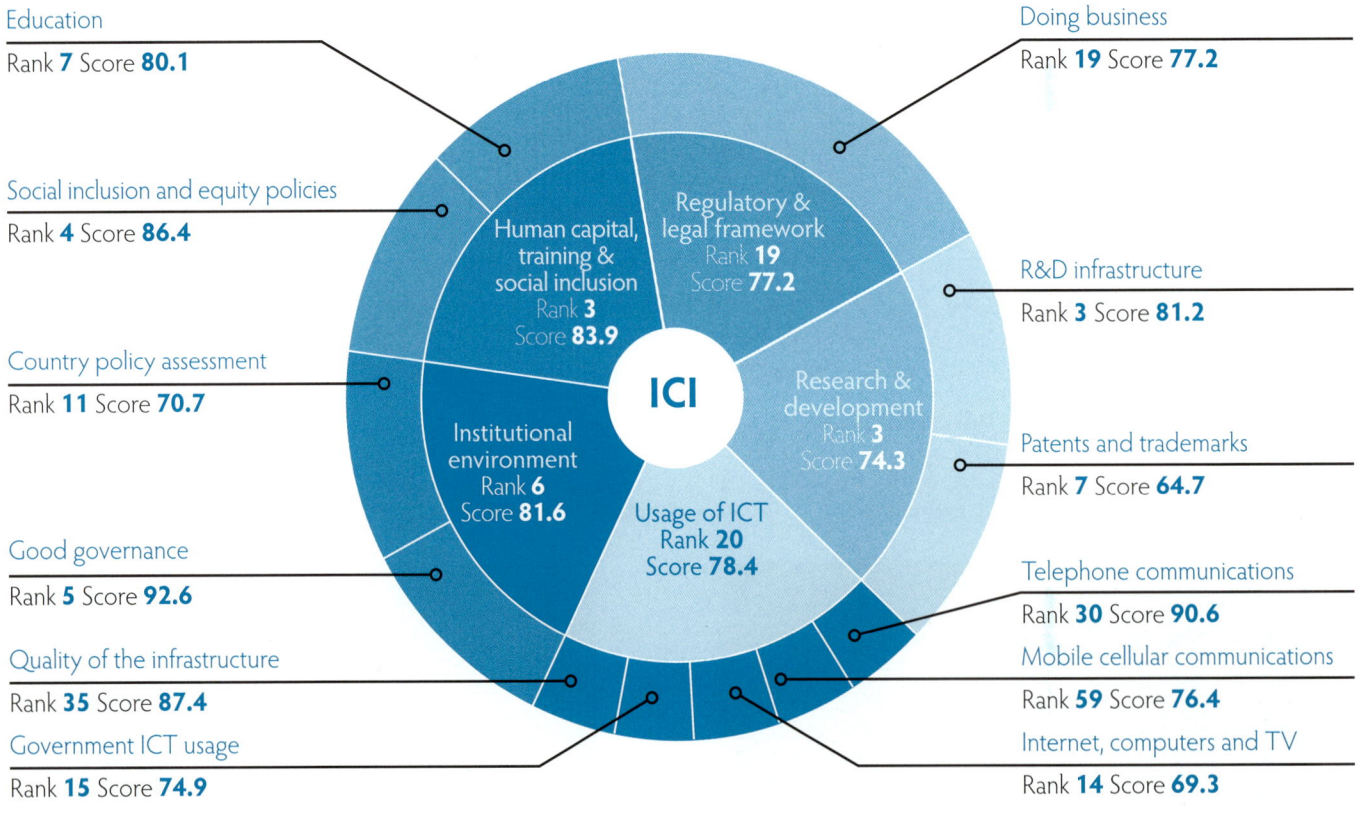

Education
Rank **7** Score **80.1**

Social inclusion and equity policies
Rank **4** Score **86.4**

Country policy assessment
Rank **11** Score **70.7**

Good governance
Rank **5** Score **92.6**

Quality of the infrastructure
Rank **35** Score **87.4**

Government ICT usage
Rank **15** Score **74.9**

Doing business
Rank **19** Score **77.2**

R&D infrastructure
Rank **3** Score **81.2**

Patents and trademarks
Rank **7** Score **64.7**

Telephone communications
Rank **30** Score **90.6**

Mobile cellular communications
Rank **59** Score **76.4**

Internet, computers and TV
Rank **14** Score **69.3**

Human capital, training & social inclusion
Rank **3** Score **83.9**

Regulatory & legal framework
Rank **19** Score **77.2**

Institutional environment
Rank **6** Score **81.6**

ICI

Research & development
Rank **3** Score **74.3**

Usage of ICT
Rank **20** Score **78.4**

Index thermometer[12]

Pillar 1 10.5% Pillar 2 10.8% Pillar 3 19.9% Pillar 4 28.7% Pillar 5 30.2% 100%

---

[1] High-income countries are those that, according to the World Bank classification, have a gross national income (GNI) per capita of $11,456 or more.
[2] Full democracies are countries whose Democracy Index scores from 1 to 10 rounded to one decimal place are of 8 to 10. [3] 2007. [4] World Bank World Development Indicators. [5] IMF World Economic Outlook. [6] 2006. [7] Estimate. [8] UNESCO, % age above 15. [9] % ages 15–64. [10] 2005. [11] 2004. [12] The index thermometer shows the relative contribution of each pillar (in %) to the total index score.

278

FINLAND

Country relative performance: **Index pillars**[13]

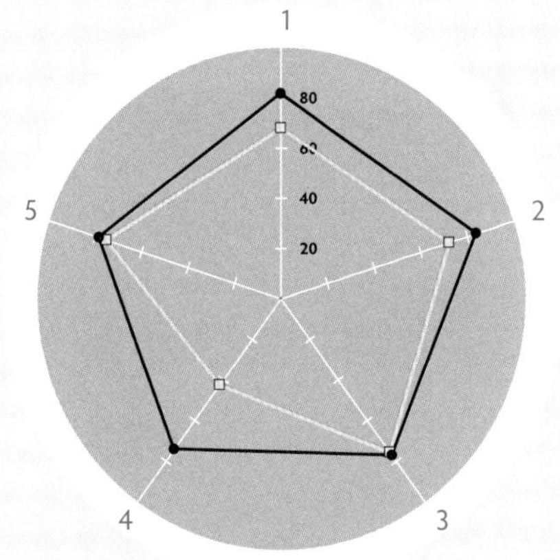

1. Institutional environment
2. Human capital, training and social inclusion
3. Regulatory and legal framework
4. Research and development
5. Usage of ICT

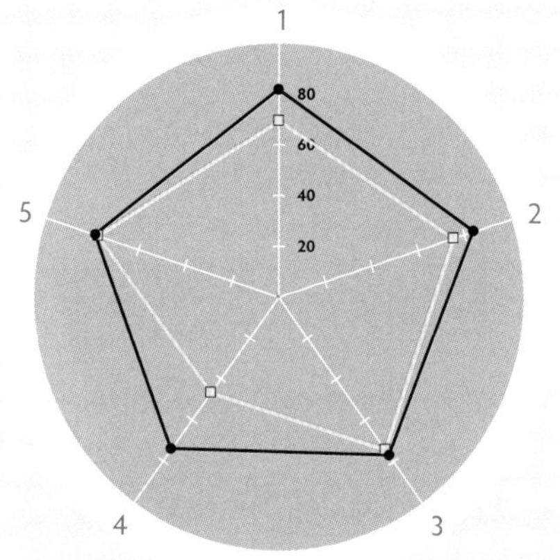

High-income ☐   Finland ●——●

Full democracies ☐   Finland ●——●

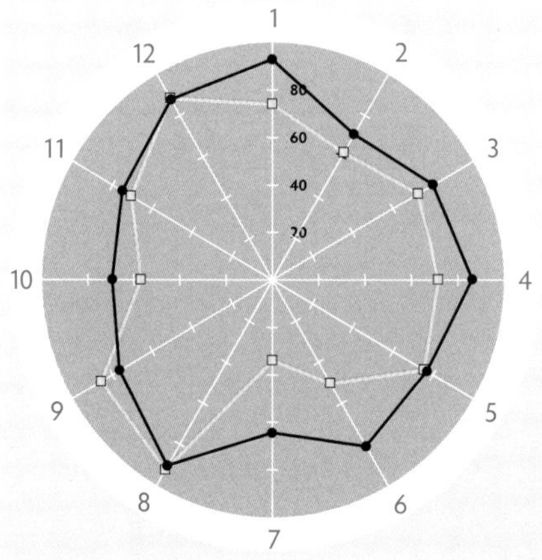

1. Good governance
2. Country policy assessment
3. Education
4. Social inclusion and equity policies
5. Doing business
6. R&D infrastructure
7. Patents and trademarks
8. Telephone communications
9. Mobile/cellular communications
10. Internet, computers and TV
11. Government ICT usage
12. Quality of the infrastructure

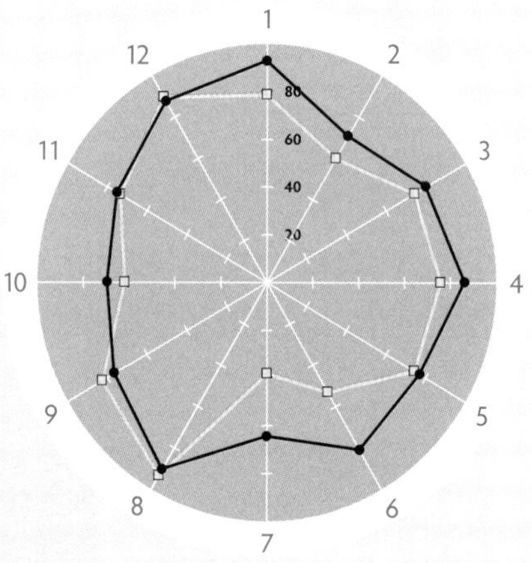

Country relative performance: **Pillar subindexes**[13]

In focus: **Top priorities for policy reform**[14]

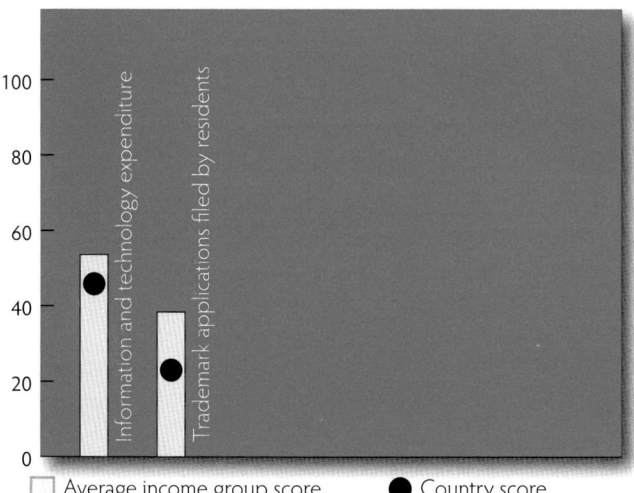

☐ Average income group score   ● Country score

In focus: **Significant indicators above income group average**[14]

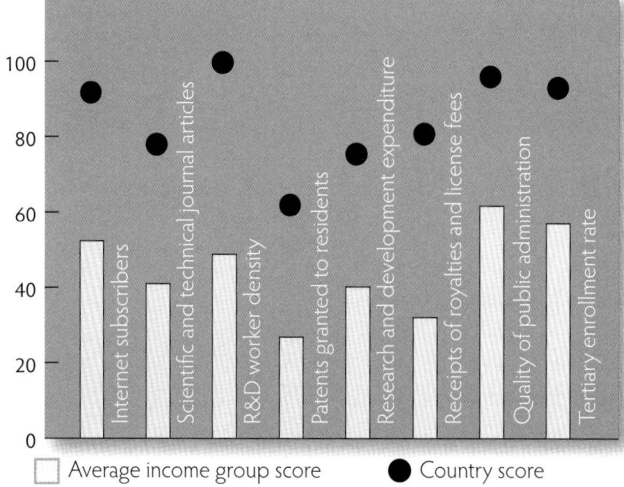

☐ Average income group score   ● Country score

---

[13] This analysis compares pillar and subindex country scores with respect to the average scores of countries grouped according to income level and political regime (see notes 1 and 2).

[14] Significant country indicators with the greatest distances above or below the income group average scores (i.e., achievements and targets for policy reform, respectively) were selected and ordered according to their Pearson correlation coefficients (highest first) with respect to the final index scores, to produce a ranking.

FRANCE

Key selected indicators

Population (in millions)[3,4]	61.7
Population density (people per sq. km)[3,4]	112
GDP (current US$ billion, PPP)[3,5]	2047
GDP per capita (current US$, PPP)[3,5]	33188
Current account balance (% of GDP)[3,5]	-1.3
Average life expectancy (years)[4,6]	80.6
Literacy rate[7,8]	99
Female labor force participation rate[4,6,9]	62.4
Unemployment rate (% of total labor force)[4,10]	9.8
CO_2 emissions (metric tons per capita)[4,11]	6.2

Income level:[1] **High-income**

Political regime:[2] **Full democracy**

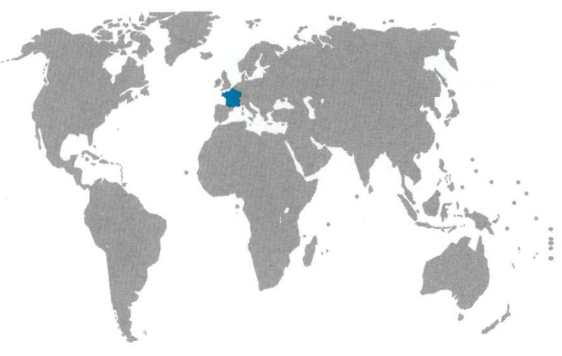

Rank
24 out of 131

Score
65.4 out of 100

Innovation Capacity Index

Pillars and pillar subindexes

Education
Rank **29** Score **70.5**

Social inclusion and equity policies
Rank **16** Score **78.4**

Country policy assessment
Rank **52** Score **55.2**

Good governance
Rank **21** Score **75.7**

Quality of the infrastructure
Rank **7** Score **98.2**

Government ICT usage
Rank **9** Score **80.4**

Doing business
Rank **64** Score **66.3**

R&D infrastructure
Rank **17** Score **55.3**

Patents and trademarks
Rank **21** Score **36.2**

Telephone communications
Rank **7** Score **96.2**

Mobile cellular communications
Rank **59** Score **76.4**

Internet, computers and TV
Rank **19** Score **65.3**

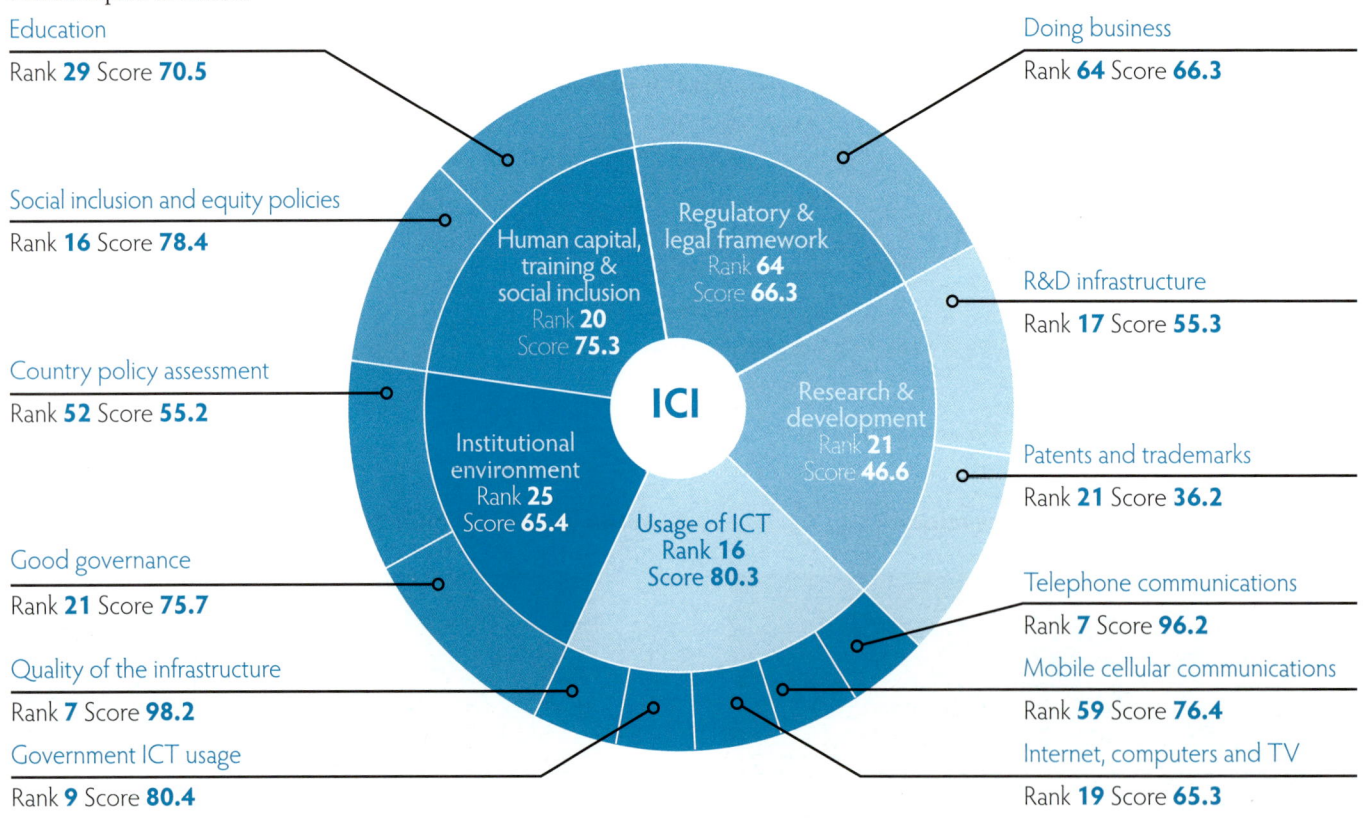

Index thermometer[12]

Pillar 1 10.0% Pillar 2 11.5% Pillar 3 20.3% Pillar 4 21.4% Pillar 5 36.8% 100%

[1] High-income countries are those that, according to the World Bank classification, have a gross national income (GNI) per capita of $11,456 or more.
[2] Full democracies are countries whose Democracy Index scores from 1 to 10 rounded to one decimal place are of 8 to 10. [3] 2007. [4] World Bank World Development Indicators. [5] IMF World Economic Outlook. [6] 2006. [7] Estimate. [8] UNESCO, % age above 15. [9] % ages 15–64. [10] 2005. [11] 2004. [12] The index thermometer shows the relative contribution of each pillar (in %) to the total index score.

280

FRANCE

Country relative performance: **Index pillars**[13]

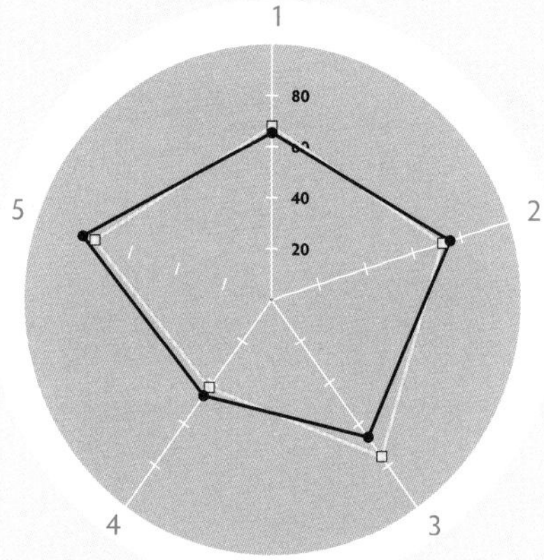

1. Institutional environment
2. Human capital, training and social inclusion
3. Regulatory and legal framework
4. Research and development
5. Usage of ICT

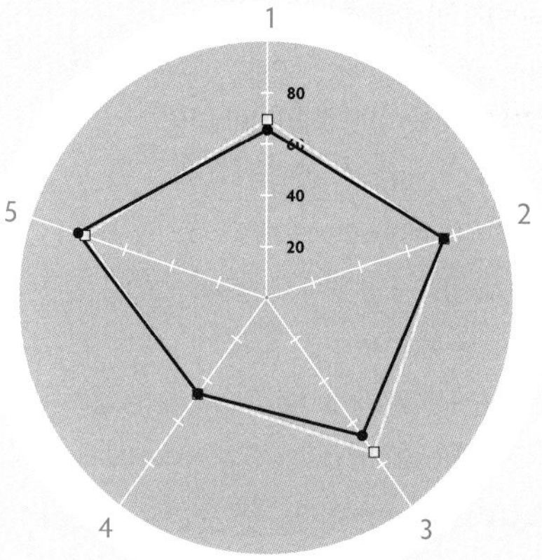

High-income ☐   France ●━━━●

Full democracies ☐   France ●━━━●

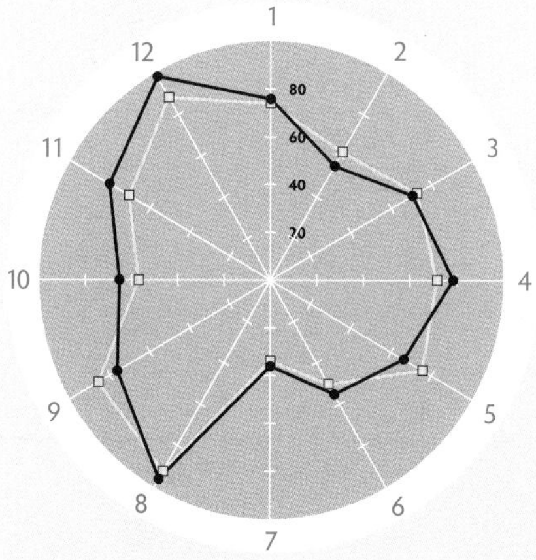

1. Good governance
2. Country policy assessment
3. Education
4. Social inclusion and equity policies
5. Doing business
6. R&D infrastructure
7. Patents and trademarks
8. Telephone communications
9. Mobile/cellular communications
10. Internet, computers and TV
11. Government ICT usage
12. Quality of the infrastructure

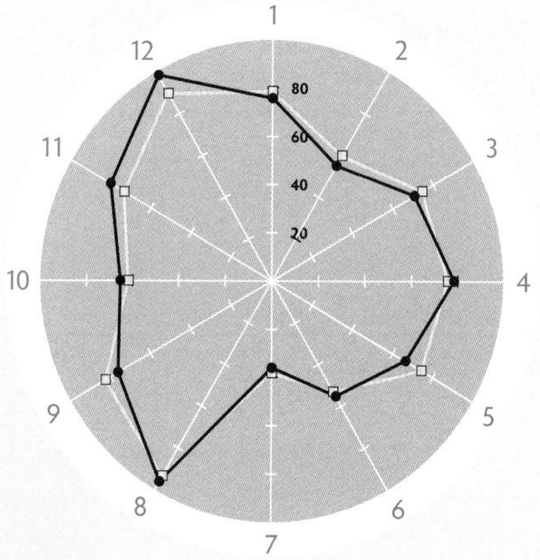

Country relative performance: **Pillar subindexes**[13]

281

In focus: **Top priorities for policy reform**[14]

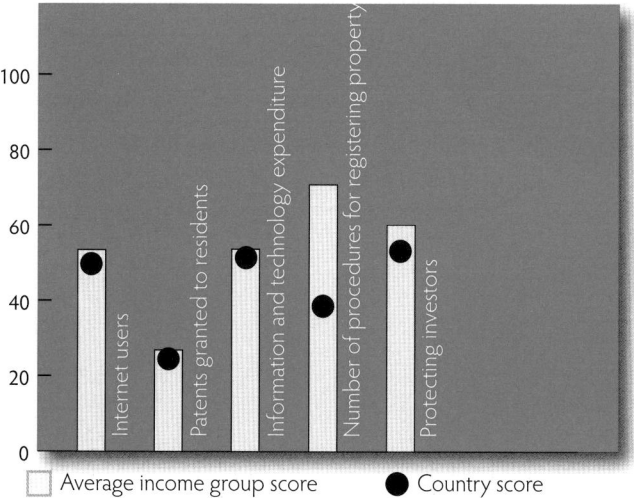

☐ Average income group score   ● Country score

In focus: **Significant indicators above income group average**[14]

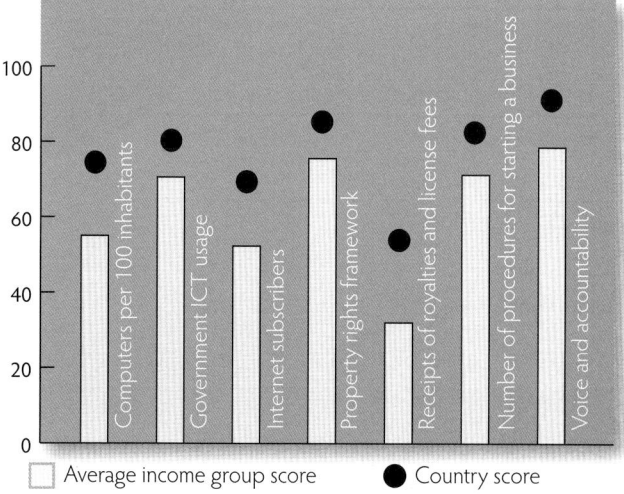

☐ Average income group score   ● Country score

[13] This analysis compares pillar and subindex country scores with respect to the average scores of countries grouped according to income level and political regime (see notes 1 and 2).

[14] Significant country indicators with the greatest distances above or below the income group average scores (i.e., achievements and targets for policy reform, respectively) were selected and ordered according to their Pearson correlation coefficients (highest first) with respect to the final index scores, to produce a ranking.

GERMANY

Key selected indicators

Population (in millions)[3,4]	82.3
Population density (people per sq. km)[3,4]	236
GDP (current US$ billion, PPP)[3,5]	2810
GDP per capita (current US$, PPP)[3,5]	34181
Current account balance (% of GDP)[5,6]	5.6
Average life expectancy (years)[4,6]	79.1
Literacy rate[7,8]	99
Female labor force participation rate[4,6,9]	68.2
Unemployment rate (% of total labor force)[4,10]	11.1
CO_2 emissions (metric tons per capita)[4,11]	9.8

Income level:[1] **High-income**

Political regime:[2] **Full democracy**

Innovation Capacity Index

Rank **20 out of 131**

Score **68.8 out of 100**

Pillars and pillar subindexes

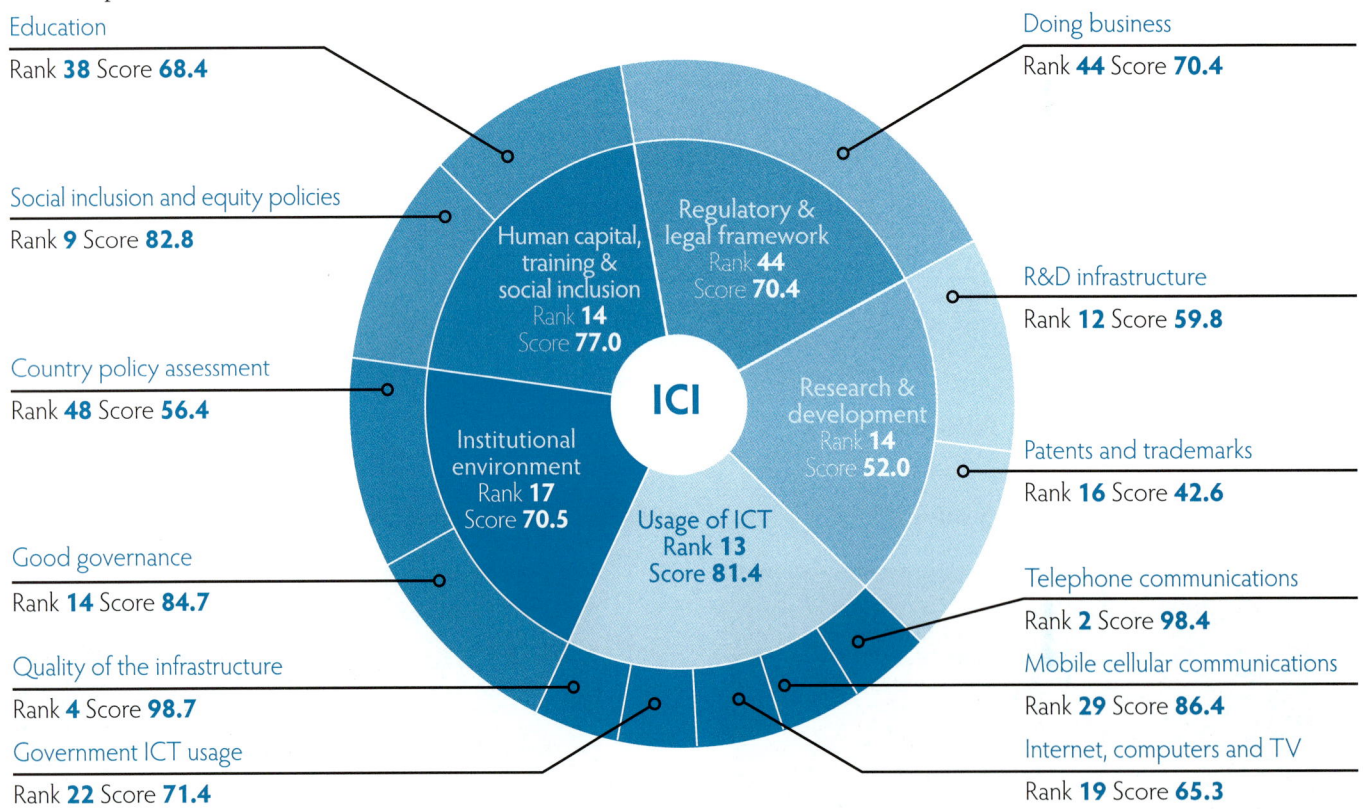

Education
Rank **38** Score **68.4**

Social inclusion and equity policies
Rank **9** Score **82.8**

Country policy assessment
Rank **48** Score **56.4**

Good governance
Rank **14** Score **84.7**

Quality of the infrastructure
Rank **4** Score **98.7**

Government ICT usage
Rank **22** Score **71.4**

Doing business
Rank **44** Score **70.4**

R&D infrastructure
Rank **12** Score **59.8**

Patents and trademarks
Rank **16** Score **42.6**

Telephone communications
Rank **2** Score **98.4**

Mobile cellular communications
Rank **29** Score **86.4**

Internet, computers and TV
Rank **19** Score **65.3**

Human capital, training & social inclusion
Rank **14** Score **77.0**

Regulatory & legal framework
Rank **44** Score **70.4**

Institutional environment
Rank **17** Score **70.5**

Research & development
Rank **14** Score **52.0**

Usage of ICT
Rank **13** Score **81.4**

ICI

Index thermometer[12]

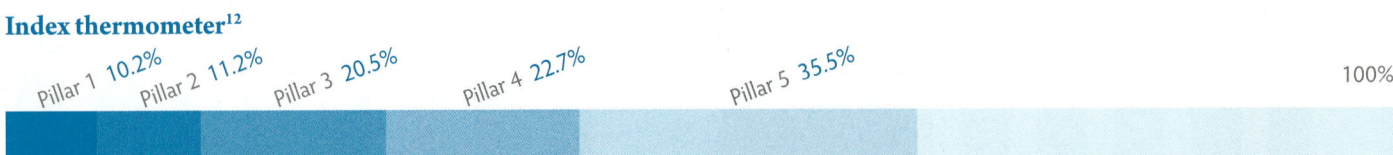

Pillar 1 10.2%   Pillar 2 11.2%   Pillar 3 20.5%   Pillar 4 22.7%   Pillar 5 35.5%   100%

[1] High-income countries are those that, according to the World Bank classification, have a gross national income (GNI) per capita of $11,456 or more.
[2] Full democracies are countries whose Democracy Index scores from 1 to 10 rounded to one decimal place are of 8 to 10. [3] 2007. [4] World Bank World Development Indicators. [5] IMF World Economic Outlook. [6] 2006. [7] Estimate. [8] UNESCO, % age above 15. [9] % ages 15–64. [10] 2005. [11] 2004. [12] The index thermometer shows the relative contribution of each pillar (in %) to the total index score.

GERMANY

Country relative performance: Index pillars[13]

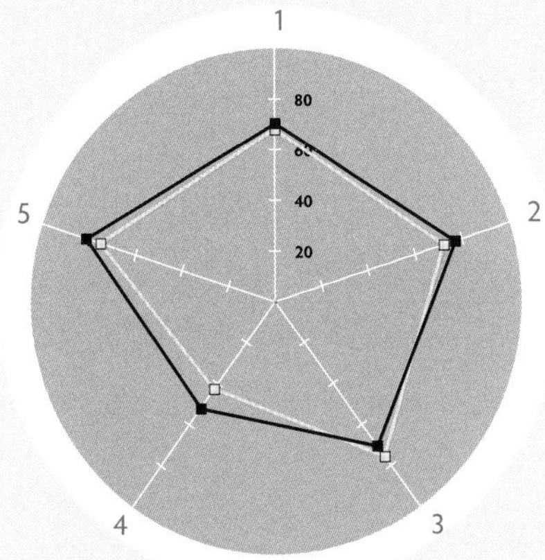

1. Institutional environment
2. Human capital, training and social inclusion
3. Regulatory and legal framework
4. Research and development
5. Usage of ICT

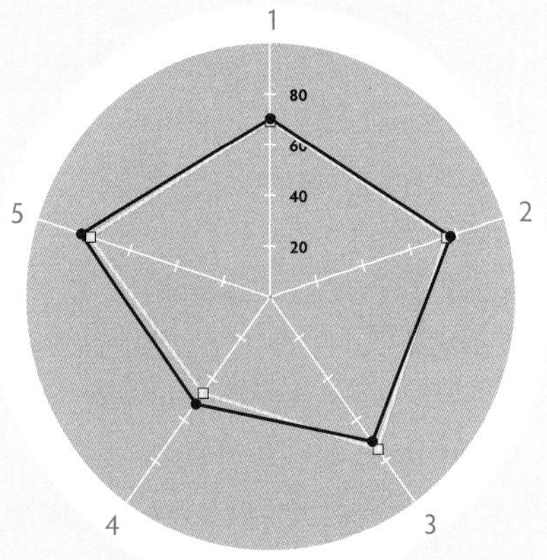

High-income ☐   Germany ●—●

Full democracies ☐   Germany ●—●

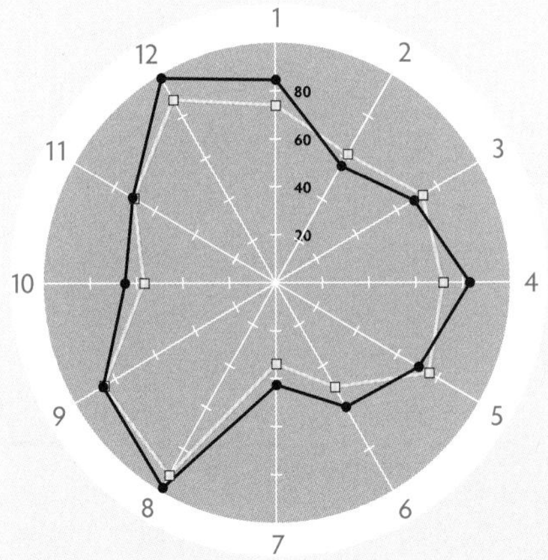

1. Good governance
2. Country policy assessment
3. Education
4. Social inclusion and equity policies
5. Doing business
6. R&D infrastructure
7. Patents and trademarks
8. Telephone communications
9. Mobile/cellular communications
10. Internet, computers and TV
11. Government ICT usage
12. Quality of the infrastructure

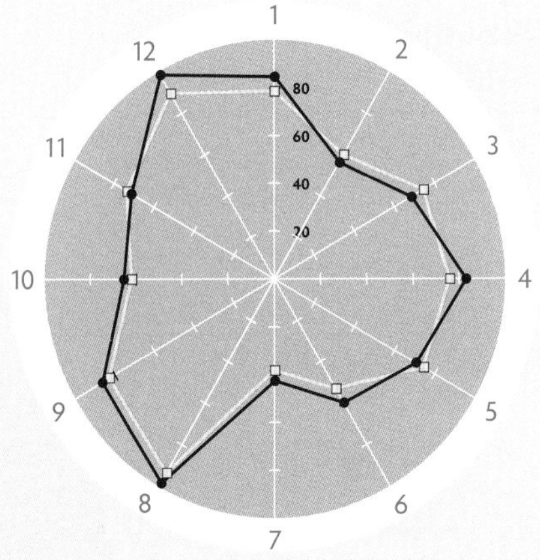

Country relative performance: Pillar subindexes[13]

In focus: Top priorities for policy reform[14]

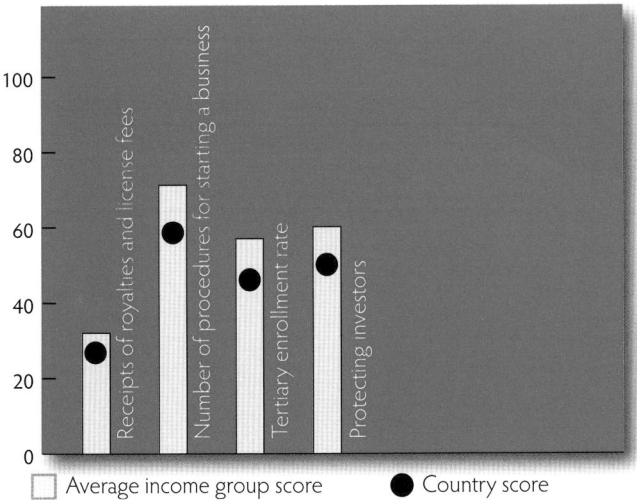

☐ Average income group score   ● Country score

In focus: Significant indicators above income group average[14]

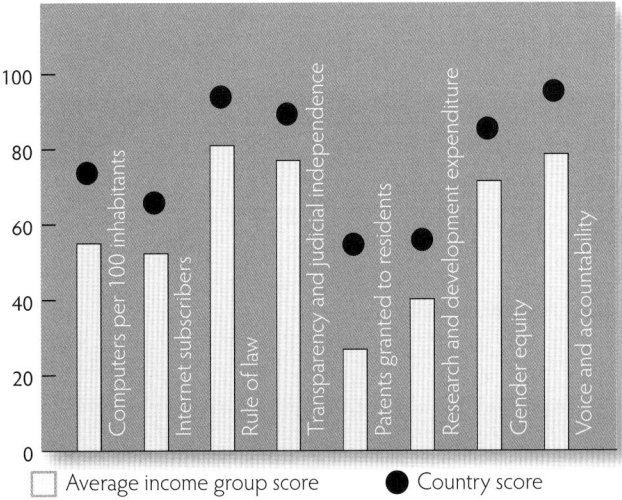

☐ Average income group score   ● Country score

[13]   This analysis compares pillar and subindex country scores with respect to the average scores of countries grouped according to income level and political regime (see notes 1 and 2).

[14]   Significant country indicators with the greatest distances above or below the income group average scores (i.e., achievements and targets for policy reform, respectively) were selected and ordered according to their Pearson correlation coefficients (highest first) with respect to the final index scores, to produce a ranking.

GHANA

Key selected indicators

Population (in millions)[3,4]	23.5
Population density (people per sq. km)[3,4]	103
GDP (current US$ billion, PPP)[3,5]	31
GDP per capita (current US$, PPP)[3,5]	1426
Current account balance (% of GDP)[3,5]	-12.8
Average life expectancy (years)[4,6]	59.7
Literacy rate[3,7]	65
Female labor force participation rate[4,6,8]	71.8
Unemployment rate (% of total labor force)[4,9]	10.1
CO_2 emissions (metric tons per capita)[4,10]	0.3

Income level:[1] **Low-income**

Political regime:[2] **Hybrid regime**

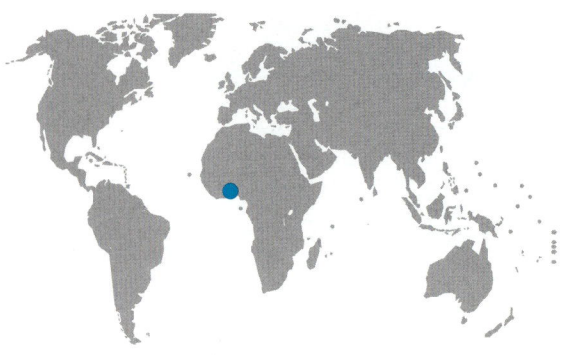

Rank	Score
77 out of 131	**46.6 out of 100**

Innovation Capacity Index

Pillars and pillar subindexes

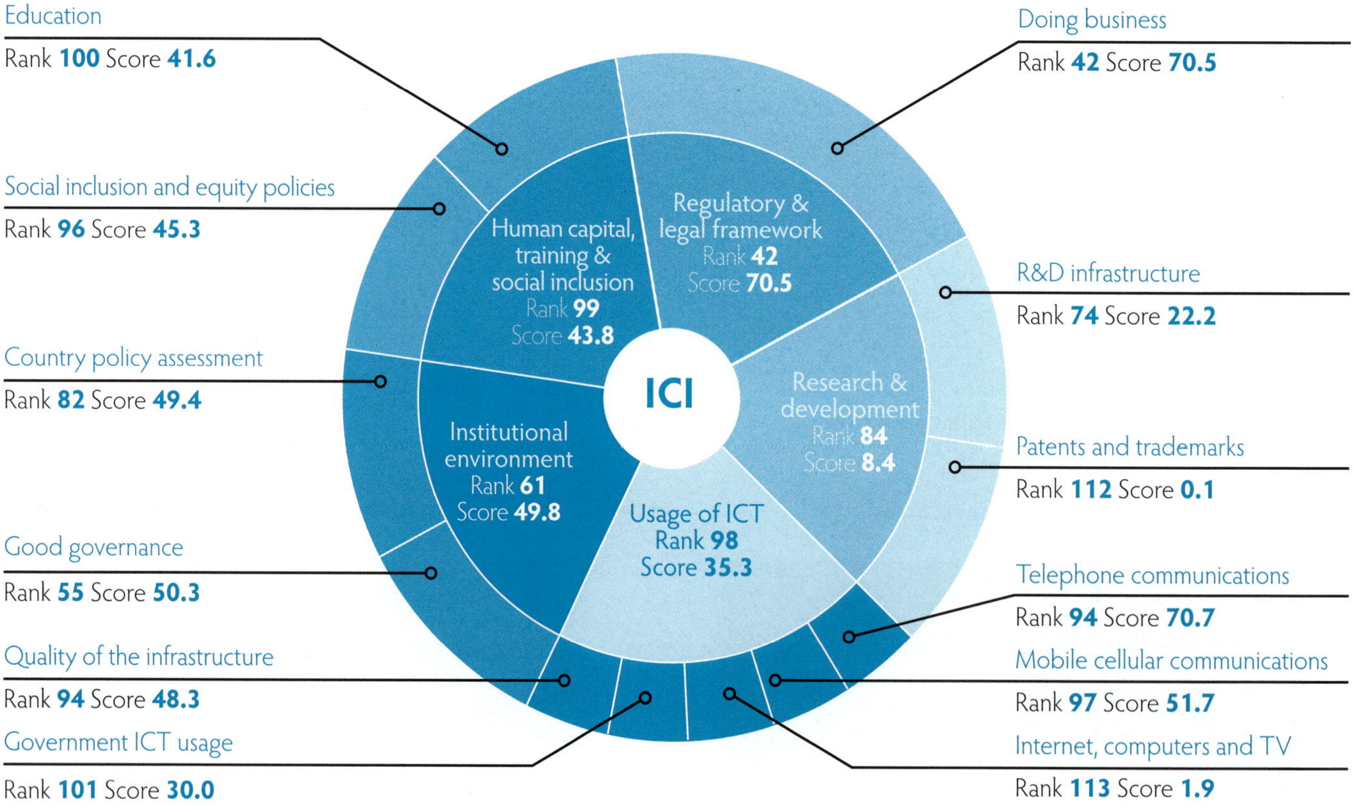

Education
Rank **100** Score **41.6**

Social inclusion and equity policies
Rank **96** Score **45.3**

Country policy assessment
Rank **82** Score **49.4**

Good governance
Rank **55** Score **50.3**

Quality of the infrastructure
Rank **94** Score **48.3**

Government ICT usage
Rank **101** Score **30.0**

Doing business
Rank **42** Score **70.5**

R&D infrastructure
Rank **74** Score **22.2**

Patents and trademarks
Rank **112** Score **0.1**

Telephone communications
Rank **94** Score **70.7**

Mobile cellular communications
Rank **97** Score **51.7**

Internet, computers and TV
Rank **113** Score **1.9**

Human capital, training & social inclusion
Rank **99**
Score **43.8**

Regulatory & legal framework
Rank **42**
Score **70.5**

Institutional environment
Rank **61**
Score **49.8**

Research & development
Rank **84**
Score **8.4**

Usage of ICT
Rank **98**
Score **35.3**

ICI

Index thermometer[11]

Pillar 1 32.1%   Pillar 2 28.2%   Pillar 3 30.3%   Pillar 4 1.8%   Pillar 5 7.6%

100%

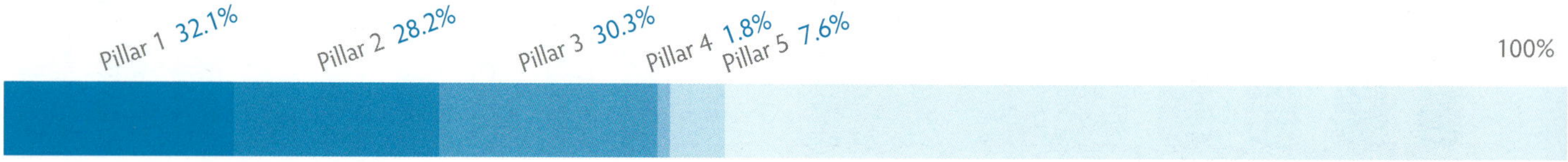

[1] Low-income countries are those that, according to the World Bank classification, have a gross national income (GNI) per capita of $935 or less. [2] Hybrid regimes are countries whose Democracy Index scores from 1 to 10 rounded to one decimal place are of 4 to 5.9. [3] 2007. [4] World Bank World Development Indicators. [5] IMF World Economic Outlook. [6] 2006. [7] UNESCO, % age above 15. [8] % ages 15–64. [9] 1999. [10] 2004. [11] The index thermometer shows the relative contribution of each pillar (in %) to the total index score.

GHANA

Country relative performance: **Index pillars**[12]

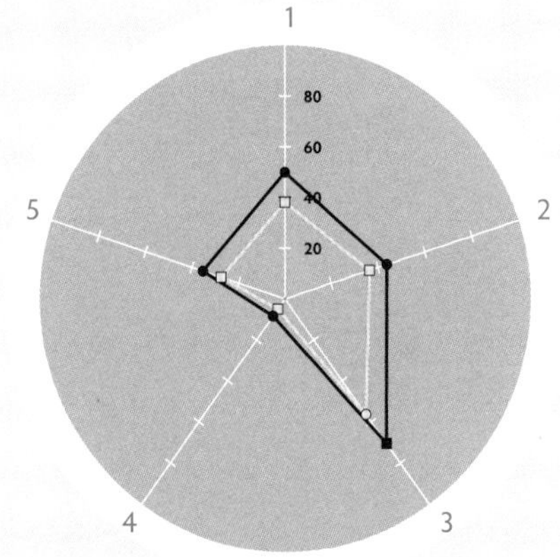

1. Institutional environment
2. Human capital, training and social inclusion
3. Regulatory and legal framework
4. Research and development
5. Usage of ICT

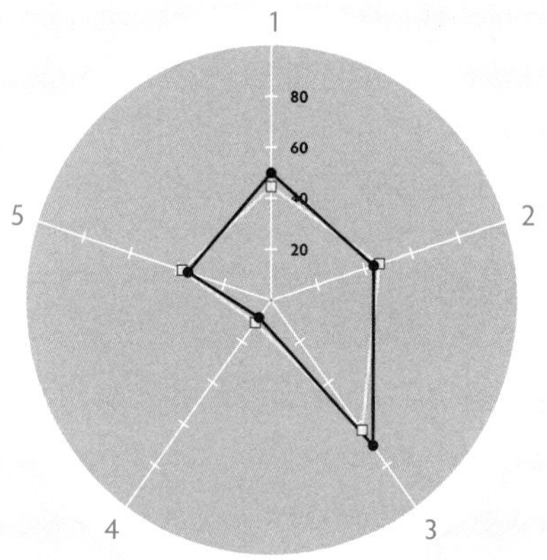

Low-income ☐ Ghana ●——

Hybrid regimes ☐ Ghana ●——

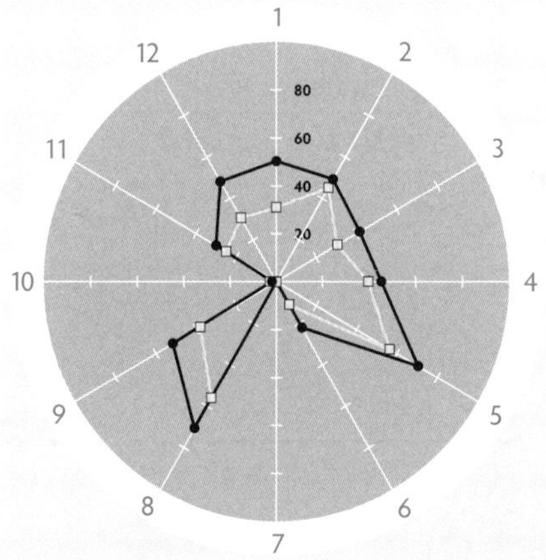

1. Good governance
2. Country policy assessment
3. Education
4. Social inclusion and equity policies
5. Doing business
6. R&D infrastructure
7. Patents and trademarks
8. Telephone communications
9. Mobile/cellular communications
10. Internet, computers and TV
11. Government ICT usage
12. Quality of the infrastructure

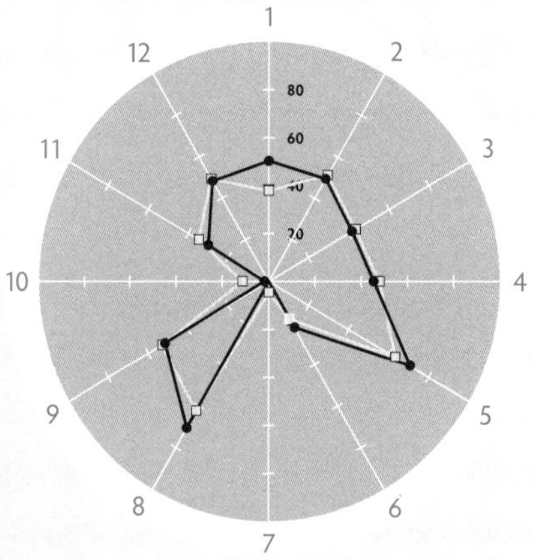

Country relative performance: **Pillar subindexes**[12]

In focus: **Top priorities for policy reform**[13]

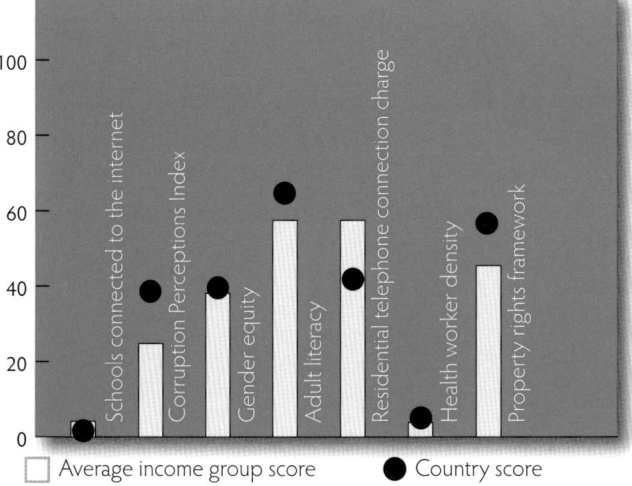

☐ Average income group score ● Country score

In focus: **Significant indicators above income group average**[13]

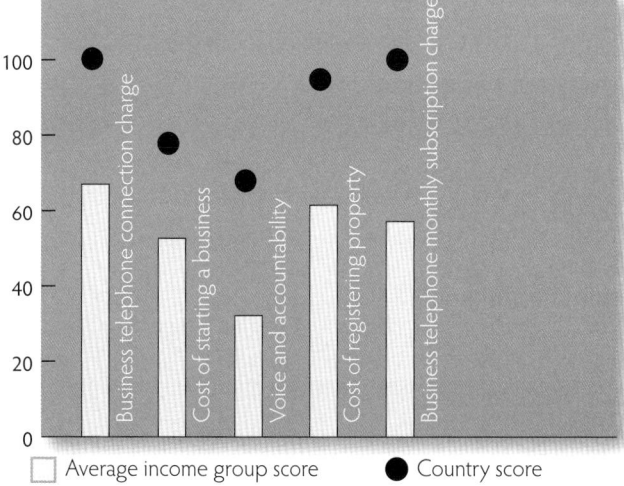

☐ Average income group score ● Country score

[12] This analysis compares pillar and subindex country scores with respect to the average scores of countries grouped according to income level and political regime (see notes 1 and 2).

[13] Significant country indicators with the greatest distances above or below the income group average scores (i.e., achievements and targets for policy reform, respectively) were selected and ordered according to their Pearson correlation coefficients (highest first) with respect to the final index scores, to produce a ranking.

GREECE

Key selected indicators

Population (in millions)[3,4]	11.2
Population density (people per sq. km)[3,4]	87
GDP (current US$ billion, PPP)[3,5]	325
GDP per capita (current US$, PPP)[3,5]	29172
Current account balance (% of GDP)[3,5]	-13.9
Average life expectancy (years)[4,6]	79.4
Literacy rate[3,7]	97
Female labor force participation rate[4,6,8]	56.9
Unemployment rate (% of total labor force)[4,9]	9.6
CO_2 emissions (metric tons per capita)[4,10]	8.7

Income level:[1] **High-income**
Political regime:[2] **Full democracy**

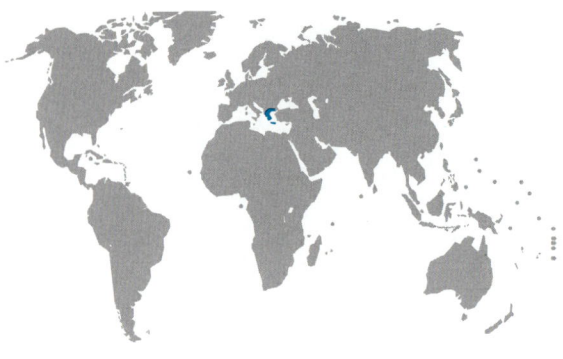

Rank **62 out of 131**   Score **50.2 out of 100**

Innovation Capacity Index

Pillars and pillar subindexes

Education
Rank **2** Score **84.2**

Social inclusion and equity policies
Rank **23** Score **71.5**

Country policy assessment
Rank **107** Score **42.4**

Good governance
Rank **46** Score **57.5**

Quality of the infrastructure
Rank **19** Score **93.1**

Government ICT usage
Rank **42** Score **57.2**

Doing business
Rank **110** Score **54.1**

R&D infrastructure
Rank **33** Score **37.4**

Patents and trademarks
Rank **41** Score **8.3**

Telephone communications
Rank **9** Score **96.0**

Mobile cellular communications
Rank **25** Score **87.9**

Internet, computers and TV
Rank **53** Score **23.8**

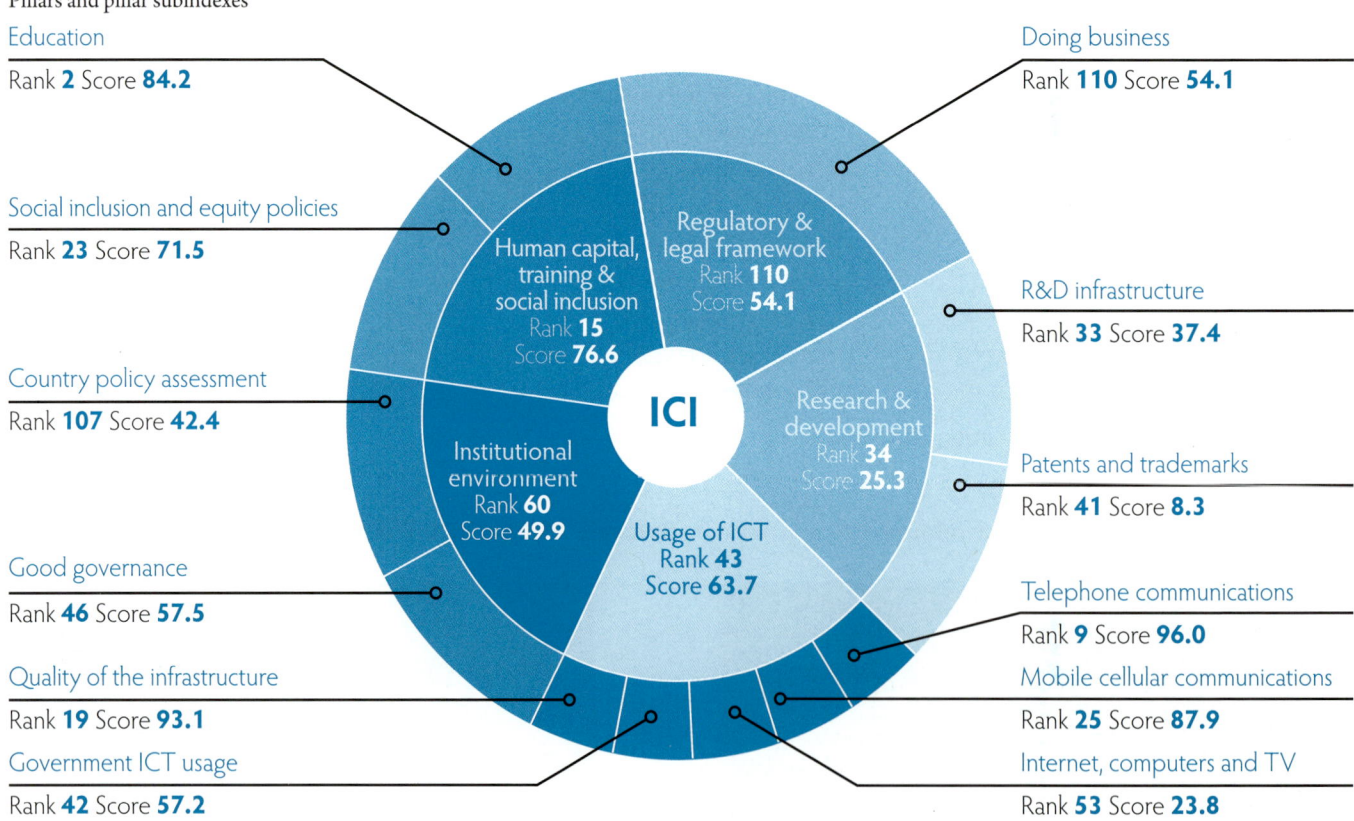

Index thermometer[11]

Pillar 1 10.0% Pillar 2 15.3% Pillar 3 21.6% Pillar 4 15.1% Pillar 5 38.1% 100%

[1] High-income countries are those that, according to the World Bank classification, have a gross national income (GNI) per capita of $11,456 or more.
[2] Full democracies are countries whose Democracy Index scores from 1 to 10 rounded to one decimal place are from 8 to 10. [3] 2007. [4] World Bank World Development Indicators. [5] IMF World Economic Outlook. [6] 2006. [7] UNESCO, % age above 15. [8] % ages 15–64. [9] 2005 [10] 2004. [11] The index thermometer shows the relative contribution of each pillar (in %) to the total index score.

GREECE

Country relative performance: **Index pillars**[12]

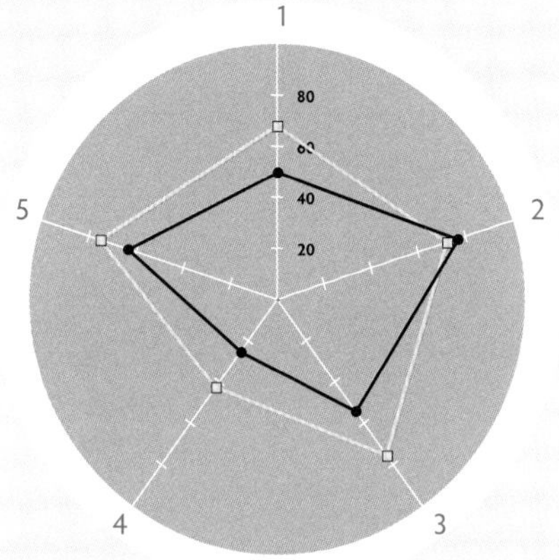

1. Institutional environment
2. Human capital, training and social inclusion
3. Regulatory and legal framework
4. Research and development
5. Usage of ICT

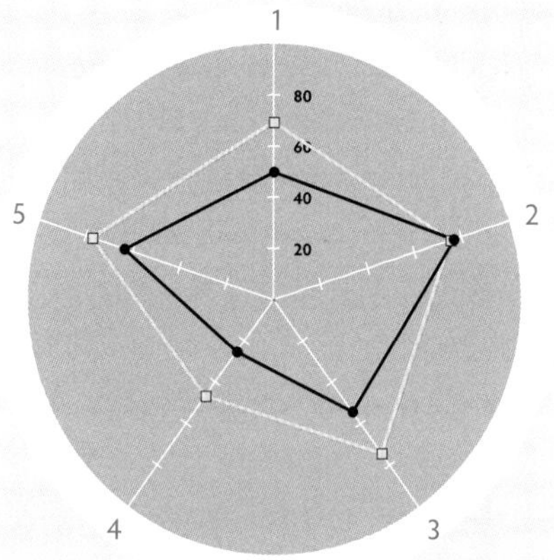

High-income ···□··· Greece —●—

Full democracies ···□··· Greece —●—

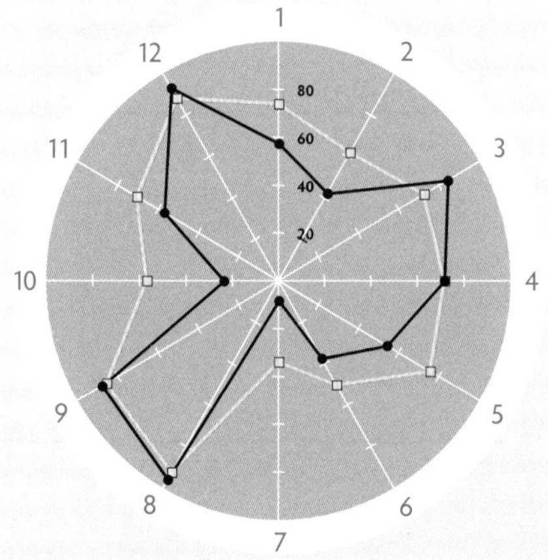

1. Good governance
2. Country policy assessment
3. Education
4. Social inclusion and equity policies
5. Doing business
6. R&D infrastructure
7. Patents and trademarks
8. Telephone communications
9. Mobile/cellular communications
10. Internet, computers and TV
11. Government ICT usage
12. Quality of the infrastructure

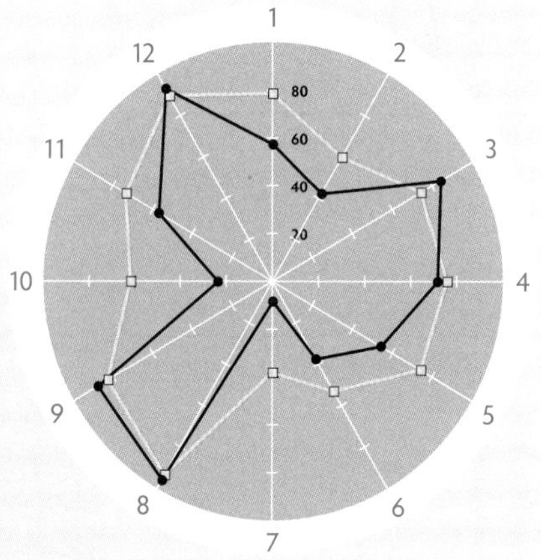

Country relative performance: **Pillar subindexes**[12]

In focus: **Top priorities for policy reform**[13]

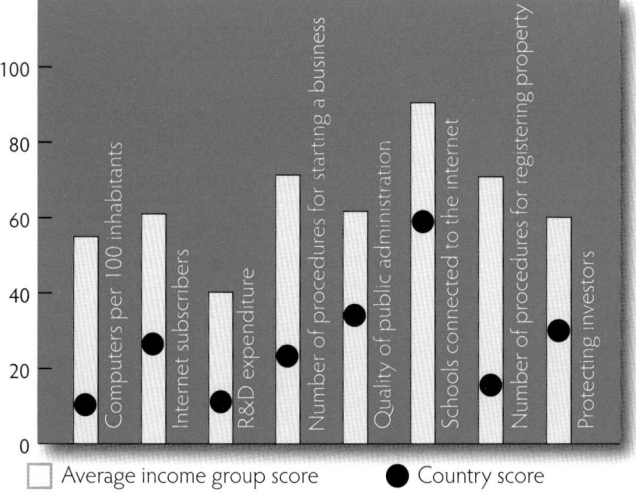

□ Average income group score ● Country score

In focus: **Significant indicators above income group average**[13]

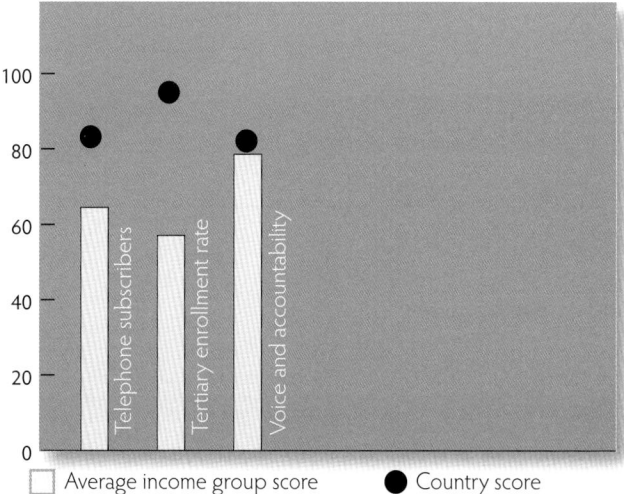

□ Average income group score ● Country score

[12] This analysis compares pillar and subindex country scores with respect to the average scores of countries grouped according to income level and political regime (see notes 1 and 2).

[13] Significant country indicators with the greatest distances above or below the income group average scores (i.e., achievements and targets for policy reform, respectively) were selected and ordered according to their Pearson correlation coefficients (highest first) with respect to the final index scores, to produce a ranking.

HUNGARY

Key selected indicators

Population (in millions)[3,4]	10.1
Population density (people per sq. km)[3,4]	112
GDP (current US$ billion, PPP)[3,5]	191
GDP per capita (current US$, PPP)[3,5]	19027
Current account balance (% of GDP)[3,5]	-5.6
Average life expectancy (years)[4,6]	73.1
Literacy rate[4,7]	99
Female labor force participation rate[4,6,8]	53.7
Unemployment rate (% of total labor force)[4,9]	7.2
CO_2 emissions (metric tons per capita)[4,10]	5.7

Income level:[1] **High-income**

Political regime:[2] **Flawed democracy**

Innovation Capacity Index

Rank **41 out of 131**

Score **55.6 out of 100**

Pillars and pillar subindexes

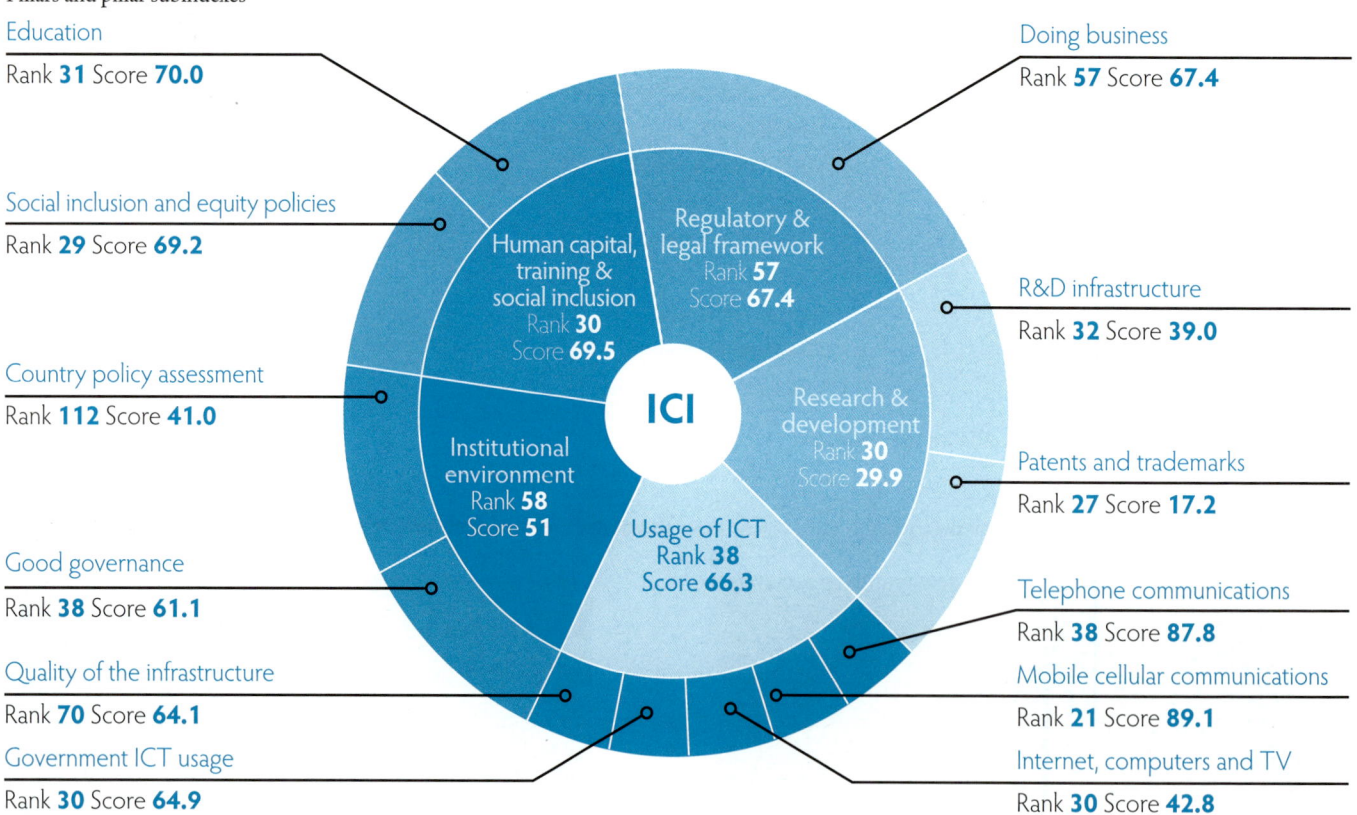

Education
Rank **31** Score **70.0**

Social inclusion and equity policies
Rank **29** Score **69.2**

Country policy assessment
Rank **112** Score **41.0**

Good governance
Rank **38** Score **61.1**

Quality of the infrastructure
Rank **70** Score **64.1**

Government ICT usage
Rank **30** Score **64.9**

Doing business
Rank **57** Score **67.4**

R&D infrastructure
Rank **32** Score **39.0**

Patents and trademarks
Rank **27** Score **17.2**

Telephone communications
Rank **38** Score **87.8**

Mobile cellular communications
Rank **21** Score **89.1**

Internet, computers and TV
Rank **30** Score **42.8**

Human capital, training & social inclusion
Rank **30** Score **69.5**

Regulatory & legal framework
Rank **57** Score **67.4**

Institutional environment
Rank **58** Score **51**

Research & development
Rank **30** Score **29.9**

Usage of ICT
Rank **38** Score **66.3**

ICI

Index thermometer[11]

Pillar 1 13.8%   Pillar 2 18.7%   Pillar 3 24.2%   Pillar 4 13.5%   Pillar 5 29.8%   100%

[1] High-income countries are those that, according to the World Bank classification, have a gross national income (GNI) per capita of $11,456 or more.
[2] Flawed democracies are countries whose Democracy Index scores from 1 to 10 rounded to one decimal place are of 6 to 7.9.   [3] 2007.   [4] World Bank World Development Indicators.   [5] IMF World Economic Outlook.   [6] 2006.   [7] UNESCO, % age above 15.   [8] % ages 15–64.   [9] 2005.   [10] 2004.   [11] The index thermometer shows the relative contribution of each pillar (in %) to the total index score.

HUNGARY

Country relative performance: **Index pillars**[12]

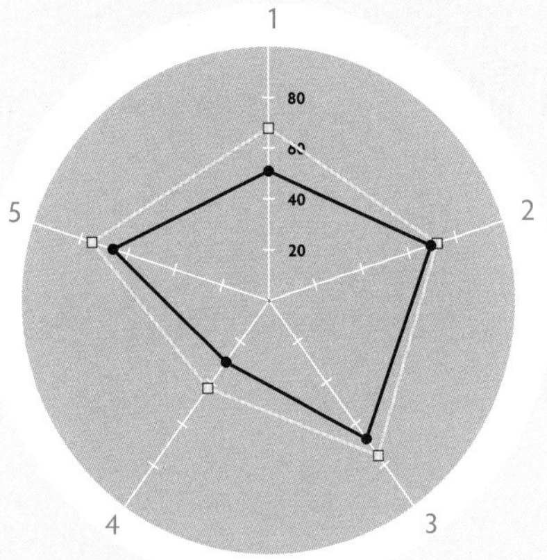

1. Institutional environment
2. Human capital, training and social inclusion
3. Regulatory and legal framework
4. Research and development
5. Usage of ICT

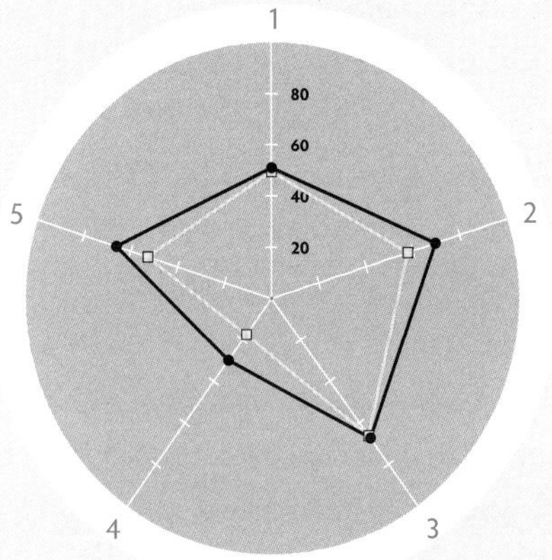

High-income · · · □ · · ·   Hungary ●━━━━●

Flawed democracies · · · □ · · ·   Hungary ●━━━━●

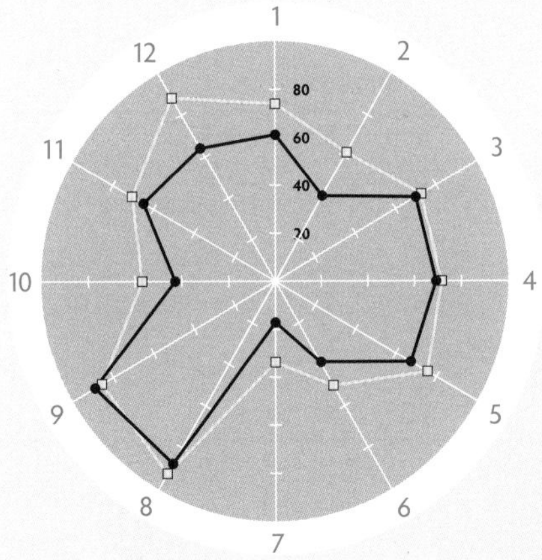

1. Good governance
2. Country policy assessment
3. Education
4. Social inclusion and equity policies
5. Doing business
6. R&D infrastructure
7. Patents and trademarks
8. Telephone communications
9. Mobile/cellular communications
10. Internet, computers and TV
11. Government ICT usage
12. Quality of the infrastructure

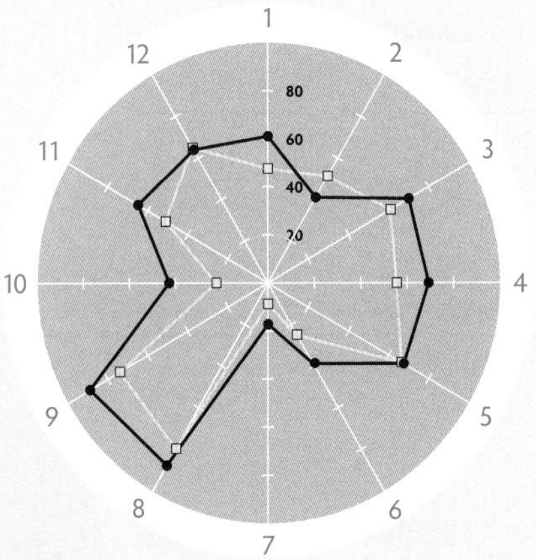

Country relative performance: **Pillar subindexes**[12]

In focus: **Top priorities for policy reform**[13]

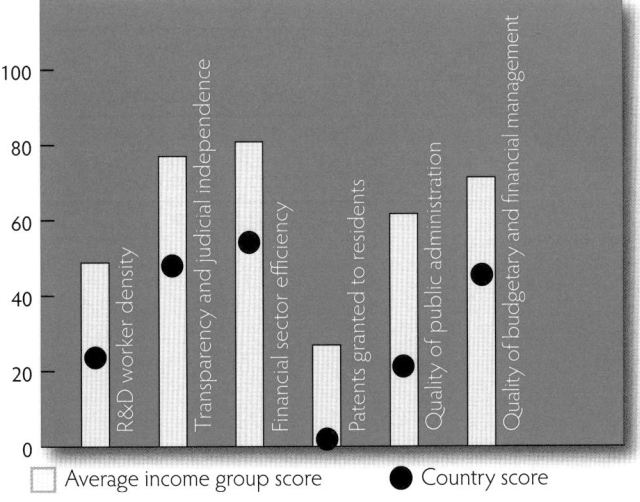

R&D worker density · Transparency and judicial independence · Financial sector efficiency · Patents granted to residents · Quality of public administration · Quality of budgetary and financial management

□ Average income group score   ● Country score

In focus: **Significant indicators above income group average**[13]

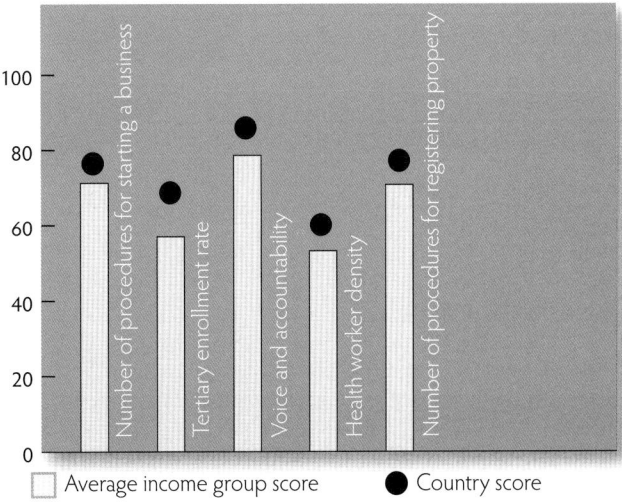

Number of procedures for starting a business · Tertiary enrollment rate · Voice and accountability · Health worker density · Number of procedures for registering property

□ Average income group score   ● Country score

---

[12] This analysis compares pillar and subindex country scores with respect to the average scores of countries grouped according to income level and political regime (see notes 1 and 2).

[13] Significant country indicators with the greatest distances above or below the income group average scores (i.e., achievements and targets for policy reform, respectively) were selected and ordered according to their Pearson correlation coefficients (highest first) with respect to the final index scores, to produce a ranking.

INDIA

Key selected indicators

Population (in millions)[3,4]	1123.3
Population density (people per sq. km)[3,4]	378
GDP (current US$ billion, PPP)[3,5]	2989
GDP per capita (current US$, PPP)[3,5]	2659
Current account balance (% of GDP)[3,5]	-1.8
Average life expectancy (years)[4,6]	64.5
Literacy rate[3,7]	66
Female labor force participation rate[4,6,8]	35.9
Unemployment rate (% of total labor force)[4,9]	5.0
CO_2 emissions (metric tons per capita)[4,9]	1.2

Income level:[1] **Lower-middle-income**
Political regime:[2] **Flawed democracy**

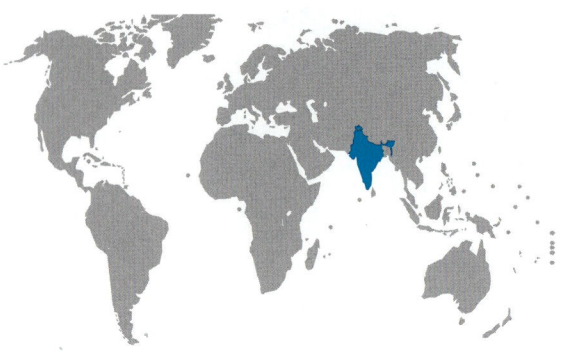

Rank	Score
85 out of 131	**45.6 out of 100**

Innovation Capacity Index

Pillars and pillar subindexes

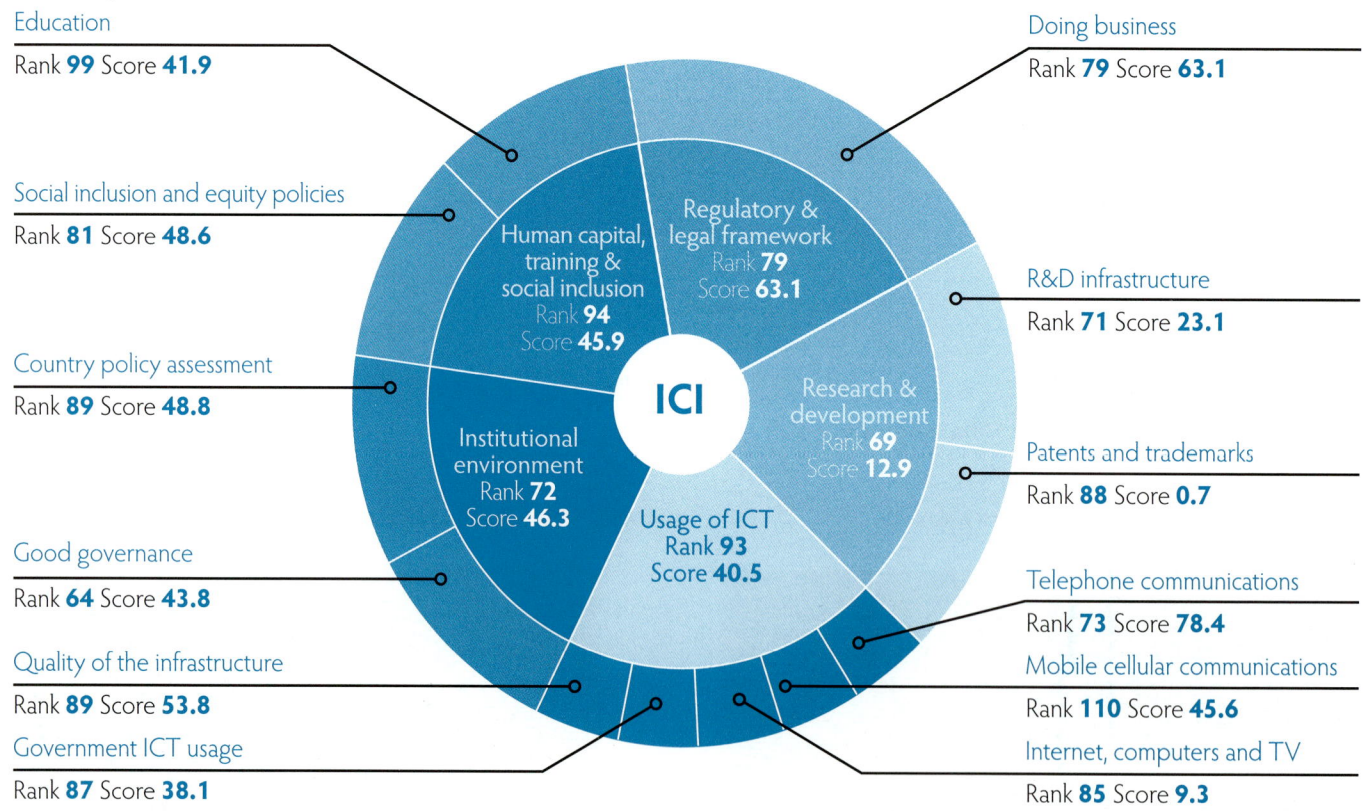

Education
Rank **99** Score **41.9**

Social inclusion and equity policies
Rank **81** Score **48.6**

Country policy assessment
Rank **89** Score **48.8**

Good governance
Rank **64** Score **43.8**

Quality of the infrastructure
Rank **89** Score **53.8**

Government ICT usage
Rank **87** Score **38.1**

Doing business
Rank **79** Score **63.1**

R&D infrastructure
Rank **71** Score **23.1**

Patents and trademarks
Rank **88** Score **0.7**

Telephone communications
Rank **73** Score **78.4**

Mobile cellular communications
Rank **110** Score **45.6**

Internet, computers and TV
Rank **85** Score **9.3**

Index thermometer[10]

Pillar 1 30.4% Pillar 2 30.2% Pillar 3 27.6% Pillar 4 2.8% Pillar 5 8.9%   100%

[1] Lower-middle-income countries are those that, according to the World Bank classification, have a gross national income (GNI) per capita of $936 to $3,705. [2] Flawed democracies are countries whose Democracy Index scores from 1 to 10 rounded to one decimal place are of 6 to 7.9. [3] 2007. [4] World Bank World Development Indicators. [5] IMF World Economic Outlook. [6] 2006. [7] UNESCO, % age above 15. [8] % ages 15–64. [9] 2004. [10] The index thermometer shows the relative contribution of each pillar (in %) to the total index score.

INDIA

Country relative performance: **Index pillars**[11]

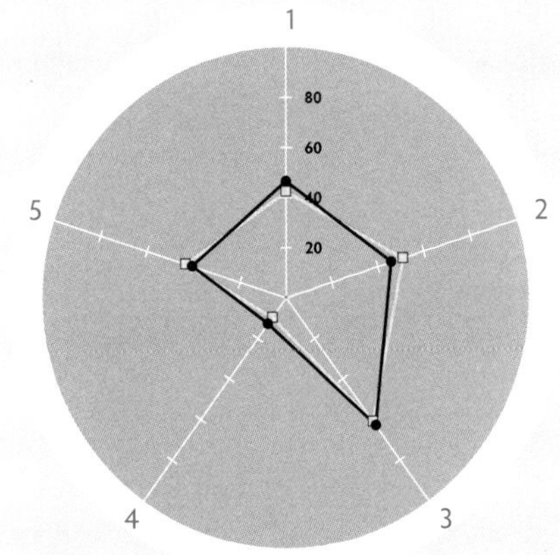

1. Institutional environment
2. Human capital, training and social inclusion
3. Regulatory and legal framework
4. Research and development
5. Usage of ICT

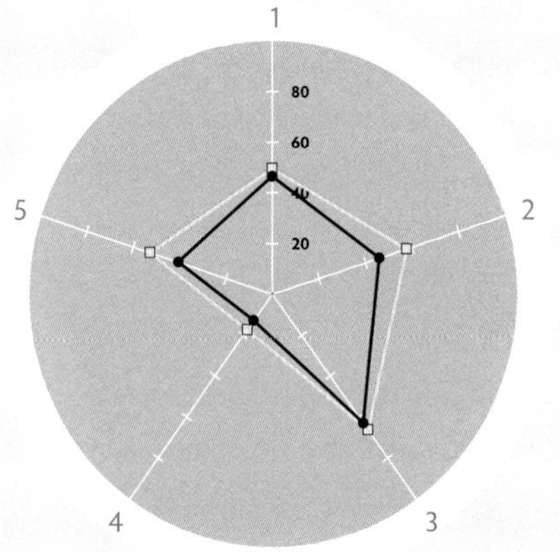

Lower-middle-income ☐   India ●—

Flawed democracies ☐   India ●—

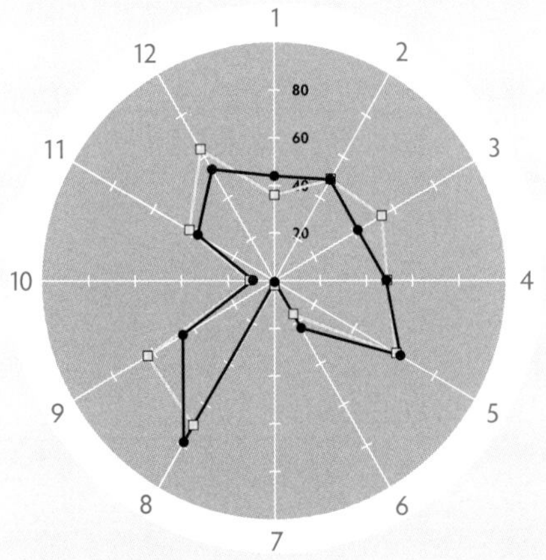

1. Good governance
2. Country policy assessment
3. Education
4. Social inclusion and equity policies
5. Doing business
6. R&D infrastructure
7. Patents and trademarks
8. Telephone communications
9. Mobile/cellular communications
10. Internet, computers and TV
11. Government ICT usage
12. Quality of the infrastructure

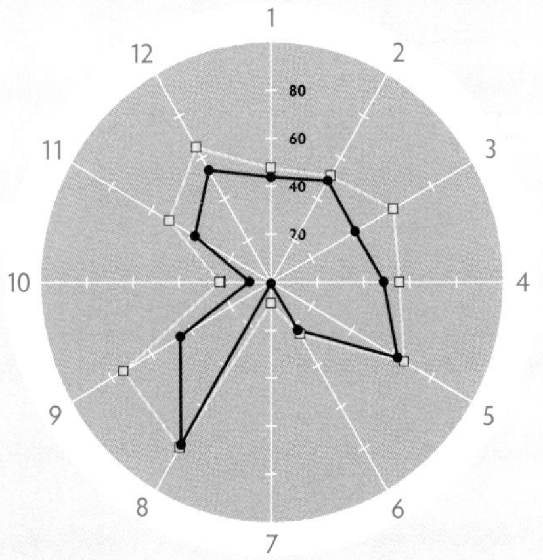

Country relative performance: **Pillar subindexes**[11]

In focus: **Top priorities for policy reform**[12]

In focus: **Significant indicators above income group average**[12]

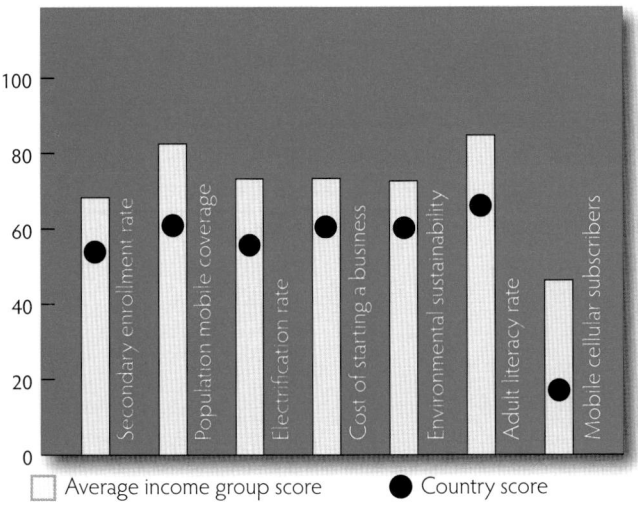

☐ Average income group score   ● Country score

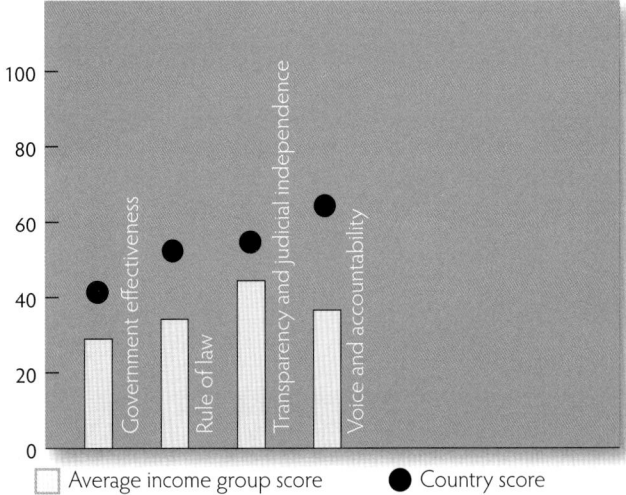

☐ Average income group score   ● Country score

---

[11] This analysis compares pillar and subindex country scores with respect to the average scores of countries grouped according to income level and political regime (see notes 1 and 2).

[12] Significant country indicators with the greatest distances above or below the income group average scores (i.e., achievements and targets for policy reform, respectively) were selected and ordered according to their Pearson correlation coefficients (highest first) with respect to the final index scores, to produce a ranking.

INDONESIA

Key selected indicators

Population (in millions)[3,4]	225.6
Population density (people per sq. km)[3,4]	125
GDP (current US$ billion, PPP)[3,5]	838
GDP per capita (current US$, PPP)[3,5]	3725
Current account balance (% of GDP)[3,5]	2.5
Average life expectancy (years)[4,6]	68.2
Literacy rate[3,7]	91
Female labor force participation rate[4,6,8]	53.3
Unemployment rate (% of total labor force)[4,6]	10.3
CO_2 emissions (metric tons per capita)[4,9]	1.7

Income level:[1] **Lower-middle-income**

Political regime:[2] **Flawed democracy**

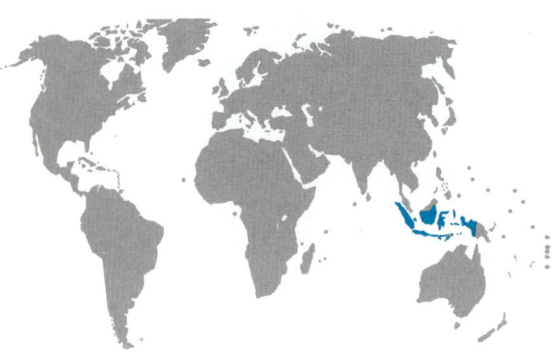

	Rank	Score
	88 out of 131	**44.9 out of 100**

Innovation Capacity Index

Pillars and pillar subindexes

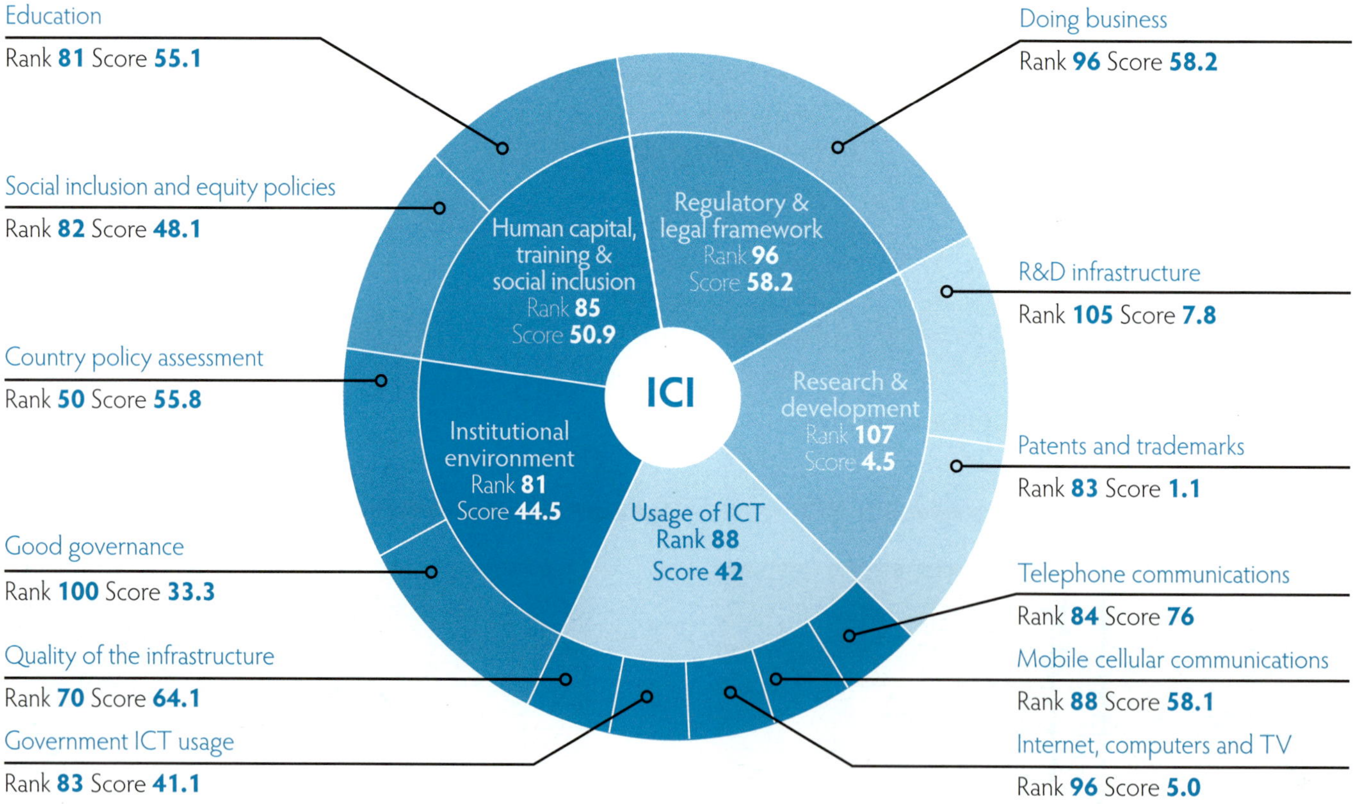

Education
Rank **81** Score **55.1**

Social inclusion and equity policies
Rank **82** Score **48.1**

Country policy assessment
Rank **50** Score **55.8**

Good governance
Rank **100** Score **33.3**

Quality of the infrastructure
Rank **70** Score **64.1**

Government ICT usage
Rank **83** Score **41.1**

Doing business
Rank **96** Score **58.2**

R&D infrastructure
Rank **105** Score **7.8**

Patents and trademarks
Rank **83** Score **1.1**

Telephone communications
Rank **84** Score **76**

Mobile cellular communications
Rank **88** Score **58.1**

Internet, computers and TV
Rank **96** Score **5.0**

(Central diagram)
ICI

Human capital, training & social inclusion
Rank **85** Score **50.9**

Regulatory & legal framework
Rank **96** Score **58.2**

Research & development
Rank **107** Score **4.5**

Institutional environment
Rank **81** Score **44.5**

Usage of ICT
Rank **88** Score **42**

Index thermometer[10]

Pillar 1 29.7% Pillar 2 34.0% Pillar 3 25.9% Pillar 4 1.0% Pillar 5 9.3%   100%

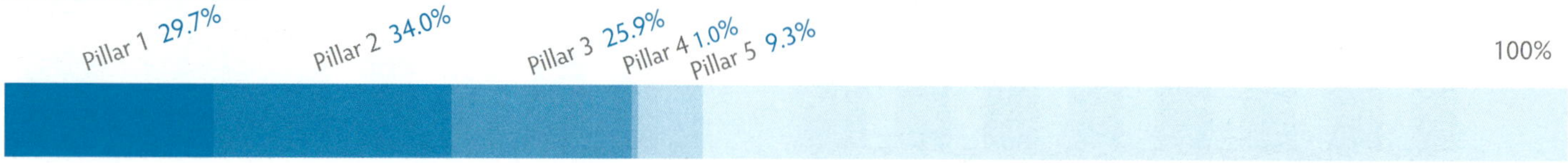

[1] Lower-middle-income countries are those that, according to the World Bank classification, have a gross national income (GNI) per capita of $936 to $3,705.   [2] Flawed democracies are countries whose Democracy Index scores from 1 to 10 rounded to one decimal place are of 6 to 7.9.   [3] 2007.   [4] World Bank World Development Indicators.   [5] IMF World Economic Outlook.   [6] 2006.   [7] UNESCO, % age above 15.   [8] % ages 15–64.   [9] 2004.   [10] The index thermometer shows the relative contribution of each pillar (in %) to the total index score.

INDONESIA

Country relative performance: **Index pillars**[11]

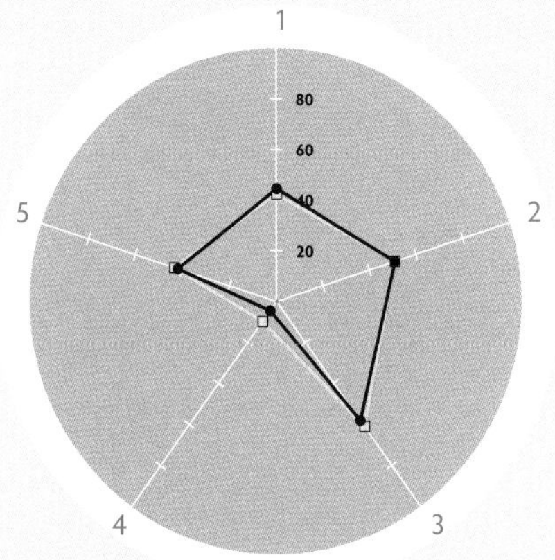

1. Institutional environment
2. Human capital, training and social inclusion
3. Regulatory and legal framework
4. Research and development
5. Usage of ICT

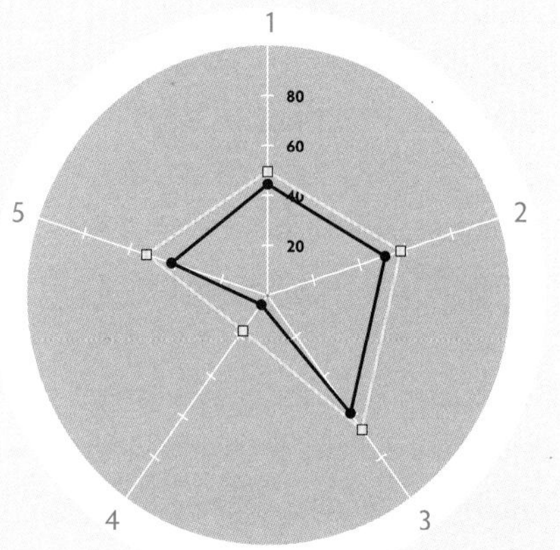

Lower-middle-income ⬚ · · · ·   Indonesia ●——

Flawed democracies ⬚ · · · ·   Indonesia ●——

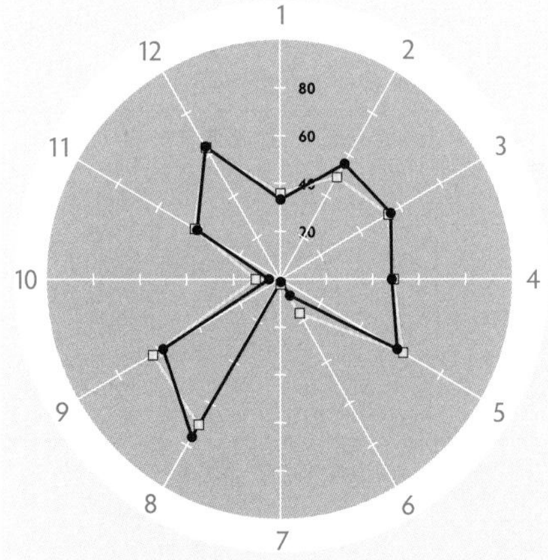

1. Good governance
2. Country policy assessment
3. Education
4. Social inclusion and equity policies
5. Doing business
6. R&D infrastructure
7. Patents and trademarks
8. Telephone communications
9. Mobile/cellular communications
10. Internet, computers and TV
11. Government ICT usage
12. Quality of the infrastructure

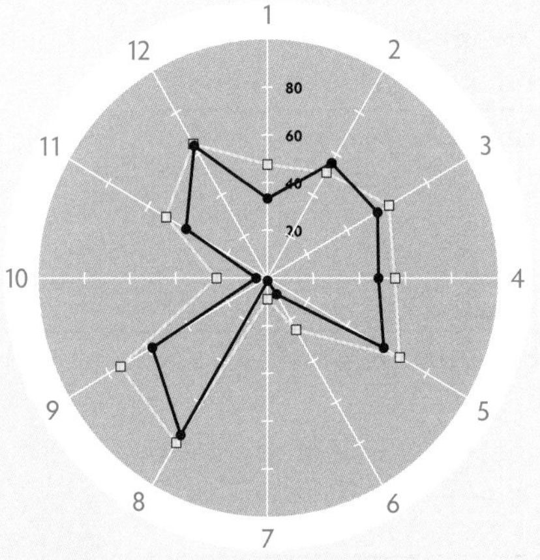

Country relative performance: **Pillar subindexes**[11]

In focus: **Top priorities for policy reform**[12]

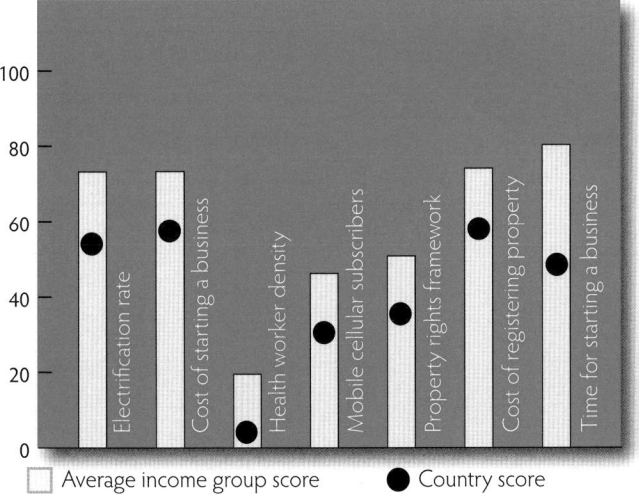

⬚ Average income group score   ● Country score

In focus: **Significant indicators above income group average**[12]

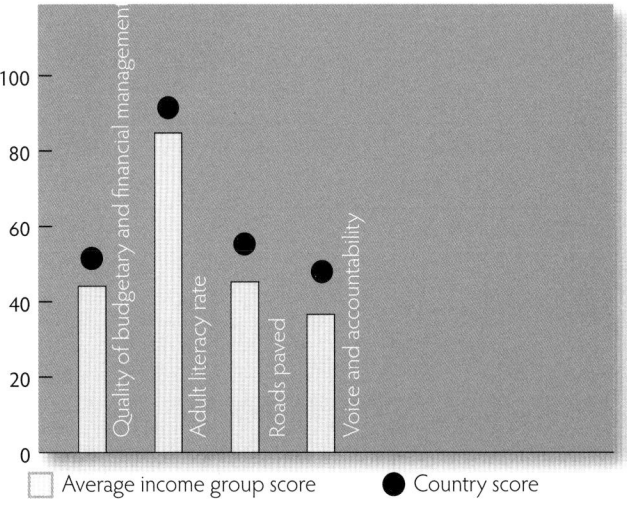

⬚ Average income group score   ● Country score

[11] This analysis compares pillar and subindex country scores with respect to the average scores of countries grouped according to income level and political regime (see notes 1 and 2).

[12] Significant country indicators with the greatest distances above or below the income group average scores (i.e., achievements and targets for policy reform, respectively) were selected and ordered according to their Pearson correlation coefficients (highest first) with respect to the final index scores, to produce a ranking.

IRAN, ISLAMIC REPUBLIC OF

Key selected indicators

Population (in millions)[3,4]	71.0
Population density (people per sq. km)[3,4]	44
GDP (current US$ billion, PPP)[3,5]	753
GDP per capita (current US$, PPP)[3,5]	10624
Current account balance (% of GDP)[3,5]	10.4
Average life expectancy (years)[4,6]	70.7
Literacy rate[3,7]	85
Female labor force participation rate[4,6,8]	41.9
Unemployment rate (% of total labor force)[4,9]	11.5
CO_2 emissions (metric tons per capita)[4,10]	6.4

Income level:[1] **Lower-middle-income**

Political regime:[2] **Authoritarian regime**

Innovation Capacity Index

Rank
84 out of 131

Score
45.7 out of 100

Pillars and pillar subindexes

Education
Rank **78** Score **56.5**

Social inclusion and equity policies
Rank **90** Score **46.4**

Country policy assessment
Rank **63** Score **52.7**

Good governance
Rank **114** Score **28.2**

Quality of the infrastructure
Rank **52** Score **78.2**

Government ICT usage
Rank **85** Score **40.7**

Doing business
Rank **95** Score **58.9**

R&D infrastructure
Rank **58** Score **28.4**

Patents and trademarks
Rank **66** Score **2.2**

Telephone communications
Rank **63** Score **81.7**

Mobile cellular communications
Rank **90** Score **57.9**

Internet, computers and TV
Rank **50** Score **24.3**

Center of wheel:

ICI

Human capital, training & social inclusion
Rank **86**
Score **50.4**

Regulatory & legal framework
Rank **95**
Score **58.9**

Institutional environment
Rank **101**
Score **39.8**

Research & development
Rank **59**
Score **16.5**

Usage of ICT
Rank **67**
Score **51.9**

Index thermometer[11]

Pillar 1 26.1%　Pillar 2 33.1%　Pillar 3 25.8%　Pillar 4 3.6%　Pillar 5 11.4%　　100%

[1] Lower-middle-income countries are those that, according to the World Bank classification, have a gross national income (GNI) per capita of $936 to $3,705.　[2] Authoritarian regimes are countries whose Democracy Index scores from 1 to 10 rounded to one decimal place are less than 4.　[3] 2007.　[4] World Bank World Development Indicators.　[5] IMF World Economic Outlook.　[6] 2006.　[7] UNESCO, % age above 15.　[8] % ages 15–64.　[9] 2005.　[10] 2004.　[11] The index thermometer shows the relative contribution of each pillar (in %) to the total index score.

IRAN, ISLAMIC REPUBLIC OF

Country relative performance: **Index pillars**[12]

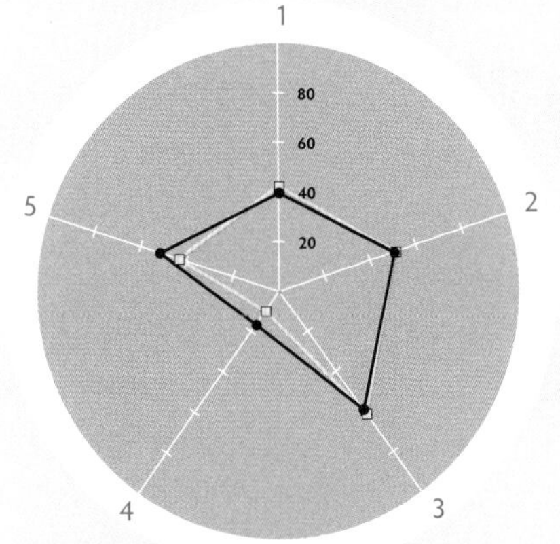

1. Institutional environment
2. Human capital, training and social inclusion
3. Regulatory and legal framework
4. Research and development
5. Usage of ICT

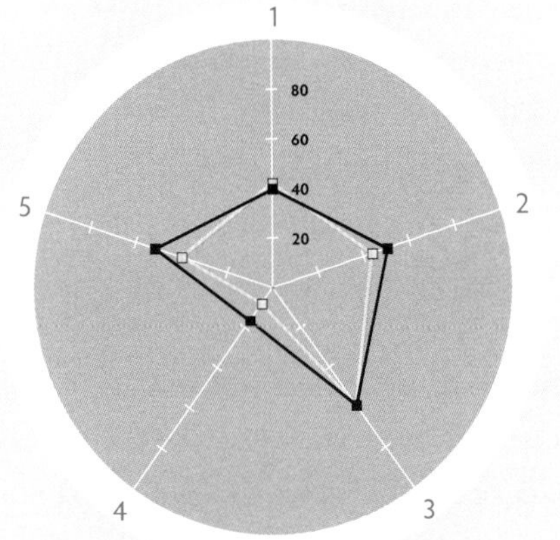

Lower-middle-income ☐   Iran ●———

Authoritarian regimes ☐   Iran ●———

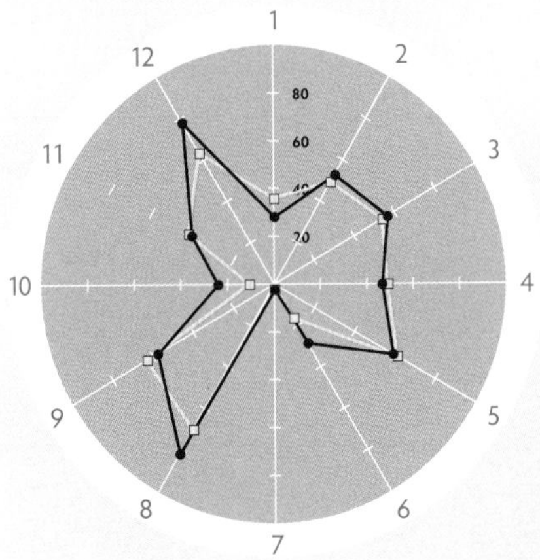

1. Good governance
2. Country policy assessment
3. Education
4. Social inclusion and equity policies
5. Doing business
6. R&D infrastructure
7. Patents and trademarks
8. Telephone communications
9. Mobile/cellular communications
10. Internet, computers and TV
11. Government ICT usage
12. Quality of the infrastructure

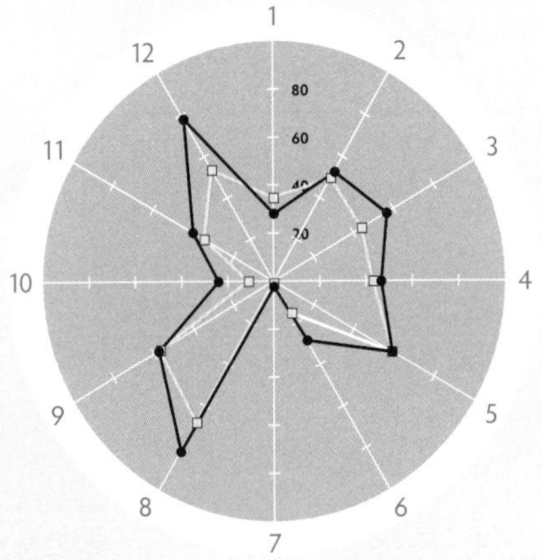

Country relative performance: **Pillar subindexes**[12]

In focus: **Top priorities for policy reform**[13]

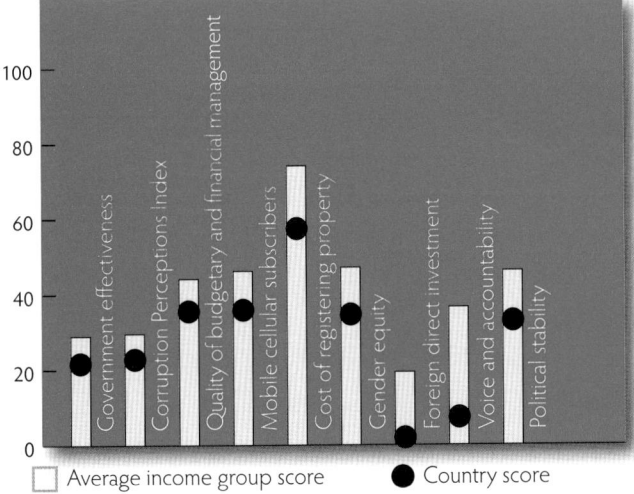

☐ Average income group score   ● Country score

In focus: **Significant indicators above income group average**[13]

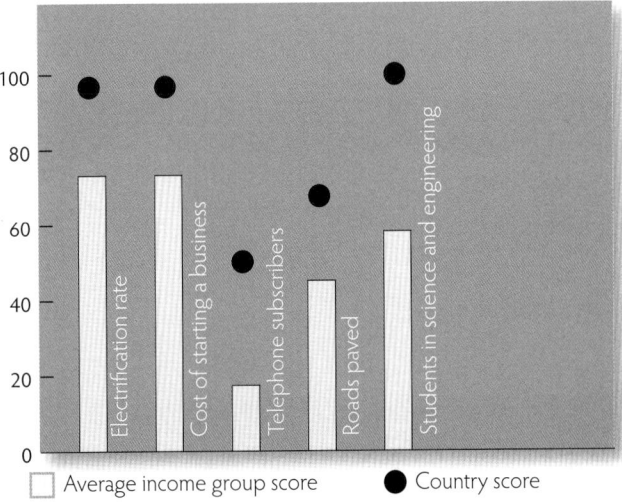

☐ Average income group score   ● Country score

[12] This analysis compares pillar and subindex country scores with respect to the average scores of countries grouped according to income level and political regime (see notes 1 and 2).

[13] Significant country indicators with the greatest distances above or below the income group average scores (i.e., achievements and targets for policy reform, respectively) were selected and ordered according to their Pearson correlation coefficients (highest first) with respect to the final index scores, to produce a ranking.

IRELAND

Key selected indicators

Population (in millions)[3,4]	4.4
Population density (people per sq. km)[3,4]	63
GDP (current US$ billion, PPP)[3,5]	186
GDP per capita (current US$, PPP)[3,5]	43144
Current account balance (% of GDP)[3,5]	-4.5
Average life expectancy (years)[4,6]	79.4
Literacy rate[7,8]	99
Female labor force participation rate[4,6,9]	63.3
Unemployment rate (% of total labor force)[4,10]	4.3
CO_2 emissions (metric tons per capita)[4,11]	10.4

Income level:[1] **High-income**

Political regime:[2] **Full democracy**

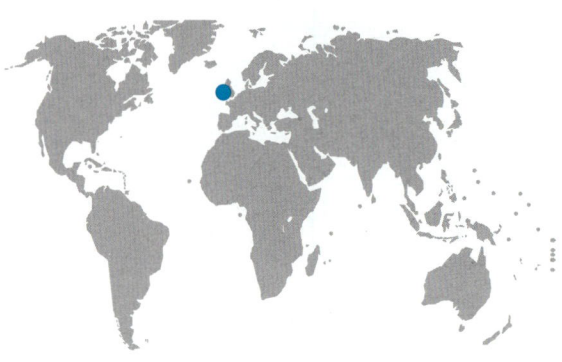

Rank	Score
18 out of 131	**70.5 out of 100**

Innovation Capacity Index

Pillars and pillar subindexes

Education
Rank **27** Score **72.3**

Social inclusion and equity policies
Rank **10** Score **81.8**

Country policy assessment
Rank **18** Score **66.7**

Good governance
Rank **15** Score **84.1**

Quality of the infrastructure
Rank **13** Score **96.6**

Government ICT usage
Rank **19** Score **73.0**

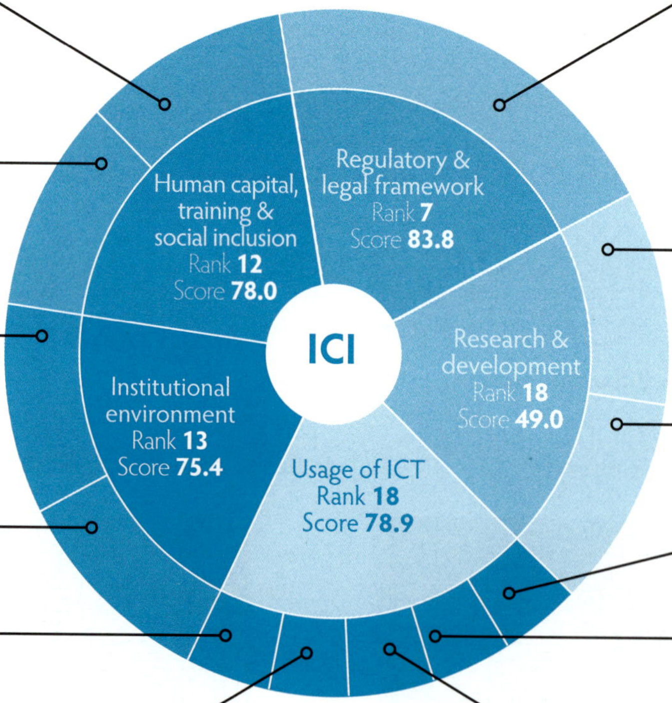

Doing business
Rank **7** Score **83.8**

R&D infrastructure
Rank **24** Score **49.7**

Patents and trademarks
Rank **12** Score **47.9**

Telephone communications
Rank **15** Score **94.2**

Mobile cellular communications
Rank **8** Score **95.0**

Internet, computers and TV
Rank **23** Score **57.6**

ICI

Human capital, training & social inclusion
Rank **12** Score **78.0**

Regulatory & legal framework
Rank **7** Score **83.8**

Institutional environment
Rank **13** Score **75.4**

Research & development
Rank **18** Score **49.0**

Usage of ICT
Rank **18** Score **78.9**

Index thermometer[12]

Pillar 1 10.7% Pillar 2 11.1% Pillar 3 23.8% Pillar 4 20.8% Pillar 5 33.6% 100%

[1] High-income countries are those that, according to the World Bank classification, have a gross national income (GNI) per capita of $11,456 or more.
[2] Full democracies are countries whose Democracy Index scores from 1 to 10 rounded to one decimal place are of 8 to 10.   [3] 2007.   [4] World Bank World Development Indicators.   [5] IMF World Economic Outlook.   [6] 2006.   [7] Estimate.   [8] UNESCO, % age above 15.   [9] % ages 15–64.   [10] 2005.   [11] 2004.   [12] The index thermometer shows the relative contribution of each pillar (in %) to the total index score.

IRELAND

Country relative performance: **Index pillars**[13]

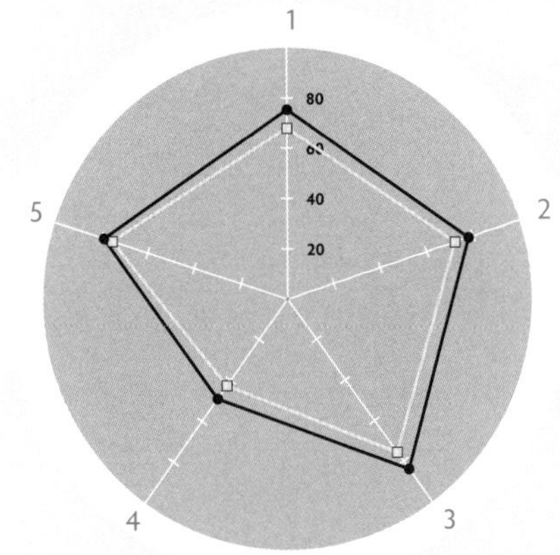

1. Institutional environment
2. Human capital, training and social inclusion
3. Regulatory and legal framework
4. Research and development
5. Usage of ICT

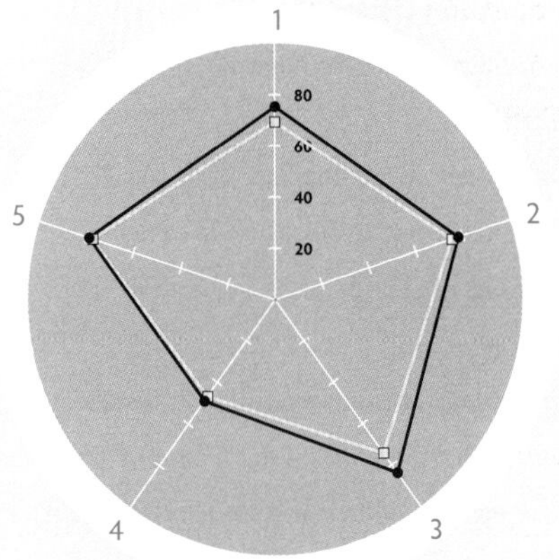

High-income ☐   Ireland ●——

Full democracies ☐   Ireland ●——

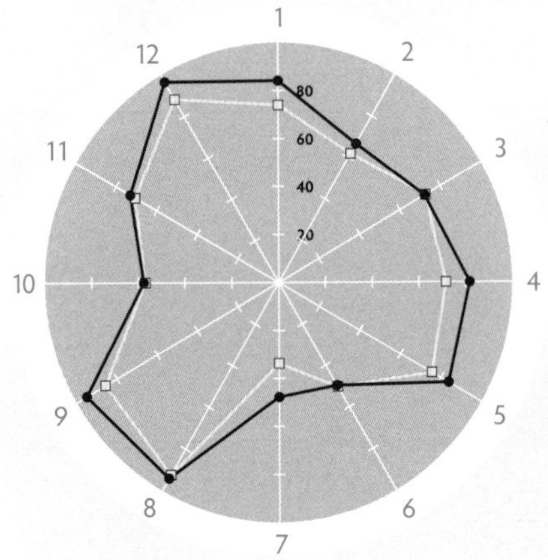

1. Good governance
2. Country policy assessment
3. Education
4. Social inclusion and equity policies
5. Doing business
6. R&D infrastructure
7. Patents and trademarks
8. Telephone communications
9. Mobile/cellular communications
10. Internet, computers and TV
11. Government ICT usage
12. Quality of the infrastructure

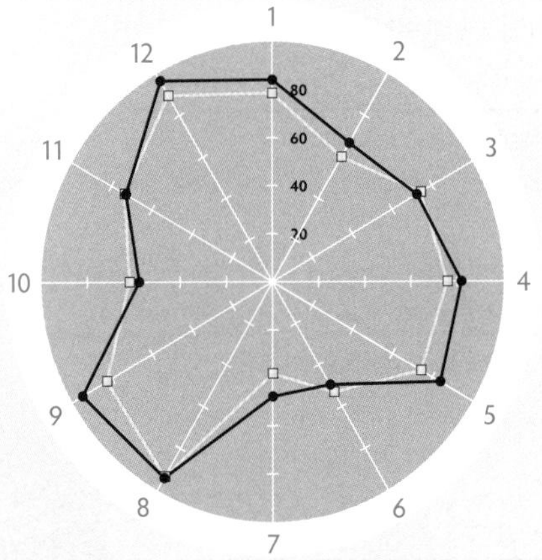

Country relative performance: **Pillar subindexes**[13]

In focus: **Top priorities for policy reform**[14]

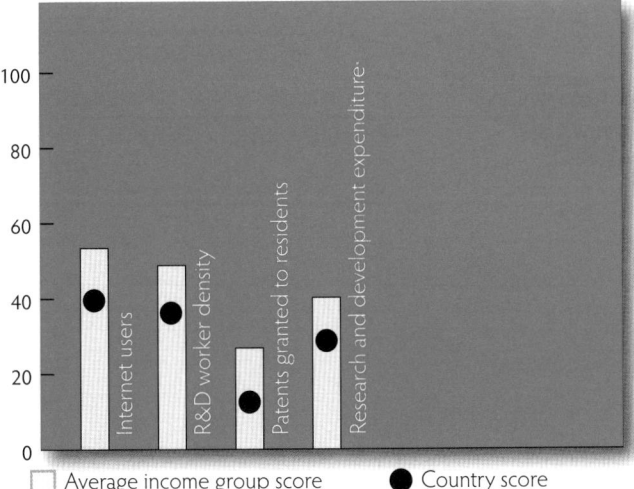

☐ Average income group score   ● Country score

In focus: **Significant indicators above income group average**[14]

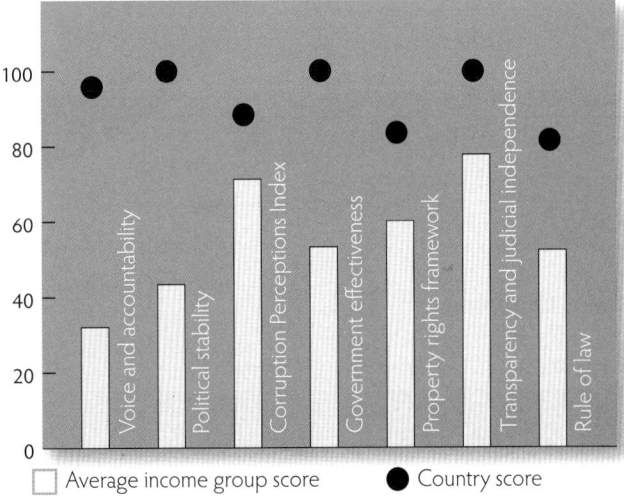

☐ Average income group score   ● Country score

[13] This analysis compares pillar and subindex country scores with respect to the average scores of countries grouped according to income level and political regime (see notes 1 and 2).

[14] Significant country indicators with the greatest distances above or below the income group average scores (i.e., achievements and targets for policy reform, respectively) were selected and ordered according to their Pearson correlation coefficients (highest first) with respect to the final index scores, to produce a ranking.

ISRAEL

Key selected indicators

Population (in millions)[3,4]	7.2
Population density (people per sq. km)[3,4]	331
GDP (current US$ billion, PPP)[3,5]	186
GDP per capita (current US$, PPP)[3,5]	25799
Current account balance (% of GDP)[3,5]	3.1
Average life expectancy (years)[4,6]	80.0
Literacy rate[3,7]	97
Female labor force participation rate[4,6,8]	59.1
Unemployment rate (% of total labor force)[4,9]	9.0
CO_2 emissions (metric tons per capita)[4,10]	10.5

Income level:[1] **High-income**

Political regime:[2] **Flawed democracy**

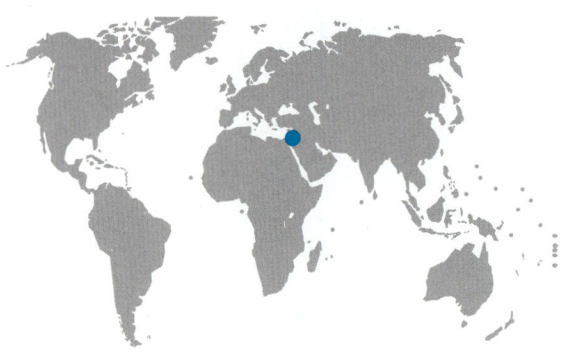

Innovation Capacity Index

Rank **21 out of 131**

Score **68.2 out of 100**

Pillars and pillar subindexes

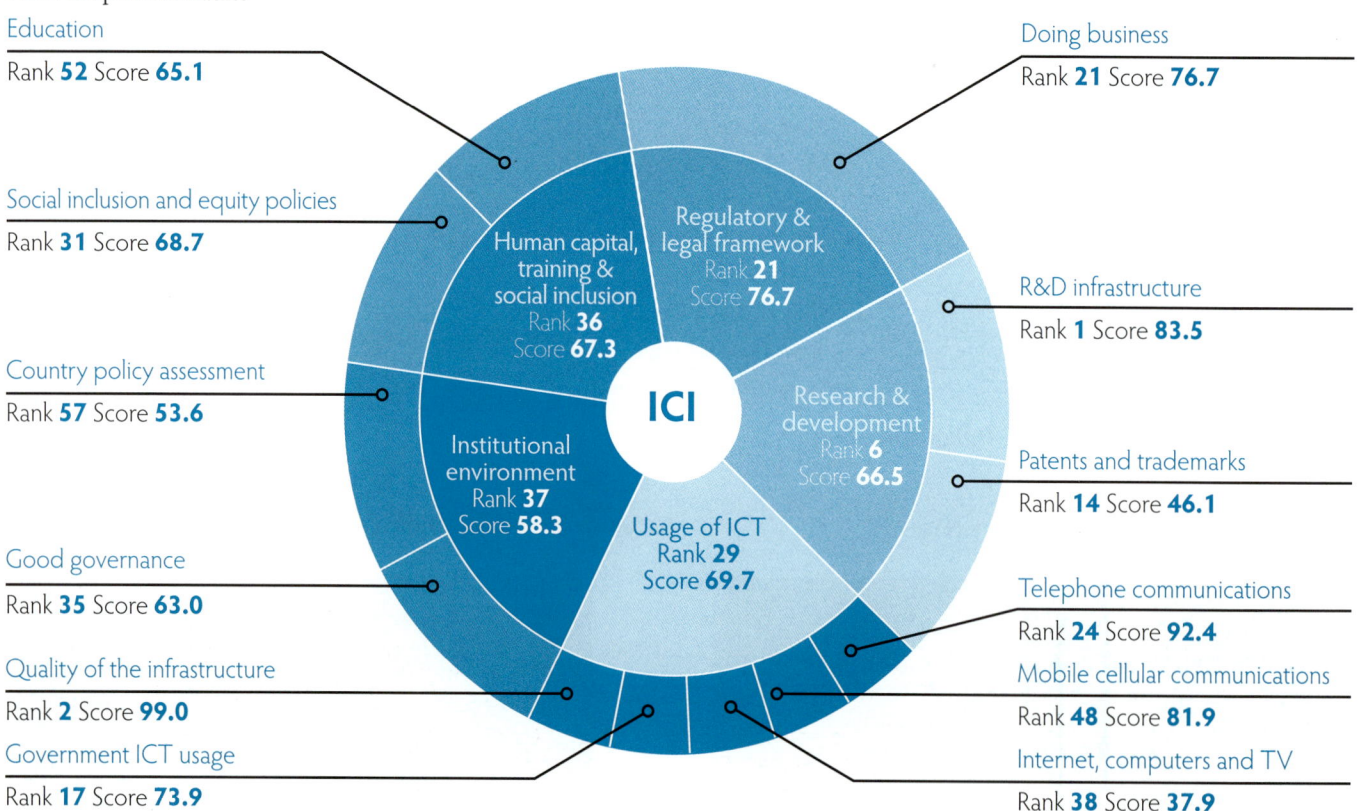

Education
Rank **52** Score **65.1**

Social inclusion and equity policies
Rank **31** Score **68.7**

Country policy assessment
Rank **57** Score **53.6**

Good governance
Rank **35** Score **63.0**

Quality of the infrastructure
Rank **2** Score **99.0**

Government ICT usage
Rank **17** Score **73.9**

Doing business
Rank **21** Score **76.7**

R&D infrastructure
Rank **1** Score **83.5**

Patents and trademarks
Rank **14** Score **46.1**

Telephone communications
Rank **24** Score **92.4**

Mobile cellular communications
Rank **48** Score **81.9**

Internet, computers and TV
Rank **38** Score **37.9**

Human capital, training & social inclusion Rank **36** Score **67.3**

Regulatory & legal framework Rank **21** Score **76.7**

Institutional environment Rank **37** Score **58.3**

Research & development Rank **6** Score **66.5**

Usage of ICT Rank **29** Score **69.7**

ICI

Index thermometer[11]

Pillar 1 12.8% Pillar 2 14.8% Pillar 3 22.5% Pillar 4 24.4% Pillar 5 25.5% 100%

298

[1] High-income countries are those that, according to the World Bank classification, have a gross national income (GNI) per capita of $11,456 or more.
[2] Flawed democracies are countries whose Democracy Index scores from 1 to 10 rounded to one decimal place are of 6 to 7.9. [3] 2007. [4] World Bank World Development Indicators. [5] IMF World Economic Outlook. [6] 2006. [7] UNESCO, % age above 15. [8] % ages 15–64. [9] 2005. [10] 2004. [11] The index thermometer shows the relative contribution of each pillar (in %) to the total index score.

ISRAEL

Country relative performance: **Index pillars**[12]

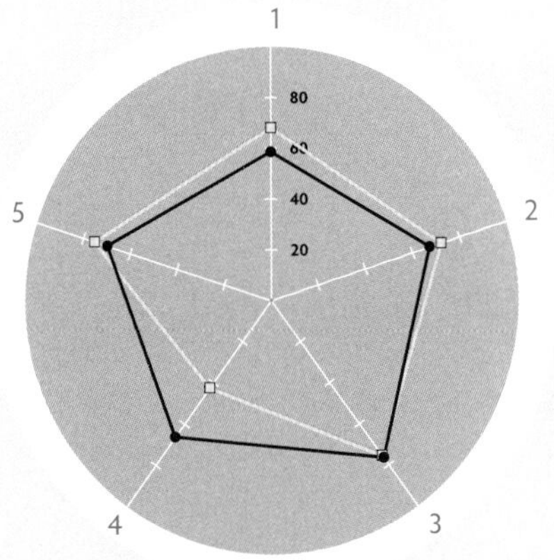

1. Institutional environment
2. Human capital, training and social inclusion
3. Regulatory and legal framework
4. Research and development
5. Usage of ICT

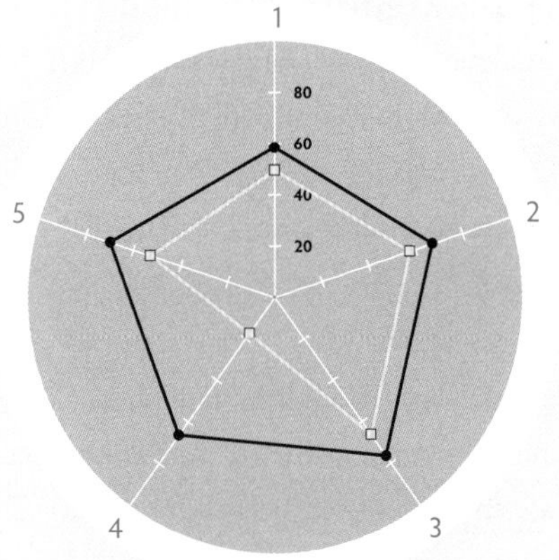

High-income   □   Israel ●

Flawed democracies   □   Israel ●

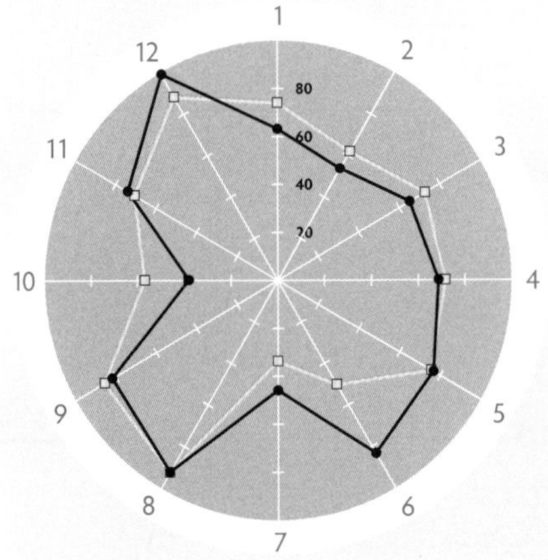

1. Good governance
2. Country policy assessment
3. Education
4. Social inclusion and equity policies
5. Doing business
6. R&D infrastructure
7. Patents and trademarks
8. Telephone communications
9. Mobile/cellular communications
10. Internet, computers and TV
11. Government ICT usage
12. Quality of the infrastructure

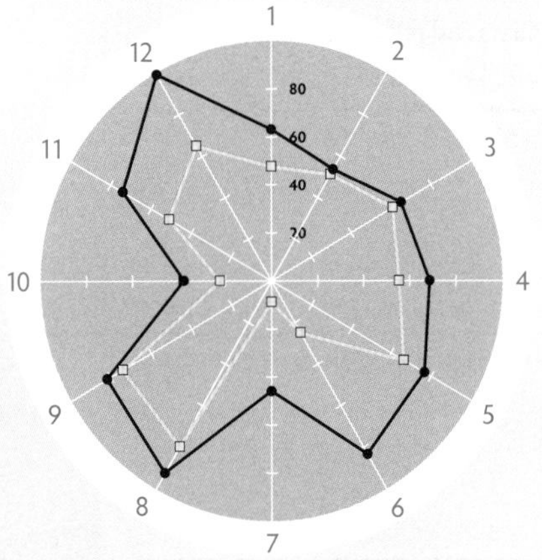

Country relative performance: **Pillar subindexes**[12]

In focus: **Top priorities for policy reform**[13]

In focus: **Significant indicators above income group average**[13]

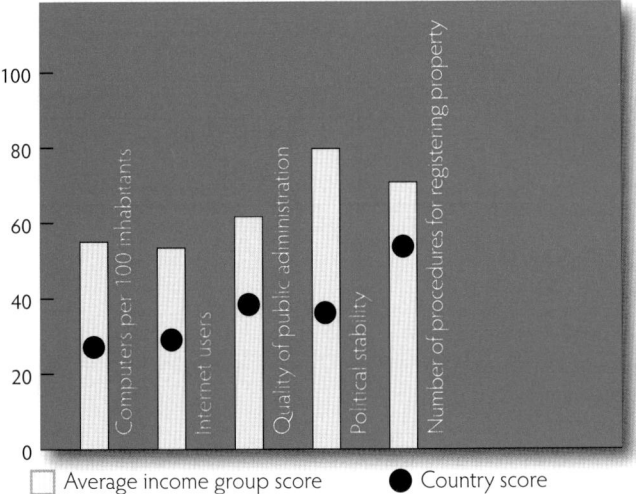

□ Average income group score   ● Country score

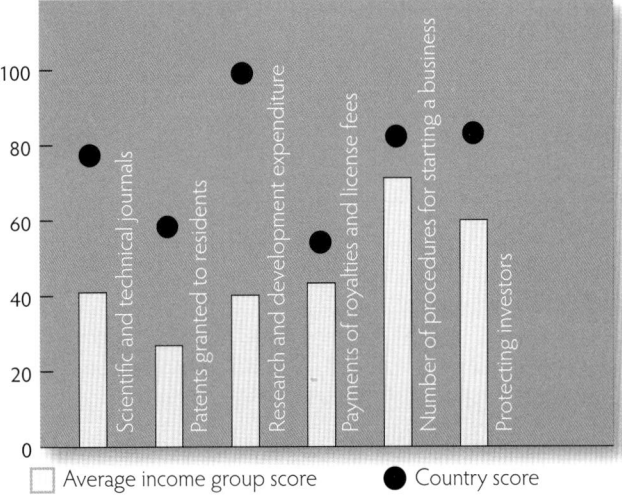

□ Average income group score   ● Country score

[12] This analysis compares pillar and subindex country scores with respect to the average scores of countries grouped according to income level and political regime (see notes 1 and 2).

[13] Significant country indicators with the greatest distances above or below the income group average scores (i.e., achievements and targets for policy reform, respectively) were selected and ordered according to their Pearson correlation coefficients (highest first) with respect to the final index scores, to produce a ranking.

299

ITALY

Key selected indicators

Population (in millions)[3,4]	59.4
Population density (people per sq. km)[3,4]	202
GDP (current US$ billion, PPP)[3,5]	1786
GDP per capita (current US$, PPP)[3,5]	30448
Current account balance (% of GDP)[3,5]	-2.2
Average life expectancy (years)[4,6]	81.1
Literacy rate[3,7]	99
Female labor force participation rate[4,6,8]	51.0
Unemployment rate (% of total labor force)[4,9]	7.7
CO_2 emissions (metric tons per capita)[4,10]	7.7

Income level:[1] **High-income**

Political regime:[2] **Full democracy**

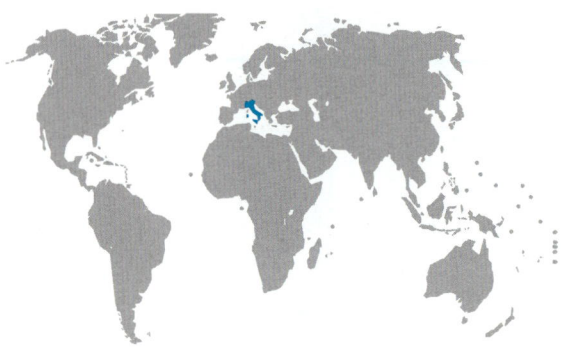

Innovation Capacity Index

Rank	Score
30 out of 131	**59.1 out of 100**

Pillars and pillar subindexes

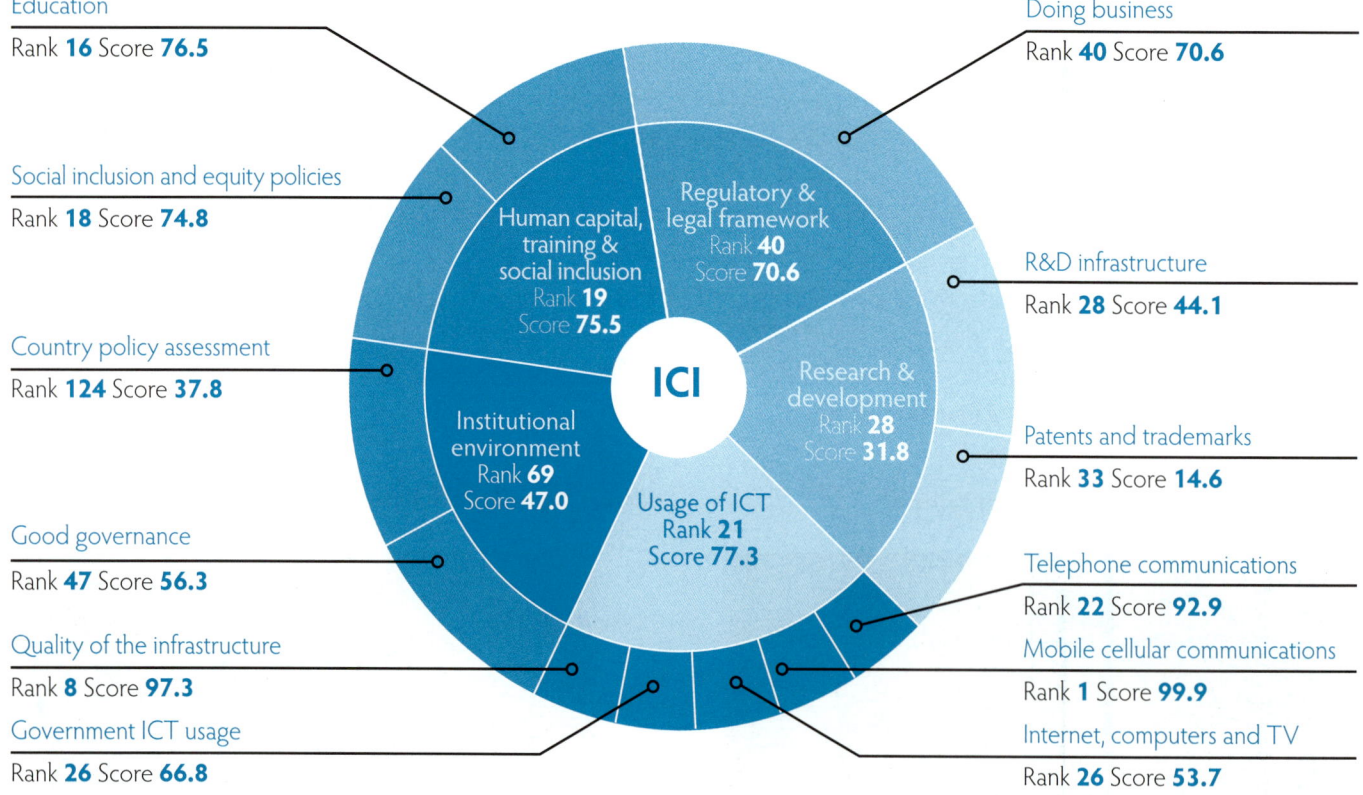

Education
Rank **16** Score **76.5**

Social inclusion and equity policies
Rank **18** Score **74.8**

Country policy assessment
Rank **124** Score **37.8**

Good governance
Rank **47** Score **56.3**

Quality of the infrastructure
Rank **8** Score **97.3**

Government ICT usage
Rank **26** Score **66.8**

Doing business
Rank **40** Score **70.6**

R&D infrastructure
Rank **28** Score **44.1**

Patents and trademarks
Rank **33** Score **14.6**

Telephone communications
Rank **22** Score **92.9**

Mobile cellular communications
Rank **1** Score **99.9**

Internet, computers and TV
Rank **26** Score **53.7**

(Inner wheel labels:)
Human capital, training & social inclusion Rank **19** Score **75.5**
Regulatory & legal framework Rank **40** Score **70.6**
Institutional environment Rank **69** Score **47.0**
Research & development Rank **28** Score **31.8**
Usage of ICT Rank **21** Score **77.3**
ICI

Index thermometer[11]

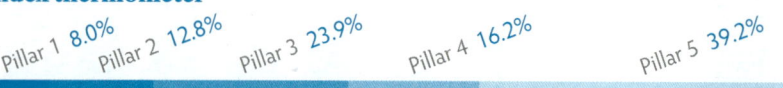

Pillar 1 8.0% Pillar 2 12.8% Pillar 3 23.9% Pillar 4 16.2% Pillar 5 39.2%   100%

[1] High-income countries are those that, according to the World Bank classification, have a gross national income (GNI) per capita of $11,456 or more.
[2] Full democracies are countries whose Democracy Index scores from 1 to 10 rounded to one decimal place are of 8 to 10. [3] 2007. [4] World Bank World Development Indicators. [5] IMF World Economic Outlook. [6] 2006. [7] UNESCO, % age above 15. [8] % ages 15–64. [9] 2005. [10] 2004. [11] The index thermometer shows the relative contribution of each pillar (in %) to the total index score.

ITALY

Country relative performance: **Index pillars**[12]

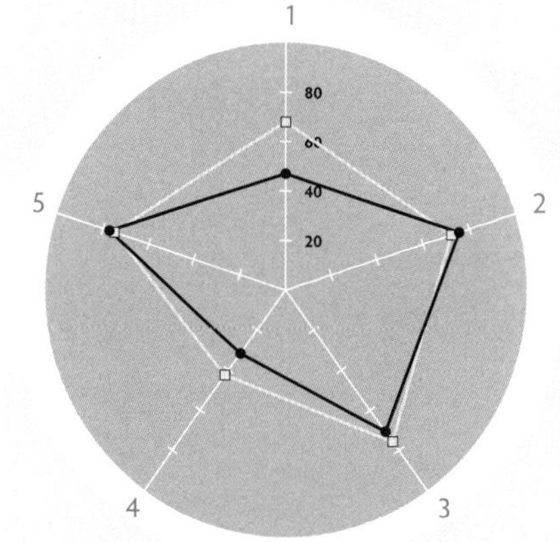

1. Institutional environment
2. Human capital, training and social inclusion
3. Regulatory and legal framework
4. Research and development
5. Usage of ICT

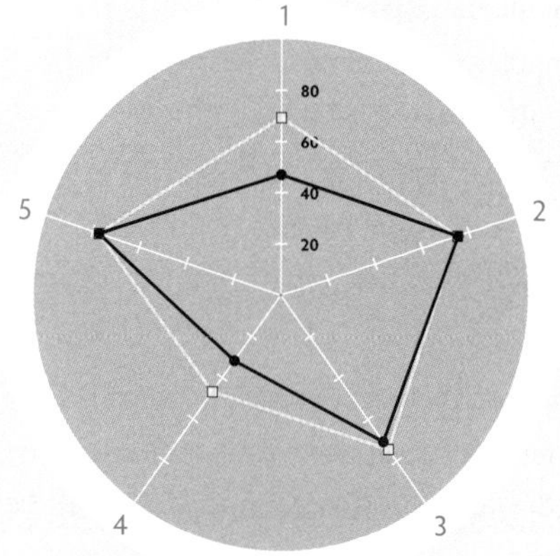

High-income   ☐   Italy ━●━

Full democracies   ☐   Italy ━●━

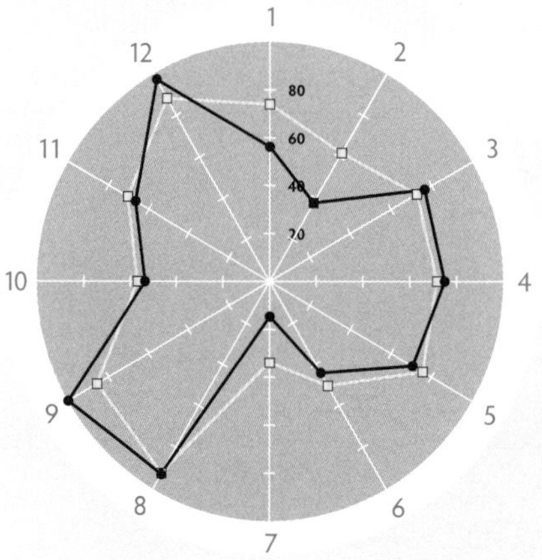

1. Good governance
2. Country policy assessment
3. Education
4. Social inclusion and equity policies
5. Doing business
6. R&D infrastructure
7. Patents and trademarks
8. Telephone communications
9. Mobile/cellular communications
10. Internet, computers and TV
11. Government ICT usage
12. Quality of the infrastructure

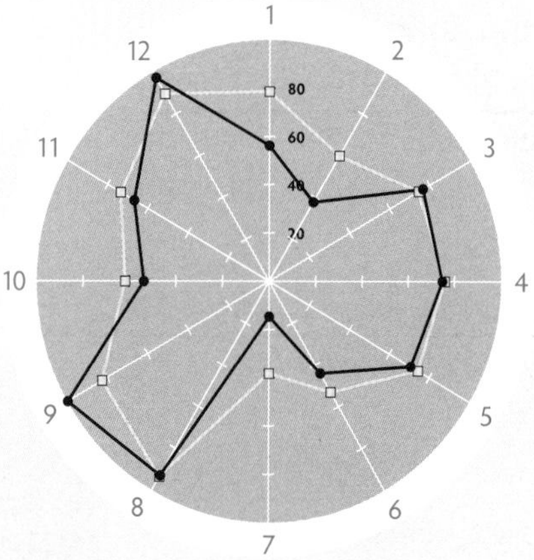

Country relative performance: **Pillar subindexes**[12]

In focus: Top priorities for policy reform[13]

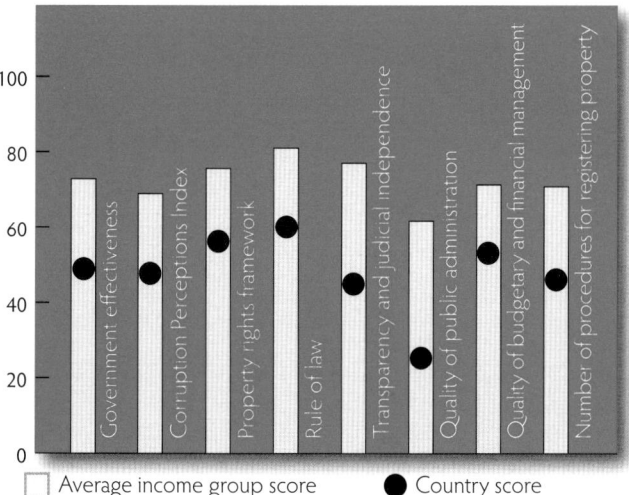

☐ Average income group score   ● Country score

In focus: Significant indicators above income group average[13]

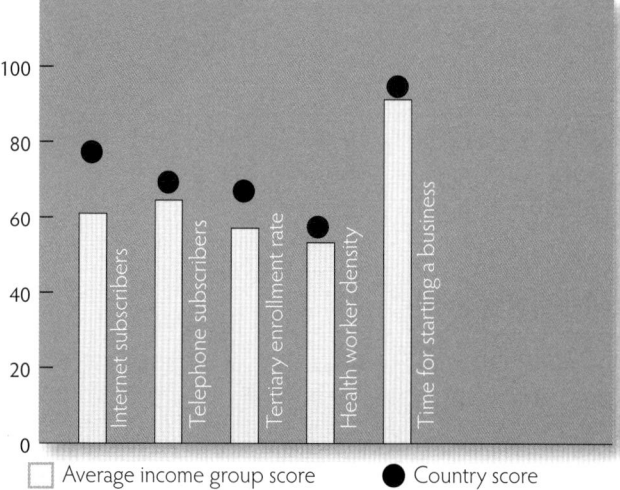

☐ Average income group score   ● Country score

301

[12] This analysis compares pillar and subindex country scores with respect to the average scores of countries grouped according to income level and political regime (see notes 1 and 2).

[13] Significant country indicators with the greatest distances above or below the income group average scores (i.e., achievements and targets for policy reform, respectively) were selected and ordered according to their Pearson correlation coefficients (highest first) with respect to the final index scores, to produce a ranking.

JAPAN

Key selected indicators

Population (in millions)[3,4]	127.8
Population density (people per sq. km)[3,4]	351
GDP (current US$ billion, PPP)[3,5]	4290
GDP per capita (current US$, PPP)[3,5]	33577
Current account balance (% of GDP)[3,5]	4.9
Average life expectancy (years)[4,6]	82.3
Literacy rate[7,8]	99
Female labor force participation rate[4,6,9]	60.6
Unemployment rate (% of total labor force)[4,10]	4.4
CO_2 emissions (metric tons per capita)[4,11]	9.8

Income level:[1] **High-income**

Political regime:[2] **Full democracy**

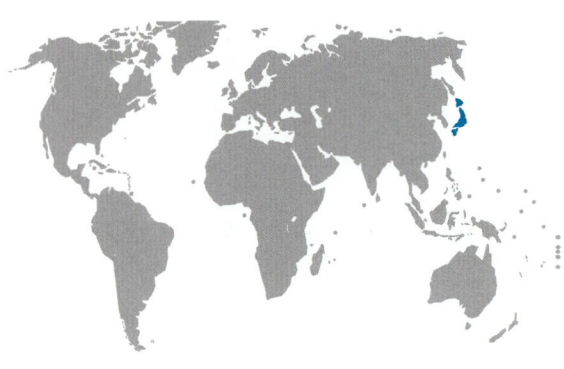

Rank	Score
15 out of 131	**72.1 out of 100**

Innovation Capacity Index

Pillars and pillar subindexes

Education
Rank **22** Score **73.8**

Social inclusion and equity policies
Rank **32** Score **68.5**

Country policy assessment
Rank **114** Score **40.7**

Good governance
Rank **20** Score **77.7**

Quality of the infrastructure
Rank **24** Score **91.4**

Government ICT usage
Rank **11** Score **77.0**

Doing business
Rank **17** Score **77.7**

R&D infrastructure
Rank **9** Score **66.7**

Patents and trademarks
Rank **4** Score **72.2**

Telephone communications
Rank **32** Score **90.3**

Mobile cellular communications
Rank **77** Score **67.2**

Internet, computers and TV
Rank **18** Score **66.0**

Central diagram (ICI):
- Human capital, training & social inclusion — Rank **29** Score **70.6**
- Regulatory & legal framework — Rank **17** Score **77.7**
- Institutional environment — Rank **35** Score **59.2**
- Research & development — Rank **4** Score **69.0**
- Usage of ICT — Rank **22** Score **76.4**

Index thermometer[12]

Pillar 1 8.2% Pillar 2 9.8% Pillar 3 21.5% Pillar 4 28.7% Pillar 5 31.8% 100%

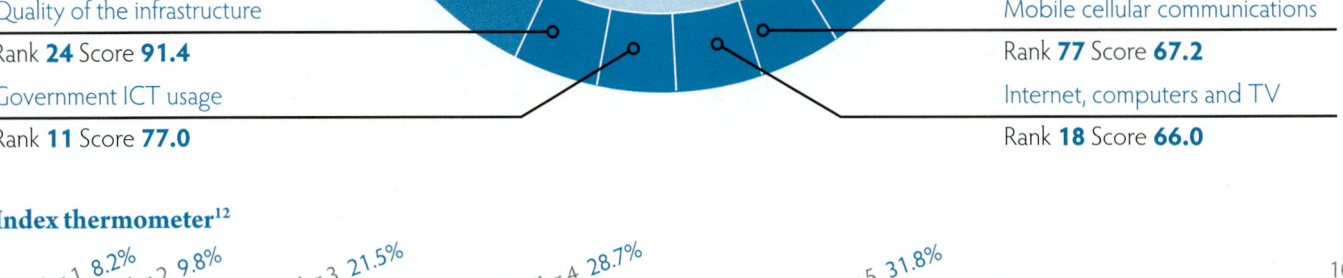

[1] High-income countries are those that, according to the World Bank classification, have a gross national income (GNI) per capita of $11,456 or more.
[2] Full democracies are countries whose Democracy Index scores from 1 to 10 rounded to one decimal place are of 8 to 10. [3] 2007. [4] World Bank World Development Indicators. [5] IMF World Economic Outlook. [6] 2006. [7] Estimate. [8] UNESCO, % age above 15. [9] % ages 15–64. [10] 2005. [11] 2004. [12] The index thermometer shows the relative contribution of each pillar (in %) to the total index score.

JAPAN

Country relative performance: **Index pillars**[13]

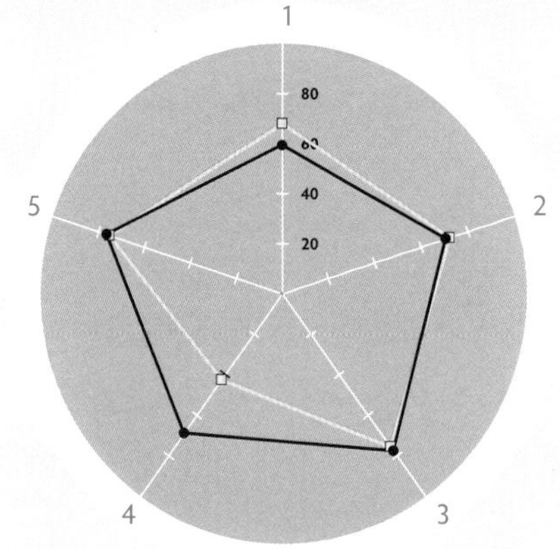

1. Institutional environment
2. Human capital, training and social inclusion
3. Regulatory and legal framework
4. Research and development
5. Usage of ICT

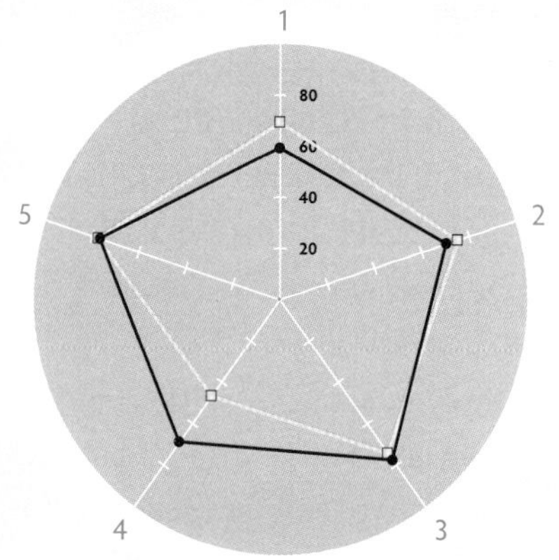

High-income □   Japan ●━━

Full democracies □   Japan ●━━

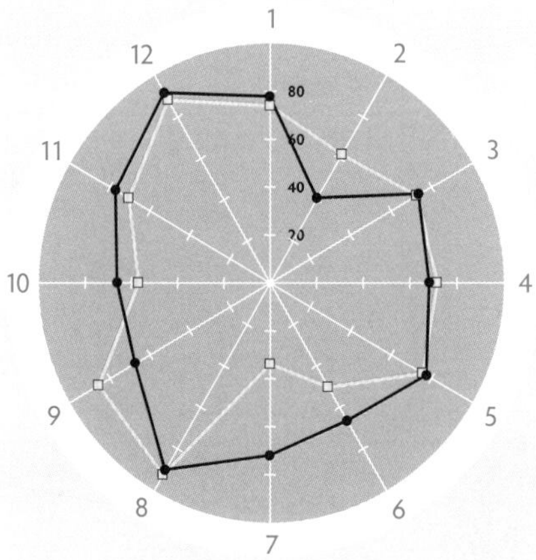

1. Good governance
2. Country policy assessment
3. Education
4. Social inclusion and equity policies
5. Doing business
6. R&D infrastructure
7. Patents and trademarks
8. Telephone communications
9. Mobile/cellular communications
10. Internet, computers and TV
11. Government ICT usage
12. Quality of the infrastructure

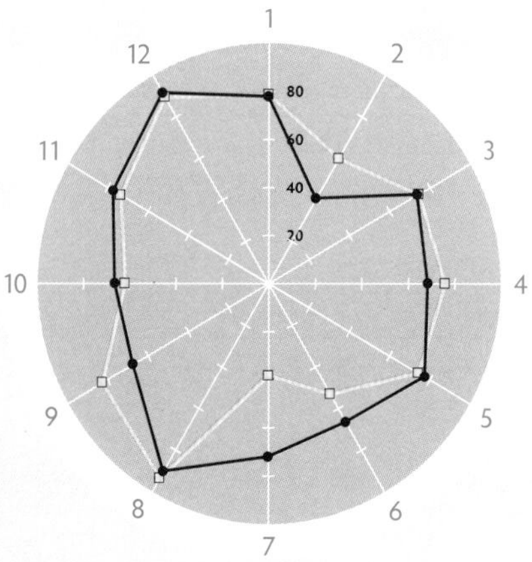

Country relative performance: **Pillar subindexes**[13]

In focus: **Top priorities for policy reform**[14]

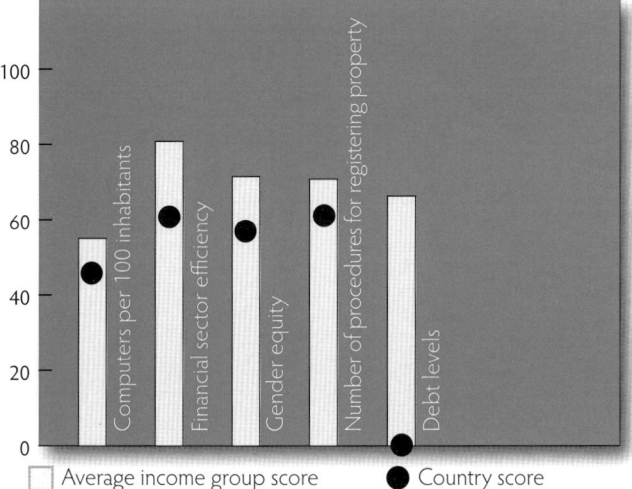

Average income group score □   ● Country score

In focus: **Significant indicators above income group average**[14]

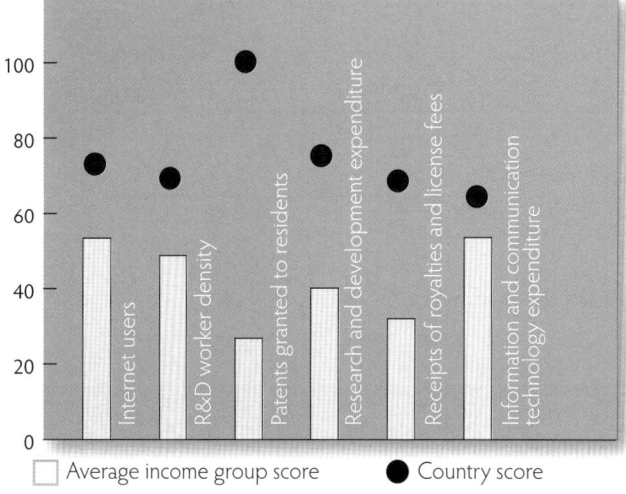

Average income group score □   ● Country score

---

[13] This analysis compares pillar and subindex country scores with respect to the average scores of countries grouped according to income level and political regime (see notes 1 and 2).

[14] Significant country indicators with the greatest distances above or below the income group average scores (i.e., achievements and targets for policy reform, respectively) were selected and ordered according to their Pearson correlation coefficients (highest first) with respect to the final index scores, to produce a ranking.

303

JORDAN

Key selected indicators

Population (in millions)[3, 4]	5.7
Population density (people per sq. km)[3, 4]	65
GDP (current US$ billion, PPP)[3, 5]	1433
GDP per capita (current US$, PPP)[3, 5]	1982
Current account balance (% of GDP)[3, 5]	-17.3
Average life expectancy (years)[4, 6]	72.2
Literacy rate[3, 7]	93
Female labor force participation rate[4, 6, 8]	29.5
Unemployment rate (% of total labor force)[4, 9]	12.4
CO_2 emissions (metric tons per capita)[4, 9]	3.1

Income level:[1] **Lower-middle-income**

Political regime:[2] **Authoritarian regime**

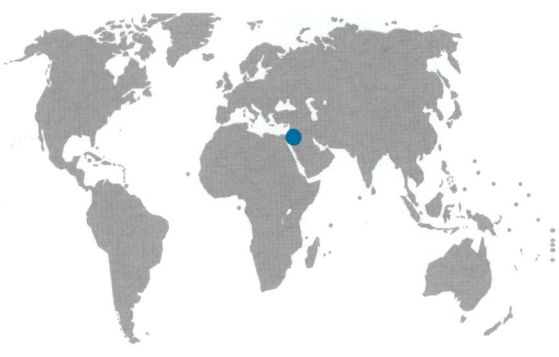

Rank	Score
44 out of 131	**53.9 out of 100**

Innovation Capacity Index

Pillars and pillar subindexes

304

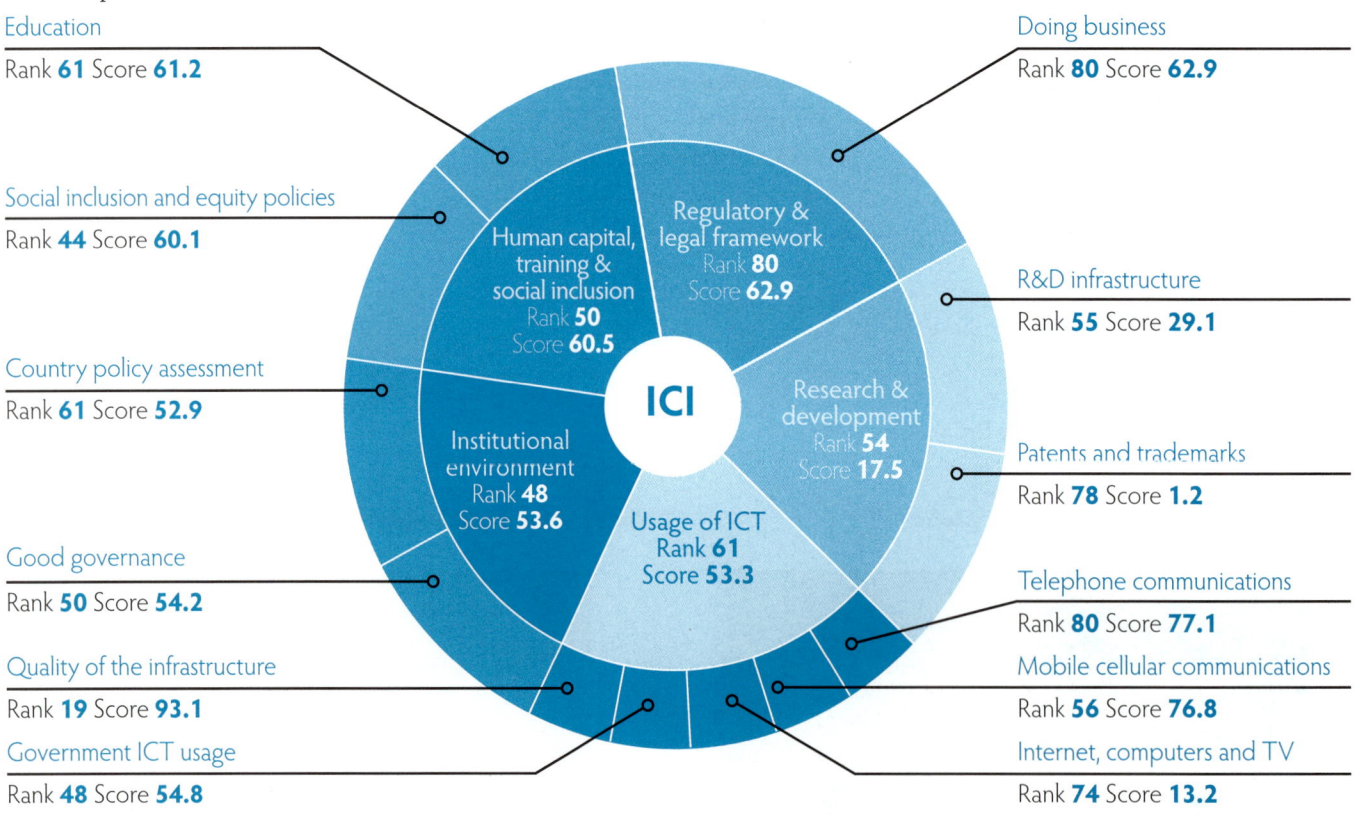

Education
Rank **61** Score **61.2**

Social inclusion and equity policies
Rank **44** Score **60.1**

Country policy assessment
Rank **61** Score **52.9**

Good governance
Rank **50** Score **54.2**

Quality of the infrastructure
Rank **19** Score **93.1**

Government ICT usage
Rank **48** Score **54.8**

Doing business
Rank **80** Score **62.9**

R&D infrastructure
Rank **55** Score **29.1**

Patents and trademarks
Rank **78** Score **1.2**

Telephone communications
Rank **80** Score **77.1**

Mobile cellular communications
Rank **56** Score **76.8**

Internet, computers and TV
Rank **74** Score **13.2**

Human capital, training & social inclusion
Rank **50** Score **60.5**

Regulatory & legal framework
Rank **80** Score **62.9**

Institutional environment
Rank **48** Score **53.6**

Research & development
Rank **54** Score **17.5**

Usage of ICT
Rank **61** Score **53.3**

ICI

Index thermometer[10]

Pillar 1 29.8% Pillar 2 33.7% Pillar 3 23.4% Pillar 4 3.2% Pillar 5 9.9% 100%

[1] Lower-middle-income countries are those that, according to the World Bank classification, have a gross national income (GNI) per capita of $936 to $3,705. [2] Authoritarian regimes are countries whose Democracy Index scores from 1 to 10 rounded to one decimal place are less than 4. [3] 2007. [4] World Bank World Development Indicators. [5] IMF World Economic Outlook. [6] 2006. [7] UNESCO, % age above 15. [8] % ages 15–64. [9] 2004. [10] The index thermometer shows the relative contribution of each pillar (in %) to the total index score.

JORDAN

Country relative performance: **Index pillars**[11]

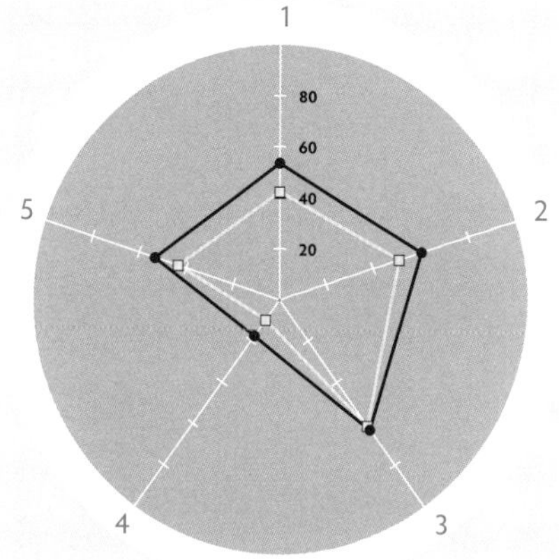

1. Institutional environment
2. Human capital, training and social inclusion
3. Regulatory and legal framework
4. Research and development
5. Usage of ICT

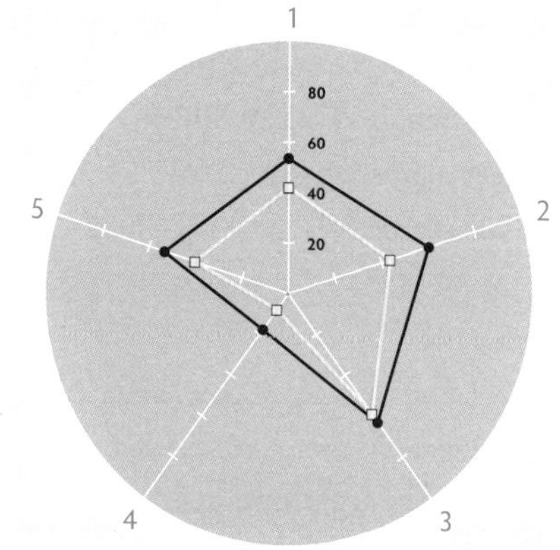

Lower-middle-income ☐   Jordan ●——

Authoritarian regimes ☐   Jordan ●——

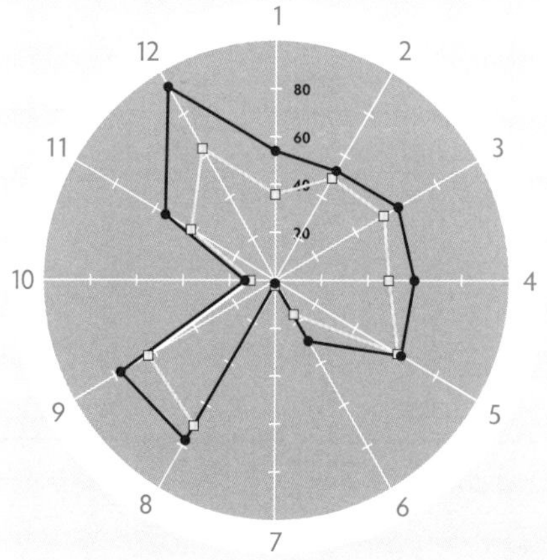

305

1. Good governance
2. Country policy assessment
3. Education
4. Social inclusion and equity policies
5. Doing business
6. R&D infrastructure
7. Patents and trademarks
8. Telephone communications
9. Mobile/cellular communications
10. Internet, computers and TV
11. Government ICT usage
12. Quality of the infrastructure

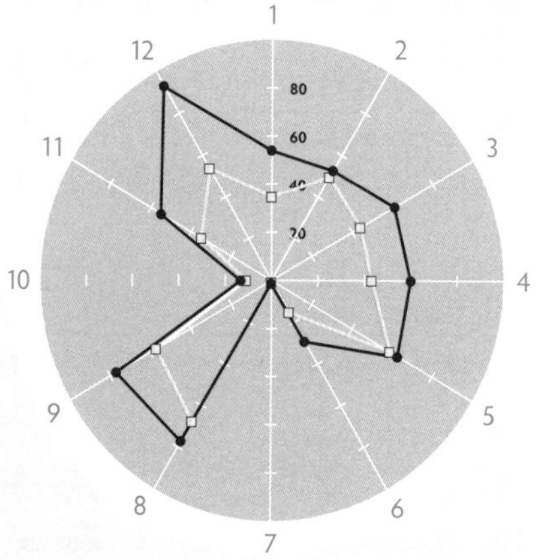

Country relative performance: **Pillar subindexes**[11]

In focus: **Top priorities for policy reform**[12]

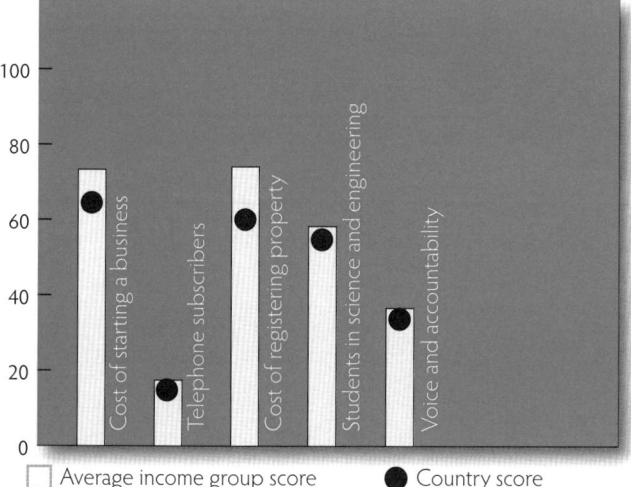

☐ Average income group score   ● Country score

In focus: **Significant indicators above income group average**[12]

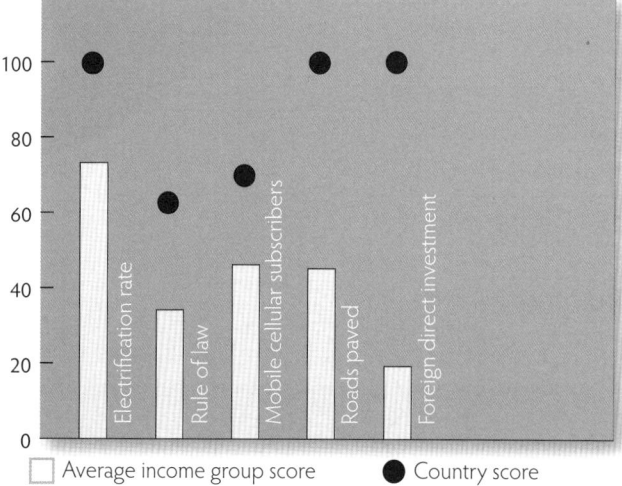

☐ Average income group score   ● Country score

---

[11] This analysis compares pillar and subindex country scores with respect to the average scores of countries grouped according to income level and political regime (see notes 1 and 2).

[12] Significant country indicators with the greatest distances above or below the income group average scores (i.e., achievements and targets for policy reform, respectively) were selected and ordered according to their Pearson correlation coefficients (highest first) with respect to the final index scores, to produce a ranking.

KAZAKHSTAN, REPUBLIC OF

Key selected indicators

Population (in millions)[3,4]	15.5
Population density (people per sq. km)[3,4]	6
GDP (current US$ billion, PPP)[3,5]	168
GDP per capita (current US$, PPP)[3,5]	11086
Current account balance (% of GDP)[3,5]	-6.6
Average life expectancy (years)[4,6]	66.2
Literacy rate[3,7]	100
Female labor force participation rate[4,6,8]	74.4
Unemployment rate (% of total labor force)[4,6]	7.8
CO_2 emissions (metric tons per capita)[4,9]	13.3

Income level:[1] **Upper-middle-income**

Political regime:[2] **Authoritarian regime**

Innovation Capacity Index

Rank	Score
57 out of 131	**51.6 out of 100**

Pillars and pillar subindexes

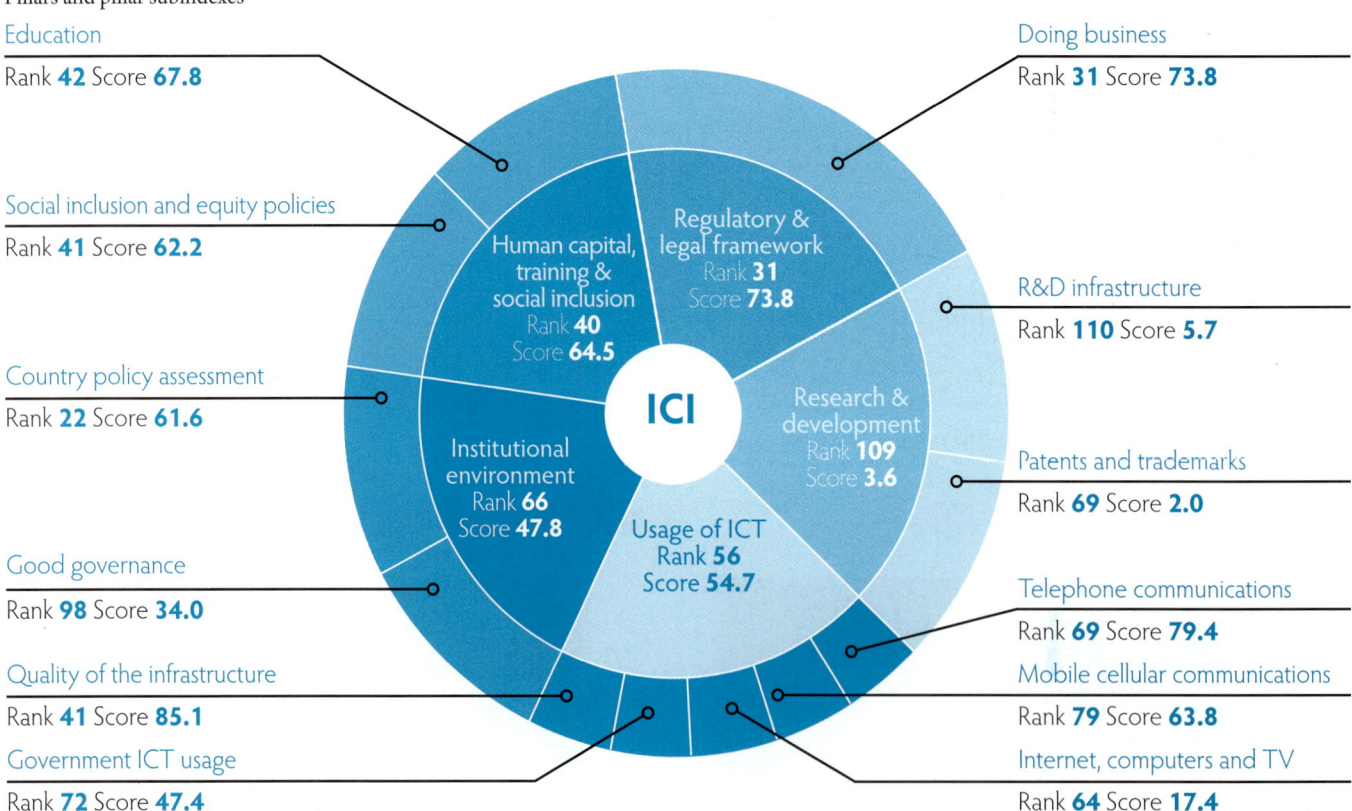

Education
Rank **42** Score **67.8**

Social inclusion and equity policies
Rank **41** Score **62.2**

Country policy assessment
Rank **22** Score **61.6**

Good governance
Rank **98** Score **34.0**

Quality of the infrastructure
Rank **41** Score **85.1**

Government ICT usage
Rank **72** Score **47.4**

Doing business
Rank **31** Score **73.8**

R&D infrastructure
Rank **110** Score **5.7**

Patents and trademarks
Rank **69** Score **2.0**

Telephone communications
Rank **69** Score **79.4**

Mobile cellular communications
Rank **79** Score **63.8**

Internet, computers and TV
Rank **64** Score **17.4**

Human capital, training & social inclusion
Rank **40**
Score **64.5**

Regulatory & legal framework
Rank **31**
Score **73.8**

Institutional environment
Rank **66**
Score **47.8**

Research & development
Rank **109**
Score **3.6**

Usage of ICT
Rank **56**
Score **54.7**

ICI

Index thermometer[10]

Pillar 1 23.2%   Pillar 2 31.3%   Pillar 3 28.6%   Pillar 4 1.1%   Pillar 5 15.9%   100%

[1] Upper-middle-income countries are those that, according to the World Bank classification, have a gross national income (GNI) per capita of $3,706 to $11,455.   [2] Authoritarian regimes are countries whose Democracy Index scores from 1 to 10 rounded to one decimal place are less than 4.   [3] 2007.
[4] World Bank World Development Indicators.   [5] IMF World Economic Outlook.   [6] 2006.   [7] UNESCO, % age above 15.   [8] % ages 15–64.   [9] 2004.   [10] The index thermometer shows the relative contribution of each pillar (in %) to the total index score.

KAZAKHSTAN, REPUBLIC OF

Country relative performance: Index pillars[11]

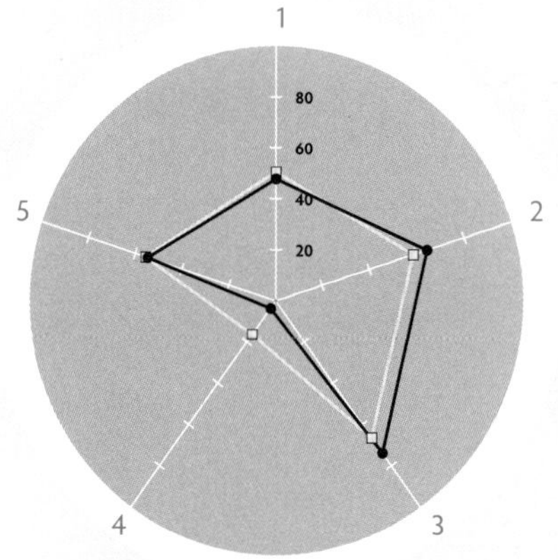

1. Institutional environment
2. Human capital, training and social inclusion
3. Regulatory and legal framework
4. Research and development
5. Usage of ICT

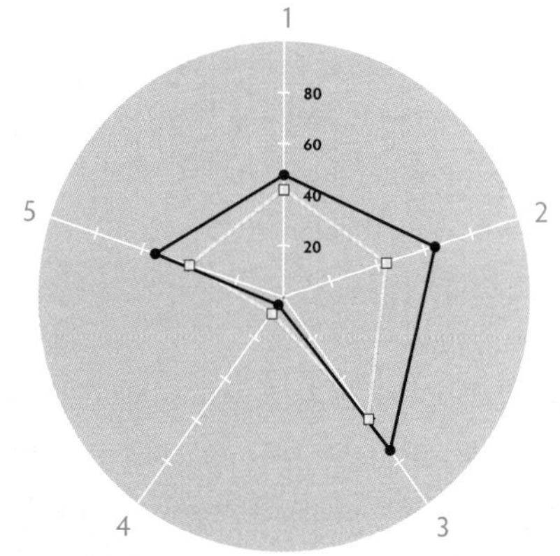

Upper-middle-income ☐　　　Kazakhstan ━●━

Authoritarian regimes ☐　　　Kazakhstan ━●━

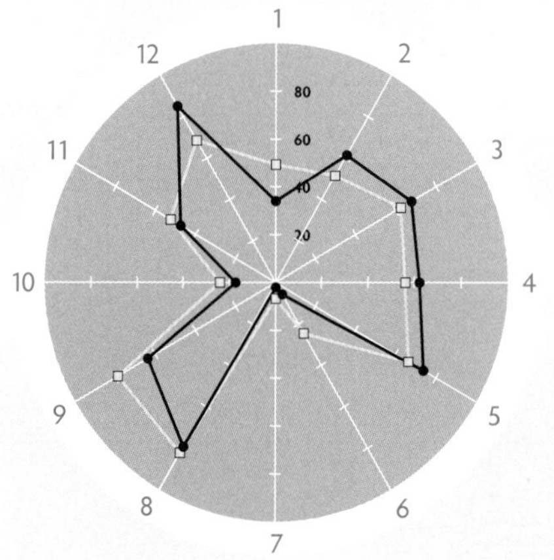

1. Good governance
2. Country policy assessment
3. Education
4. Social inclusion and equity policies
5. Doing business
6. R&D infrastructure
7. Patents and trademarks
8. Telephone communications
9. Mobile/cellular communications
10. Internet, computers and TV
11. Government ICT usage
12. Quality of the infrastructure

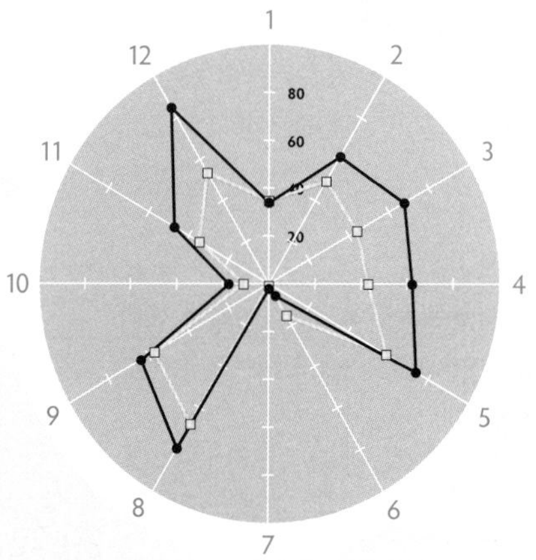

Country relative performance: Pillar subindexes[11]

307

In focus: Top priorities for policy reform[12]

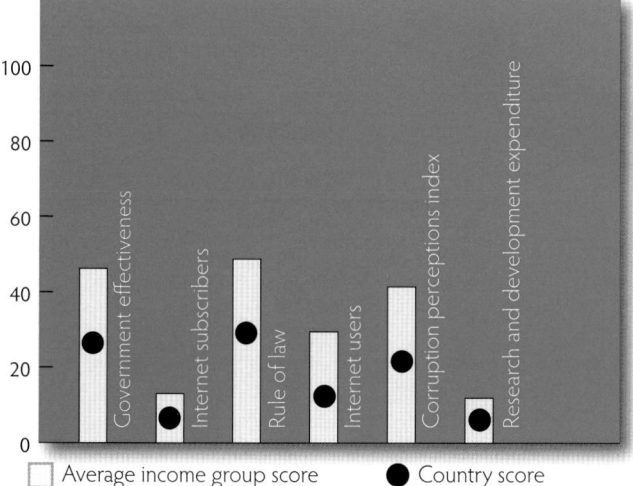

☐ Average income group score　● Country score

In focus: Significant indicators above income group average[12]

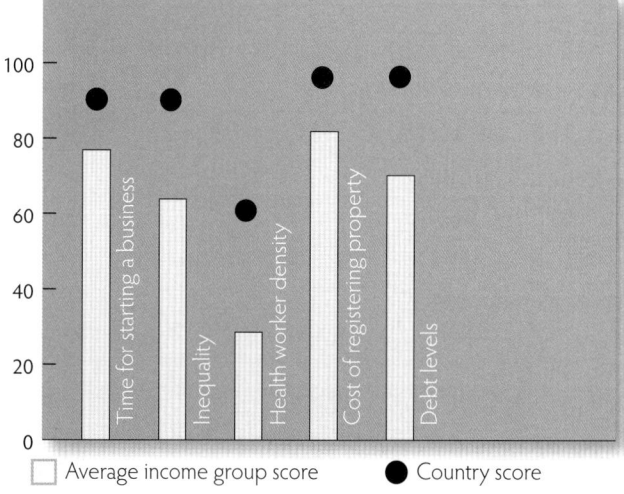

☐ Average income group score　● Country score

[11] This analysis compares pillar and subindex country scores with respect to the average scores of countries grouped according to income level and political regime (see notes 1 and 2).

[12] Significant country indicators with the greatest distances above or below the income group average scores (i.e., achievements and targets for policy reform, respectively) were selected and ordered according to their Pearson correlation coefficients (highest first) with respect to the final index scores, to produce a ranking.

KENYA

Key selected indicators

Population (in millions)[3,4]	37.5
Population density (people per sq. km)[3,4]	66
GDP (current US$ billion, PPP)[3,5]	59
GDP per capita (current US$, PPP)[3,5]	1699
Current account balance (% of GDP)[3,5]	-3.5
Average life expectancy (years)[4,6]	53.4
Literacy rate[7,8]	74
Female labor force participation rate[4,6,9]	71.6
Unemployment rate (% of total labor force)[4,10]	9.8
CO_2 emissions (metric tons per capita)[4,11]	0.3

Income level:[1] **Low-income**

Political regime:[2] **Hybrid regime**

Innovation Capacity Index

Rank	Score
95 out of 131	**43.3 out of 100**

Pillars and pillar subindexes

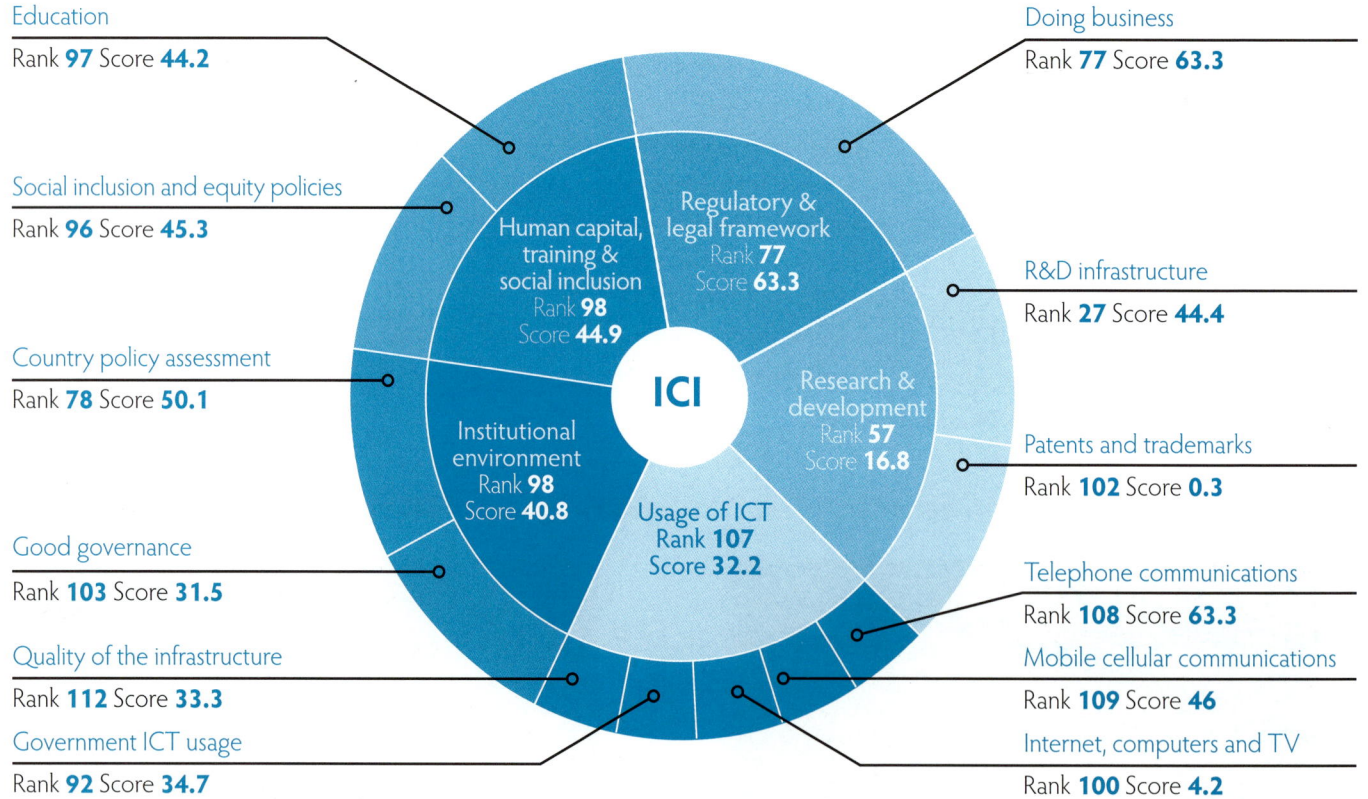

Education
Rank **97** Score **44.2**

Social inclusion and equity policies
Rank **96** Score **45.3**

Country policy assessment
Rank **78** Score **50.1**

Good governance
Rank **103** Score **31.5**

Quality of the infrastructure
Rank **112** Score **33.3**

Government ICT usage
Rank **92** Score **34.7**

Doing business
Rank **77** Score **63.3**

R&D infrastructure
Rank **27** Score **44.4**

Patents and trademarks
Rank **102** Score **0.3**

Telephone communications
Rank **108** Score **63.3**

Mobile cellular communications
Rank **109** Score **46**

Internet, computers and TV
Rank **100** Score **4.2**

Human capital, training & social inclusion
Rank **98** Score **44.9**

Regulatory & legal framework
Rank **77** Score **63.3**

ICI

Institutional environment
Rank **98** Score **40.8**

Research & development
Rank **57** Score **16.8**

Usage of ICT
Rank **107** Score **32.2**

Index thermometer[12]

Pillar 1 28.3% Pillar 2 31.1% Pillar 3 29.3% Pillar 4 3.9% Pillar 5 7.4%

100%

[1] Low-income countries are those that, according to the World Bank classification, have a gross national income (GNI) per capita of $935 or less. [2] Hybrid regimes are countries whose Democracy Index scores from 1 to 10 rounded to one decimal place are of 4 to 5.9. [3] 2007. [4] World Bank World Development Indicators. [5] IMF World Economic Outlook. [6] 2006. [7] 2000. [8] UNESCO, % age above 15. [9] % ages 15–64. [10] 1999. [11] 2004. [12] The index thermometer shows the relative contribution of each pillar (in %) to the total index score.

KENYA

Country relative performance: **Index pillars**[13]

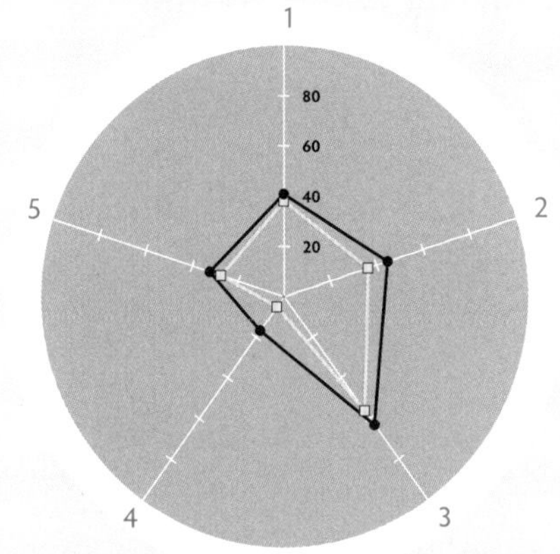

1. Institutional environment
2. Human capital, training and social inclusion
3. Regulatory and legal framework
4. Research and development
5. Usage of ICT

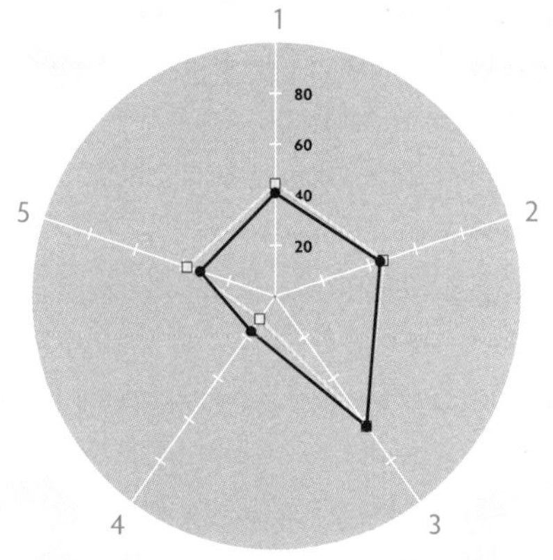

Low-income ☐   Kenya ●——

Hybrid regimes ☐   Kenya ●——

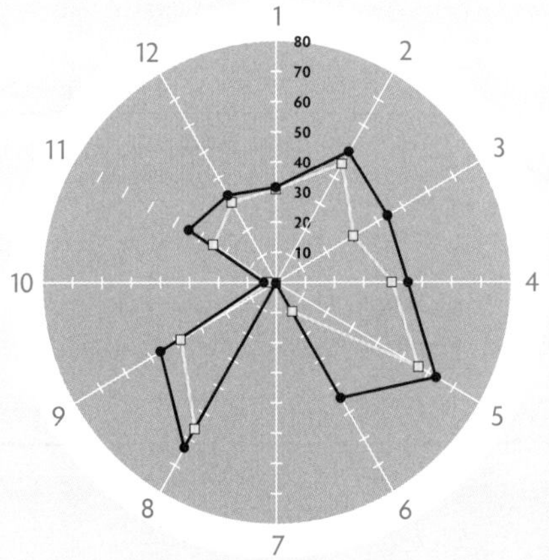

1. Good governance
2. Country policy assessment
3. Education
4. Social inclusion and equity policies
5. Doing business
6. R&D infrastructure
7. Patents and trademarks
8. Telephone communications
9. Mobile/cellular communications
10. Internet, computers and TV
11. Government ICT usage
12. Quality of the infrastructure

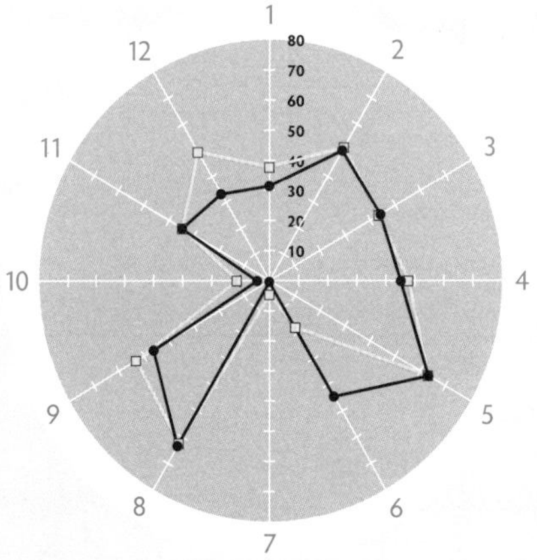

Country relative performance: **Pillar subindexes**[13]

In focus: **Top priorities for policy reform**[14]

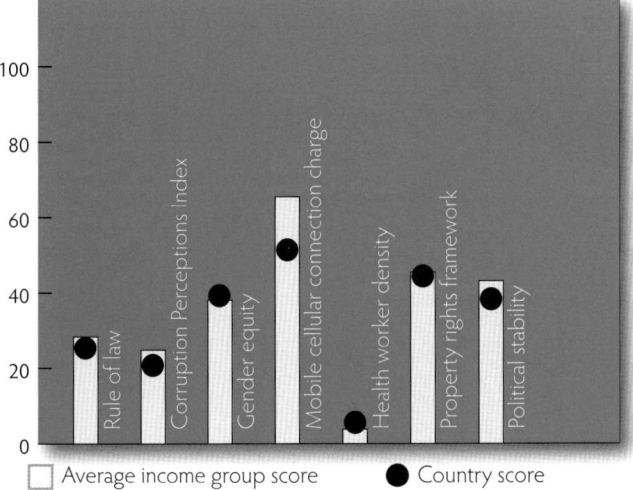

☐ Average income group score   ● Country score

In focus: **Significant indicators above income group average**[14]

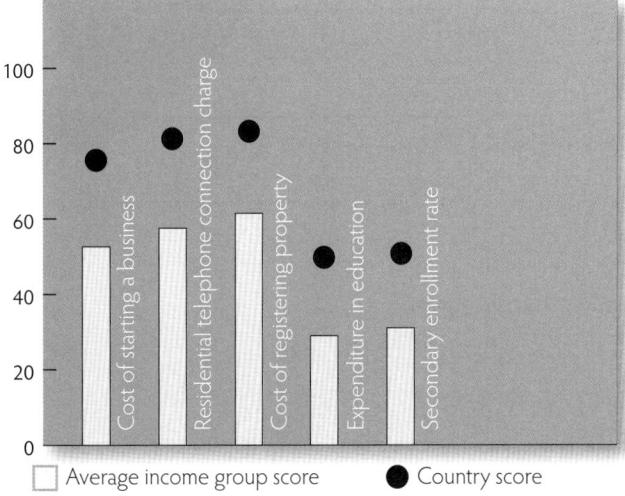

☐ Average income group score   ● Country score

[13] This analysis compares pillar and subindex country scores with respect to the average scores of countries grouped according to income level and political regime (see notes 1 and 2).

[14] Significant country indicators with the greatest distances above or below the income group average scores (i.e., achievements and targets for policy reform, respectively) were selected and ordered according to their Pearson correlation coefficients (highest first) with respect to the final index scores, to produce a ranking.

KOREA, REPUBLIC OF

Key selected indicators

Population (in millions)[3, 4]	48.5
Population density (people per sq. km)[3, 4]	492
GDP (current US$ billion, PPP)[3, 5]	1201
GDP per capita (current US$, PPP)[3, 5]	24783
Current account balance (% of GDP)[3, 5]	0.6
Average life expectancy (years)[4, 6]	78.5
Literacy rate[7, 8]	99
Female labor force participation rate[4, 6, 9]	54.3
Unemployment rate (% of total labor force)[4, 10]	3.7
CO_2 emissions (metric tons per capita)[4, 11]	9.7

Income level:[1] **High-income**

Political regime:[2] **Full democracy**

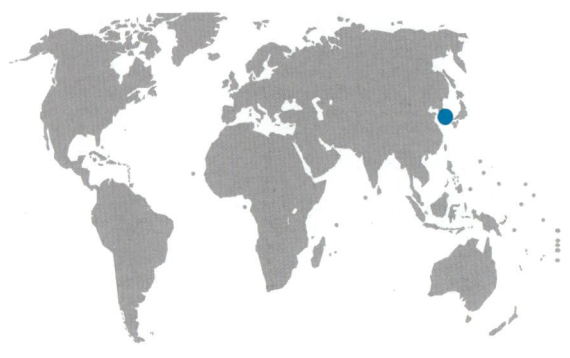

Innovation Capacity Index

Rank	Score
19 out of 131	**70 out of 100**

Pillars and pillar subindexes

Education
Rank **5** Score **81.4**

Social inclusion and equity policies
Rank **48** Score **59.2**

Country policy assessment
Rank **29** Score **60.1**

Good governance
Rank **30** Score **65.7**

Quality of the infrastructure
Rank **22** Score **91.8**

Government ICT usage
Rank **6** Score **83.2**

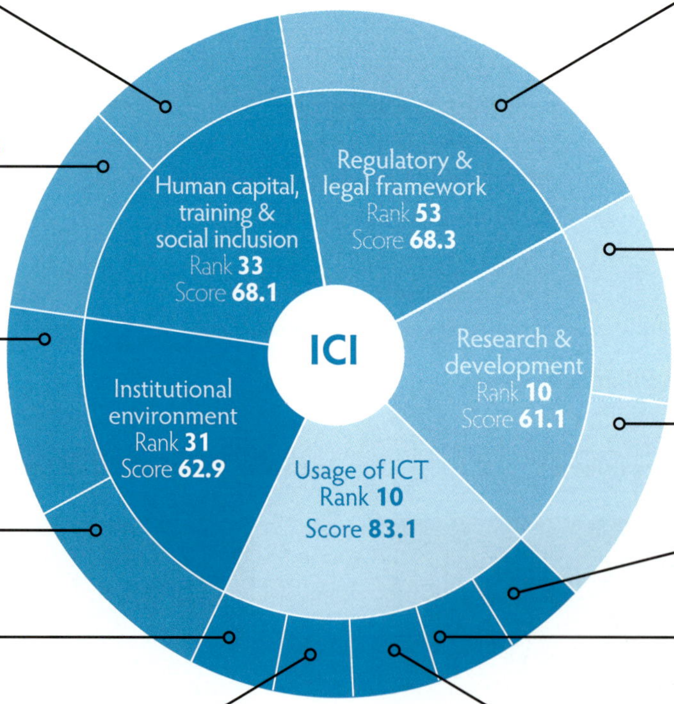

Doing business
Rank **53** Score **68.3**

R&D infrastructure
Rank **7** Score **68.9**

Patents and trademarks
Rank **10** Score **50.1**

Telephone communications
Rank **14** Score **94.5**

Mobile cellular communications
Rank **12** Score **91.4**

Internet, computers and TV
Rank **13** Score **69.7**

Center diagram labels:
- ICI
- Human capital, training & social inclusion — Rank **33** Score **68.1**
- Regulatory & legal framework — Rank **53** Score **68.3**
- Institutional environment — Rank **31** Score **62.9**
- Research & development — Rank **10** Score **61.1**
- Usage of ICT — Rank **10** Score **83.1**

Index thermometer[12]

Pillar 1 9.0%   Pillar 2 9.7%   Pillar 3 19.5%   Pillar 4 26.2%   Pillar 5 35.6%   100%

---

[1] High-income countries are those that, according to the World Bank classification, have a gross national income (GNI) per capita of $11,456 or more.
[2] Full democracies are countries whose Democracy Index scores from 1 to 10 rounded to one decimal place are of 8 to 10.   [3] 2007.   [4] World Bank World Development Indicators.   [5] IMF World Economic Outlook.   [6] 2006.   [7] Estimate.   [8] UNESCO, % age above 15.   [9] % ages 15–64.   [10] 2005.   [11] 2004.   [12] The index thermometer shows the relative contribution of each pillar (in %) to the total index score.

KOREA, REPUBLIC OF

Country relative performance: **Index pillars**[13]

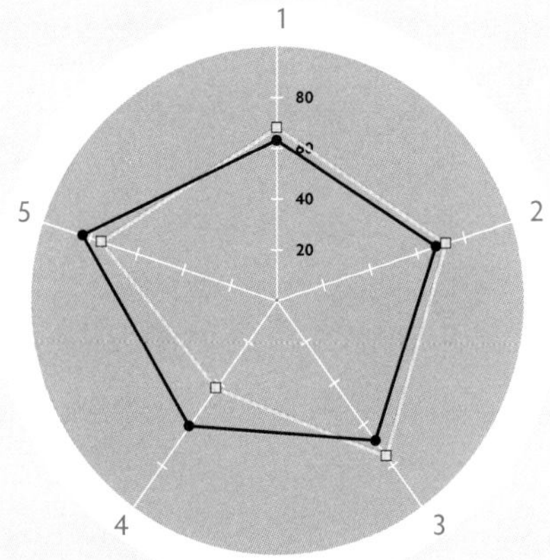

1. Institutional environment
2. Human capital, training and social inclusion
3. Regulatory and legal framework
4. Research and development
5. Usage of ICT

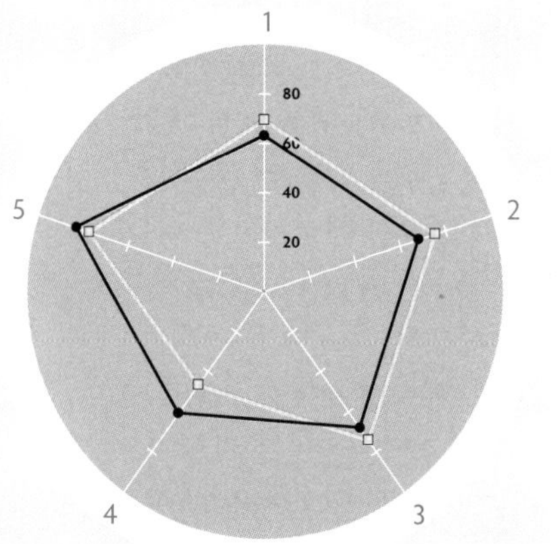

High-income □   Korea ●——

Full democracies □   Korea ●——

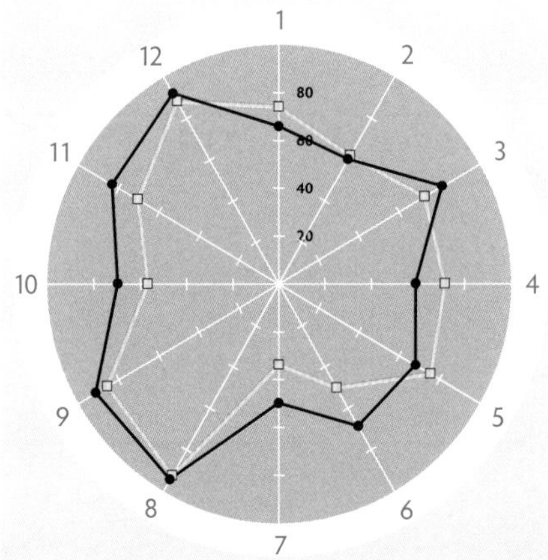

1. Good governance
2. Country policy assessment
3. Education
4. Social inclusion and equity policies
5. Doing business
6. R&D infrastructure
7. Patents and trademarks
8. Telephone communications
9. Mobile/cellular communications
10. Internet, computers and TV
11. Government ICT usage
12. Quality of the infrastructure

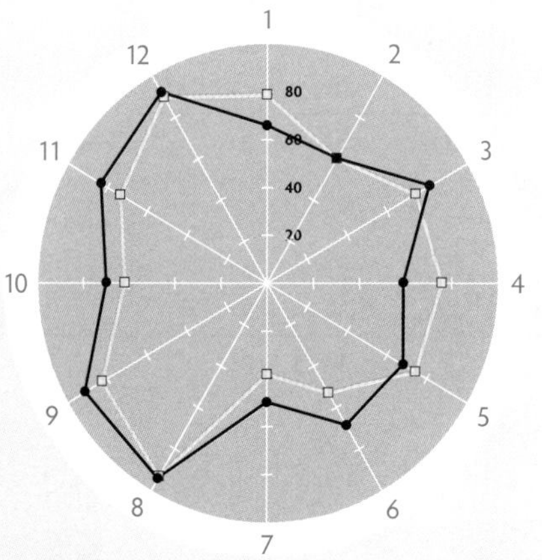

Country relative performance: **Pillar subindexes**[13]

311

In focus: **Top priorities for policy reform**[14]

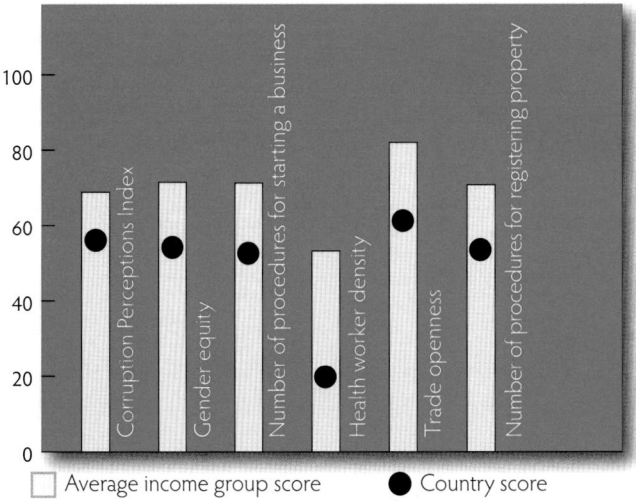

□ Average income group score   ● Country score

In focus: **Significant indicators above income group average**[14]

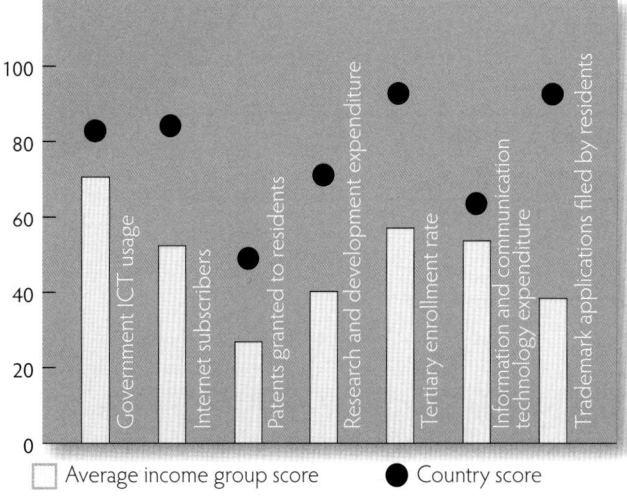

□ Average income group score   ● Country score

[13] This analysis compares pillar and subindex country scores with respect to the average scores of countries grouped according to income level and political regime (see notes 1 and 2).

[14] Significant country indicators with the greatest distances above or below the income group average scores (i.e., achievements and targets for policy reform, respectively) were selected and ordered according to their Pearson correlation coefficients (highest first) with respect to the final index scores, to produce a ranking.

KUWAIT

Key selected indicators

Population (in millions)[3,4]	2.7
Population density (people per sq. km)[3,4]	149
GDP (current US$ billion, PPP)[3,5]	130
GDP per capita (current US$, PPP)[3,5]	39306
Current account balance (% of GDP)[3,5]	47.4
Average life expectancy (years)[4,6]	77.7
Literacy rate[3,7]	94
Female labor force participation rate[4,6,8]	51.4
Unemployment rate (% of total labor force)[4,9]	1.7
CO_2 emissions (metric tons per capita)[4,9]	40.4

Income level:[1] **High-income**

Political regime:[2] **Authoritarian regime**

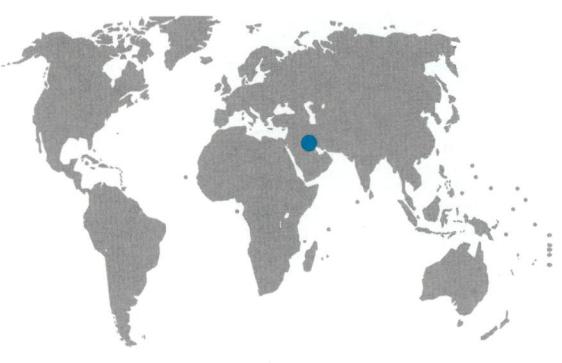

Innovation Capacity Index

Rank	Score
64 out of 131	**50.1 out of 100**

Pillars and pillar subindexes

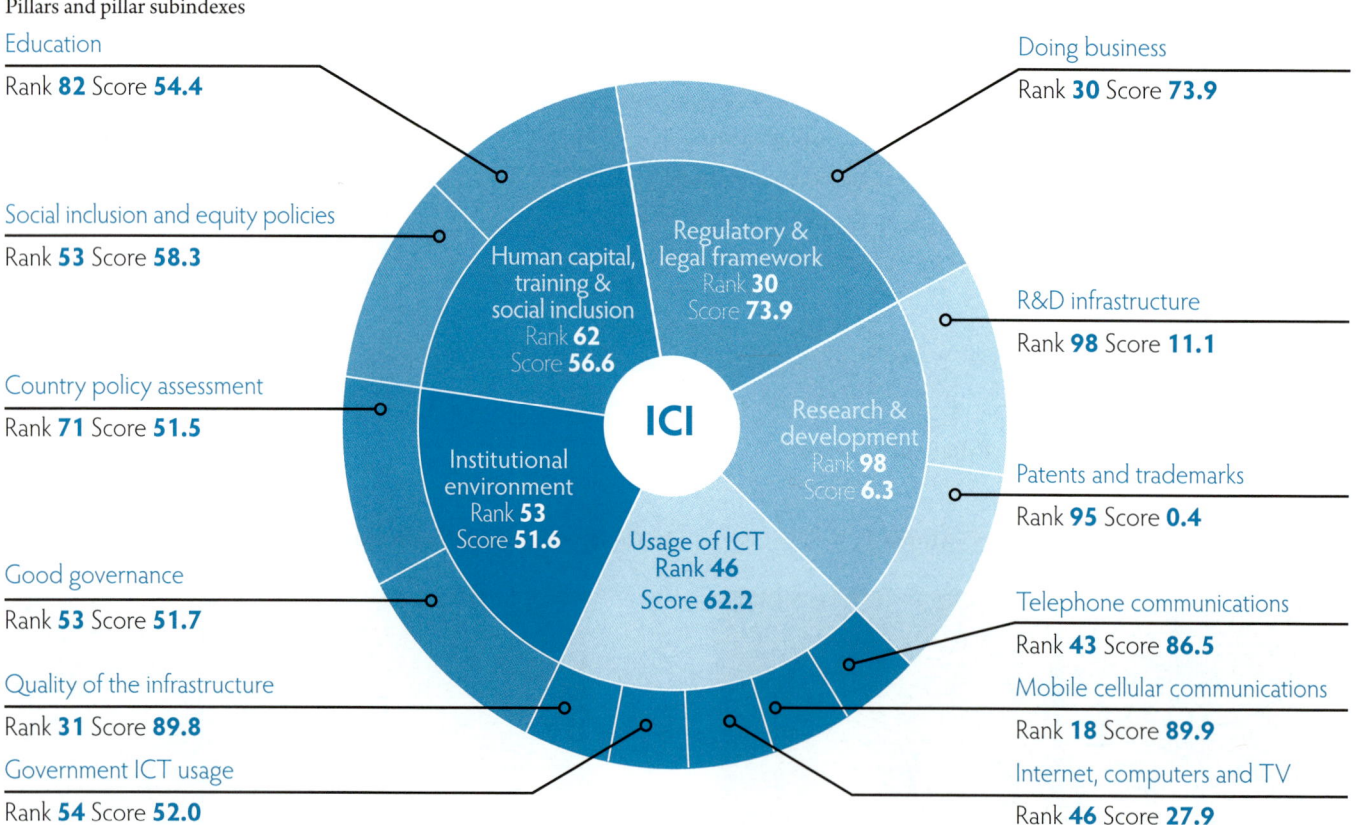

Education
Rank **82** Score **54.4**

Social inclusion and equity policies
Rank **53** Score **58.3**

Country policy assessment
Rank **71** Score **51.5**

Good governance
Rank **53** Score **51.7**

Quality of the infrastructure
Rank **31** Score **89.8**

Government ICT usage
Rank **54** Score **52.0**

Doing business
Rank **30** Score **73.9**

R&D infrastructure
Rank **98** Score **11.1**

Patents and trademarks
Rank **95** Score **0.4**

Telephone communications
Rank **43** Score **86.5**

Mobile cellular communications
Rank **18** Score **89.9**

Internet, computers and TV
Rank **46** Score **27.9**

Human capital, training & social inclusion
Rank 62
Score **56.6**

Regulatory & legal framework
Rank **30**
Score **73.9**

Institutional environment
Rank **53**
Score **51.6**

Research & development
Rank **98**
Score **6.3**

Usage of ICT
Rank **46**
Score **62.2**

ICI

Index thermometer[10]

Pillar 1 20.6%   Pillar 2 22.6%   Pillar 3 29.5%   Pillar 4 2.5%   Pillar 5 24.8%

100%

---

[1] High-income countries are those that, according to the World Bank classification, have a gross national income (GNI) per capita of $11,456 or more.
[2] Authoritarian regimes are countries whose Democracy Index scores from 1 to 10 rounded to one decimal place are less than 4.   [3] 2007.   [4] World Bank World Development Indicators.   [5] IMF World Economic Outlook.   [6] 2006.   [7] UNESCO, % age above 15.   [8] % ages 15–64.   [9] 2004.   [10] The index thermometer shows the relative contribution of each pillar (in %) to the total index score.

KUWAIT

Country relative performance: Index pillars[11]

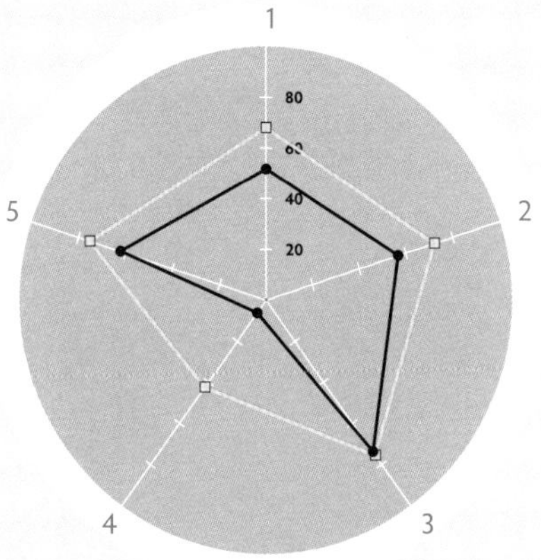

1. Institutional environment
2. Human capital, training and social inclusion
3. Regulatory and legal framework
4. Research and development
5. Usage of ICT

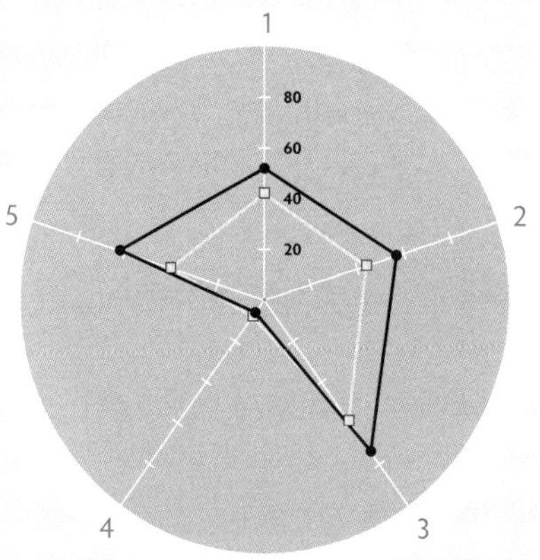

High-income ☐   Kuwait ●

Authoritarian regimes ☐   Kuwait ●

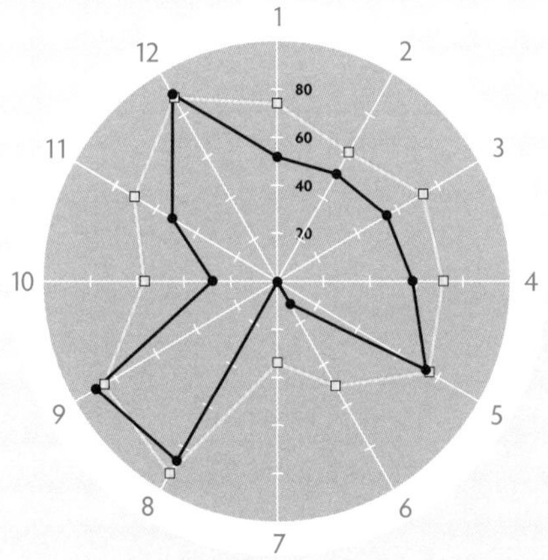

1. Good governance
2. Country policy assessment
3. Education
4. Social inclusion and equity policies
5. Doing business
6. R&D infrastructure
7. Patents and trademarks
8. Telephone communications
9. Mobile/cellular communications
10. Internet, computers and TV
11. Government ICT usage
12. Quality of the infrastructure

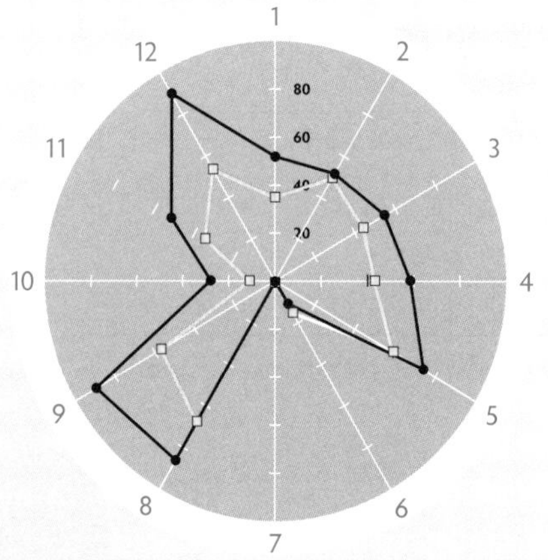

313

Country relative performance: Pillar subindexes[11]

In focus: Top priorities for policy reform[12]

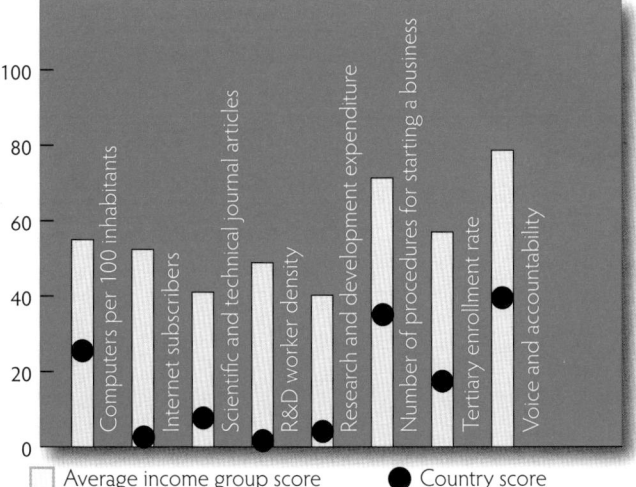

☐ Average income group score   ● Country score

In focus: Significant indicators above income group average[12]

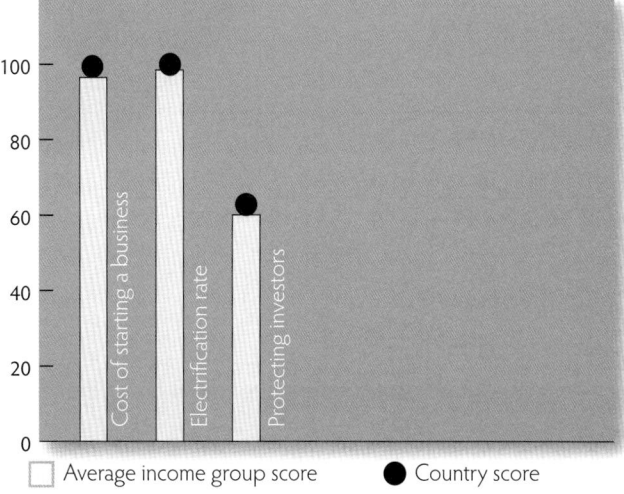

☐ Average income group score   ● Country score

---

[11]   This analysis compares pillar and subindex country scores with respect to the average scores of countries grouped according to income level and political regime (see notes 1 and 2).

[12]   Significant country indicators with the greatest distances above or below the income group average scores (i.e., achievements and targets for policy reform, respectively) were selected and ordered according to their Pearson correlation coefficients (highest first) with respect to the final index scores, to produce a ranking.

LATVIA, REPUBLIC OF

Key selected indicators

Population (in millions)[3,4]	2.3
Population density (people per sq. km)[3,4]	37
GDP (current US$ billion, PPP)[3,5]	40
GDP per capita (current US$, PPP)[3,5]	17416
Current account balance (% of GDP)[3,5]	-23.3
Average life expectancy (years)[4,6]	70.9
Literacy rate[3,7]	100
Female labor force participation rate[4,6,8]	63.1
Unemployment rate (% of total labor force)[4,9]	8.7
CO_2 emissions (metric tons per capita)[4,10]	3.1

Income level:[1] **Upper-middle-income**

Political regime:[2] **Flawed democracy**

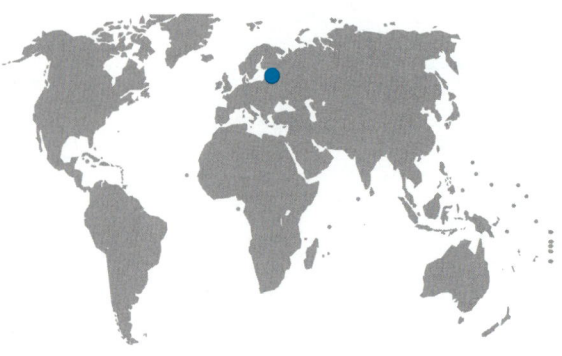

Innovation Capacity Index

Rank
27 out of 131

Score
60.5 out of 100

Pillars and pillar subindexes

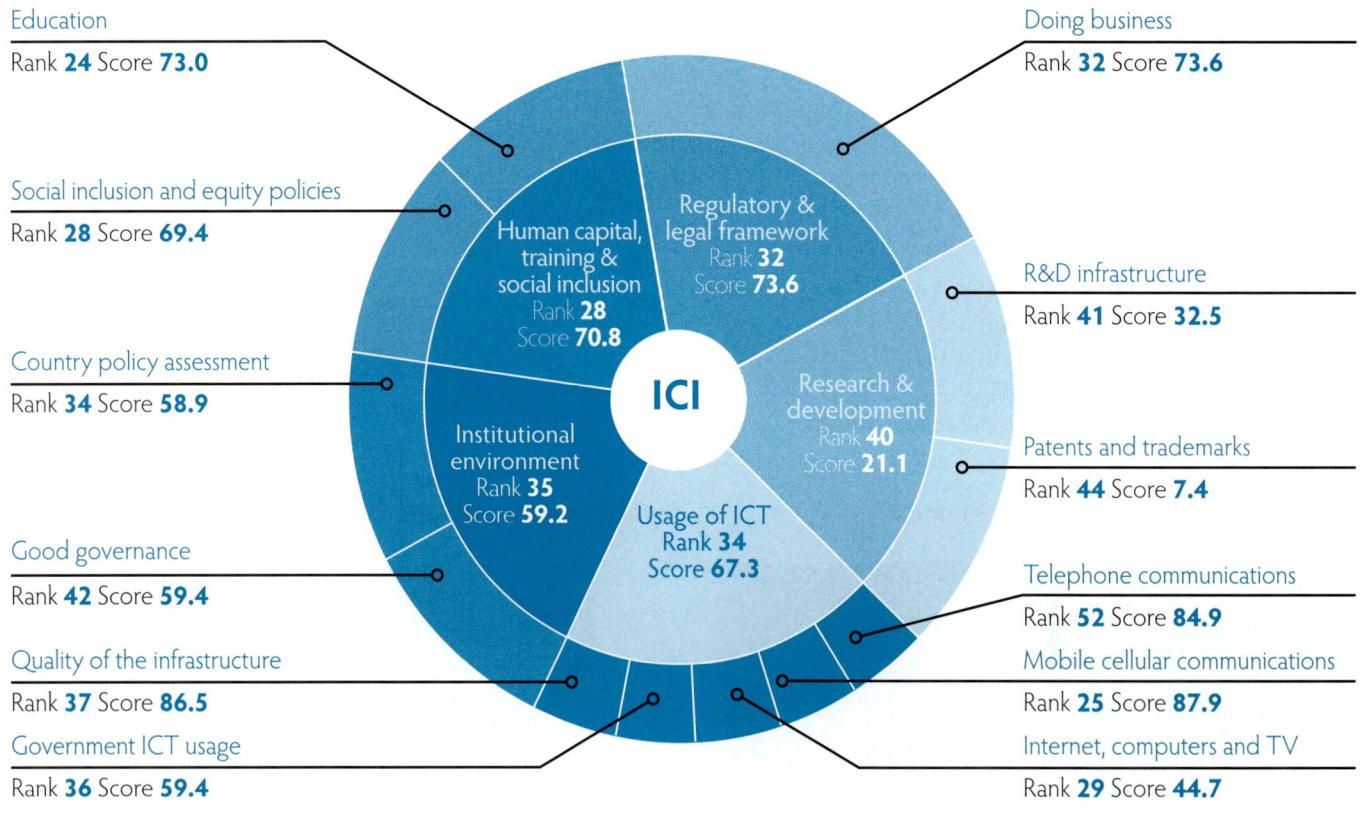

Education
Rank **24** Score **73.0**

Social inclusion and equity policies
Rank **28** Score **69.4**

Country policy assessment
Rank **34** Score **58.9**

Good governance
Rank **42** Score **59.4**

Quality of the infrastructure
Rank **37** Score **86.5**

Government ICT usage
Rank **36** Score **59.4**

Doing business
Rank **32** Score **73.6**

R&D infrastructure
Rank **41** Score **32.5**

Patents and trademarks
Rank **44** Score **7.4**

Telephone communications
Rank **52** Score **84.9**

Mobile cellular communications
Rank **25** Score **87.9**

Internet, computers and TV
Rank **29** Score **44.7**

Human capital, training & social inclusion
Rank **28** Score **70.8**

Regulatory & legal framework
Rank **32** Score **73.6**

Research & development
Rank **40** Score **21.1**

Institutional environment
Rank **35** Score **59.2**

Usage of ICT
Rank **34** Score **67.3**

ICI

Index thermometer[11]

Pillar 1 24.5% Pillar 2 29.3% Pillar 3 24.3% Pillar 4 5.2% Pillar 5 16.7% 100%

[1] Upper-middle-income countries are those that, according to the World Bank classification, have a gross national income (GNI) per capita of $3,706 to $11,455. [2] Flawed democracies are countries whose Democracy Index scores from 1 to 10 rounded to one decimal place are of 6 to 7.9. [3] 2007. [4] World Bank World Development Indicators. [5] IMF World Economic Outlook. [6] 2006. [7] UNESCO, % age above 15. [8] % ages 15–64. [9] 2005. [10] 2004. [11] The index thermometer shows the relative contribution of each pillar (in %) to the total index score.

LATVIA, REPUBLIC OF

Country relative performance: **Index pillars**[12]

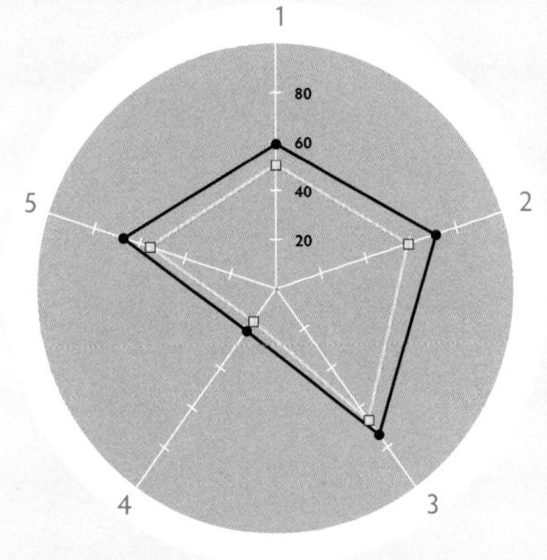

1. Institutional environment
2. Human capital, training and social inclusion
3. Regulatory and legal framework
4. Research and development
5. Usage of ICT

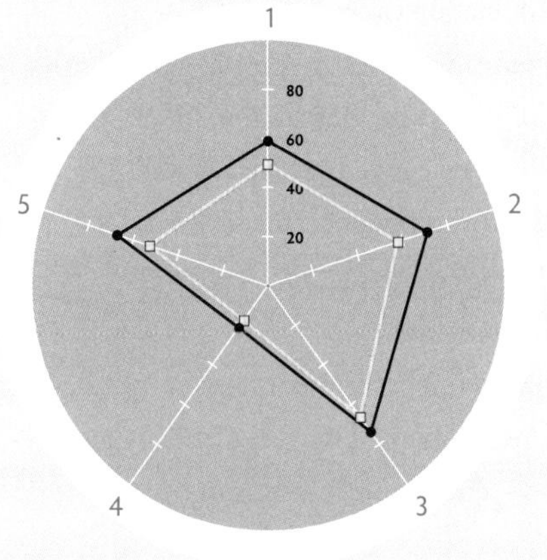

Upper-middle-income ▫   Latvia ●—●

Flawed democracies ▫   Latvia ●—●

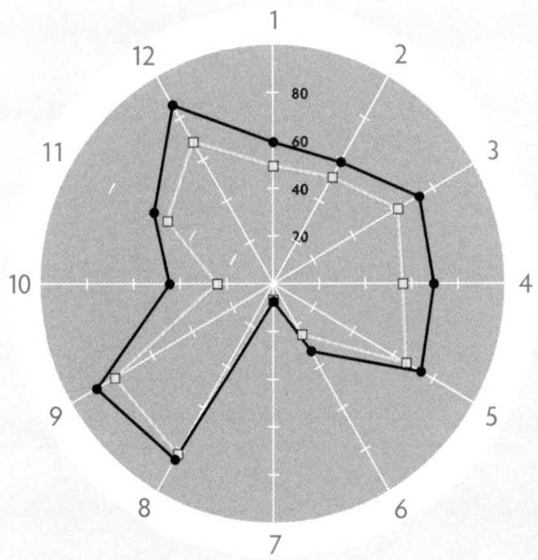

1. Good governance
2. Country policy assessment
3. Education
4. Social inclusion and equity policies
5. Doing business
6. R&D infrastructure
7. Patents and trademarks
8. Telephone communications
9. Mobile/cellular communications
10. Internet, computers and TV
11. Government ICT usage
12. Quality of the infrastructure

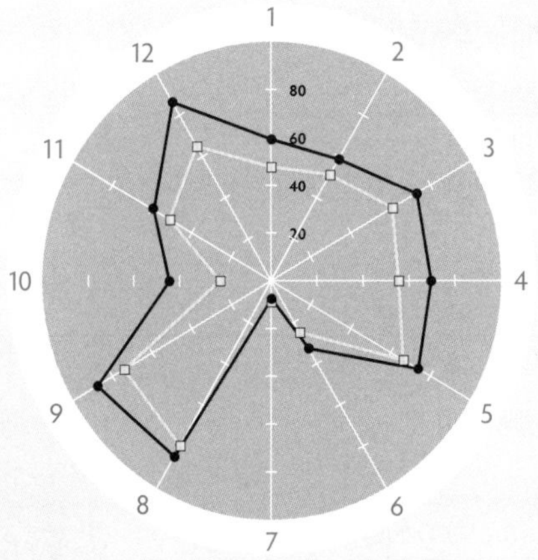

Country relative performance: **Pillar subindexes**[12]

In focus: **Top priorities for policy reform**[13]

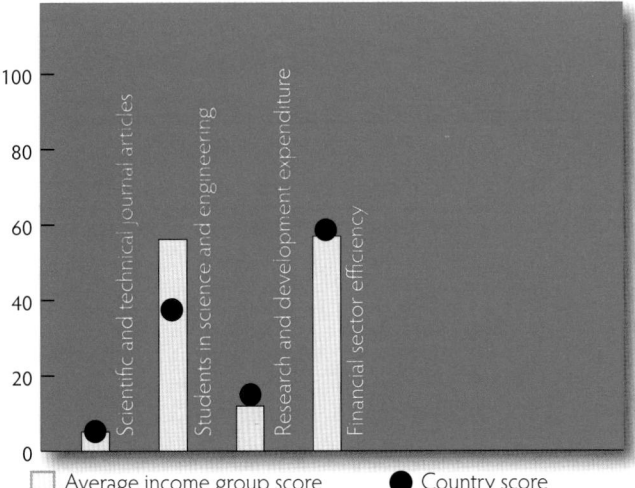

□ Average income group score   ● Country score

In focus: **Significant indicators above income group average**[13]

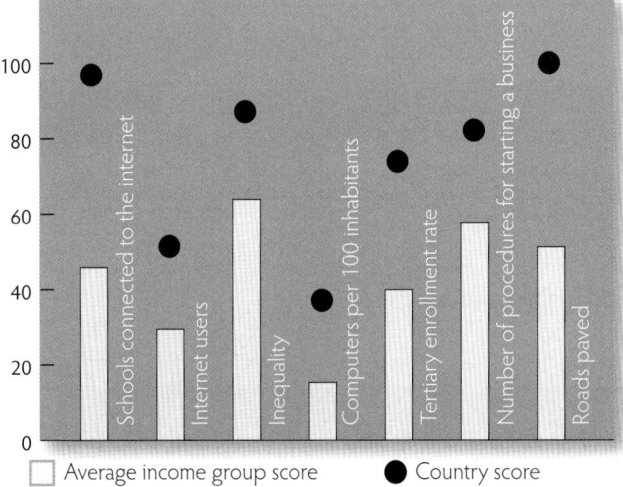

□ Average income group score   ● Country score

[12] This analysis compares pillar and subindex country scores with respect to the average scores of countries grouped according to income level and political regime (see notes 1 and 2).

[13] Significant country indicators with the greatest distances above or below the income group average scores (i.e., achievements and targets for policy reform, respectively) were selected and ordered according to their Pearson correlation coefficients (highest first) with respect to the final index scores, to produce a ranking.

LITHUANIA, REPUBLIC OF

Key selected indicators

Population (in millions)[3, 4]	3.4
Population density (people per sq. km)[3, 4]	54
GDP (current US$ billion, PPP)[3, 5]	60
GDP per capita (current US$, PPP)[3, 5]	17661
Current account balance (% of GDP)[3, 5]	-13.0
Average life expectancy (years)[4, 6]	71.0
Literacy rate[3, 7]	100
Female labor force participation rate[4, 6, 8]	66.4
Unemployment rate (% of total labor force)[4, 9]	8.3
CO_2 emissions (metric tons per capita)[4, 10]	3.9

Income level:[1] **Upper-middle-income**

Political regime:[2] **Flawed democracy**

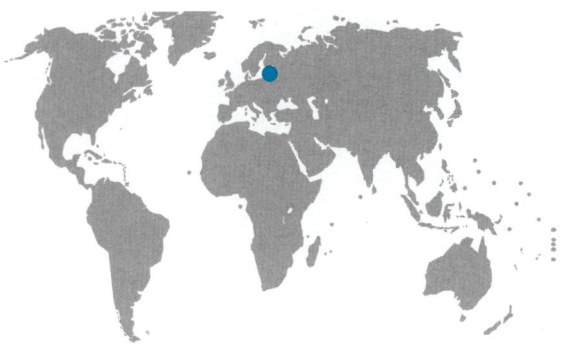

Innovation Capacity Index

Rank	Score
26 out of 131	**60.7 out of 100**

Pillars and pillar subindexes

Education
Rank **12** Score **77.5**

Social inclusion and equity policies
Rank **26** Score **69.8**

Country policy assessment
Rank **36** Score **57.8**

Good governance
Rank **43** Score **58.8**

Quality of the infrastructure
Rank **44** Score **83.7**

Government ICT usage
Rank **28** Score **66.2**

Doing business
Rank **29** Score **74.1**

R&D infrastructure
Rank **42** Score **32.4**

Patents and trademarks
Rank **50** Score **5.8**

Telephone communications
Rank **47** Score **86.2**

Mobile cellular communications
Rank **4** Score **97.7**

Internet, computers and TV
Rank **39** Score **36.5**

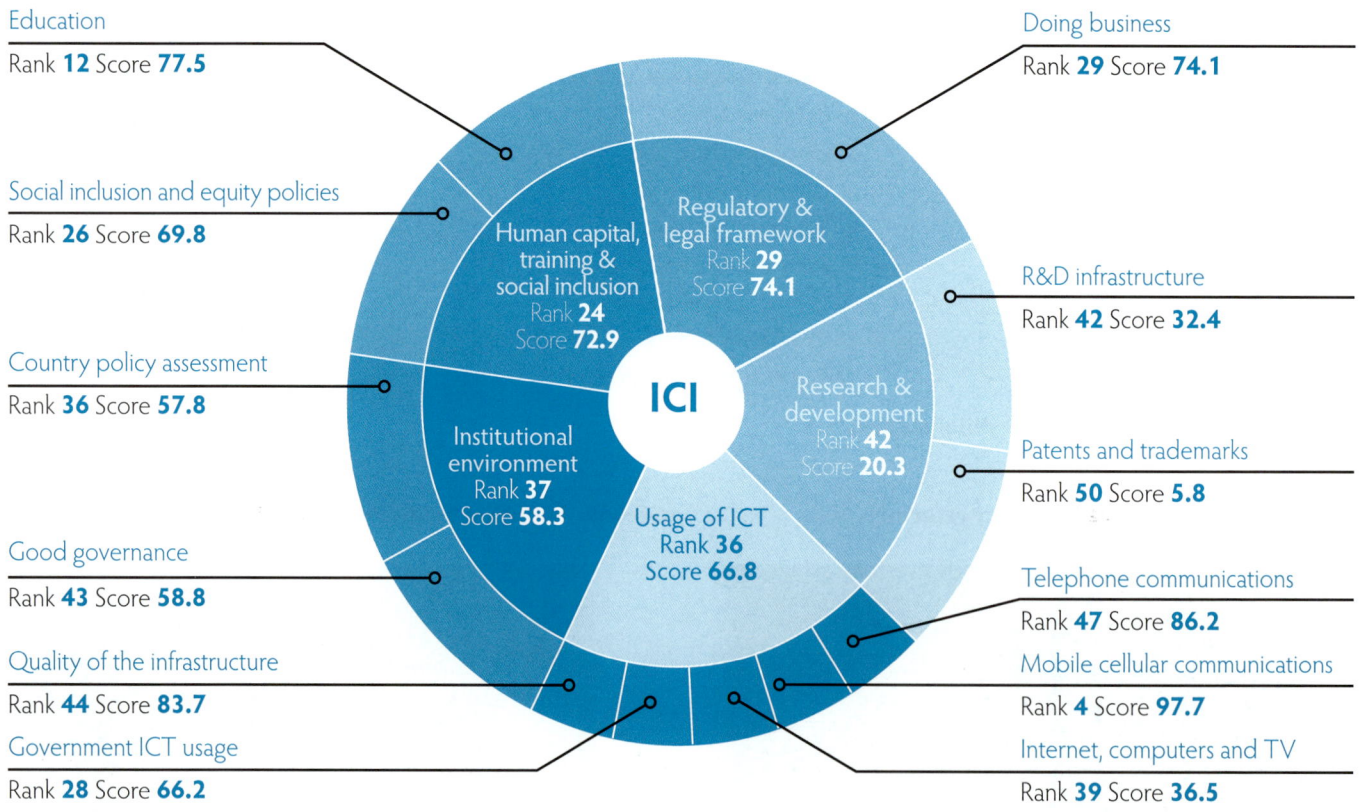

Human capital, training & social inclusion
Rank **24** Score **72.9**

Regulatory & legal framework
Rank **29** Score **74.1**

Institutional environment
Rank **37** Score **58.3**

Research & development
Rank **42** Score **20.3**

Usage of ICT
Rank **36** Score **66.8**

ICI

Index thermometer[11]

Pillar 1 24.0% Pillar 2 30.0% Pillar 3 24.4% Pillar 4 5.0% Pillar 5 16.5% 100%

[1] Upper-middle-income countries are those that, according to the World Bank classification, have a gross national income (GNI) per capita of $3,706 to $11,455. [2] Flawed democracies are countries whose Democracy Index scores from 1 to 10 rounded to one decimal place are of 6 to 7.9. [3] 2007. [4] World Bank World Development Indicators. [5] IMF World Economic Outlook. [6] 2006. [7] UNESCO, % age above 15. [8] % ages 15–64. [9] 2005. [10] 2004. [11] The index thermometer shows the relative contribution of each pillar (in %) to the total index score.

LITHUANIA, REPUBLIC OF

Country relative performance: **Index pillars**[12]

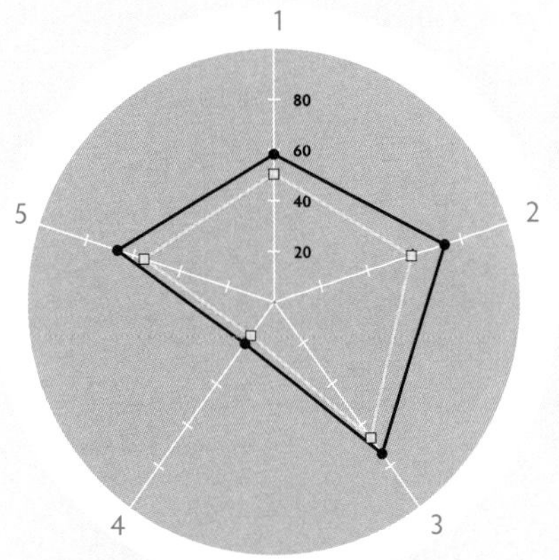

1. Institutional environment
2. Human capital, training and social inclusion
3. Regulatory and legal framework
4. Research and development
5. Usage of ICT

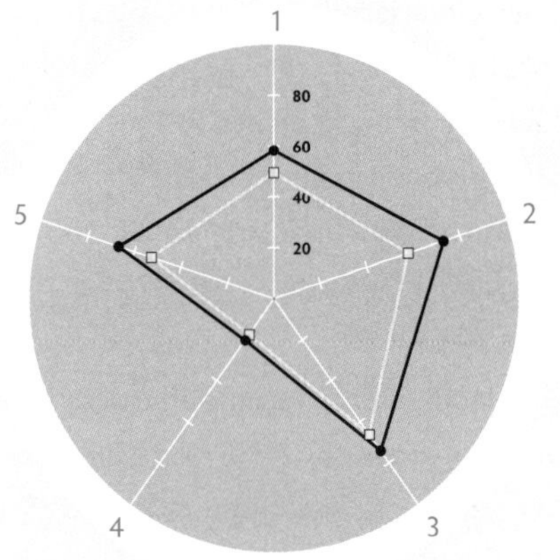

Upper-middle-income ☐   Lithuania ●——

Flawed democracies ☐   Lithuania ●——

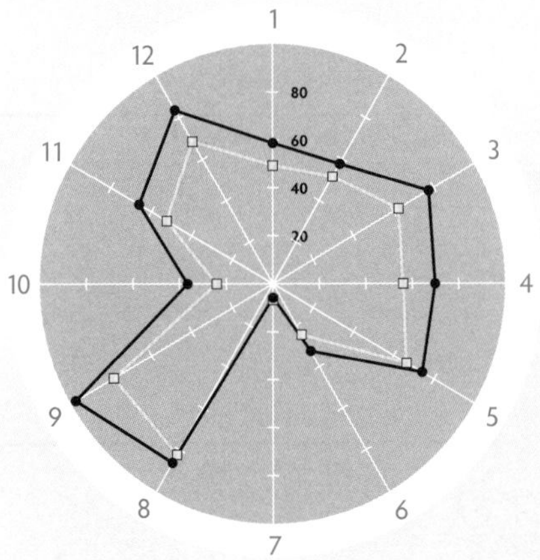

1. Good governance
2. Country policy assessment
3. Education
4. Social inclusion and equity policies
5. Doing business
6. R&D infrastructure
7. Patents and trademarks
8. Telephone communications
9. Mobile/cellular communications
10. Internet, computers and TV
11. Government ICT usage
12. Quality of the infrastructure

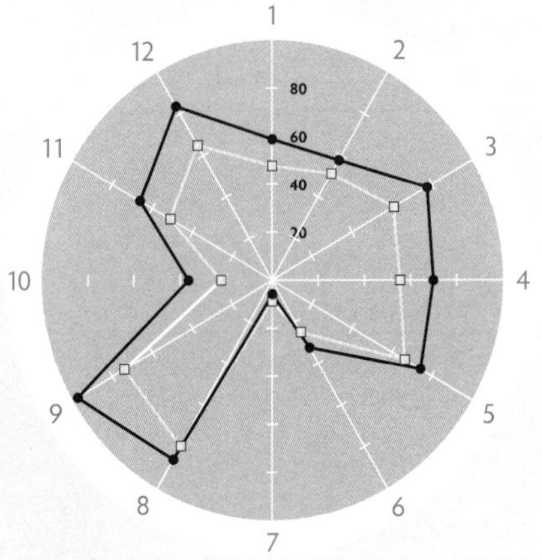

Country relative performance: **Pillar subindexes**[12]

In focus: **Top priorities for policy reform**[13]

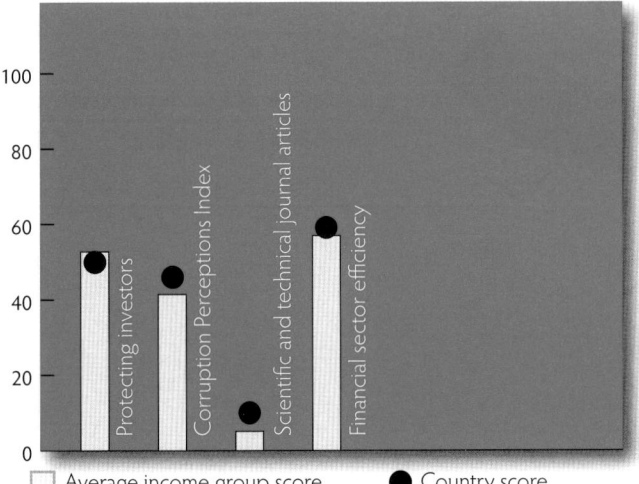

☐ Average income group score   ● Country score

In focus: **Significant indicators above income group average**[13]

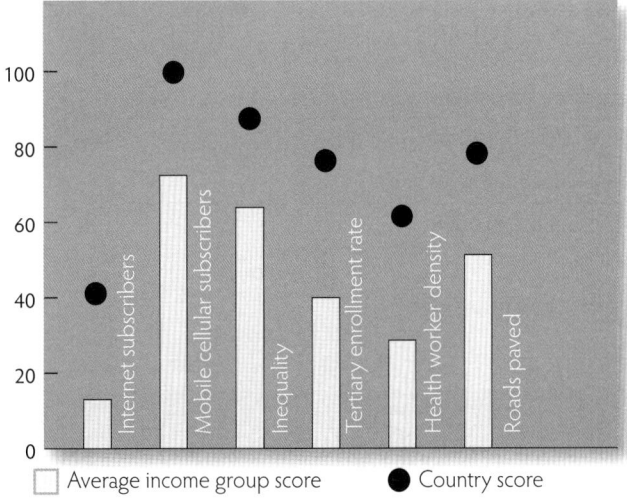

☐ Average income group score   ● Country score

[12] This analysis compares pillar and subindex country scores with respect to the average scores of countries grouped according to income level and political regime (see notes 1 and 2).

[13] Significant country indicators with the greatest distances above or below the income group average scores (i.e., achievements and targets for policy reform, respectively) were selected and ordered according to their Pearson correlation coefficients (highest first) with respect to the final index scores, to produce a ranking.

MALAYSIA

Key selected indicators

Population (in millions)[3,4]	26.5
Population density (people per sq. km)[3,4]	81
GDP (current US$ billion, PPP)[3,5]	357
GDP per capita (current US$, PPP)[3,5]	13315
Current account balance (% of GDP)[3,5]	14.0
Average life expectancy (years)[4,6]	74.0
Literacy rate[3,7]	92
Female labor force participation rate[4,6,8]	48.5
Unemployment rate (% of total labor force)[4,9]	3.5
CO$_2$ emissions (metric tons per capita)[4,9]	7.0

Income level:[1] **Upper-middle-income**
Political regime:[2] **Flawed democracy**

Innovation Capacity Index

Rank **34 out of 131** Score **57.3 out of 100**

Pillars and pillar subindexes

Education — Rank **76** Score **56.9**
Social inclusion and equity policies — Rank **68** Score **53.2**
Country policy assessment — Rank **35** Score **58.7**
Good governance — Rank **45** Score **58.0**
Quality of the infrastructure — Rank **21** Score **92.2**
Government ICT usage — Rank **34** Score **60.6**

Doing business — Rank **14** Score **80.1**
R&D infrastructure — Rank **49** Score **30.3**
Patents and trademarks — Rank **41** Score **8.3**
Telephone communications — Rank **54** Score **84.6**
Mobile cellular communications — Rank **47** Score **82.2**
Internet, computers and TV — Rank **37** Score **38.7**

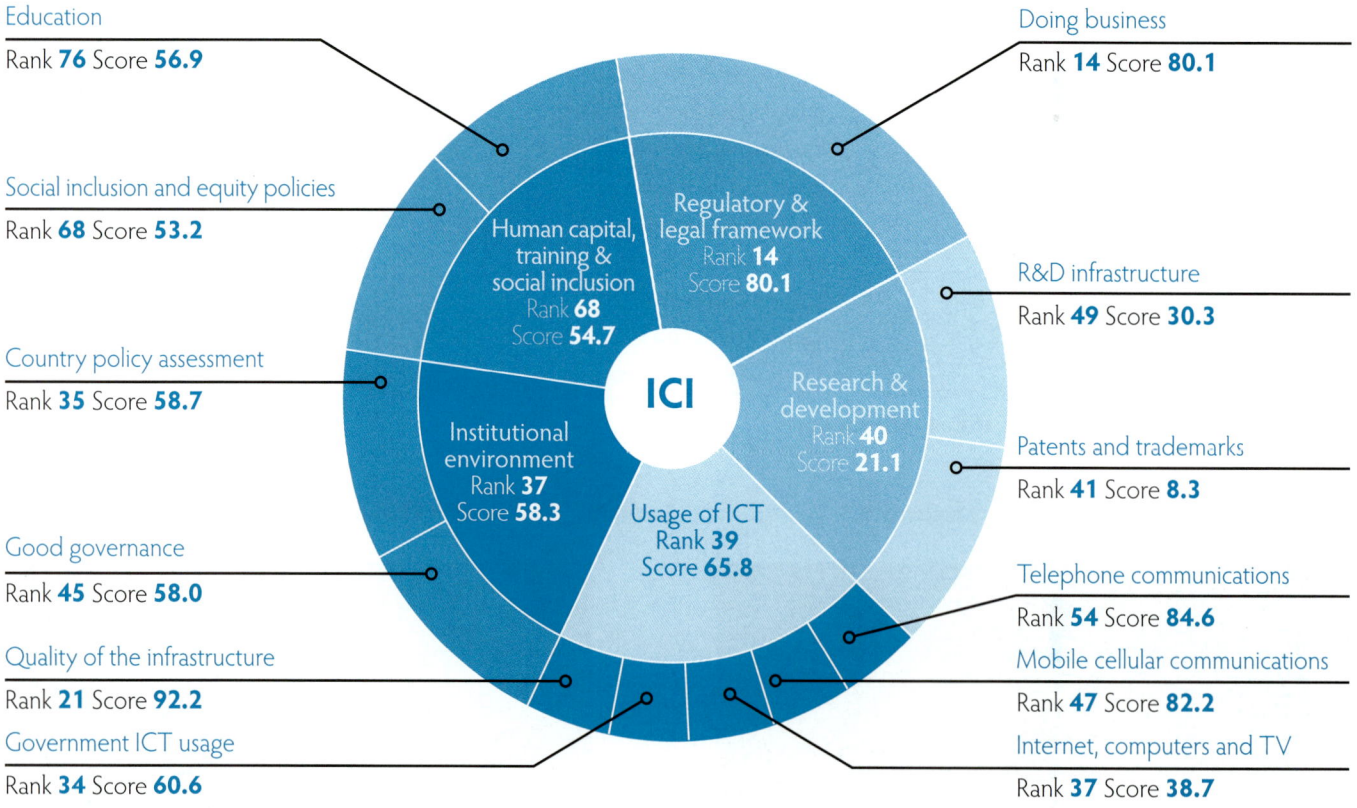

Index thermometer[10]

Pillar 1 25.5% Pillar 2 23.9% Pillar 3 27.9% Pillar 4 5.5% Pillar 5 17.2% 100%

[1] Upper-middle-income countries are those that, according to the World Bank classification, have a gross national income (GNI) per capita of $3,706 to $11,455. [2] Flawed democracies are countries whose Democracy Index scores from 1 to 10 rounded to one decimal place are of 6 to 7.9. [3] 2007. [4] World Bank World Development Indicators. [5] IMF World Economic Outlook. [6] 2006. [7] UNESCO, % age above 15. [8] % ages 15–64. [9] 2005. [10] The index thermometer shows the relative contribution of each pillar (in %) to the total index score.

MALAYSIA

Country relative performance: **Index pillars**[11]

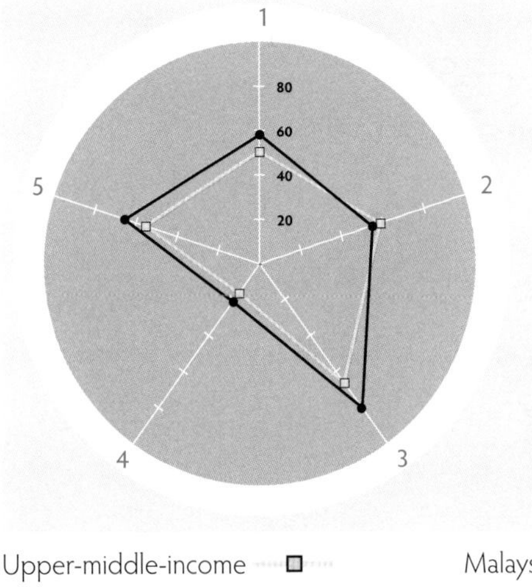

1. Institutional environment
2. Human capital, training and social inclusion
3. Regulatory and legal framework
4. Research and development
5. Usage of ICT

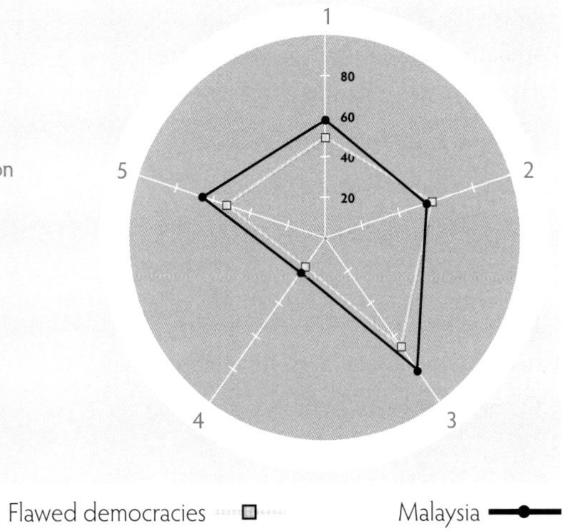

Upper-middle-income ☐   Malaysia ●

Flawed democracies ☐   Malaysia ●

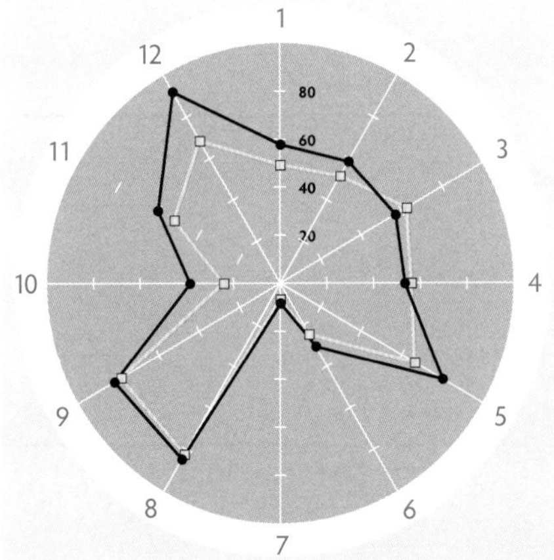

1. Good governance
2. Country policy assessment
3. Education
4. Social inclusion and equity policies
5. Doing business
6. R&D infrastructure
7. Patents and trademarks
8. Telephone communications
9. Mobile/cellular communications
10. Internet, computers and TV
11. Government ICT usage
12. Quality of the infrastructure *

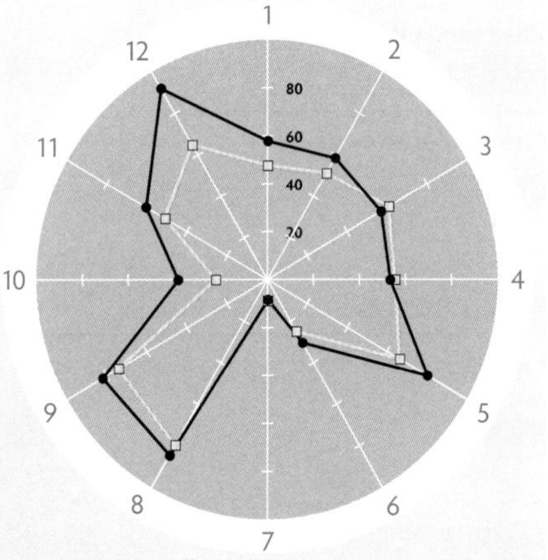

Country relative performance: **Pillar subindexes**[11]

In focus: **Top priorities for policy reform**[12]

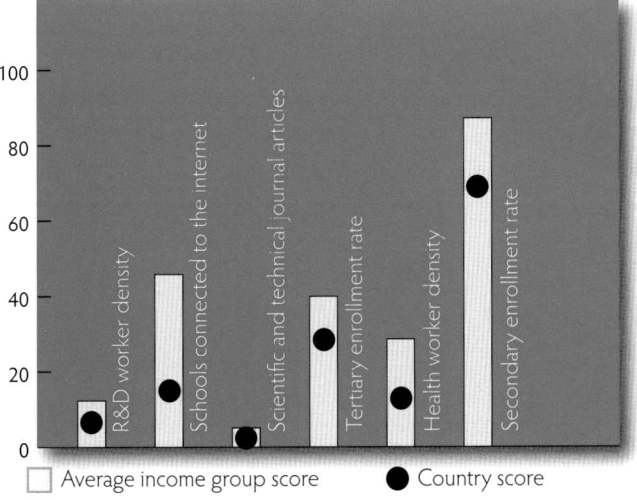

☐ Average income group score   ● Country score

In focus: **Significant indicators above income group average**[12]

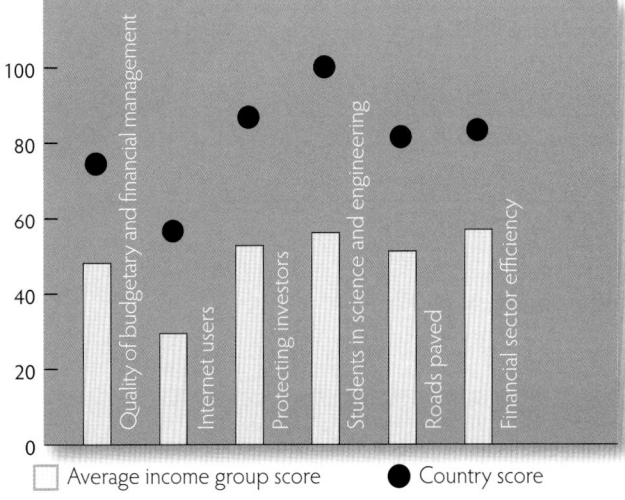

☐ Average income group score   ● Country score

[11] This analysis compares pillar and subindex country scores with respect to the average scores of countries grouped according to income level and political regime (see notes 1 and 2).

[12] Significant country indicators with the greatest distances above or below the income group average scores (i.e., achievements and targets for policy reform, respectively) were selected and ordered according to their Pearson correlation coefficients (highest first) with respect to the final index scores, to produce a ranking.

MEXICO

Key selected indicators

Population (in millions)[3,4]	105.3
Population density (people per sq. km)[3,4]	54
GDP (current US$ billion, PPP)[3,5]	1346
GDP per capita (current US$, PPP)[3,5]	12775
Current account balance (% of GDP)[3,5]	-0.8
Average life expectancy (years)[4,6]	74.5
Literacy rate[3,7]	92
Female labor force participation rate[4,6,8]	43.0
Unemployment rate (% of total labor force)[4,9]	3.5
CO_2 emissions (metric tons per capita)[4,10]	4.3

Income level:[1] **Upper-middle-income**

Political regime:[2] **Flawed democracy**

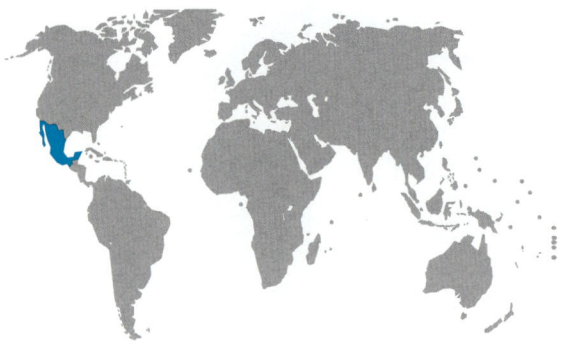

Rank	Score
61 out of 131	**50.5 out of 100**

Innovation Capacity Index

Pillars and pillar subindexes

Education
Rank **70** Score **58.5**

Social inclusion and equity policies
Rank **57** Score **55.9**

Country policy assessment
Rank **87** Score **48.9**

Good governance
Rank **71** Score **42.3**

Quality of the infrastructure
Rank **86** Score **55.8**

Government ICT usage
Rank **37** Score **58.9**

Doing business
Rank **47** Score **69.9**

R&D infrastructure
Rank **50** Score **30.2**

Patents and trademarks
Rank **54** Score **4.7**

Telephone communications
Rank **60** Score **83.0**

Mobile cellular communications
Rank **61** Score **76.2**

Internet, computers and TV
Rank **59** Score **19.0**

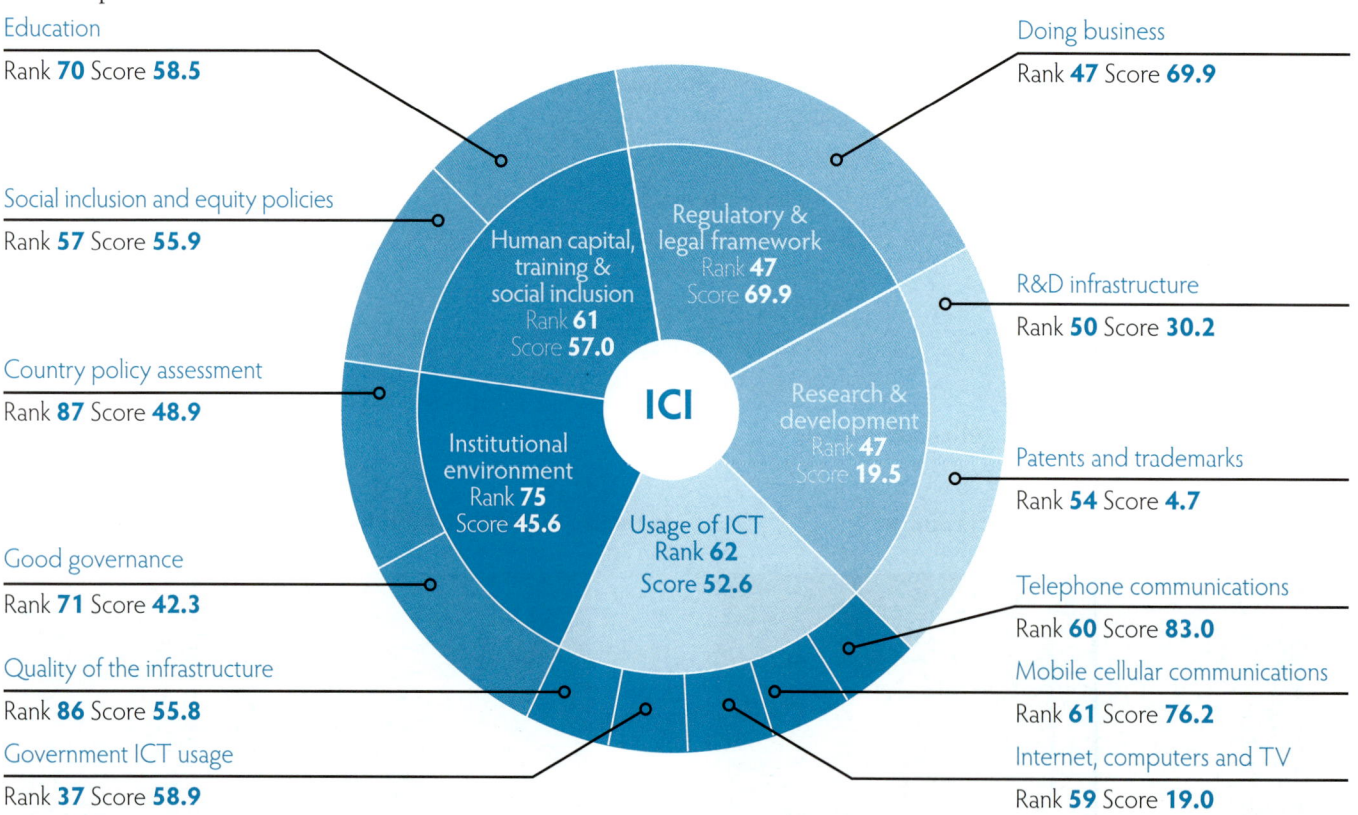

ICI

Human capital, training & social inclusion
Rank **61** Score **57.0**

Regulatory & legal framework
Rank **47** Score **69.9**

Institutional environment
Rank **75** Score **45.6**

Research & development
Rank **47** Score **19.5**

Usage of ICT
Rank **62** Score **52.6**

Index thermometer[11]

Pillar 1 22.6% Pillar 2 28.2% Pillar 3 27.7% Pillar 4 5.8% Pillar 5 15.6%

100%

[1] Upper-middle-income countries are those that, according to the World Bank classification, have a gross national income (GNI) per capita of $3,706 to $11,455. [2] Flawed democracies are countries whose Democracy Index scores from 1 to 10 rounded to one decimal place are of 6 to 7.9. [3] 2007. [4] World Bank World Development Indicators. [5] IMF World Economic Outlook. [6] 2006. [7] UNESCO, % age above 15. [8] % ages 15–64. [9] 2005. [10] 2004 . [11] The index thermometer shows the relative contribution of each pillar (in %) to the total index score.

MEXICO

Country relative performance: **Index pillars**[12]

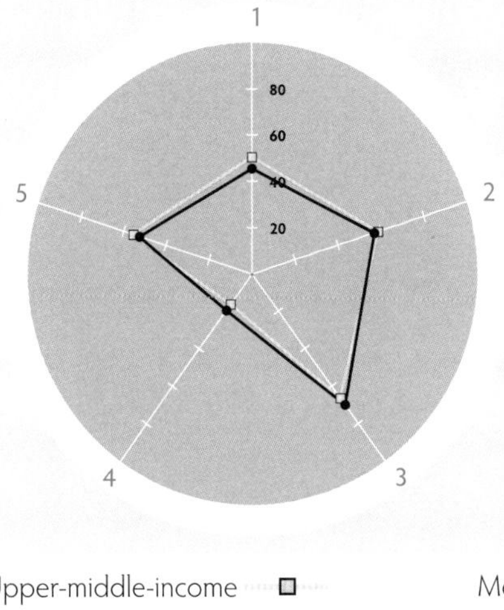

1. Institutional environment
2. Human capital, training and social inclusion
3. Regulatory and legal framework
4. Research and development
5. Usage of ICT

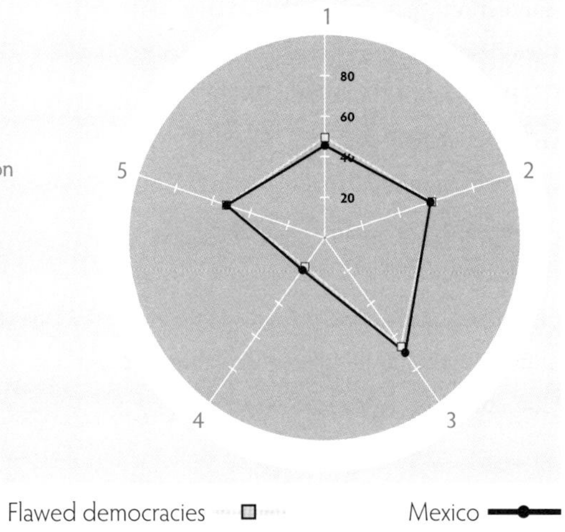

Upper-middle-income ▫   Mexico ●—

Flawed democracies ▫   Mexico ●—

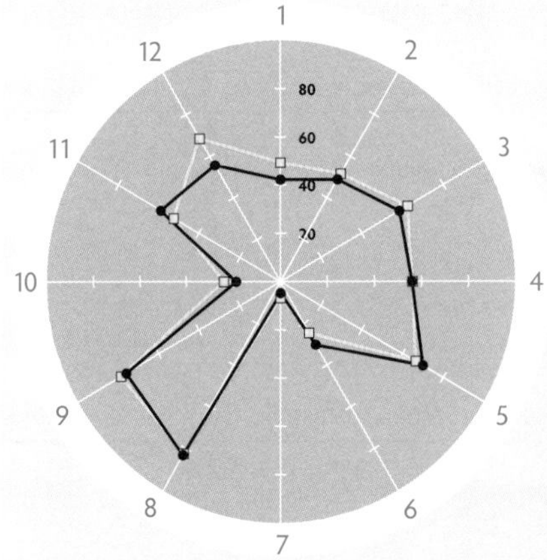

1. Good governance
2. Country policy assessment
3. Education
4. Social inclusion and equity policies
5. Doing business
6. R&D infrastructure
7. Patents and trademarks
8. Telephone communications
9. Mobile/cellular communications
10. Internet, computers and TV
11. Government ICT usage
12. Quality of the infrastructure

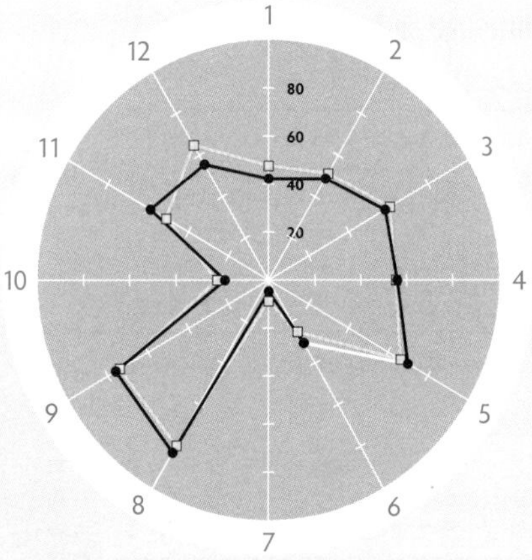

Country relative performance: **Pillar subindexes**[12]

In focus: **Top priorities for policy reform**[13]

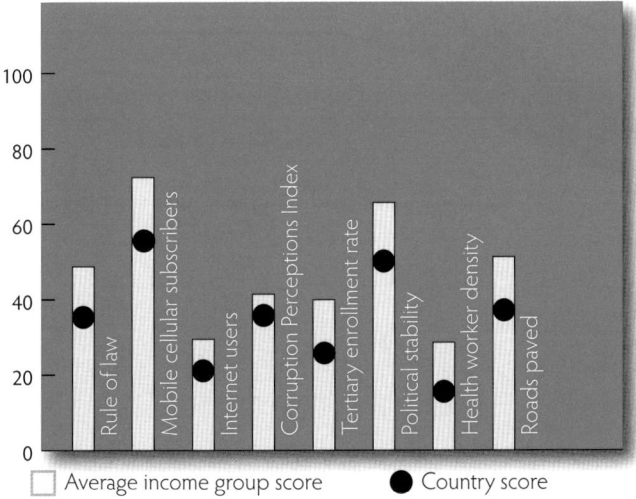

☐ Average income group score   ● Country score

In focus: **Significant indicators above income group average**[13]

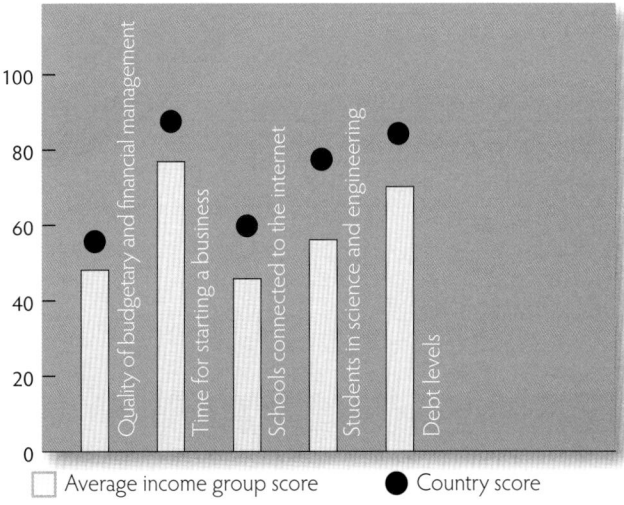

☐ Average income group score   ● Country score

---

[12] This analysis compares pillar and subindex country scores with respect to the average scores of countries grouped according to income level and political regime (see notes 1 and 2).

[13] Significant country indicators with the greatest distances above or below the income group average scores (i.e., achievements and targets for policy reform, respectively) were selected and ordered according to their Pearson correlation coefficients (highest first) with respect to the final index scores, to produce a ranking.

NETHERLANDS

Key selected indicators

Population (in millions)[3,4]	16.4
Population density (people per sq. km)[3,4]	484
GDP (current US$ billion, PPP)[3,5]	640
GDP per capita (current US$, PPP)[3,5]	38486
Current account balance (% of GDP)[3,5]	6.6
Average life expectancy (years)[4,6]	79.7
Literacy rate[7,8]	99
Female labor force participation rate[4,6,9]	70.1
Unemployment rate (% of total labor force)[4,10]	5.2
CO_2 emissions (metric tons per capita)[4,11]	8.7

Income level:[1] **High-income**

Political regime:[2] **Full democracy**

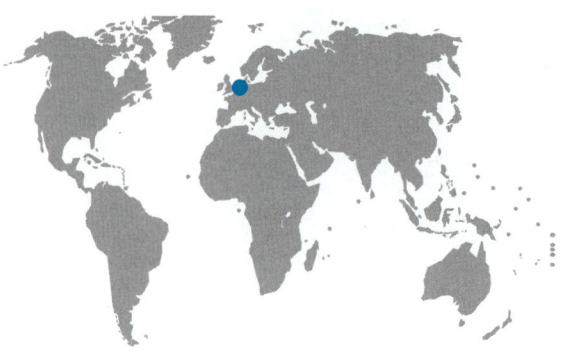

Rank	Score
5 out of 131	**76.6 out of 100**

Innovation Capacity Index

Pillars and pillar subindexes

Education
Rank **19** Score **75.2**

Social inclusion and equity policies
Rank **3** Score **86.6**

Country policy assessment
Rank **20** Score **65.1**

Good governance
Rank **7** Score **89.3**

Quality of the infrastructure
Rank **17** Score **95.6**

Government ICT usage
Rank **5** Score **86.3**

Doing business
Rank **28** Score **74.2**

R&D infrastructure
Rank **19** Score **53.9**

Patents and trademarks
Rank **5** Score **71.1**

Telephone communications
Rank **17** Score **93.4**

Mobile cellular communications
Rank **30** Score **86.2**

Internet, computers and TV
Rank **1** Score **95.9**

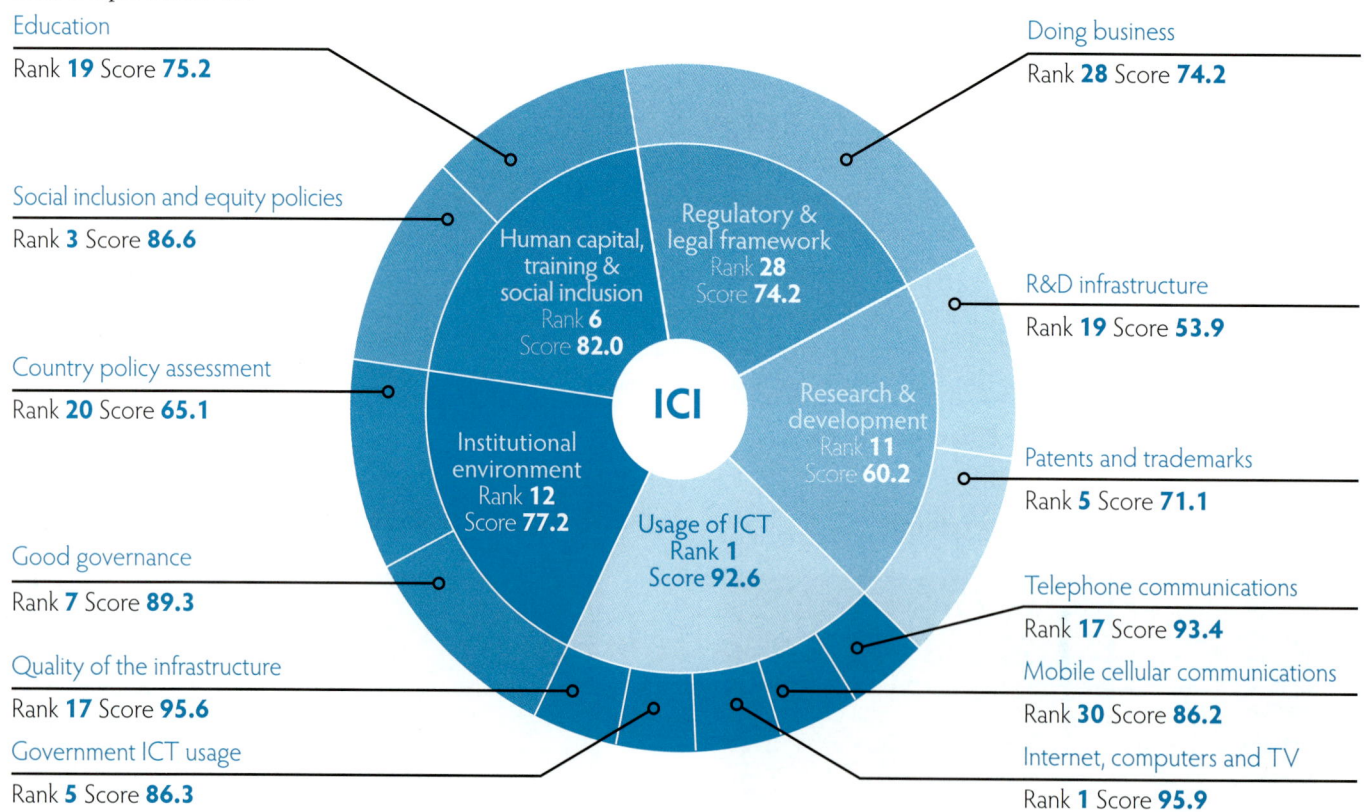

ICI

Human capital, training & social inclusion
Rank **6** Score **82.0**

Regulatory & legal framework
Rank **28** Score **74.2**

Research & development
Rank **11** Score **60.2**

Institutional environment
Rank **12** Score **77.2**

Usage of ICT
Rank **1** Score **92.6**

Index thermometer[12]

Pillar 1 10.1%   Pillar 2 10.7%   Pillar 3 19.4%   Pillar 4 23.6%   Pillar 5 36.3%   100%

[1] High-income countries are those that, according to the World Bank classification, have a gross national income (GNI) per capita of $11,456 or more.
[2] Full democracies are countries whose Democracy Index scores from 1 to 10 rounded to one decimal place are of 8 to 10.   [3] 2007.   [4] World Bank World Development Indicators.   [5] IMF World Economic Outlook.   [6] 2006.   [7] Estimate.   [8] UNESCO, % age above 15.   [9] % ages 15–64.   [10] 2005.   [11] 2004.   [12] The index thermometer shows the relative contribution of each pillar (in %) to the total index score.

NETHERLANDS

Country relative performance: **Index pillars**[13]

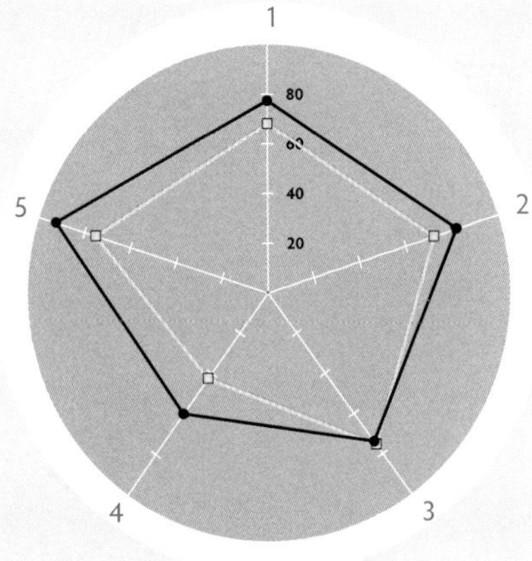

1. Institutional environment
2. Human capital, training and social inclusion
3. Regulatory and legal framework
4. Research and development
5. Usage of ICT

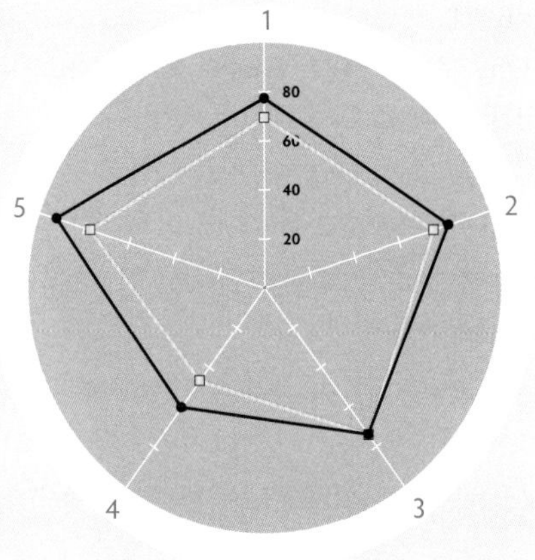

High-income ·····□····   Netherlands ●━━━

Full democracies ·····□····   Netherlands ●━━━

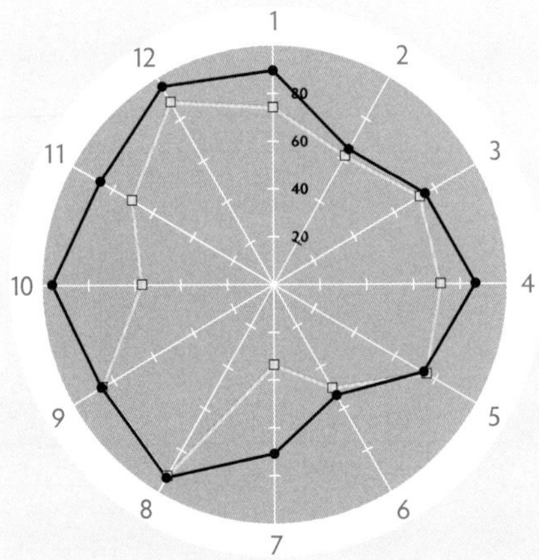

1. Good governance
2. Country policy assessment
3. Education
4. Social inclusion and equity policies
5. Doing business
6. R&D infrastructure
7. Patents and trademarks
8. Telephone communications
9. Mobile/cellular communications
10. Internet, computers and TV
11. Government ICT usage
12. Quality of the infrastructure

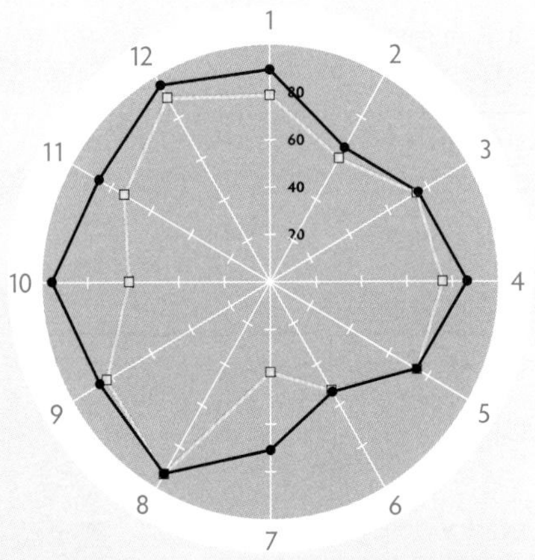

Country relative performance: **Pillar subindexes**[13]

In focus: **Top priorities for policy reform**[14]

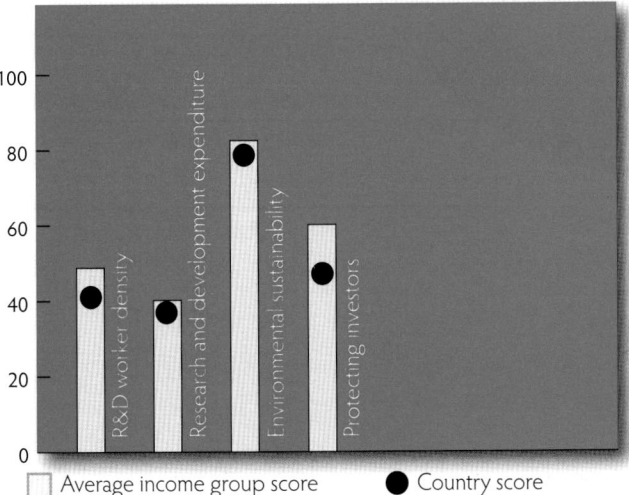

□ Average income group score   ● Country score

In focus: **Significant indicators above income group average**[14]

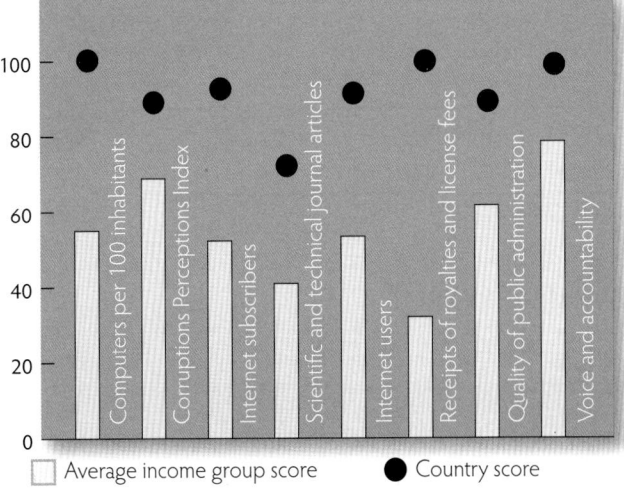

□ Average income group score   ● Country score

[13] This analysis compares pillar and subindex country scores with respect to the average scores of countries grouped according to income level and political regime (see notes 1 and 2).

[14] Significant country indicators with the greatest distances above or below the income group average scores (i.e., achievements and targets for policy reform, respectively) were selected and ordered according to their Pearson correlation coefficients (highest first) with respect to the final index scores, to produce a ranking.

NEW ZEALAND

Key selected indicators

Population (in millions)[3,4]	4.2
Population density (people per sq. km)[3,4]	16
GDP (current US$ billion, PPP)[3,5]	112
GDP per capita (current US$, PPP)[3,5]	26379
Current account balance (% of GDP)[3,5]	-8.1
Average life expectancy (years)[4,6]	79.9
Literacy rate[7,8]	99
Female labor force participation rate[4,6,9]	71.8
Unemployment rate (% of total labor force)[4,10]	3.7
CO$_2$ emissions (metric tons per capita)[4,11]	7.7

Income level:[1] **High-income**

Political regime:[2] **Full democracy**

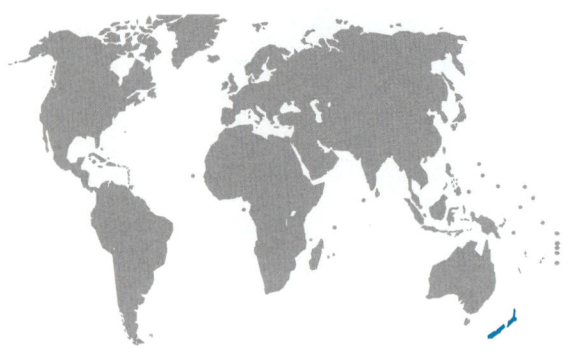

Innovation Capacity Index

Rank **10 out of 131**

Score **73.4 out of 100**

Pillars and pillar subindexes

Education
Rank **9** Score **79.6**

Social inclusion and equity policies
Rank **14** Score **78.9**

Country policy assessment
Rank **10** Score **71.4**

Good governance
Rank **4** Score **92.7**

Quality of the infrastructure
Rank **39** Score **85.6**

Government ICT usage
Rank **17** Score **73.9**

Doing business
Rank **1** Score **96.2**

R&D infrastructure
Rank **22** Score **52.0**

Patents and trademarks
Rank **18** Score **37.9**

Telephone communications
Rank **27** Score **91.8**

Mobile cellular communications
Rank **28** Score **86.6**

Internet, computers and TV
Rank **11** Score **71.6**

Human capital, training & social inclusion
Rank **10**
Score **79.2**

Regulatory & legal framework
Rank **1**
Score **96.2**

ICI

Institutional environment
Rank **5**
Score **82.1**

Research & development
Rank **22**
Score **46.1**

Usage of ICT
Rank **15**
Score **80.8**

Index thermometer[12]

Pillar 1 11.2% Pillar 2 10.8% Pillar 3 26.2% Pillar 4 18.8% Pillar 5 33% 100%

[1] High-income countries are those that, according to the World Bank classification, have a gross national income (GNI) per capita of $11,456 or more.
[2] Full democracies are countries whose Democracy Index scores from 1 to 10 rounded to one decimal place are of 8 to 10. [3] 2007. [4] World Bank World Development Indicators. [5] IMF World Economic Outlook. [6] 2006. [7] Estimate. [8] UNESCO, % age above 15. [9] % ages 15–64. [10] 2005. [11] 2004. [12] The index thermometer shows the relative contribution of each pillar (in %) to the total index score.

NEW ZEALAND

Country relative performance: **Index pillars**[13]

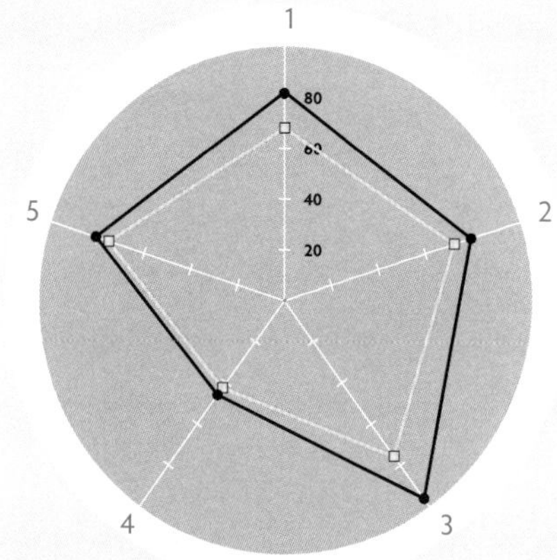

1. Institutional environment
2. Human capital, training and social inclusion
3. Regulatory and legal framework
4. Research and development
5. Usage of ICT

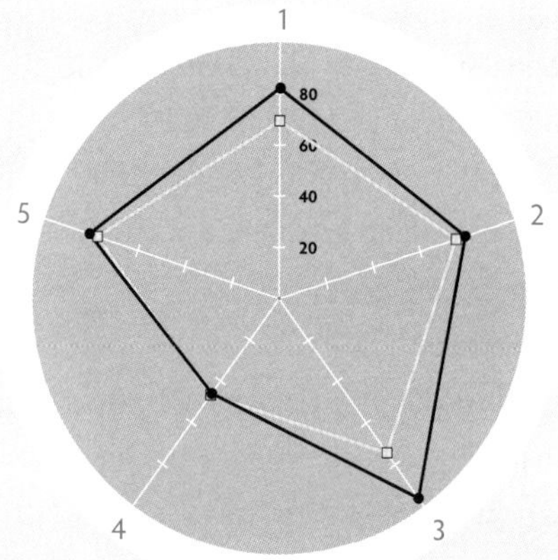

High-income ▫ · · · New Zealand ●━━

Full democracies ▫ · · · New Zealand ●━━

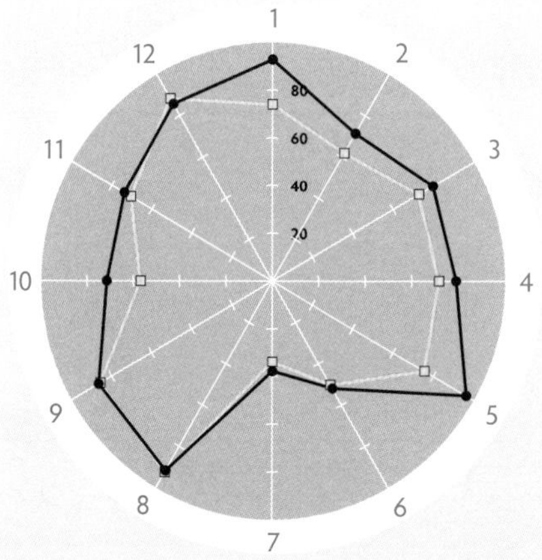

1. Good governance
2. Country policy assessment
3. Education
4. Social inclusion and equity policies
5. Doing business
6. R&D infrastructure
7. Patents and trademarks
8. Telephone communications
9. Mobile/cellular communications
10. Internet, computers and TV
11. Government ICT usage
12. Quality of the infrastructure

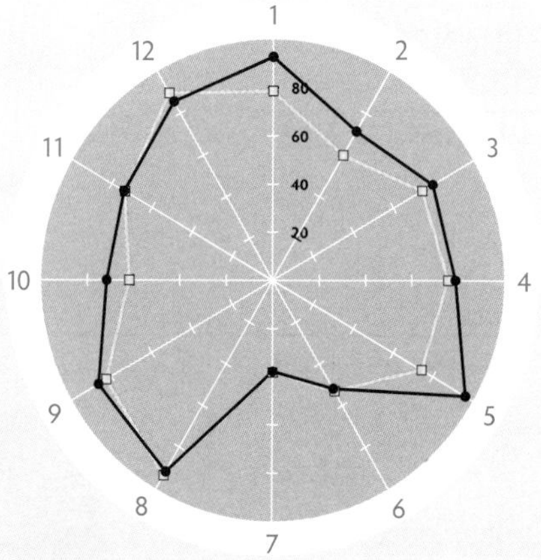

Country relative performance: **Pillar subindexes**[13]

In focus: **Top priorities for policy reform**[14]

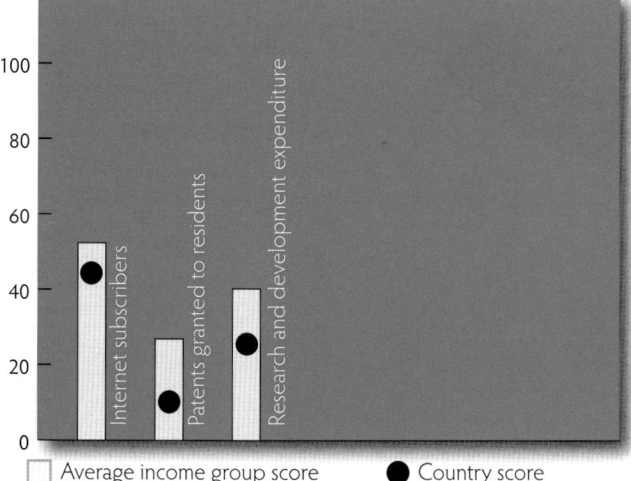

☐ Average income group score   ● Country score

In focus: **Significant indicators above income group average**[14]

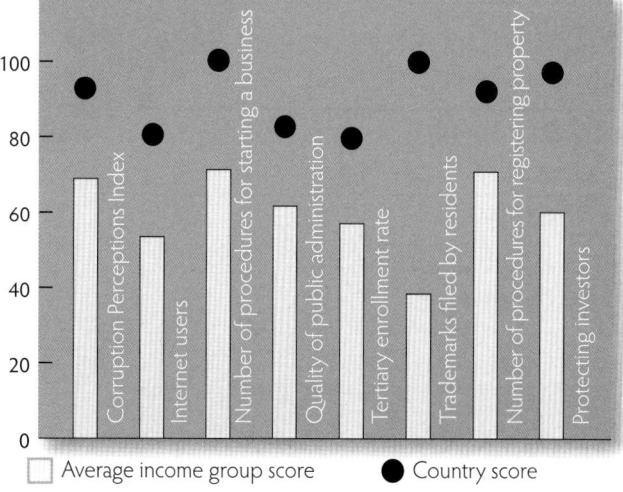

☐ Average income group score   ● Country score

[13] This analysis compares pillar and subindex country scores with respect to the average scores of countries grouped according to income level and political regime (see notes 1 and 2).

[14] Significant country indicators with the greatest distances above or below the income group average scores (i.e., achievements and targets for policy reform, respectively) were selected and ordered according to their Pearson correlation coefficients (highest first) with respect to the final index scores, to produce a ranking.

NIGERIA

Key selected indicators

Population (in millions)[3, 4]	148.0
Population density (people per sq. km)[3, 4]	162
GDP (current US$ billion, PPP)[3, 5]	293
GDP per capita (current US$, PPP)[3, 5]	2035
Current account balance (% of GDP)[3, 5]	0.7
Average life expectancy (years)[4, 6]	46.8
Literacy rate[3, 7]	72
Female labor force participation rate[4, 6, 8]	46.7
Unemployment rate (% of total labor force)[9, 10]	4.9
CO_2 emissions (metric tons per capita)[4, 11]	0.8

Income level:[1] **Low-income**

Political regime:[2] **Authoritarian regime**

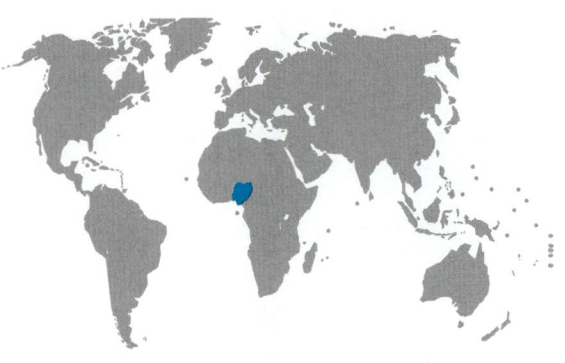

Innovation Capacity Index

Rank	Score
104 out of 131	**40.2 out of 100**

Pillars and pillar subindexes

Education
Rank **101** Score **40.6**

Social inclusion and equity policies
Rank **109** Score **40.9**

Country policy assessment
Rank **46** Score **56.5**

Good governance
Rank **109** Score **28.9**

Quality of the infrastructure
Rank **105** Score **40.3**

Government ICT usage
Rank **100** Score **30.6**

Doing business
Rank **105** Score **56.2**

R&D infrastructure
Rank **99** Score **10.4**

Patents and trademarks
Rank **106** Score **0.2**

Telephone communications
Rank **99** Score **68.0**

Mobile cellular communications
Rank **102** Score **48.2**

Internet, computers and TV
Rank **103** Score **4.0**

Central circle chart:
- ICI
- Human capital, training & social inclusion — Rank **107** Score **40.8**
- Regulatory & legal framework — Rank **105** Score **56.2**
- Research & development — Rank **106** Score **4.6**
- Usage of ICT — Rank **100** Score **34.0**
- Institutional environment — Rank **87** Score **42.8**

Index thermometer[12]

Pillar 1 31.9% Pillar 2 30.5% Pillar 3 28.0% Pillar 4 1.1% Pillar 5 8.5% 100%

326

[1] Low-income countries are those that, according to the World Bank classification, have a gross national income (GNI) per capita of $935 or less. [2] Authoritarian regimes are countries whose Democracy Index scores from 1 to 10 rounded to one decimal place are less than 4. [3] 2007. [4] World Bank World Development Indicators. [5] IMF World Economic Outlook. [6] 2006. [7] UNESCO, % age above 15. [8] % ages 15–64. [9] 2007 est. [10] CIA World Factbook [11] 2004. [12] The index thermometer shows the relative contribution of each pillar (in %) to the total index score.

NIGERIA

Country relative performance: **Index pillars**[13]

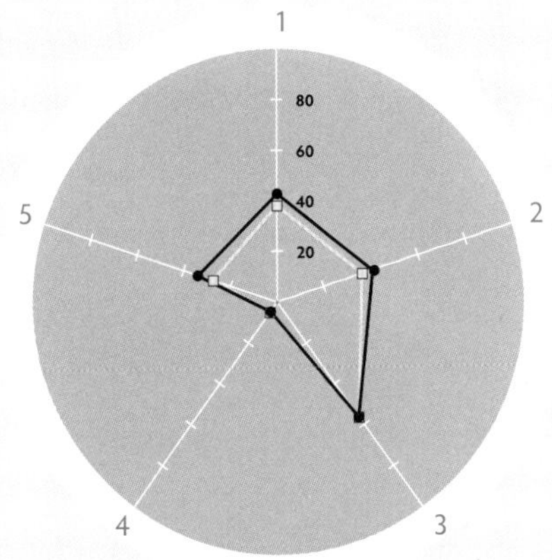

1. Institutional environment
2. Human capital, training and social inclusion
3. Regulatory and legal framework
4. Research and development
5. Usage of ICT

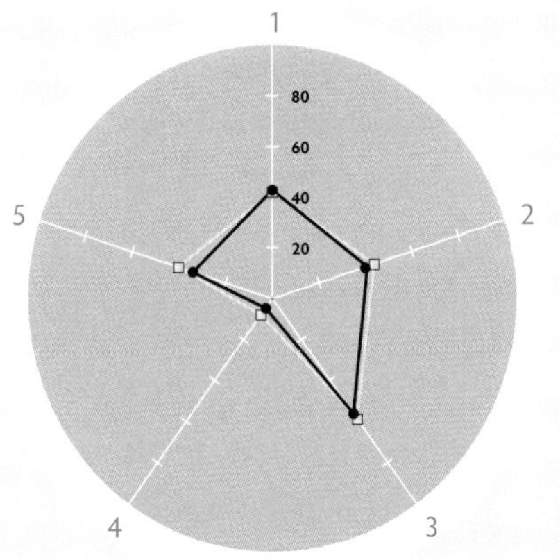

Low-income □   Nigeria ●—

Authoritarian regimes □   Nigeria ●—

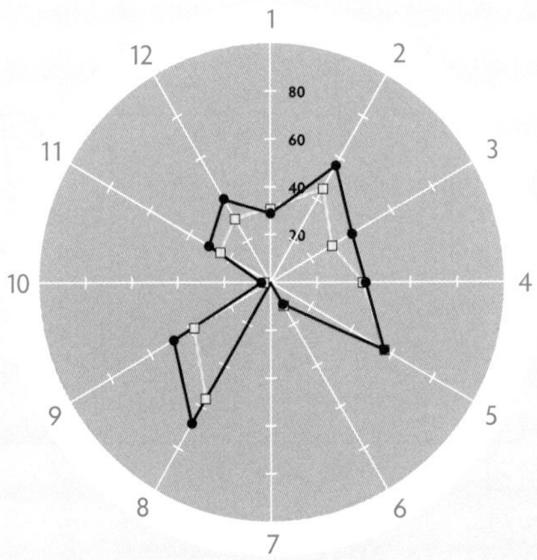

1. Good governance
2. Country policy assessment
3. Education
4. Social inclusion and equity policies
5. Doing business
6. R&D infrastructure
7. Patents and trademarks
8. Telephone communications
9. Mobile/cellular communications
10. Internet, computers and TV
11. Government ICT usage
12. Quality of the infrastructure

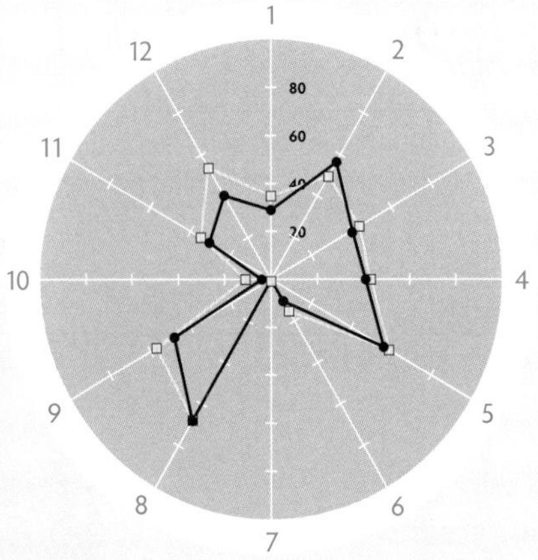

Country relative performance: **Pillar subindexes**[13]

In focus: **Top priorities for policy reform**[14]

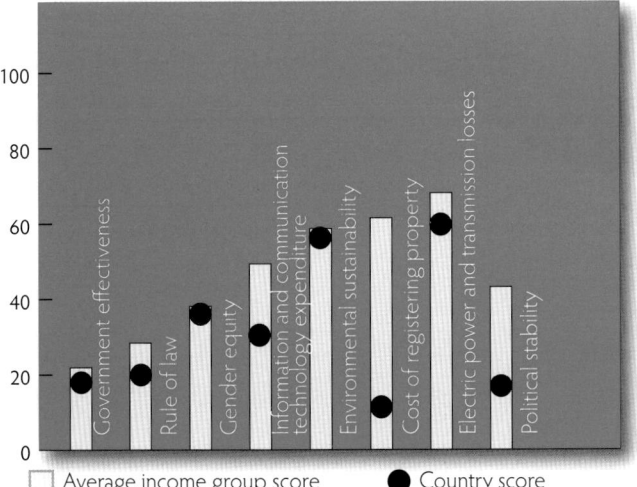

□ Average income group score   ● Country score

In focus: **Significant indicators above income group average**[14]

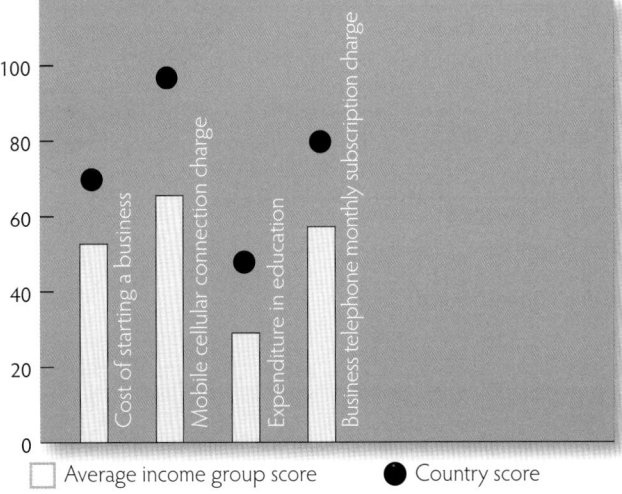

□ Average income group score   ● Country score

---

[13] This analysis compares pillar and subindex country scores with respect to the average scores of countries grouped according to income level and political regime (see notes 1 and 2).

[14] Significant country indicators with the greatest distances above or below the income group average scores (i.e., achievements and targets for policy reform, respectively) were selected and ordered according to their Pearson correlation coefficients (highest first) with respect to the final index scores, to produce a ranking.

327

NORWAY

Key selected indicators

Population (in millions)[3, 4]	4.7
Population density (people per sq. km)[3, 4]	15
GDP (current US$ billion, PPP)[3, 5]	247
GDP per capita (current US$, PPP)[3, 5]	53037
Current account balance (% of GDP)[3, 5]	16.3
Average life expectancy (years)[4, 6]	80.3
Literacy rate[7, 8]	99
Female labor force participation rate[4, 6, 8]	77.4
Unemployment rate (% of total labor force)[4, 10]	4.6
CO_2 emissions (metric tons per capita)[4, 11]	19.1

Income level:[1] **High-income**

Political regime:[2] **Full democracy**

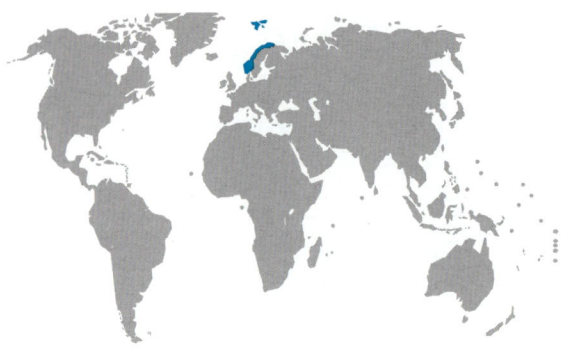

Innovation Capacity Index

Rank	Score
9 out of 131	**73.5 out of 100**

Pillars and pillar subindexes

Education
Rank **4** Score **82.4**

Social inclusion and equity policies
Rank **1** Score **93.2**

Country policy assessment
Rank **6** Score **73.5**

Good governance
Rank **12** Score **87.6**

Quality of the infrastructure
Rank **32** Score **89.7**

Government ICT usage
Rank **3** Score **89.2**

Doing business
Rank **8** Score **81.9**

R&D infrastructure
Rank **18** Score **54.1**

Patents and trademarks
Rank **19** Score **37.4**

Telephone communications
Rank **21** Score **93.1**

Mobile cellular communications
Rank **49** Score **81.6**

Internet, computers and TV
Rank **5** Score **83.4**

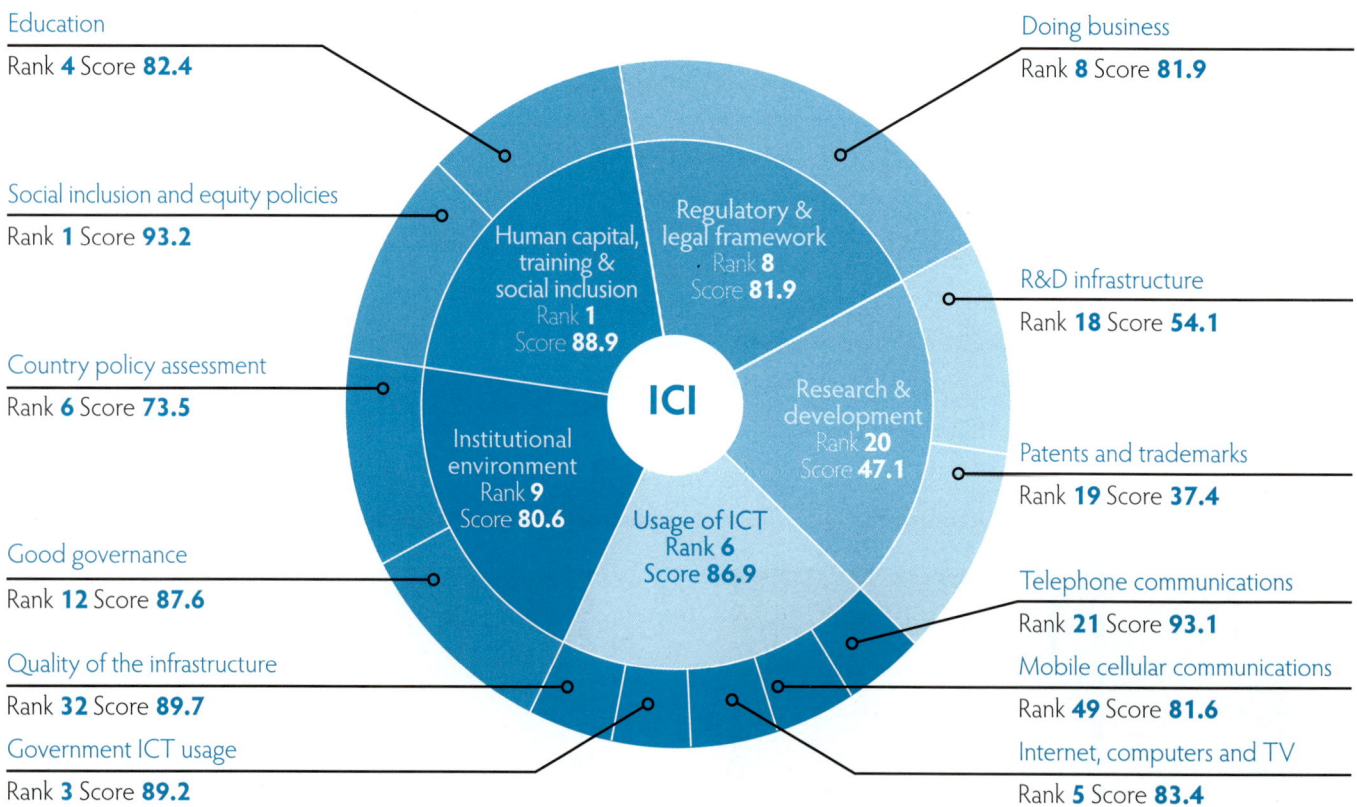

Human capital, training & social inclusion
Rank **1**
Score **88.9**

Regulatory & legal framework
Rank **8**
Score **81.9**

Institutional environment
Rank **9**
Score **80.6**

ICI

Research & development
Rank **20**
Score **47.1**

Usage of ICT
Rank **6**
Score **86.9**

Index thermometer[12]

Pillar 1 11.0% Pillar 2 12.1% Pillar 3 22.3% Pillar 4 19.2% Pillar 5 35.5% 100%

[1] High-income countries are those that, according to the World Bank classification, have a gross national income (GNI) per capita of $11,456 or more.
[2] Full democracies are countries whose Democracy Index scores from 1 to 10 rounded to one decimal place are of 8 to 10. [3] 2007. [4] World Bank World Development Indicators. [5] IMF World Economic Outlook. [6] 2006. [7] Estimate. [8] UNESCO, % age above 15. [9] % ages 15–64. [10] 2005. [11] 2004. [12] The index thermometer shows the relative contribution of each pillar (in %) to the total index score.

NORWAY

Country relative performance: **Index pillars**[13]

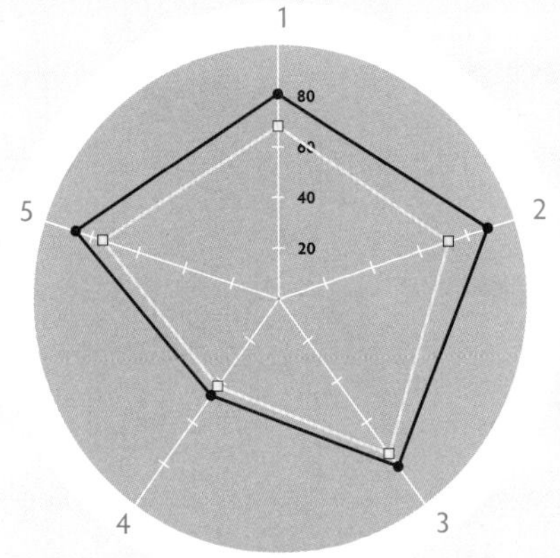

1. Institutional environment
2. Human capital, training and social inclusion
3. Regulatory and legal framework
4. Research and development
5. Usage of ICT

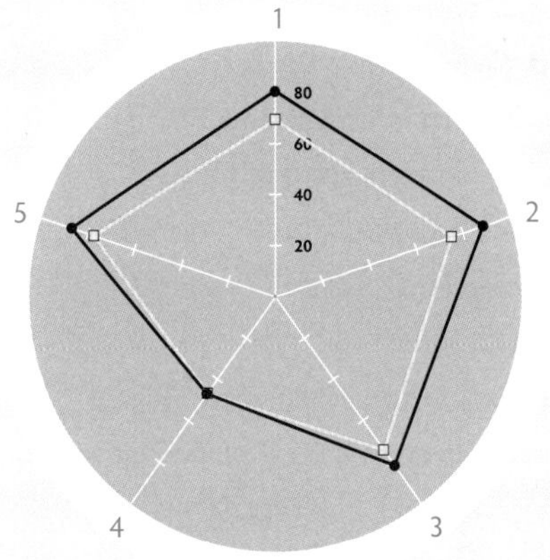

High-income ········ ▫   Norway ●━━━●

Full democracies ········ ▫   Norway ●━━━●

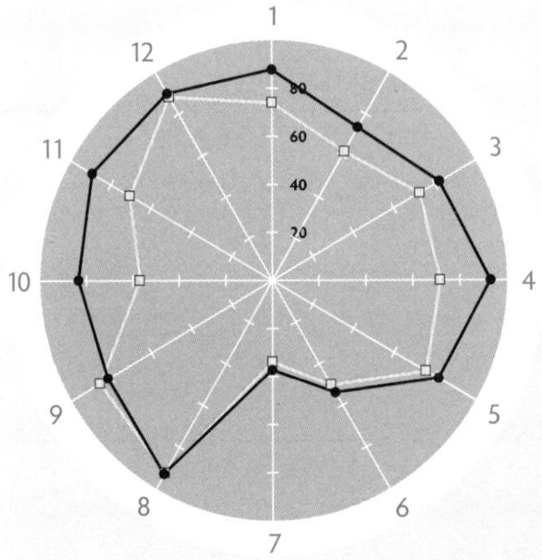

1. Good governance
2. Country policy assessment
3. Education
4. Social inclusion and equity policies
5. Doing business
6. R&D infrastructure
7. Patents and trademarks
8. Telephone communications
9. Mobile/cellular communications
10. Internet, computers and TV
11. Government ICT usage
12. Quality of the infrastructure

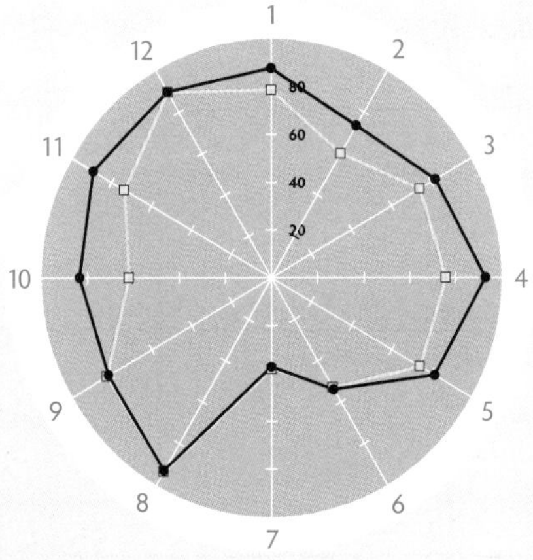

Country relative performance: **Pillar subindexes**[13]

In focus: **Top priorities for policy reform**[14]

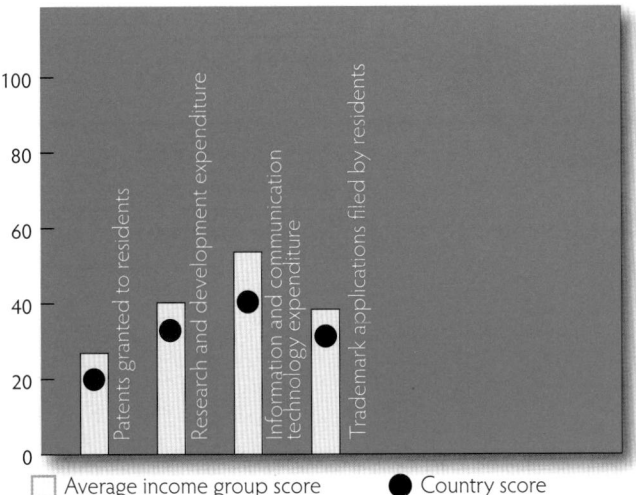

☐ Average income group score   ● Country score

In focus: **Significant indicators above income group average**[14]

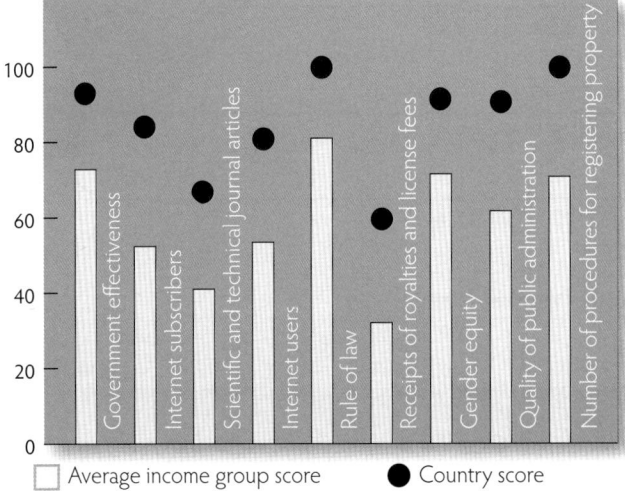

☐ Average income group score   ● Country score

---

[13] This analysis compares pillar and subindex country scores with respect to the average scores of countries grouped according to income level and political regime (see notes 1 and 2).

[14] Significant country indicators with the greatest distances above or below the income group average scores (i.e., achievements and targets for policy reform, respectively) were selected and ordered according to their Pearson correlation coefficients (highest first) with respect to the final index scores, to produce a ranking.

PAKISTAN

Key selected indicators

Population (in millions)[3,4]	162.4
Population density (people per sq. km)[3,4]	211
GDP (current US$ billion, PPP)[3,5]	410
GDP per capita (current US$, PPP)[3,5]	2592
Current account balance (% of GDP)[3,5]	-4.9
Average life expectancy (years)[4,6]	65.2
Literacy rate[3,7]	55
Female labor force participation rate[4,6,8]	34.3
Unemployment rate (% of total labor force)[4,9]	7.7
CO_2 emissions (metric tons per capita)[4,10]	0.8

Income level:[1] **Low-income**

Political regime:[2] **Hybrid regime**

Innovation Capacity Index

Rank	Score
97 out of 131	**42.7 out of 100**

Pillars and pillar subindexes

Education
Rank **113** Score **32.4**

Social inclusion and equity policies
Rank **88** Score **46.6**

Country policy assessment
Rank **84** Score **49.1**

Good governance
Rank **120** Score **26.1**

Quality of the infrastructure
Rank **78** Score **59.3**

Government ICT usage
Rank **97** Score **31.6**

Doing business
Rank **50** Score **69.1**

R&D infrastructure
Rank **75** Score **21.8**

Patents and trademarks
Rank **94** Score **0.5**

Telephone communications
Rank **81** Score **76.3**

Mobile cellular communications
Rank **81** Score **62.3**

Internet, computers and TV
Rank **94** Score **5.6**

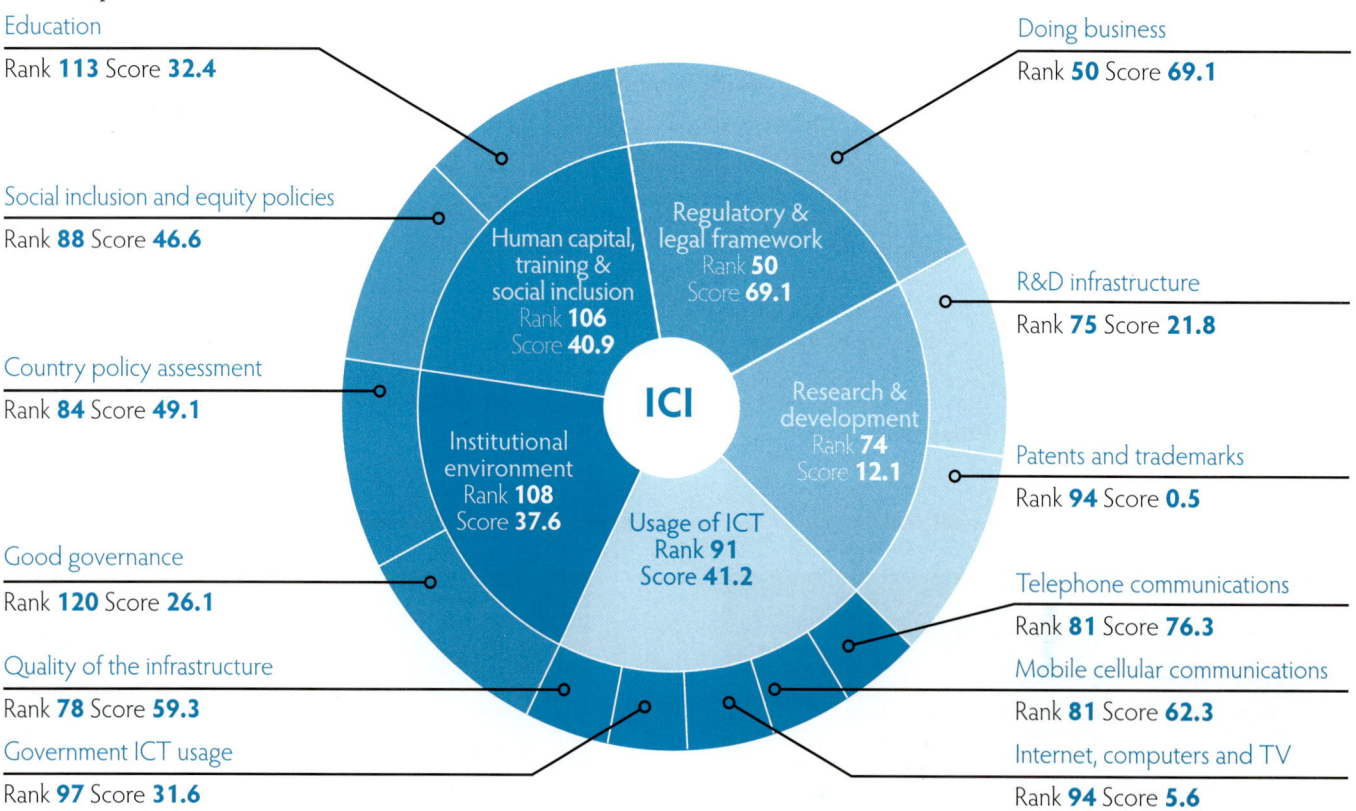

Human capital, training & social inclusion
Rank **106** Score **40.9**

Regulatory & legal framework
Rank **50** Score **69.1**

ICI

Research & development
Rank **74** Score **12.1**

Institutional environment
Rank **108** Score **37.6**

Usage of ICT
Rank **91** Score **41.2**

Index thermometer[11]

Pillar 1 26.4% Pillar 2 28.7% Pillar 3 32.3% Pillar 4 2.8% Pillar 5 9.7%

100%

[1] Low-income countries are those that, according to the World Bank classification, have a gross national income (GNI) per capita of $935 or less. [2] Hybrid regimes are countries whose Democracy Index scores from 1 to 10 rounded to one decimal place are of 4 to 5.9. [3] 2007. [4] World Bank World Development Indicators. [5] IMF World Economic Outlook. [6] 2006. [7] UNESCO, % age above 15. [8] % ages 15–64. [9] 2005. [10] 2004. [11] The index thermometer shows the relative contribution of each pillar (in %) to the total index score.

PAKISTAN

Country relative performance: **Index pillars**[12]

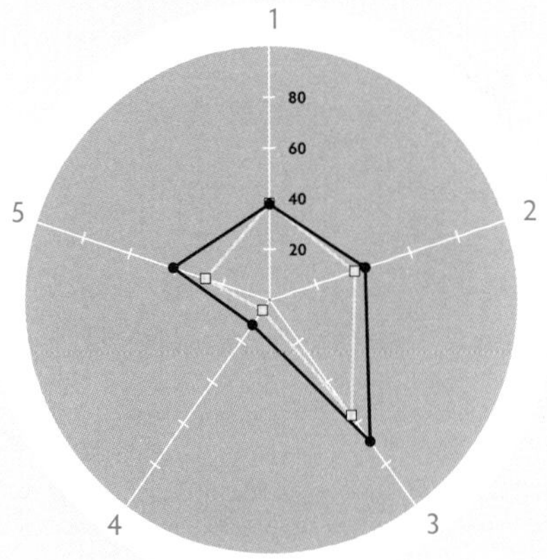

1. Institutional environment
2. Human capital, training and social inclusion
3. Regulatory and legal framework
4. Research and development
5. Usage of ICT

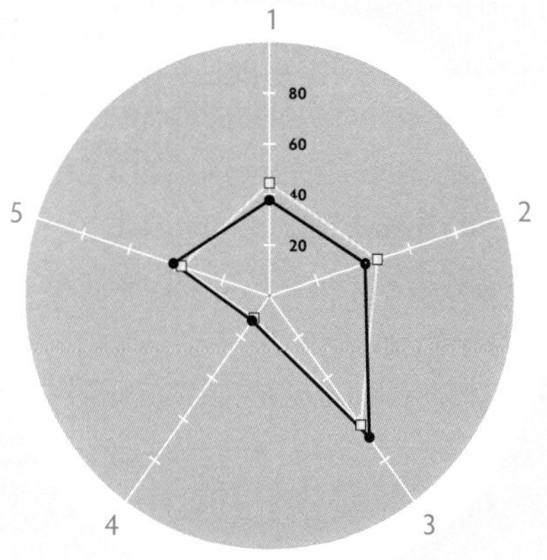

Low-income ☐   Pakistan ●━━

Hybrid regimes ☐   Pakistan ●━━

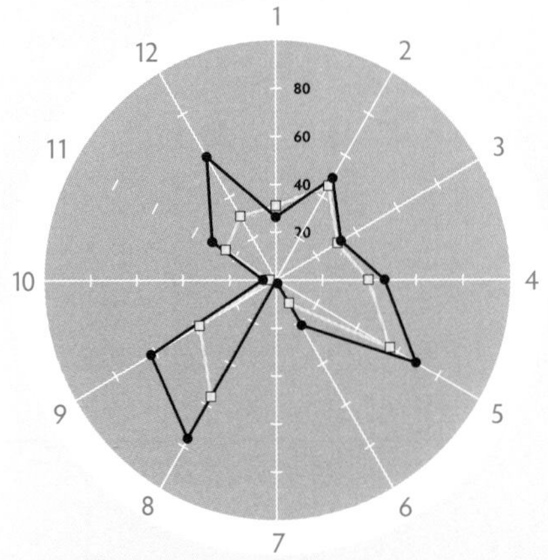

1. Good governance
2. Country policy assessment
3. Education
4. Social inclusion and equity policies
5. Doing business
6. R&D infrastructure
7. Patents and trademarks
8. Telephone communications
9. Mobile/cellular communications
10. Internet, computers and TV
11. Government ICT usage
12. Quality of the infrastructure

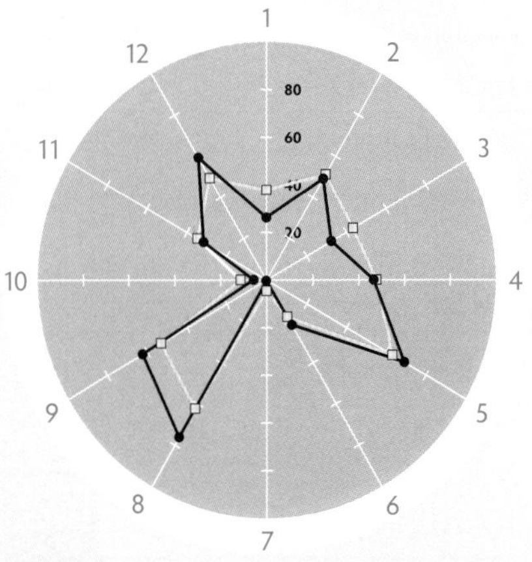

Country relative performance: **Pillar subindexes**[12]

In focus: **Top priorities for policy reform**[13]

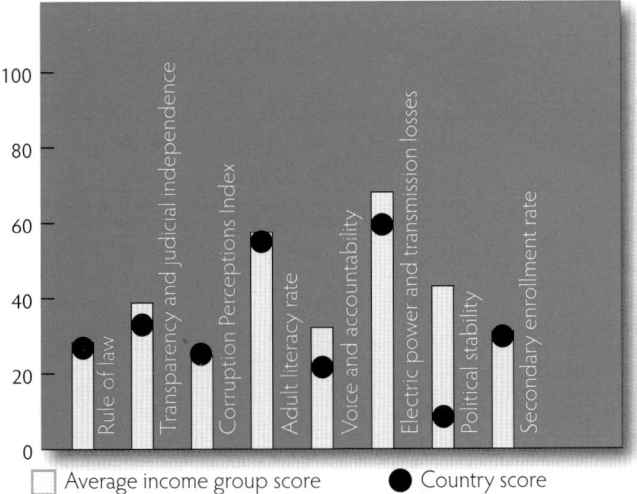

☐ Average income group score   ● Country score

In focus: **Significant indicators above income group average**[13]

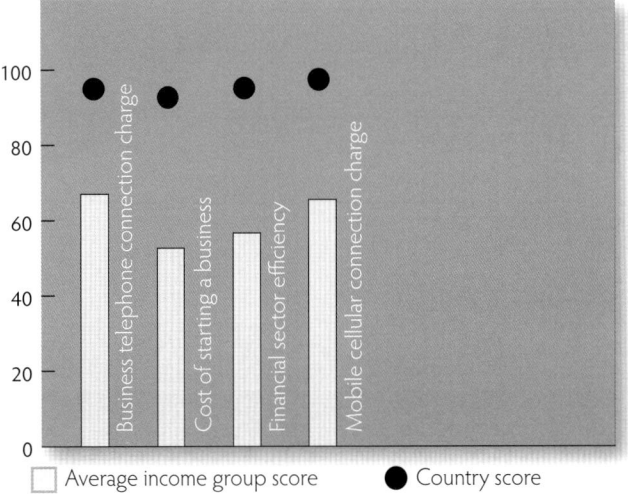

☐ Average income group score   ● Country score

---

[12]   This analysis compares pillar and subindex country scores with respect to the average scores of countries grouped according to income level and political regime (see notes 1 and 2).

[13]   Significant country indicators with the greatest distances above or below the income group average scores (i.e., achievements and targets for policy reform, respectively) were selected and ordered according to their Pearson correlation coefficients (highest first) with respect to the final index scores, to produce a ranking.

PERU

Key selected indicators

Population (in millions)[3,4]	27.9
Population density (people per sq. km)[3,4]	22
GDP (current US$ billion, PPP)[3,5]	219
GDP per capita (current US$, PPP)[3,5]	7803
Current account balance (% of GDP)[3,5]	1.6
Average life expectancy (years)[4,6]	71.1
Literacy rate[3,7]	90
Female labor force participation rate[4,6,8]	62.1
Unemployment rate (% of total labor force)[4,9]	11.4
CO_2 emissions (metric tons per capita)[4,10]	1.2

Income level:[1] **Lower-middle-income**

Political regime:[2] **Flawed democracy**

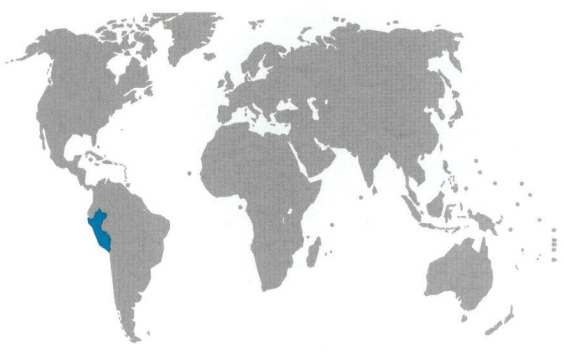

Rank	Score
60 out of 131	**50.6 out of 100**

Innovation Capacity Index

Pillars and pillar subindexes

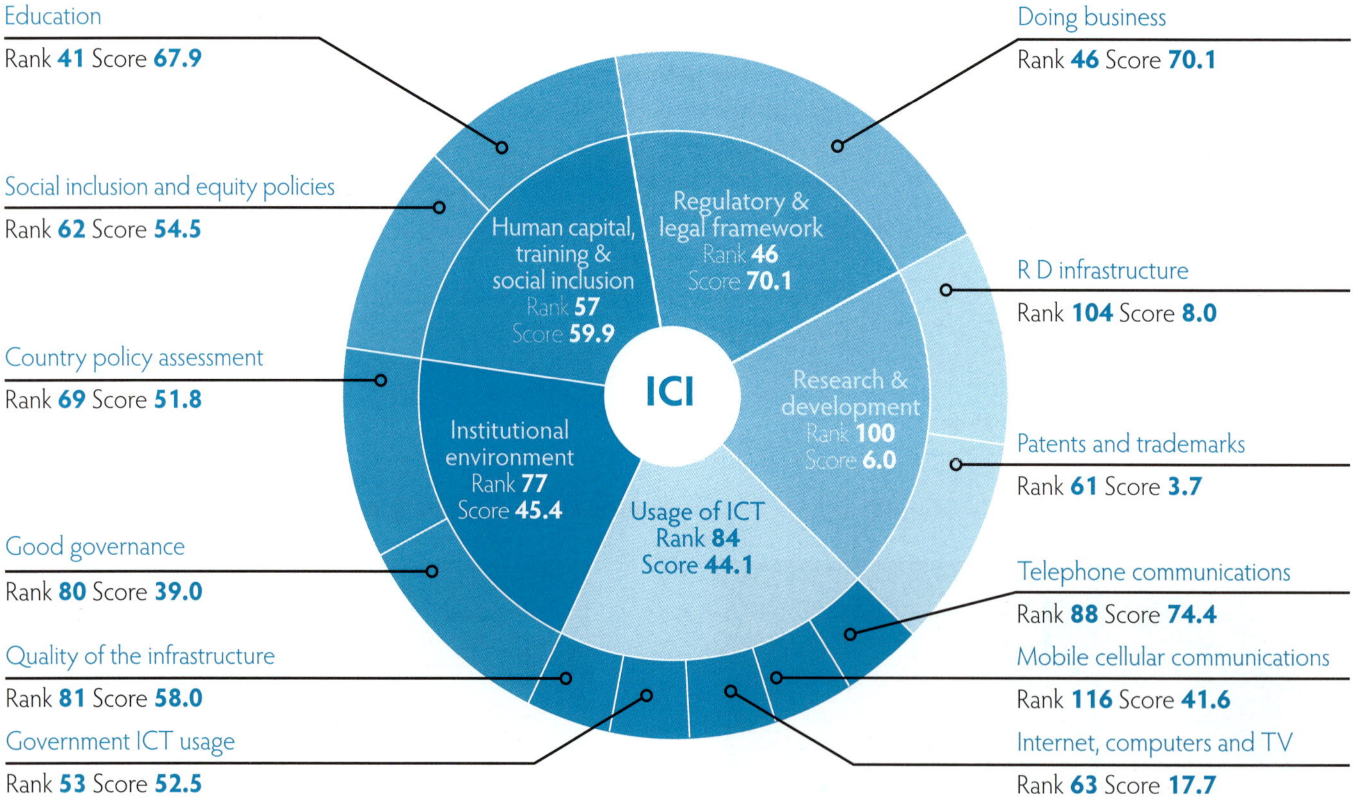

Education
Rank **41** Score **67.9**

Social inclusion and equity policies
Rank **62** Score **54.5**

Country policy assessment
Rank **69** Score **51.8**

Good governance
Rank **80** Score **39.0**

Quality of the infrastructure
Rank **81** Score **58.0**

Government ICT usage
Rank **53** Score **52.5**

Doing business
Rank **46** Score **70.1**

R D infrastructure
Rank **104** Score **8.0**

Patents and trademarks
Rank **61** Score **3.7**

Telephone communications
Rank **88** Score **74.4**

Mobile cellular communications
Rank **116** Score **41.6**

Internet, computers and TV
Rank **63** Score **17.7**

Central diagram:

ICI

- Human capital, training & social inclusion — Rank **57** Score **59.9**
- Regulatory & legal framework — Rank **46** Score **70.1**
- Research & development — Rank **100** Score **6.0**
- Usage of ICT — Rank **84** Score **44.1**
- Institutional environment — Rank **77** Score **45.4**

Index thermometer[11]

Pillar 1 26.9%   Pillar 2 35.5%   Pillar 3 27.7%   Pillar 4 1.2%   Pillar 5 8.7%   100%

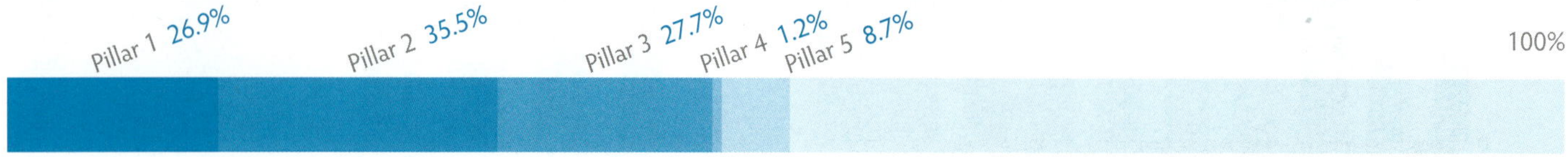

[1] Lower-middle-income countries are those that, according to the World Bank classification, have a gross national income (GNI) per capita of $936 to $3,705.   [2] Flawed democracies are countries whose Democracy Index scores from 1 to 10 rounded to one decimal place are of 6 to 7.9.   [3] 2007.   [4] World Bank World Development Indicators.   [5] IMF World Economic Outlook.   [6] 2006.   [7] UNESCO, % age above 15.   [8] % ages 15–64.   [9] 2005.   [10] 2004.   [11] The index thermometer shows the relative contribution of each pillar (in %) to the total index score.

332

PERU

Country relative performance: **Index pillars**[12]

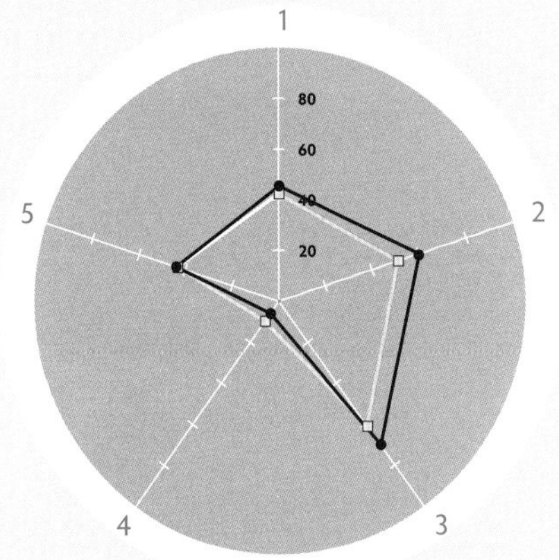

1. Institutional environment
2. Human capital, training and social inclusion
3. Regulatory and legal framework
4. Research and development
5. Usage of ICT

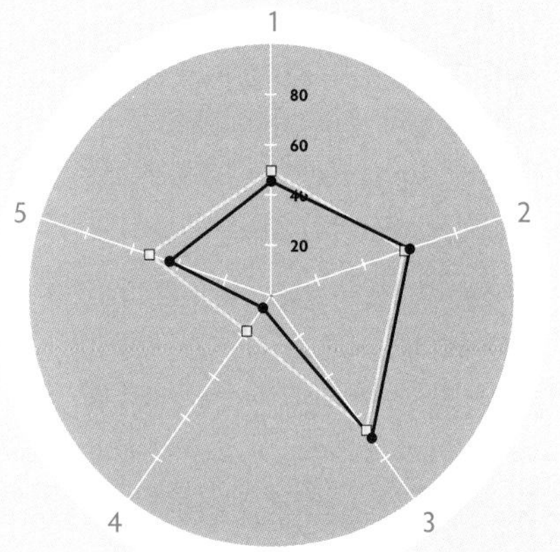

Lower-middle-income ☐ ......   Peru ●——●

Flawed democracies ☐ ......   Peru ●——●

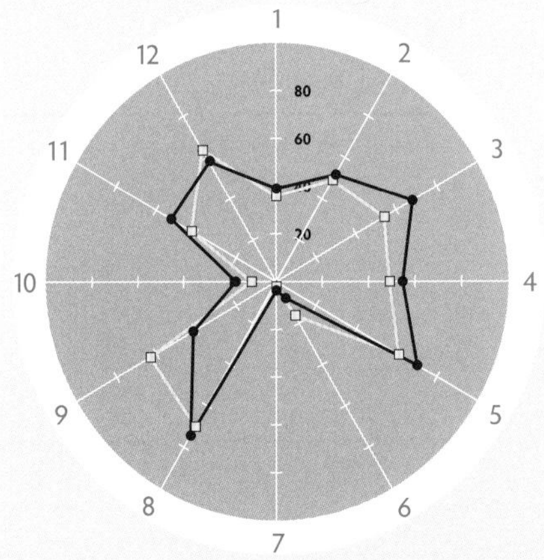

1. Good governance
2. Country policy assessment
3. Education
4. Social inclusion and equity policies
5. Doing business
6. R&D infrastructure
7. Patents and trademarks
8. Telephone communications
9. Mobile/cellular communications
10. Internet, computers and TV
11. Government ICT usage
12. Quality of the infrastructure

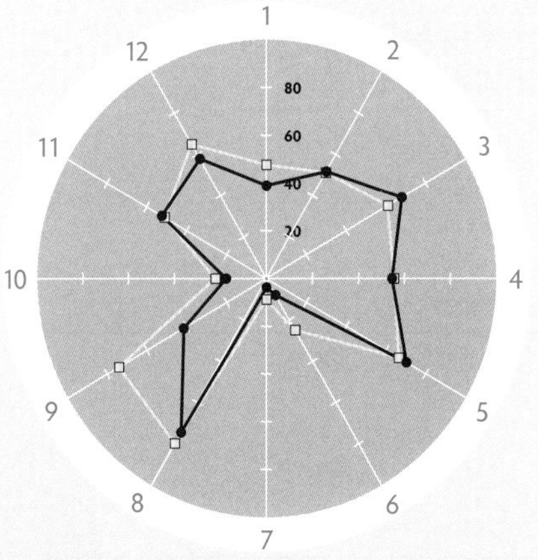

Country relative performance: **Pillar subindexes**[12]

In focus: **Top priorities for policy reform**[13]

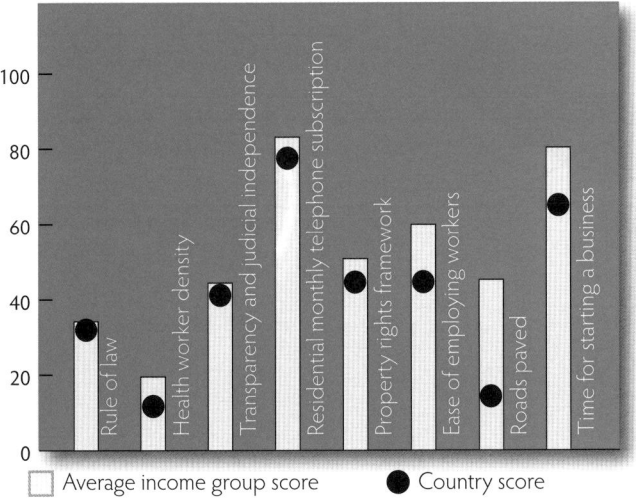

☐ Average income group score   ● Country score

In focus: **Significant indicators above income group average**[13]

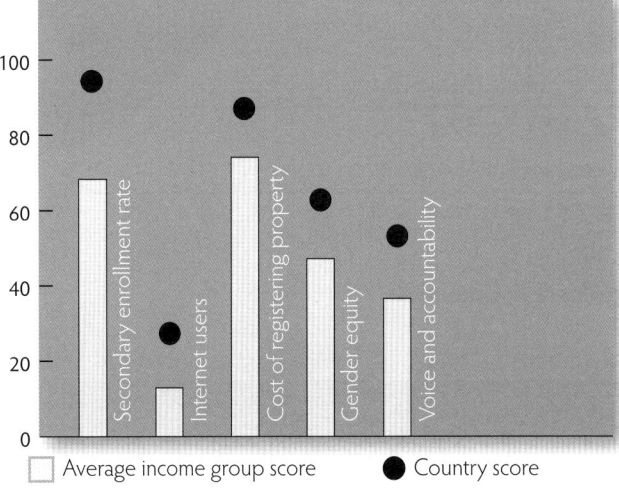

☐ Average income group score   ● Country score

[12] This analysis compares pillar and subindex country scores with respect to the average scores of countries grouped according to income level and political regime (see notes 1 and 2).

[13] Significant country indicators with the greatest distances above or below the income group average scores (i.e., achievements and targets for policy reform, respectively) were selected and ordered according to their Pearson correlation coefficients (highest first) with respect to the final index scores, to produce a ranking.

PHILIPPINES

Key selected indicators

Population (in millions)[3,4]	87.9
Population density (people per sq. km)[3,4]	295
GDP (current US$ billion, PPP)[3,5]	300
GDP per capita (current US$, PPP)[3,5]	3378
Current account balance (% of GDP)[3,5]	4.4
Average life expectancy (years)[4,6]	71.4
Literacy rate[3,7]	93
Female labor force participation rate[4,6,8]	57.6
Unemployment rate (% of total labor force)[4,9]	7.4
CO_2 emissions (metric tons per capita)[4,10]	1.0

Income level:[1] **Lower-middle-income**

Political regime:[2] **Flawed democracy**

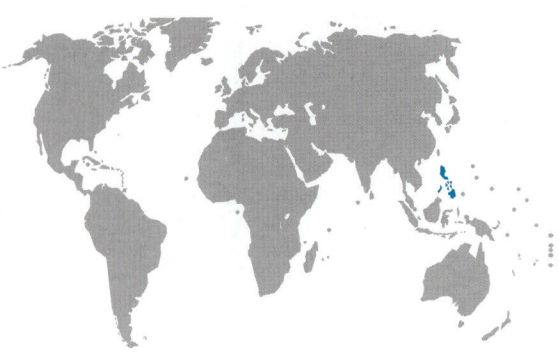

Innovation Capacity Index

	Rank	Score
	75 out of 131	**47.0 out of 100**

Pillars and pillar subindexes

Education
Rank **56** Score **63.7**

Social inclusion and equity policies
Rank **50** Score **58.9**

Country policy assessment
Rank **99** Score **45.2**

Good governance
Rank **99** Score **33.9**

Quality of the infrastructure
Rank **82** Score **57.8**

Government ICT usage
Rank **62** Score **50.0**

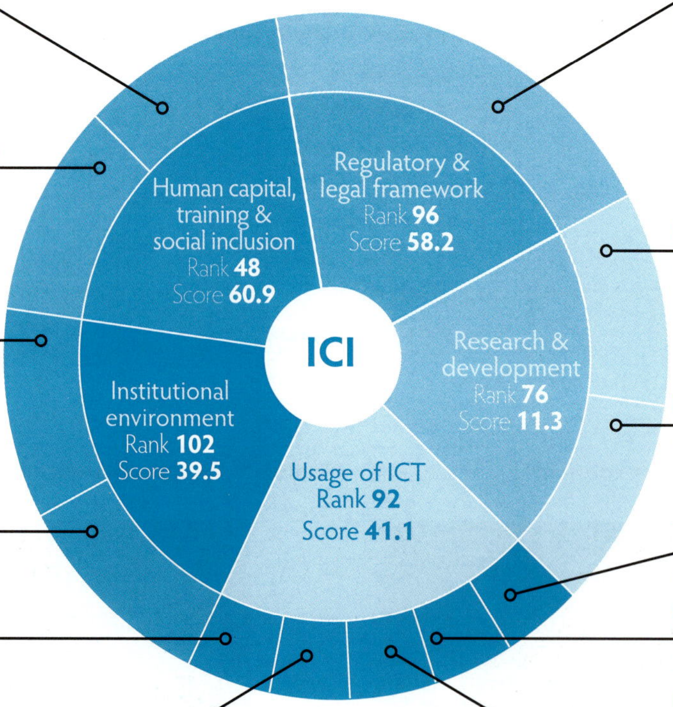

Doing business
Rank **96** Score **58.2**

R&D infrastructure
Rank **82** Score **18.4**

Patents and trademarks
Rank **78** Score **1.2**

Telephone communications
Rank **112** Score **58.4**

Mobile cellular communications
Rank **72** Score **69.8**

Internet, computers and TV
Rank **89** Score **8.2**

Circular chart:

ICI

Human capital, training & social inclusion
Rank **48** Score **60.9**

Regulatory & legal framework
Rank **96** Score **58.2**

Institutional environment
Rank **102** Score **39.5**

Research & development
Rank **76** Score **11.3**

Usage of ICT
Rank **92** Score **41.1**

Index thermometer[11]

Pillar 1 25.2%   Pillar 2 38.8%   Pillar 3 24.8%   Pillar 4 2.4%   Pillar 5 8.7%   100%

---

[1] Lower-middle-income countries are those that, according to the World Bank classification, have a gross national income (GNI) per capita of $936 to $3,705.   [2] Flawed democracies are countries whose Democracy Index scores from 1 to 10 rounded to one decimal place are of 6 to 7.9.   [3] 2007.   [4] World Bank World Development Indicators.   [5] IMF World Economic Outlook.   [6] 2006.   [7] UNESCO, % age above 15.   [8] % ages 15–64.   [9] 2005.   [10] 2004.   [11] The index thermometer shows the relative contribution of each pillar (in %) to the total index score.

PHILIPPINES

Country relative performance: **Index pillars**[12]

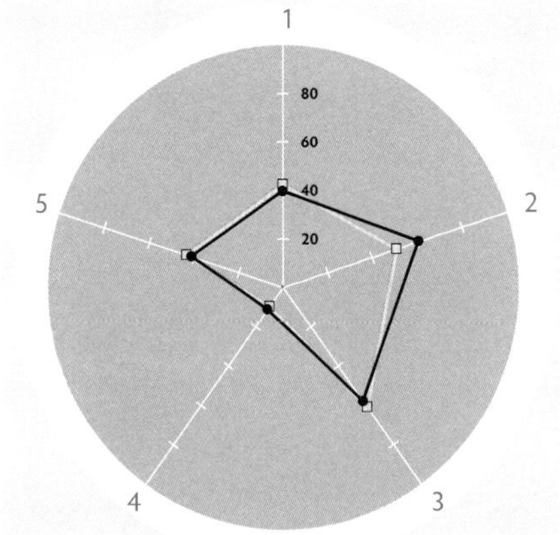

1. Institutional environment
2. Human capital, training and social inclusion
3. Regulatory and legal framework
4. Research and development
5. Usage of ICT

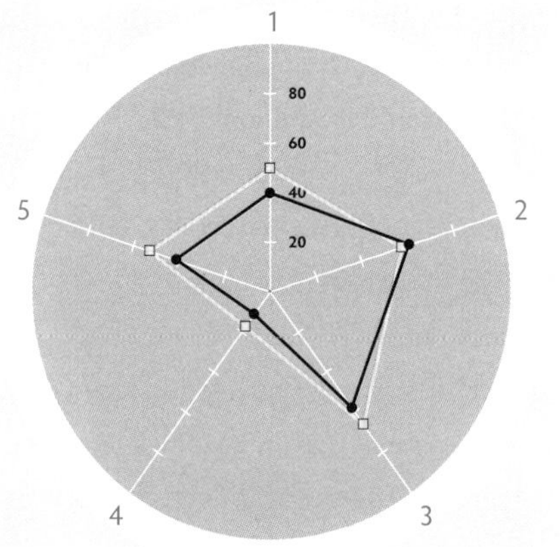

Lower-middle-income ☐   Philippines ●——

Flawed democracies ☐   Philippines ●——

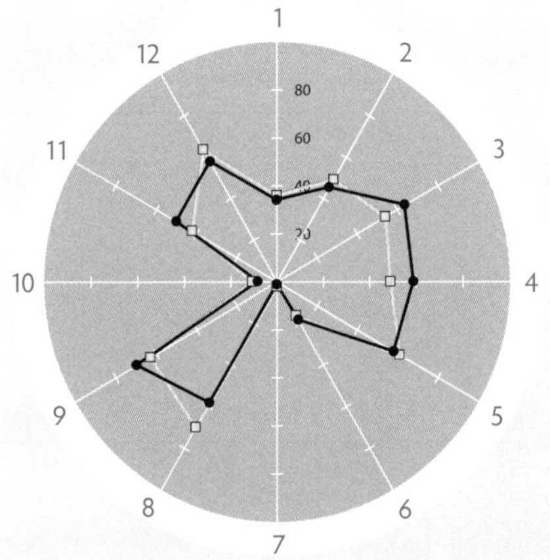

1. Good governance
2. Country policy assessment
3. Education
4. Social inclusion and equity policies
5. Doing business
6. R&D infrastructure
7. Patents and trademarks
8. Telephone communications
9. Mobile/cellular communications
10. Internet, computers and TV
11. Government ICT usage
12. Quality of the infrastructure

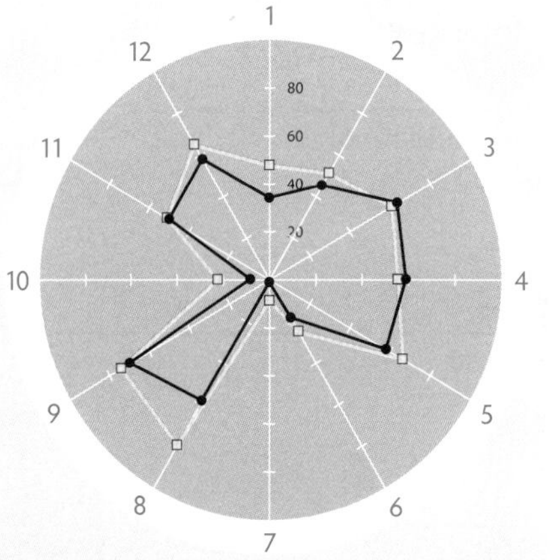

335

Country relative performance: **Pillar subindexes**[12]

In focus: **Top priorities for policy reform**[13]

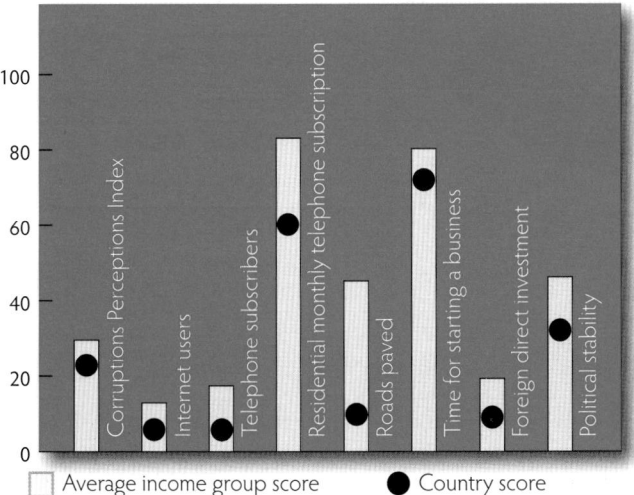

☐ Average income group score   ● Country score

In focus: **Significant indicators above income group average**[13]

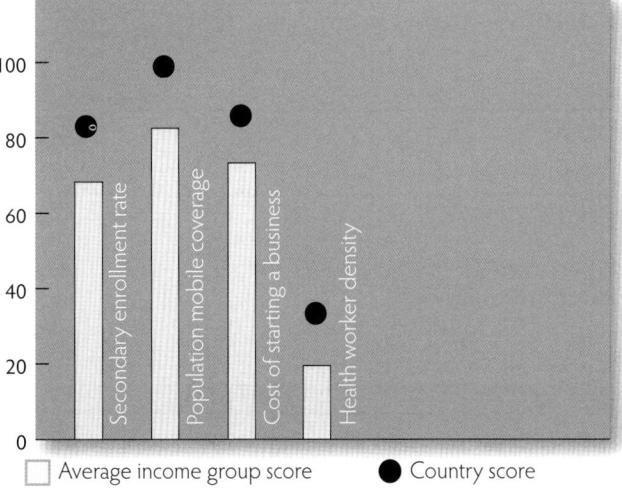

☐ Average income group score   ● Country score

[12] This analysis compares pillar and subindex country scores with respect to the average scores of countries grouped according to income level and political regime (see notes 1 and 2).

[13] Significant country indicators with the greatest distances above or below the income group average scores (i.e., achievements and targets for policy reform, respectively) were selected and ordered according to their Pearson correlation coefficients (highest first) with respect to the final index scores, to produce a ranking.

POLAND

Key selected indicators

Population (in millions)[3, 4]	38.1
Population density (people per sq. km)[3, 4]	124
GDP (current US$ billion, PPP)[3, 5]	621
GDP per capita (current US$, PPP)[3, 5]	16311
Current account balance (% of GDP)[3, 5]	-3.7
Average life expectancy (years)[4, 6]	75.1
Literacy rate[3, 7]	99
Female labor force participation rate[4, 6, 8]	57.3
Unemployment rate (% of total labor force)[4, 9]	17.7
CO_2 emissions (metric tons per capita)[4, 10]	8.0

Income level:[1] **Upper-middle-income**

Political regime:[2] **Flawed democracy**

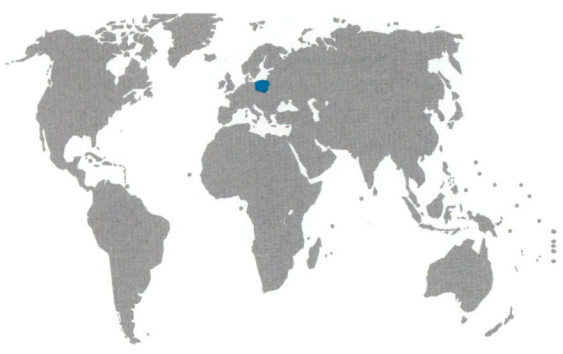

Innovation Capacity Index

Rank
40 out of 131

Score
55.7 out of 100

Pillars and pillar subindexes

Education
Rank **46** Score **67.2**

Social inclusion and equity policies
Rank **39** Score **65.3**

Country policy assessment
Rank **84** Score **49.1**

Good governance
Rank **51** Score **54.0**

Quality of the infrastructure
Rank **51** Score **78.6**

Government ICT usage
Rank **33** Score **61.3**

Doing business
Rank **59** Score **66.9**

R&D infrastructure
Rank **34** Score **36.2**

Patents and trademarks
Rank **46** Score **6.9**

Telephone communications
Rank **43** Score **86.5**

Mobile cellular communications
Rank **22** Score **88.9**

Internet, computers and TV
Rank **42** Score **31.0**

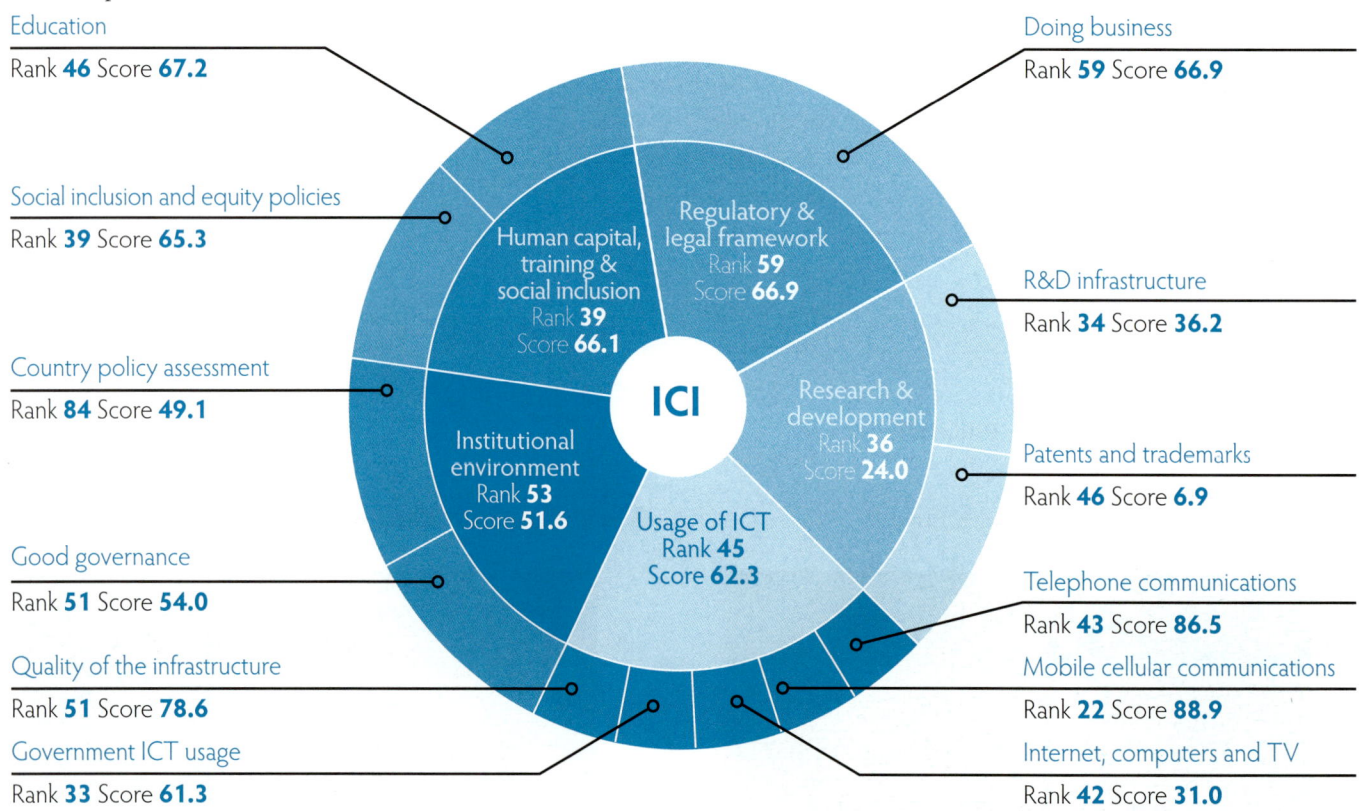

Index thermometer[11]

Pillar 1 23.1% Pillar 2 29.6% Pillar 3 24.0% Pillar 4 6.5% Pillar 5 16.8%   100%

336

[1] Upper-middle-income countries are those that, according to the World Bank classification, have a gross national income (GNI) per capita of $3,706 to $11,455. [2] Flawed democracies are countries whose Democracy Index scores from 1 to 10 rounded to one decimal place are of 6 to 7.9. [3] 2007. [4] World Bank World Development Indicators. [5] IMF World Economic Outlook. [6] 2006. [7] UNESCO, % age above 15. [8] % ages 15–64. [9] 2005. [10] 2004. [11] The index thermometer shows the relative contribution of each pillar (in %) to the total index score.

POLAND

Country relative performance: **Index pillars**[12]

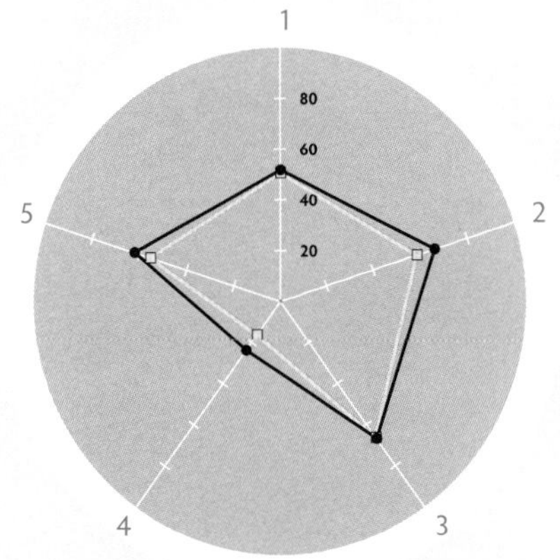

1. Institutional environment
2. Human capital, training and social inclusion
3. Regulatory and legal framework
4. Research and development
5. Usage of ICT

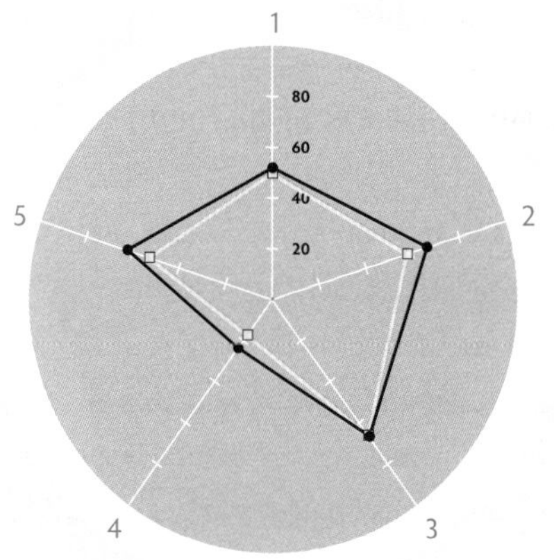

Upper-middle-income ☐   Poland ●——

Flawed democracies ☐   Poland ●——

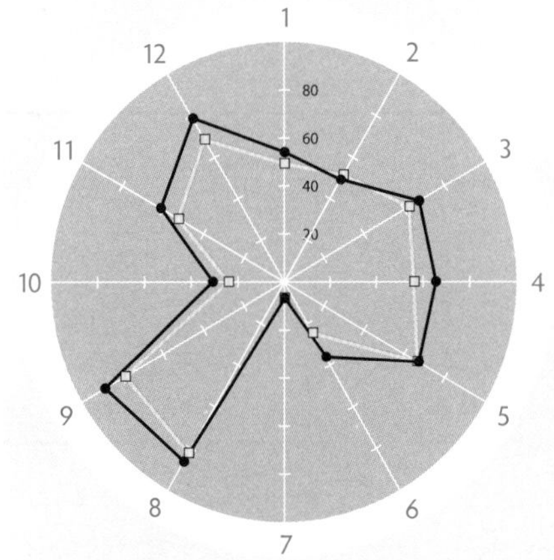

1. Good governance
2. Country policy assessment
3. Education
4. Social inclusion and equity policies
5. Doing business
6. R&D infrastructure
7. Patents and trademarks
8. Telephone communications
9. Mobile/cellular communications
10. Internet, computers and TV
11. Government ICT usage
12. Quality of the infrastructure

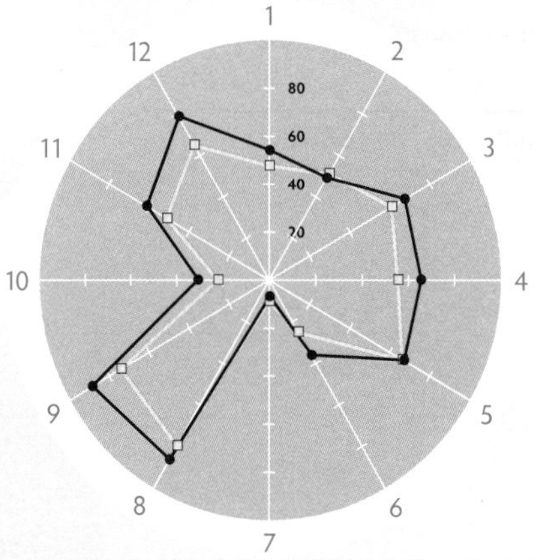

Country relative performance: **Pillar subindexes**[12]

In focus: **Top priorities for policy reform**[13]

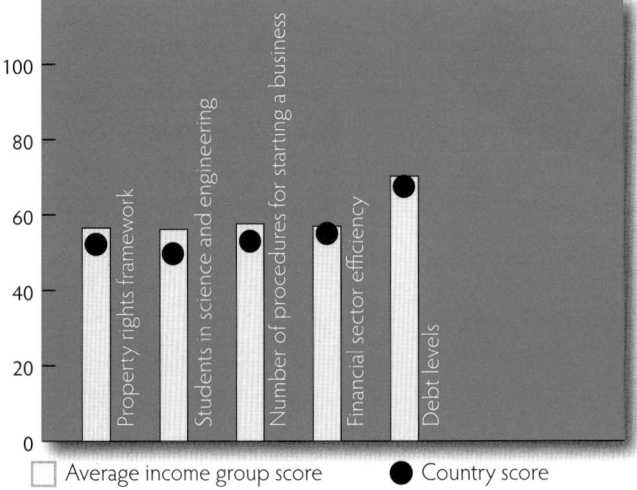

☐ Average income group score   ● Country score

In focus: **Significant indicators above income group average**[13]

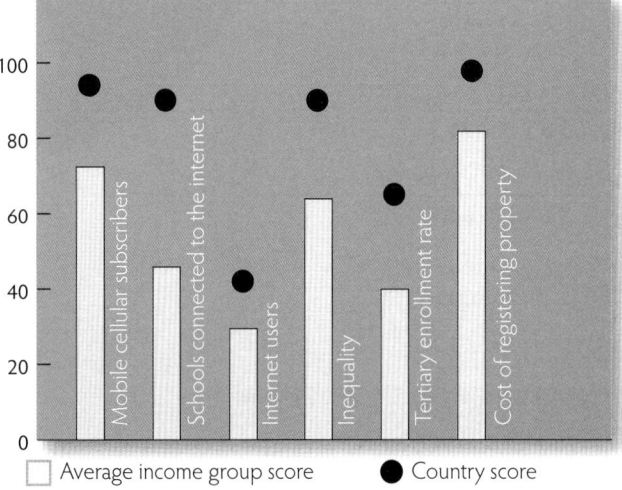

☐ Average income group score   ● Country score

[12] This analysis compares pillar and subindex country scores with respect to the average scores of countries grouped according to income level and political regime (see notes 1 and 2).

[13] Significant country indicators with the greatest distances above or below the income group average scores (i.e., achievements and targets for policy reform, respectively) were selected and ordered according to their Pearson correlation coefficients (highest first) with respect to the final index scores, to produce a ranking.

PORTUGAL

Key selected indicators

Population (in millions)[3,4]	10.6
Population density (people per sq. km)[3,4]	116
GDP (current US$ billion, PPP)[3,5]	231
GDP per capita (current US$, PPP)[3,5]	21701
Current account balance (% of GDP)[3,5]	-9.4
Average life expectancy (years)[4,6]	78.4
Literacy rate[3,7]	95
Female labor force participation rate[4,6,8]	68.5
Unemployment rate (% of total labor force)[4,9]	7.6
CO_2 emissions (metric tons per capita)[4,10]	5.6

Income level:[1] **High-income**

Political regime:[2] **Full democracy**

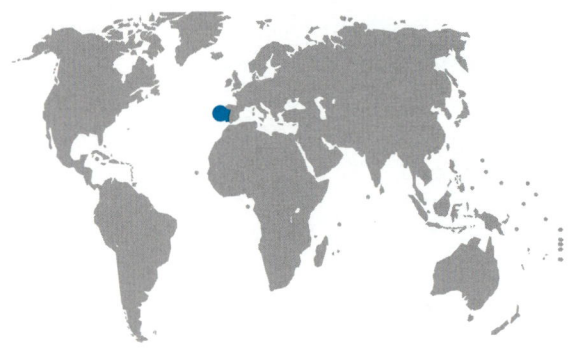

Innovation Capacity Index

Rank **35 out of 131**

Score **57.2 out of 100**

Pillars and pillar subindexes

Education
Rank **35** Score **68.8**

Social inclusion and equity policies
Rank **22** Score **72.3**

Country policy assessment
Rank **75** Score **51.2**

Good governance
Rank **27** Score **69.2**

Quality of the infrastructure
Rank **26** Score **91.3**

Government ICT usage
Rank **31** Score **64.8**

Human capital, training & social inclusion
Rank **27**
Score **70.9**

Regulatory & legal framework
Rank **34**
Score **73.1**

ICI

Institutional environment
Rank **33**
Score **60.2**

Research & development
Rank **29**
Score **30.7**

Usage of ICT
Rank **32**
Score **67.7**

Doing business
Rank **34** Score **73.1**

R&D infrastructure
Rank **29** Score **43.1**

Patents and trademarks
Rank **35** Score **13.3**

Telephone communications
Rank **33** Score **90.1**

Mobile cellular communications
Rank **5** Score **97.3**

Internet, computers and TV
Rank **40** Score **32.6**

Index thermometer[11]

Pillar 1 10.5%   Pillar 2 12.4%   Pillar 3 25.5%   Pillar 4 16.1%   Pillar 5 35.5%   100%

[1] High-income countries are those that, according to the World Bank classification, have a gross national income (GNI) per capita of $11,456 or more.
[2] Full democracies are countries whose Democracy Index scores from 1 to 10 rounded to one decimal place are of 8 to 10.   [3] 2007.   [4] World Bank World Development Indicators.   [5] IMF World Economic Outlook.   [6] 2006.   [7] UNESCO, % age above 15.   [8] % ages 15–64.   [9] 2005.   [10] 2004.   [11] The index thermometer shows the relative contribution of each pillar (in %) to the total index score.

PORTUGAL

Country relative performance: **Index pillars**[12]

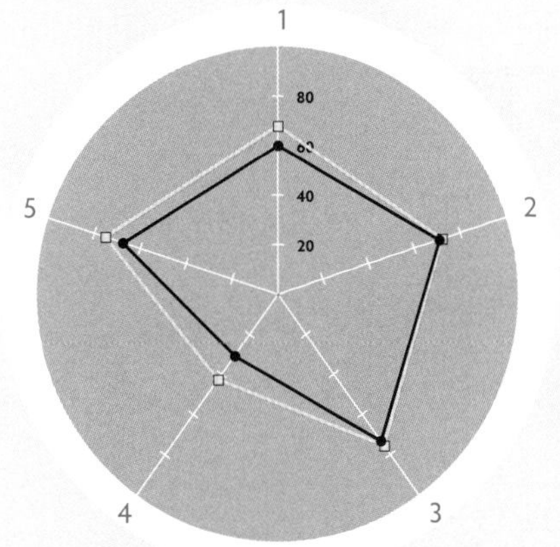

1. Institutional environment
2. Human capital, training and social inclusion
3. Regulatory and legal framework
4. Research and development
5. Usage of ICT

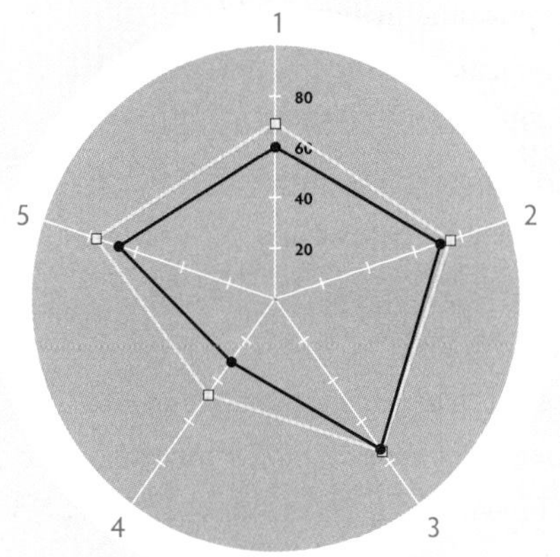

High-income ☐   Portugal ●━━

Full democracies ☐   Portugal ●━━

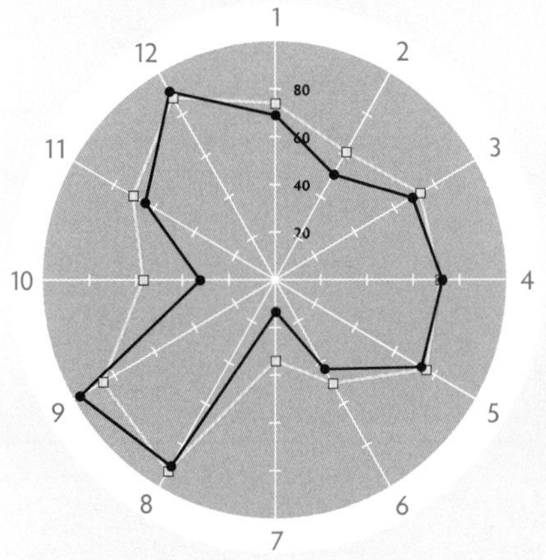

1. Good governance
2. Country policy assessment
3. Education
4. Social inclusion and equity policies
5. Doing business
6. R&D infrastructure
7. Patents and trademarks
8. Telephone communications
9. Mobile/cellular communications
10. Internet, computers and TV
11. Government ICT usage
12. Quality of the infrastructure

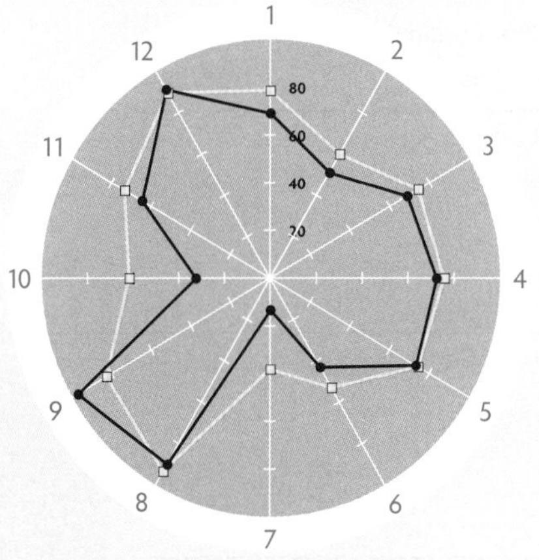

Country relative performance: **Pillar subindexes**[12]

In focus: **Top priorities for policy reform**[13]

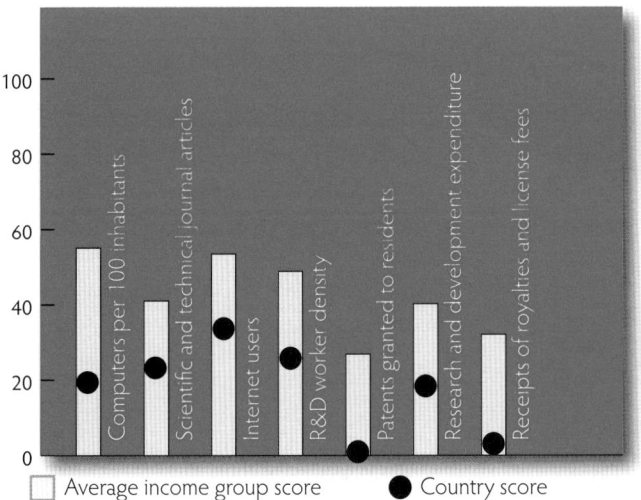

Computers per 100 inhabitants
Scientific and technical journal articles
Internet users
R&D worker density
Patents granted to residents
Research and development expenditure
Receipts of royalties and license fees

☐ Average income group score   ● Country score

In focus: **Significant indicators above income group average**[13]

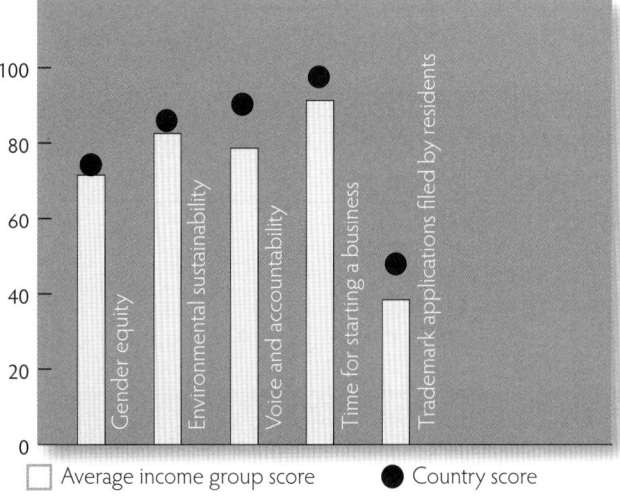

Gender equity
Environmental sustainability
Voice and accountability
Time for starting a business
Trademark applications filed by residents

☐ Average income group score   ● Country score

[12] This analysis compares pillar and subindex country scores with respect to the average scores of countries grouped according to income level and political regime (see notes 1 and 2).

[13] Significant country indicators with the greatest distances above or below the income group average scores (i.e., achievements and targets for policy reform, respectively) were selected and ordered according to their Pearson correlation coefficients (highest first) with respect to the final index scores, to produce a ranking.

ROMANIA

Key selected indicators

Population (in millions)[3,4]	21.5
Population density (people per sq. km)[3,4]	94
GDP (current US$ billion, PPP)[3,5]	246
GDP per capita (current US$, PPP)[3,5]	11387
Current account balance (% of GDP)[3,5]	-13.9
Average life expectancy (years)[4,6]	72.2
Literacy rate[3,7]	98
Female labor force participation rate[4,6,8]	54.4
Unemployment rate (% of total labor force)[4,9]	7.2
CO_2 emissions (metric tons per capita)[4,10]	4.2

Income level:[1] **Upper-middle-income**

Political regime:[2] **Flawed democracy**

Innovation Capacity Index

Rank	Score
47 out of 131	**53.1 out of 100**

Pillars and pillar subindexes

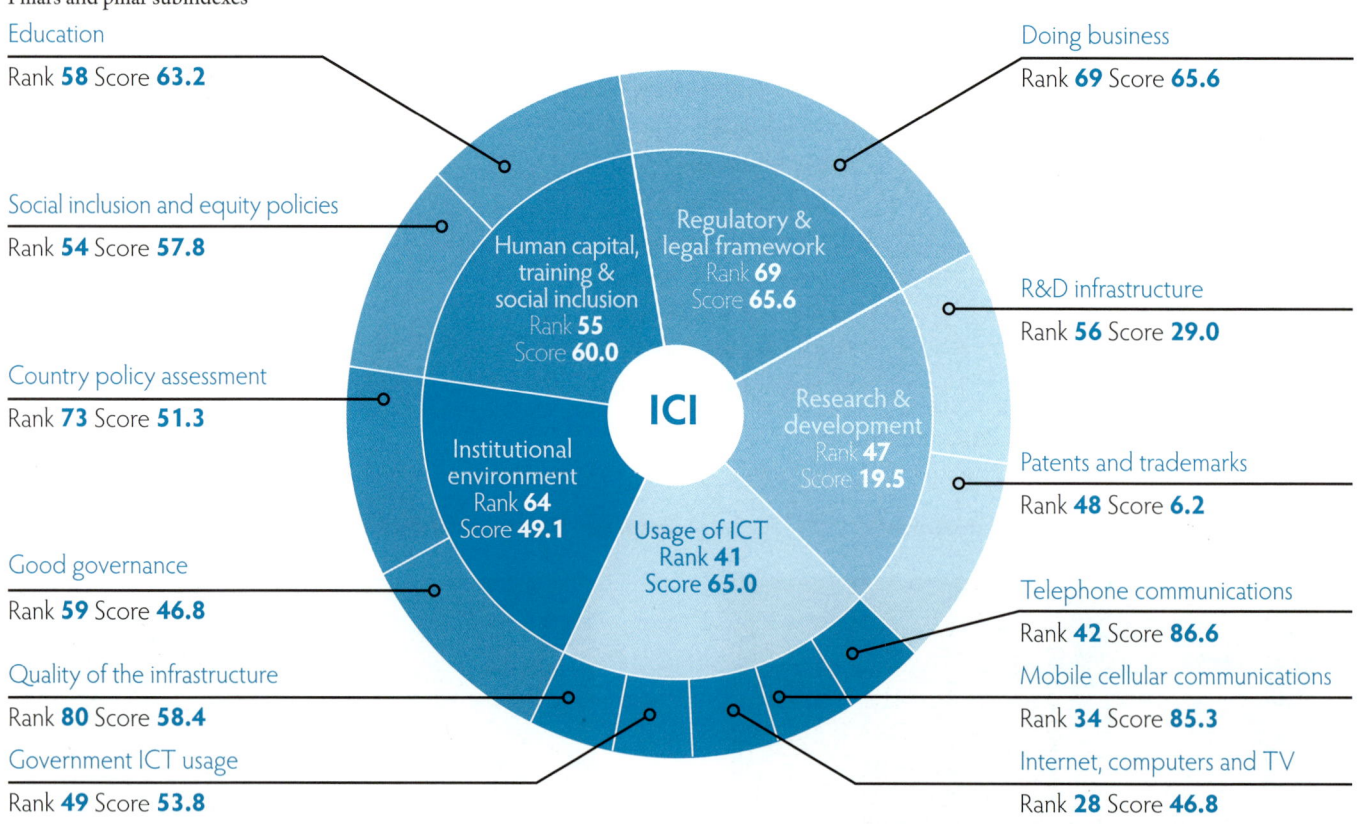

Education
Rank **58** Score **63.2**

Social inclusion and equity policies
Rank **54** Score **57.8**

Country policy assessment
Rank **73** Score **51.3**

Good governance
Rank **59** Score **46.8**

Quality of the infrastructure
Rank **80** Score **58.4**

Government ICT usage
Rank **49** Score **53.8**

Doing business
Rank **69** Score **65.6**

R&D infrastructure
Rank **56** Score **29.0**

Patents and trademarks
Rank **48** Score **6.2**

Telephone communications
Rank **42** Score **86.6**

Mobile cellular communications
Rank **34** Score **85.3**

Internet, computers and TV
Rank **28** Score **46.8**

Human capital, training & social inclusion Rank **55** Score **60.0**

Regulatory & legal framework Rank **69** Score **65.6**

Institutional environment Rank **64** Score **49.1**

Research & development Rank **47** Score **19.5**

Usage of ICT Rank **41** Score **65.0**

ICI

Index thermometer[11]

Pillar 1 23.1% Pillar 2 28.3% Pillar 3 24.7% Pillar 4 5.5% Pillar 5 18.4%   100%

[1] Upper-middle-income countries are those that, according to the World Bank classification, have a gross national income (GNI) per capita of $3,706 to $11,455. [2] Flawed democracies are countries whose Democracy Index scores from 1 to 10 rounded to one decimal place are of 6 to 7.9. [3] 2007. [4] World Bank World Development Indicators. [5] IMF World Economic Outlook. [6] 2006. [7] UNESCO, % age above 15. [8] % ages 15–64. [9] 2005. [10] 2004. [11] The index thermometer shows the relative contribution of each pillar (in %) to the total index score.

ROMANIA

Country relative performance: **Index pillars**[12]

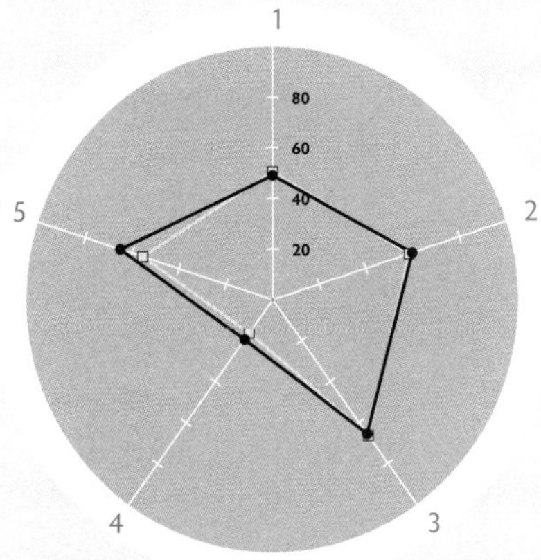

1. Institutional environment
2. Human capital, training and social inclusion
3. Regulatory and legal framework
4. Research and development
5. Usage of ICT

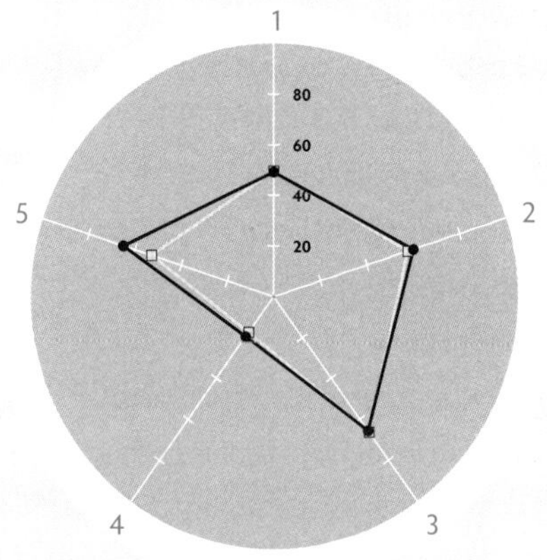

Upper-middle-income ☐   Romania ●——

Flawed democracies ☐   Romania ●——

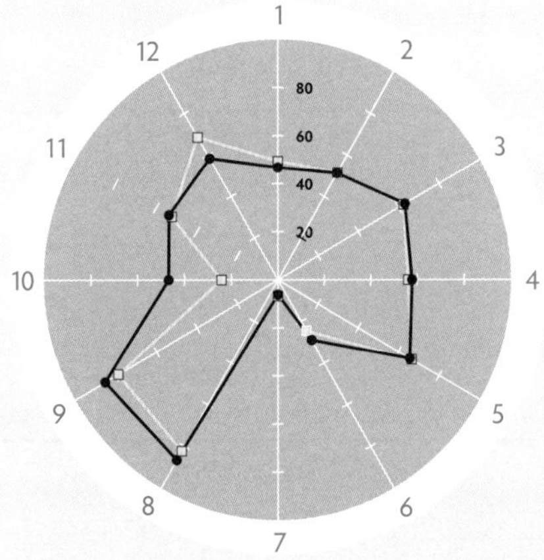

1. Good governance
2. Country policy assessment
3. Education
4. Social inclusion and equity policies
5. Doing business
6. R&D infrastructure
7. Patents and trademarks
8. Telephone communications
9. Mobile/cellular communications
10. Internet, computers and TV
11. Government ICT usage
12. Quality of the infrastructure

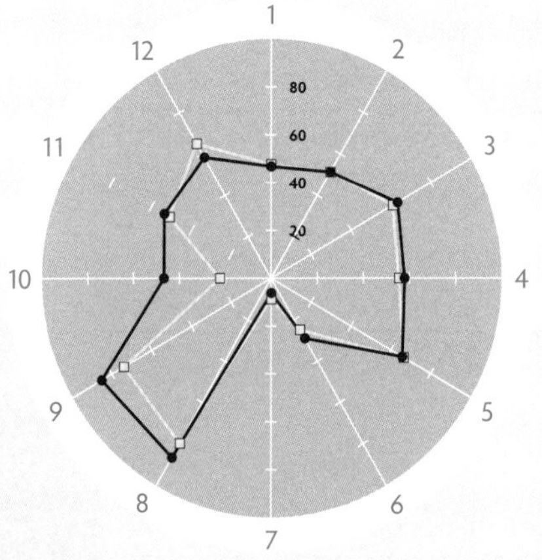

Country relative performance: **Pillar subindexes**[12]

In focus: **Top priorities for policy reform**[13]

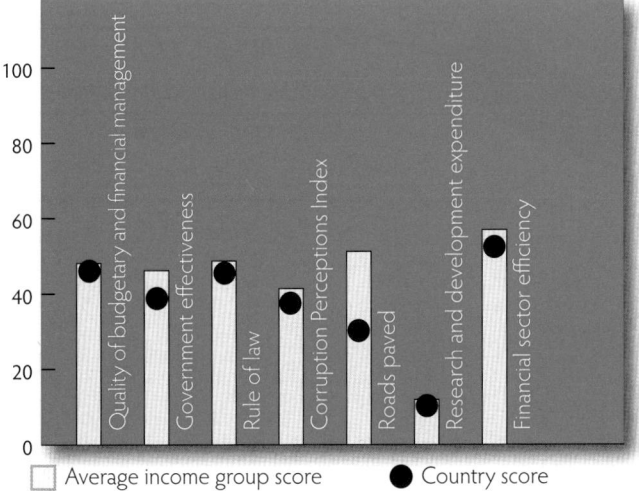

☐ Average income group score   ● Country score

In focus: **Significant indicators above income group average**[13]

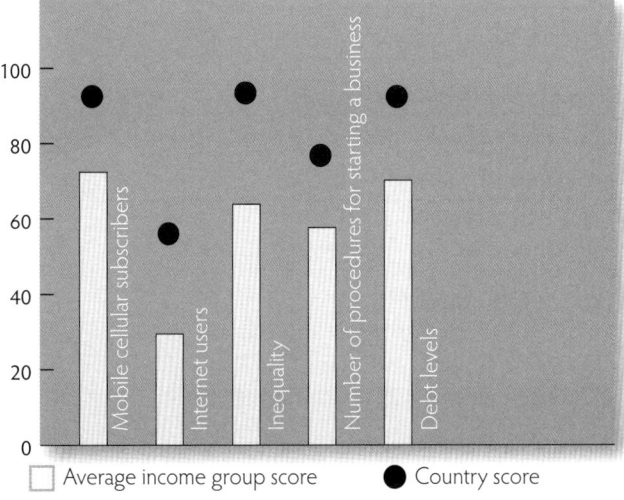

☐ Average income group score   ● Country score

[12] This analysis compares pillar and subindex country scores with respect to the average scores of countries grouped according to income level and political regime (see notes 1 and 2).

[13] Significant country indicators with the greatest distances above or below the income group average scores (i.e., achievements and targets for policy reform, respectively) were selected and ordered according to their Pearson correlation coefficients (highest first) with respect to the final index scores, to produce a ranking.

RUSSIAN FEDERATION

Key selected indicators

Population (in millions)[3,4]	141.6
Population density (people per sq. km)[3,4]	9
GDP (current US$ billion, PPP)[3,5]	2088
GDP per capita (current US$, PPP)[3,5]	14,692
Current account balance (% of GDP)[3,5]	5.9
Average life expectancy (years)[4,6]	65.6
Literacy rate[3,7]	100
Female labor force participation rate[4,6,8]	67.2
Unemployment rate (% of total labor force)[4,9]	7.9
CO_2 emissions (metric tons per capita)[4,9]	10.6

Income level:[1] **Upper-middle-income**

Political regime:[2] **Hybrid regime**

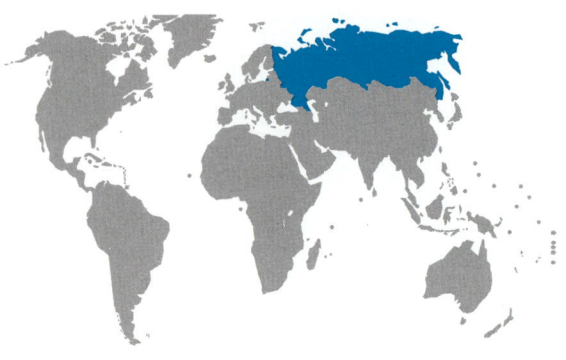

342

Innovation Capacity Index

Rank **49 out of 131**

Score **52.8 out of 100**

Pillars and pillar subindexes

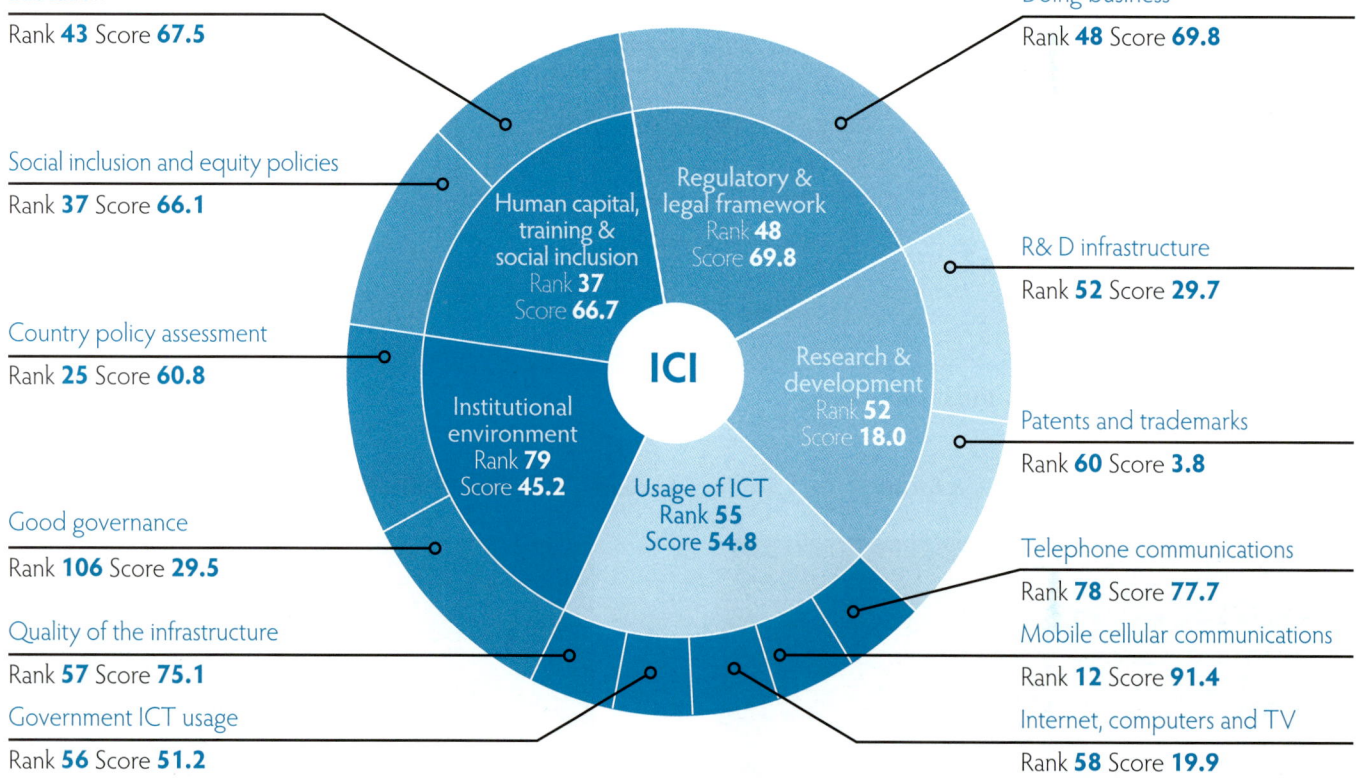

Education
Rank **43** Score **67.5**

Social inclusion and equity policies
Rank **37** Score **66.1**

Country policy assessment
Rank **25** Score **60.8**

Good governance
Rank **106** Score **29.5**

Quality of the infrastructure
Rank **57** Score **75.1**

Government ICT usage
Rank **56** Score **51.2**

Doing business
Rank **48** Score **69.8**

R&D infrastructure
Rank **52** Score **29.7**

Patents and trademarks
Rank **60** Score **3.8**

Telephone communications
Rank **78** Score **77.7**

Mobile cellular communications
Rank **12** Score **91.4**

Internet, computers and TV
Rank **58** Score **19.9**

Within the chart:

Human capital, training & social inclusion — Rank **37** Score **66.7**

Regulatory & legal framework — Rank **48** Score **69.8**

Institutional environment — Rank **79** Score **45.2**

Research & development — Rank **52** Score **18.0**

Usage of ICT — Rank **55** Score **54.8**

ICI

Index thermometer[10]

Pillar 1 21.4% Pillar 2 31.5% Pillar 3 26.4% Pillar 4 5.1% Pillar 5 15.6%   100%

---

[1] Upper-middle-income countries are those that, according to the World Bank classification, have a gross national income (GNI) per capita of $3,706 to $11,455. [2] Hybrid regimes are countries whose Democracy Index scores from 1 to 10 rounded to one decimal place are of 4 to 5.9. [3] 2007. [4] World Bank World Development Indicators. [5] IMF World Economic Outlook. [6] 2006. [7] UNESCO, % age above 15. [8] % ages 15–64. [9] 2004. [10] The index thermometer shows the relative contribution of each pillar (in %) to the total index score.

RUSSIAN FEDERATION

Country relative performance: **Index pillars**[11]

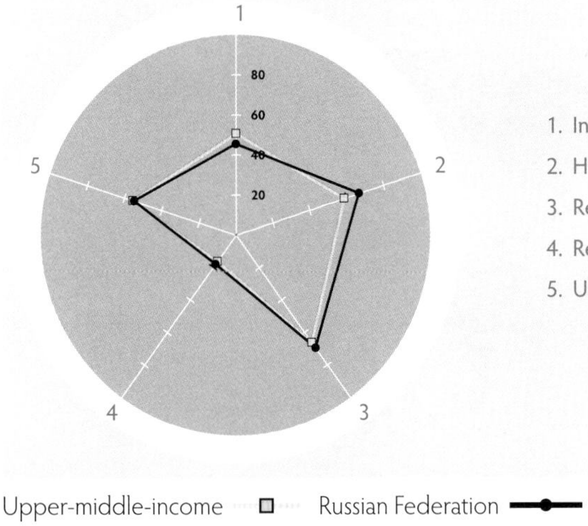

1. Institutional environment
2. Human capital, training and social inclusion
3. Regulatory and legal framework
4. Research and development
5. Usage of ICT

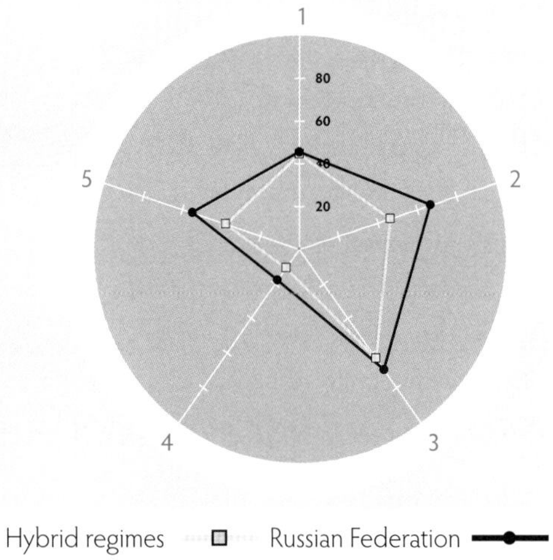

Upper-middle-income ☐ Russian Federation ●—

Hybrid regimes ☐ Russian Federation ●—

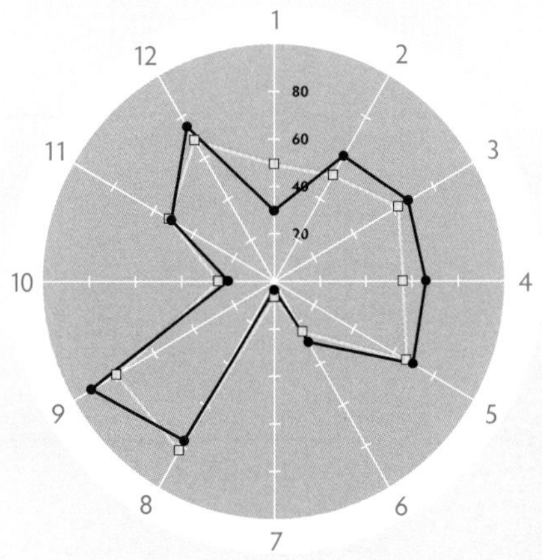

1. Good governance
2. Country policy assessment
3. Education
4. Social inclusion and equity policies
5. Doing business
6. R&D infrastructure
7. Patents and trademarks
8. Telephone communications
9. Mobile/cellular communications
10. Internet, computers and TV
11. Government ICT usage
12. Quality of the infrastructure

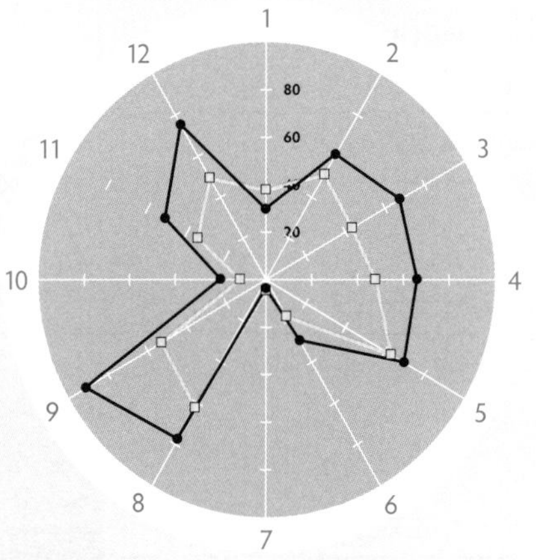

Country relative performance: **Pillar subindexes**[11]

In focus: **Top priorities for policy reform**[12]

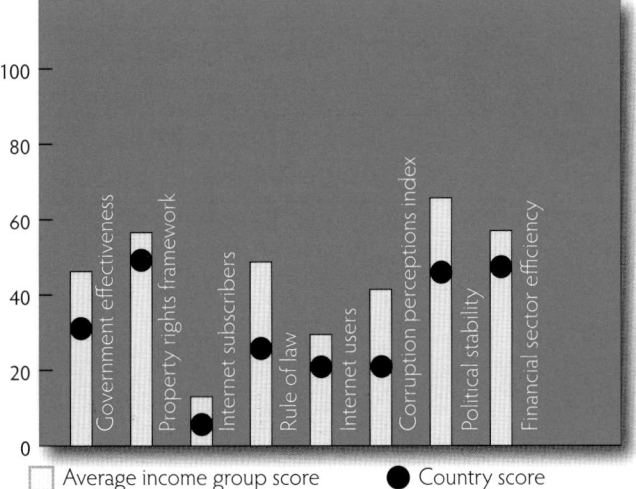

☐ Average income group score   ● Country score

In focus: **Significant indicators above income group average**[12]

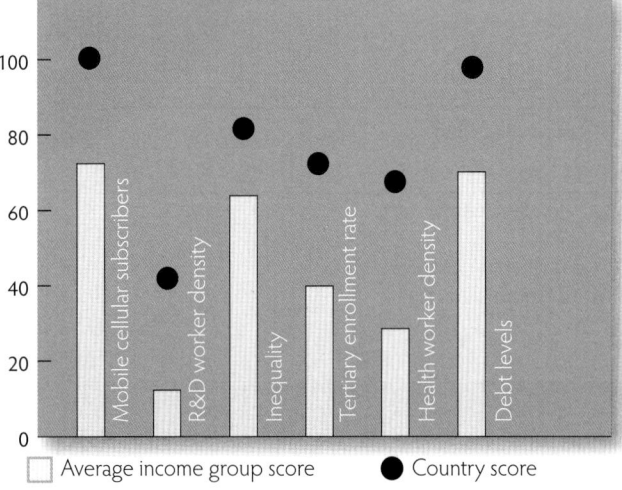

☐ Average income group score   ● Country score

[11] This analysis compares pillar and subindex country scores with respect to the average scores of countries grouped according to income level and political regime (see notes 1 and 2).

[12] Significant country indicators with the greatest distances above or below the income group average scores (i.e., achievements and targets for policy reform, respectively) were selected and ordered according to their Pearson correlation coefficients (highest first) with respect to the final index scores, to produce a ranking.

SAUDI ARABIA

Key selected indicators

Population (in millions)[3,4]	24.2
Population density (people per sq. km)[3,4]	12
GDP (current US$ billion, PPP)[3,5]	565
GDP per capita (current US$, PPP)[3,5]	23243
Current account balance (% of GDP)[3,5]	26.8
Average life expectancy (years)[3,4]	85.0
Literacy rate[6,7]	83
Female labor force participation rate[4,8,9]	19.1
Unemployment rate (% of total labor force)[4,8]	6.2
CO_2 emissions (metric tons per capita)[4,6]	13.7

Income level:[1] **High-income**
Political regime:[2] **Authoritarian regime**

Innovation Capacity Index

Rank	Score
55 out of 131	**51.9 out of 100**

Pillars and pillar subindexes

Education
Rank **15** Score **77.0**

Social inclusion and equity policies
Rank **120** Score **37.4**

Country policy assessment
Rank **8** Score **72.9**

Good governance
Rank **72** Score **42.2**

Quality of the infrastructure
Rank **62** Score **70.7**

Government ICT usage
Rank **64** Score **49.4**

Doing business
Rank **15** Score **79.4**

R&D infrastructure
Rank **57** Score **28.9**

Patents and trademarks
Rank **95** Score **0.4**

Telephone communications
Rank **51** Score **85.3**

Mobile cellular communications
Rank **20** Score **89.2**

Internet, computers and TV
Rank **56** Score **20.3**

Central diagram (ICI):
- Human capital, training & social inclusion — Rank **67** Score **55.0**
- Regulatory & legal framework — Rank **15** Score **79.4**
- Research & development — Rank **77** Score **11.1**
- Institutional environment — Rank **42** Score **57.6**
- Usage of ICT — Rank **50** Score **56.4**

Index thermometer[10]

Pillar 1 22.2% Pillar 2 21.2% Pillar 3 30.6% Pillar 4 4.3% Pillar 5 21.8%   100%

344

[1] High-income countries are those that, according to the World Bank classification, have a gross national income (GNI) per capita of $11,456 or more.
[2] Authoritarian regimes are countries whose Democracy Index scores from 1 to 10 rounded to one decimal place are less than 4.   [3] 2007.   [4] World Bank World Development Indicators.   [5] IMF World Economic Outlook.   [6] 2004.   [7] UNESCO, % age above 15.   [8] 2006.   [9] % ages 15–64.   [10] The index thermometer shows the relative contribution of each pillar (in %) to the total index score.

SAUDI ARABIA

Country relative performance: **Index pillars**[11]

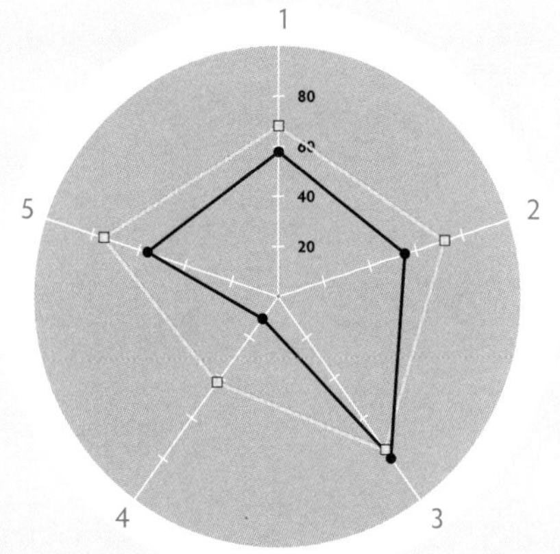

1. Institutional environment
2. Human capital, training and social inclusion
3. Regulatory and legal framework
4. Research and development
5. Usage of ICT

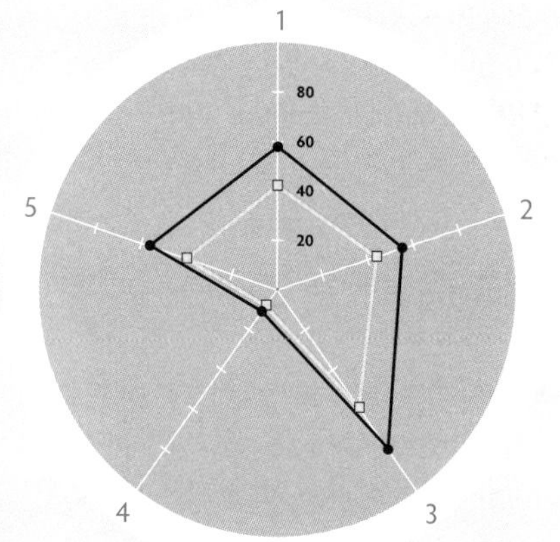

High-income ⬜   Saudi Arabia ●

Authoritarian regimes ⬜   Saudi Arabia ●

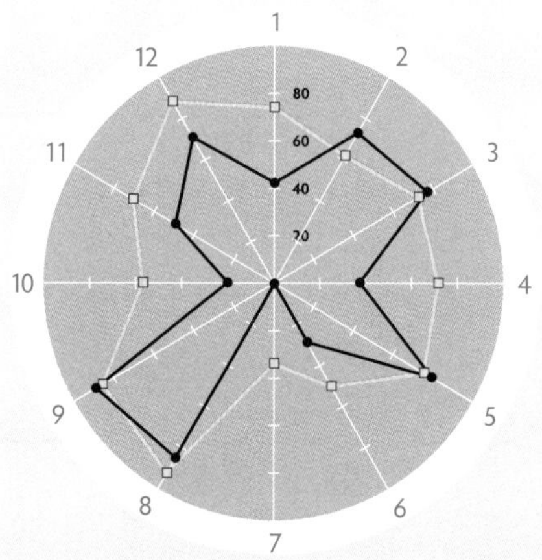

1. Good governance
2. Country policy assessment
3. Education
4. Social inclusion and equity policies
5. Doing business
6. R&D infrastructure
7. Patents and trademarks
8. Telephone communications
9. Mobile/cellular communications
10. Internet, computers and TV
11. Government ICT usage
12. Quality of the infrastructure

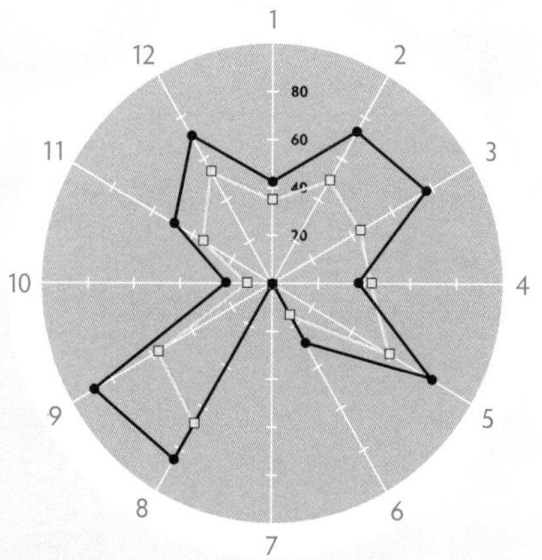

Country relative performance: **Pillar subindexes**[11]

In focus: **Top priorities for policy reform**[12]

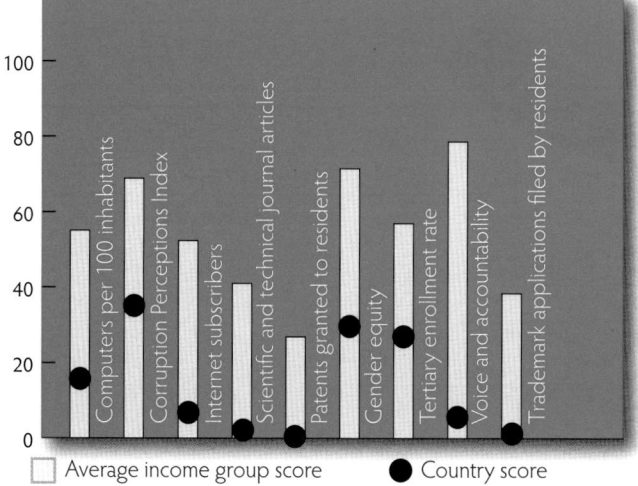

⬜ Average income group score   ● Country score

In focus: **Significant indicators above income group average**[12]

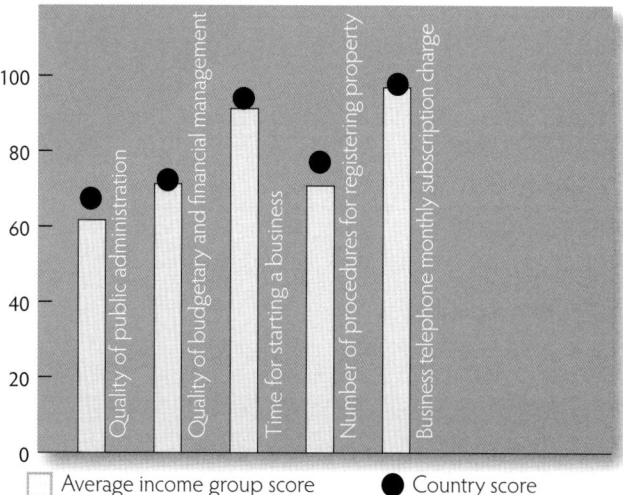

⬜ Average income group score   ● Country score

[11] This analysis compares pillar and subindex country scores with respect to the average scores of countries grouped according to income level and political regime (see notes 1 and 2).

[12] Significant country indicators with the greatest distances above or below the income group average scores (i.e., achievements and targets for policy reform, respectively) were selected and ordered according to their Pearson correlation coefficients (highest first) with respect to the final index scores, to produce a ranking.

SINGAPORE

Key selected indicators

Population (in millions)[3,4]	4.6
Population density (people per sq. km)[3,4]	6660
GDP (current US$ billion, PPP)[3,5]	228
GDP per capita (current US$, PPP)[3,5]	49714
Current account balance (% of GDP)[3,5]	24.3
Average life expectancy (years)[4,6]	79.9
Literacy rate[3,7]	94
Female labor force participation rate[4,6,8]	56.7
Unemployment rate (% of total labor force)[4,9]	4.2
CO_2 emissions (metric tons per capita)[4,10]	12.5

Income level:[1] **High-income**

Political regime:[2] **Hybrid regime**

Rank	Score
6 out of 131	**76.5 out of 100**

Innovation Capacity Index

Pillars and pillar subindexes

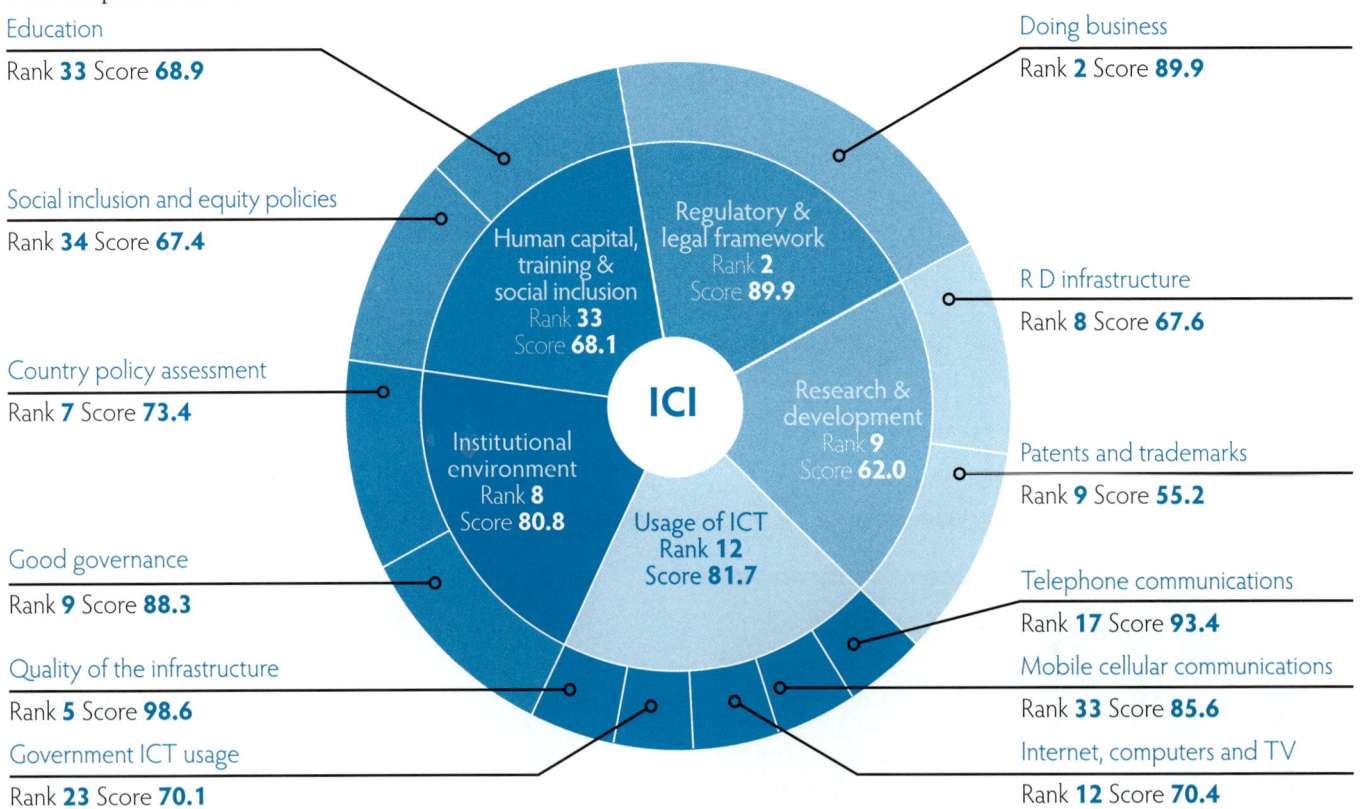

Education
Rank **33** Score **68.9**

Social inclusion and equity policies
Rank **34** Score **67.4**

Country policy assessment
Rank **7** Score **73.4**

Good governance
Rank **9** Score **88.3**

Quality of the infrastructure
Rank **5** Score **98.6**

Government ICT usage
Rank **23** Score **70.1**

Doing business
Rank **2** Score **89.9**

R D infrastructure
Rank **8** Score **67.6**

Patents and trademarks
Rank **9** Score **55.2**

Telephone communications
Rank **17** Score **93.4**

Mobile cellular communications
Rank **33** Score **85.6**

Internet, computers and TV
Rank **12** Score **70.4**

(Wheel:) ICI — Human capital, training & social inclusion Rank **33** Score **68.1**; Regulatory & legal framework Rank **2** Score **89.9**; Research & development Rank **9** Score **62.0**; Usage of ICT Rank **12** Score **81.7**; Institutional environment Rank **8** Score **80.8**

Index thermometer[11]

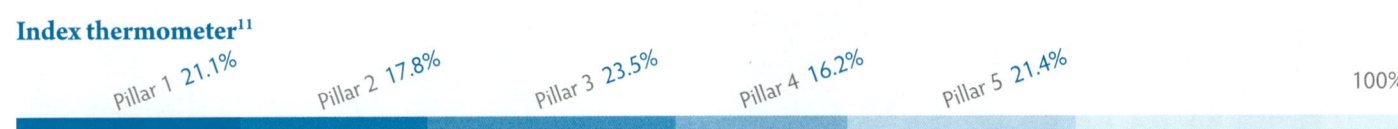

Pillar 1 21.1%   Pillar 2 17.8%   Pillar 3 23.5%   Pillar 4 16.2%   Pillar 5 21.4%   100%

---

[1] High-income countries are those that, according to the World Bank classification, have a gross national income (GNI) per capita of $11,456 or more.
[2] Hybrid regimes are countries whose Democracy Index scores from 1 to 10 rounded to one decimal place are of 4 to 5.9. [3] 2007. [4] World Bank World Development Indicators. [5] IMF World Economic Outlook. [6] 2006. [7] UNESCO, % age above 15. [8] % ages 15–64. [9] 2005. [10] 2004. [11] The index thermometer shows the relative contribution of each pillar (in %) to the total index score.

SINGAPORE

Country relative performance: **Index pillars**[12]

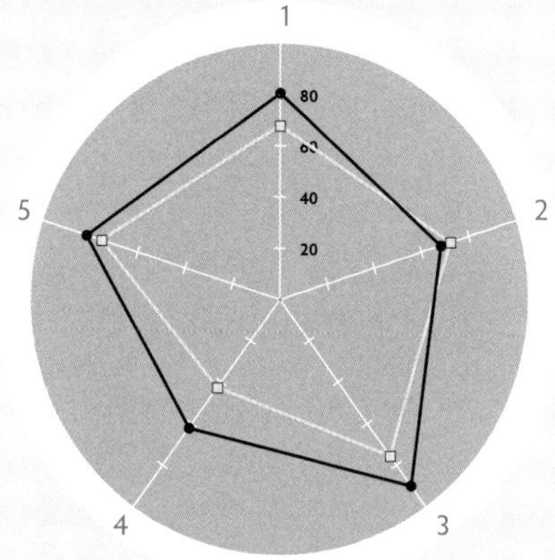

1. Institutional environment
2. Human capital, training and social inclusion
3. Regulatory and legal framework
4. Research and development
5. Usage of ICT

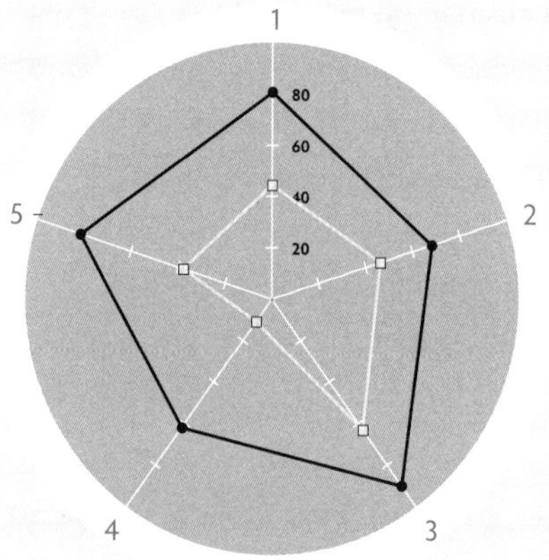

High-income ⬜   Singapore ●━━

Hybrid regimes ⬜   Singapore ●━━

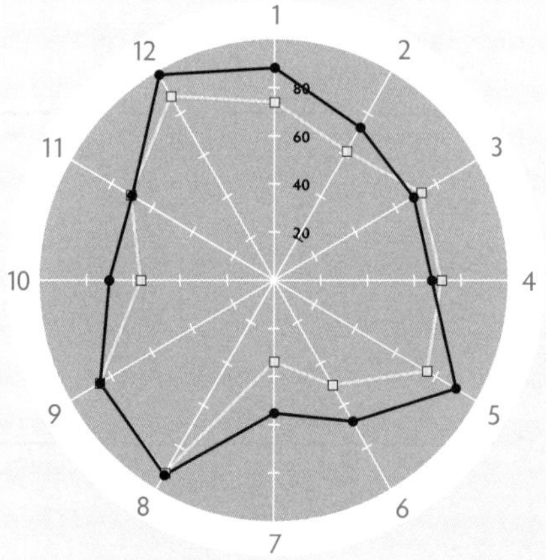

1. Good governance
2. Country policy assessment
3. Education
4. Social inclusion and equity policies
5. Doing business
6. R&D infrastructure
7. Patents and trademarks
8. Telephone communications
9. Mobile/cellular communications
10. Internet, computers and TV
11. Government ICT usage
12. Quality of the infrastructure

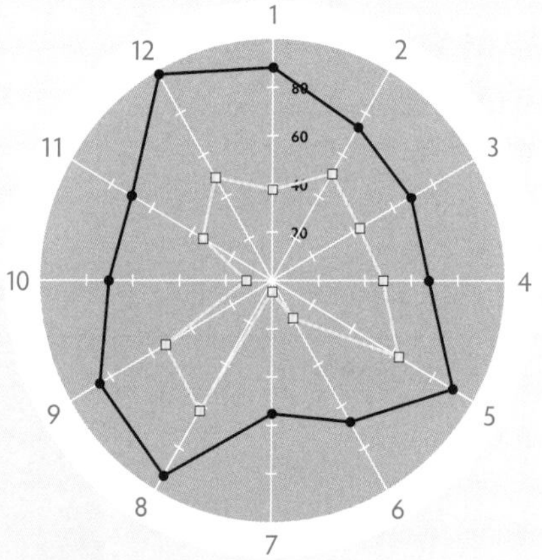

Country relative performance: **Pillar subindexes**[12]

In focus: **Top priorities for policy reform**[13]

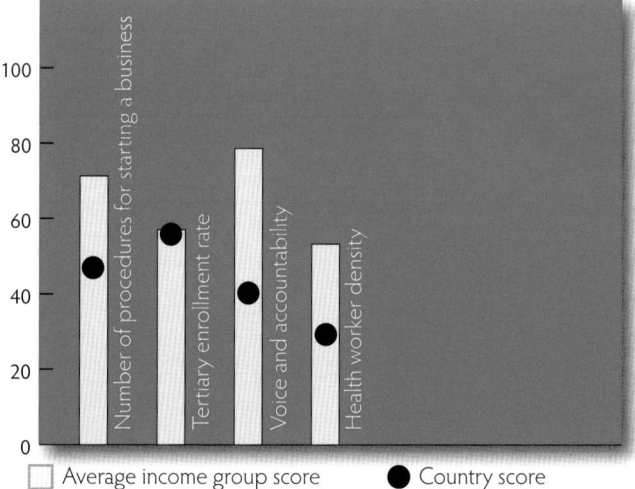

☐ Average income group score   ● Country score

In focus: **Significant indicators above income group average**[13]

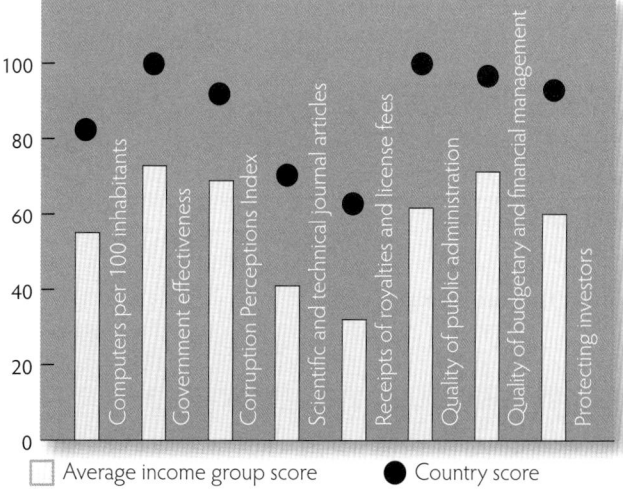

☐ Average income group score   ● Country score

---

[12] This analysis compares pillar and subindex country scores with respect to the average scores of countries grouped according to income level and political regime (see notes 1 and 2).

[13] Significant country indicators with the greatest distances above or below the income group average scores (i.e., achievements and targets for policy reform, respectively) were selected and ordered according to their Pearson correlation coefficients (highest first) with respect to the final index scores, to produce a ranking.

SLOVENIA, REPUBLIC OF

Key selected indicators

Population (in millions)[3,4]	2.0
Population density (people per sq. km)[3,4]	100
GDP (current US$ billion, PPP)[3,5]	55
GDP per capita (current US$, PPP)[3,5]	27205
Current account balance (% of GDP)[3,5]	-4.8
Average life expectancy (years)[4,6]	77.7
Literacy rate[3,7]	100
Female labor force participation rate[4,6,8]	67.0
Unemployment rate (% of total labor force)[4,9]	5.8
CO_2 emissions (metric tons per capita)[4,10]	8.1

Income level:[1] **High-income**

Political regime:[2] **Full democracy**

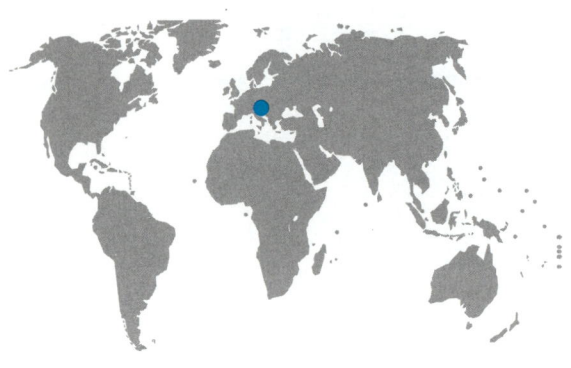

Rank	Score
31 out of 131	**58.6 out of 100**

Innovation Capacity Index

Pillars and pillar subindexes

Education
Rank **26** Score **72.9**

Social inclusion and equity policies
Rank **27** Score **69.7**

Country policy assessment
Rank **39** Score **57.2**

Good governance
Rank **26** Score **69.9**

Quality of the infrastructure
Rank **12** Score **96.7**

Government ICT usage
Rank **26** Score **66.8**

Doing business
Rank **85** Score **62.0**

R&D infrastructure
Rank **23** Score **50.2**

Patents and trademarks
Rank **31** Score **15.5**

Telephone communications
Rank **26** Score **92**

Mobile cellular communications
Rank **51** Score **79.4**

Internet, computers and TV
Rank **25** Score **55.2**

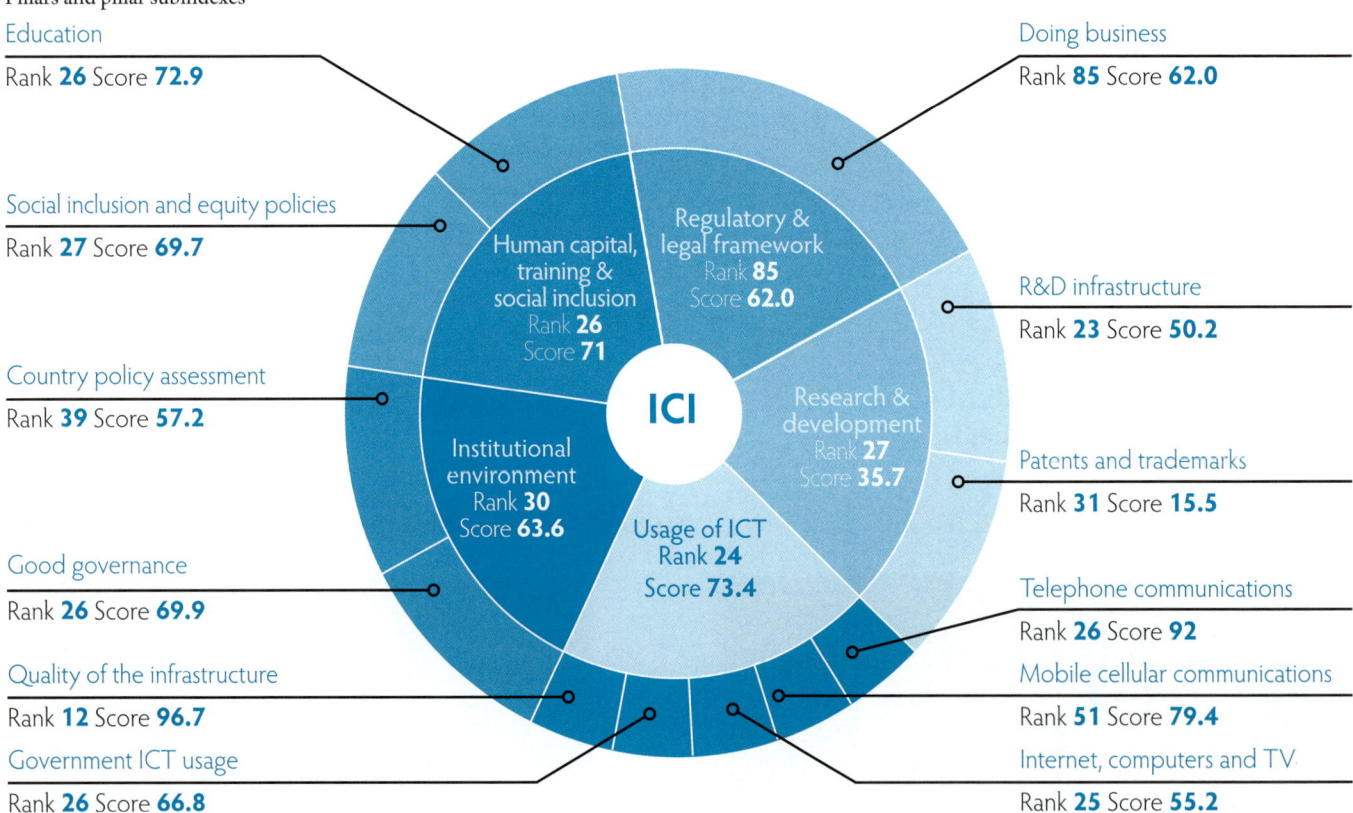

Human capital, training & social inclusion
Rank **26** Score **71**

Regulatory & legal framework
Rank **85** Score **62.0**

Institutional environment
Rank **30** Score **63.6**

Research & development
Rank **27** Score **35.7**

Usage of ICT
Rank **24** Score **73.4**

ICI

Index thermometer[11]

Pillar 1 10.9% Pillar 2 12.1% Pillar 3 21.2% Pillar 4 18.3% Pillar 5 37.6% 100%

[1] High-income countries are those that, according to the World Bank classification, have a gross national income (GNI) per capita of $11,456 or more.
[2] Full democracies are countries whose Democracy Index scores from 1 to 10 rounded to one decimal place are of 8 to 10. [3] 2007. [4] World Bank World Development Indicators. [5] IMF World Economic Outlook. [6] 2006. [7] UNESCO, % age above 15. [8] % ages 15–64. [9] 2005. [10] 2004. [11] The index thermometer shows the relative contribution of each pillar (in %) to the total index score.

SLOVENIA, REPUBLIC OF

Country relative performance: **Index pillars**[12]

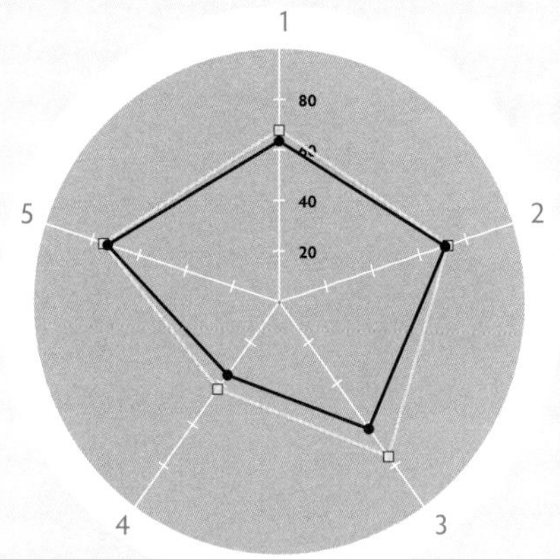

1. Institutional environment
2. Human capital, training and social inclusion
3. Regulatory and legal framework
4. Research and development
5. Usage of ICT

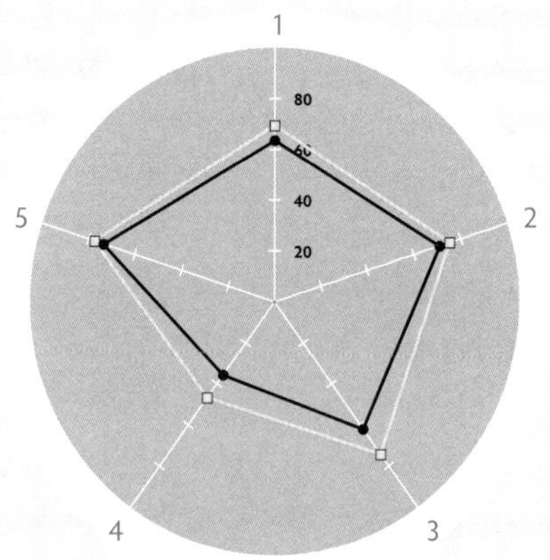

High-income ⊡   Slovenia ●─

Full democracies ⊡   Slovenia ●─

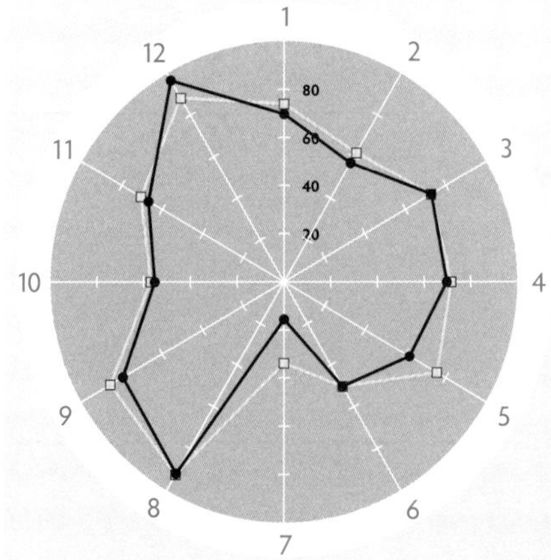

1. Good governance
2. Country policy assessment
3. Education
4. Social inclusion and equity policies
5. Doing business
6. R&D infrastructure
7. Patents and trademarks
8. Telephone communications
9. Mobile/cellular communications
10. Internet, computers and TV
11. Government ICT usage
12. Quality of the infrastructure

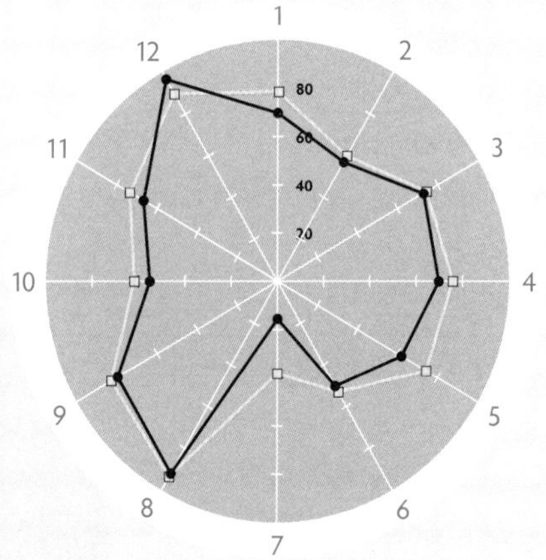

Country relative performance: **Pillar subindexes**[12]

In focus: **Top priorities for policy reform**[13]

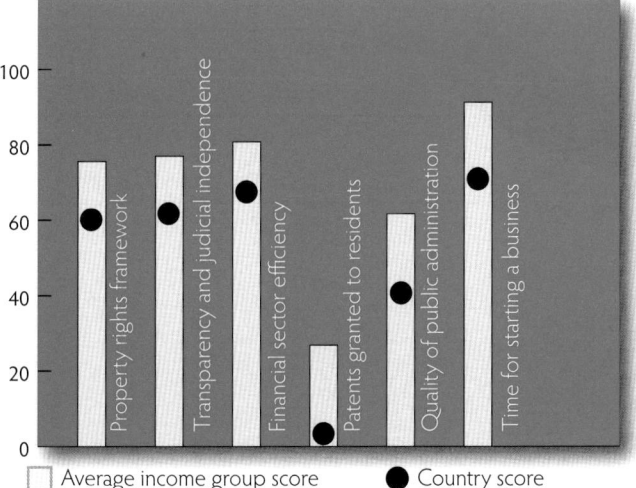

☐ Average income group score   ● Country score

In focus: **Significant indicators above income group average**[13]

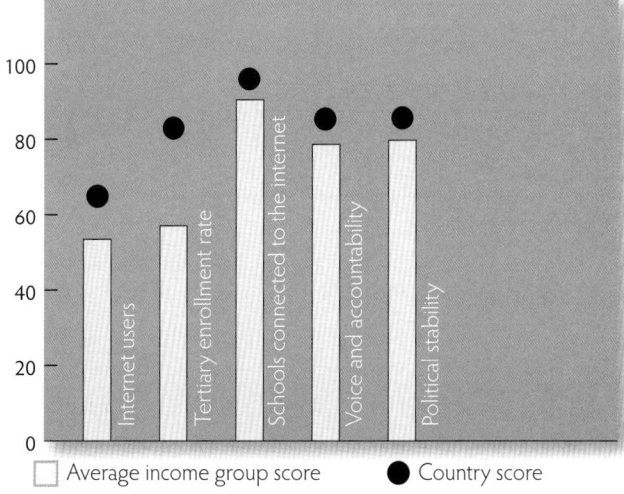

☐ Average income group score   ● Country score

---

[12] This analysis compares pillar and subindex country scores with respect to the average scores of countries grouped according to income level and political regime (see notes 1 and 2).

[13] Significant country indicators with the greatest distances above or below the income group average scores (i.e., achievements and targets for policy reform, respectively) were selected and ordered according to their Pearson correlation coefficients (highest first) with respect to the final index scores, to produce a ranking.

SOUTH AFRICA

Key selected indicators

Population (in millions)[3,4]	47.6
Population density (people per sq. km)[3,4]	39
GDP (current US$ billion, PPP)[3,5]	467
GDP per capita (current US$, PPP)[3,5]	9761
Current account balance (% of GDP)[3,5]	-7.3
Average life expectancy (years)[4,6]	50.7
Literacy rate[3,7]	88
Female labor force participation rate[4,6,8]	49.3
Unemployment rate (% of total labor force)[4,9]	26.7
CO_2 emissions (metric tons per capita)[4,10]	9.4

Income level:[1] **Upper-middle-income**

Political regime:[2] **Flawed democracy**

Innovation Capacity Index

Rank	Score
46 out of 131	**53.3 out of 100**

Pillars and pillar subindexes

Education
Rank **63** Score **60.5**

Social inclusion and equity policies
Rank **92** Score **46.0**

Country policy assessment
Rank **41** Score **56.9**

Good governance
Rank **40** Score **59.7**

Quality of the infrastructure
Rank **75** Score **60.2**

Government ICT usage
Rank **56** Score **51.2**

Doing business
Rank **21** Score **76.7**

R&D infrastructure
Rank **47** Score **30.7**

Patents and trademarks
Rank **52** Score **5.4**

Telephone communications
Rank **68** Score **79.5**

Mobile cellular communications
Rank **45** Score **82.3**

Internet, computers and TV
Rank **82** Score **11.1**

Human capital, training & social inclusion Rank **78** Score **51.8**

Regulatory & legal framework Rank **21** Score **76.7**

ICI

Institutional environment Rank **37** Score **58.3**

Research & development Rank **43** Score **20.2**

Usage of ICT Rank **71** Score **49.6**

Index thermometer[11]

Pillar 1 27.3% Pillar 2 24.3% Pillar 3 28.7% Pillar 4 5.7% Pillar 5 14.0% 100%

[1] Upper-middle-income countries are those that, according to the World Bank classification, have a gross national income (GNI) per capita of $3,706 to $11,455. [2] Flawed democracies are countries whose Democracy Index scores from 1 to 10 rounded to one decimal place are of 6 to 7.9. [3] 2007. [4] World Bank World Development Indicators. [5] IMF World Economic Outlook. [6] 2006. [7] UNESCO, % age above 15. [8] % ages 15–64. [9] 2005. [10] 2004. [11] The index thermometer shows the relative contribution of each pillar (in %) to the total index score.

SOUTH AFRICA

Country relative performance: **Index pillars**[12]

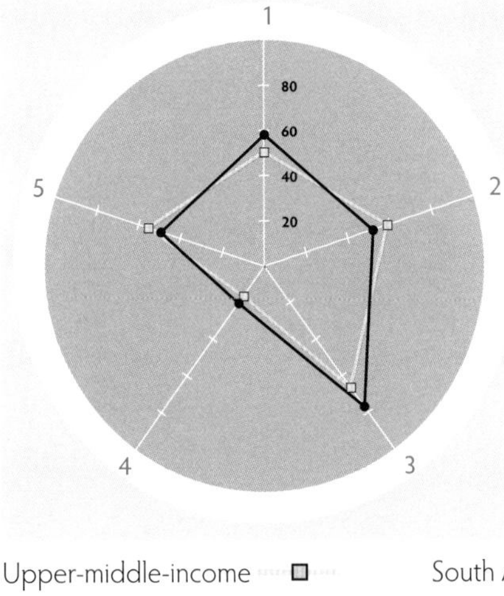

1. Institutional environment
2. Human capital, training and social inclusion
3. Regulatory and legal framework
4. Research and development
5. Usage of ICT

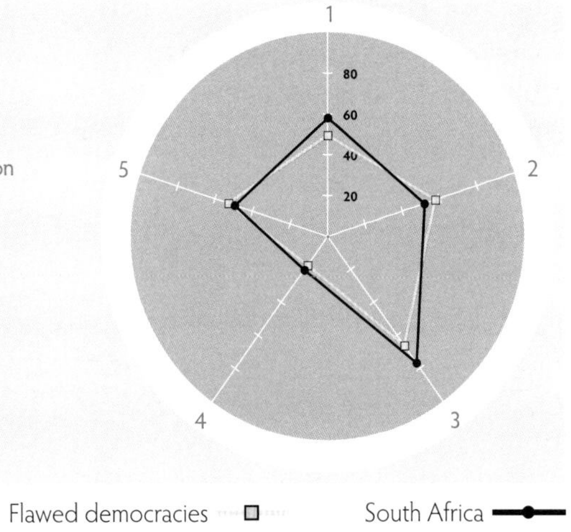

Upper-middle-income ⬜   South Africa ●━━●

Flawed democracies ⬜   South Africa ●━━●

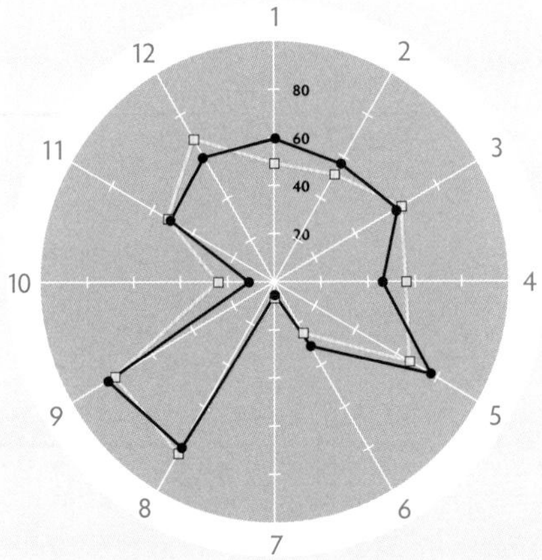

1. Good governance
2. Country policy assessment
3. Education
4. Social inclusion and equity policies
5. Doing business
6. R&D infrastructure
7. Patents and trademarks
8. Telephone communications
9. Mobile/cellular communications
10. Internet, computers and TV
11. Government ICT usage
12. Quality of the infrastructure

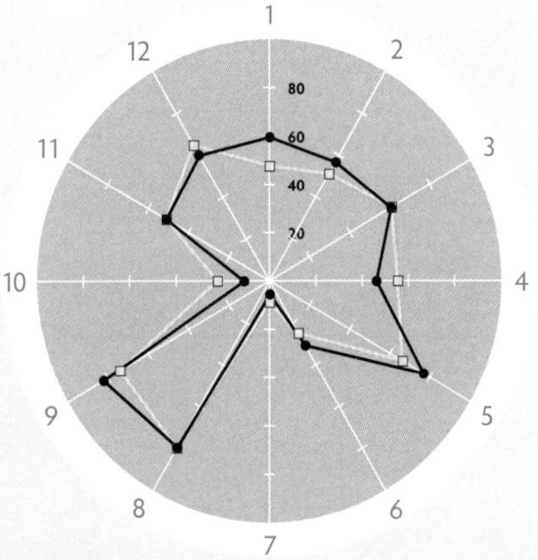

Country relative performance: **Pillar subindexes**[12]

In focus: **Top priorities for policy reform**[13]

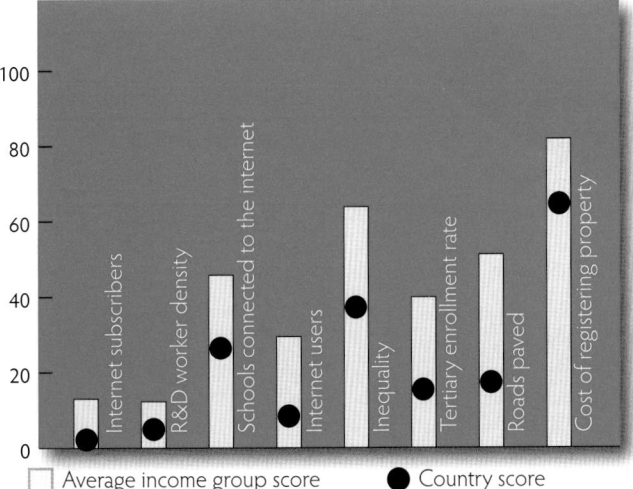

⬜ Average income group score   ● Country score

In focus: **Significant indicators above income group average**[13]

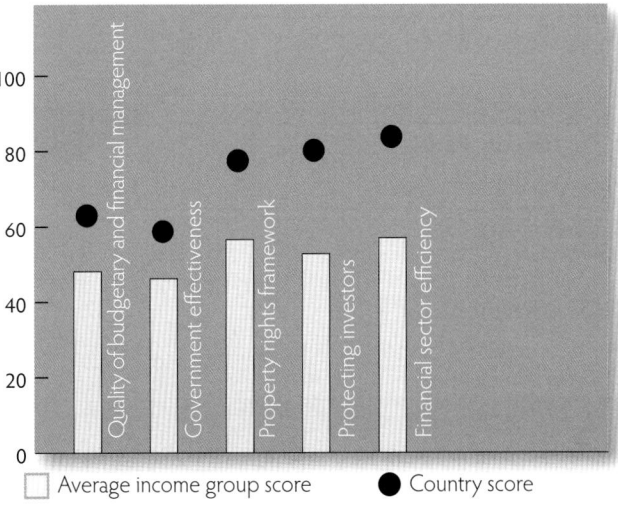

⬜ Average income group score   ● Country score

[12] This analysis compares pillar and subindex country scores with respect to the average scores of countries grouped according to income level and political regime (see notes 1 and 2).

[13] Significant country indicators with the greatest distances above or below the income group average scores (i.e., achievements and targets for policy reform, respectively) were selected and ordered according to their Pearson correlation coefficients (highest first) with respect to the final index scores, to produce a ranking.

SPAIN

Key selected indicators

Population (in millions)[3,4]	44.9
Population density (people per sq. km)[3,4]	90
GDP (current US$ billion, PPP)[3,5]	1352
GDP per capita (current US$, PPP)[3,5]	30120
Current account balance (% of GDP)[3,5]	-10.1
Average life expectancy (years)[4,6]	80.8
Literacy rate[3,7]	97
Female labor force participation rate[4,6,8]	57.8
Unemployment rate (% of total labor force)[4,9]	9.2
CO_2 emissions (metric tons per capita)[4,10]	7.7

Income level:[1] **High-income**

Political regime:[2] **Full democracy**

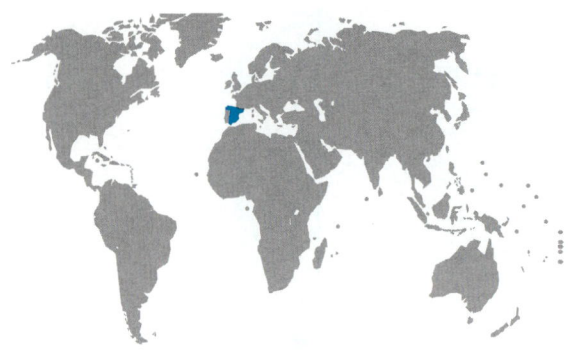

Innovation Capacity Index

Rank	Score
28 out of 131	**60.3 out of 100**

Pillars and pillar subindexes

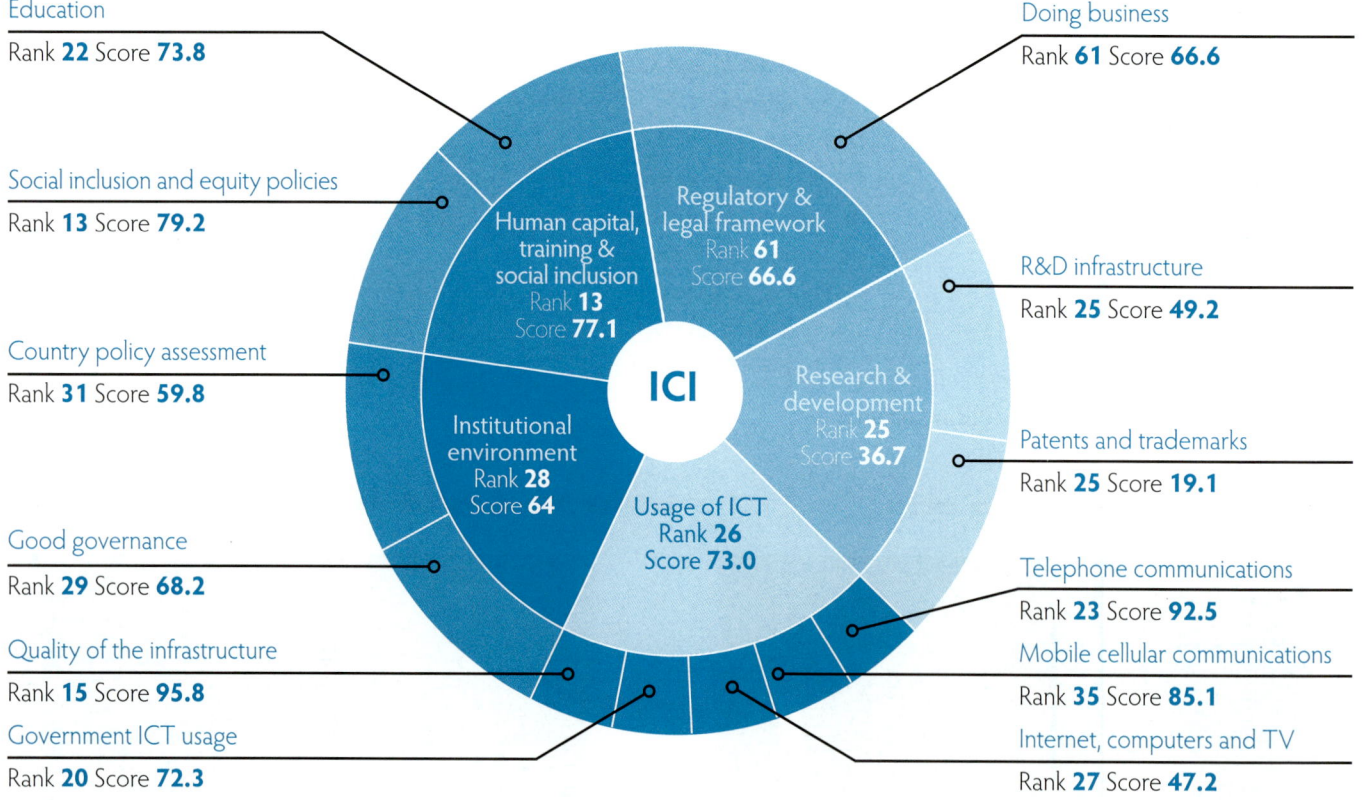

Education
Rank **22** Score **73.8**

Social inclusion and equity policies
Rank **13** Score **79.2**

Country policy assessment
Rank **31** Score **59.8**

Good governance
Rank **29** Score **68.2**

Quality of the infrastructure
Rank **15** Score **95.8**

Government ICT usage
Rank **20** Score **72.3**

Doing business
Rank **61** Score **66.6**

R&D infrastructure
Rank **25** Score **49.2**

Patents and trademarks
Rank **25** Score **19.1**

Telephone communications
Rank **23** Score **92.5**

Mobile cellular communications
Rank **35** Score **85.1**

Internet, computers and TV
Rank **27** Score **47.2**

Human capital, training & social inclusion
Rank **13** Score **77.1**

Regulatory & legal framework
Rank **61** Score **66.6**

Institutional environment
Rank **28** Score **64**

ICI

Research & development
Rank **25** Score **36.7**

Usage of ICT
Rank **26** Score **73.0**

Index thermometer[11]

Pillar 1 10.6% Pillar 2 12.8% Pillar 3 22.1% Pillar 4 18.2% Pillar 5 36.3% 100%

[1] High-income countries are those that, according to the World Bank classification, have a gross national income (GNI) per capita of $11,456 or more. [2] Full democracies are countries whose Democracy Index scores from 1 to 10 rounded to one decimal place are of 8 to 10. [3] 2007. [4] World Bank World Development Indicators. [5] IMF World Economic Outlook. [6] 2006. [7] UNESCO, % age above 15. [8] % ages 15–64. [9] 2005. [10] 2004. [11] The index thermometer shows the relative contribution of each pillar (in %) to the total index score.

SPAIN

Country relative performance: **Index pillars**[12]

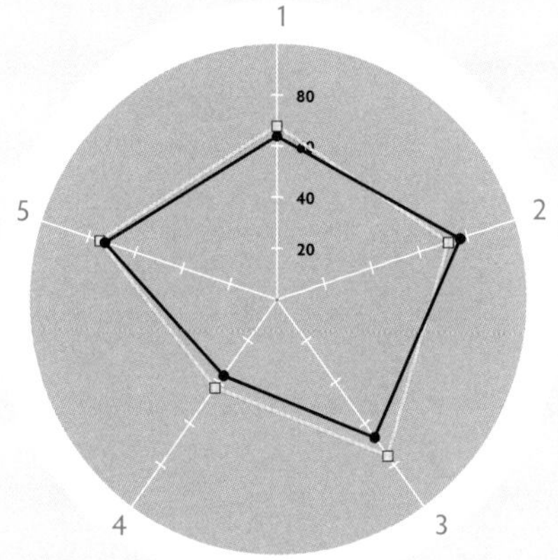

1. Institutional environment
2. Human capital, training and social inclusion
3. Regulatory and legal framework
4. Research and development
5. Usage of ICT

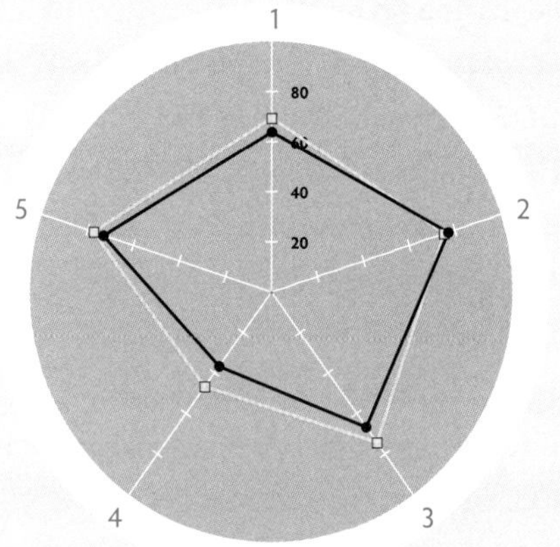

High-income ☐ ⋯⋯ Spain ●━━━

Full democracies ☐ ⋯⋯ Spain ●━━━

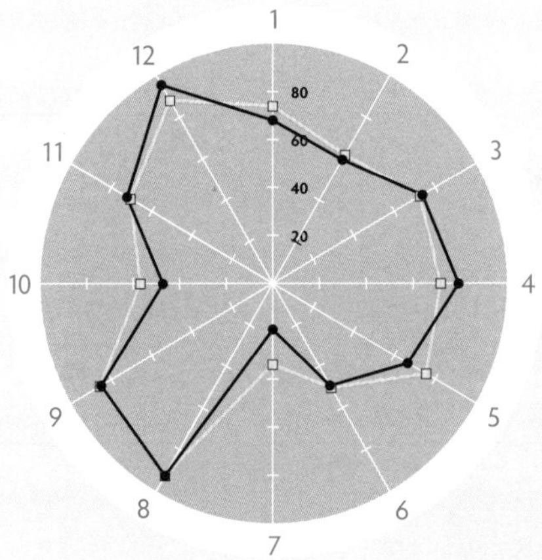

1. Good governance
2. Country policy assessment
3. Education
4. Social inclusion and equity policies
5. Doing business
6. R&D infrastructure
7. Patents and trademarks
8. Telephone communications
9. Mobile/cellular communications
10. Internet, computers and TV
11. Government ICT usage
12. Quality of the infrastructure

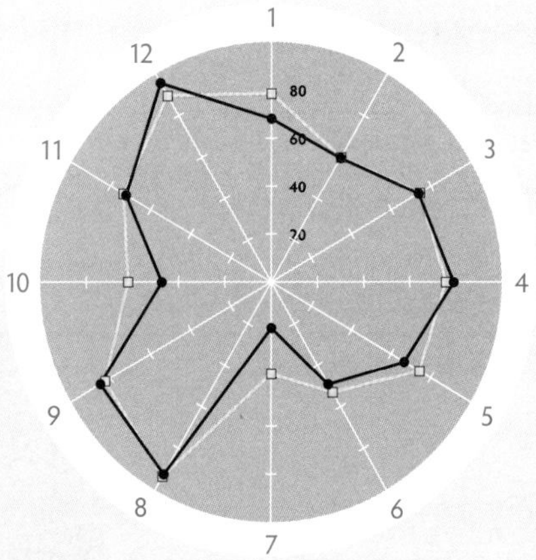

Country relative performance: **Pillar subindexes**[12]

In focus: **Top priorities for policy reform**[13]

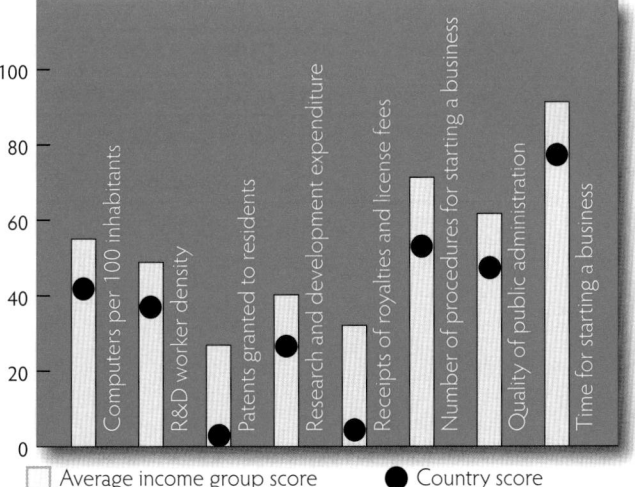

☐ Average income group score ● Country score

In focus: **Significant indicators above income group average**[13]

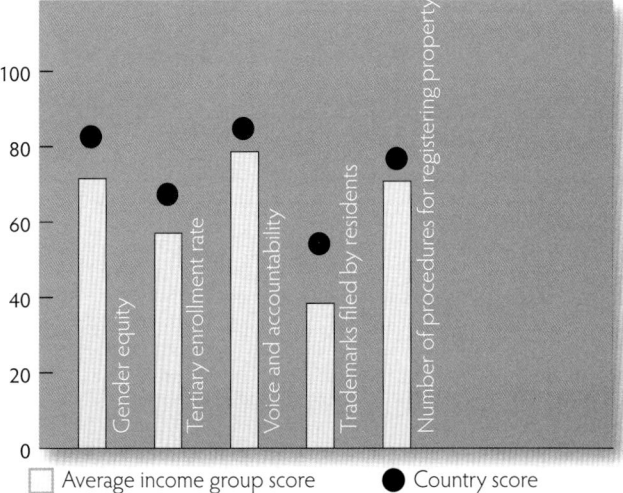

☐ Average income group score ● Country score

[12] This analysis compares pillar and subindex country scores with respect to the average scores of countries grouped according to income level and political regime (see notes 1 and 2).

[13] Significant country indicators with the greatest distances above or below the income group average scores (i.e., achievements and targets for policy reform, respectively) were selected and ordered according to their Pearson correlation coefficients (highest first) with respect to the final index scores, to produce a ranking.

SRI LANKA

Key selected indicators

Population (in millions)[3,4]	19.9
Population density (people per sq. km)[3,4]	309
GDP (current US$ billion, PPP)[3,5]	81
GDP per capita (current US$, PPP)[3,5]	4079
Current account balance (% of GDP)[3,5]	-4.6
Average life expectancy (years)[4,6]	75.0
Literacy rate[3,7]	92
Female labor force participation rate[4,6,8]	38.1
Unemployment rate (% of total labor force)[4,9]	7.6
CO_2 emissions (metric tons per capita)[4,10]	0.6

Income level:[1] **Lower-middle-income**

Political regime:[2] **Flawed democracy**

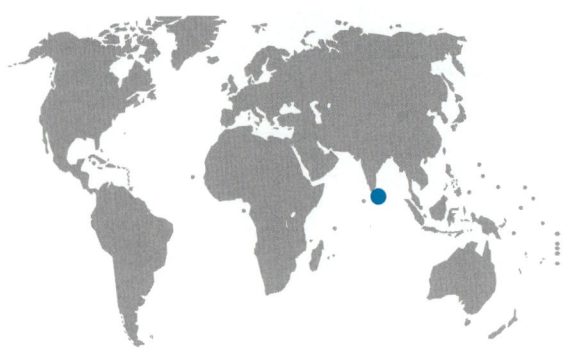

Innovation Capacity Index

Rank
86 out of 131

Score
45.5 out of 100

Pillars and pillar subindexes

Education
Rank **65** Score **59.9**

Social inclusion and equity policies
Rank **84** Score **47.8**

Country policy assessment
Rank **122** Score **39.1**

Good governance
Rank **87** Score **36.6**

Quality of the infrastructure
Rank **58** Score **74.5**

Government ICT usage
Rank **82** Score **42.4**

Doing business
Rank **55** Score **67.8**

R&D infrastructure
Rank **92** Score **13.0**

Patents and trademarks
Rank **72** Score **1.8**

Telephone communications
Rank **110** Score **62.0**

Mobile cellular communications
Rank **82** Score **62.2**

Internet, computers and TV
Rank **98** Score **4.6**

ICI

Human capital, training & social inclusion
Rank **73** Score **52.7**

Regulatory & legal framework
Rank **55** Score **67.8**

Institutional environment
Rank **107** Score **37.8**

Research & development
Rank **94** Score **7.4**

Usage of ICT
Rank **93** Score **40.5**

Index thermometer[11]

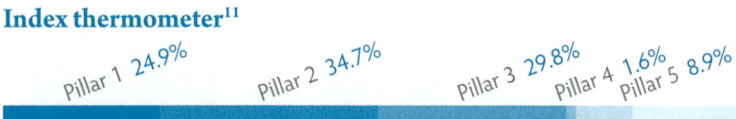

Pillar 1 24.9%   Pillar 2 34.7%   Pillar 3 29.8%   Pillar 4 1.6%   Pillar 5 8.9%   100%

[1] Lower-middle-income countries are those that, according to the World Bank classification, have a gross national income (GNI) per capita of $936 to $3,705. [2] Flawed democracies are countries whose Democracy Index scores from 1 to 10 rounded to one decimal place are of 6 to 7.9. [3] 2007. [4] World Bank World Development Indicators. [5] IMF World Economic Outlook. [6] 2006. [7] UNESCO, % age above 15. [8] % ages 15–64. [9] 2005. [10] 2004. [11] The index thermometer shows the relative contribution of each pillar (in %) to the total index score.

SRI LANKA

Country relative performance: **Index pillars**[12]

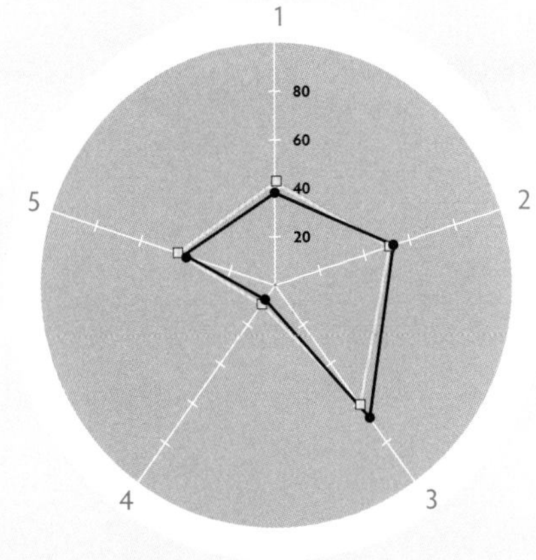

1. Institutional environment
2. Human capital, training and social inclusion
3. Regulatory and legal framework
4. Research and development
5. Usage of ICT

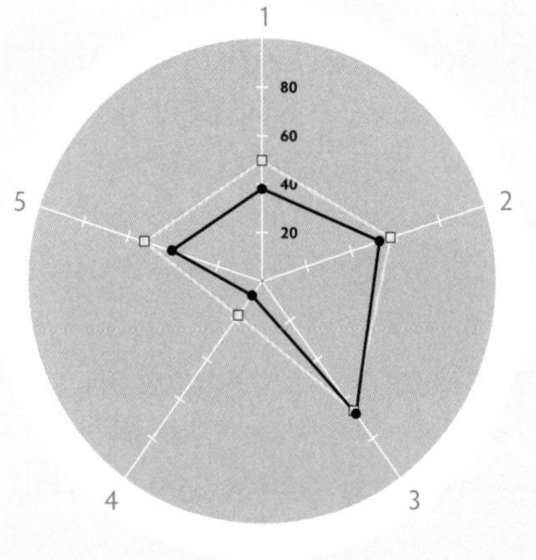

Lower-middle-income ▢   Sri Lanka ●——

Flawed democracies ▢   Sri Lanka ●——

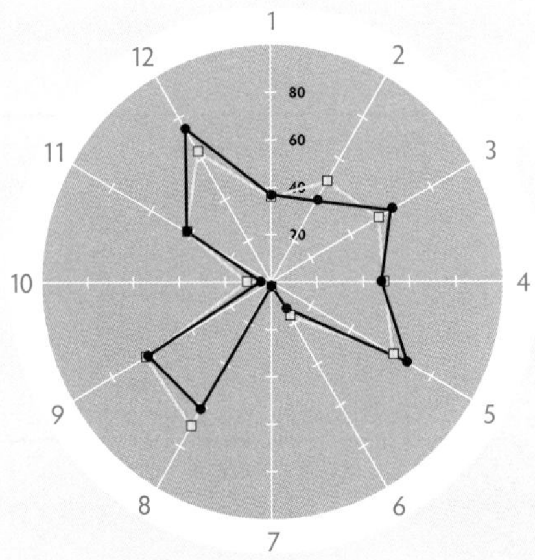

1. Good governance
2. Country policy assessment
3. Education
4. Social inclusion and equity policies
5. Doing business
6. R&D infrastructure
7. Patents and trademarks
8. Telephone communications
9. Mobile/cellular communications
10. Internet, computers and TV
11. Government ICT usage
12. Quality of the infrastructure

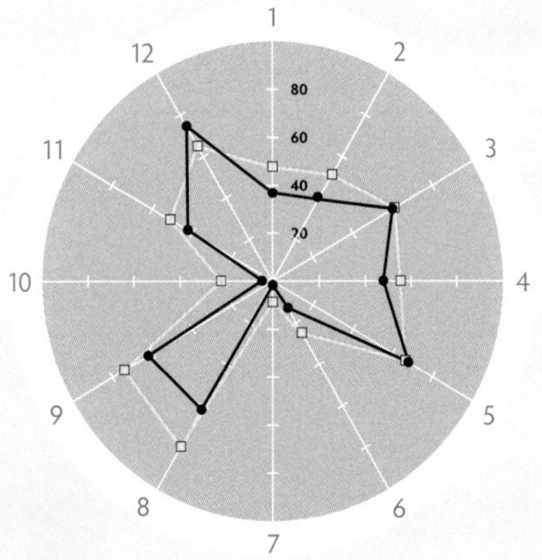

Country relative performance: **Pillar subindexes**[12]

In focus: **Top priorities for policy reform**[13]

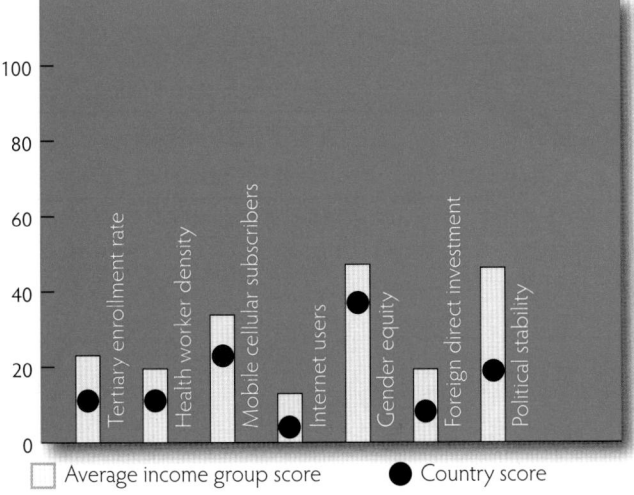

☐ Average income group score   ● Country score

In focus: **Significant indicators above income group average**[13]

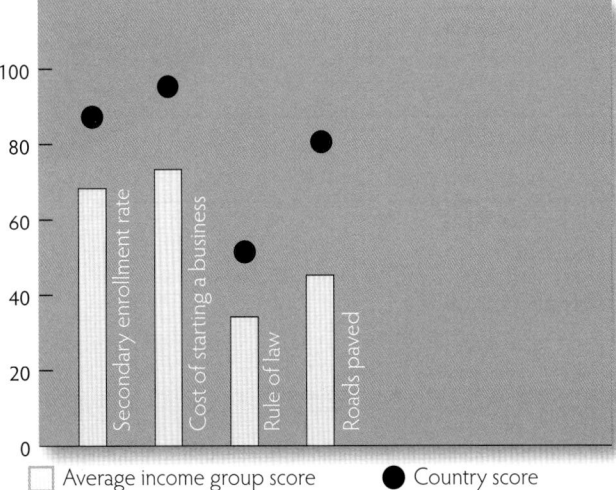

☐ Average income group score   ● Country score

[12] This analysis compares pillar and subindex country scores with respect to the average scores of countries grouped according to income level and political regime (see notes 1 and 2).

[13] Significant country indicators with the greatest distances above or below the income group average scores (i.e., achievements and targets for policy reform, respectively) were selected and ordered according to their Pearson correlation coefficients (highest first) with respect to the final index scores, to produce a ranking.

SWEDEN

Key selected indicators

Population (in millions)[3,4]	9.1
Population density (people per sq. km)[3,4]	22
GDP (current US$ billion, PPP)[3,5]	335
GDP per capita (current US$, PPP)[3,5]	36494
Current account balance (% of GDP)[3,5]	8.3
Average life expectancy (years)[4,6]	80.8
Literacy rate[7,8]	99
Female labor force participation rate[4,6,9]	74.7
Unemployment rate (% of total labor force)[4,10]	7.7
CO_2 emissions (metric tons per capita)[4,11]	5.9

Income level:[1] **High-income**

Political regime:[2] **Full democracy**

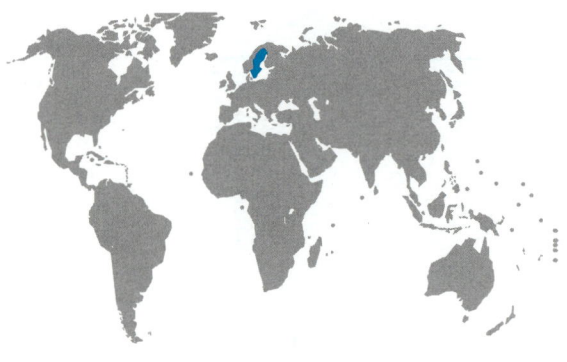

Innovation Capacity Index

Rank	Score
1 out of 131	**82.2 out of 100**

Pillars and pillar subindexes

Education
Rank **21** Score **74.3**

Social inclusion and equity policies
Rank **2** Score **89.5**

Country policy assessment
Rank **13** Score **69.7**

Good governance
Rank **2** Score **93.4**

Quality of the infrastructure
Rank **59** Score **74.2**

Government ICT usage
Rank **1** Score **91.6**

Doing business
Rank **11** Score **80.8**

R&D infrastructure
Rank **2** Score **82.4**

Patents and trademarks
Rank **6** Score **66.1**

Telephone communications
Rank **6** Score **97.2**

Mobile cellular communications
Rank **36** Score **85.0**

Internet, computers and TV
Rank **2** Score **91.0**

Chart center: **ICI**

- Human capital, training & social inclusion — Rank **4** Score **83.4**
- Regulatory & legal framework — Rank **11** Score **80.8**
- Research & development — Rank **2** Score **75.6**
- Institutional environment — Rank **6** Score **81.6**
- Usage of ICT — Rank **2** Score **89.6**

Index thermometer[12]

Pillar 1 9.9% Pillar 2 10.1% Pillar 3 19.7% Pillar 4 27.6% Pillar 5 32.7% 100%

356

[1] High-income countries are those that, according to the World Bank classification, have a gross national income (GNI) per capita of $11,456 or more. [2] Full democracies are countries whose Democracy Index scores from 1 to 10 rounded to one decimal place are of 8 to 10. [3] 2007. [4] World Bank World Development Indicators. [5] IMF World Economic Outlook. [6] 2006. [7] Estimate. [8] UNESCO, % age above 15. [9] % ages 15–64. [10] 2005. [11] 2004. [12] The index thermometer shows the relative contribution of each pillar (in %) to the total index score.

SWEDEN

Country relative performance: **Index pillars**[13]

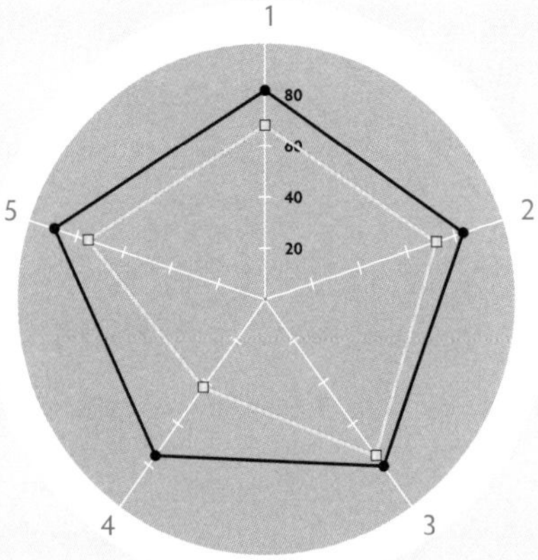

1. Institutional environment
2. Human capital, training and social inclusion
3. Regulatory and legal framework
4. Research and development
5. Usage of ICT

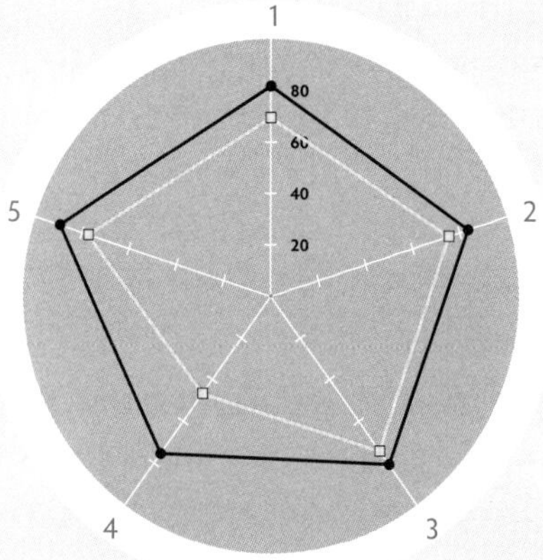

High-income ·····□·····   Sweden ——●——

Full democracies ·····□·····   Sweden ——●——

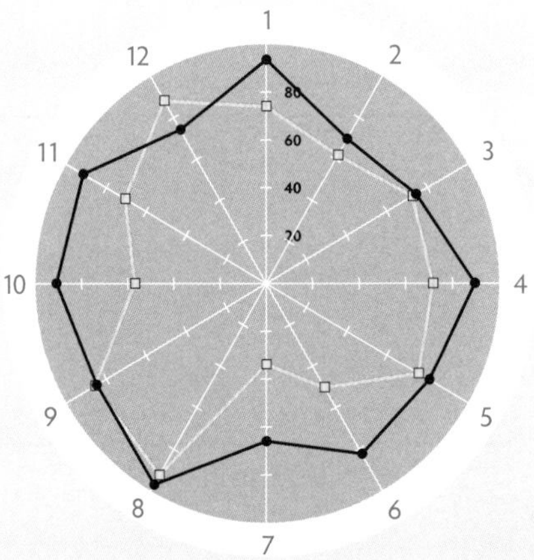

1. Good governance
2. Country policy assessment
3. Education
4. Social inclusion and equity policies
5. Doing business
6. R&D infrastructure
7. Patents and trademarks
8. Telephone communications
9. Mobile/cellular communications
10. Internet, computers and TV
11. Government ICT usage
12. Quality of the infrastructure

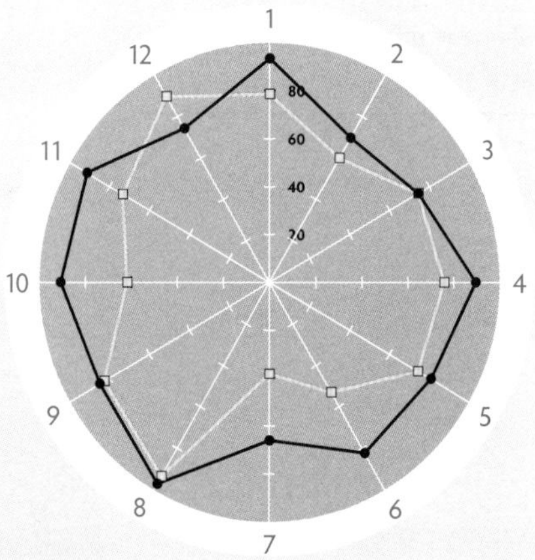

Country relative performance: **Pillar subindexes**[13]

In focus: **Top priorities for policy reform**[14]

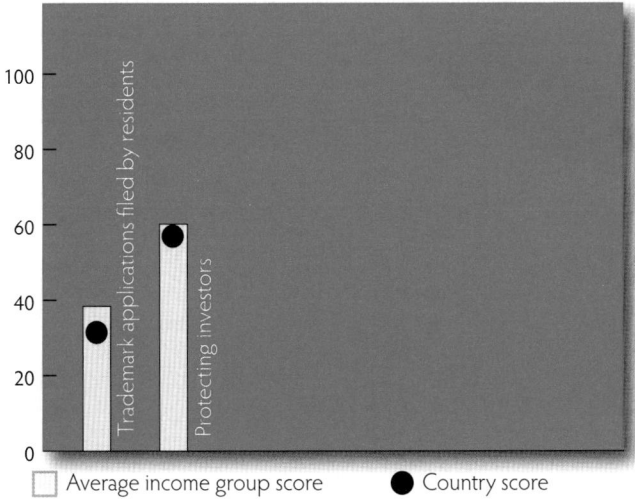

☐ Average income group score   ● Country score

In focus: **Significant indicators above income group average**[14]

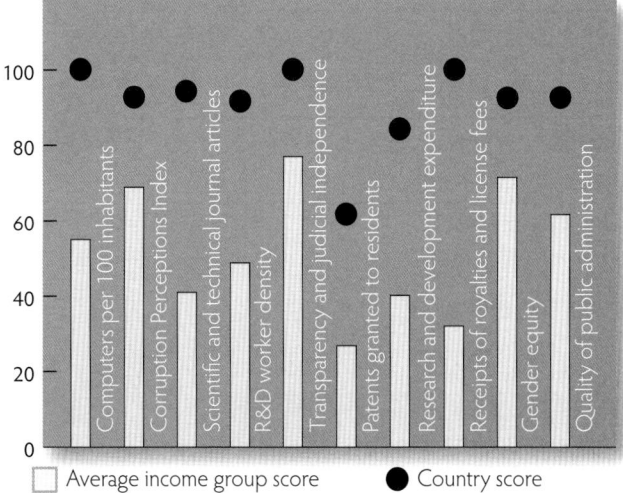

☐ Average income group score   ● Country score

357

[13] This analysis compares pillar and subindex country scores with respect to the average scores of countries grouped according to income level and political regime (see notes 1 and 2).

[14] Significant country indicators with the greatest distances above or below the income group average scores (i.e., achievements and targets for policy reform, respectively) were selected and ordered according to their Pearson correlation coefficients (highest first) with respect to the final index scores, to produce a ranking.

SWITZERLAND

Key selected indicators

Population (in millions)[3,4]	7.6
Population density (people per sq. km)[3,4]	189
GDP (current US$ billion, PPP)[3,5]	300
GDP per capita (current US$, PPP)[3,5]	41128
Current account balance (% of GDP)[3,5]	17.2
Average life expectancy (years)[4,6]	81.5
Literacy rate[7,8]	99
Female labor force participation rate[4,6,9]	76.0
Unemployment rate (% of total labor force)[4,10]	4.4
CO_2 emissions (metric tons per capita)[4,11]	5.5

Income level:[1] **High-income**

Political regime:[2] **Full democracy**

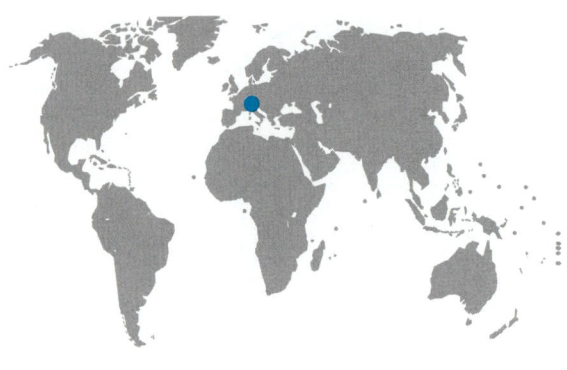

Innovation Capacity Index

Rank **4 out of 131**   Score **77 out of 100**

Pillars and pillar subindexes

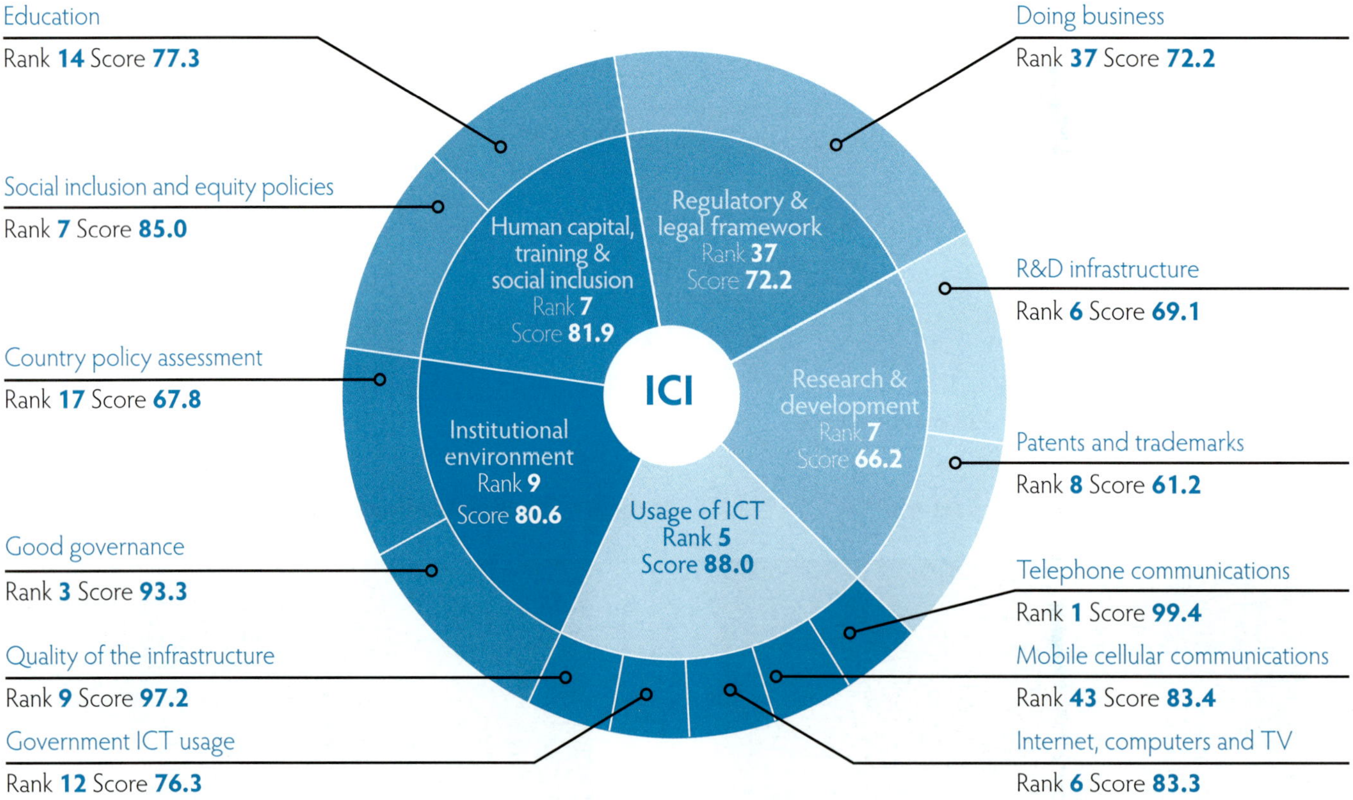

Education
Rank **14** Score **77.3**

Social inclusion and equity policies
Rank **7** Score **85.0**

Country policy assessment
Rank **17** Score **67.8**

Good governance
Rank **3** Score **93.3**

Quality of the infrastructure
Rank **9** Score **97.2**

Government ICT usage
Rank **12** Score **76.3**

Doing business
Rank **37** Score **72.2**

R&D infrastructure
Rank **6** Score **69.1**

Patents and trademarks
Rank **8** Score **61.2**

Telephone communications
Rank **1** Score **99.4**

Mobile cellular communications
Rank **43** Score **83.4**

Internet, computers and TV
Rank **6** Score **83.3**

ICI circle diagram:
- Human capital, training & social inclusion — Rank **7** Score **81.9**
- Regulatory & legal framework — Rank **37** Score **72.2**
- Research & development — Rank **7** Score **66.2**
- Institutional environment — Rank **9** Score **80.6**
- Usage of ICT — Rank **5** Score **88.0**

Index thermometer[12]

Pillar 1 10.5%   Pillar 2 10.6%   Pillar 3 18.8%   Pillar 4 25.8%   Pillar 5 34.3%   100%

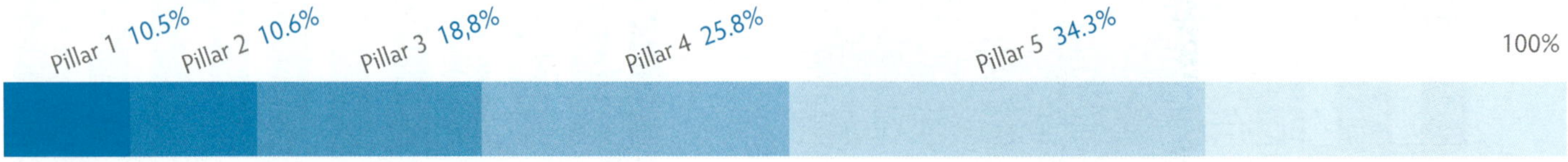

[1] High-income countries are those that, according to the World Bank classification, have a gross national income (GNI) per capita of $11,456 or more.
[2] Full democracies are countries whose Democracy Index scores from 1 to 10 rounded to one decimal place are of 8 to 10.   [3] 2007.   [4] World Bank World Development Indicators.   [5] IMF World Economic Outlook.   [6] 2006.   [7] Estimate.   [8] UNESCO, % age above 15.   [9] % ages 15–64.   [10] 2005.   [11] 2004.   [12] The index thermometer shows the relative contribution of each pillar (in %) to the total index score.

SWITZERLAND

Country relative performance: **Index pillars**[13]

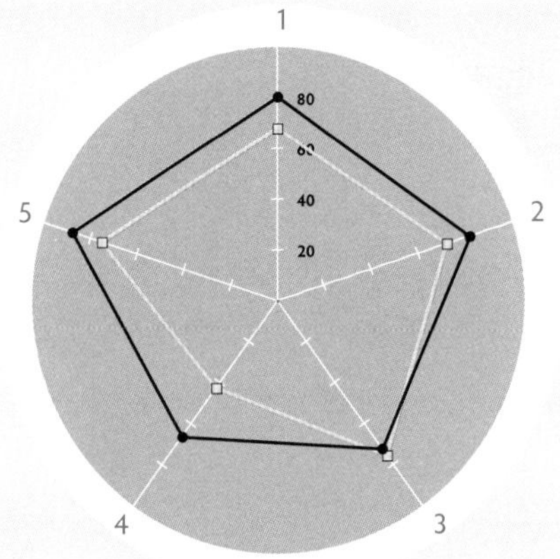

1. Institutional environment
2. Human capital, training and social inclusion
3. Regulatory and legal framework
4. Research and development
5. Usage of ICT

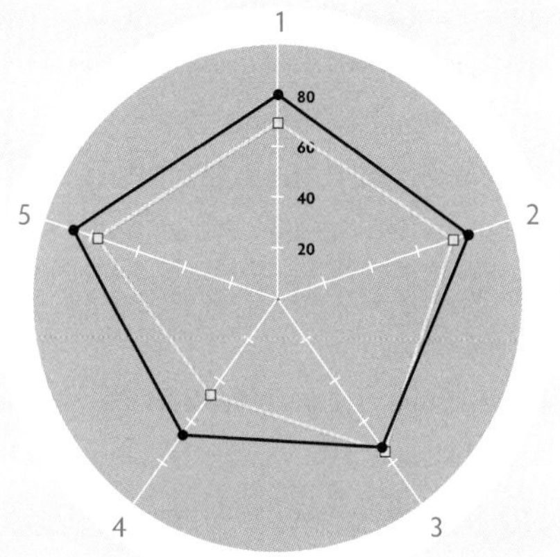

High-income ☐   Switzerland ●—

Full democracies ☐   Switzerland ●—

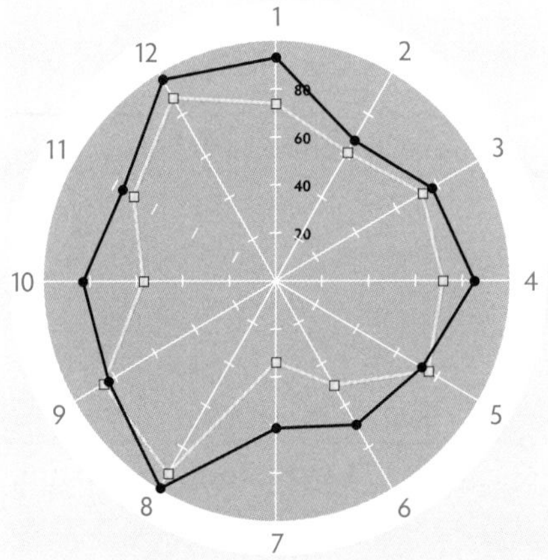

1. Good governance
2. Country policy assessment
3. Education
4. Social inclusion and equity policies
5. Doing business
6. R&D infrastructure
7. Patents and trademarks
8. Telephone communications
9. Mobile/cellular communications
10. Internet, computers and TV
11. Government ICT usage
12. Quality of the infrastructure

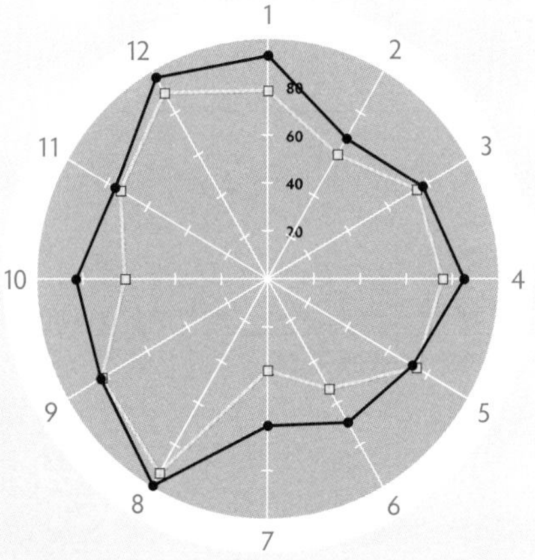

Country relative performance: **Pillar subindexes**[13]

In focus: **Top priorities for policy reform**[14]

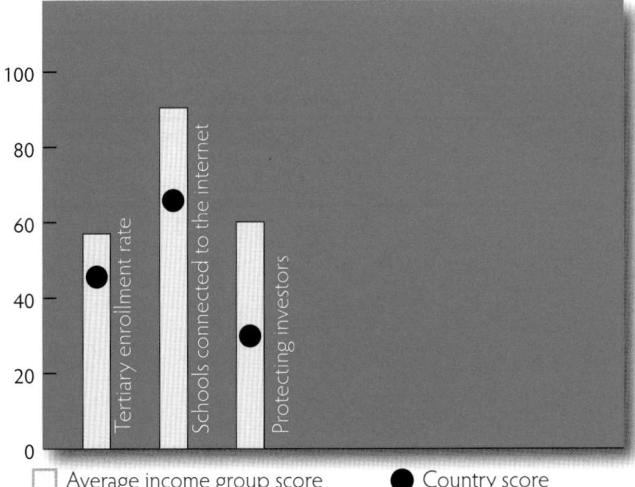

☐ Average income group score   ● Country score

In focus: **Significant indicators above income group average**[14]

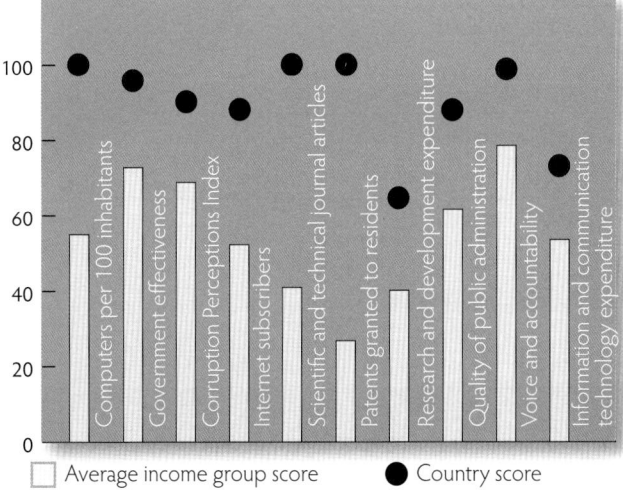

☐ Average income group score   ● Country score

[13] This analysis compares pillar and subindex country scores with respect to the average scores of countries grouped according to income level and political regime (see notes 1 and 2).

[14] Significant country indicators with the greatest distances above or below the income group average scores (i.e., achievements and targets for policy reform, respectively) were selected and ordered according to their Pearson correlation coefficients (highest first) with respect to the final index scores, to produce a ranking.

TAIWAN

Key selected indicators

Population (in millions)[3,4]	22.9
Population density (people per sq. km)[3,4]	635
GDP (current US$ billion, PPP)[3,5]	695
GDP per capita (current US$, PPP)[3,5]	30126
Current account balance (% of GDP)[3,5]	8.3
Average life expectancy (years)[4,6]	77.5
Literacy rate[3,4,7]	98
Female labor force participation rate[4,6,7]	49.0
Unemployment rate (% of total labor force)[3,4]	3.9
CO_2 emissions (metric tons per capita)[8,9,10]	12.5

Income level:[1] **High-income**

Political regime:[2] **Flawed democracy**

Rank	Score
13 out of 131	**72.9 out of 100**

Innovation Capacity Index

Pillars and pillar subindexes

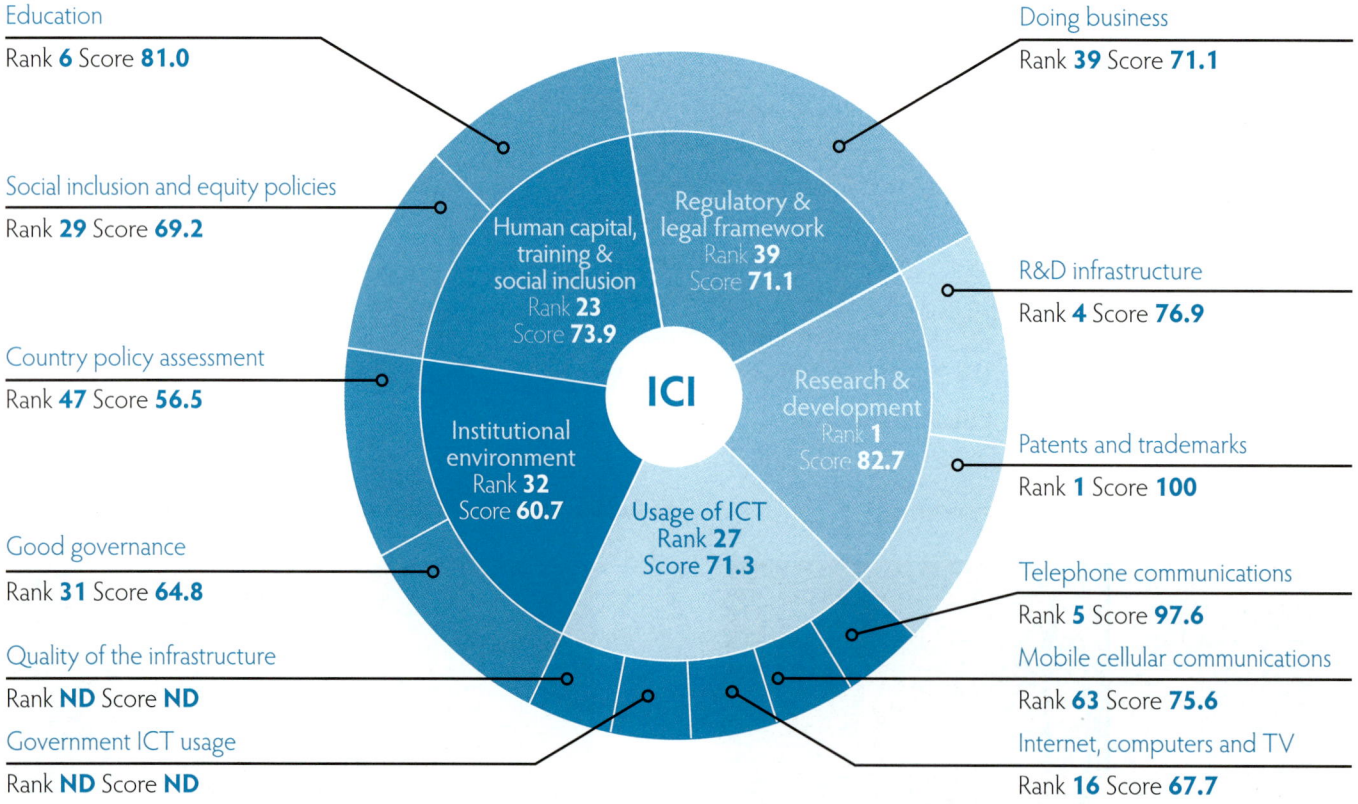

Education
Rank **6** Score **81.0**

Social inclusion and equity policies
Rank **29** Score **69.2**

Country policy assessment
Rank **47** Score **56.5**

Good governance
Rank **31** Score **64.8**

Quality of the infrastructure
Rank **ND** Score **ND**

Government ICT usage
Rank **ND** Score **ND**

Doing business
Rank **39** Score **71.1**

R&D infrastructure
Rank **4** Score **76.9**

Patents and trademarks
Rank **1** Score **100**

Telephone communications
Rank **5** Score **97.6**

Mobile cellular communications
Rank **63** Score **75.6**

Internet, computers and TV
Rank **16** Score **67.7**

Human capital, training & social inclusion
Rank **23** Score **73.9**

Regulatory & legal framework
Rank **39** Score **71.1**

Institutional environment
Rank **32** Score **60.7**

Research & development
Rank **1** Score **82.7**

Usage of ICT
Rank **27** Score **71.3**

ICI

Index thermometer[11]

Pillar 1 12.5%   Pillar 2 15.2%   Pillar 3 19.5%   Pillar 4 28.3%   Pillar 5 24.5%   100%

[1] High-income countries are those that, according to the World Bank classification, have a gross national income (GNI) per capita of $11,456 or more.
[2] Flawed democracies are countries whose Democracy Index scores from 1 to 10 rounded to one decimal place are of 6 to 7.9.   [3] 2007.   [4] National sources.
[5] IMF World Economic Outlook.   [6] 2006.   [7] % age above 15.   [8] 2005.   [9] Energy Information Administration, US Government.
[10] Estimate.   [11] The index thermometer shows the relative contribution of each pillar (in %) to the total index score.

TAIWAN

Country relative performance: **Index pillars**[12]

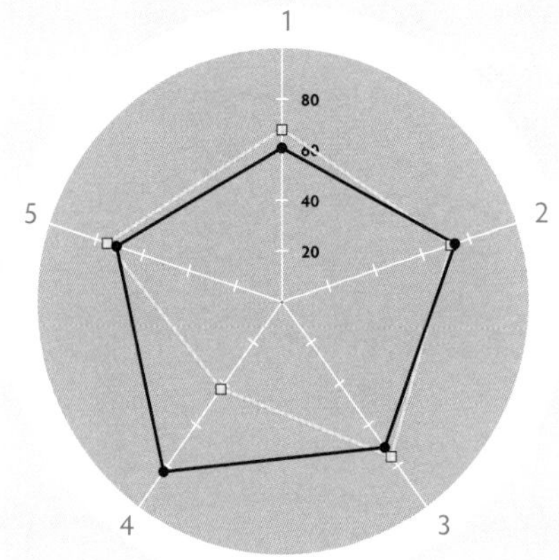

1. Institutional environment
2. Human capital, training and social inclusion
3. Regulatory and legal framework
4. Research and development
5. Usage of ICT

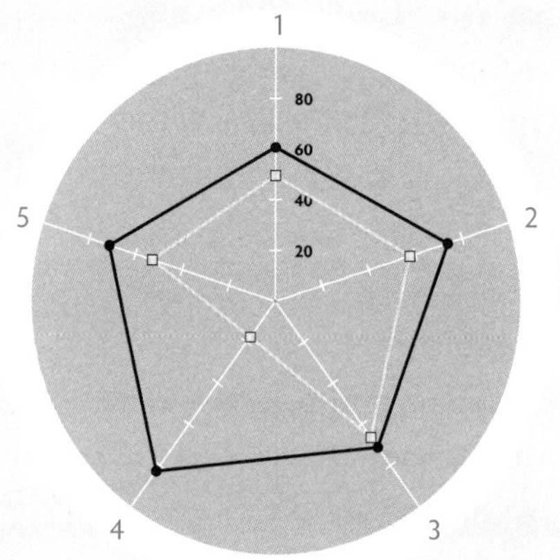

High-income ☐   Taiwan ●——

Flawed democracies ☐   Taiwan ●——

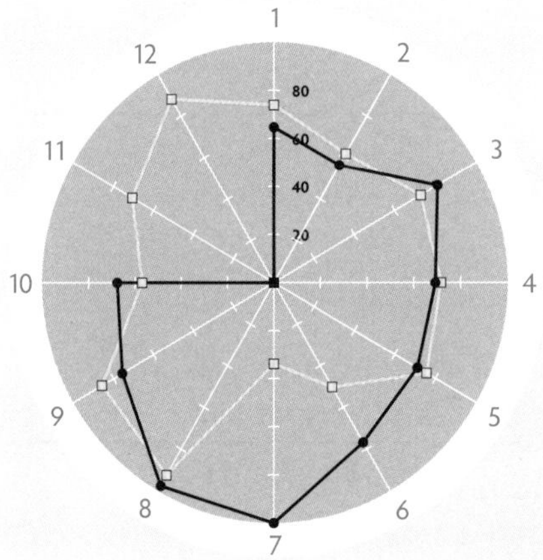

1. Good governance
2. Country policy assessment
3. Education
4. Social inclusion and equity policies
5. Doing business
6. R&D infrastructure
7. Patents and trademarks
8. Telephone communications
9. Mobile/cellular communications
10. Internet, computers and TV
11. Government ICT usage
12. Quality of the infrastructure

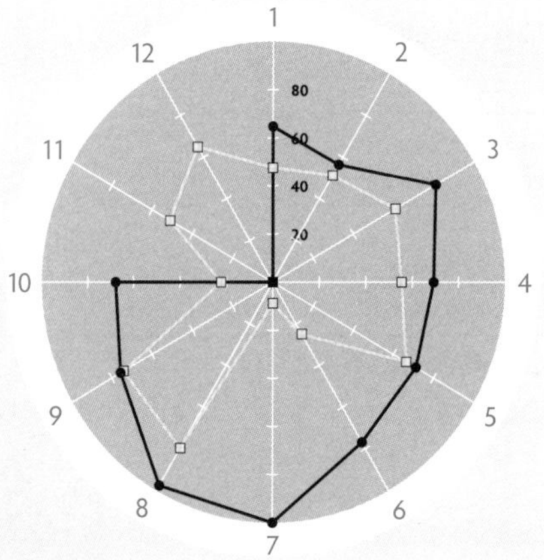

Country relative performance: **Pillar subindexes**[12]

In focus: **Top priorities for policy reform**[13]

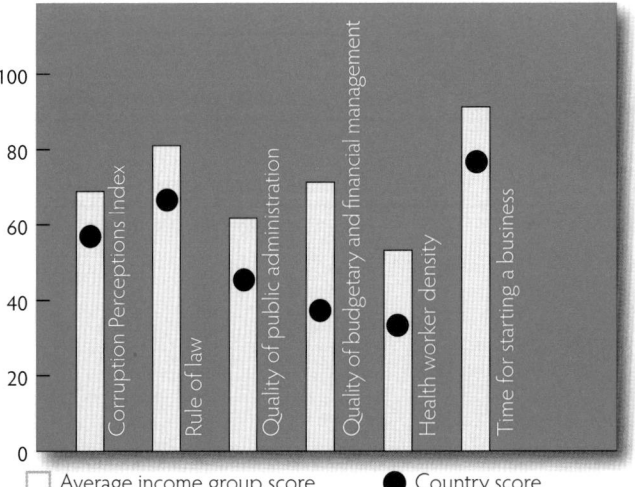

☐ Average income group score   ● Country score

In focus: **Significant indicators above income group average**[13]

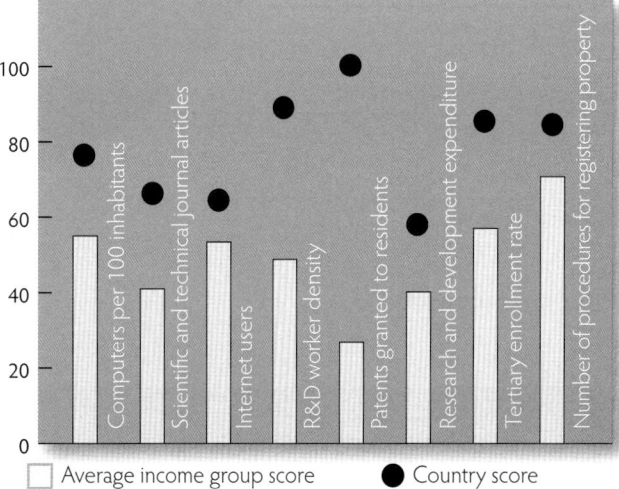

☐ Average income group score   ● Country score

[12] This analysis compares pillar and subindex country scores with respect to the average scores of countries grouped according to income level and political regime (see notes 1 and 2).

[13] Significant country indicators with the greatest distances above or below the income group average scores (i.e., achievements and targets for policy reform, respectively) were selected and ordered according to their Pearson correlation coefficients (highest first) with respect to the final index scores, to produce a ranking.

THAILAND

Key selected indicators

Population (in millions)[3,4]	63.8
Population density (people per sq. km)[3,4]	125
GDP (current US$ billion, PPP)[3,5]	519
GDP per capita (current US$, PPP)[3,5]	7900
Current account balance (% of GDP)[3,5]	6.1
Average life expectancy (years)[4,6]	70.2
Literacy rate[3,7]	94
Female labor force participation rate[4,6,8]	72.2
Unemployment rate (% of total labor force)[4,9]	1.3
CO_2 emissions (metric tons per capita)[4,10]	4.3

Income level:[1] **Lower-middle-income**

Political regime:[2] **Flawed democracy**

Innovation Capacity Index

Rank 43 out of 131   **Score** 54.6 out of 100

Pillars and pillar subindexes

Education — Rank 54 Score 64.6

Social inclusion and equity policies — Rank 63 Score 54.4

Country policy assessment — Rank 43 Score 56.8

Good governance — Rank 69 Score 42.6

Quality of the infrastructure — Rank 15 Score 95.8

Government ICT usage — Rank 60 Score 50.3

Doing business — Rank 20 Score 77.1

R&D infrastructure — Rank 83 Score 18.2

Patents and trademarks — Rank 48 Score 6.2

Telephone communications — Rank 66 Score 79.7

Mobile cellular communications — Rank 53 Score 78.3

Internet, computers and TV — Rank 72 Score 14.1

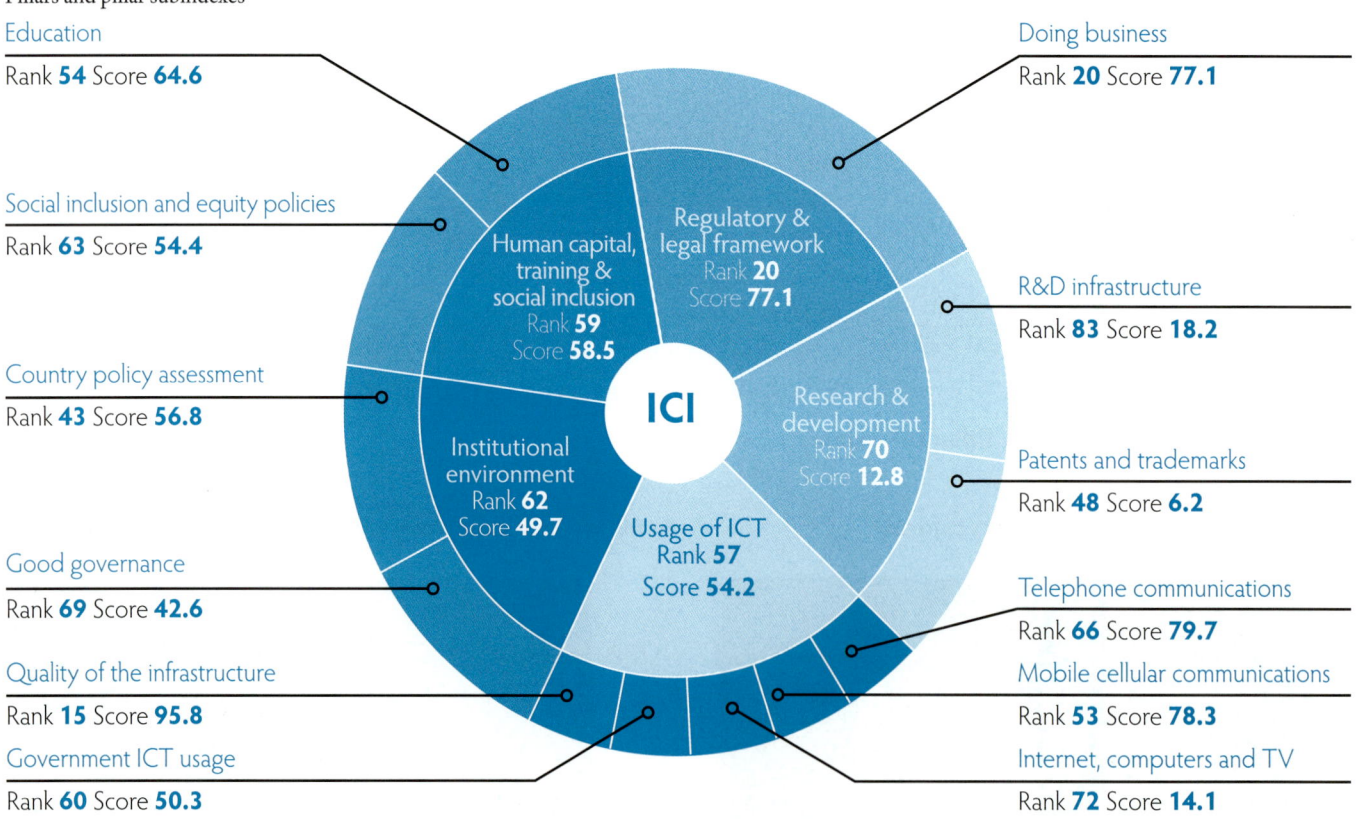

362

Index thermometer[11]

Pillar 1 27.3% Pillar 2 32.1% Pillar 3 28.2% Pillar 4 2.3% Pillar 5 9.9% 100%

[1] Lower-middle-income countries are those that, according to the World Bank classification, have a gross national income (GNI) per capita of $936 to $3,705. [2] Flawed democracies are countries whose Democracy Index scores from 1 to 10 rounded to one decimal place are of 6 to 7.9. [3] 2007. [4] World Bank World Development Indicators. [5] IMF World Economic Outlook. [6] 2006. [7] UNESCO, % age above 15. [8] % ages 15–64. [9] 2005. [10] 2004. [11] The index thermometer shows the relative contribution of each pillar (in %) to the total index score.

THAILAND

Country relative performance: Index pillars[12]

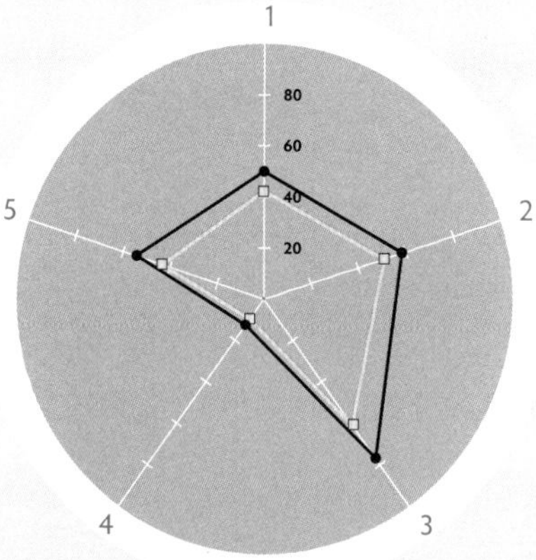

1. Institutional environment
2. Human capital, training and social inclusion
3. Regulatory and legal framework
4. Research and development
5. Usage of ICT

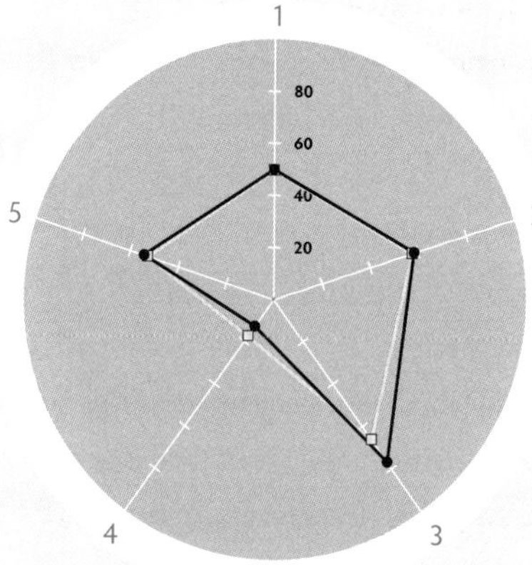

Lower-middle-income ☐ ⋯⋯⋯   Thailand ●━━━

Flawed democracies ☐ ⋯⋯⋯   Thailand ●━━━

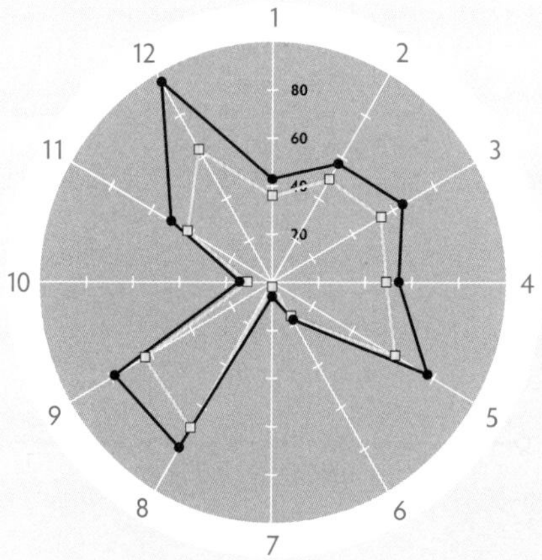

1. Good governance
2. Country policy assessment
3. Education
4. Social inclusion and equity policies
5. Doing business
6. R&D infrastructure
7. Patents and trademarks
8. Telephone communications
9. Mobile/cellular communications
10. Internet, computers and TV
11. Government ICT usage
12. Quality of the infrastructure

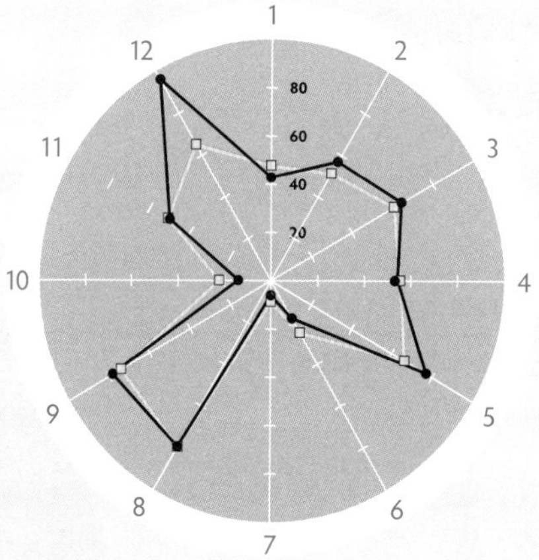

Country relative performance: Pillar subindexes[12]

363

In focus: **Top priorities for policy reform**[13]

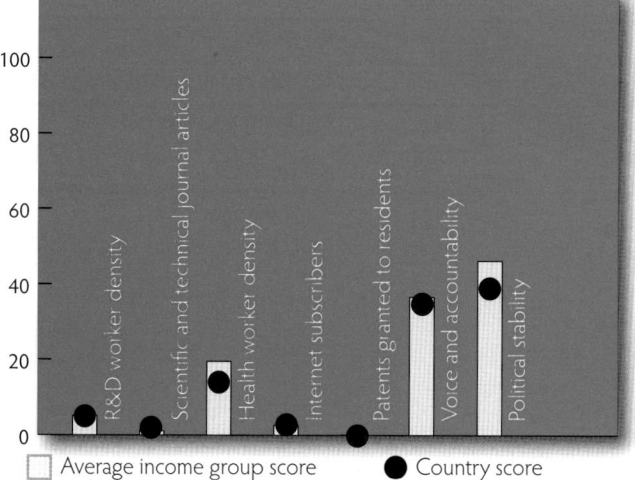

☐ Average income group score   ● Country score

In focus: **Significant indicators above income group average**[13]

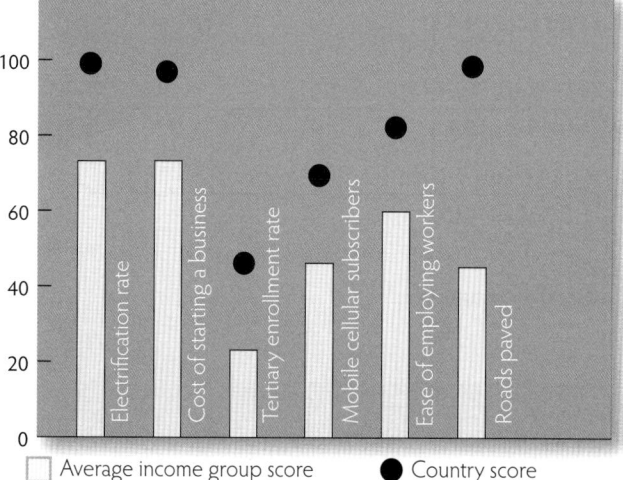

☐ Average income group score   ● Country score

[12] This analysis compares pillar and subindex country scores with respect to the average scores of countries grouped according to income level and political regime (see notes 1 and 2).

[13] Significant country indicators with the greatest distances above or below the income group average scores (i.e., achievements and targets for policy reform, respectively) were selected and ordered according to their Pearson correlation coefficients (highest first) with respect to the final index scores, to produce a ranking.

TURKEY

Key selected indicators

Population (in millions)[3,4]	73.9
Population density (people per sq. km)[3,4]	96
GDP (current US$ billion, PPP)[3,5]	888
GDP per capita (current US$, PPP)[3,5]	12888
Current account balance (% of GDP)[3,5]	-5.7
Average life expectancy (years)[4,6]	71.5
Literacy rate[3,7]	89
Female labor force participation rate[4,6,8]	28.8
Unemployment rate (% of total labor force)[4,9]	10.3
CO_2 emissions (metric tons per capita)[4,10]	3.2

Income level:[1] **Upper-middle-income**

Political regime:[2] **Hybrid regime**

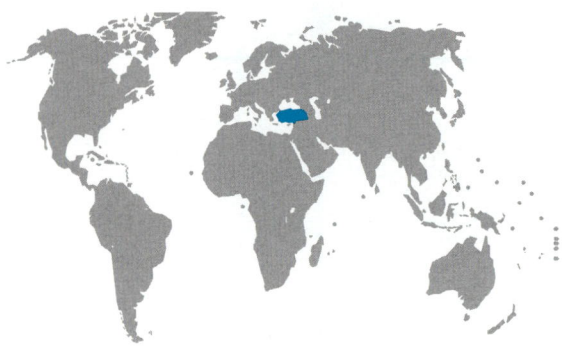

Innovation Capacity Index

Rank	Score
59 out of 131	**50.8 out of 100**

Pillars and pillar subindexes

Education
Rank **75** Score **57.0**

Social inclusion and equity policies
Rank **86** Score **47.6**

Country policy assessment
Rank **54** Score **53.9**

Good governance
Rank **56** Score **47.6**

Quality of the infrastructure
Rank **87** Score **55.6**

Government ICT usage
Rank **70** Score **48.3**

Doing business
Rank **33** Score **73.3**

R&D infrastructure
Rank **66** Score **26.4**

Patents and trademarks
Rank **45** Score **7.1**

Telephone communications
Rank **49** Score **86.1**

Mobile cellular communications
Rank **50** Score **81.2**

Internet, computers and TV
Rank **65** Score **17.1**

ICI

Human capital, training & social inclusion
Rank **82**
Score **51.4**

Regulatory & legal framework
Rank **33**
Score **73.3**

Research & development
Rank **50**
Score **18.4**

Institutional environment
Rank **59**
Score **50.7**

Usage of ICT
Rank **64**
Score **52.2**

Index thermometer[11]

Pillar 1 25% Pillar 2 25.3% Pillar 3 28.9% Pillar 4 5.4% Pillar 5 15.4%

100%

[1] Upper-middle-income countries are those that, according to the World Bank classification, have a gross national income (GNI) per capita of $3,706 to $11,455. [2] Hybrid regimes are countries whose Democracy Index scores from 1 to 10 rounded to one decimal place are of 4 to 5.9. [3] 2007. [4] World Bank World Development Indicators. [5] IMF World Economic Outlook. [6] 2006. [7] UNESCO, % age above 15. [8] % ages 15–64. [9] 2005. [10] 2004. [11] The index thermometer shows the relative contribution of each pillar (in %) to the total index score.

TURKEY

Country relative performance: **Index pillars**[12]

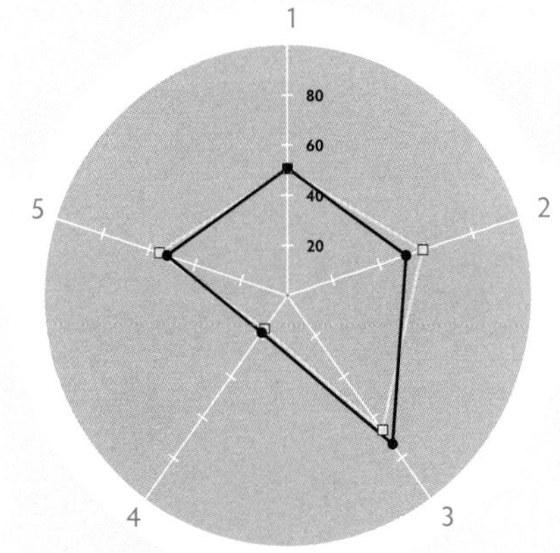

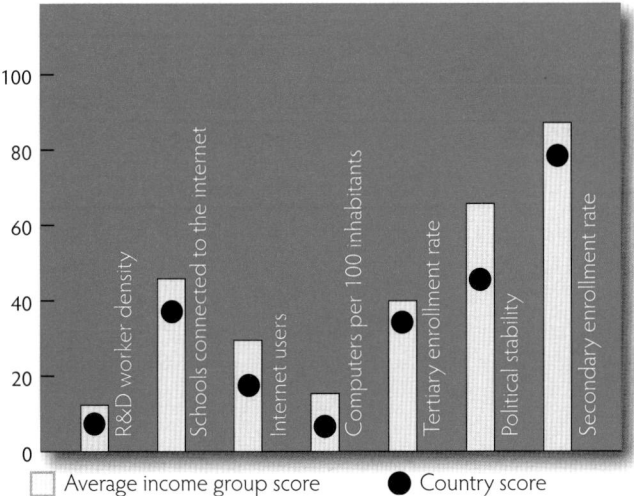

1. Institutional environment
2. Human capital, training and social inclusion
3. Regulatory and legal framework
4. Research and development
5. Usage of ICT

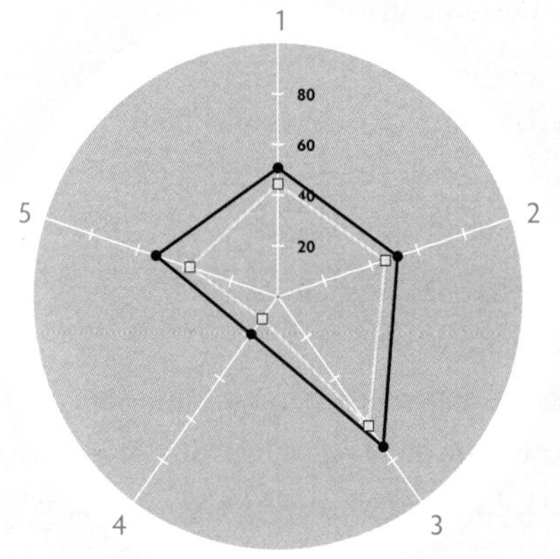

Upper-middle-income □   Turkey ●━━

Hybrid regimes □   Turkey ●━━

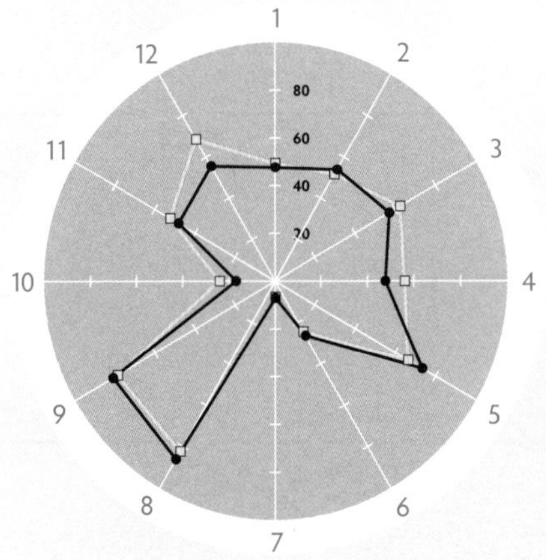

1. Good governance
2. Country policy assessment
3. Education
4. Social inclusion and equity policies
5. Doing business
6. R&D infrastructure
7. Patents and trademarks
8. Telephone communications
9. Mobile/cellular communications
10. Internet, computers and TV
11. Government ICT usage
12. Quality of the infrastructure

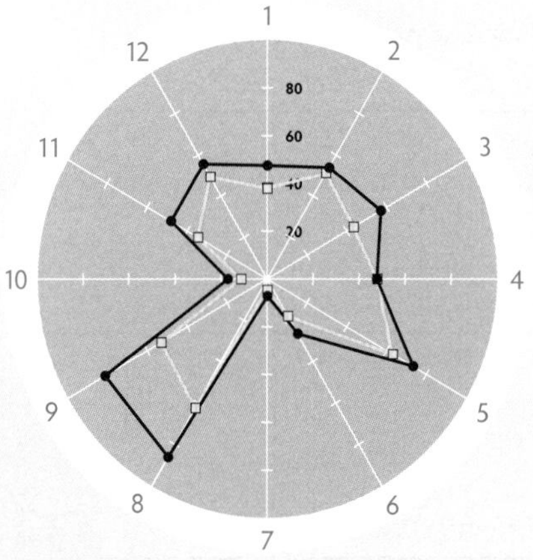

Country relative performance: **Pillar subindexes**[12]

In focus: **Top priorities for policy reform**[13]

In focus: **Significant indicators above income group average**[13]

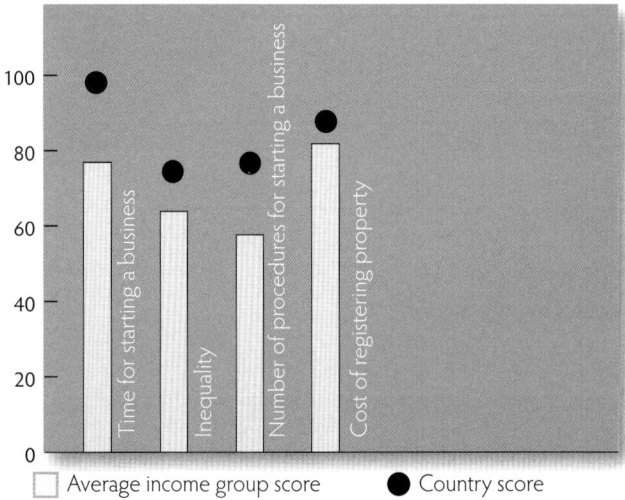

□ Average income group score   ● Country score

□ Average income group score   ● Country score

[12] This analysis compares pillar and subindex country scores with respect to the average scores of countries grouped according to income level and political regime (see notes 1 and 2).

[13] Significant country indicators with the greatest distances above or below the income group average scores (i.e., achievements and targets for policy reform, respectively) were selected and ordered according to their Pearson correlation coefficients (highest first) with respect to the final index scores, to produce a ranking.

UKRAINE

Key selected indicators

Population (in millions)[3,4]	46.4
Population density (people per sq. km)[3,4]	80
GDP (current US$ billion, PPP)[3,5]	320
GDP per capita (current US$, PPP)[3,5]	6941
Current account balance (% of GDP)[3,5]	-4.2
Average life expectancy (years)[4,6]	68.0
Literacy rate[3,7]	100
Female labor force participation rate[4,6,8]	63.0
Unemployment rate (% of total labor force)[4,9]	7.2
CO_2 emissions (metric tons per capita)[4,10]	6.9

Income level:[1] **Lower-middle-income**

Political regime:[2] **Flawed democracy**

Innovation Capacity Index

Rank	Score
54 out of 131	**52.0 out of 100**

Pillars and pillar subindexes

Education
Rank **20** Score **75.1**

Social inclusion and equity policies
Rank **43** Score **60.8**

Country policy assessment
Rank **60** Score **53.0**

Good governance
Rank **89** Score **35.8**

Quality of the infrastructure
Rank **34** Score **88.7**

Government ICT usage
Rank **41** Score **57.3**

Doing business
Rank **108** Score **55.4**

R&D infrastructure
Rank **34** Score **36.2**

Patents and trademarks
Rank **57** Score **4.3**

Telephone communications
Rank **89** Score **74.3**

Mobile cellular communications
Rank **7** Score **95.4**

Internet, computers and TV
Rank **61** Score **18.1**

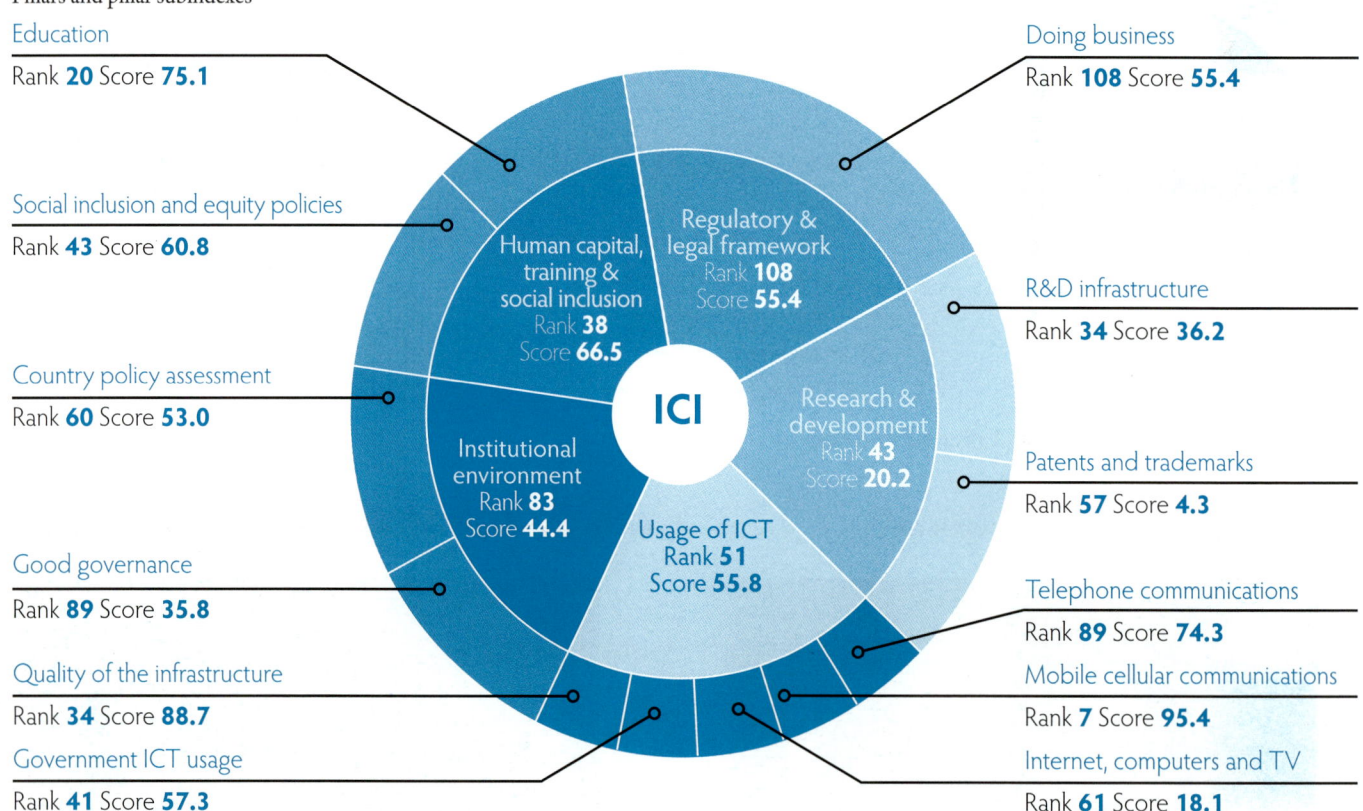

Index thermometer[11]

Pillar 1 25.6% Pillar 2 38.4% Pillar 3 21.3% Pillar 4 3.9% Pillar 5 10.7% 100%

[1] Lower-middle-income countries are those that, according to the World Bank classification, have a gross national income (GNI) per capita of $936 to $3,705. [2] Flawed democracies are countries whose Democracy Index scores from 1 to 10 rounded to one decimal place are of 6 to 7.9. [3] 2007. [4] World Bank World Development Indicators. [5] IMF World Economic Outlook. [6] 2006. [7] UNESCO, % age above 15. [8] % ages 15–64. [9] 2005. [10] 2004. [11] The index thermometer shows the relative contribution of each pillar (in %) to the total index score.

UKRAINE

Country relative performance: **Index pillars**[12]

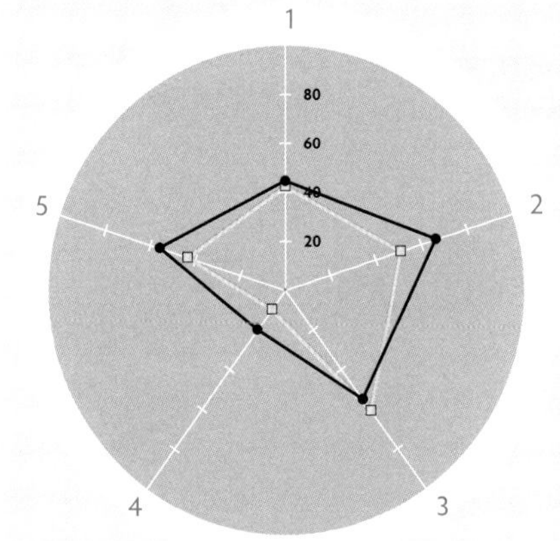

1. Institutional environment
2. Human capital, training and social inclusion
3. Regulatory and legal framework
4. Research and development
5. Usage of ICT

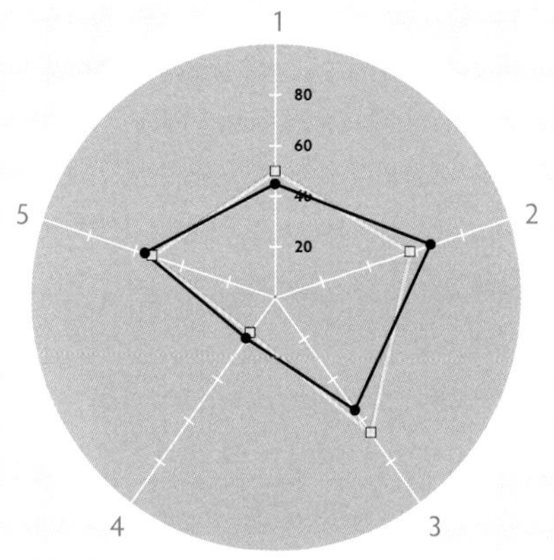

Lower-middle-income ☐   Ukraine ●

Flawed democracies ☐   Ukraine ●

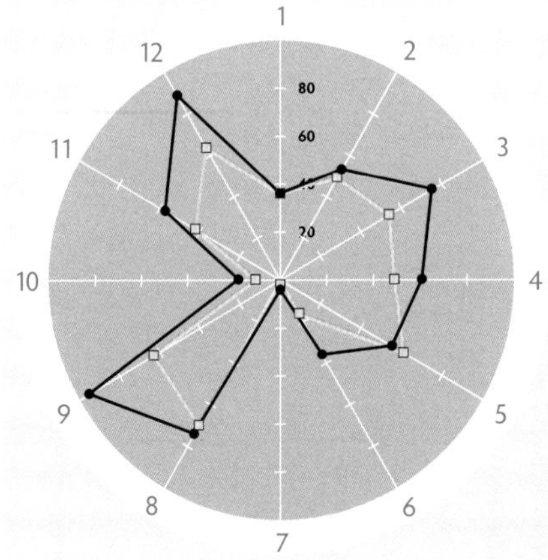

1. Good governance
2. Country policy assessment
3. Education
4. Social inclusion and equity policies
5. Doing business
6. R&D infrastructure
7. Patents and trademarks
8. Telephone communications
9. Mobile/cellular communications
10. Internet, computers and TV
11. Government ICT usage
12. Quality of the infrastructure

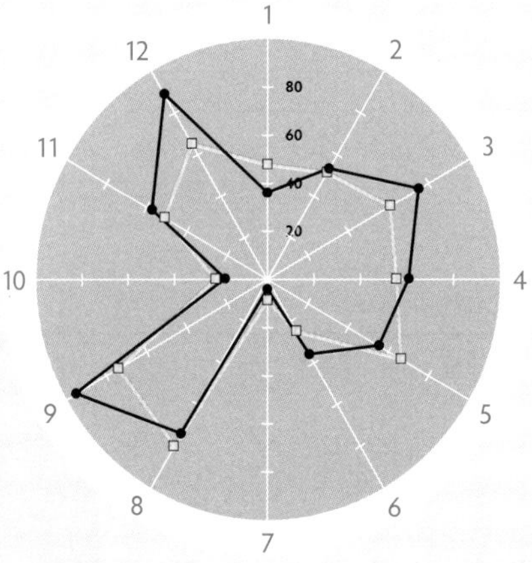

Country relative performance: **Pillar subindexes**[12]

In focus: **Top priorities for policy reform**[13]

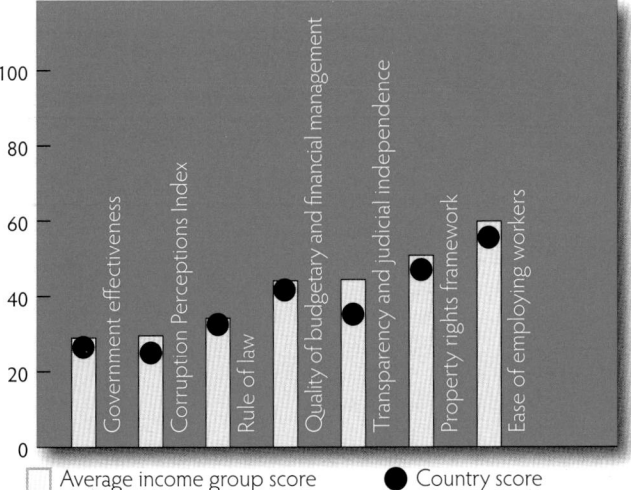

☐ Average income group score   ● Country score

In focus: **Significant indicators above income group average**[13]

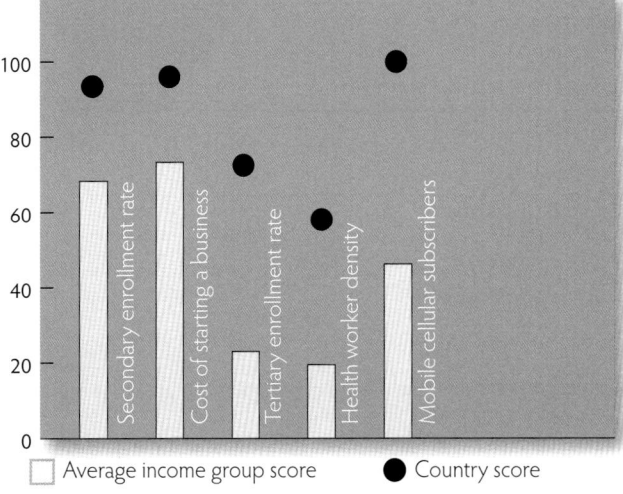

☐ Average income group score   ● Country score

[12] This analysis compares pillar and subindex country scores with respect to the average scores of countries grouped according to income level and political regime (see notes 1 and 2).

[13] Significant country indicators with the greatest distances above or below the income group average scores (i.e., achievements and targets for policy reform, respectively) were selected and ordered according to their Pearson correlation coefficients (highest first) with respect to the final index scores, to produce a ranking.

UNITED KINGDOM

Key selected indicators

Population (in millions)[3,4]	61.0
Population density (people per sq. km)[3,4]	252
GDP (current US$ billion, PPP)[3,5]	2137
GDP per capita (current US$, PPP)[3,5]	35134
Current account balance (% of GDP)[3,5]	-4.9
Average life expectancy (years)[4,6]	79.1
Literacy rate[7,8]	99
Female labor force participation rate[4,6,9]	69.5
Unemployment rate (% of total labor force)[4,10]	4.6
CO_2 emissions (metric tons per capita)[4,11]	9.8

Income level:[1] **High-income**

Political regime:[2] **Full democracy**

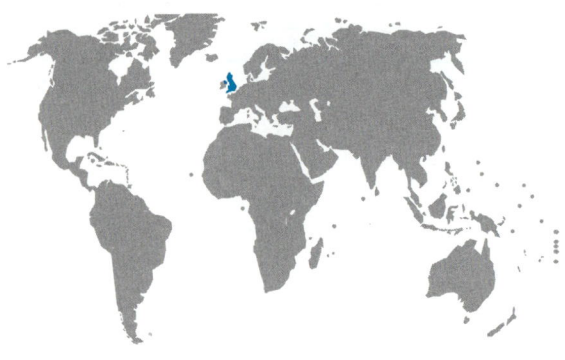

Rank	Score
8 out of 131	**74.6 out of 100**

Innovation Capacity Index

Pillars and pillar subindexes

Education
Rank **24** Score **73.0**

Social inclusion and equity policies
Rank **14** Score **78.9**

Country policy assessment
Rank **32** Score **59.7**

Good governance
Rank **17** Score **81.3**

Quality of the infrastructure
Rank **13** Score **96.6**

Government ICT usage
Rank **10** Score **78.7**

Doing business
Rank **5** Score **87.3**

R&D infrastructure
Rank **16** Score **57.3**

Patents and trademarks
Rank **13** Score **47.4**

Telephone communications
Rank **13** Score **94.8**

Mobile cellular communications
Rank **10** Score **93.0**

Internet, computers and TV
Rank **7** Score **81.9**

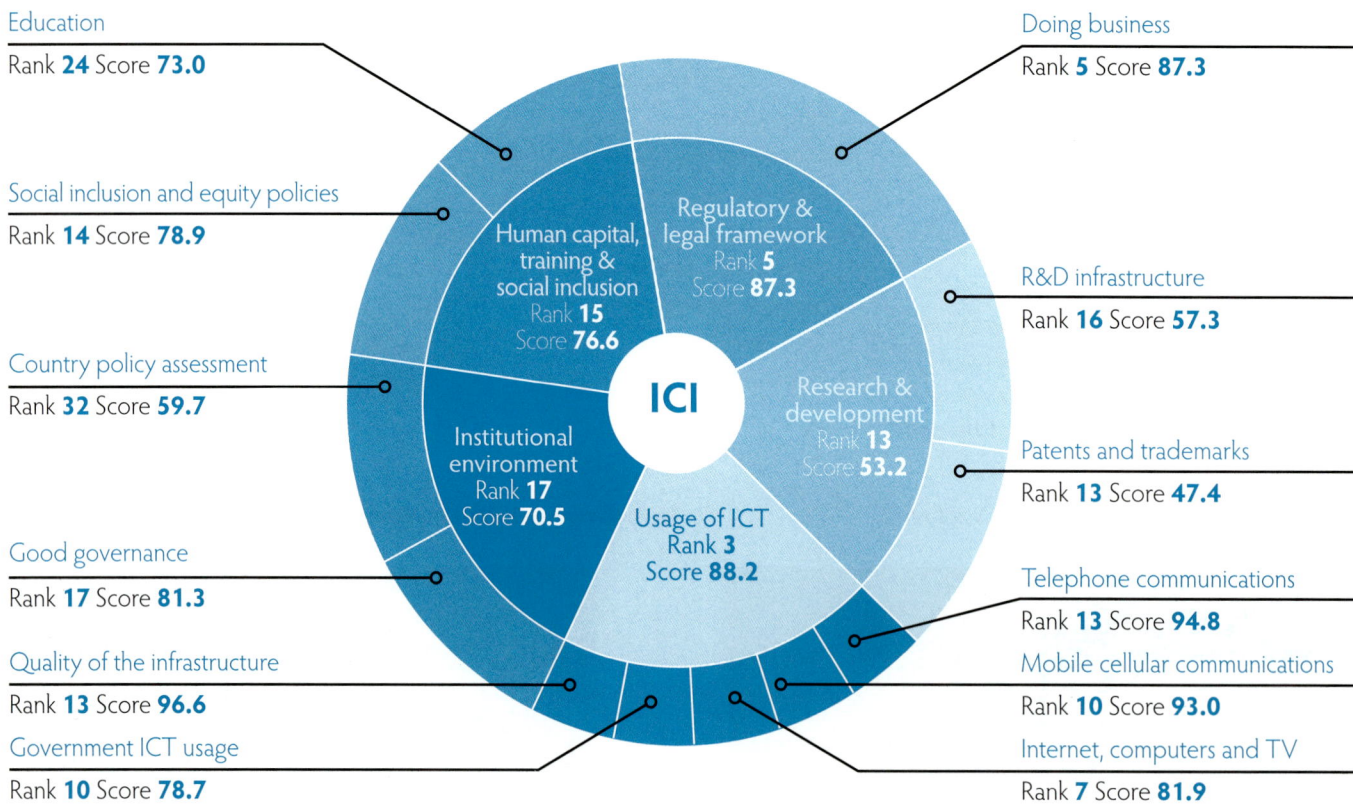

Human capital, training & social inclusion
Rank **15** Score **76.6**

Regulatory & legal framework
Rank **5** Score **87.3**

ICI

Institutional environment
Rank **17** Score **70.5**

Research & development
Rank **13** Score **53.2**

Usage of ICT
Rank **3** Score **88.2**

Index thermometer[12]

Pillar 1 9.5% Pillar 2 10.3% Pillar 3 23.4% Pillar 4 21.4% Pillar 5 35.5% 100%

[1] High-income countries are those that, according to the World Bank classification, have a gross national income (GNI) per capita of $11,456 or more.
[2] Full democracies are countries whose Democracy Index scores from 1 to 10 rounded to one decimal place are of 8 to 10. [3] 2007. [4] World Bank World Development Indicators. [5] IMF World Economic Outlook. [6] 2006. [7] Estimate. [8] UNESCO, % age above 15. [9] % ages 15–64. [10] 2005. [11] 2004. [12] The index thermometer shows the relative contribution of each pillar (in %) to the total index score.

UNITED KINGDOM

Country relative performance: **Index pillars**[13]

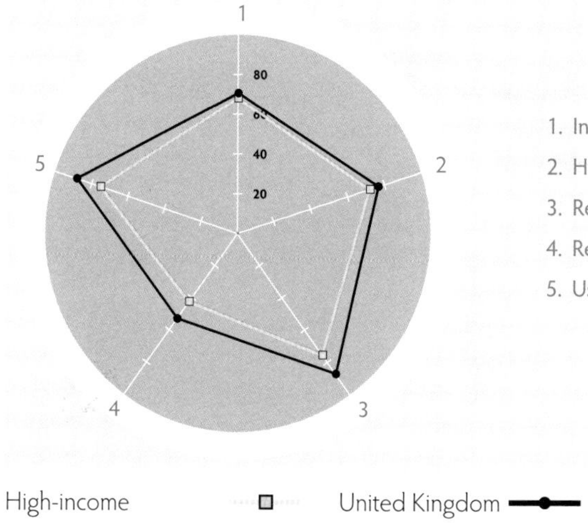

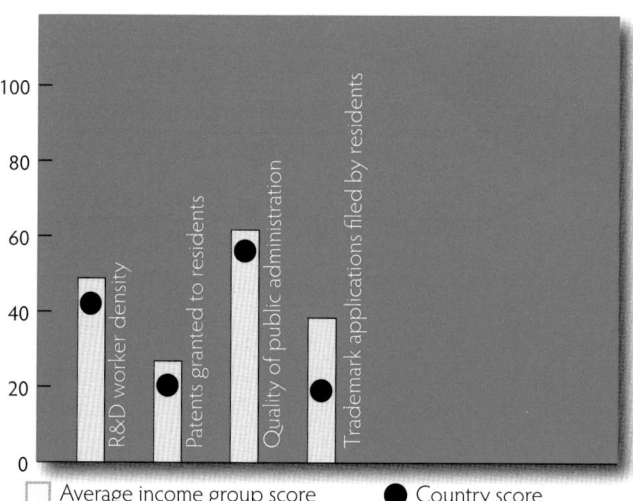

1. Institutional environment
2. Human capital, training and social inclusion
3. Regulatory and legal framework
4. Research and development
5. Usage of ICT

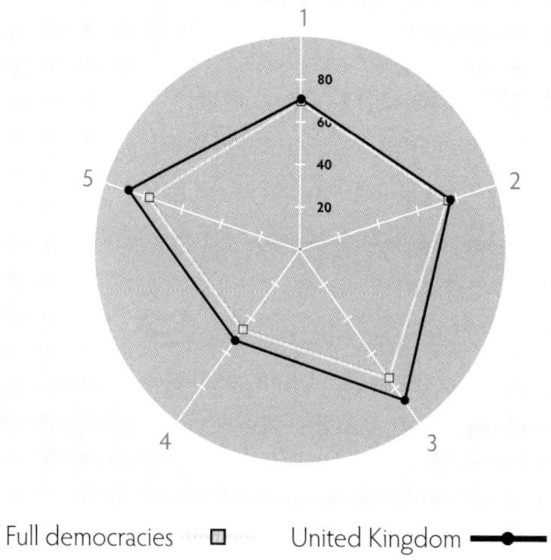

High-income ▫   United Kingdom ●

Full democracies ▫   United Kingdom ●

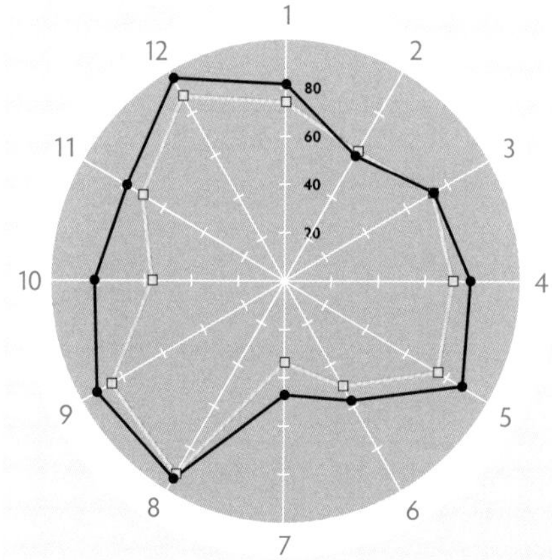

1. Good governance
2. Country policy assessment
3. Education
4. Social inclusion and equity policies
5. Doing business
6. R&D infrastructure
7. Patents and trademarks
8. Telephone communications
9. Mobile/cellular communications
10. Internet, computers and TV
11. Government ICT usage
12. Quality of the infrastructure

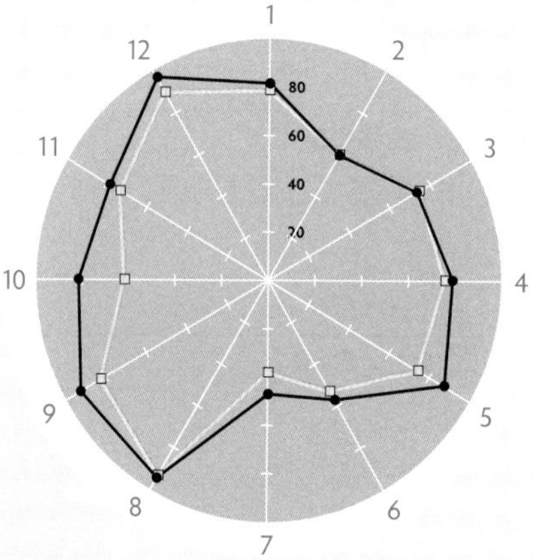

Country relative performance: **Pillar subindexes**[13]

In focus: **Top priorities for policy reform**[14]

In focus: **Significant indicators above income group average**[14]

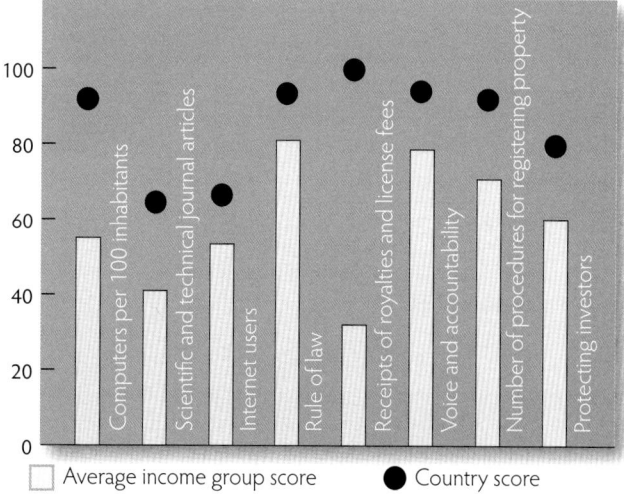

□ Average income group score   ● Country score

[13] This analysis compares pillar and subindex country scores with respect to the average scores of countries grouped according to income level and political regime (see notes 1 and 2).

[14] Significant country indicators with the greatest distances above or below the income group average scores (i.e., achievements and targets for policy reform, respectively) were selected and ordered according to their Pearson correlation coefficients (highest first) with respect to the final index scores, to produce a ranking.

UNITED STATES

Key selected indicators

Population (in millions)[3,4]	301.6
Population density (people per sq. km)[3,4]	33
GDP (current US$ billion, PPP)[3,5]	13844
GDP per capita (current US$, PPP)[3,5]	45845
Current account balance (% of GDP)[3,5]	-5.3
Average life expectancy (years)[4,6]	77.8
Literacy rate[7,8]	99
Female labor force participation rate[4,6,9]	70
Unemployment rate (% of total labor force)[4,10]	5.1
CO_2 emissions (metric tons per capita)[4,11]	20.6

Income level:[1] **High-income**

Political regime:[2] **Full democracy**

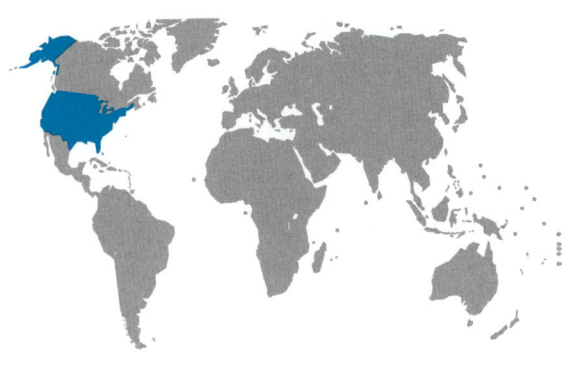

Rank	Score
3 out of 131	**77.5 out of 100**

Innovation Capacity Index

Pillars and pillar subindexes

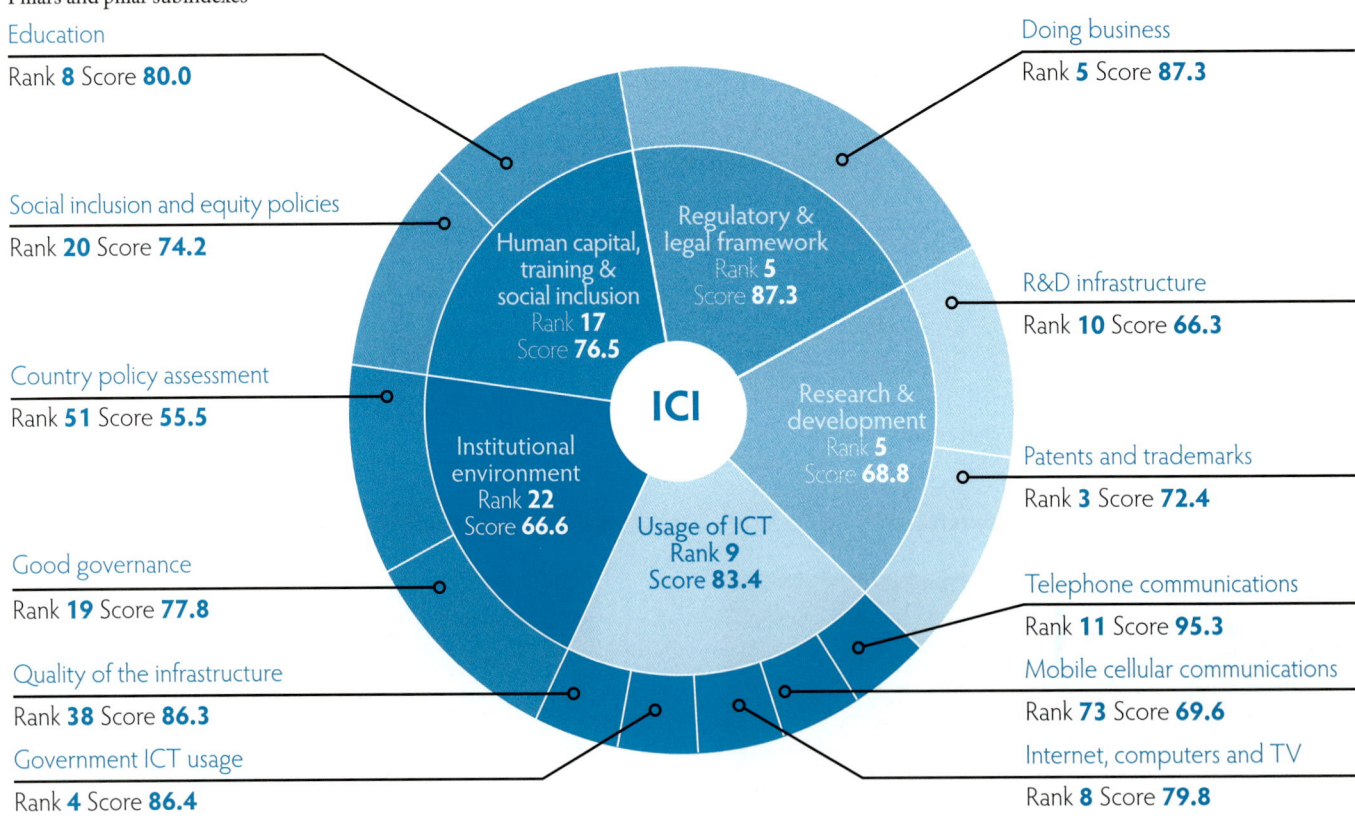

Education
Rank **8** Score **80.0**

Social inclusion and equity policies
Rank **20** Score **74.2**

Country policy assessment
Rank **51** Score **55.5**

Good governance
Rank **19** Score **77.8**

Quality of the infrastructure
Rank **38** Score **86.3**

Government ICT usage
Rank **4** Score **86.4**

Doing business
Rank **5** Score **87.3**

R&D infrastructure
Rank **10** Score **66.3**

Patents and trademarks
Rank **3** Score **72.4**

Telephone communications
Rank **11** Score **95.3**

Mobile cellular communications
Rank **73** Score **69.6**

Internet, computers and TV
Rank **8** Score **79.8**

Human capital, training & social inclusion
Rank **17** Score **76.5**

Regulatory & legal framework
Rank **5** Score **87.3**

Institutional environment
Rank **22** Score **66.6**

Research & development
Rank **5** Score **68.8**

Usage of ICT
Rank **9** Score **83.4**

ICI

370

Index thermometer[12]

Pillar 1 8.5% Pillar 2 9.9% Pillar 3 22.5% Pillar 4 26.7% Pillar 5 32.3% 100%

[1] High-income countries are those that, according to the World Bank classification, have a gross national income (GNI) per capita of $11,456 or more.
[2] Full democracies are countries whose Democracy Index scores from 1 to 10 rounded to one decimal place are of 8 to 10. [3] 2007. [4] World Bank World Development Indicators. [5] IMF World Economic Outlook. [6] 2006. [7] Estimate. [8] UNESCO, % age above 15. [9] % ages 15–64. [10] 2005. [11] 2004. [12] The index thermometer shows the relative contribution of each pillar (in %) to the total index score.

UNITED STATES

Country relative performance: **Index pillars**[13]

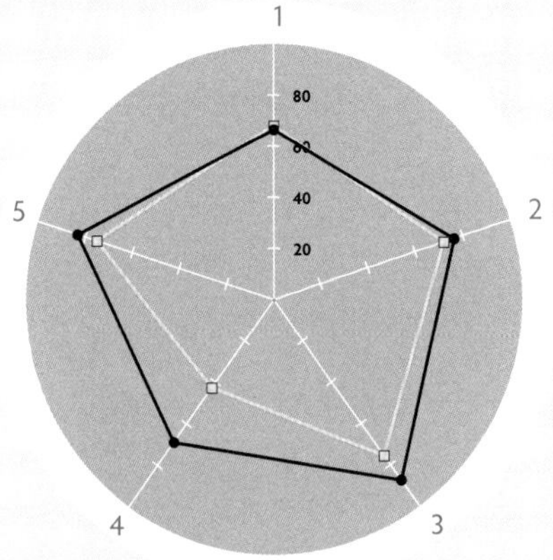

1. Institutional environment
2. Human capital, training and social inclusion
3. Regulatory and legal framework
4. Research and development
5. Usage of ICT

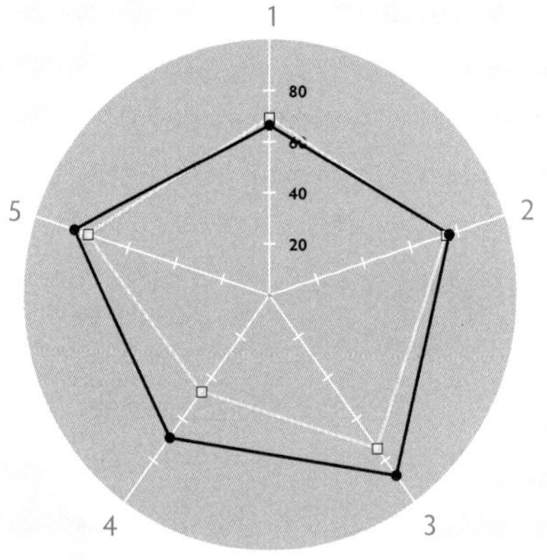

High-income ☐   United States ●

Full democracies ☐   United States ●

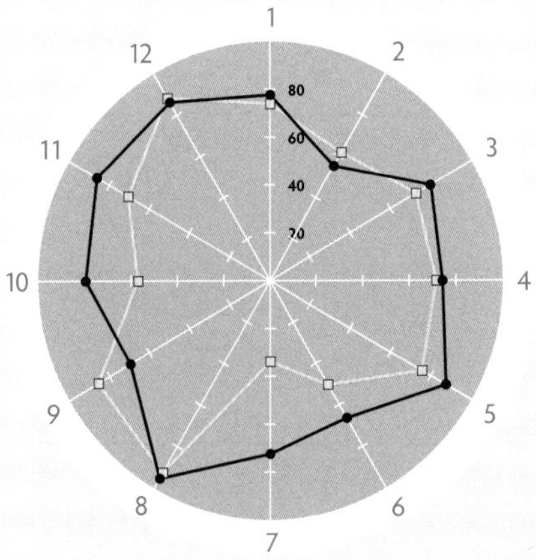

1. Good governance
2. Country policy assessment
3. Education
4. Social inclusion and equity policies
5. Doing business
6. R&D infrastructure
7. Patents and trademarks
8. Telephone communications
9. Mobile/cellular communications
10. Internet, computers and TV
11. Government ICT usage
12. Quality of the infrastructure

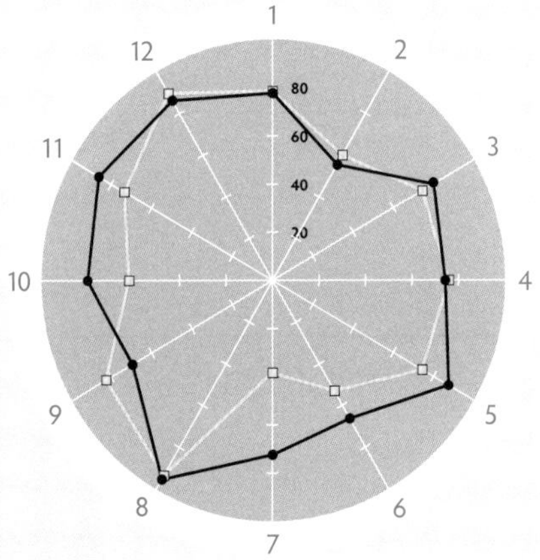

Country relative performance: **Pillar subindexes**[13]

In focus: **Top priorities for policy reform**[14]

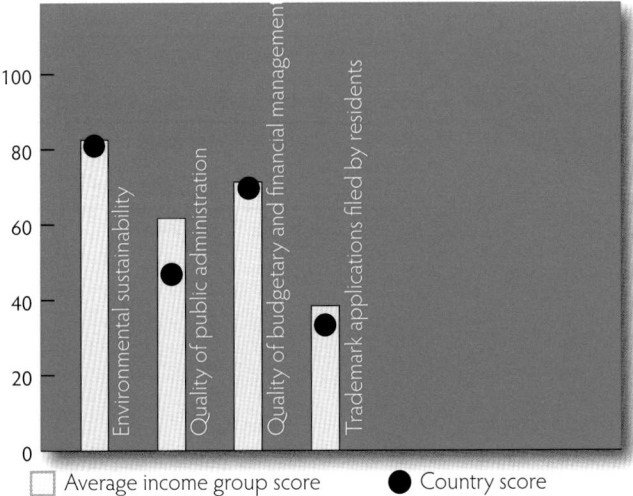

☐ Average income group score   ● Country score

In focus: **Significant indicators above income group average**[14]

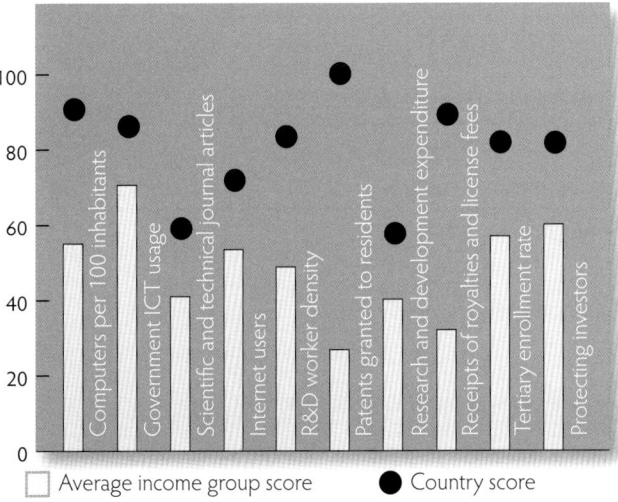

☐ Average income group score   ● Country score

---

[13] This analysis compares pillar and subindex country scores with respect to the average scores of countries grouped according to income level and political regime (see notes 1 and 2).

[14] Significant country indicators with the greatest distances above or below the income group average scores (i.e., achievements and targets for policy reform, respectively) were selected and ordered according to their Pearson correlation coefficients (highest first) with respect to the final index scores, to produce a ranking.

URUGUAY

Key selected indicators

Population (in millions)[3,4]	3.3
Population density (people per sq. km)[3,4]	19
GDP (current US$ billion, PPP)[3,5]	37
GDP per capita (current US$, PPP)[3,5]	11621
Current account balance (% of GDP)[3,5]	-0.8
Average life expectancy (years)[4,6]	75.7
Literacy rate[3,7]	98
Female labor force participation rate[4,6,8]	66.8
Unemployment rate (% of total labor force)[4,9]	12.2
CO_2 emissions (metric tons per capita)[4,10]	1.7

Income level:[1] **Upper-middle-income**

Political regime:[2] **Full democracy**

Innovation Capacity Index

Rank **49 out of 131**   Score **52.8 out of 100**

Pillars and pillar subindexes

Education
Rank **53** Score **65.0**

Social inclusion and equity policies
Rank **56** Score **57.1**

Country policy assessment
Rank **94** Score **47.1**

Good governance
Rank **28** Score **68.7**

Quality of the infrastructure
Rank **88** Score **55.4**

Government ICT usage
Rank **46** Score **56.5**

Doing business
Rank **85** Score **62.0**

R&D infrastructure
Rank **78** Score **21.0**

Patents and trademarks
Rank **31** Score **15.5**

Telephone communications
Rank **41** Score **86.7**

Mobile cellular communications
Rank **74** Score **69.5**

Internet, computers and TV
Rank **52** Score **24.0**

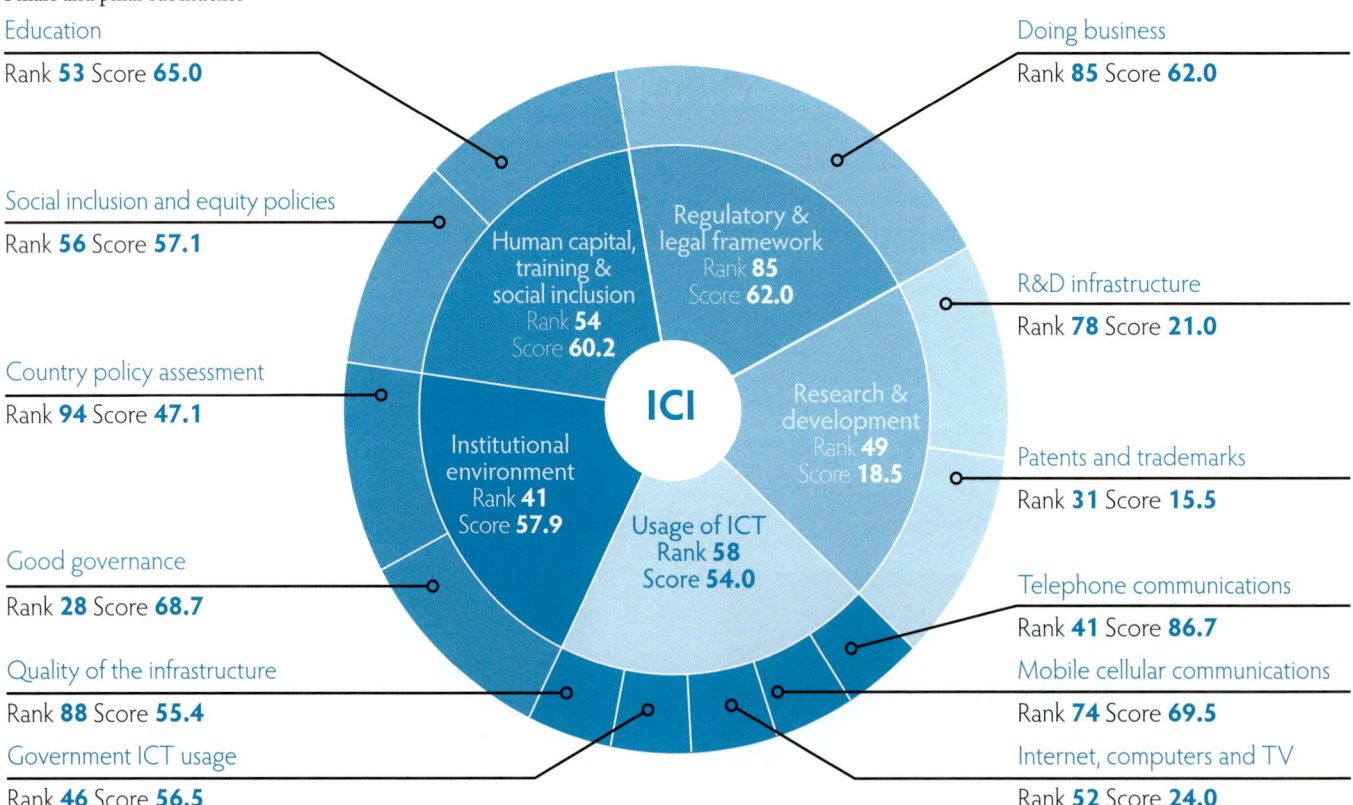

Human capital, training & social inclusion
Rank **54** Score **60.2**

Regulatory & legal framework
Rank **85** Score **62.0**

Institutional environment
Rank **41** Score **57.9**

ICI

Research & development
Rank **49** Score **18.5**

Usage of ICT
Rank **58** Score **54.0**

372

Index thermometer[11]

Pillar 1 23.5%   Pillar 2 28.5%   Pillar 3 27.4%   Pillar 4 5.3%   Pillar 5 15.3%   100%

[1] Upper-middle-income countries are those that, according to the World Bank classification, have a gross national income (GNI) per capita of $3,706 to $11,455. [2] Full democracies are countries whose Democracy Index scores from 1 to 10 rounded to one decimal place are of 8 to 10. [3] 2007. [4] World Bank World Development Indicators. [5] IMF World Economic Outlook. [6] 2006. [7] UNESCO, % age above 15. [8] % ages 15–64. [9] 2005. [10] 2004. [11] The index thermometer shows the relative contribution of each pillar (in %) to the total index score.

URUGUAY

Country relative performance: **Index pillars**[12]

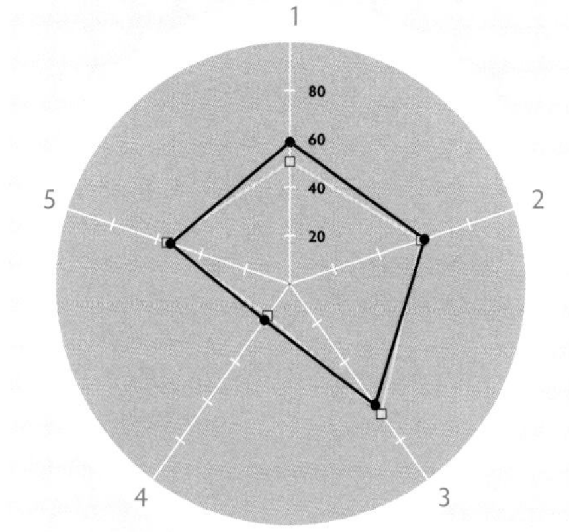

1. Institutional environment
2. Human capital, training and social inclusion
3. Regulatory and legal framework
4. Research and development
5. Usage of ICT

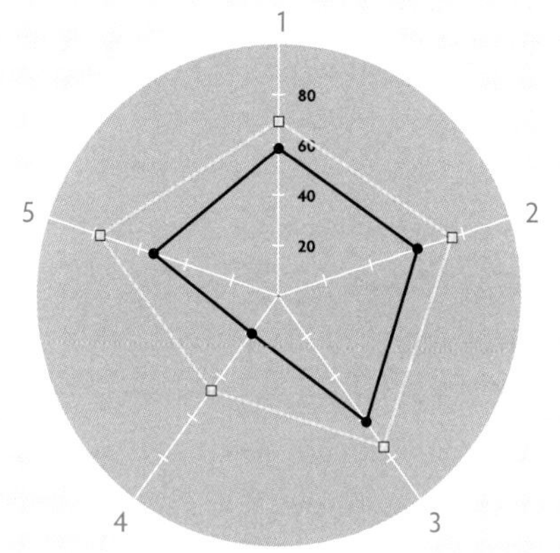

Upper-middle-income ▢   Uruguay ●━━

Full democracies ▢   Uruguay ●━━

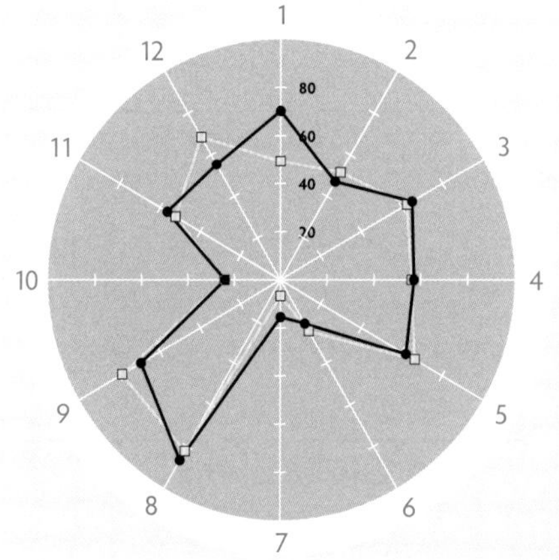

1. Good governance
2. Country policy assessment
3. Education
4. Social inclusion and equity policies
5. Doing business
6. R&D infrastructure
7. Patents and trademarks
8. Telephone communications
9. Mobile/cellular communications
10. Internet, computers and TV
11. Government ICT usage
12. Quality of the infrastructure

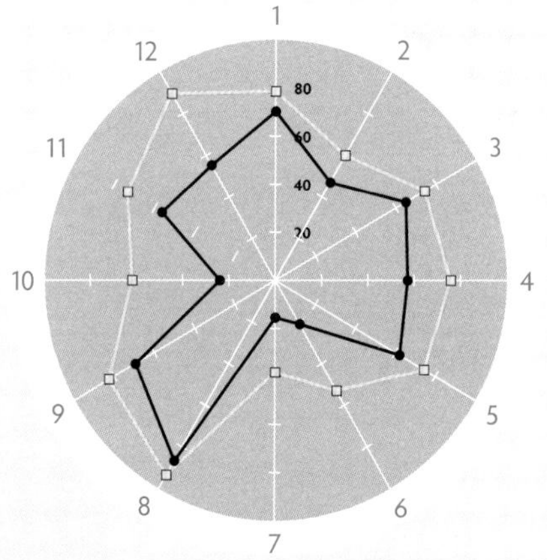

Country relative performance: **Pillar subindexes**[12]

In focus: **Top priorities for policy reform**[13]

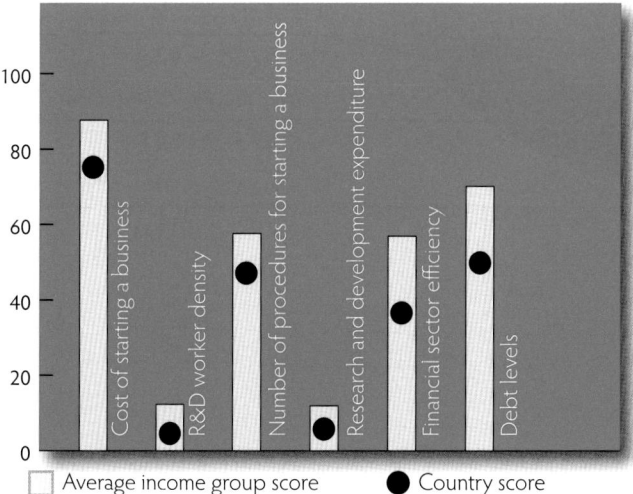

▢ Average income group score   ● Country score

In focus: **Significant indicators above income group average**[13]

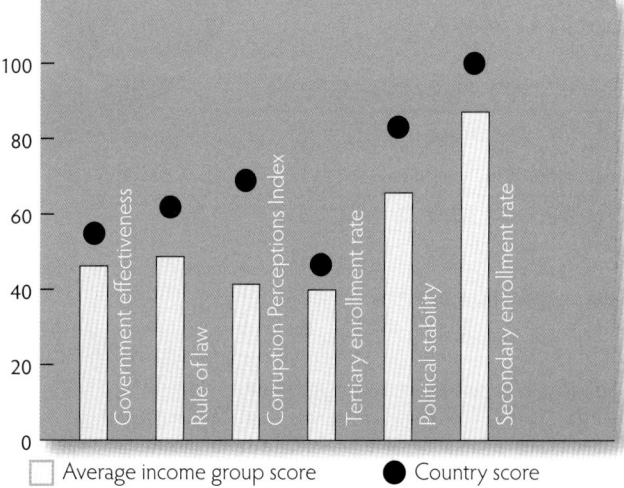

▢ Average income group score   ● Country score

---

[12] This analysis compares pillar and subindex country scores with respect to the average scores of countries grouped according to income level and political regime (see notes 1 and 2).

[13] Significant country indicators with the greatest distances above or below the income group average scores (i.e., achievements and targets for policy reform, respectively) were selected and ordered according to their Pearson correlation coefficients (highest first) with respect to the final index scores, to produce a ranking.

VENEZUELA

Key selected indicators

Population (in millions)[3,4]	27.5
Population density (people per sq. km)[3,4]	31
GDP (current US$ billion, PPP)[3,5]	335
GDP per capita (current US$, PPP)[3,5]	12166
Current account balance (% of GDP)[3,5]	9.8
Average life expectancy (years)[4,6]	74.4
Literacy rate[7,8]	93
Female labor force participation rate[4,6,9]	63.5
Unemployment rate (% of total labor force)[4,10]	15.0
CO_2 emissions (metric tons per capita)[4,10]	6.6

Income level:[1] **Upper-middle-income**

Political regime:[2] **Hybrid regime**

Innovation Capacity Index

Rank **102 out of 131**

Score **40.9 out of 100**

Pillars and pillar subindexes

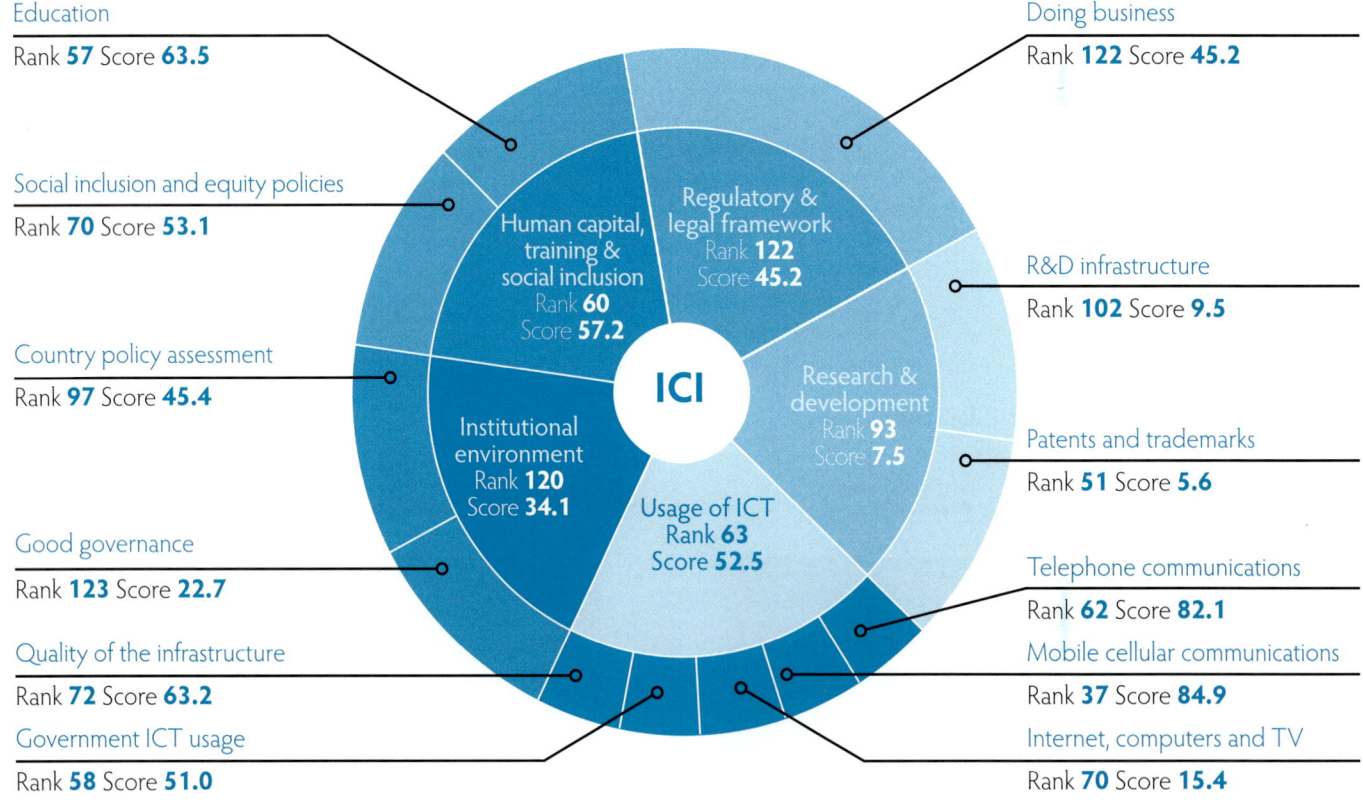

Education
Rank **57** Score **63.5**

Social inclusion and equity policies
Rank **70** Score **53.1**

Country policy assessment
Rank **97** Score **45.4**

Good governance
Rank **123** Score **22.7**

Quality of the infrastructure
Rank **72** Score **63.2**

Government ICT usage
Rank **58** Score **51.0**

Doing business
Rank **122** Score **45.2**

R&D infrastructure
Rank **102** Score **9.5**

Patents and trademarks
Rank **51** Score **5.6**

Telephone communications
Rank **62** Score **82.1**

Mobile cellular communications
Rank **37** Score **84.9**

Internet, computers and TV
Rank **70** Score **15.4**

Human capital, training & social inclusion
Rank **60**
Score **57.2**

Regulatory & legal framework
Rank **122**
Score **45.2**

Institutional environment
Rank **120**
Score **34.1**

Research & development
Rank **93**
Score **7.5**

Usage of ICT
Rank **63**
Score **52.5**

ICI

Index thermometer[11]

Pillar 1 20.8%   Pillar 2 35.0%   Pillar 3 22.1%   Pillar 4 2.8%   Pillar 5 19.3%

100%

374

[1] Upper-middle-income countries are those that, according to the World Bank classification, have a gross national income (GNI) per capita of $3,706 to $11,455.   [2] Hybrid regimes are countries whose Democracy Index scores from 1 to 10 rounded to one decimal place are of 4 to 5.9.   [3] 2007.   [4] World Bank World Development Indicators.   [5] IMF World Economic Outlook.   [6] 2006.   [7] 2001.   [8] UNESCO, % age above 15.   [9] % ages 15–64.   [10] 2004.   [11] The index thermometer shows the relative contribution of each pillar (in %) to the total index score.

VENEZUELA

Country relative performance: **Index pillars**[12]

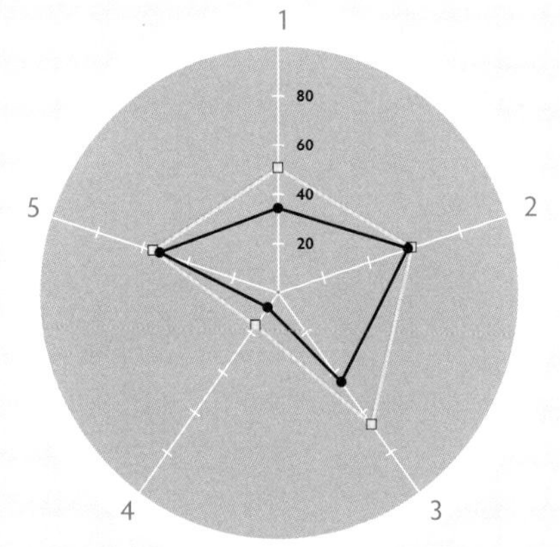

1. Institutional environment
2. Human capital, training and social inclusion
3. Regulatory and legal framework
4. Research and development
5. Usage of ICT

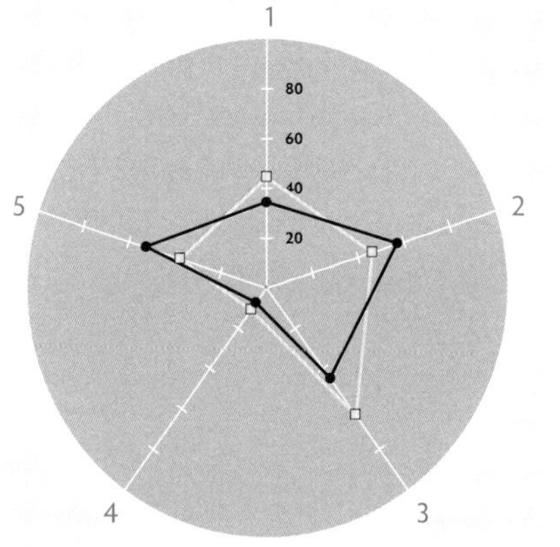

Upper-middle-income ☐   Venezuela ●━━

Hybrid regimes ☐   Venezeuela ●━━

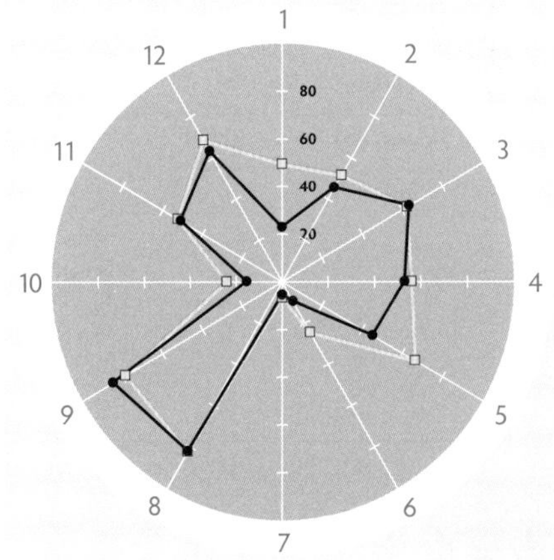

1. Good governance
2. Country policy assessment
3. Education
4. Social inclusion and equity policies
5. Doing business
6. R&D infrastructure
7. Patents and trademarks
8. Telephone communications
9. Mobile/cellular communications
10. Internet, computers and TV
11. Government ICT usage
12. Quality of the infrastructure

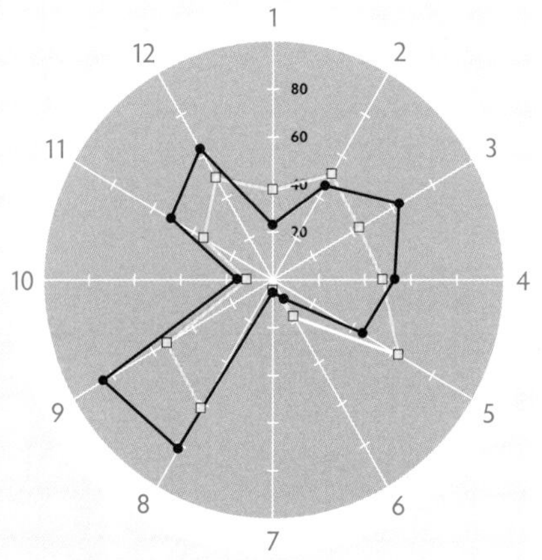

Country relative performance: **Pillar subindexes**[12]

In focus: **Top priorities for policy reform**[13]

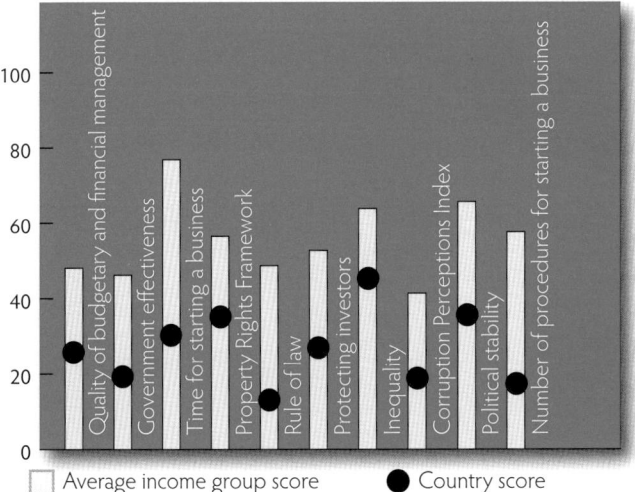

☐ Average income group score   ● Country score

In focus: **Significant indicators above income group average**[13]

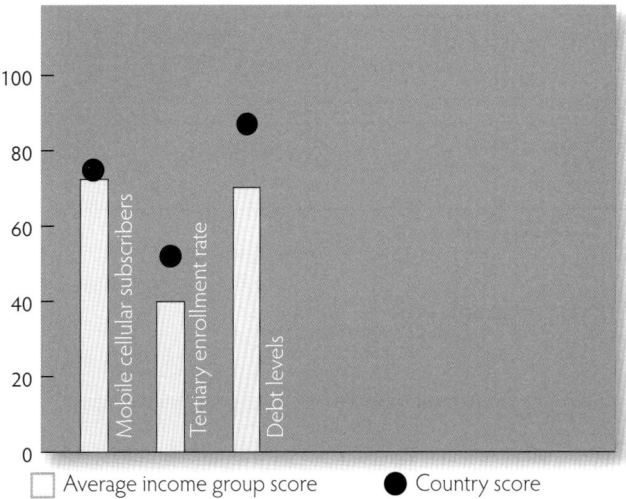

☐ Average income group score   ● Country score

---

[12] This analysis compares pillar and subindex country scores with respect to the average scores of countries grouped according to income level and political regime (see notes 1 and 2).

[13] Significant country indicators with the greatest distances above or below the income group average scores (i.e., achievements and targets for policy reform, respectively) were selected and ordered according to their Pearson correlation coefficients (highest first) with respect to the final index scores, to produce a ranking.

VIETNAM

Key selected indicators

Population (in millions)[3, 4]	85.1
Population density (people per sq. km)[3, 4]	275
GDP (current US$ billion, PPP)[3, 5]	221
GDP per capita (current US$, PPP)[3, 5]	2587
Current account balance (% of GDP)[3, 5]	-9.6
Average life expectancy (years)[4, 6]	70.8
Literacy rate[7, 8]	90
Female labor force participation rate[4, 6, 9]	77.2
Unemployment rate (% of total labor force)[4, 10]	2.1
CO_2 emissions (metric tons per capita)[4, 10]	1.2

Income level:[1] **Low-income**

Political regime:[2] **Authoritarian regime**

Rank	Score
78 out of 131	**46.4 out of 100**

Innovation Capacity Index

Pillars and pillar subindexes

Education
Rank **95** Score **45.4**

Social inclusion and equity policies
Rank **57** Score **55.9**

Country policy assessment
Rank **81** Score **49.8**

Good governance
Rank **93** Score **35.1**

Quality of the infrastructure
Rank **69** Score **64.5**

Government ICT usage
Rank **79** Score **45.6**

Doing business
Rank **89** Score **61.1**

R&D infrastructure
Rank **81** Score **19.0**

Patents and trademarks
Rank **78** Score **1.2**

Telephone communications
Rank **61** Score **82.8**

Mobile cellular communications
Rank **98** Score **51.1**

Internet, computers and TV
Rank **69** Score **15.6**

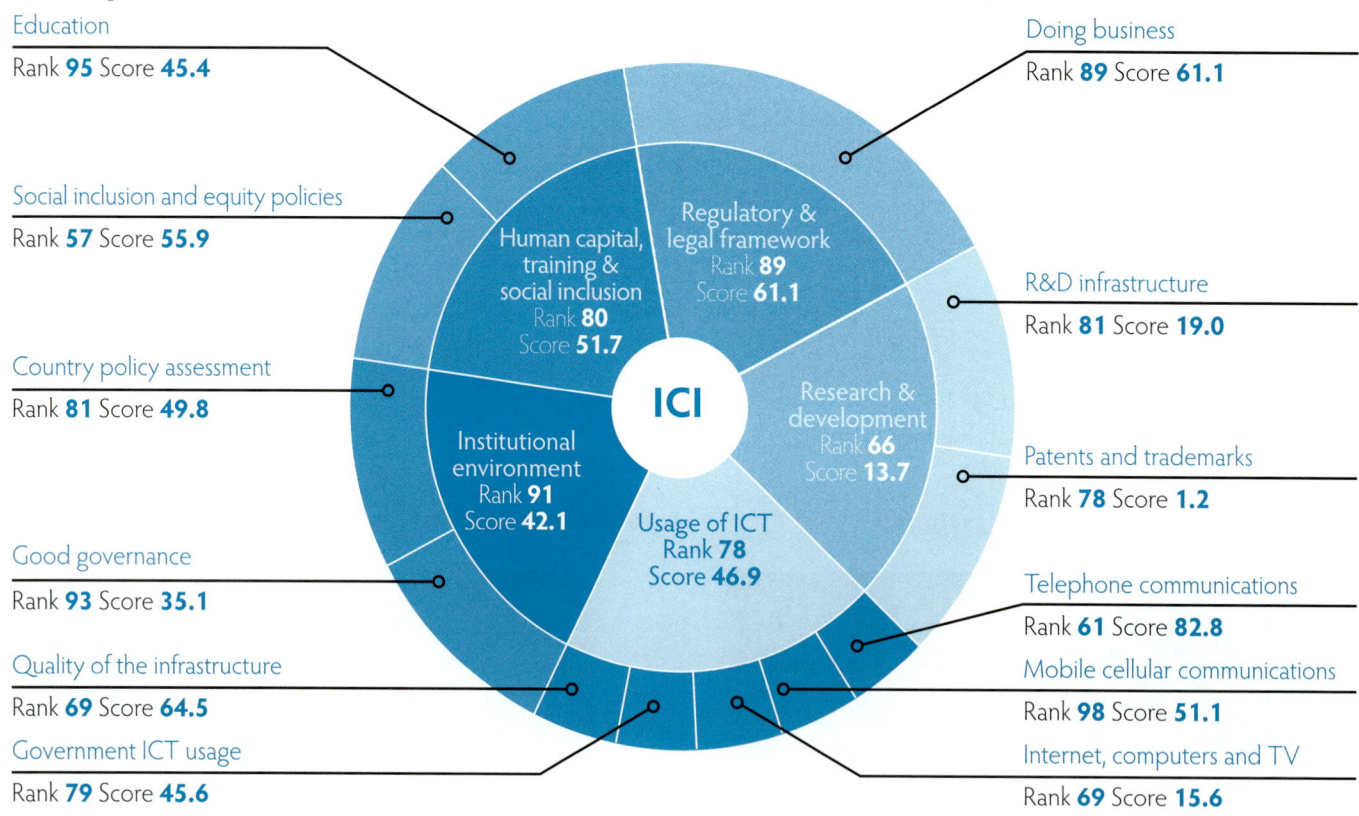

Index thermometer[11]

Pillar 1 27.2% Pillar 2 33.4% Pillar 3 26.3% Pillar 4 2.9% Pillar 5 10.1% 100%

[1] Low-income countries are those that, according to the World Bank classification, have a gross national income (GNI) per capita of $935 or less.
[2] Authoritarian regimes are countries whose Democracy Index scores from 1 to 10 rounded to one decimal place are less than 4. [3] 2007. [4] World Bank World Development Indicators. [5] IMF World Economic Outlook. [6] 2006. [7] 1999. [8] UNESCO, % age above 15. [9] % ages 15–64. [10] 2004. [11] The index thermometer shows the relative contribution of each pillar (in %) to the total index score.

VIETNAM

Country relative performance: **Index pillars**[12]

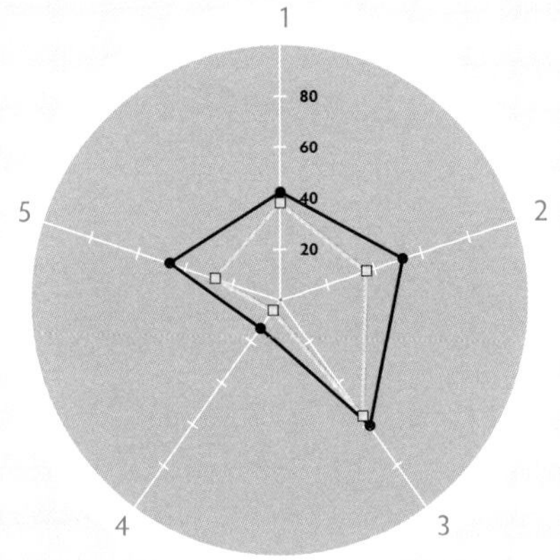

1. Institutional environment
2. Human capital, training and social inclusion
3. Regulatory and legal framework
4. Research and development
5. Usage of ICT

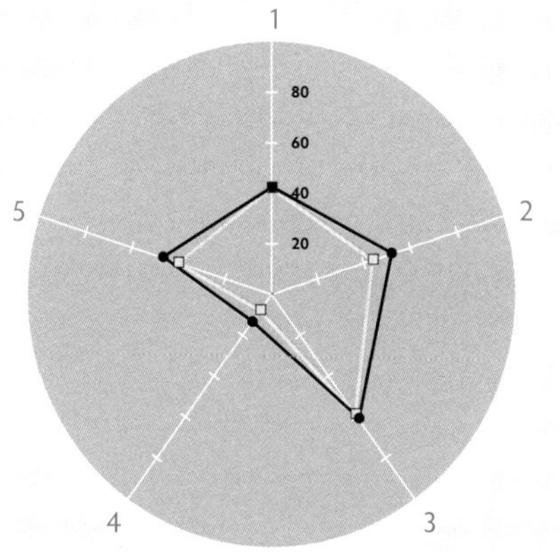

Low-income ☐   Vietnam ●———

Authoritarian regimes ☐   Vietnam ●———

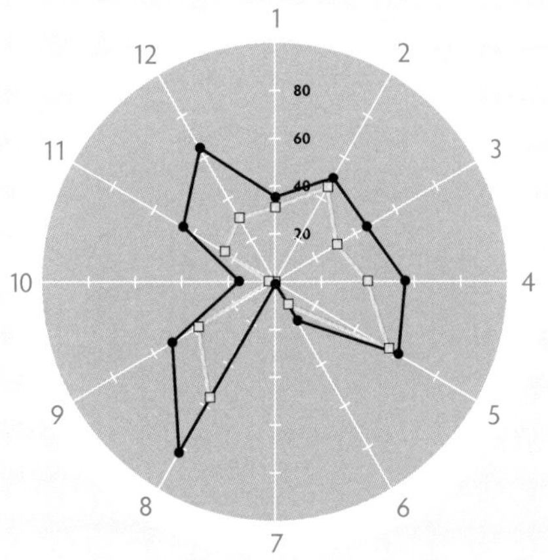

1. Good governance
2. Country policy assessment
3. Education
4. Social inclusion and equity policies
5. Doing business
6. R&D infrastructure
7. Patents and trademarks
8. Telephone communications
9. Mobile/cellular communications
10. Internet, computers and TV
11. Government ICT usage
12. Quality of the infrastructure

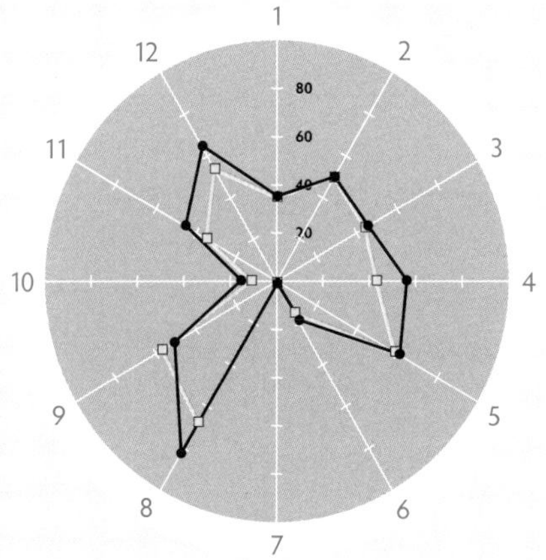

Country relative performance: **Pillar subindexes**[12]

In focus: **Top priorities for policy reform**[13]

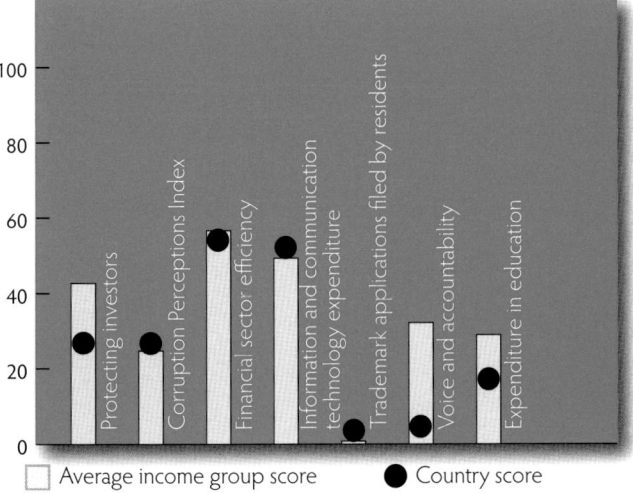

☐ Average income group score   ● Country score

In focus: **Significant indicators above income group average**[13]

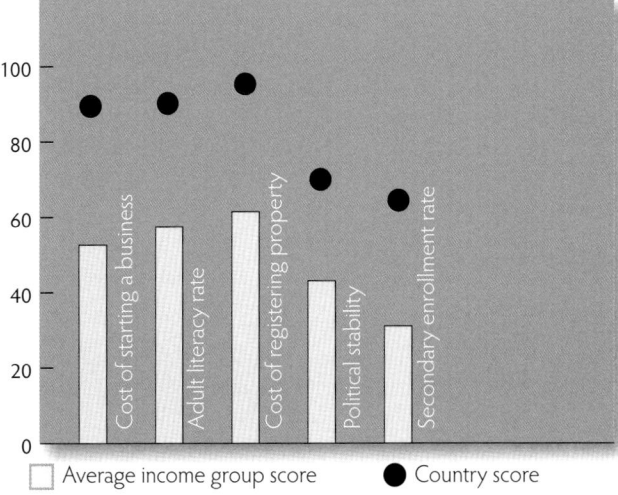

☐ Average income group score   ● Country score

---

[12] This analysis compares pillar and subindex country scores with respect to the average scores of countries grouped according to income level and political regime (see notes 1 and 2).

[13] Significant country indicators with the greatest distances above or below the income group average scores (i.e., achievements and targets for policy reform, respectively) were selected and ordered according to their Pearson correlation coefficients (highest first) with respect to the final index scores, to produce a ranking.

About the Authors

Laura Altinger serves as Economic Advisor in the Office of the Director, Environment, Housing and Land Management Division at the United Nations Economic Commission for Europe (UNECE), and focuses on climate change and green economy issues. Previously, as Research Director, she played a key role in creating the *Humanitarian Response Index*, endorsed by Kofi Annan. In 2006, she worked as Associate Director at the World Economic Forum, contributing to a number of publications and underlying indexes, including the *Global Competitiveness Report 2006–2007*, the *Global Gender Gap Report 2006*, the *Latin America Review 2006*, and the *Global Information Technology Report 2006*. She served six years at the United Nations, preparing various editions of the Economic Survey of Europe and analytical studies on environmental issues for the Aarhus Convention Secretariat. She has also held positions at the European Commission in Kenya and the Council of the European Union. Laura holds a PhD from the London Business School, an MA from the University of Cambridge, an MSc from Bristol University and a Graduate Diploma of Law from the College of Law, London. She recently co-founded the London Business School Carbon Club, and acts as its co-President.

Ester Basri is a Senior Analyst in the OECD Directorate for Science, Technology and Industry. She is responsible for the Working Group on Research Institutions and Human Resources (RIHR), which analyzes institutional, regulatory, and management issues facing governments as they aim to strengthen the capabilities of their public research institutions. Ester is the coordinator of the Science, Technology and Industry Outlook publication and has worked on a number of horizontal projects within the OECD including the Tertiary Education Review. She is co-leading the work on human capital for the OECD's Innovation Strategy. Prior to joining the OECD in 2006, Ester managed an academic research centre at the University of Western Sydney, was the manager of the Innovation Analysis Section at the Australian Department of Industry and was part of the Prime Minister's Mapping Science and Innovation Taskforce at the Education and Science Ministry. She holds a PhD from the Australian National University.

Sarah Box is an economist in the OECD's Directorate for Science, Technology and Industry. Her current role involves analyzing issues related to human capital and innovation, as well as public research organizations. She also contributes to the Directorate's regular examination of trends, prospects, and policy directions in science, technology, and industry. Prior to joining the OECD, Sarah began her career as an analyst at the New Zealand Treasury, where she undertook research on such topics as economic geography and economic integration. She also provided policy advice on telecommunications and regional, industry, and economic development. Following this, she worked as a Senior Research Economist for the Australian Government Productivity Commission, where she co-authored analytical reports providing policy advice on microeconomic issues. She holds Master of Commerce and Bachelor of Commerce (Honours) degrees in economics from the University of Auckland, New Zealand.

Simon Commander is Managing Partner of Altura Advisers and Senior Adviser at the European Bank for Reconstruction and Development in London. Between 1999 and 2008, he was a faculty member and Director of the Centre for New and Emerging Markets at London Business School. He holds a BA from Oxford University and a PhD from Cambridge University. He previously worked for over a decade at the World Bank in Washington, DC, in research, training and operations, while serving in a range of university posts. He has published widely in peer-reviewed research journals as well as books.

Alexander Ebner is Professor of Socio-Economics at Goethe University in Frankfurt am Main, Germany. He has also held teaching and research positions at Jacobs University Bremen, the University of Erfurt in Germany, and the Grenoble École de Management in France. Dr. Ebner maintains research affiliations with the University of California at Berkeley and the Institute of Southeast Asian Studies in Singapore. His main research interests focus on innovation, governance, and international development. He is the author of *Embedded Entrepreneurship* (Routledge) and co-edited the volume *Institutions of the Market* (Oxford University Press). He received his PhD in economics and political science from Goethe University.

Anil K. Gupta is a professor at the Centre for Management in Agriculture of the Indian Institute of Management, and Coordinator of the Society for Research and Initiatives for Sustainable Technologies and Institutions (SRISTI) and the Honey Bee Network of India. Dr. Gupta has devoted himself to analyzing the indigenous knowledge of farmers, artisans, and pastoralists, building bridges to science-based knowledge, and to ensuring that grassroots innovators are both encouraged and given credit for the results of their initiative and creativity. For his unique work in this area, Professor Gupta was elected at a young age to India's National Academy of Agricultural Sciences and received a Pew Conservation Scholar Award from University of Michigan. He was judged one of the 50 most influential people in the field of intellectual property rights in the world in 2003 and was accorded the Padma Shri National Award by the President of India for distinguished achievements in the field of management education. Prof. Gupta earned an MA in Biochemical Genetics in 1974 from Haryana Agricultural University, Haryana, and his PhD in management from Kurukshetra University (India) in 1986.

Markus Haacker is a growth and development economist based in London. From 1999–2008, he worked at the International Monetary Fund, mainly at the African Department. Since 2008, he has been an Honorary Lecturer at the London School of Hygiene and Tropical Medicine, and as a consultant to the World Bank. His work on the macroeconomic aspects of information and communication technologies includes contributions to the International Monetary Fund's *World Economic Outlook and the Global Information Technology Report* published by the World Economic Forum. Other notable work includes several publications on macroeconomic and fiscal aspects of HIV/AIDS and the response to HIV/AIDS, notably the "Macroeconomics of HIV/AIDS," published by the IMF. Dr. Haacker is chairman of the German publishing house ARCO. He holds a PhD in Economics from the London School of Economics.

Alan Hughes is Margaret Thatcher Professor of Enterprise Studies at the Judge Business School and Director of the Centre for Business Research at the University of Cambridge, where he is also a Fellow of Sidney Sussex College. He is the Director of the UK Innovation Research Centre, a joint venture between Cambridge and Imperial College, London. He has worked extensively on the role of universities in innovation and on the nature of knowledge exchange patterns between universities and the science base. His work in this area with colleagues at the Centre for Business Research, Cambridge, and at the Industrial Performance Center at MIT has been published in the report by Cosh, Hughes and Lester *UK PLC: Just How Innovative Are We?* He is currently completing an analysis on university-industry links at national and regional levels (University-Industry Knowledge Exchange: Demand Pull, Supply Push and the Public Space Role of Higher Education Institutions in the UK Regions). In 2004 he was appointed by the Prime Minister of the UK to membership of the UK's senior policy advisory body, the Council for Science and Technology.

Adam B. Jaffe is Fred C. Hecht Professor in Economics and Dean of the Faculty of Arts and Sciences at Brandeis University. He was previously Professor and Chair of the Economics Department, having come to Brandeis in 1994 from Harvard University. A graduate of Massachusetts Institute of Technology (BS in Chemistry, 1976; MS in Technology and Policy, 1978) and Harvard University (Ph.D. in Economics, 1985), Jaffe is the author of two books: *Patents, Citations and Innovations: A Window on the Knowledge Economy* (with Manuel Trajtenberg, 2002), and *Innovation and Its Discontents: How Our Broken Patent System is Endangering Innovation and Progress and What to Do About It* (with Josh Lerner, 2004). He was a co-founder of the Innovation Policy and the Economy group of the National Bureau of Economic Research and served as senior staff economist at the President's Council of Economic Advisers (1990–91). Research interests include intellectual property, science and technology policy, and innovation related to environment and energy.

Daniel Kaufmann is a Senior Fellow in the Global Economy and Development program at the Brookings Institution. Most recently, he served as Director in the World Bank Institute, where he pioneered new approaches to measure and analyze governance and corruption, helping countries formulate action programs. Well known for his writing on governance, corruption, and development, Kaufmann and his colleagues have pioneered new approaches to the diagnosis and analysis of country governance. At the World Bank, he also held senior positions focused on finance, regulation, anti-corruption, and capacity building for Latin America. After working as a se-

nior economist in Africa, he served as lead economist both in economies-in-transition and in the World Bank's research department. In the early 1990s, Kaufmann was the first Chief of Mission of the World Bank to the Ukraine, and then held a visiting position at Harvard University, prior to resuming his career at the World Bank. His research on economic development, governance, the unofficial economy, macro-economics, investment, corruption, privatization, and urban and labor economics has been published in leading journals. Kaufmann is a Chilean national who received his MA and Ph.D. in Economics at Harvard, and a BA in Economics and Statistics from the Hebrew University of Jerusalem.

Mohsen Khalil is a joint Director at the World Bank and the International Finance Corporation (IFC) of the Global Information and Communication Technologies Department. He oversees the World Bank Groups activities concerning telecommunications and information technologies world-wide, advising governments on sector reforms, regulatory frameworks, and institutional capacity building, in addition to supporting private investments in developing countries. Dr. Khalil held the post of Director of IFC's Central Asia, Middle East and North Africa Department, and as Chief Investment Officer in the Telecommunications, Transport, and Utilities Department. He was a Professor of Business at the American University of Beirut, and served as Chief Advisor to the Lebanese Minister of Post and Telecommunications, the Board Director of Lebanon's Autonomous Fund for Housing, and various governments and major corporations in the Middle East. He received an MSc in Electrical Engineering from the University of Wisconsin, Madison, an MS from MIT Sloan School of Management, and a PhD in Electrical Engineering from the University of Southern California.

Josh Lerner is the Jacob H. Schiff Professor of Investment Banking at Harvard Business School, with a joint appointment in the Finance and Entrepreneurial Management Units. He graduated from Yale College with a Special Divisional Major that combined physics with the history of technology. Before obtaining his Ph.D. in economics at Harvard, he worked for several years on issues concerning technological innovation and public policy, at the Brookings Institution, for a public-private task force in Chicago, and on Capitol Hill. Dr.

Lerner focuses on the world of alternative investments, with a particular emphasis on venture capital and private equity, and how public policies can boost entrepreneurship and technological and financial innovation. He is leading an international team of scholars in a multi-year study of the future of alternative investments for the World Economic Forum. He is the author (with Adam Jaffe) of *Innovation and Its Discontents,* a textbook *Venture Capital and Private Equity: A Casebook,* and many articles published in academic journals.

Augusto López-Claros is Honorary Professor at the European Business School in Frankfurt. He is also the founder of EFD–Global Consulting Network, an international consultancy specializing in economic, financial, and development issues. From 2003 to 2006, he was Chief Economist and Director of the Global Competitiveness Program at the World Economic Forum in Geneva, where he led the effort to expand the international profile of the Forum's work on issues of economic growth and productivity. At the Forum, he served as Editor of the Forum's *Global Competitiveness Report* and a number of other publications exploring issues of growth and development in various regions of the world and the impact of innovation, technology, and gender on economic growth. Before joining the Forum, he was Executive Director and Senior International Economist with Lehman Brothers International (London), and Resident Representative of the International Monetary Fund in the Russian Federation (Moscow) from 1992 to 1995. Prior to joining the IMF, he was Professor of Economics at the University of Chile in Santiago. He has written and lectured extensively in the United States, Europe, Latin America, Africa, and Asia on a broad range of subjects, including aspects of economic reform in transition economies, economic integration, the role of technology and innovation in advancing the development process, interdependence and cooperation, governance, gender, and the role of international organizations. Dr. López-Claros received his PhD in Economics from Duke University and a diploma in Mathematical Statistics from Cambridge University.

Yasmina N. Mata is a consultant with EFD—Global Consulting Network, formerly a researcher with the Center for Biological Research (CIB), which forms part of the National Research Council, in Madrid. She collaborates with the Extractive Metallurgy Research Group of Complutense University. She is pursuing a career as an independent consultant, offering scientific, academic, and research knowledge in areas of innovation not traditionally linked to natural sciences, and bridging different scientific and academic disciplines. She has published a number of papers in specialized journals about the biosorption processes of heavy and precious metals with biomass. Dr. Mata received two undergraduate degrees in Biology and a PhD in Science from Complutense University.

Ellen Olafsen is an Operations Officer at infoDev, a multi-donor partnership for technology-enabled innovation hosted by the World Bank. Her responsibilities include managing infoDev's networks on business incubation in Africa, Asia, Eastern Europe, Latin America, and the Middle East, which together comprise almost 200 business incubators in 80 developing countries. She also leads a team of business incubation experts that help least-developed countries plan and operate their business incubators effectively. Previously, Ellen worked for the Grassroots Business Initiative of the International Finance Corporation, and for the Development Gateway, where she led global communities of practice on microfinance and the knowledge economy. Ellen has an MBA in International Finance and an MA in International Affairs, focused on small and medium enterprise development in Africa. She received the Hall of Nations Award for scholarly promise from American University.

Hernán Rincón is President of Microsoft Latin America, responsible for both the long-term business and people strategy for 46 countries and territories in Latin America and the Caribbean, including hiring and retaining talented individuals and senior executives. He oversees all sales, marketing, and services operations and, as an ambassador for Microsoft, is engaged in the Corporate Citizenship programs that enable jobs, opportunities, and local innovation to support the economic and social development of Latin America. Earlier, he served as the Sales and Marketing Vice President for Microsoft Latin America. Prior to joining Microsoft, Mr. Rincón was President and CEO of Ferag Americas (a Swiss company specializing in state-of-the-art, high-tech solutions for the print media industry) and held several executive positions at Unisys global headquarters in Pennsylvania, USA. After completing a BA in Mathematics and Computer Science from the State University of New York, he received an MA from Harvard's John F. Kennedy School of Government and an MA in Science from the Andes University, in Colombia. As Citizenship Ambassador, Mr. Rincón is closely involved with the education and technology access initiatives which are helping the children and families in Latin America to realize their full potential.

Andrew Stirling is Science Director at SPRU (Science and Technology Policy Research, University of Sussex), and co-directs the Economic and Social Research Council STEPS Centre. Formerly a Director of Greenpeace International, he has since collaborated with a range of government, industry, and public interest organisations. His research interests focus on technological risk, innovation policy, scientific uncertainty, and democratic governance in a number of sectors, including energy, chemicals, nuclear systems, medicine, and food. Dr. Stirling has been involved in developing participatory appraisal methods, as well as general frameworks for implementing the Precautionary Principle and analyzing diversity, flexibility, and resilience in technology and research. He has served on a number of policy advisory committees, including the UK Government Advisory Committee on Toxic Substances and GM Science Review Panel, the European Commission Expert Group on Science and Governance, the Department for the Environment, Food, and Rural Affairs' Science Advisory Council, and the Sciencewise Panel of the Department for Business, Innovation and Skills.

Florian Täube is Assistant Professor of Growth Management at the Strascheg Institute of Innovation and Entrepreneurship (SIIE) at the European Business School (EBS) in Germany. Before joining EBS in 2008, Dr. Täube was with the Innovation and Entrepreneurship Group at the Tanaka Business School, Imperial College, London. His research interests lie at the intersection of entrepreneurship, economic geography, organization theory, and international business. Using mainly qualitative methods, he focuses his studies on the internationalization of project-based industries, particularly those related to IT, film, pharmaceuticals, and construction. He has had a long-standing interest is India, where he works with the evolution of the Bangalore IT cluster, and on a collaborative proj-

ect on the organization, networks and growth strategies of the Indian film and pharmaceutical industries. During his doctoral studies, he was a Visiting Scholar to the Indian Institute of Science and The Wharton School. Dr. Täube has published articles in the *Journal of International Management and Environment and Planning*. He holds a PhD in Economics from Johann Wolfgang Goethe-University, Frankfurt, Germany.

Acknowledgments

No publication of this magnitude and complexity can be accomplished without the assistance of many others.

I would like to take this opportunity to thank most warmly our editor, Nancy Ackerman, of AmadeaEditing (http://www.amadeaediting.com) in Toronto, for her careful editing of all the text components of this volume, for the writing of the Executive Summaries, and for her unflagging assistance with the overall management of the publication process. Her professionalism has been an invaluable asset in putting together this first edition of *The Innovation for Development Report*.

The responsibility for the layout of the volume has rested on the capable shoulders of designer René Steiner of Steiner Graphics (http://www.steinergraphics.com) in Geneva, who has patiently endured the many stages of this painstaking work with remarkable willingness and good cheer.

Finally, may I express my deep gratitude to two other individuals: writer Steve Pulley, for his conscientious and good-humored proofreading of the Report and for many helpful suggestions from start to finish; and Alan Yoshioka, of AY's edit: (http://www.aysedit.com) for compiling the first-rate Index.

Index